D1422951

Fundamentals of Electronic Devices and Circuits

FIFTH EDITION

DAVID A. BELL

OXFORD
UNIVERSITY PRESS

OXFORD

UNIVERSITY PRESS

70 Wynford Drive, Don Mills, Ontario M3C 1J9
www.oup.com/ca

Oxford University Press is a department of the University of Oxford.
It furthers the University's objective of excellence in research, scholarship,
and education by publishing worldwide in

Oxford New York
Auckland Cape Town Dar es Salaam Hong Kong Karachi
Kuala Lumpur Madrid Melbourne Mexico City Nairobi
New Delhi Shanghai Taipei Toronto

With offices in
Argentina Austria Brazil Chile Czech Republic France Greece
Guatemala Hungary Italy Japan Poland Portugal Singapore
South Korea Switzerland Thailand Turkey Ukraine Vietnam

Oxford is a trade mark of Oxford University Press
in the UK and in certain other countries

Published in Canada
by Oxford University Press

Library and Archives Canada Cataloguing in Publication

Bell, David A., 1930–
Fundamentals of electronic devices and circuits / David A. Bell.—5th ed.

Includes index.
Previous ed. published under title: Electronic devices and circuits.
ISBN 978-0-19-542523-9

1. Semiconductors—Textbooks. 2. Electronic circuits—Textbooks.
3. Electronic apparatus and appliances—Textbooks. I. Bell, David A., 1930–.
Electronic devices and circuits. II. Title.

TK7871.85.B44 2007   621.3815   C2007-904822-6

Cover image: NASA
Cover design: Sherill Chapman

1 2 3 4 – 11 10 09 08

This book is printed on permanent (acid-free) paper ∞.

Printed in Canada

PREFACE

This book is intended for use as an electronics technology course text for colleges and universities, and as a reference text for practising professionals.

The objectives of the book are to provide clear explanations of the operation of all important electronics devices generally available today, and to show how each device is used in circuits. I am convinced that an understanding of devices and circuits is most easily achieved by learning how to design circuits. *Practical circuit design is usually quite simple, much simpler than some methods of circuit analysis.*

After a discussion of device operation, characteristics, and parameters, typical circuits using the device are explained, then circuit design and analysis are treated. The text includes many practical examples, using parameters from device manufacturers' data sheets. The circuit design procedure most often involves determining appropriate current and voltage levels, and then applying Ohm's law and the capacitor impedance equation. Most equations are derived, so that the student knows exactly what is going on. Instead of rigorous analysis methods, practical approximations are employed wherever possible.

Conventional current direction is used because it is the direction normally employed by device and integrated circuit manufacturers, and because every device graphic symbol uses an arrowhead that indicates conventional current direction.

I am always grateful for suggestions that might improve my presentation of the material or for additional topics that should be treated.

David Bell

CONTENTS

CHAPTER 1
Basic Semiconductor and *pn*-Junction Theory

CONTENTS

Objectives

You will be able to:

1   Describe an atom, and sketch a diagram to show its components.
2   Explain energy levels and energy bands relating to electrons, and define valence band and conduction band.
3   Describe how electric current flow occurs by electron motion and by hole transfer.
4   Explain conventional current direction and direction of electron flow.
5   Sketch a diagram to show the relationship between covalently bonded atoms.
6   Discuss the differences between conductors, insulators, and semiconductors.

7   Explain how *n*-type and *p*-type semiconductor materials are created, and discuss the differences between the two types.
8   Sketch a *pn*-junction, and explain the origin of the junction depletion region.
9   Draw diagrams to show the effects of forward biasing and reverse biasing a *pn*-junction.
10 Sketch the current/voltage characteristics for forward-biased and reverse-biased *pn*-junctions.
11 Discuss temperature effects on conductors, insulators, semiconductors, and *pn*-junctions.

INTRODUCTION

An electronic device controls the movement of electrons. The study of electronic devices requires a basic understanding of the relationship between electrons and the other components of an atom. The movement of electrons within a solid and the bonding forces between atoms can then be investigated. This leads to a knowledge of the differences between conductors, insulators, and semiconductors, and to an understanding of *p*-type and *n*-type semiconductor material.

Junctions of *p*-type and *n*-type material (*pn*-junctions) are basic to all but a very few semiconductor devices. Forces act upon electrons that are adjacent to a *pn*-junction, and these forces are altered by the presence of an external bias voltage.

1-1 ATOMIC THEORY

The Atom

The atom can be thought of as consisting of a central *nucleus* surrounded by orbiting *electrons* (see Fig. 1-1). It may be compared to a planet with orbiting satellites. Just as satellites are held in orbit by the attractive force of gravity due to the mass of the planet, so each electron is held in orbit by an *electrostatic force of attraction* between it and the nucleus.

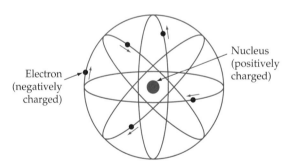

Figure 1-1 The atom consists of a central nucleus surrounded by orbiting electrons. The electrons have a negative charge, and the nucleus contains protons that are positively charged.

Each electron has a negative electrical charge of 1.602×10^{-19} Coulombs (C), and some particles within the nucleus have a positive charge of the same magnitude. Because opposite charges attract, a force of attraction exists between the oppositely charged electron and nucleus. Compared to the mass of the nucleus, electrons are relatively tiny particles of almost negligible mass. In fact, they can be considered to be little particles of negative electricity having no mass at all.

The nucleus of an atom (Fig. 1-2) is largely a cluster of two types of particles, *protons* and *neutrons.* Protons have a positive electrical charge, equal in magnitude (but opposite in polarity) to the negative charge on an electron. A neutron has no charge at all. Protons and neutrons each have masses about 1800 times the mass of an electron. For a given atom, the number of protons in the nucleus normally equals the number of orbiting electrons.

Figure 1-2 The nucleus of an atom is largely a cluster of protons and neutrons.

Because the protons and orbital electrons are equal in number and equal and opposite in charge, they neutralize each other electrically. For this reason, all atoms are normally electrically neutral. If an atom loses an electron, it has lost some negative charge. Consequently, it becomes positively charged and is referred to as a *positive ion* (see Fig. 1-3a). Similarly, if an atom gains an additional electron, it becomes negatively charged and is termed a *negative ion* (Fig. 1-3b).

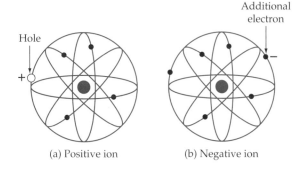

(a) Positive ion   (b) Negative ion

Figure 1-3 Positive and negative ions are created when an atom gains or loses an electron.

The differences between atoms consist largely of dissimilar numbers and arrangements of the three basic types of particle. However, all electrons are identical, as are all protons and all neutrons. An electron from one atom could replace an electron in any other atom. Different materials are made up of different types of atoms or different combinations of several types of atoms.

The number of protons in an atom is referred to as the *atomic number* of the atom. The *atomic weight* is approximately equal to the total number of protons and neutrons in the nucleus of the atom. The atom of the semiconductor material *silicon* has 14 protons and 14 neutrons in its nucleus, as well as 14 orbital electrons. Therefore, the atomic number for silicon is 14, and its atomic weight is approximately 28.

Electron Orbits and Energy Levels

Atoms may be conveniently represented by the two-dimensional diagrams shown in Fig. 1-4. It has been found that electrons can occupy only certain orbital rings or *shells* at fixed distances from the nucleus and that each shell can contain only a particular number of electrons. The electrons in the outer shell determine the electrical (and chemical) characteristics of each particular type of atom. These electrons are usually referred to as *valence electrons*. An atom may have its outer shell, or *valence shell*, completely filled or only partially filled.

The atoms represented in Fig. 1-4 are those of two important semiconductor materials, *silicon (Si)* and *germanium (Ge)*. It is seen that each of these atoms has four electrons in a valence shell that can contain a maximum of eight. So, the valence shells have four electrons and four *holes*. A *hole* is defined as an absence of an electron in a shell where one could exist. Even though the valence shells

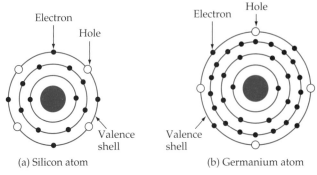

Electron

Hole

Valence shell

(a) Silicon atom

Electron   Hole

Valence shell

(b) Germanium atom

Figure 1-4 Two-dimensional representations of silicon and germanium atoms. The outer shells have four electrons and four holes.

of silicon and germanium have four holes, both types of atoms are electrically neutral because the total number of orbiting (negatively charged) electrons equals the total number of (positively charged) protons in the nucleus.

The closer an electron is to the nucleus, the stronger are the forces that bind it to the atom. Each shell has an *energy level* associated with it that represents the amount of energy required to extract an electron from the atom. Since the electrons in the valence shell are farthest from the nucleus, they require the least amount of energy to extract them (see Fig. 1-5a). Conversely, those electrons closest to the nucleus require the greatest energy application to extract them from the atom.

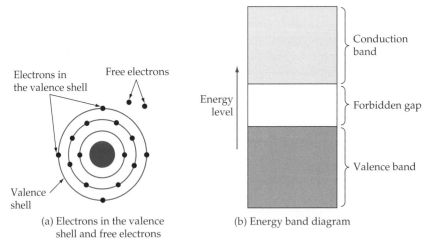

Electrons in the valence shell   Free electrons

Valence shell

(a) Electrons in the valence shell and free electrons

Energy level

Conduction band

Forbidden gap

Valence band

(b) Energy band diagram

Figure 1-5 The energy level of electrons within a solid is shown by the energy band diagram. Electrons in orbit around a nucleus are in the valence band. Electrons which have broken away from an atom are in the conduction band.

The energy levels of the orbiting electrons are measured in *electron volts* (eV). An electron volt is defined as the amount of energy required to move one electron through a voltage difference of one volt.

Energy Bands

Up to this point the discussion has concerned a system of electrons in one isolated atom. The electrons of an isolated atom are acted upon only by the forces within that atom. However, when atoms are brought closer together, as in a solid, the electrons come under the influence of forces from other atoms. Under these circumstances, the energy levels that may be occupied by electrons merge into bands of energy levels. There are two distinct energy bands in which electrons may exist: the *valence band* and the *conduction band*. Separating these two bands is an *energy gap*, termed the *forbidden gap*, in which electrons cannot normally exist. The valence band, conduction band, and forbidden gap are shown diagrammatically in Fig. 1-5b.

Electrons in the conduction band of energy levels have become disconnected from atoms and are drifting around in the material. Conduction band electrons may be easily moved by the application of relatively small amounts of energy. Much larger amounts of energy must be applied to move an electron in the valence band of energy levels. Electrons in the valence band are usually in orbit around a nucleus. Depending upon the particular material, the forbidden gap may be large, small, or non-existent. The distinction between conductors, insulators, and semiconductors is largely concerned with the relative widths of the forbidden gap.

It is important to note that *the energy band diagram is simply a graphic representation of the energy levels associated with electrons.* To repeat, the electrons in the valence band are actually in orbit around the nucleus of an atom; those in the conduction band are drifting in the spaces between atoms.

Section 1-1 Review

1-1.1 Define nucleus, electron, electronic charge, proton, neutron, shell, positive ion, and negative ion.

1-1.2 What is meant by atomic number and atomic weight? State the atomic number and atomic weight for silicon.

1-1.3 Define conduction band, valence band, and forbidden gap.

1-2 CONDUCTION IN SOLIDS

Electron Motion and Hole Transfer

Conduction occurs in any material when an applied voltage causes electrons in the material to move in a particular direction. This may be due to one or both of two processes: *electron motion* and *hole transfer.* In electron motion, free electrons in the conduction band are moved under the influence of the applied electric field, thus creating an electric current (see Fig. 1-6). Since electrons have a negative charge, they are repelled from the negative terminal of the applied voltage and attracted toward the positive terminal. Hole transfer involves electrons which are still attached to atoms (those in the valence band).

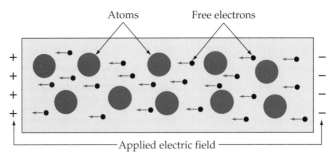

Figure 1-6   Free electrons are easily moved by an applied voltage to create an electric current.

If some of the electron positions in the valence shell of an atom are not occupied by electrons, there are holes where electrons could exist. When sufficient energy is applied, an electron may be made to jump from one atom to a hole in another atom. When it jumps, the electron leaves a hole behind it, and consequently the hole has moved in a direction opposite to that of the electron.

Figure 1-7 illustrates a situation where there are no free electrons. However, those electrons in orbit around atoms experience a force of attraction to the positive terminal of the applied voltage, and repulsion from the negative terminal. This force can cause an electron to jump from one atom to another, moving toward the positive terminal.

Figure 1-7a shows an electron jumping from atom Y to atom X (toward the positive terminal of the applied voltage). When this occurs, the hole in the

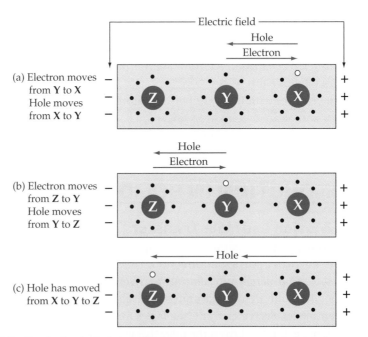

Figure 1-7   Conduction by hole transfer. An electron can be made to jump from one atom to another atom under the influence of an applied voltage. This causes the hole to move in the opposite direction.

valence shell of atom X is filled, and a hole is left in the valence shell of atom Y (Fig. 1-7b). If an electron now jumps from atom Z to fill the hole in Y, a hole is left in the valence shell of Z (Fig. 1-7c). The hole has moved from atom X to atom Y to atom Z. A flow of current (electron motion) has occurred, and this may be said to be due to *hole movement,* or *hole transfer.*

Holes may be thought of as *positively charged particles,* and as such, they move through an electric field in a direction opposite to that of electrons. (Positive particles are attracted toward the negative terminal of an applied voltage.) In the circumstance illustrated in Fig. 1-7 where there are few free electrons, it is more convenient to think of hole movement than of electrons jumping from atom to atom.

Since the flow of electric current is constituted by the movement of electrons and holes, electrons and holes are referred to as *charge carriers.* Each time a hole moves, an electron must be supplied with enough energy to enable it to escape from its atom. Free electrons can be moved with less application of energy than holes because they are already disconnected from their atoms. For this reason, electrons have *greater mobility* than holes.

Conventional Current and Electron Flow

In the early years of electrical experimentation it was believed that a positive charge represented an increased amount of electricity and that a negative charge was a reduced quantity. Consequently, current was assumed to flow from positive to negative. This is a convention that remains in use today even though current is now known to be a movement of electrons from negative to positive (see Fig. 1-8).

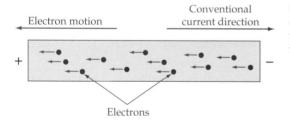

Electron motion

Conventional current direction

+   −

Electrons

Figure 1-8 Conventional current direction is from positive to negative. Electron flow is from negative to positive.

Current flow from positive to negative is referred to as the conventional current direction.

Electron flow from negative to positive is known as the direction of electron flow.

It is important to understand both electron flow and conventional current direction. The operation of electronic devices is explained in terms of electron movement. However, *every graphic symbol used to represent an electronic device has an arrowhead which indicates conventional current direction* (see the diode symbol in Fig. 1-9). Consequently, electronic circuits are most easily explained by using current flow from positive to negative.

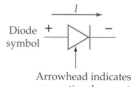

Diode symbol

Arrowhead indicates conventional current direction

Figure 1-9 All graphic symbols for electronic devices have arrowheads that indicate conventional current direction.

Section 1-2 Review

1-2.1 Draw sketches to show the process of current flow by electron motion and by hole transfer.

1-2.2 Explain the difference between conventional current direction and direction of electron flow.

1-3 CONDUCTORS, SEMICONDUCTORS, AND INSULATORS

Bonding Forces between Atoms

Whether a material is a conductor, a semiconductor, or an insulator depends largely upon what happens to the outer-shell electrons when the atoms bond themselves together to form a solid. In the case of copper, the easily detached valence electrons are given up by the atoms. As illustrated in Fig. 1-10, this creates a great mass of free electrons (or *electron gas*) drifting about in the space between the copper atoms. The electrons are easily moved under the influence of an applied voltage to create a current flow. The bonding force that holds atoms together in a conductor is known as *metallic bonding*.

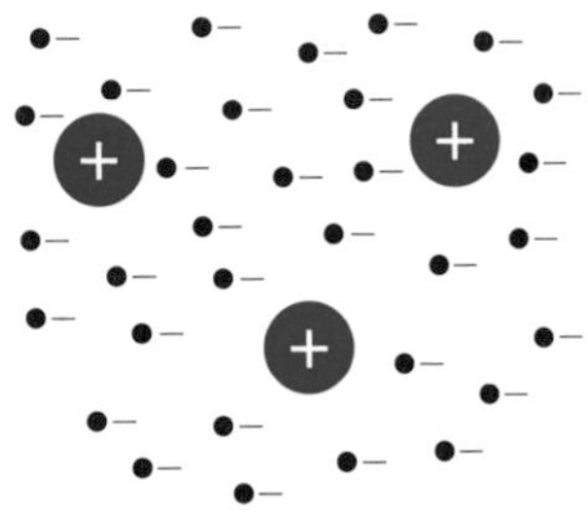

Semiconductor atoms normally have four outer-shell electrons and four holes, and they are so close together that the outer-shell electrons behave as if they were orbiting in the valence shells of two atoms. In this way each valence-shell electron fills one of the holes in the valence shell of an adjacent atom. This produces a bonding force referred to as *covalent bonding*. As shown in Fig. 1-11, it would appear that there are no holes and no free electrons drifting about in the semiconductor material. In fact, some of the electrons are so weakly attached to their atoms that they can be made to break away to create a current flow when a voltage is applied.

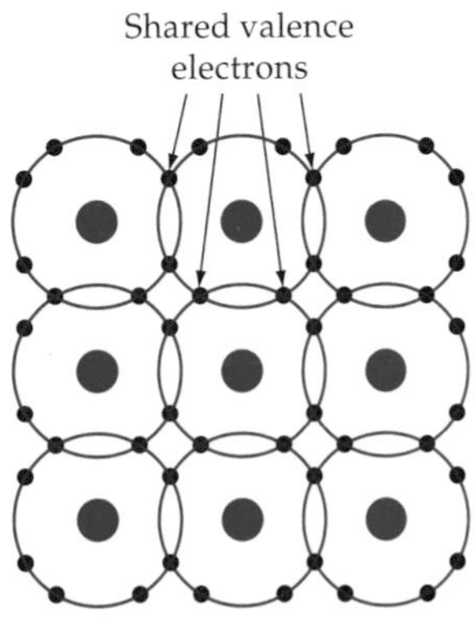

In some insulating materials the atoms bond together in a similar way to semiconductor atoms (*covalent bonding*). But the valence-shell electrons are so strongly attached to the atoms in an insulator that no charge carriers are available for current flow. In other types of insulating materials, some atoms give up outer-shell electrons, which are accepted into the orbit of nearby atoms. Because the atoms are *ionized*, this is termed *ionic bonding* (see Fig. 1-12). Here again, all of the electrons are very strongly attached to atoms, and the possibility of current flow is virtually zero.

Energy Bands in Different Materials

The energy band diagrams in Fig. 1-13 show that insulators have a wide forbidden gap, semiconductors have a narrow forbidden gap, and conductors have no forbidden gap at all.

Figure 1-10 In metallic bonding a mass of free electrons is drifting around in the space between atoms.

Shared valence electrons

Figure 1-11 Covalent bonding in semiconductors produces very few free electrons or holes for current flow.

Negative ion   Positive ion

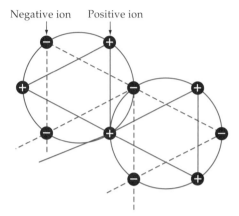

Figure 1-12 Ionic bonding occurs in some insulating materials. There are no holes or free electrons to facilitate current flow.

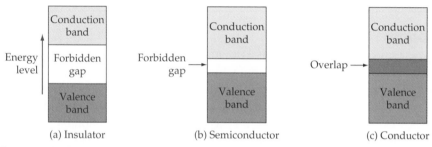

(a) Insulator   (b) Semiconductor   (c) Conductor

Figure 1-13 Energy band diagrams for insulators, semiconductors, and conductors.

In the case of insulators, there are practically no electrons in the conduction band, and the valence band is filled. Moreover, the forbidden gap is so wide (Fig. 1-13a) that it would require the application of relatively large amounts of energy to cause an electron to cross from the valence band to the conduction band. Therefore, when a voltage is applied to an insulator, conduction cannot normally take place either by electron motion or hole transfer.

For semiconductors at a temperature of absolute zero ($-273°C$), the valence band is usually full and there may be no electrons in the conduction band. As shown in Fig. 1-13b, the forbidden gap in a semiconductor is very much narrower than that in an insulator, and the application of small amounts of energy can raise electrons from the valence band to the conduction band. Sufficient thermal energy for this purpose is available when the semiconductor is at normal room temperature. If a voltage is applied to the semiconductor, conduction can occur by both electron movement in the conduction band and by hole transfer in the valence band.

In the case of a conductor (Fig. 1-13c) there is no forbidden gap, and the valence and conduction energy bands overlap. For this reason, very large numbers of electrons are available for conduction, even at extremely low temperatures.

Typical resistance values for a one-centimetre-cube sample are the following: conductor, 10^{-6} Ω; semiconductor 10 Ω; insulator 10^{14} Ω.

Section 1-3 Review

1-3.1 Briefly describe the type of atomic bonding that occurs in (a) insulators and (b) conductors.

1-3.2 Draw a sketch to show the bonding of atoms in semiconductor material. Briefly explain.

1-3.3 Sketch and explain the energy band diagrams for conductors, insulators, and semiconductors.

1-4 *n*-TYPE AND *p*-TYPE SEMICONDUCTORS

Doping

Pure semiconductor material is known as *intrinsic* material. Before intrinsic material can be used in the manufacture of a device, *impurity atoms* must be added to improve its conductivity. The process of adding the atoms is termed *doping*. Two different types of doping are possible: *donor doping* and *acceptor doping*. Donor doping generates free electrons in the conduction band (that is, electrons that are not tied to an atom). Acceptor doping produces valence-band holes, or a shortage of valence electrons in the material. After doping, the semiconductor material is known as *extrinsic* material.

n-Type Material

In donor doping, illustrated in Fig. 1-14, impurity atoms that have five electrons and three holes in their valence shells are added to the undoped material.

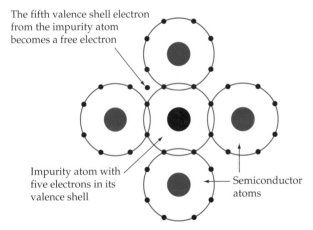

Figure 1-14 In donor doping, impurity atoms with five valence-shell electrons and three holes are added to the semiconductor material.

The impurity atoms form covalent bonds with the silicon or germanium atoms. Because the semiconductor atoms have only four electrons and four holes in their valence shells, there is one extra valence-shell electron for each impurity atom added. Each additional electron produced in this way enters the conduction band as a free electron. Because the free electrons have *negative charges,* donor-doped semiconductor is known as *n-type material.*

Free electrons in the conduction band are easily moved around under the influence of an electric field. Consequently, conduction takes place largely by electron motion in donor-doped semiconductor material. The doped material remains electrically neutral (neither positively charged nor negatively charged) because the total number of electrons (including the free electrons) is still equal to the total number of protons in the atomic nuclei.

The term *donor doping* comes from the fact that an electron is *donated* to the conduction band by each impurity atom. Typical donor materials are *antimony, phosphorus,* and *arsenic.* Since these atoms each have five valence electrons, they are referred to as *pentavalent atoms.*

p-Type Material

The impurity atoms used for acceptor doping (see Fig. 1-15) have outer shells containing three electrons and five holes. Suitable atoms with three valence electrons (*trivalent atoms*) are *boron, aluminum,* and *gallium.* These atoms form covalent bonds with the semiconductor atoms, but the bonds lack one electron for a complete outer shell of eight. In Fig. 1-15 the impurity atom illustrated has only three valence electrons; so a hole exists in its bond with the surrounding atoms. It is seen that in acceptor doping, holes are introduced into the valence band so that conduction may occur by the process of hole transfer.

Since holes can be said to have a *positive charge,* acceptor-doped semiconductor material is referred to as *p-type material.* Like the *n*-type, *p*-type material

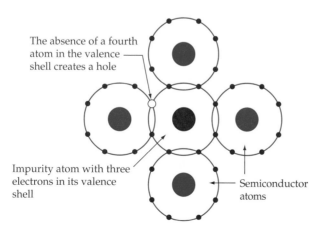

Figure 1-15   Acceptor doping requires the use of impurity atoms that have three electrons and five holes in the valence shells. This creates holes in the semiconductor material.

remains electrically neutral, because the total number of orbital electrons in each impurity atom is equal to the total number of protons in its atomic nucleus. The term *acceptor doping* comes from the fact that holes can *accept* a free electron.

Majority and Minority Charge Carriers

In undoped semiconductor material at room temperature there are a number of free electrons and holes. That is because thermal energy causes some electrons to break the bonds with their atoms and enter the conduction band. This process, which creates pairs of holes and electrons, is appropriately termed *hole-electron pair generation.* The opposite effect, called *recombination,* occurs when an electron *falls into* a hole in the valence band. Because there are many more electrons than holes in *n*-type material, electrons are referred to as the *majority charge carriers,* and holes as the *minority carriers* in *n*-type material. In *p*-type material, holes are the majority carriers and electrons are minority carriers.

Effects of Heat and Light

When a conductor is heated, the atoms (which are in fixed locations) tend to vibrate, and the vibration impedes the movement of the surrounding mass of electrons (see Fig. 1-10). This causes a reduction in the flow of the electrons that constitute the electric current. The reduced current flow means that the conductor's resistance has increased. So a conductor has a *positive temperature coefficient (PTC)* of resistance, a resistance that increases with increasing temperature. This is illustrated in Fig. 1-16a.

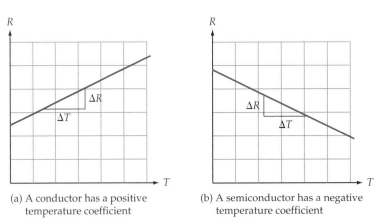

(a) A conductor has a positive temperature coefficient

(b) A semiconductor has a negative temperature coefficient

Figure 1-16 The resistances of conductors and semiconductors are affected differently by temperature change.

When undoped semiconductor material is at a temperature of absolute zero (−273°C), all electrons are in normal orbit around the atoms and there are virtually no free electrons in the conduction band and no holes in the valence band. Consequently, at −273°C a semiconductor behaves as an insulator.

When the temperature of the semiconductor is raised, electrons break away from their atoms and move from the valence band to the conduction band. This produces holes in the valence band and free electrons in the conduction band, allowing conduction to occur by electron movement and by hole transfer. Increasing the application of thermal energy generates an increasing number of hole-electron pairs.

As in the case of a conductor, thermal vibration of atoms occurs in a semiconductor. However, there are very few electrons to be impeded in a semiconductor compared to the large numbers in a conductor. The thermal generation of electrons is the dominating factor, and the semiconductor current increases as the temperature rises. This represents a decrease in resistance with rising temperature, or a *negative temperature coefficient* (*NTC*) (Fig. 1-16b). Heavily doped semiconductor material is an exception in that it behaves more like a conductor than a semiconductor.

Just as thermal energy can cause electrons to break their atomic bonds, so hole-electron pairs can be generated by energy applied to a semiconductor in the form of light. A material that has few free electrons available for conduction when not illuminated is said to have a high *dark resistance*. When the semiconductor material is illuminated, its resistance decreases and may become comparable to that of a conductor.

Section 1-4 Review
1-4.1 Draw a sketch to illustrate donor doping.
1-4.2 Draw a sketch to illustrate acceptor doping.
1-4.3 Define *n*-type material, *p*-type material, minority charge carriers, majority charge carriers, positive temperature coefficient, negative temperature coefficient, and dark resistance.

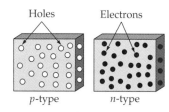

Figure 1-17 *p*-type and *n*-type semiconductor materials.

1-5 THE *pn*-JUNCTION

Junction of *p*-Type and *n*-Type

Two blocks of semiconductor material are represented in Fig. 1-17: one block is *p*-type material, and the other is *n*-type. The small circles in the *p*-type material represent holes, which are the majority charge carriers in *p*-type. The dots in the *n*-type material represent the majority charge carrier free electrons within that material. Normally, the holes are uniformly distributed throughout the volume of the *p*-type semiconductor and the electrons are uniformly distributed in the *n*-type.

In Fig. 1-18, *p*-type and *n*-type semiconductor materials are shown side by side, representing a *pn-junction*. Since holes and electrons are close together at the junction, some free electrons from the *n*-side are attracted across the junction to fill adjacent holes on the *p*-side. They are said to *diffuse* across the junction from a region of high carrier concentration to one of low concentration.

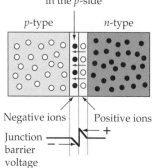

Figure 1-18 At a *pn*-junction, electrons cross from the *n*-side to fill holes in a layer on the *p*-side close to the junction.

The free electrons crossing the junction create negative ions on the p-side by giving some atoms one more electron than their total number of protons. The electrons also leave positive ions (atoms with one fewer electron than the number of protons) behind them on the n-side.

Barrier Voltage

The n-type and p-type materials are both electrically neutral before the charge carriers diffuse across the junction. When negative ions are created on the p-side, the portion of the p-side close to the junction acquires a negative voltage (Fig. 1-18). Similarly, the positive ions created on the n-side give the n-side a positive voltage close to the junction. The negative voltage on the p-side tends to repel additional electrons crossing from the n-side. Also (thinking of the holes as positive particles), the positive voltage on the n-side tends to repel any movement of holes from the p-side. So, the initial diffusion of charge carriers creates a *barrier voltage* at the junction, which is negative on the p-side and positive on the n-side. The transfer of charge carriers and the resultant creation of the barrier voltage occur when the pn-junctions are formed during the manufacturing process (see Chapter 7).

The magnitude of the barrier voltage at a pn-junction can be calculated from a knowledge of the doping densities, electronic charge, and junction temperature. Typical barrier voltages at 25°C are 0.3 V for germanium junctions and 0.7 V for silicon.

It has been explained that the barrier voltage at the junction opposes both the flow of electrons from the n-side and the flow of holes from the p-side. Because electrons are the majority charge carriers in the n-type material, and holes are the majority charge carriers in the p-type, it is seen that *the barrier voltage opposes the flow of majority carriers* across the pn-junction (see Fig. 1-19). Any free electrons generated by thermal energy on the p-side of the junction are attracted across the positive barrier to the n-side. Similarly, thermally generated holes on the n-side are attracted to the p-side through the negative barrier presented to them at the junction. Electrons on the p-side and holes on the n-side are minority charge carriers. Therefore, *the barrier voltage assists the flow of minority carriers across the junction* (Fig. 1-19).

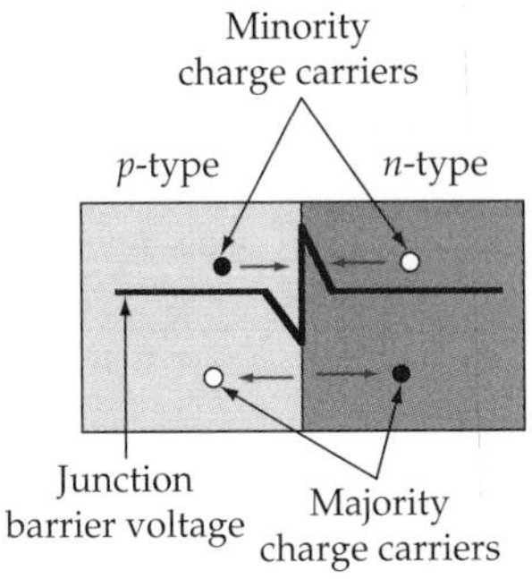

Minority
charge carriers

p-type n-type

Junction
barrier voltage   Majority
charge carriers

Figure 1-19 The barrier voltage at a *pn*-junction assists the flow of minority charge carriers and opposes the flow of majority charge carriers.

Depletion Region

The movement of charge carriers across the junction leaves a layer on each side that is depleted of charge carriers. This is the *depletion region* shown in Fig. 1-20a. On the n-side, the depletion region consists of donor impurity atoms that, having lost the free electron associated with them, have become positively charged. The depletion region on the p-side is made up of acceptor impurity atoms that have become negatively charged by losing the hole associated with them. (The hole has been filled by an electron.)

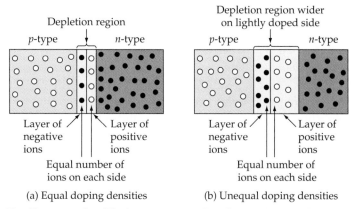

Figure 1-20 Charge carrier diffusion across a *pn*-junction creates a region that is depleted of charge carriers and that penetrates deepest into the more lightly doped side.

On each side of the junction, equal numbers of impurity atoms are involved in the depletion region. If the two blocks of semiconductor material have equal doping densities, the depletion layers on each side have equal widths (Fig. 1-20a). If the *p*-side is more heavily doped than the *n*-side, as illustrated in Fig. 1-20b, the depletion region penetrates more deeply into the *n*-side in order to include an equal number of impurity atoms on each side of the junction. Conversely, when the *n*-side is more heavily doped, the depletion region penetrates deeper into the *p*-type material.

Summary

- A region depleted of charge carriers spreads across both sides of a *pn*-junction, penetrating farther into the less doped side.
- The depletion region contains an equal number of ionized atoms on opposite sides of the junction.
- A barrier voltage is created by the charge carrier depletion effect, positive on the *n*-side and negative on the *p*-side.
- The barrier voltage opposes majority charge carrier flow and assists the flow of minority charge carriers across the junction.

Section 1-5 Review

1-5.1 Sketch a *pn*-junction showing the depletion region. Briefly explain how the depletion region is created.

1-5.2 Explain the origin of the barrier voltage at a *pn*-junction. Discuss the effect of the barrier voltage on minority and majority charge carriers.

1-6 BIASED JUNCTIONS

Reverse-Biased Junction

When an external bias voltage is applied to a *pn*-junction, positive to the *n*-side and negative to the *p*-side, electrons from the *n*-side are attracted to the positive terminal, and holes from the *p*-side are attracted to the negative terminal. As shown in Fig. 1-21, holes on the *p*-side of the junction are attracted away from the junction and electrons are attracted away from the junction on the *n*-side. This causes the depletion region to be widened and the barrier voltage to be increased, as illustrated. With the barrier voltage increase, there is no possibility of a majority charge carrier current flow across the junction, and the junction is said to be *reverse-biased*. Because there is only a very small reverse current, a reverse-biased *pn*-junction can be said to have a high resistance.

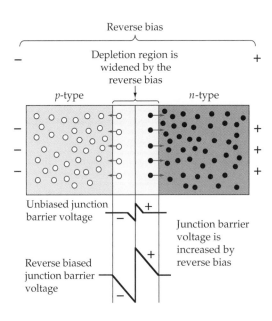

Figure 1-21 A reverse bias applied to a *pn*-junction (positive on the *n*-side, negative on the *p*-side) causes the depletion region to widen and increases the barrier voltage. Only a very small reverse current flows across the junction.

Although there is no possibility that a majority charge carrier current can flow across a reverse-biased junction, minority carriers generated on each side can still cross the junction. Electrons in the *p*-side are attracted across the junction to the positive voltage on the *n*-side. Holes on the *n*-side may flow across to the negative voltage on the *p*-side. This is shown by the junction reverse characteristic, or the graph of reverse current (I_R) versus reverse voltage (V_R) (Fig. 1-22). Since only a very small reverse-bias voltage is necessary to direct all available minority carriers across the junction, further increases in bias voltage do not increase the current level. This current is referred to as a *reverse saturation current*.

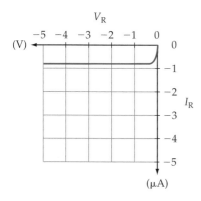

Figure 1-22 Current-versus-voltage characteristic for a reverse-biased *pn*-junction.

The reverse saturation current is normally a very small quantity, ranging from nanoamps to microamps, depending on the junction area, temperature, and semiconductor material.

Forward-Biased Junction

Consider the effect of an external bias voltage applied with the polarity shown in Fig. 1-23: positive on the *p*-side and negative on the *n*-side. The holes on the *p*-side, being positively charged particles, are repelled from the positive terminal and driven toward the junction. Similarly, the electrons on the *n*-side are repelled from the negative terminal toward the junction. The result is that the width of the depletion region and the barrier potential are both reduced.

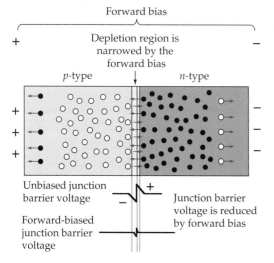

Figure 1-23 Forward biasing a *pn*-junction (positive on the *p*-side, negative on the *n*-side) narrows the depletion region, reduces the barrier voltage, and causes a relatively large current to flow across the junction.

When the applied bias voltage is progressively increased from zero, the barrier voltage gets smaller until it effectively disappears and charge carriers easily flow across the junction. Electrons from the *n*-side are now attracted across to the positive bias terminal on the *p*-side, and holes from the *p*-side flow across to the negative terminal on the *n*-side (thinking of holes as positively charged particles). A majority carrier current flows, and the junction is said to be *forward-biased*.

The graph in Fig. 1-24 shows the forward current (I_F) plotted against forward voltage (V_F) for typical germanium and silicon *pn*-junctions. In each case, the graph is known as the *forward characteristic* of the junction. It is seen that there is very little forward current until V_F exceeds the junction barrier voltage (0.3 V for germanium, 0.7 V for silicon). When V_F is increased from zero toward the knee of the characteristic, the barrier voltage is progressively overcome, allowing more majority charge carriers to flow across the junction. Above the knee of the characteristic, I_F increases almost linearly with increase in V_F. The level of current that can be made to flow across a forward-biased *pn*-junction largely depends on the area of the junction.

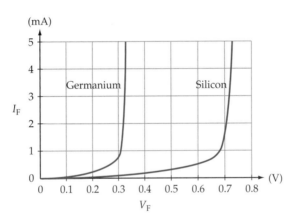

Figure 1-24 *pn*-junction forward characteristics. Germanium junctions are forward-biased at approximately 0.3 V. A silicon junction requires approximately 0.7 V for forward bias.

Junction Temperature Effects

As already discussed, the reverse saturation current (I_R) at a *pn*-junction consists of minority charge carriers. When the temperature of semiconductor material is raised, more and more electrons break away from their atoms. This generates additional minority charge carriers, causing I_R to increase as the junction temperature rises. It can be shown that the reverse current approximately doubles with each 10°C increase in temperature, and that for a given junction, there is a definite I_R level for each temperature level (Fig. 1-25a).

The junction forward voltage drop (V_F) is also affected by temperature. The horizontal line (at 3 mA) on Fig. 1-25b shows that, if I_F is held constant while the junction temperature is changing, the forward voltage decreases with rising junction temperature. This means that V_F has a negative temperature coefficient. It is found that the temperature coefficient for the forward voltage of a *pn*-junction is approximately −1.8 mV/°C for a silicon junction and −2.02 mV/°C for germanium. A figure of −2 mV/°C is normally used as an approximation.

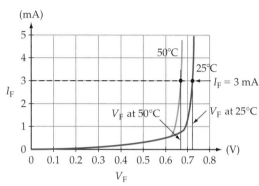
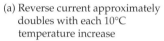
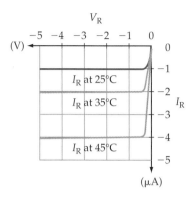

(a) Reverse current approximately
doubles with each 10°C
temperature increase

(b) Forward voltage-drop changes by
approximately −2 mV/°C

Figure 1-25 Junction reverse current and forward voltage drop are affected by temperature change.

Summary

- A *pn*-junction is reverse-biased when a voltage is applied positive to the *n*-side and negative to the *p*-side.
- A reverse-biased junction has a wide depletion region.
- A very small *minority charge carrier* current (I_R) flows across a reverse-biased junction.
- The junction reverse characteristic is the graph of I_R versus V_R.
- The *reverse saturation current* (I_R) tends to be constant regardless of the reverse bias voltage (V_R).
- The reverse current approximately doubles with every 10°C rise in junction temperature.
- A *pn*-junction is *forward-biased* when a voltage is applied positive to the *p*-side and negative to the *n*-side.
- A forward-biased junction has a narrow depletion region.
- A *majority charge carrier* current (I_F) flows across a forward-biased junction.
- The junction forward characteristic is the graph of I_F versus V_F.
- The *forward current* (I_F) depends on the level of forward bias voltage (V_F).
- Typical junction forward voltage drops are 0.3 V for germanium and 0.7 V for silicon.
- The junction forward voltage drop changes by approximately −2 mV/°C as the junction temperature increases.

Section 1-6 Review

1-6.1 Sketch typical voltage/current characteristics for a forward-biased silicon *pn*-junction. Briefly explain.

1-6.2 Sketch typical voltage/current characteristics for a reverse-biased *pn*-junction. Briefly explain.

1-6.3 Discuss the effects of temperature change on forward- and reverse-biased *pn*-junctions.

Review Questions

Section 1-1

1-1 Describe the atom, and draw a two-dimensional diagram to illustrate your description. Compare the atom to a planet with orbiting satellites.

1-2 Define nucleus, electron, electronic charge, proton, neutron, shell, positive ion, and negative ion.

1-3 Sketch two-dimensional diagrams of silicon and germanium atoms. Describe the valence shell of each atom.

1-4 Explain atomic number and atomic weight. State the atomic number and atomic weight for silicon.

1-5 Explain what is meant by energy levels and energy bands. Sketch an energy band diagram, and define conduction band, valence band, and forbidden gap.

Section 1-2

1-6 Draw a sketch to show the process of current flow by electron motion. Briefly explain.

1-7 Draw sketches to show the process of current flow by hole transfer. Which have greater mobility, electrons or holes? Explain why.

1-8 Define conventional current direction and direction of electron flow. State why each is important.

Section 1-3

1-9 Name the three kinds of bonds that hold atoms together in a solid. What kind of bonding might be found in (a) conductors, (b) insulators, and (c) semiconductors?

1-10 Draw sketches to illustrate metallic bonding and ionic bonding. Explain what happens in each case.

1-11 Draw a sketch to illustrate covalent bonding. Explain the bonding process.

1-12 Draw energy band diagrams for conductors, insulators, and semiconductors. Explain the reasons for the differences between the diagrams.

Section 1-4

1-13 Define acceptor doping, and draw a sketch to illustrate the process. Explain.

1-14 Define donor doping, and draw a sketch to illustrate the process. Explain.

1-15 State the names given to acceptor-doped material and donor-doped material. Explain.

1-16 What is meant by majority charge carriers and minority charge carriers? Which are majority carriers and why in (a) donor-doped material, and (b) acceptor-doped material?

1-17 Explain what happens to resistance with increasing temperature in the case of (a) a conductor, (b) a semiconductor, and (c) a heavily doped semiconductor. What do you think would happen to the resistance of an insulator with increasing temperature? Why?

1-18 Explain hole-electron pair generation and recombination.

1-19 Discuss the effects that light can produce on semiconductor materials.

1-20 Define *n*-type material, *p*-type material, majority charge carriers, minority charge carriers, positive temperature coefficient, negative temperature coefficient, and dark resistance.

Section 1-5

1-21 Using illustrations, explain how the depletion region and barrier voltage are produced at a *pn*-junction. List the characteristics of the depletion region.

1-22 Draw a sketch to show the depletion region and barrier voltage at a *pn*-junction with unequal doping of each side. Briefly explain.

Section 1-6

1-23 A bias is applied to a *pn*-junction, positive to the *p*-side, negative to the *n*-side. Show, by a series of sketches, the effect of this bias on depletion region width, barrier voltage, minority carriers, and majority carriers. Briefly explain the effect in each case.

1-24 Repeat Question 1-23 for a bias applied negative to the *p*-side and positive to the *n*-side.

1-25 Sketch the voltage/current characteristics for a *pn*-junction (a) with forward bias and (b) with reverse bias. Show how temperature change affects the characteristics.

1-26 State typical values of barrier voltage for silicon and germanium junctions. Discuss the resistances of forward-biased and reverse-biased *pn*-junctions.

1-27 State typical reverse saturation current levels for *pn*-junctions. Explain the origin of reverse saturation current.

CHAPTER 2
Semiconductor Diodes

CONTENTS

Objectives

You will be able to:

1   Sketch a diode circuit symbol, identify the terminals, and discuss diode circuit behaviour.
2   Explain diode forward and reverse characteristics.
3   List important diode parameters and determine parameter values from the characteristics.
4   Sketch approximate diode characteristics and dc equivalent circuits, and apply them to circuit analysis.
5   Draw dc load lines on diode characteristics to precisely analyze diode circuits.

6   Determine maximum diode power dissipations and forward voltages at various temperatures.
7   Sketch diode ac equivalent circuits and calculate switching times.
8   Determine diode parameter values from device data sheets.
9   Test diodes, and plot the forward and reverse characteristics.
10 Sketch a Zener diode circuit symbol, identify the terminals, and explain the characteristics.
11 Determine Zener diode parameter values from device characteristics and from data sheets.
12 Analyze basic Zener diode circuits.

INTRODUCTION

The term *diode* refers to a two-electrode, or two-terminal, device. A semiconductor diode is simply a *pn*-junction with a connecting lead on each side. *A diode is a one-way device, offering a low resistance when forward-biased, and behaving almost as an open switch when reverse-biased.* An approximately constant voltage drop occurs across a forward-biased diode, and this simplifies diode circuit analysis. Diode forward and reverse characteristics are graphs of corresponding current and voltage levels. For precise circuit analysis, dc load lines are drawn on the diode forward characteristic.

Some diodes are low-current devices for use in switching circuits. High-current diodes are most often used as rectifiers for ac to dc conversion. Zener diodes are operated in reverse breakdown because they have a very stable breakdown voltage.

2-1 *pn*-JUNCTION DIODE

As explained in Sections 1-5 and 1-6, a *pn*-junction permits substantial current flow when forward-biased, and blocks current when reverse-biased. Thus, it can be used as a switch: *on* when forward-biased and *off* when biased in reverse. A *pn*-junction provided with copper wire connecting leads becomes an electronic device known as a *diode* (see Fig. 2-1).

The circuit symbol (or *graphic symbol*) for a diode is an arrowhead and bar (Fig. 2-2). The arrowhead indicates the conventional direction of current flow when the diode is forward-biased (from the positive terminal through the device to the negative terminal). The *p*-side of the diode is always the positive terminal for forward bias and is termed the *anode*. The *n*-side, called the *cathode*, is the negative terminal when the device is forward-biased.

A *pn*-junction diode can be destroyed if a high level of forward current overheats the device. It can also be destroyed if a large reverse voltage causes the

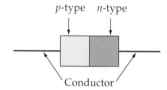

Figure 2-1 A semiconductor diode is a *pn*-junction with conductors on each side of the junction for connecting the device to a circuit.

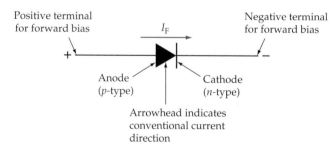

Figure 2-2 Diode circuit symbol. Current flows in the arrowhead direction when the diode is forward-biased: positive (+) on the anode and negative (−) on the cathode.

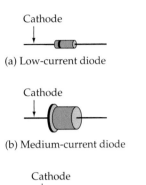

(a) Low-current diode

(b) Medium-current diode

(c) High-current diode

Figure 2-3 The size and appearance of a diode depend upon the level of forward current that the device is designed to pass.

junction to break down. The maximum forward current and reverse voltage for diodes are specified on the manufacturer's data sheets (see Section 2-7). In general, physically large diodes pass the largest currents and survive the largest reverse voltages. Small diodes are limited to low current levels and low reverse voltages.

Figure 2-3 shows the appearance of low-, medium-, and high-current diodes. Since the body of the low-current device in Fig. 2-3a may be only 0.3 cm long, the cathode is usually identified by a coloured band. This type of diode is usually capable of passing a maximum forward current of approximately 100 mA. It can also survive about 75 V reverse bias without breaking down, and its reverse current is usually less than 1 μA at 25°C.

The medium-current diode shown in Fig. 2-3b can usually pass a forward current of about 400 mA and survive over 200 V reverse bias. The anode and cathode terminals may be indicated by a diode symbol on the side of the device.

Low-current and medium-current diodes are usually mounted by soldering the connecting leads to terminals. Power dissipated in the device is then carried away by air convection and by heat conduction along the connecting leads.

High-current diodes, or *power diodes* (see Fig. 2-3c), generate a lot of heat. So air convection would be completely inadequate. Such devices are designed to be connected mechanically to a metal heat sink. Power diodes can pass forward currents of many amperes and can survive several hundred volts of reverse bias.

2-2 CHARACTERISTICS AND PARAMETERS

Forward and Reverse Characteristics

The semiconductor diode is essentially a *pn*-junction, and its characteristics are those discussed in Chapter 1. Figures 2-4 and 2-5 show typical forward and reverse characteristics for low-current silicon and germanium diodes. From the silicon diode characteristics in Fig. 2-4, it is seen that the forward current (I_F) remains very low (less than 100 μA microamps) until the diode forward-bias voltage (V_F) exceeds approximately 0.7 V. At V_F levels greater than 0.7 V, I_F increases almost linearly.

Because the diode reverse current (I_R) is very much smaller than its forward current, the reverse characteristics are plotted with expanded current scales. For a silicon diode, I_R is normally less than 100 nA, and it is almost completely independent of the reverse-bias voltage. As already explained in Chapter 1, I_R is largely a minority charge carrier *reverse saturation current*. A small increase in I_R can occur with increasing reverse-bias voltage, as a result of minority charge carriers leaking along the junction surface. For a diode with the characteristics

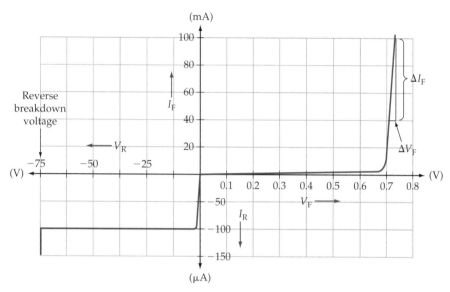

Figure 2-4 Typical forward and reverse characteristics for a silicon diode. There is a substantial forward current (I_F) when the forward voltage (V_F) exceeds approximately 0.7 V.

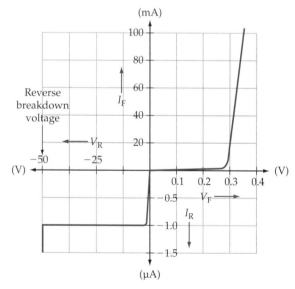

Figure 2-5 Typical forward and reverse characteristics for a germanium diode. Substantial forward current (I_F) flows when the forward voltage (V_F) exceeds approximately 0.3 V.

in Fig. 2-4, the reverse current is usually less than 1/10 000 of the lowest normal forward current level. Therefore, I_R is quite negligible when compared to I_F, and a reverse-biased diode may be treated almost as an open switch. This is investigated further in Example 2-1.

Example 2-1

Calculate the forward and reverse resistances offered by a silicon diode with the characteristics in Fig. 2.4 at $I_F = 100$ mA, and at $V_R = 50$ V.

Solution

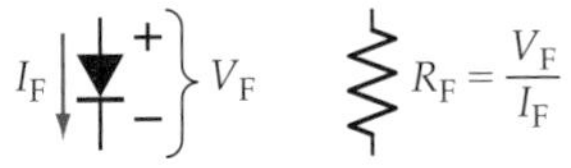

(a) Forward resistance

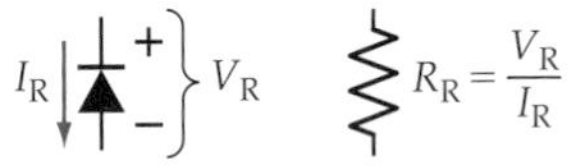

(b) Reverse resistance

Figure 2-6 Determination of diode forward and reverse resistance.

At $I_F = 100$ mA, $V_F \approx 0.75$ V

$$R_F = \frac{V_F}{I_F} = \frac{0.75 \text{ V}}{100 \text{ mA}} \quad \text{(see Fig. 2-6a)}$$

$$= 7.5 \ \Omega$$

At $V_R = 50$ V, $I_R \approx 100$ nA

$$R_R = \frac{V_R}{I_R} = \frac{50 \text{ V}}{100 \text{ nA}} \quad \text{(see Fig. 2-6b)}$$

$$= 500 \text{ M}\Omega$$

When the diode reverse voltage (V_R) is sufficiently increased, the device goes into *reverse breakdown*. For the characteristics shown in Fig. 2-4, this occurs where $V_R = 75$ V. Reverse breakdown can destroy a diode unless the current is limited by a suitable series-connected resistor. Reverse breakdown is usefully applied in Zener diodes, which are introduced in Section 2-9.

The characteristics of a germanium diode are similar to those of a silicon diode but with some important differences (see Fig. 2-5). The forward voltage drop of a germanium diode is typically 0.3 V, compared to 0.7 V for silicon. For a germanium device, the reverse saturation current at 25°C may be about 1 μA, which is much larger than the reverse current for a silicon diode. Finally, the reverse breakdown voltage for germanium devices is likely to be substantially lower than that for silicon devices.

The lower forward voltage drop for germanium diodes can be a distinct advantage. However, the lower reverse current and higher reverse breakdown voltage of silicon diodes make them preferable to germanium devices for most applications.

Diode Parameters

The diode parameters of greatest interest are

V_F	forward voltage drop
I_R	reverse saturation current
V_{BR}	reverse breakdown voltage
r_d	dynamic resistance
$I_{F(max)}$	maximum forward current

The values of these parameters are normally listed on the diode data sheet provided by device manufacturers (see Section 2-7). Some of the parameters can also be determined directly from the diode characteristics. For the silicon diode characteristics in Fig. 2-4, $V_F \approx 0.7$ V, $I_R = 100$ nA, and $V_{BR} = 75$ V.

The forward resistance calculated in Example 2-1 is a *static quantity*. It is the constant resistance (or dc resistance) of the diode at a particular constant forward current. The *dynamic resistance* of the diode is the resistance offered to changing levels of forward voltage. The dynamic resistance, also known as the *incremental resistance* or *ac resistance*, is the reciprocal of the slope of the forward characteristics beyond the knee. Referring to Fig. 2-4 and Fig. 2-7,

$$r_d = \frac{\Delta V_F}{\Delta I_F} \tag{2-1}$$

The dynamic resistance can also be calculated from the rule-of-thumb equation

$$r_d' = \frac{26 \text{ mV}}{I_F} \tag{2-2}$$

where I_F is the dc forward current. Thus, for example, the dynamic resistance for a diode passing a 1 mA forward current is $r_d' = 26 \text{ mV}/1 \text{ mA} = 26 \ \Omega$.

Equation 2-2 shows that the diode dynamic resistance changes with the level of dc forward current. Since this is not shown in Figs. 2-4 and 2-5, the characteristics are approximations of the actual device characteristics. It should also be noted that Equation 2-2 gives the ac resistance only for the junction. It does not include the dc resistance of the semiconductor material, which might be as large as 2 Ω depending on the design of the device. The resistance derived from the slope of the device characteristic does include the semiconductor dc resistance. So r_d (from the characteristic) should be slightly larger than r_d' calculated from Equation 2-2.

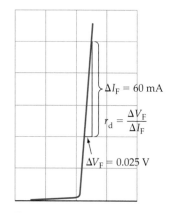

Figure 2-7 Determination of the dynamic resistance (r_d) of a diode from the forward characteristic.

Example 2-2

Determine the dynamic resistance at a forward current of 70 mA for the diode characteristics given in Fig. 2-4. Using Equation 2-2, estimate the diode dynamic resistance.

Solution

In Fig. 2-4 at $I_F = 70$ mA,

$$\Delta I_F = 60 \text{ mA} \quad \text{and} \quad \Delta V_F \approx 0.025 \text{ V}$$

Eq. 2-1:
$$r_d = \frac{\Delta V_F}{\Delta I_F} = \frac{0.025 \text{ V}}{60 \text{ mA}} \quad \text{(see Fig. 2-7)}$$

$$= 0.42 \ \Omega$$

Eq. 2-2:
$$r_d' = \frac{26 \text{ mV}}{I_F} = \frac{26 \text{ mV}}{70 \text{ mA}}$$

$$= 0.37 \ \Omega$$

Practice Problems

2-2.1 Calculate the resistances offered by a diode with the characteristics in Fig. 2.5 at 30 V reverse bias and at 60 mA of forward current.

2-2.2 Determine the dynamic resistance at a 50 mA forward current for a diode with the characteristics in Fig. 2-5. Use Eq. 2-2 to estimate the diode dynamic resistance.

2-3 DIODE APPROXIMATIONS

Ideal Diodes and Practical Diodes

As already explained, a diode is essentially a one-way device, offering a low resistance when forward-biased and a high resistance when biased in reverse. An *ideal diode* (or perfect diode) would have zero forward resistance and zero forward voltage drop. It would also have an infinitely high reverse resistance, which would result in zero reverse current. Figure 2-8a shows the current/voltage characteristics of an ideal diode.

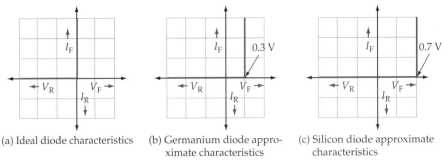

(a) Ideal diode characteristics   (b) Germanium diode approximate characteristics   (c) Silicon diode approximate characteristics

Figure 2-8   An ideal diode has $V_F = 0$ and $I_R = 0$. Practical diodes can be treated as near-ideal devices if the forward voltage drop is taken into account.

Although an ideal diode does not exist, there are many applications where diodes can be assumed to be near-ideal devices. In circuits with supply voltages much larger than the diode forward voltage drop, V_F can be assumed to be constant without introducing any serious error. Also, the diode reverse current is normally so much smaller than the forward current that the reverse current can be ignored. These assumptions lead to the near-ideal, or approximate, characteristics for silicon and germanium diodes shown in Fig. 2-8b and c. Example 2-3 investigates a situation where the diode V_F is assumed to be constant.

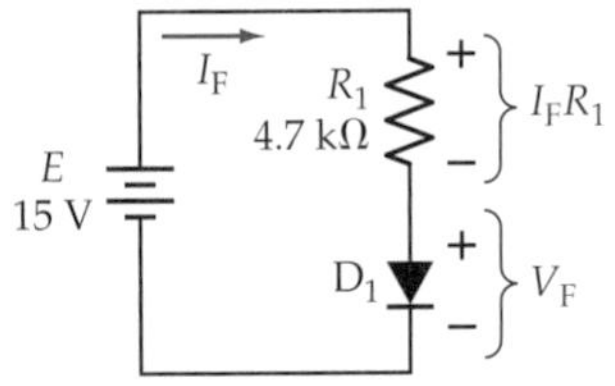

Example 2-3

A silicon diode is used in the circuit shown in Fig. 2-9. Calculate the diode current.

Figure 2-9   Circuit for Example 2-3.

Solution

$$E = I_F R_1 + V_F$$

or

$$I_F = \frac{E - V_F}{R_1} = \frac{15\text{ V} - 0.7\text{ V}}{4.7\text{ k}\Omega}$$

$$= 3.04\text{ mA}$$

Piecewise Linear Characteristic

When the forward characteristic of a diode is not available, a straight-line approximation called the *piecewise linear characteristic* may be employed. To construct the piecewise linear characteristic, V_F is first marked on the horizontal axis, as shown in Fig. 2-10. Then, from V_F, a straight line is drawn with a slope equal to the diode dynamic resistance. Example 2-4 demonstrates the process.

Example 2-4

Construct the piecewise linear characteristic for a silicon diode that has a $0.25\ \Omega$ dynamic resistance and a 200 mA maximum forward current.

Solution

Plot point A on the horizontal axis at

$$V_F = 0.7\text{ V} \qquad (\text{see Fig. 2-10})$$

$$\Delta V_F = \Delta I_F \times r_d = 200\text{ mA} \times 0.25\ \Omega$$

$$= 0.05\text{ V}$$

Plot point B (on Fig. 2-10) at

$$I_F = 200\text{ mA} \quad \text{and} \quad V_F = (0.7\text{ V} + 0.05\text{ V})$$

Draw the characteristic through points A and B.

DC Equivalent Circuits

An *equivalent circuit* for a device is a circuit that represents the device behaviour. Usually, the equivalent circuit is made up of a number of components, such as resistors and voltage cells. A diode equivalent circuit may be substituted for the device when investigating a circuit containing the diode. Equivalent circuits may also be used as device *models* for computer analysis.

In Example 2-3 a forward-biased diode is assumed to have a constant forward voltage drop (V_F) and negligible series resistance. In this case the diode equivalent circuit is assumed to be a voltage cell with a voltage V_F (see Fig. 2-11a). This simple dc equivalent circuit is quite suitable for a great many diode applications.

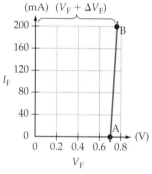

Figure 2-10   Diode piecewise linear characteristic, or straight-line approximation of the diode forward characteristic.

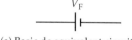

(a) Basic dc equivalent circuit

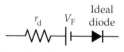

(b) Complete dc equivalent circuit

Figure 2-11   DC equivalent circuits for a junction diode.

A more accurate equivalent circuit includes the diode dynamic resistance (r_d) in series with the voltage cell, as shown in Fig. 2-11b. This takes account of the small variations in V_F that occur with change in forward current. An ideal diode is also included to show that current flows only in one direction. The equivalent circuit without r_d assumes that the diode has the approximate characteristics illustrated in Fig. 2-8b or c. With r_d included, the equivalent circuit represents a diode with the type of piecewise linear characteristic shown in Fig. 2-10. Consequently, the circuit in Fig. 2-11b is termed the *piecewise linear equivalent circuit*.

Example 2-5

Calculate I_F for the diode circuit in Fig. 2-12a assuming that the diode has $V_F = 0.7$ V and $r_d = 0$. Then recalculate the current taking $r_d = 0.25\ \Omega$.

Solution

Substituting V_F as the diode equivalent circuit (Fig. 2-12b),

$$I_F = \frac{E - V_F}{R_1} = \frac{1.5\ \text{V} - 0.7\ \text{V}}{10\ \Omega}$$

$$= 80\ \text{mA}$$

Substituting V_F and r_d as the diode equivalent circuit (Fig. 2-12c),

$$I_F = \frac{E - V_F}{R_1 + r_d} = \frac{1.5\ \text{V} - 0.7\ \text{V}}{10\ \Omega + 0.25\ \Omega}$$

$$= 78\ \text{mA}$$

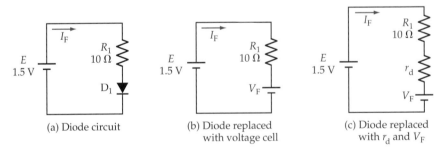

| (a) Diode circuit | (b) Diode replaced with voltage cell | (c) Diode replaced with r_d and V_F |

Figure 2-12 Diode circuits for Example 2-5.

Practice Problems

2-3.1 Calculate the new level of diode current in the circuit in Fig. 2-9 when D_1 is replaced with two series-connected silicon diodes.

2-3.2 A germanium diode has a maximum forward current of 100 mA and a $0.5\ \Omega$ dynamic resistance. Construct the piecewise linear characteristics for this diode on Fig. 2-10.

2-3.3 Calculate the circuit current when the diode in Problem 2-3.2 is forward-biased in series with a 15 Ω resistor and a 3 V battery.

2-4 DC LOAD LINE ANALYSIS

DC Load Line

Figure 2-13(a) shows a diode in series with a 100 Ω resistor (R_1) and a supply voltage (E). The polarity of E is such that the diode is forward-biased, so that there is a diode forward current (I_F). As already discussed, the circuit current can be determined approximately by assuming a constant diode forward voltage drop (V_F). When the precise levels of the diode current and voltage must be calculated, *graphical analysis* (also termed *dc load line analysis*) is employed.

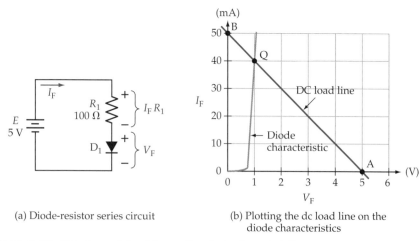

(a) Diode-resistor series circuit

(b) Plotting the dc load line on the diode characteristics

Figure 2-13   Drawing a dc load line on the diode characteristic.

For graphical analysis, a dc load line is drawn on the diode forward characteristics (Fig. 2-13b). This is a straight line that illustrates all dc conditions that could exist within the circuit. Because the load line is always straight, it can be constructed by plotting any two corresponding current and voltage points and then drawing a straight line through them. To determine two points on the load line, an equation relating voltage, current, and resistance is first derived for the circuit. From Fig. 2-13a,

$$E = (I_F R_1) + V_F \qquad (2\text{-}3)$$

Any two convenient levels of I_F can be substituted into Equation 2-3 to calculate corresponding V_F levels, or vice versa. As demonstrated in Example 2-6, it is convenient to calculate V_F when $I_F = 0$, and to determine I_F when $V_F = 0$.

Example 2-6

Draw the dc load line for the circuit in Fig. 2-13a on the diode forward characteristic given in Fig. 2-13b.

Solution

Substitute $I_F = 0$ into Eq. 2-3,

$$E = (I_F R_1) + V_F = 0 + V_F$$

or　　　　　　　$V_F = E = 5\text{ V}$

Plot point A on the diode characteristic at

$$I_F = 0 \text{ and } V_F = 5\text{ V}$$

Now substitute $V_F = 0$ into Eq. 2-3,

$$E = (I_F R_1) + 0$$

giving　　　　　　　$I_F = \dfrac{E}{R_1} = \dfrac{5\text{ V}}{100\text{ }\Omega}$

$$= 50\text{ mA}$$

Plot point B on the diode characteristic at

$$I_F = 50\text{ mA and } V_F = 0$$

Draw the dc load line through points A and B.

Q-Point

The relationship between the diode forward voltage and current in the circuit in Fig. 2-13a is defined by the device characteristic. Consequently, there is only one point on the dc load line where the diode voltage and current are compatible with the circuit conditions. That is *point Q*, termed the *quiescent point* or *dc bias point*, where the load line intersects the characteristic. This may be checked by substituting the levels of I_F and V_F at point Q into Equation 2-3. From the Q point on Fig. 2-13b, $I_F = 40$ mA and $V_F = 1$ V. Equation 2-3 states that $E = (I_F R_1) + V_F$. Therefore,

$$E = (40\text{ mA} \times 100\text{ }\Omega) + 1\text{ V}$$

$$= 5\text{ V}$$

So, with $E = 5$ V and $R_1 = 100$ Ω, the only levels of I_F and V_F that can satisfy Equation 2-3 on the diode characteristics in Fig. 2-13b are 40 mA and 1 V.

Note that, although 0 and 5 V were used for V_F when the dc load line was drawn in Example 2-6, no functioning semiconductor diode would have a 5 V forward voltage drop. This is simply a convenient theoretical level for plotting the dc load line.

Calculating Load Resistance and Supply Voltage

In a diode series circuit (see Fig. 2-14a), resistor R_1 dictates the slope of the dc load line, and supply voltage E determines point A on the load line. So the circuit conditions can be altered by changing either R_1 or E.

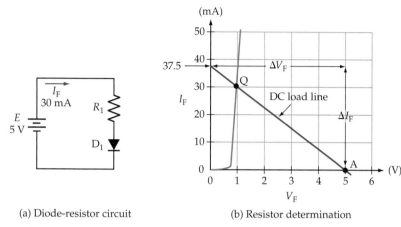

(a) Diode-resistor circuit   (b) Resistor determination

Figure 2-14 Determination of the required circuit series resistance R_1 from the slope of the dc load line.

When designing a diode circuit, one may be required to use a given supply voltage and set up a specified forward current. In this case, points A and Q are first plotted and the load line is drawn. Resistor R_1 is then calculated from the slope of the load line. The problem could also occur in another way. For example, R_1 and the required I_F are known, and the supply voltage is to be determined. This problem is solved by plotting point Q and drawing the load line with slope $1/R_1$. The supply voltage is then read at point A.

Example 2-7

Using the device characteristics in Fig. 2-14b, determine the required load resistance for the circuit in Fig. 2-14a to give $I_F = 30$ mA.

Solution

From Eq. 2-3, $V_F = E - (I_F R_1)$

Substituting $I_F = 0$, $V_F = E - 0 = 5$ V

Plot point A on the diode characteristic in Fig. 2-14b at

$$I_F = 0 \text{ and } V_F = 5 \text{ V}$$

Now plot point Q in Fig. 2-14b at

$$I_F = 30 \text{ mA}$$

Draw the dc load line through points A and Q.
From the load line,

$$R_1 = \frac{\Delta V_F}{\Delta I_F} = \frac{5 \text{ V}}{37.5 \text{ mA}}$$

$$= 133 \ \Omega$$

Example 2-8

Determine a new supply voltage for the circuit in Fig. 2-14a to give a 50 mA diode forward current when $R_1 = 100\ \Omega$.

Solution

Plot point Q on the diode characteristic in Fig. 2-15 at

$$I_F = 50\ \text{mA}$$

From the characteristic, read

$$V_F = 1.1\ \text{V}$$

From Eq. 2-3, $\qquad V_F = E - (I_F\,R_1)$

When I_F changes from 50 mA to 0,

$$\Delta I_F = 50\ \text{mA} \qquad (\text{see Fig. 2-15})$$

and $\qquad\qquad \Delta V_F = I_F\,R_1 = 50\ \text{mA} \times 100\ \Omega \qquad (\text{see Fig. 2-15})$

$$= 5\ \text{V}$$

The new supply voltage is

$$E = V_F + \Delta V_F = 1.1\ \text{V} + 5\ \text{V}$$

$$= 6.1\ \text{V}$$

Point A may now be plotted (on Fig. 2-15) at $I_F = 0$ and $E = 6.1$ V, and the new dc load line may be drawn through points A and Q.

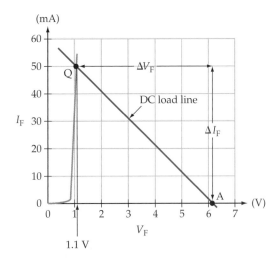

Figure 2-15 Determination of the required supply voltage for a diode-resistor circuit with a given resistor and a specified load current.

Practice Problems

2-4.1 A diode with the characteristics in Fig. 2-15 is connected in series with a 3 V supply and a 100 Ω resistor. Draw the dc load line for the circuit, and determine the forward current level.

2-4.2 Calculate a new load resistance for the circuit in Problem 2-4.1 to produce a 40 mA diode forward current.

2-4.3 A diode with the characteristics in Fig. 2-15 is to have a 35 mA forward current when connected in series with a 70 Ω resistor. Determine the required supply voltage.

2-5 TEMPERATURE EFFECTS

Diode Power Dissipation

The power dissipation in a diode is simply calculated as the device terminal voltage multiplied by the current level:

$$P = V_F I_F \qquad\qquad (2\text{-}4)$$

Device manufacturers specify a maximum power dissipation for each type of diode. If the specified level is exceeded, the device will overheat and may short-circuit or open-circuit. The maximum power that may be dissipated in a diode (or any other electronic device) is normally specified for an ambient temperature of 25°C or, sometimes, for a 25°C case temperature. When the temperature exceeds this level, the device maximum power dissipation must be derated.

Figure 2-16 shows the type of power-versus-temperature graph provided on device data sheets. The maximum power dissipation for any temperature is

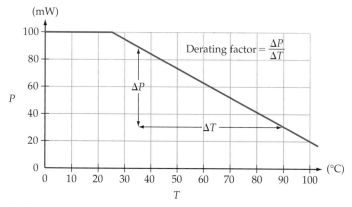

Figure 2-16 A power-versus-temperature graph shows how the maximum power dissipation of a device must be derated with increasing temperature.

simply read from the graph; then the maximum forward current level is calculated from Equation 2-4. Instead of a power-versus-temperature graph, the rectifier diode data sheet A-2 in Appendix A has a current-versus-temperature graph, which directly gives the maximum forward current at any temperature in the range of operation.

As an alternative to power or current graphs, a *derating factor* is often listed on device data sheets. The derating factor defines the slope of the power-versus-temperature graph (see Fig. 2-16); it can be employed to draw the graph or used directly without reference to the graph. The equation for the maximum power dissipation when the temperature changes involves the specified power at the specified temperature (P_1 at T_1), the derating factor (D), and the temperature change (ΔT).

$$P_2 = (P_1 \text{ at } T_1) - (D \times \Delta T) \tag{2-5}$$

Example 2-9

A diode with 700 mW maximum power dissipation at 25°C has a 5 mW/°C derating factor. If the forward voltage drop remains constant at 0.7 V, calculate the maximum forward current at 25°C and at 65°C.

Solution

At 25°C:

From Eq. 2-4, $I_F = \dfrac{P}{V_F} = \dfrac{700 \text{ mW}}{0.7 \text{ V}}$

$= 1 \text{ A}$

At 65°C:

Eq. 2-5: $P_2 = (P_1 \text{ at } T_1) - (D \times \Delta T)$

$= 700 \text{ mW} - [5 \text{ mW/°C} \times (65°C - 25°C)]$

$= 500 \text{ mW}$

From Eq. 2-4, $I_F = \dfrac{P_2}{V_F} = \dfrac{500 \text{ mW}}{0.7 \text{ V}} I_F$

$= 714 \text{ mA}$

Forward Voltage Drop

Sometimes it is important to know the precise level of a diode forward voltage drop. In these cases, the graphical analysis techniques discussed in Section 2-4 can be used. However, as explained in Section 1-6 and illustrated in Fig. 2-17,

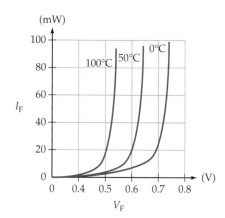

Figure 2-17   The forward voltage drop across a diode decreases by approximately 2 mV/°C as the device temperature increases.

the voltage drop across a forward-biased *pn*-junction changes with temperature by approximately -1.8 mV/°C for a silicon device and by -2.02 mV/°C for germanium. A diode forward voltage drop at any temperature can be calculated from a knowledge of V_F at the starting temperature (V_{F1} at T_1), the temperature change (ΔT), and the voltage/temperature coefficient (ΔV_F/°C).

$$V_{F2} = (V_{F1} \text{ at } T_1) + [\Delta T \,(\Delta V_F/°C)] \qquad (2\text{-}6)$$

Dynamic Resistance

Equation 2-2 for calculating the dynamic resistance of a forward-biased diode is correct only for a junction temperature of 25°C. For higher or lower temperatures, the equation must be modified to

$$r_d' = \frac{26 \text{ mV}}{I_F}\left(\frac{T + 273°C}{298°C}\right) \qquad (2\text{-}7)$$

where T is the junction temperature in degrees Celsius.

Example 2-10

A silicon diode with a 0.7 V forward voltage drop at 25°C is to be operated with a constant forward current up to a temperature of 100°C. Calculate the diode V_F at 100°C. Also, determine the junction dynamic resistance at 25°C and at 100°C if the forward current is 26 mA.

Solution

Eq. 2-6:
$$V_{F2} = (V_{F1} \text{ at } T_1) + [\Delta T \,(\Delta V_F/°C)]$$

$$= 0.7 \text{ V} + [(100°C - 25°C)(-1.8 \text{ mV}/°C)]$$

$$= 0.565 \text{ V (at } 100°C)$$

At 25°C:

Eq. 2-7:
$$r'_d = \frac{26 \text{ mV}}{I_F}\left[\frac{T + 273°C}{298°C}\right] = \frac{26 \text{ mV}}{26 \text{ mA}}\left[\frac{25°C + 273°C}{298°C}\right]$$

$$= 1 \, \Omega$$

At 100°C:

$$r'_d = \frac{26 \text{ mV}}{26 \text{ mA}}\left[\frac{100°C + 273°C}{298°C}\right]$$

$$= 1.25 \, \Omega$$

Practice Problems

2-5.1 A diode with a constant 0.65 V forward drop has the power-versus-temperature graph in Fig. 2-16. Calculate the maximum forward current that may be passed at 25°C and at 80°C.

2-5.2 Calculate the maximum and minimum levels of V_F for a germanium diode with $V_F = 0.3$ V at 25°C when operated over a temperature range of 10°C to 80°C. Determine the device dynamic resistances at the temperature extremes if I_F is 20 mA.

2-5.3 A diode with a 1 W maximum power dissipation at 25°C has a 4 mW/°C derating factor. Calculate the maximum power that may be dissipated in the diode when its temperature is 80°C.

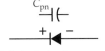

(a) A depletion layer capacitance occurs at a reverse-biased diode

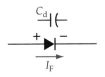

(b) A diffusion capacitance is present at a forward-biased diode

Figure 2-18 The capacitance of a diode depends upon the polarity of the applied voltage and on the current level. The diffusion capacitance is much larger than the depletion layer capacitance.

2-6 DIODE AC MODELS

Junction Capacitances

The depletion region of a *pn*-junction (see Section 1-5) is a layer depleted of charge carriers situated between two blocks of low-resistance material. Since this is also the description of a capacitor, the junction depletion region clearly has a capacitance. The *depletion layer capacitance* (C_{pn}) may be calculated from the equation for a parallel-plate capacitor if the junction dimensions are known. Typically, C_{pn} is 4 pF for a low-current diode.

The depletion layer capacitance is essentially the capacitance of a reverse-biased *pn*-junction (Fig. 2-18a). Consider the forward-biased junction in Fig. 2-18b. If the applied voltage is suddenly reversed, forward current I_F ceases immediately, leaving some majority charge carriers in the depletion region. These charge carriers must flow back out of the depletion region, which is widened when the junction is reverse-biased. The result is that, when a forward-biased junction is suddenly reversed, there is a reverse current, which is large at first and which slowly decreases to the level of the reverse saturation current. The effect may be likened to the discharging of a capacitor, and so it is represented by a capacitance known as the *diffusion capacitance* (C_d).

It can be shown that C_d is proportional to the forward current. This is to be expected, since the number of charge carriers in the depletion region must be directly proportional to I_F. For a low-current diode with a 10 mA forward current, a typical value of diffusion capacitance is about 1 nF.

AC Equivalent Circuits (Reverse-Biased and Forward-Biased)

A reverse-biased diode can be simply represented by the high reverse resistance R_R in parallel with the depletion layer capacitance C_{pn} (see Fig. 2-19a). The equivalent circuit (or model) for a forward-biased diode consists of the dynamic resistance r_d in series with a voltage cell representing V_F, as discussed in Section 2-3. To allow for the effect of the diffusion capacitance, C_d is included in parallel to give the complete equivalent circuit shown in Fig. 2-19b.

The complete equivalent circuit for the forward-biased diode may be modified into an *ac equivalent circuit*, which can be used for diodes that are maintained in a forward-biased condition while subjected to small variations in I_F and V_F. The ac equivalent circuit is created simply by removing the voltage cell representing V_F from the complete equivalent circuit (Fig. 2-19c).

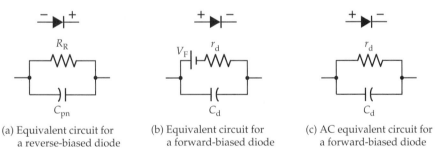

(a) Equivalent circuit for a reverse-biased diode   (b) Equivalent circuit for a forward-biased diode   (c) AC equivalent circuit for a forward-biased diode

Figure 2-19   Equivalent circuits (or models) for reverse-biased and forward-biased diodes.

Reverse Recovery Time

In many applications, diodes must switch rapidly between forward and reverse bias. Most diodes switch very quickly into the forward-biased condition; however, there is a longer *turnoff* time owing to the junction diffusion capacitance.

Figure 2-20 illustrates the effect of a voltage pulse on the diode forward current. When the pulse switches from positive to negative, the diode conducts in reverse instead of switching *off* sharply (see Fig. 2-20a). The reverse current (I_R) initially equals the forward current (I_F); then it gradually decreases toward zero. The high level of reverse current occurs because at the instant of reverse bias there are charge carriers crossing the junction depletion region, and these must be removed. (This is the same effect that produces diffusion capacitance.) The *reverse recovery time* (t_{rr}) is the time required for the current to decrease to the reverse saturation current level. Typical values of t_{rr} for switching diodes

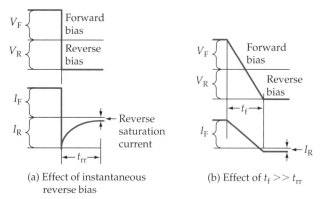

(a) Effect of instantaneous reverse bias

(b) Effect of $t_f \gg t_{rr}$

Figure 2-20   The minimum on-off switching time of a diode is limited by the reverse recovery time (t_{rr}). The reverse saturation current can be minimized by using a pulse with $t_f \gg t_{rr}$.

range from 4 ns to 50 ns. The diode reverse current can be kept to a minimum if the *fall time* (t_f) of the applied voltage pulse is much larger than the diode reverse recovery time. This is illustrated in Fig. 2-20b. Typically

$$t_{f(min)} = 10\, t_{rr} \qquad\qquad (2\text{-}8)$$

Example 2-11

Calculate the minimum fall times for voltage pulses applied to a circuit using 1N915 and 1N917 diodes to keep the diode reverse current to a minimum.

Solution

From diode data sheet A-1 in Appendix A,

$$t_{rr} = 10 \text{ ns for the 1N915} \quad \text{and} \quad t_{rr} = 3 \text{ ns for the 1N917}$$

For the 1N915:

Eq. 2-8: $\qquad\qquad t_{f(min)} = 10\ t_{rr} = 10 \times 10 \text{ ns}$

$$= 100 \text{ ns}$$

For the 1N917:

Eq. 2-8: $\qquad\qquad t_{f(min)} = 10\ t_{rr} = 10 \times 3 \text{ ns}$

$$= 30 \text{ ns}$$

Practice Problems

2-6.1 Determine the maximum reverse recovery time for satisfactory operation of a diode with an applied voltage pulse that has a 0.5 μs fall time.

2-6.2 Estimate a suitable minimum fall time for a pulse that switches a diode from *on* to *off* if the reverse recovery time of the diode is 15 ns.

2-7 DIODE SPECIFICATIONS

Diode Data Sheets

To select a suitable diode for a particular application, the *data sheets*, or *specifications*, provided by device manufacturers must be consulted. Portions of typical diode data sheets are shown in Fig. 2-21 and as data sheets 1 to 3 in Appendix A.

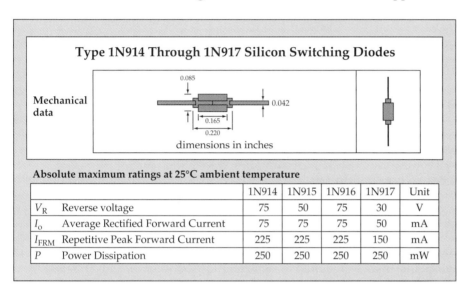

Type 1N914 Through 1N917 Silicon Switching Diodes

Mechanical data

dimensions in inches

Absolute maximum ratings at 25°C ambient temperature

		1N914	1N915	1N916	1N917	Unit
V_R	Reverse voltage	75	50	75	30	V
I_o	Average Rectified Forward Current	75	75	75	50	mA
I_{FRM}	Repetitive Peak Forward Current	225	225	225	150	mA
P	Power Dissipation	250	250	250	250	mW

Figure 2-21 Part of data sheet for type 1N914 to 1N917 diodes.

Most data sheets start with the device type number at the top of the page, such as '1N914 through 1N917' or '1N5391 through 1N5399'. The 1 (one) in the type number signifies a one-junction device: a diode. This is followed by a short descriptive title, for example, *silicon switching diode* or *silicon rectifier*. *Mechanical data* are also given, usually in the form of an illustration showing the shape and dimensions of the package. The *maximum ratings* at 25°C are then listed (see Fig. 2-21).

The maximum ratings are the maximum voltages, currents, and so on, that can be applied without destroying the device. It is very important that these ratings not be exceeded; otherwise the diode is quite likely to fail. *For reliability, the maximum ratings should not even be approached.* If a diode is to survive a 50 V reverse bias, a device that has a 75 V peak reverse voltage should be selected. If the diode peak forward current is to be 100 mA, a device that can handle 150 mA should be used. It is also important to note that the maximum ratings must be adjusted downward for operation at temperatures greater than 25°C (see Section 2-5).

A list of other electrical characteristics for the device normally follows the maximum ratings. An understanding of all the parameters specified on a data sheet will not be achieved until the data sheets have been consulted frequently. However, some of the most important parameters are considered below.

V_R or V_{RRM} *Peak reverse voltage* (also termed *peak inverse voltage* and *dc blocking voltage*). This is the maximum reverse voltage that may be applied across the diode.

I_o or $I_{F(AV)}$	*Steady-state forward current.* The maximum current that may be passed continuously through the diode.
I_{FSM}	*Non-repetitive peak surge current.* This current may be passed for a specified time. The non-repetitive surge current is very much higher than the normal maximum forward current. It is a current that may be allowed to flow briefly when a circuit is first switched on.
I_{FRM}	*Repetitive peak surge current.* Peak current that may be repeated over and over again, for example, during each cycle of a rectified waveform.
V_F	*Static forward voltage drop.* The maximum forward volt drop for a given forward current and device temperature.
P	*Continuous power dissipation at 25°C.* The maximum power that the device can safely dissipate continuously in free air. This rating must be downgraded at higher temperatures (see Section 2-5).

Low-Power Diodes

The data sheet portion in Fig. 2-21 identifies the 1N914 to 1N917 devices as *switching diodes*. The average rectified forward current is listed as 75 mA (except for the 1N917). Maximum reverse voltage ranges from 30 V to 75 V. Thus, these diodes are intended for relatively low-current, low-voltage applications, in which they may be required to switch rapidly between *on* and *off* states.

Rectifier Diodes

The low-power rectifier data sheets (see Appendix A, data sheets A-2 and A-3) show that the 1N4000 range of rectifiers can pass an average forward current of 1 A, and that the 1N5390 range can pass 1.5 A. Both types have maximum reverse voltages ranging from 50 V to 1000 V. Unlike data sheet A-1, for switching diodes, the rectifier data sheet does not list the reverse recovery time. Rectifier diodes are generally intended for low-frequency applications (60 Hz to perhaps 400 Hz) in which switching time is not important.

Example 2-12

Referring to Fig. 2-21, determine the following quantities for a 1N915 diode: peak reverse voltage, steady-stage forward current, peak repetitive forward current and power dissipation.

Solution

For the 1N915:

$$PIV = V_R = 50 \text{ V}$$

$$I_o = 75 \text{ mA}$$

$$I_{FRM} = 225 \text{ mA}$$

$$P = 250 \text{ mW}$$

Practice Problems

2-7.1 Referring to Appendix A, data sheet A-3, determine the following quantities for a 1N5397 diode: peak reverse voltage, steady-stage forward current, and non-repetitive forward current.

2-7.2 A rectifier has to pass an average current of 600 mA and survive a repetitive reverse voltage of 75 V. Choose a suitable diode from Appendix A, data sheets A-1 to A-3.

2-8 DIODE TESTING

Ohmmeter Tests

Diode failure is usually the result of passing excessive forward current through the device, or application of excessive reverse voltage. Both situations can result in devices that offer either an open circuit or a short circuit. Several methods are available for testing diodes. One of the simplest and quickest tests can be made by using an ohmmeter to measure the forward and reverse resistance (see Fig. 2-22a). The diode should offer a low resistance when forward-biased and a high resistance when reverse-biased. A diode is short-circuited when it displays a low resistance for both forward and reverse bias, and open-circuited if a high resistance is measured with both bias polarities. In the case of an analog ohmmeter with a 1.5 V battery, the low resistance indicated is normally about half the selected range. It is important to note that *when some multifunction instruments are used as ohmmeters, the voltage polarity at the terminals may*

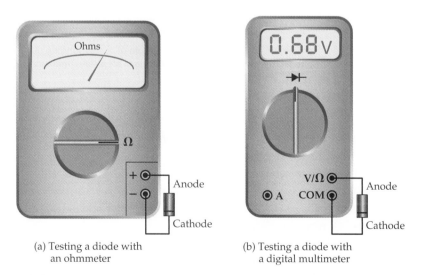

(a) Testing a diode with   (b) Testing a diode with
    an ohmmeter   a digital multimeter

Figure 2-22 Analog or digital multimeters may be used for diode testing. When forward-biased, the diode should display a low resistance, or the typical forward voltage drop. When reverse biased, it should indicate a high reverse resistance.

not be the same as the polarity marked on the instrument. A voltmeter can be used to check the ohmmeter terminal polarity.

Use of a Digital Meter

Many portable multi-function digital instruments have a diode-testing facility which displays the diode forward voltage when the terminals are connected *positive* to the anode and *negative* to the cathode (see Fig. 2-22b). The meter function switch should be set to the diode symbol, as illustrated. When reverse-connected, a functioning diode produces either an OL display or an indication of the meter internal voltage.

Plotting Diode Characteristics

The forward characteristics of a diode can be obtained by use of the circuit illustrated in Fig. 2-23a. The diode voltage is set at a series of convenient levels, and the corresponding current levels are measured and recorded. The characteristics are then plotted from the resulting table of quantities. The reverse characteristics can be derived in the same way, except that a very sensitive microammeter is required in order to measure the reverse current (Fig. 2-23b). The microammeter must be connected directly in series with the diode, as shown; otherwise the voltmeter current may introduce a serious error.

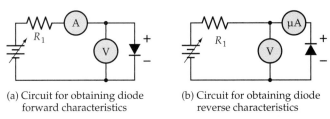

(a) Circuit for obtaining diode   (b) Circuit for obtaining diode
     forward characteristics   reverse characteristics

Figure 2-23 Diode characteristics can be plotted from a table of corresponding current and voltage measurements, obtained by varying the applied voltage in steps and measuring *V* and *I* at each step.

Figure 2-24 shows a method of using an *XY recorder* for drawing the forward characteristics of a diode. The resistor voltage (V_{R1}) is directly proportional to the diode forward current (I_F). So V_{R1} is applied to the vertical input terminals of the XY recorder, as illustrated. The diode forward voltage (V_F) goes to the horizontal input terminals. When the power supply voltage is slowly increased from zero, the diode forward characteristic is traced out by the pen on the XY recorder.

If R_1 in Fig. 2-24 is a 1 kΩ resistor, there is a 1 V drop across it for every 1 mA of diode current. Therefore, with the vertical scale of the XY recorder set to 1 V/cm, the (vertical) current coordinate of the graph is 1 mA/cm. A convenient scale for the (horizontal) voltage coordinate is 0.1 V/cm. Diode characteristics

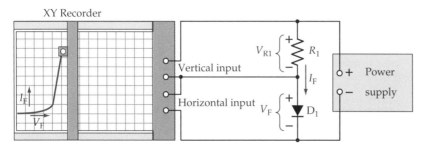

Figure 2-24 An XY recorder can be used for directly plotting diode characteristics.

can also be investigated by means of a *curve tracer,* and by computer graphic analysis software.

Example 2-13

The arrangement shown in Fig. 2-24 is to be used to plot the characteristics of a 1N914 diode. Select a suitable resistor for R_1, and determine appropriate V/cm scales for the XY recorder.

Solution

From Fig. 2-21, $$I_o = 75 \text{ mA} \quad \text{(for the 1N914)}$$

Select a vertical scale of 5 mA/cm for the characteristic so that

$$\text{vertical scale length (for } I_F) = \frac{75 \text{ mA}}{5 \text{ mA/cm}}$$

$$= 15 \text{ cm}$$

Select a vertical scale of 1 V/cm for the XY recorder, so that 15 cm represents 15 V as well as 75 mA.

$$R_1 = \frac{15 \text{ V}}{75 \text{ mA}}$$

$$= 200 \ \Omega$$

and
$$P_{R1(max)} = (I_{F(max)})^2 \, R_1 = (75 \text{ mA})^2 \times 200 \ \Omega$$

$$= 1.1 \text{ W}$$

V_F might be as large as 0.8 V, so select a horizontal scale of 0.1 V/cm so that

$$\text{horizontal scale length (for } V_F) = \frac{0.8 \text{ V}}{0.1 \text{V/cm}}$$

$$= 8 \text{ cm}$$

Practice Problems

2-8.1 Plot the forward characteristics for a silicon diode from the following experimental data:

V_F (V)	0.6	0.62	0.64	0.66	0.68	0.7	0.72	0.74	0.76
I_F (mA)	1	1.4	1.8	2.5	10	50	90	130	170

2-8.2 An XY recorder is used (as in Fig. 2-24) to plot silicon diode forward characteristics to approximately fill a 10 cm by 10 cm square. If the maximum forward current is to be 100 mA, select suitable V/cm scales for the recorder and a suitable resistance for R_1.

2-9 ZENER DIODES

Junction Breakdown

When a junction diode is reverse-biased, there is normally only a very small reverse saturation current: I_S on the reverse characteristic in Fig. 2-25a. When the reverse voltage is sufficiently increased, the junction *breaks down* and a large reverse current flows. If the reverse current is limited by means of a suitable series-connected resistor (R_1 in the circuit in Fig. 2-25b), the power dissipation in the diode can be kept to a level that will not destroy the device. In this case, the diode may be operated continuously in reverse breakdown. The reverse current returns to its normal level when the voltage is reduced below the reverse breakdown level.

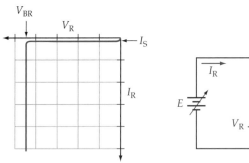

(a) Diode reverse characteristic   (b) Diode-resistor circuit

Figure 2-25 A diode can be operated in reverse breakdown if the current is limited by means of a series-connected resistor. Zener diodes are designed for operation in reverse breakdown.

Diodes designed for operation in reverse breakdown are found to have a breakdown voltage that remains extremely stable over a wide range of current levels. This property gives the *breakdown diode* many useful applications as a voltage reference source. There are two mechanisms that cause breakdown in a reverse biased *pn*-junction. With a very narrow depletion region, the electric field strength (volts/width) produced by a reverse bias voltage can be very high. The high-intensity electric field causes electrons to break away from their atoms, thus converting the depletion region from an insulating material into a conductor. This is *ionization by electric field*, also called Zener breakdown, and it usually occurs with reverse bias voltages less than 5 V.

In cases where the depletion region is too wide for Zener breakdown, the electrons in the reverse saturation current can be given sufficient energy to cause other electrons to break free when they strike atoms within the depletion region. This is termed *ionization by collision*. The electrons released in this way collide with other atoms to produce more free electrons in an *avalanche* effect. *Avalanche breakdown* is normally produced by reverse voltage levels above 5 V. Although *Zener* and *avalanche* are two different types of breakdown, the name *Zener diode* is commonly applied to all breakdown diodes.

Circuit Symbol and Package

The circuit symbol for a Zener diode in Fig. 2-26a is the same as that for an ordinary diode but with the cathode bar approximately in the shape of a letter Z. The arrowhead on the symbol still points in the (conventional) direction of forward current when the device is forward-biased. As illustrated, for operation in reverse bias, the voltage drop (V_Z) is positive (+) on the cathode and negative (−) on the anode.

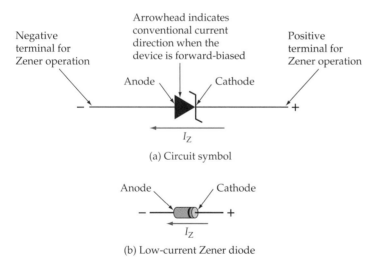

(a) Circuit symbol

(b) Low-current Zener diode

Figure 2-26 Zener diode circuit symbol and low-current Zener diode package.

Low-power Zener diodes are available in a variety of packages. For the device package shown in Fig. 2-26b, the coloured band identifies the cathode terminal, as in the case of an ordinary low-current diode. High-current Zener diodes are also available in the type of package that allows for mounting on a heat sink.

Characteristics and Parameters

The typical characteristics of a Zener diode are shown in detail in Fig. 2-27. Note that the forward characteristic is simply that of an ordinary forward-biased diode. Some important points on the reverse characteristic are

V_Z Zener breakdown voltage
I_{ZT} Test current for measuring V_Z
I_{ZK} Reverse current near the knee of the characteristic, the minimum reverse current to sustain breakdown
I_{ZM} Maximum Zener current, limited by the maximum power dissipation

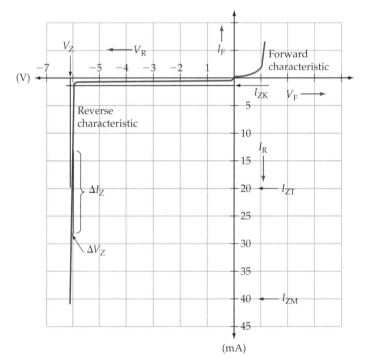

Figure 2-27   Typical characteristics for a Zener diode. The most important parameters are the Zener voltage V_Z, the knee current I_{ZK}, the test current I_{ZT}, and the maximum current I_{ZM}.

The *dynamic impedance* (Z_Z) is another important parameter that may be derived from the characteristics.

$$Z_Z = \frac{\Delta V_Z}{\Delta I_Z} \tag{2-9}$$

As illustrated in Fig. 2-27, Z_Z defines how V_Z changes with variations in diode reverse current. When measured at I_{ZT}, the dynamic impedance is designated (Z_{ZT}). The dynamic impedance measured at the knee of the characteristic (Z_{ZK}) is substantially larger than Z_{ZT}.

The Zener diode may be operated at any (reverse) current level between I_{ZK} and I_{ZM}. For greatest voltage stability, the diode is normally operated at the test current (I_{ZT}). Many low-power Zener diodes have a test current specified as 20 mA; however, some devices have lower test currents.

Data Sheet

A portion of a data sheet for low-power Zener diodes with voltages ranging from 3.3 V to 12 V is shown in Fig. 2-28. (See also data sheet A-4 for 2.4 V to 110 V Zener diodes, in Appendix A.) Note in Fig. 2-28 that the V_Z tolerance is stated as ±5% or ±10%. This means that the devices can be purchased with either a ±5% or ±10% tolerance on V_Z. For a 1N753 with a ±10% tolerance, the actual V_Z is 6.2 V ±10%, or 5.58 V to 6.82 V. The Zener voltage remains stable at whatever it happens to be within this range.

Type 1N746 Through 1N759 Silicon Zener Diodes

3.3 V to 12 V (±5% or ±10%)
P_D = 400 mW (derate linearly above 50°C at 3.2 mW/°C)

Electrical characteristics at 25°C ambient temperature

Type Number	Nominal Zener voltage V_Z (V)	Test current I_Z (mA)	Zener impedance Z_{ZT} (Ω)	Leakage current I_R (μA)	Reverse temperature coefficient α_Z (%/°C)
IN746	3.3	20	28	10	−0.062
IN747	3.6	20	24	10	−0.055
IN753	6.2	20	7	0.1	+0.022
IN755	7.5	20	6	0.1	+0.045
IN757	9.1	20	10	0.1	+0.056
IN759	12.0	20	30	0.1	+0.060

Figure 2-28 Portions of a data sheet for low-power Zener diodes.

The data sheet also lists the dynamic impedance (Z_{ZT}); reverse leakage current (I_R), which is the reverse current before breakdown; and the temperature coefficient (α_Z) for the V_Z of each device. The Zener voltages at any temperature can be calculated as follows:

$$V_{Z2} = (V_{Z1} \text{ at } T_1) + [\Delta T \alpha_Z V_Z/100] \tag{2-10}$$

Temperature-compensated Zener diodes are also available with extremely low temperature coefficients.

Low-power Zener diodes are generally limited to a maximum power dissipation of 400 mW (P_D in Fig. 2-28). Higher-power devices are available. All of the power dissipations must be derated with temperature increase, exactly as explained in Section 2-5. When the maximum Zener current is not listed on the device data sheet, it may be calculated from the power dissipation equation.

$$P_D = V_Z I_{ZM} \tag{2-11}$$

Example 2-14

Calculate the maximum current that may be allowed to flow through a 1N755 Zener diode at device temperatures of 50°C and 100°C.

Solution

From data sheet A-4 in Appendix A, it can be seen that for the 1N755 Zener diode,

$$V_Z = 7.5 \text{ V}, \quad P_D = 400 \text{ mW at } 50°\text{C, and the derating factor} = 3.2 \text{ mW/°C}$$

At 50°C:

From Eq. 2-11, $I_{ZM} = \dfrac{P_D}{V_Z} = \dfrac{400 \text{ mV}}{7.5 \text{ V}}$

$$= 53.3 \text{ mA}$$

At 100°C:

Eq. 2-5: $P_2 = (P_1 \text{ at } T_1) - [\Delta T \times (\textit{derating factor})]$

$$= 400 \text{ mW} - [(100°\text{C} - 50°\text{C}) \times 3.2 \text{ mW/°C}]$$

$$= 240 \text{ mW}$$

From Eq. 2-11, $I_{ZM} = \dfrac{P_2}{V_Z} = \dfrac{240 \text{ mW}}{7.5 \text{ V}}$

$$= 32 \text{ mA}$$

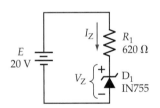

Figure 2-29 Circuit for Example 2-15.

Example 2-15

For the Zener diode circuit in Fig. 2-29, $E = 20$ V, and $R_1 = 620 \ \Omega$. The Zener diode is a 1N755. Calculate the diode current and power dissipation.

Solution

From the data sheet in Fig. 2-28, it is seen that for the 1N755 Zener diode, $V_Z = 7.5$ V

$$V_{R1} = E - V_Z = 20 \text{ V} - 7.5 \text{ V}$$

$$= 12.5 \text{ V}$$

$$I_Z = I_{R1} = \frac{V_{R1}}{R_1} = \frac{12.5 \text{ V}}{620 \ \Omega}$$

$$= 20.16 \text{ mA}$$

$$P_D = V_Z I_Z = 7.5 \text{ V} \times 20.16 \text{ mA}$$

$$= 151 \text{ mW}$$

Equivalent Circuit

The dc equivalent circuit for a Zener diode is simply a voltage cell with a voltage V_Z, as in Fig. 2-30a. This is the complete equivalent circuit for the device for all dc calculations. For the ac equivalent circuit (Fig. 2-30b), the dynamic impedance is included in series with the voltages cell. The ac equivalent circuit is used in situations where the Zener current is varied by small amounts. It must be understood that these equivalent circuits apply only when the Zener diode is maintained in reverse breakdown. If the device becomes forward-biased, then the equivalent circuit for a forward-biased diode must be used.

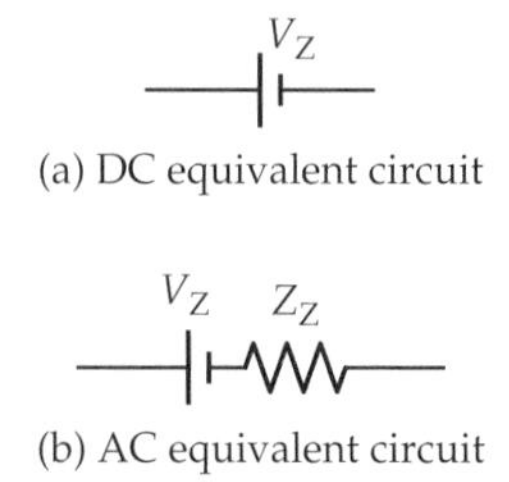

(a) DC equivalent circuit

(b) AC equivalent circuit

Figure 2-30 DC and AC equivalent circuits for a Zener diode.

Example 2-16

A Zener diode with $V_Z = 4.3$ V has Z_Z equal to 22 Ω when $I_Z = 20$ mA. Calculate the upper and lower limits of V_Z when I_Z changes by ± 5 mA.

Solution

$$\Delta V_Z = \pm(\Delta I_Z \times Z_Z) = \pm(5 \text{ mA} \times 22 \text{ }\Omega)$$

$$= \pm 110 \text{ mV}$$

$$V_{Z(max)} = V_Z + \Delta V_Z = 4.3 \text{ V} + 110 \text{ mV}$$

$$= 4.41 \text{ V}$$

$$V_{Z(min)} = V_Z - \Delta V_Z = 4.3 \text{ V} - 110 \text{ mV}$$

$$= 4.19 \text{ V}$$

Practice Problems

2-9.1 Calculate the maximum current for a 1N753 Zener diode at device temperatures of 50°C and 100°C.

2-9.2 A 1N749 Zener diode is connected in series with an 820 Ω resistor and a 12 V supply. Calculate the diode current and power dissipation.

Review Questions

Section 2-1

2-1 Sketch the symbol for a semiconductor diode, labelling the anode and cathode and showing the polarity and current direction for forward bias. Show the direction of movement of charge carriers when the device is (a) forward-biased and (b) reverse-biased.

2-2 Draw sketches to show the appearance of low-current and medium-current diodes, and show how the cathode is identified in each case.

Section 2-2

2-3 Sketch typical forward and reverse characteristics for a germanium diode and for a silicon diode. Discuss the characteristics, and compare silicon and germanium diodes.

2-4 For diodes, define forward voltage drop, maximum forward current, dynamic resistance, reverse saturation current, and reverse breakdown voltage.

2-5 Show how the diode dynamic resistance can be determined from the forward characteristics. Write an equation for calculating the dynamic resistance from the dc forward current.

Section 2-3

2-6 Sketch the characteristics for an ideal diode and the approximate characteristics for practical diodes. Briefly explain each characteristic.

2-7 Draw the dc equivalent circuit for a diode and the piecewise linear equivalent circuit. Discuss the application of each.

Section 2-4

2-8 Explain the purpose of a dc load line. Write the equation for drawing a dc load line for a series circuit consisting of a supply voltage (E), a resistor (R_1), and a diode (D_1).

2-9 Define the Q-point in a diode circuit, and explain how it is related to the diode characteristics and the dc load line.

Section 2-5

2-10 Sketch and explain a power-versus-temperature graph for a diode. Define the power-derating factor for a diode.

2-11 Discuss how temperature change affects diode forward voltage drop.

Section 2-6

2-12 Explain the origins of depletion layer capacitance and diffusion capacitance, and discuss the importance of each.

2-13 Sketch the complete equivalent circuits for forward-biased and reverse-biased diodes. Sketch the ac equivalent circuit for a forward-biased diode. Briefly explain each circuit.

2-14 Define reverse recovery time. Sketch waveforms to show the effect of reverse recovery time on a diode switched rapidly from on to off. Explain each waveform.

Section 2-7

2-15 Discuss the major differences between switching diodes and rectifier diodes.

2-16 Define the following diode quantities: peak reverse voltage, repetitive peak surge current, and steady-state forward current.

Section 2-8

2-17 Describe how an ohmmeter and a digital multimeter may be used for testing diodes.

2-18 Sketch circuits for obtaining diode forward and reverse characteristics by measuring corresponding current and voltage levels. Explain.

2-19 Draw a sketch to show how the forward characteristics of a diode may be plotted on an XY recorder.

Section 2-9

2-20 Discuss the different types of junction breakdown that can occur in a reverse-biased diode. Sketch the circuit symbol for a Zener diode, and briefly explain Zener diode operation.

2-21 Sketch typical characteristics for a Zener diode. Explain the shape of the characteristics, and identify the important points.

2-22 Sketch the equivalent circuit for a Zener diode. Briefly explain.

Problems

Section 2-2

2-1 Calculate the static forward resistance for the characteristics in Fig. 2-31 at a 200 mA forward current. Determine the reverse resistance at a 75 V reverse voltage.

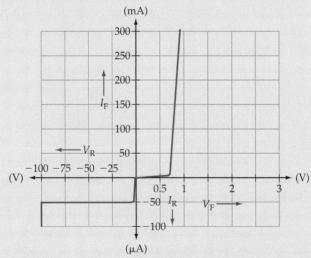

Figure 2-31 Diode characteristics for Problems 2-1, 2-2, and 2-9.

2-2 Determine the dynamic resistance at a 150 mA forward current for a diode with the characteristics shown in Fig. 2-31.

2-3 Calculate the static forward resistance for the diode characteristics shown in Fig. 2-32 at a 25 mA forward current. Also, determine the dynamic resistance for the device at $I_F = 25$ mA, using the characteristic and using Eq. 2-2.

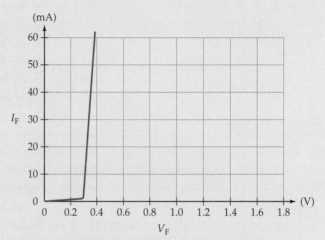

Figure 2-32   Diode characteristics for Problems 2-3, 2-10, and 2-11.

Section 2-3

2-4   Calculate the forward current in a circuit consisting of a germanium diode connected in series with a 9 V battery and a 3.3 kΩ resistor.

2-5   A silicon diode in series with a 2.7 kΩ resistor and a battery is to have an I_F of 1.96 mA. Calculate the battery voltage.

2-6   Draw the piecewise linear characteristics for a silicon diode with a 0.6 Ω dynamic resistance and a 75 mA maximum forward current.

2-7   Draw a straight-line approximation of the forward characteristic for a silicon diode that has a 0.5 Ω dynamic resistance and a maximum forward current of 300 mA.

Section 2-4

2-8   A diode with the characteristics in Fig. 2-13b is to pass a 35 mA current from a 5.5 V supply. Draw the dc load line and calculate the required series resistance value. Determine the new current level when the supply is reduced to 3.5 V.

2-9   A diode with the forward characteristic in Fig. 2-31 is connected in series with a 30 Ω resistance and a 6 V supply. Determine the diode current, and find the new current when the resistance is changed to 20 Ω.

2-10   A diode which has the characteristics shown in Fig. 2-32 is to pass a 20 mA forward current when the supply is 12 V. Determine the value of resistance that must be connected in series with the diode.

2-11   Calculate the new supply voltage for the circuit in Problem 2-10 to give a 15 mA current level.

Section 2-5

2-12   A diode with a maximum power dissipation of 1000 mW at 25°C is to pass an average forward current of 500 mA. The forward voltage drop for the device is 0.8 V, and the power derating factor is 10 mW/°C. Calculate the maximum temperature at which the diode may be safely operated.

2-13   The diode specified in Problem 2-12 is to be operated at a temperature of 75°C. Calculate the maximum level of average forward current that the diode can safely pass.

2-14   Draw the power dissipation-versus-temperature graph for the diode specified in Problem 2-12.

2-15   A diode with a 0.9 V forward drop has a 1.5 W maximum power dissipation at 25°C. If the device derating factor is 7.5 mW/°C, calculate the maximum forward current level at 25°C and at 75°C. Assume that V_F remains constant.

2-16   Estimate the forward voltage drop at 75°C of the silicon diode in Problem 2-15. Determine the junction dynamic resistances at the temperature extremes if I_F is 20 mA.

2-17   A 5 V supply is applied via a 150 Ω resistor to a silicon diode and a germanium diode that are connected in series. Determine the diode current at 25°C and at 100°C.

Section 2-6

2-18   Calculate the minimum fall time for a voltage pulse applied to a diode with a reverse recovery time of (a) 6 ns and (b) 50 ns.

2-19   A diode has an applied voltage with a 200 ns fall time. Determine the diode maximum reverse recovery time for satisfactory operation.

2-20   Calculate a suitable minimum fall time for a switching waveform applied to a diode that has a 12 ns reverse recovery time.

2-21   Calculate the minimum fall time for a voltage pulse applied to a circuit with a 1N914 diode.

Section 2-7

2-22   A diode connected in series with a 560 Ω resistor has a supply voltage that alternates between peak levels of +150 V and −150 V. Select a suitable device from the data sheets A-1 and A-2 in Appendix A.

2-23   Referring to data sheet A-3 in Appendix A, determine the peak reverse voltage and average rectified forward current for a 1N5398 diode.

2-24   A rectifier diode has to pass an average current of 55 mA and survive a 40 V peak reverse voltage. Select a suitable device from data sheets A-1 to A-3 in Appendix A.

Section 2-8

2-25   The arrangement shown in Fig. 2-24 is to be used to plot the forward characteristics of a 1N917 diode. Determine a resistance value for R_1, and select appropriate V/cm scales for the horizontal and vertical inputs of the XY recorder. The graph should be approximately 20 cm × 20 cm.

2-26   Plot the forward characteristics of a diode from the following experimental data:

V_F (V)	0.24	0.26	0.28	0.29	0.30	0.31	0.32	0.33	0.34
I_F (mA)	0.05	0.07	0.09	0.5	0.9	20	40	60	80

Section 2-9

2-27 Determine the maximum current that may be used with a 1N757 Zener diode at temperatures of 25°C and 80°C.

2-28 A 1N750 Zener diode is connected in series with an 470 Ω resistor and a 10 V supply voltage. Calculate the diode current and power dissipation.

Practice Problem Answers

2-2.1	5.5 Ω, 30 MΩ
2-2.2	0.5 Ω, 0.5 Ω
2-3.1	2.89 mA
2-3.2	(Point A: 0 mA, 0.3 V), (Point B: 100 mA, 0.35 V)
2-3.3	174 mA
2-4.1	20 mA
2-4.2	50 Ω
2-4.3	3.45 V
2-5.1	154 mA, 61.5 mA
2-5.2	267 mV, 421 mV, 1.23 Ω, 1.54 Ω
2-5.3	780 mV
2-6.1	50 ns
2-6.2	150 ns
2-7.1	600 V, 1.5 A, 50 A
2-7.2	1N5392
2-8.2	Vertical 1 V/cm, Horizontal 0.1 V/cm, 100 Ω
2-9.1	64.5 mA, 38.7 mA
2-9.2	40.4 mW

CHAPTER 3
Diode Applications

CONTENTS

Introduction

Objectives

You will be able to:

1 Sketch various diode half-wave and full-wave rectifier circuits and their input and output waveforms. Explain the operation of each circuit.

2 Sketch basic dc power supply circuits using rectifiers and capacitor filters. Discuss the operation and performance of each circuit.

3 Design dc power supply circuits and analyze them to determine diode current and voltage levels, output ripple voltage, line and load effects, and line and load regulation.

4 Sketch Zener diode voltage regulator circuits, and explain their operation.

5 Design and analyze Zener diode voltage regulator circuits.

6 Draw diagrams for series and shunt clipping circuits and sketch their input and output waveforms. Explain the operation of each circuit.

7 Design and analyze series and shunt clipping circuits.

8 Draw diagrams for clamping circuits and dc voltage multipliers. Sketch the input and output waveforms, and explain the circuit operation.

9 Design and analyze clamping circuits and dc voltage multipliers.

10 Sketch diode AND and OR gate circuits, and explain their operation. Calculate circuit current and voltage levels.

INTRODUCTION

One of the most important applications of diodes is rectification: conversion of a sinusoidal ac waveform into single-polarity half cycles. Rectification may be performed by half-wave or full-wave rectifier circuits. A dc power supply converts a sinusoidal ac supply to dc by rectification and filtering. The filtering process normally involves nothing more than the use of a large *reservoir capacitor*, which charges to the peak input voltage level to produce the dc output. The capacitor partially discharges between peaks of the rectified waveform, and this results in a *ripple* voltage on the output. Power supplies are specified according to the dc output voltage, load current, and ripple voltage. Power supply performance is defined in terms of the output voltage stability when the input voltage or the load current changes.

Other important diode applications include clipping, clamping, dc voltage multiplication, and logic circuits. Diode clipping circuits are used for *clipping off* an unwanted portion of a waveform. Clamping circuits change the dc voltage level of a waveform without affecting the wave shape. DC voltage multipliers are applied to change the level of a dc voltage source to a desired higher level. Logic circuits produce a *high* or *low* output voltage, depending upon the voltage levels at several input terminals.

3-1 HALF-WAVE RECTIFICATION

Positive Half-Wave Rectifier

A diode *positive half-wave rectifier* circuit is shown in Fig. 3-1a. An alternating input voltage is applied via a transformer (T_1) to a single diode connected in series with a load resistor R_L. The transformer is normally necessary to *dc-isolate* the rectifier circuit from the ac supply. The diode is forward-biased during the positive half cycles of the input waveform, and reverse-biased during the negative half cycles. Substantial current flows through R_L only during the positive half cycles of the input. For the duration of the negative half cycles, the diode behaves almost as an open switch. The output voltage waveform developed across R_L is a series of positive half cycles of alternating voltage with intervening very small negative voltage levels produced by the diode reverse saturation current.

When the diode is forward-biased (see Fig. 3-1b), the voltage drop across it is V_F, and the output voltage is (input voltage) $-V_F$. So, the peak output voltage is

$$V_{po} = V_{pi} - V_F \qquad (3\text{-}1)$$

Note that $V_{pi} = 1.414\ V_i$, where V_i is the rms level of the sinusoidal input voltage to the rectifier circuit (from the transformer output).

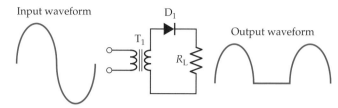

(a) Transformer-coupled half-wave rectifier circuit

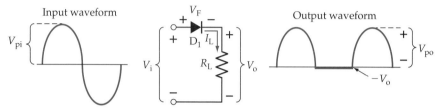

(b) Half-wave rectifier circuit with input and output waveforms

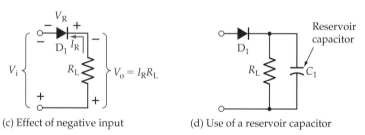

(c) Effect of negative input   (d) Use of a reservoir capacitor

Figure 3.1   Positive half-wave rectifier circuit. The diode is forward-biased during the positive half-cycle of the applied waveform and reverse-biased during the negative half-cycle.

The diode peak forward current is

$$I_P = \frac{V_{po}}{R_L} \tag{3-2}$$

During the negative half-cycle of the input (Fig. 3-1c), the reverse-biased diode offers a very high resistance. So there is only a very small reverse current (I_R), giving an output voltage

$$-V_o = -I_R R_L \tag{3-3}$$

While the diode is reverse-biased, the peak voltage of the negative half-cycle of the input is applied to its terminals. Thus the *peak reverse voltage*, or *peak inverse voltage (PIV)*, applied to the diode is

$$V_R = PIV = V_{pi} \tag{3-4}$$

The average and rms values of the half-wave rectified waveform can be determined as $V_{o(ave)} = 0.318\, V_{po}$ and $V_{o(rms)} = 0.5\, V_{po}$. However, most rectifier circuits use a *reservoir capacitor* at the output terminals to smooth the rectified voltage wave into direct voltage (see Fig. 3-1d). It is very important to note that *the presence of the capacitor changes the output waveform and that it substantially affects the load current and voltage and the diode current and voltage.* This is discussed in Sections 3-2 and 3-3.

Example 3-1

A diode with $V_F = 0.7$ V is connected as a half-wave rectifier. The load resistance is 500 Ω, and the (rms) ac input is 22 V. Determine the peak output voltage, the peak load current, and the diode peak reverse voltage.

Solution

$$V_{pi} = 1.414\, V_i = 1.414 \times 22 \text{ V}$$

$$= 31.1 \text{ V}$$

Eq. 3-1:
$$V_{po} = V_{pi} - V_F = 31.1 \text{ V} - 0.7 \text{ V}$$

$$= 30.4 \text{ V}$$

Eq. 3-2:
$$I_p = \frac{V_{po}}{R_L} = \frac{30.4 \text{ V}}{500 \ \Omega}$$

$$= 60.8 \text{ mA}$$

Eq. 3-4:
$$\text{PIV} = V_{pi} = 31.1 \text{ V}$$

Negative Half-Wave Rectifier

Figure 3-2a shows the effect of reversing the diode polarity in the circuit of Fig. 3-1. The negative half-cycle of the ac input waveform (instead of the positive half-cycle) is passed to the load resistor. Consequently, the peak output voltage and current are negative quantities.

Figure 3-2b shows a positive half-wave rectifier circuit with the positive output terminal grounded. This is possible because, as already discussed, the ac input to the rectifier circuit is normally derived from the secondary of a transformer. Either of the two output terminals may be grounded so long as there is no other grounded point in the circuit. Now, when the transformer output waveform is at its peak positive level, the load waveform is actually a peak negative quantity, as shown, because output *terminal B* is negative with respect to the grounded *terminal A*. The diode is reverse-biased during the negative half-cycle of the transformer output, and so the load voltage is zero. Thus, a negative half-wave rectified waveform can be generated simply by grounding the positive output terminal of a positive rectifier circuit. It is important to note that one of the rectifier circuit output terminals may be grounded *only when the transformer is present to provide complete dc isolation* from the ac supply. If a variable transformer (*autotransformer* or *variac*) is used without a power supply transformer, a 1:1 *isolation transformer* must be substituted for the power supply transformer to provide the required dc isolation (see Fig. 3-2c).

(a) Negative rectification by reversing the diode polarity

(b) Negative rectification by grounding the positive output terminal

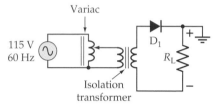

(c) An isolation transformer should be used when the power supply transformer is not available

Figure 3-2 Negative half-wave rectification is performed when the diode is forward-biased during the negative half-cycle of the input. Negative output half-cycles can also be derived from a positive half-wave rectifier if the positive output terminal is grounded. Rectifier circuit should be transformer-coupled for dc isolation from the input.

Practice Problems

3-1.1 A half-wave rectifier circuit has a 15 V ac input and a 330 Ω load resistance. Calculate the peak output voltage, the peak load current, and the diode maximum reverse voltage.

3-1.2 A half-wave rectifier (as in Fig. 3-1) produces a 40 mA peak load current through a 1.2 kΩ resistor. If the diode is silicon, calculate the rms input voltage and the diode PIV.

3-2 FULL-WAVE RECTIFICATION

Two-Diode Full-Wave Rectifier

The *full-wave rectifier* circuit in Fig. 3-3 uses two diodes, and its input voltage is supplied from a transformer (T_1) with a centre-tapped secondary winding. The circuit is essentially a combination of two half-wave rectifier circuits, each supplied from one half of the transformer secondary.

When the transformer output voltage is positive at the top, as illustrated in Fig. 3-4a, the anode of D_1 is positive, and the centre tap of the transformer is connected to the cathode of D_1 by R_L. Consequently, D_1 is forward-biased, and load current (I_L) flows from the top of the transformer secondary through D_1, through R_L from top to bottom, and back to the transformer centre tap. During this time, the polarity of the voltage from the bottom half of the transformer secondary causes diode D_2 to be reverse-biased, as illustrated.

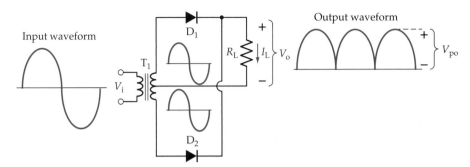

Figure 3-3 Full-wave rectification can be performed by two diodes used with a transformer that has a centre-tapped secondary. The diodes are connected to conduct during positive half-cycles of the transformer output to give a positive output waveform.

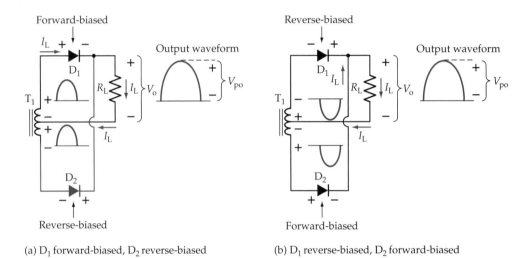

(a) D_1 forward-biased, D_2 reverse-biased

(b) D_1 reverse-biased, D_2 forward-biased

Figure 3-4 In a two-diode full-wave rectifier circuit, diodes D_1 and D_2 conduct alternately. When D_1 is forward-biased, D_2 is reverse-biased, and vice versa.

For the duration of the negative half-cycle of the transformer output, the polarity of the transformer secondary voltage causes D_1 to be reverse-biased and D_2 to be forward-biased (see Fig. 3-4b). I_L flows from the bottom terminal of the transformer secondary through diode D_2, through R_L from top to bottom, and back to the transformer centre tap. The output waveform is the combination of the two half-cycles, that is, a continuous series of positive half-cycles of sinusoidal waveform. This is *positive full-wave rectification*.

Figure 3-5 shows that if the polarity of the diodes is reversed, the output waveform is a series of sinusoidal negative half-cycles: negative full-wave rectification. The centre tap of the transformer is normally grounded, as shown in Fig. 3-5, and so the only way to obtain a negative output from this type of circuit is to reverse the diode polarity.

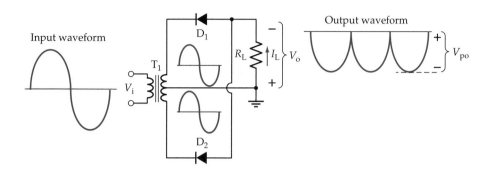

Input waveform

Output waveform

V_i

T_1

D_1

R_L

I_L V_o

D_2

V_{po}

Figure 3-5 Two-diode negative full-wave rectification is produced by connecting the diodes to conduct during negative half-cycles of the transformer output.

Bridge Rectifier

The centre-tapped transformer used in the circuit of Fig. 3-3 is usually more expensive and requires more space than additional diodes. So a *bridge rectifier* is the circuit most frequently used for full-wave rectification.

The bridge rectifier circuit in Fig. 3-6 is seen to consist of four diodes connected with their arrowhead symbols all pointing toward the positive output terminal of the circuit. Diodes D_1 and D_2 are series-connected, as are D_3 and D_4. The AC input terminals are the junction of D_1 and D_2 and the junction of D_3 and D_4. The positive output terminal is at the cathodes of D_1 and D_3, and the negative output is at the anodes of D_2 and D_4.

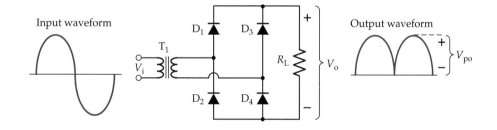

Input waveform

Output waveform

V_i

T_1

D_1 D_3

R_L V_o

D_2 D_4

V_{po}

Figure 3-6 A full-wave bridge rectifier circuit uses four diodes and a transformer with a single secondary winding.

During the positive half-cycle of input voltage, diodes D_1 and D_4 are in series with R_L, as illustrated in Figs. 3-7a and b. Load current (I_L) flows from the positive input terminal through D_1 to R_L, and then through R_L and D_4 back to the negative input terminal. Note that the direction of the load current through R_L is from top to bottom. During this time, the positive input terminal is applied to the cathode of D_2 and the negative output is at the D_2 anode (see Fig. 3-7a). So D_2 is reverse-biased during the positive half-cycle of the input. Similarly, D_3 has the negative input at its anode and the positive output at its cathode during the positive input half-cycle, causing D_3 to be reverse-biased.

Figures 3-7c and d show that diodes D_2 and D_3 are forward-biased during the negative half-cycle of the input waveform, while D_1 and D_4 are reverse-biased. Although the input terminal polarity is reversed, I_L again flows through R_L from top to bottom, via D_3 and D_2.

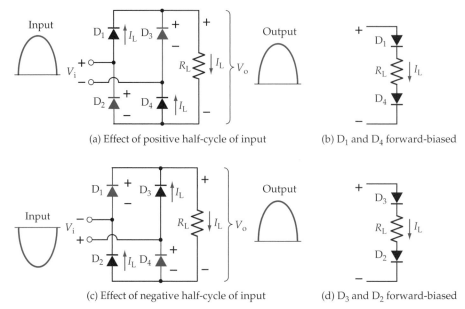

(a) Effect of positive half-cycle of input

(b) D_1 and D_4 forward-biased

(c) Effect of negative half-cycle of input

(d) D_3 and D_2 forward-biased

Figure 3-7   In a bridge rectifier circuit diodes D_1 and D_4 are forward-biased during the positive half cycle of the input, while D_2 and D_3 are reverse-biased. During the input negative half-cycle, D_2 and D_3 are forward-biased and D_1 and D_4 are reversed. In each case current flows through load resistor R_L in the same direction (from top to bottom).

It is seen that during both half-cycles of the input, the output terminal polarity is always positive at the top of R_L and negative at the bottom. Both positive and negative half-cycles of the input are passed to the output. The negative half-cycles are inverted, so that the output is a continuous series of positive half-cycles of sinusoidal voltage.

A full-wave bridge rectifier circuit always requires that the input be derived from a transformer that provides dc isolation from the supply. The circuit will not function correctly if one of its input terminals is grounded. However, given the required dc isolation between supply and output, either output terminal may be grounded to provide a positive or negative output voltage.

The bridge rectifier has two forward-biased diodes in series with the supply voltage and the load. Because each diode has a forward voltage drop (V_F), the peak output voltage is

$$V_{po} = V_{pi} - 2V_F \qquad (3\text{-}5)$$

The average and rms values of the full-wave rectified waveform can be determined as follows: $V_{o(ave)} = 0.637 \, V_{po}$ and $V_{o(rms)} = 0.707 \, V_{po}$. However, once again it should be noted that most rectifier circuits use a *reservoir capacitor* to smooth the rectified voltage wave into direct voltage, and *the presence of the capacitor changes the output waveform and substantially affects the load current and voltage and the diode current and voltage.* See Sections 3-2 and 3-3.

Example 3-2

Determine the peak output voltage and current for the bridge rectifier circuit in Fig. 3-7 when $V_i = 30$ V, $R_L = 300$ Ω, and the diodes have $V_F = 0.7$ V.

Solution

$$V_{pi} = 1.414\ V_i = 1.414 \times 30\ \text{V}$$

$$= 42.42\ \text{V}$$

Eq. 3-5:

$$V_{po} = V_{pi} - 2\ V_F = 42.42\ \text{V} - (2 \times 0.7\ \text{V})$$

$$= 41\ \text{V}$$

Eq. 3-2:

$$I_p = \frac{V_{po}}{R_L} = \frac{41\ \text{V}}{300\ \Omega}$$

$$= 137\ \text{mA}$$

More Bridge Rectifier Circuits

Figure 3-8 shows two common methods of drawing a bridge rectifier circuit. Although they both look more complex than the circuit in Fig. 3-7, they are exactly the same circuit. The cathodes of D_1 and D_3 in all three circuits are connected to the positive output terminal, and the anodes of D_2 and D_4 are connected to the negative output terminal. The ac input is applied to the junction of D_1 and D_2 and to the junction of D_3 and D_4.

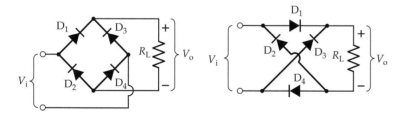

Figure 3-8 Two additional ways of drawing a bridge rectifier circuit diagram. Both of these circuits have the diodes connected exactly as in Fig. 3-6.

Practice Problems

3-2.1 Determine the peak load voltage, peak current, and power dissipation in a 470 load resistor connected to a bridge rectifier circuit that has a 24 V ac input. The rectifier diodes are germanium.

3-2.2 A bridge rectifier with silicon diodes and a 680 Ω load resistor has an 18 V peak output. Calculate the power dissipated in the load resistor and the rms input voltage.

3-3 HALF-WAVE RECTIFIER POWER SUPPLY

Capacitor Filter Circuit

When a sinusoidal alternating voltage is rectified (see Sections 3-1 and 3-2), the resulting output waveform is a series of positive (or negative) half-cycles of the input; it is *not* direct voltage. To convert to direct voltage (dc voltage), a *smoothing circuit* or *filter* must be employed. Figure 3-9a shows a half-wave rectifier circuit with a single capacitor filter (C_1) and a load resistor (R_L), and Fig. 3-9b shows the output waveform. The capacitor, termed a *reservoir capacitor*, is charged almost to the peak level of the circuit input voltage when the diode is forward-biased. This occurs at V_{pi}, as illustrated in Fig. 3-9c, giving a peak capacitor voltage:

$$V_C = V_{pi} - V_F \qquad (3\text{-}6)$$

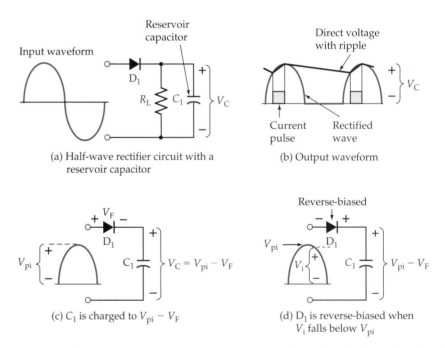

(a) Half-wave rectifier circuit with a reservoir capacitor

(b) Output waveform

(c) C_1 is charged to $V_{pi} - V_F$

(d) D_1 is reverse-biased when V_i falls below V_{pi}

Figure 3-9 A reservoir capacitor smooths the output from a rectifier circuit by charging to the peak output voltage and retaining most of its charge between peaks.

When the instantaneous level of input voltage (at the diode anode) falls below V_{pi}, the diode becomes reverse-biased because the capacitor voltage (V_C) (at the diode cathode) remains close to ($V_{pi} - V_F$) (see Fig. 3-9d). With the diode reverse-biased, there is no capacitor charging current, and the capacitor begins to discharge through the load resistor (R_L). So V_C falls slowly, as shown by the capacitor voltage waveform in Fig. 3-9b.

The diode remains reverse-biased (as V_C decreases) throughout the rest of the input positive half-cycle, the negative half-cycle, and the first part of the positive half-cycle again until the instantaneous level of V_i becomes greater than V_C once more. At this point, current flows through the diode to recharge the capacitor, causing the capacitor voltage to return to $(V_{pi} - V_F)$. The charge and discharge of the capacitor cause the small increase and decrease in the capacitor voltage, which is also the circuit output voltage. It is seen that the circuit output is a direct voltage with a small ripple voltage waveform superimposed (Fig. 3-9b).

Ripple Amplitude and Capacitance

The amplitude of the ripple voltage is affected by the load current, the reservoir capacitor value, and the capacitor discharge time. The discharge time depends upon the frequency of the ripple waveform, which is the same as the ac input frequency in the case of a half-wave rectifier. With a constant load current, the ripple amplitude is inversely proportional to the capacitance: the largest capacitance produces the smallest ripple.

The ripple amplitude can be calculated from the capacitor value, the load current, and the capacitor discharge time. Consider the circuit output voltage waveform illustrated in Fig. 3-10a. The waveform quantities are as follows:

E_{ave} average DC output voltage
$E_{o(max)}$ maximum output voltage
$E_{o(min)}$ minimum output voltage
V_r ripple voltage peak-to-peak amplitude
T time period of the ac input waveform
t_1 capacitor discharge time
t_2 capacitor charge time
θ_1 phase angle of the input wave from zero to $E_{o(min)}$
θ_2 phase angle of the input wave from $E_{o(min)}$ to $E_{o(max)}$

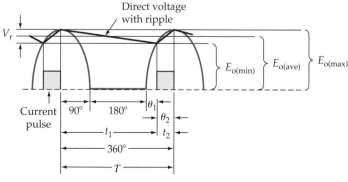

(a) Capacitor waveform amplitudes, angles, and times

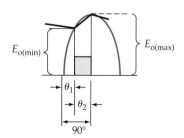

(b) Relationship between $E_{o(min)}$ and $E_{o(max)}$

Figure 3-10 The capacitance value for a reservoir capacitor can be calculated from the load current, ripple voltage, and input frequency.

Figure 3-10b shows that, because the input wave is sinusoidal,

$$E_{o(min)} = E_{o(max)} \sin \theta_1$$

which gives

$$\theta_1 = \sin^{-1} \frac{E_{o(min)}}{E_{o(max)}} \tag{3-7}$$

Also from Fig. 3-10b,

$$\theta_2 = 90° - \theta_1 \tag{3-8}$$

The time period is

$$T = \frac{1}{f}$$

where f is the frequency of the ac input waveform.

The input waveform goes through a 360° phase angle during time T, which gives the time per degree as

$$t/\text{degree} = \frac{T}{360°}$$

So

$$t_2 = \frac{\theta_2 T}{360°} \tag{3-9}$$

and

$$t_1 = T - t_2 \tag{3-10}$$

Taking the current as a constant quantity,

$$C_1 = \frac{I_L t_1}{V_r} \tag{3-11}$$

Approximate Calculations

Equations 3-9 to 3-11 can be used to analyze a filter circuit or to calculated a capacitance value for a specified ripple amplitude. The assumption that the load current is constant (for Eq. 3-11) is an approximation, because the load current changes by a small amount as the output voltage falls (due to ripple). Normally, the change in the load current is so small that it has no significant effect on the calculation. Another way of simplifying the calculations is to take the discharge time (t_1) as approximately equal to the input waveform time period (Fig. 3-10a).

$$t_1 \approx T \tag{3-12}$$

This implies that the capacitor charging time (t_2) is so much smaller than t_1 that it can be neglected, which is a reasonable assumption where the ripple voltage is small (about 10% of V_o). Using Eq. 3-12 when calculating capacitance gives a slightly larger value than a more precise calculation. This is acceptable because a larger capacitor will produce a smaller ripple voltage.

Example 3-3

Determine the peak-to-peak ripple voltage for a half-wave rectifier and filter circuit which has a 680 μF reservoir capacitor, an average output of 28 V, and a 200 Ω load resistance.

Solution

$$I_L = \frac{E_{o(ave)}}{R_L} = \frac{28 \text{ V}}{200 \text{ Ω}}$$

$$= 140 \text{ mA}$$

$$T = \frac{1}{f} = \frac{1}{60 \text{ Hz}}$$

$$= 16.7 \text{ ms}$$

Eq. 3-12:

$$t_1 \approx T = 16.7 \text{ ms}$$

From Eq. 3-11,

$$V_r = \frac{I_L t_1}{C_1} = \frac{140 \text{ mA} \times 16.7 \text{ ms}}{680 \text{ μF}}$$

$$\approx 3.4 \text{ V}$$

Capacitor Selection

When a capacitance value is determined for a specified ripple, a suitable capacitor has to be chosen from a manufacturer's list of available standard values. The calculated capacitance for a reservoir capacitor is always the lowest that can give a specified ripple voltage amplitude. If a larger capacitance than calculated is used, the ripple voltage will be lower than the specified maximum. So a larger standard-value capacitor is always chosen. The standard-value capacitors listed in Appendix B-2 are usually available with ±20% tolerance. In the case of capacitors greater than 10 μF, the tolerance is often listed as −10% +50%. This means that a 100 μF capacitor might have a capacitance as low as 90 μF or as high as 150 μF. In dc power supply circuits, a higher capacitance value is always acceptable.

The voltage levels that a capacitor will be subjected to must be taken into consideration. The maximum voltage that may be safely applied to a capacitor is stated in terms of its dc *working voltage*. This can be quite small for large-value capacitors. For example, some 10 μF capacitors have 6.3 V working voltages. If the specified working voltage is exceeded, the capacitor dielectric may break down.

Capacitor Polarity

It is very important that polarized capacitors be connected with the correct polarity. The positive terminal should be connected to the more positive of the two points in the circuit where the capacitor is to be installed (see Fig. 3-11). The straight bar on the capacitor graphic symbol represents the positive terminal. *Polarized capacitors can sometimes explode when incorrectly connected.*

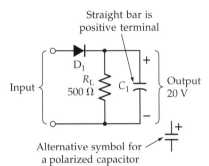

Straight bar is positive terminal

Input

D_1

R_L 500 Ω C_1

Output 20 V

Alternative symbol for a polarized capacitor

Figure 3-11 It is very important that polarized capacitors be connected with the correct polarity.

Example 3-4

A half-wave rectifier dc power supply is to provide 20 V to a 500 Ω load (as in Fig. 3-11). The peak-to-peak ripple voltage is not to exceed 10% of the average output voltage, and the ac input frequency is 60 Hz. Calculate the required reservoir capacitance.

Solution

$$T = \frac{1}{f} = \frac{1}{60 \text{ Hz}}$$

$$= 16.7 \text{ ms}$$

Eq. 3-12:

$$t_1 \approx T = 16.7 \text{ ms}$$

$$V_r = 10\% \text{ of } E_{o(ave)}$$

$$= 10\% \text{ of } 20 \text{ V}$$

$$= 2 \text{ V}$$

$$I_L = \frac{E_{o(ave)}}{R_L} = \frac{20 \text{ V}}{500 \text{ Ω}}$$

$$= 40 \text{ mA}$$

Eq. 3-11:

$$C_1 = \frac{I_L t_1}{V_r} = \frac{40 \text{ mA} \times 16.7 \text{ ms}}{2 \text{ V}}$$

$$= 334 \text{ μF (use 330 μF standard value)}$$

Example 3-4 uses the approximate method of calculating the capacitor value (assuming $t_1 \approx T$). A more precise determination of C_1 can be made by first calculating t_1 and t_2.

Example 3-5

Determine the times t_1 and t_2 for the half-wave rectifier power supply in Ex. 3-4, and recalculate C_1.

Solution

$$V_r = 10\% \text{ of } E_{o(ave)} = 10\% \text{ of } 20 \text{ V}$$

$$= 2 \text{ V}$$

$$E_{o(min)} = E_{o(ave)} - 0.5\,V_r = 20 \text{ V} - (0.5 \times 2 \text{ V})$$

$$= 19 \text{ V}$$

$$E_{o(max)} = E_{o(ave)} + 0.5\,V_r = 20 \text{ V} + (0.5 \times 2 \text{ V})$$

$$= 21 \text{ V}$$

Eq. 3-7:
$$\theta_1 = \sin^{-1}\frac{E_{o(min)}}{E_{o(max)}} = \sin^{-1}\frac{19 \text{ V}}{21 \text{ V}}$$

$$\approx 65°$$

Eq. 3-8:
$$\theta_2 = 90° - \theta_1 = 90° - 65°$$

$$= 25°$$

Eq. 3-9:
$$t_2 = \frac{\theta_2 T}{360°} = \frac{25° \times 16.7 \text{ ms}}{360°}$$

$$= 1.16 \text{ ms}$$

Eq. 3-10:
$$t_1 = T - t_2 = 16.7 \text{ ms} - 1.16 \text{ ms}$$

$$= 15.54 \text{ ms}$$

Eq. 3-11:
$$C_1 = \frac{I_L t_1}{V_r} = \frac{40 \text{ mA} \times 15.54 \text{ ms}}{2 \text{ V}}$$

$$= 310 \text{ }\mu\text{F}$$

If the approximate and precisely determined reservoir capacitance values are compared, it is seen that the approximate method gives a larger capacitance than actually required. Consequently, when the approximate method is used, it is acceptable to choose a standard capacitance value slightly smaller than the value calculated.

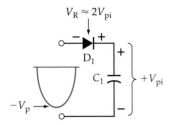

Figure 3-12 The diode reverse voltage in a half-wave rectifier circuit with a reservoir capacitor is twice the peak of the input voltage.

Diode Specification

Rectifier diodes must be specified in terms of the currents and voltages that they are subjected to. The diodes selected must be able to survive higher levels than the calculated maximums for a given circuit. Consider Fig. 3-12, which illustrates the situation when the ac input wave is at its negative peak voltage ($-V_p$). The capacitor has already been charged up to approximately the positive peak level of the input ($+V_p$). Consequently, the diode has $-V_p$ at its anode and $+V_p$ at its cathode, and so the diode peak reverse voltage is

$$V_R \approx 2 V_p \qquad (3\text{-}13)$$

The average forward rectified current ($I_{F(av)}$) that the diode must pass is equal to the dc output current:

$$I_{F(av)} = I_L \qquad (3\text{-}14)$$

As explained, the diode in a half-wave rectifier circuit with a reservoir capacitor does not conduct continuously, but repeatedly passes pulses of current to recharge the capacitor each time the diode becomes forward-biased. This is illustrated in Fig. 3-13 (and in Fig. 3-10). The current pulse is known as the *repetitive surge current* and is designated I_{FRM}. The repetitive surge current is actually likely to be spike-shaped, but it can be thought of as rectangular for calculation purposes. Since the average input current to the rectifier circuit must equal the average load current (I_L), I_{FRM} averaged over time period T equals I_L (see Fig. 3-13):

$$I_L = \frac{I_{FRM}\, t_2}{(t_1 + t_2)}$$

giving

$$I_{FRM} = \frac{I_L\,(t_1 + t_2)}{t_2} \qquad (3\text{-}15)$$

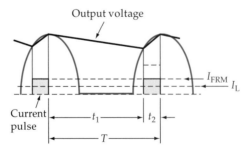

Output voltage

Figure 3-13 A repetitive surge current flows to recharge the capacitor in a rectifier circuit with a reservoir capacitor. The output load current is the average value of the repetitive surge current.

The repetitive surge current is an important quantity that *must* be known when a rectifier diode is being selected for a particular application. Unlike the case of the reservoir capacitor, there is no approximate method for finding a suitable value for I_{FRM}. In order to calculate the value, the quantities t_1 and t_2 must first be determined by using Equations 3-7 to 3-10 (as in Example 3-5).

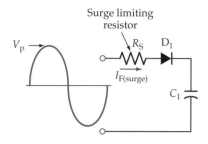

Surge limiting
resistor

V_P

R_S D_1

$I_{F(surge)}$

C_1

Figure 3-14 A surge current flows at the instant of switch-on of the ac input to a rectifier circuit with a reservoir capacitor. Diodes must be protected by the use of a surge-limiting resistor.

Figure 3-14 shows a half-wave rectifier circuit with a resistor (R_S) connected in series with the diode. This is a low-resistance component known as a *surge limiting resistor*. As its name suggests, the purpose of R_S is to limit any surge current that might pass through the diode. The highest surge current occurs when the ac supply is first switched on to the rectifier circuit. Before switch-on, the reservoir capacitor normally contains no charge, and so it behaves as a short-circuit at the instant of switch-on. If switch-on occurs at the instant when the ac input is at its peak level, the surge current is

$$I_{F(surge)} = \frac{V_P}{R_S}$$

For a diode with a specified maximum *non-repetitive surge current* (I_{FSM}), the surge limiting resistor is calculated as

$$R_S = \frac{V_P}{I_{FSM}} \qquad (3\text{-}16)$$

It should be noted that, because a transformer is normally used at the input to a rectifier circuit, and because its secondary winding might have a resistance greater than the required value of R_S, an additional surge limiting resistor might not be required.

Example 3-6

Specify the diode for the half-wave rectifier power supply circuit in Examples 3-4 and 3-5. Select a suitable device from Appendix A, and calculate the required surge-limiting resistance.

Solution

$$V_P = E_{o(max)} + V_F = 21 \text{ V} + 0.7 \text{ V}$$

$$= 21.7 \text{ V}$$

Eq. 3-13: $$V_R \approx 2\,V_P = 2 \times 21.7 \text{ V}$$

$$= 43.4 \text{ V}$$

Eq. 3-14: $I_{F(av)} = I_L = 40\text{ mA}$

Eq. 3-15: $I_{FRM} = \dfrac{I_L(t_1 + t_2)}{t_2} = \dfrac{40\text{ mA} \times 16.7\text{ ms}}{1.16\text{ ms}}$

$= 576\text{ mA}$

In data sheet A-1 in Appendix A, for the 1N4001, $V_R = 50$ V, $I_{F(av)} = I_o = 1$ A, and $I_{FRM} = I_{Fm(rep)} = 10$ A. So its specification is better than required for this application. Any of the 1N4002 to 1N4007 diodes could be used, but those with higher reverse voltage are more expensive than the 1N4001.

For the 1N4001: $I_{F(surge)} = 30$ A

Eq. 3-16: $R_S = \dfrac{V_p}{I_{FSM}} = \dfrac{21.7\text{ V}}{30\text{ A}}$

$\approx 0.7\ \Omega$

Transformer Selection

A power supply transformer is normally defined in terms of its rms input and output voltage and current. The input is usually the 115 V, 60 Hz supply, and the transformer peak output voltage is calculated by adding the rectifier voltage drop to the power supply peak output. The peak voltage is then converted to rms to give the secondary value.

$$V_{s(rms)} = 0.707\,(E_{o(max)} + V_F) \tag{3-17}$$

The rms secondary current determination is not quite as straightforward as the voltage calculation. For a half-wave rectifier circuit with a resistive load, the relationship can be shown to be $I_{rms} = 2.2\,I_{L(dc)}$. However, as explained, the reservoir capacitor causes a peak repetitive current to be passed through the rectifiers from the transformer, and this complicates the determination of the transformer secondary current. For half-wave rectifiers with capacitor filters, manufacturers of power supply transformers recommend the following relationships:

$$I_{L(dc)} = 0.28\,I_{s(rms)}$$

which gives $$I_{s(rms)} = 3.6\,I_{L(dc)} \tag{3-18}$$

The transformer primary current is

$$I_{p(rms)} = \dfrac{V_{s(rms)} \times I_{s(rms)}}{V_{p(rms)}} \tag{3-19}$$

Example 3-7
Specify the transformer for the half-wave rectifier power supply circuit in Examples 3-4 and 3-5.

Solution

Eq. 3-17: $V_{s(rms)} = 0.707\,(E_{o(max)} + V_F) = 0.707\,(21\,V + 0.7\,V)$

$= 15.3\,V$

Eq. 3-18: $I_{s(rms)} = 3.6\,I_{L(dc)} = 3.6 \times 40\,mA$

$= 144\,mA$

$V_{p(rms)} = 115\,V, 60\,Hz$

Eq. 3-19: $I_{p(rms)} = \dfrac{V_{s(rms)} \times I_{s(rms)}}{V_{p(rms)}} = \dfrac{15.3\,V \times 144\,mA}{115\,V}$

$= 19.2\,mA$

Practice Problems
3-3.1 A dc power supply consisting of a half-wave rectifier and a reservoir capacitor has to supply 12 V to a 200 Ω load. Determine the capacitance required if the maximum ripple voltage is to be ±5% of the average output and the input frequency is 60 Hz.

3-3.2 For Problem 3-3.1, recalculate the capacitance approximately by assuming that the capacitor discharge time is very much larger than the charging time.

3-3.3 For the circuit in Problem 3-3.1, specify the diode, select a suitable device from the available data sheets, determine the required surge limiting resistance, and specify a suitable transformer.

3-4 FULL-WAVE RECTIFIER POWER SUPPLY

Like half-wave rectifiers, full-wave rectifiers require filter circuits in order to convert the output waveform to direct voltage. Figure 3-15 shows a full-wave rectifier circuit with a reservoir capacitor and a surge-limiting resistor. These components operate exactly as explained for the half-wave rectifier circuit, with a few important exceptions.

The capacitor-smoothed full-wave rectifier waveforms are shown in detail in Fig. 3-16. Equation 3-7, derived for the half-wave rectifier circuit, still applies for determining the angle θ_1.

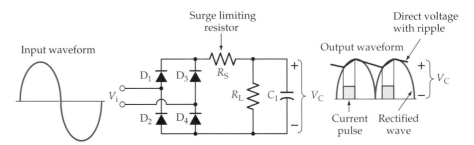

Figure 3-15 Bridge rectifier circuit with a reservoir capacitor to smooth the output voltage and a surge limiting resistor to protect the diodes.

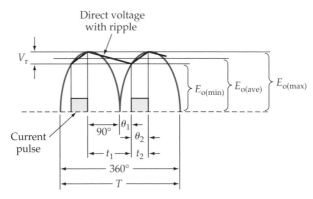

Figure 3-16 Ripple voltage, waveform angles, and time periods for a full-wave rectifier circuit with a reservoir capacitor.

Eq. 3-7:
$$\theta_1 = \sin^{-1} \frac{E_{o(min)}}{E_{o(max)}}$$

θ_2 can be determined from

Eq. 3-8:
$$\theta_2 = 90° - \theta_1$$

and to calculate time t_2,

Eq. 3-9:
$$t_2 = \frac{\theta_2 T}{360°}$$

A comparison of Figs. 3-16 and 3-10 shows that the capacitor discharge time t_1 for the half-wave rectifier circuit is approximately equal to the waveform time period T, while for the full-wave rectifier t_1 approximately equals $T/2$. More precisely,

$$t_1 = \left(\frac{T}{2}\right) - t_2 \qquad \text{(3-20)}$$

By using the correct value of t_1, the reservoir capacitance for a full-wave rectifier circuit can be calculated from

Eq. 3-11:
$$C = \frac{I_L t_1}{V_r}$$

Similarly (with the correct value of t_1 in Fig. 3-16), the repetitive current (I_{FRM}) can be determined from

Eq. 3-15:
$$I_{FRM} = \frac{I_L(t_1 + t_2)}{t_2}$$

The average forward current passed by the bridge rectifier circuit is equal to the load current. But each pair of diodes supplies current for no more than a half-cycle of the input wave. The other pair conducts during the other half-cycle. So, as illustrated in Fig. 3-17, the average forward current for each diode is half of the load current.

$$I_{F(av)} = \frac{I_L}{2} \tag{3-21}$$

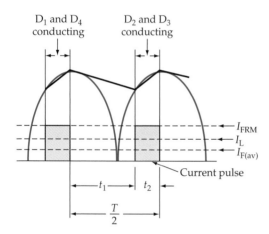

Figure 3-17 Because the diodes in a bridge rectifier conduct during alternate half-cycles, the average forward current ($I_{F(av)}$) in each diode is equal to half the load current (I_L).

Another difference between the half-wave and full-wave rectifier power supply circuits concerns the reverse voltage applied to the diodes. Consider Fig. 3-18. When the instantaneous input voltage is $+V_p$, as illustrated, V_p is

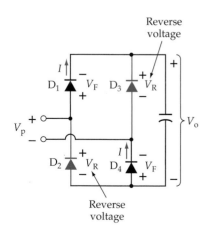

Figure 3-18 The diode reverse voltage in a bridge rectifier circuit with a reservoir capacitor is equal to the peak of the input voltage.

applied across forward-biased diode D_1 in series with reverse-biased diode D_3. Therefore, the reverse voltage across D_3 is

$$V_R = V_p - V_F$$

or

$$V_R \approx V_p$$

Examination of the circuit in Eq. 3-18 shows that $V_R \approx V_p$ applies to each diode when it is reverse-biased.

As in the case of the half-wave circuit, capacitor and ripple calculations can be simplified by assuming that time t_2 is very much smaller than t_1. This gives the approximation that the capacitor discharge time is equal to half the input waveform time period.

$$t_1 \approx \frac{T}{2} \tag{3-22}$$

Example 3-8

The full-wave rectifier dc power supply in Fig. 3-19 is to supply 20 V to a 500 Ω load. The peak-to-peak ripple voltage is not to exceed 10% of the average output voltage, and the ac input frequency is 60 Hz. Accurately calculate the required reservoir capacitor value. (Note that the specifications for this example are similar to those for the half-wave circuit in Examples 3-4 and 3-5.)

Solution

From Example 3-5, $V_r = 2$ V, $T = 16.7$ ms, $t_2 = 1.16$ ms, and $I_L = 40$ mA

Eq. 3-20:

$$t_1 = \frac{T}{2} - t_2 = \frac{16.7 \text{ ms}}{2} - 1.16 \text{ ms}$$

$$= 7.19 \text{ ms}$$

Eq. 3-11:

$$C_1 = \frac{I_L t_1}{V_r} = \frac{40 \text{ mA} \times 7.19 \text{ ms}}{2 \text{ V}}$$

$$= 144 \ \mu\text{F (use 150 } \mu\text{F standard value)}$$

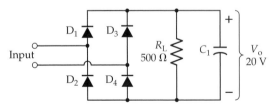

Figure 3-19 Full-wave rectifier power supply for Example 3-8.

Example 3-9

Assuming that t_2 is very much smaller than t_1, for the full-wave rectifier circuit in Example 3-8, recalculate the required reservoir capacitor value.

Solution

From Example 3-5, $V_r = 2$ V, $T = 16.7$ ms, and $I_L = 40$ mA

Eq. 3-22:
$$t_1 \approx \frac{T}{2} = \frac{16.7 \text{ ms}}{2}$$

$$= 8.35 \text{ ms}$$

Eq. 3-11:
$$C_1 = \frac{I_L t_1}{V_r} = \frac{40 \text{ mA} \times 8.35 \text{ ms}}{2 \text{ V}}$$

$$= 167 \text{ μF (use 150 μF standard value)}$$

Example 3-10

Specify the diodes for the bridge rectifier circuit in Ex. 3-8. Select a suitable device and calculate the surge-limiting resistance.

Solution

$$V_p = E_{o(max)} + (2 \, V_F) = 21 \text{ V} + (2 \times 0.7 \text{ V})$$

$$= 22.4 \text{ V}$$

$$V_R \approx V_p = 22.4 \text{ V}$$

Eq. 3-21:
$$I_{F(ave)} = \frac{I_L}{2} = \frac{40 \text{ mA}}{2}$$

$$= 20 \text{ mA}$$

Eq. 3-15:
$$I_{FRM} = \frac{I_L(t_1 + t_2)}{t_2} = \frac{40 \text{ mA} \times 8.35 \text{ ms}}{1.16 \text{ ms}}$$

$$= 288 \text{ mA}$$

From data sheet A-2 in Appendix A, it is seen that the 1N4001 is suitable.
     For the 1N4001, $I_{F(surge)} = 30$ A

Eq. 3-16:
$$R_S = \frac{V_p}{I_{FSM}} = \frac{22.4 \text{ V}}{30 \text{ A}}$$

$$\approx 0.75 \, \Omega$$

Transformer Selection

The transformer specification for a full-wave bridge rectifier power supply is determined similarly to that for a half-wave circuit, with some exceptions. Two diode voltage drops are involved in calculating the secondary rms voltage.

$$V_{s(rms)} = 0.707 \, (E_{o(max)} + 2V_F) \qquad \text{(3-23)}$$

For full-wave rectifiers with capacitor filters, power supply transformers manufacturers recommend

$$I_{L(dc)} = 0.62\, I_{s(rms)}$$

which gives

$$I_{s(rms)} = 1.6\, I_{L(dc)} \qquad \text{(3-24)}$$

As with all transformers, the primary current is

Eq. 3-19:

$$I_{p(rms)} = \frac{V_{s(rms)} \times I_{s(rms)}}{V_{p(rms)}}$$

Example 3-11

Specify the transformer for the full-wave rectifier power supply circuit in Examples 3-8 and 3-9.

Solution

Eq. 3-23:

$$V_{s(rms)} = 0.707(E_{o(max)} + 2V_F) = 0.707(21\ \text{V} + 1.4\ \text{V})$$

$$= 15.8\ \text{V}$$

Eq. 3-24:

$$I_{s(rms)} = 1.6\, I_{L(dc)} = 1.6 \times 40\ \text{mA}$$

$$= 64\ \text{mA}$$

$$V_{p(rms)} = 115\ \text{V, 60 Hz}$$

Eq. 3-19:

$$I_{p(rms)} = \frac{V_{s(rms)} \times I_{s(rms)}}{V_{p(rms)}} = \frac{15.8\ \text{V} \times 64\ \text{mA}}{115\ \text{V}}$$

$$= 8.8\ \text{mA}$$

Practice Problems

3-4.1 A dc power supply consisting of a bridge rectifier and a reservoir capacitor has to supply 15 V to a 150 Ω load. Determine the capacitance required if the maximum ripple voltage is to be ±5% of the average output and the input frequency is 60 Hz.

3-4.2 For Problem 3-4.1, recalculate the capacitance approximately by assuming that the capacitor discharge time is very much larger than the charging time.

3-4.3 For the circuit in Problem 3-4.1, specify the diodes, select a suitable device from the available data sheets, determine the required surge limiting resistance, and specify a suitable transformer.

3-5 POWER SUPPLY PERFORMANCE AND TESTING

Source Effect

The ac supply to the input of a transformer in a dc power supply does not always remain constant. A ±10% variation in the ac *source voltage* (V_S) (also termed the *line voltage*) is not unusual. When the source voltage varies, there is some variation in the dc output voltage from a power supply (see Fig. 3-20). This output voltage change (ΔE_o) due to a change in the input is termed the *source effect*. If the output varies by 100 mV when the source voltage changes by ±10%, the source effect is 100 mV. Another way of stating this output change is to express ΔE_o as a percentage of the dc output voltage (E_o). In this case, the term *line regulation* is used:

$$\text{Source effect} = \Delta E_o \text{ for a 10\% change in } V_S \qquad \text{(3-25)}$$

$$\text{Line regulation} = \frac{(\Delta E_o \text{ for a 10\% change in } V_S) \times 100\%}{E_o} \qquad \text{(3-26)}$$

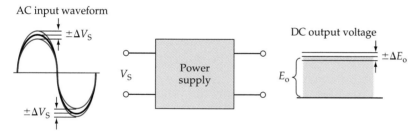

Figure 3-20 Power supply source effect is the output voltage change that occurs when the input voltage changes by ±10%.

Load Effect

Power supply output voltage is also affected by changes in load current (I_L). The output voltage decreases when I_L is increased and rises when I_L is reduced. The *load effect* defines how the output voltage changes when the load current is increased from zero to its specified maximum level ($I_{L(max)}$). If the load current change (ΔI_L) produces a voltage change (ΔE_o) of 100 mV, the load effect is 100 mV. As for the source effect, the load effect can also be expressed as a percentage of the output voltage. This is termed the *load regulation*.

$$\text{Load effect} = \Delta E_o \text{ for } \Delta I_{L(max)} \qquad \text{(3-27)}$$

$$\text{Load regulation} = \frac{(\Delta E_o \text{ for } \Delta I_{L(max)}) \times 100\%}{E_o} \qquad \text{(3-28)}$$

Test Procedure

Figure 3-21 shows an arrangement for dc power supply testing. The *variable voltage transformer* (*variac*) provides for adjustment of the ac input voltage; ac voltmeter V_1 measures the transformer output; dc voltmeter V_2 measures the power supply dc output voltage; and the oscilloscope displays the output waveform.

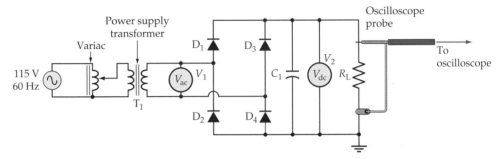

Figure 3-21   Circuit for testing a dc power supply. The ac input voltage is adjusted by means of the variac, and the power supply input and output voltages are monitored by voltmeters V_1 and V_2. The oscilloscope displays the ripple voltage waveform.

The test procedure is as follows:

- Set the variac for zero output, then connect the ac supply and switch on.
- Slowly adjust the variac to set the power supply output to the desired level (E_o).
- Use the oscilloscope to measure the amplitude of the output ripple voltage (V_r).
- Adjust the variac to decrease the power supply input voltage by 10%. Measure the dc output voltage change (the source effect, $\Delta E_{o(source)}$).
- Adjust the variac to return the input voltage to its normal level. Disconnect R_L, and measure the resultant dc output voltage change (the load effect, $\Delta E_{o(load)}$).

Example 3-12

The output voltage of a dc power supply changes from 20 V to 19.7 V when the load is increased from zero to maximum. The voltage also increases to 20.2 V when the ac supply increases by 10%. Calculate the load and source effects and the load and line regulations.

Solution

Eq. 3-27:   Load effect $= \Delta E_o$ for $\Delta I_{L(max)} = 20\ V - 19.7\ V$

$$= 300\ mV$$

Eq. 3-28:   Load regulation $= \dfrac{(\Delta E_o \text{ for } \Delta I_{L(max)}) \times 100\%}{E_o} = \dfrac{300 \text{ mV} \times 100\%}{20 \text{ V}}$

$= 1.5\%$

Eq. 3-25:   Source effect $= \Delta E_o$ for 10% change in $V_S = 20.2$ V $- 20$ V

$= 200$ mV

Eq. 3-26:   Line regulation $= \dfrac{(\Delta E_o \text{ for } 10\% \text{ change in } V_S) \times 100\%}{E_o}$

$= \dfrac{200 \text{ mV} \times 100\%}{20 \text{ V}}$

$= 1\%$

Short Circuits and Open Circuits

All power supplies normally have a fuse on the input side of the supply trans-former, as shown in Fig. 3-22, and a blown fuse usually indicates a power sup-ply fault. The symptoms are zero output voltage with the input connected and switched on. The fuse is checked and found to be blown (melted). This is usu-ally due to an overload or demand for an excessive supply current; this in turn is most often due to a short circuit, either at the output terminals or in the power supply circuit. In a laboratory situation the overload might be caused, for example, by a fault in a circuit being tested. If this appears to be the case, the circuit should be disconnected from the power supply, the fuse replaced, and the power supply switched on again and checked for the correct output voltage. Obviously, the circuit fault must be located and repaired before the circuit is reconnected to the power supply.

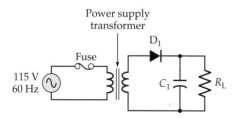

Figure 3-22   All dc power supplies normally have a fuse on the supply side of the transformer. When an overload occurs, the fuse will melt, thus interrupting the ac supply.

When the problem appears to be in the power supply circuit, *the ac supply should be switched off and the supply cord disconnected before anyone works on the circuit* (Fig. 3-23a). It should also be kept in mind that *charged capacitors can have a terminal voltage even when the supply is disconnected*. Small capacitors may be temporarily short-circuited with an insulated wire to ensure dis-charge. Large-value capacitors should be discharged via a suitable resistor (see Fig. 3-23b).

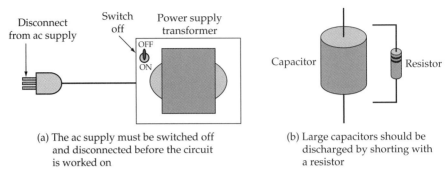

(a) The ac supply must be switched off and disconnected before the circuit is worked on

(b) Large capacitors should be discharged by shorting with a resistor

Figure 3-23 Electricity can be lethal. The ac supply should be switched off and disconnected, and large capacitors should be discharged before a circuit is worked on.

Possible Faults and Remedies

- Incorrect connections in the case of a newly constructed power supply. The circuit connections should be checked very carefully.
- One or more diodes have become shorted or open-circuited as the result of an overload at the output terminals. This may be obvious when the diodes look *burned*. Faulty diodes must be replaced.
- The capacitor has developed a short-circuit or an open-circuit, possibly from being connected with the wrong polarity. If the capacitor is shorted, it will also have damaged the diodes. The capacitor and diodes will have to be replaced.
- The transformer windings have developed a short circuit or an open circuit. The transformer must be replaced.

Using an Oscilloscope

Diagnosing a circuit problem usually requires the use of an oscilloscope to investigate the waveforms at various points in the circuit, as illustrated in Fig. 3-24. The output voltage and ripple waveform should first be checked (Fig. 3-24a). The ripple should be zero without R_L connected at the output, and there should be an acceptable ripple voltage with R_L connected. An unsmoothed rectified waveform indicates that the reservoir capacitor is damaged or open-circuited. With a full-wave or bridge rectifier, an excessive level of ripple voltage suggests that one or more of the diodes is open-circuited. A zero output voltage with a half-wave rectifier circuit might mean that the diode is open-circuited or that there is no input voltage (from the transformer). Similarly, with a full-wave rectifier and a bridge rectifier, a zero output voltage could mean that all the diodes are open-circuited or that the problem lies with the rectifier circuit input.

Oscilloscope probes have two input terminals, one of which is grounded, and care must be taken to *avoid grounding more than one point in the circuit*. For example, consider the bridge rectifier circuit in Fig. 3-24b. If the negative output terminal is grounded and a probe is connected across the transformer secondary as shown, diode D_4 is short-circuited by the two ground connections.

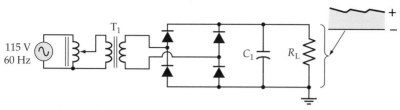

(a) Waveform at power supply output

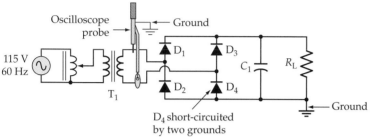

(b) More than one circuit ground point must be avoided

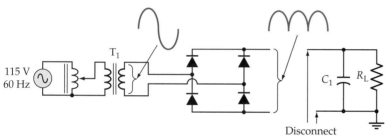

(c) Waveforms at rectifier circuit input and output

Figure 3-24   An oscilloscope is used to investigate the voltage waveforms at various points in a power supply circuit. Care must be taken to avoid more than one ground point.

To display the waveform at the transformer secondary, first make sure there are no other grounded points in the power supply circuit or in any circuit connected to its output. Before the rectifier circuit input and output waveforms are investigated, the reservoir capacitor and load must be disconnected (Fig. 3-24c). The rectifier output waveform is changed by the presence of the capacitor, and the transformer output can be checked only when there are no other ground points, as already explained.

Practice Problem

3-5.1 A dc power supply output drops from 15 V to 14.95 V when the ac source voltage falls by 10%. The output also falls from 15 V to 14.9 V when the load current is increased from zero to maximum. Calculate the source effect, load effect, line regulation, and load regulation.

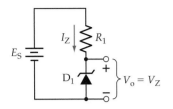

Figure 3-25 Zener diode used as a reference voltage source, or voltage regulator.

3-6 ZENER DIODE VOLTAGE REGULATORS

Regulator Circuit with No Load

The most important application of Zener diodes (discussed in Section 2-9) is in dc voltage regulator circuits. These can be the simple regulator circuit shown in Fig. 3-25 or the more complex regulators discussed in Chapter 18. The circuit in Fig. 3-25 is usually employed as a voltage reference source that supplies only a very low current (much lower than I_Z) to the output. Resistor R_1 in Fig. 3-25 limits the Zener diode current to the desired level. I_Z is calculated as follows:

$$I_Z = \frac{E_S - V_Z}{R_1} \qquad (3\text{-}29)$$

The Zener current may be just greater than the diode knee current (I_{ZK}). However, for the most stable reference voltage, I_Z should be selected as I_{ZT} (the specified test current). Example 3-13 demonstrates the circuit design procedure.

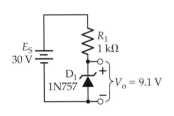

Figure 3-26 Zener diode circuit designed in Example 3-13.

Example 3-13

A 9 V reference source is to use a series-connected Zener diode and resistor connected to a 30 V supply (see Fig. 3-26). Select suitable components, and calculate the circuit current when the supply voltage drops to 27 V.

Solution

Data sheet A-4 in Appendix A, for Zener diodes, shows that the most suitable device is a 1N757, which has $V_Z = 9.1$ V and $I_{ZT} = 20$ mA.

With $E_S = 30$ V:

From Eq. 3-29, $R_1 = \dfrac{E_S - V_Z}{I_Z} = \dfrac{30 \text{ V} - 9.1 \text{ V}}{20 \text{ mA}}$

$= 1.05$ kΩ (use 1 kΩ—see Table B in Appendix B)

P_D for R_1: $P_{R1} = I_1^2 R_1 = (20 \text{ mA})^2 \times 1 \text{ k}\Omega$

$= 0.4$ W

When $E_S = 27$ V:

$$I_Z = \frac{E_S - V_Z}{R_1} = \frac{27 \text{ V} - 9.1 \text{ V}}{1 \text{ k}\Omega}$$

$= 17.9$ mA

Loaded Regulator

When a Zener diode regulator has to supply a load current (I_L), as shown in Fig. 3-27, the total supply current (flowing through resistor R_1) is the sum of I_L and I_Z. Care must be taken to ensure that the minimum Zener diode current is large enough to keep the diode in reverse breakdown. Typically, $I_{Z(min)} = 5\,\text{mA}$ for a Zener diode with an I_{ZT} of 20 mA. The circuit current equation is

$$I_Z + I_L = \frac{E_S - V_Z}{R_1} \tag{3-30}$$

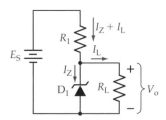

Figure 3-27   Zener diode voltage regulator circuit supplying a load current (I_L). The diode must be able to pass a maximum current of ($I_L + I_Z$).

In some cases, the load current in the type of circuit shown in Fig. 3-27 may be reduced to zero. Because the voltage drop across R_1 remains constant, the supply current remains constant:

$$I_{R1} = I_Z + I_L$$

All of this current flows through the Zener diode when R_L is disconnected. The circuit design must ensure that the total current does not exceed the maximum Zener diode current.

Example 3-14 demonstrates the design process for a loaded voltage reference source that uses a low-power Zener diode. It should be noted that high-power Zener diodes are available for higher-current applications.

Example 3-14

Design a 6 V dc reference source to operate from a 16 V supply (see Fig. 3-28). The circuit is to use a low-power Zener diode (as in data sheet A-4 in Appendix A) and is to produce the maximum possible load current. Calculate the maximum load current that can be taken from the circuit.

Solution

From the Zener diode data sheet A-4 in Appendix A, it is seen that the 1N753 has $V_Z = 6.2\,\text{V}$ and $P_D = 400\,\text{mW}$.

From Eq. 2-11,
$$I_{ZM} = \frac{P_D}{V_Z} = \frac{400\,\text{mW}}{6.2\,\text{V}}$$

$$= 64.5\,\text{mA}$$

$$I_{L(max)} + I_{Z(min)} = I_{ZM} = 64.5\,\text{mA}$$

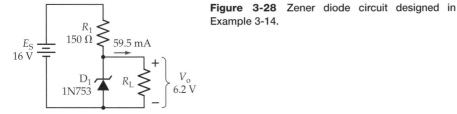

Figure 3-28 Zener diode circuit designed in Example 3-14.

From Eq. 3-29, $R_1 = \dfrac{E_S - V_Z}{I_{ZM}} = \dfrac{16 \text{ V} - 6.2 \text{ V}}{64.5 \text{ mA}}$

$$= 152 \ \Omega \text{ (use 150 } \Omega \text{ standard value)}$$

$$P_{R1} = I_1^2 R_1 = (64.5 \text{ mA})^2 \times 150 \ \Omega$$

$$= 0.62 \text{ W}$$

Select $I_{Z(min)} = 5 \text{ mA}$

$$I_{L(max)} = I_{ZM} - I_{Z(min)} = 64.5 \text{ mA} - 5 \text{ mA}$$

$$= 59.5 \text{ mA}$$

Regulator Performance

The performance of a Zener diode voltage regulator may be expressed in terms of the source and load effects and the line and load regulations, exactly as discussed in Section 3-5. Equations 3-25 to 3-28 may be applied. If there is an input ripple voltage, it will be severely attenuated. The *ripple rejection ratio* is the ratio of the output to input ripple amplitudes.

To assess the performance of a Zener diode voltage regulator, the ac equivalent circuit is first drawn by replacing the diode with its dynamic impedance (Z_Z), as shown in Fig. 3-29a. The complete ac equivalent circuit is seen to be a simple voltage divider. When the input voltage changes by ΔE_S, the output voltage change is

$$\Delta V_o = \frac{\Delta E_S \times Z_Z}{R_1 + Z_Z} \tag{3-31}$$

Equation 3-31 assumes that there is no load connected to the regulator output. When there is a load, R_L appears in parallel with Z_Z in the ac equivalent circuit (see Fig. 3-29b). The equation for the output voltage change now becomes

$$\Delta V_o = \frac{\Delta E_S \times (Z_Z \| R_L)}{R_1 + (Z_Z \| R_L)} \tag{3-32}$$

The regulator source effect can be determined from either Equation 3-31 or Equation 3-32, as appropriate. The equations can also be used for calculating the ripple rejection ratio. The input ripple amplitude (V_{ri}) and the output ripple (V_{ro}) are substituted for the input and output voltages in Equations 3-31 and 3-32. Thus, Equation 3-31 can be modified to give a ripple rejection ratio equation,

$$\frac{V_{ro}}{V_{ri}} = \frac{Z_Z}{R_1 + Z_Z} \tag{3-33}$$

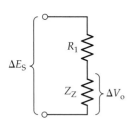

(a) Regulator without a load

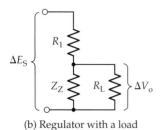

(b) Regulator with a load

Figure 3-29 AC equivalent circuits for Zener diode voltage regulator.

For a loaded regulator, Eq. 3-32 gives

$$\frac{V_{ro}}{V_{ri}} = \frac{Z_Z \| R_L}{R_1 + (Z_Z \| R_L)} \qquad (3\text{-}34)$$

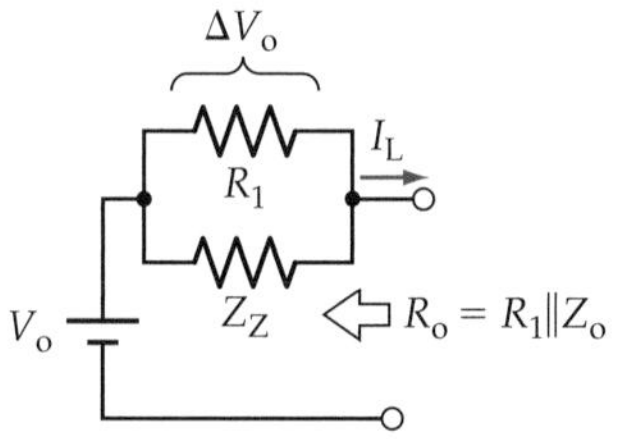

To determine the load effect of the Zener diode voltage regulator, the circuit output resistance has to be calculated. The regulator Thevenin equivalent circuit in Fig. 3-30 shows that, assuming a zero source resistance, the circuit output resistance is

$$R_o = Z_Z \| R_1 \qquad (3\text{-}35)$$

When the load current changes by ΔI_L, the output voltage change is

$$\Delta V_o = \Delta I_L (Z_Z \| R_1) \qquad (3\text{-}36)$$

Figure 3-30   Thevenin equivalent circuit for a Zener diode voltage regulator.

Example 3-15

Calculate the line regulation, load regulation, and ripple rejection ratio for the voltage regulator in Example 3-14.

Solution

From data sheet A-4 in Appendix A, it is seen that the 1N753 has $Z_Z = 7\ \Omega$.

Source effect:

$$\Delta E_S = 10\% \text{ of } E_S = 10\% \text{ of } 16\ \text{V}$$

$$= 1.6\ \text{V}$$

$$R_L = \frac{V_o}{I_L} = \frac{6.2\ \text{V}}{59.5\ \text{mA}}$$

$$= 104\ \Omega$$

Eq. 3-32: $\quad \Delta V_o = \dfrac{\Delta E_S \times (Z_Z \| R_L)}{R_1 + (Z_Z \| R_L)} = \dfrac{1.6\ \text{V} \times (7\ \Omega \| 104\ \Omega)}{150\ \Omega + (7\ \Omega \| 104\ \Omega)}$

$$= 67\ \text{mV}$$

Eq. 3-26: $\quad$ Line regulation $= \dfrac{(\Delta E_o \text{ for } 10\% \text{ change in } E_S) \times 100\%}{E_o}$

$$= \frac{67\ \text{mV} \times 100\%}{6.2\ \text{V}}$$

$$= 1.08\%$$

Load effect:

Eq. 3-36:
$$\Delta V_o = \Delta I_L (Z_Z \| R_1) = 59.5 \text{ mA} \times (7\ \Omega \| 150\ \Omega)$$
$$= 398 \text{ mV}$$

Eq. 3-28: Load regulation $= \dfrac{(\Delta E_o \text{ for } \Delta I_{L(max)}) \times 100\%}{E_o} = \dfrac{398 \text{ mV} \times 100\%}{6.2 \text{ V}}$
$$= 6.4\%$$

Ripple rejection:

Eq. 3-34:
$$\frac{V_{ro}}{V_{ri}} = \frac{Z_Z \| R_L}{R_1 + (Z_Z \| R_L)} = \frac{7\ \Omega \| 104\ \Omega}{150\ \Omega + (7\ \Omega \| 104\ \Omega)}$$
$$= 4.19 \times 10^{-2}$$

Practice Problems

3-6.1 Design a 12 V dc reference source (consisting of a Zener diode and series-connected resistor) to operate from a 25 V supply. Determine the effect on the diode current when the supply drops to 22 V.

3-6.2 An 8 V dc reference source is to be designed to produce the maximum possible output current from a low-power Zener diode. The supply voltage is 20 V. Design the circuit and determine the maximum load current.

3-6.3 Calculate the line effect, load effect, and ripple rejection for the circuit designed in Problem 3-6.2.

3-7 SERIES CLIPPING CIRCUITS

Series Clipper

The function of a *clipper* (or *limiter*) is to clip off an unwanted portion of a waveform. This is sometimes necessary to protect a device or circuit that might be destroyed by a large-amplitude signal (negative or positive).

A half-wave rectifier can be described as a clipper because it passes only the positive (or negative) half-cycle of an alternating waveform and clips off the other half-cycle. In fact, a diode *series clipper* is simply a half-wave rectifier circuit.

Figure 3-31a shows a *negative series clipper* circuit with a square wave input symmetrical above and below ground level. While the input is positive, D_1 is forward-biased and the positive half-cycle is passed to the output.

$$V_o = E - V_F \tag{3-37}$$

During the negative half-cycle of the input, the diode is reverse-biased. Consequently, the output remains at zero and the negative half-cycle is effectively clipped off.

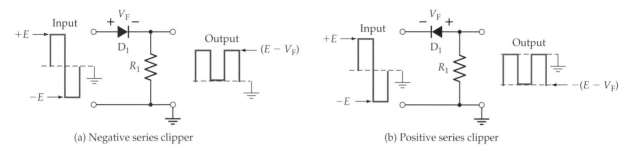

(a) Negative series clipper   (b) Positive series clipper

Figure 3-31   Diode series clipping circuits clip off unwanted portions of input waveforms. These are essentially half-wave rectifier circuits.

The *zero level* output from a series clipper circuit is not exactly zero. The reverse saturation current (I_R) of the diode produces a voltage drop across resistor R_1:

$$V_o = -I_R R_1 \qquad\qquad (3\text{-}38)$$

This voltage drop is almost always so small that it can be ignored.

If the diode in Fig. 3-31a is reconnected with reversed polarity, as shown in Fig. 3-31b, the positive half-cycles are clipped off and the circuit becomes a positive series clipper. The input waveforms to a clipper may be square, sinusoidal, or any other shape.

The output terminals of series clippers are usually connected to circuits that have a high input resistance; so the (current-limiting) resistor R_1 is selected to pass an acceptable minimum current through the diode. This current must be sufficient to operated the diode beyond the *knee* of its forward characteristic (see Fig. 2-4). Usually, a current of 1 mA is adequate. The resistor value is calculated from the output voltage and the selected current level. The remaining part of the design process is specifying the diode.

Example 3-16

The negative series clipper in Fig. 3-32 has a ±9 V input and zero load current. Determine a suitable resistance for R_1, and specify the diode forward current and reverse voltage. Use a silicon diode.

Solution

$$V_o = E - V_F$$
$$= 9\text{ V} - 0.7\text{ V}$$
$$= 8.3\text{ V}$$

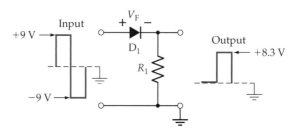

Figure 3-32 Series clipper circuit for Example 3-16.

Select $I_F = 1$ mA

$$R_1 = \frac{V_o}{I_F} = \frac{8.3 \text{ V}}{1 \text{ mA}}$$

$$= 8.3 \text{ k}\Omega \text{ (use 8.2 k}\Omega \text{ standard value)}$$

Diode specification:

$$V_R = E = 9 \text{ V}$$

$$I_F = 1 \text{ mA}$$

Current Level Selection

In most electronic circuits it is best to select the smallest possible level of current. One reason is to keep the total supply current to a minimum. High supply currents require larger, more expensive power supplies than low current levels. Where a battery supply is used, batteries last longer with low supply currents. Another reason to keep circuit current levels low whenever possible is to minimize component power dissipation. Low power dissipation allows the smallest components to be used and avoids circuit heating problems.

Series Noise Clipper

Digital signal waveforms sometimes have unwanted lower-level *noise* voltages. The noise can be removed by a *series noise clipping circuit*, which simply consists of two inverse-parallel connected diodes, as illustrated in Fig. 3-33. When the noise amplitude is much smaller than the diode forward voltage

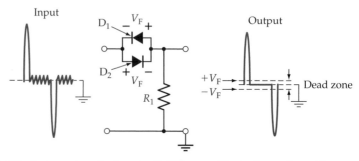

Figure 3-33 Series noise clipping circuit. Noise amplitudes lower than $\pm V_F$ cannot pass to the output.

drop (V_F), and the signal amplitude is larger than V_F, the signals are passed and the noise is blocked. The signal peak output voltage is $E - V_F$.

A *dead zone* of $\pm V_F$ exists around ground level in the output waveform, indicating that inputs must exceed $\pm V_F$ to pass to the output. For noise voltage amplitudes that approach $\pm V_F$, two diodes may be connected in series to give a larger dead zone.

Practice Problems

3-7.1 A positive series clipping circuit, as in Fig. 3-31b, has a ± 7 V input and zero load current. Calculate a suitable resistance for R_1, and specify the diode.

3-7.2 A ± 6 V square wave is applied via a series clipping circuit to a device that cannot survive $+6$ V. The device input current is 100 μA when its input voltage is negative. Select a suitable clipping circuit, determine the required resistance, and specify the diode.

3-8 SHUNT CLIPPING CIRCUITS

Shunt Clipper

A *positive shunt clipper circuit* is illustrated in Fig. 3-34a. Here the diode is connected in *shunt* (or parallel) with the output terminals. When the input is negative, the diode is reverse-biased and there is only a small voltage drop across R_1, due to

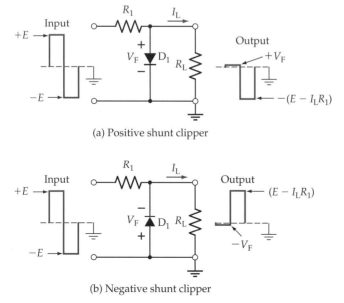

(a) Positive shunt clipper

(b) Negative shunt clipper

Figure 3-34 Shunt clipping circuits. The output voltage cannot exceed the diode voltage drop (V_F) when the input voltage forward-biases the diode.

load current I_L. This means that the circuit output voltage (V_o) is approximately equal to the negative input peak ($-E$). When the input is $+E$, D_1 is forward-biased and the output voltage equals the diode voltage drop ($+V_F$). Thus, the positive half of the waveform is clipped off. As illustrated, the upper and lower levels of the output of a positive shunt clipper are approximately $+V_F$ and $-E$.

A *negative shunt clipper circuit* is exactly the same as a positive shunt clipper with the diode polarity reversed (see Fig. 3-34b). The negative half-cycle of the waveform is clipped off.

The load current on a shunt clipper produces a voltage drop ($I_L R_1$) across the resistor; this might be insignificant where the load current is very low:

$$V_o = E - (I_L R_1) \qquad (3\text{-}39)$$

As in the case of series clipping circuits, shunt clippers may be used with square, sinusoidal, or other input waveforms.

Example 3-17

The negative shunt clipper in Fig. 3-35 has a ± 5 V input, and is to produce a $+4.5\ V$ minimum output when the load current is 2 mA. Determine a suitable resistance for R_1, and specify the diode forward current and reverse voltage.

Solution

When the diode is reverse-biased:

Eq. 3-39: $\qquad\qquad V_o = E - (I_L R_1)$

or $\qquad\qquad R_1 = \dfrac{E - V_o}{I_L} = \dfrac{5\ V - 4.5\ V}{2\ mA}$

$$= 250\ \Omega\ (\text{use } 220\ \Omega)$$

Diode reverse voltage:

$$V_R = E = 5\ V$$

When the diode is forward-biased:

$$I_F = \dfrac{E - V_F}{R_1} = \dfrac{5\ V - 0.7\ V}{220\ \Omega}$$

$$= 19.5\ mA$$

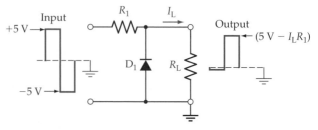

Figure 3-35 Negative shunt clipper for Example 3-17.

Shunt Noise Clipper

The shunt noise clipper shown in Fig. 3-36 removes noise riding on the peaks of an input waveform. The signal amplitude must be greater than the diode forward voltage drop. The output waveform is clipped at $\pm V_F$, so that the noise is not passed to the output. This type of clipper is used with pulse signals where the pulse amplitude is not important. In this case, information is contained in the pulse width or simply in its presence or absence.

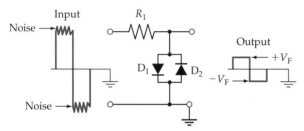

Figure 3-36   Shunt noise clipper circuit. Input amplitudes greater than $\pm V_F$ are clipped off.

Biased Shunt Clipper

Figure 3-37 shows a clipping circuit that uses two diodes that have different bias voltages. The cathode of D_1 is connected to a $+2$ V bias (V_{B1}), and the anode of D_2 has a -2 V bias (V_{B2}). While the input waveform amplitude is less than $\pm(V_B + V_F)$, neither diode is forward-biased, and the input is simply passed to the output. When the positive input is greater than ($V_{B1} + V_F$), D_1 becomes forward-biased, and the output cannot exceed this voltage. Similarly, when the negative input goes below ($-V_{B2} - V_F$), D_2 is forward-biased and the output is limited to $-(V_{B2} + V_F)$.

$$V_o = \pm(V_B + V_F) \tag{3-40}$$

For obvious reasons, the circuit is termed a *biased shunt clipper.* Biased shunt clippers are used to protect circuits or devices from (positive/negative) input voltages that must not exceed specified levels.

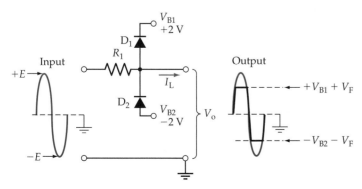

Figure 3-37   Biased shunt clipper circuit. Input amplitudes greater than $\pm(V_B + V_F)$ are clipped off.

The voltage across resistor R_1 is $(E - V_o)$ in Fig. 3-37, and the resistor current is the sum of the load current and the diode forward current $(I_L + I_F)$. As in other diode circuits, a minimum level of I_F is selected, and the resistor value is calculated as

$$R_1 = \frac{E - V_o}{I_L + I_F} \tag{3-41}$$

Example 3-18

The biased shunt clipper in Fig. 3-37 has a ± 9 V input, and its output is to be limited to ± 2.7 V. Determine a suitable resistance for R_1 if the clipper output current is to be ± 1 mA.

Solution

Eq. 3-40: $V_o = \pm(V_B + V_F)$

giving $V_B = \pm(V_o - V_F)$

$$= \pm(2.7 \text{ V} - 0.7 \text{ V})$$

$$= \pm 2 \text{ V}$$

Select the diode forward current as

$$I_F = 1 \text{ mA}$$

Eq. 3-41: $R_1 = \dfrac{E - V_o}{I_L + I_F} = \dfrac{9 \text{ V} - 2.7 \text{ V}}{1 \text{ mA} + 1 \text{ mA}}$

$$= 3.15 \text{ k}\Omega \quad \text{(use 2.7 k}\Omega \text{ to give } I_F > 1 \text{ mA)}$$

Zener Diode Shunt Clipper

A *Zener diode shunt clipper* produces the same kind of result as a biased shunt clipper without the need for bias voltages. The clipper circuit in Fig. 3-38 has two back-to-back series-connected Zener diodes. When the input voltage is positive and has sufficient amplitude, D_1 is forward-biased and D_2 is biased into reverse breakdown. At this time, the output voltage is limited to $(V_F + V_{Z2})$. A negative input voltage produces a maximum negative output of $-(V_F + V_{Z1})$. With equal-voltage Zener diodes, the maximum output voltage is

$$V_o = \pm(V_F + V_Z) \tag{3-42}$$

The resistor voltage is $(E - V_o)$, and the resistor current is $(I_L + I_Z)$. A minimum level of I_Z (greater than the device knee current) is selected, and the resistor value is calculated as

$$R_1 = \frac{E - V_o}{I_L + I_Z} \tag{3-43}$$

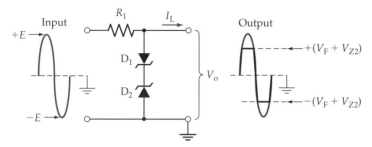

Figure 3-38   Zener diode shunt clipper circuit. Input amplitudes greater than $\pm(V_Z + V_F)$ are clipped off.

Example 3-19

A Zener diode shunt clipper, as in Fig. 3-38, is to be connected between a ±20 V square wave signal and a circuit that cannot accept inputs greater than ±5 V. Select suitable Zener diodes, and determine R_1. The clipper output current is to be ±1 mA.

Solution

From Eq. 3-42, $\qquad V_Z = V_o - V_F$

$$= 5\text{ V} - 0.7\text{ V}$$

$$= 4.3\text{ V}$$

From data sheet A-4 in Appendix A, it is seen that the 1N749 has $V_Z = 4.3$ V. Use 1N749 Zener diodes.

$\qquad$ Select $I_{Z(min)} = 5$ mA

Eq. 3-43: $\qquad R_1 = \dfrac{E - V_o}{I_L + I_Z} = \dfrac{20\text{ V} - 5\text{ V}}{1\text{ mA} + 5\text{ mA}}$

$$= 1.86\text{ k}\Omega \quad \text{(use 1.8 k}\Omega \text{ to give } I_Z > 5\text{ mA)}$$

Practice Problems

3-8.1 A positive shunt clipper, as in Fig. 3-34a, has a ±7 V input. The negative output voltage is to be 6.5 V when the load current is 1.5 mA. Calculate the required resistance for R_1, and specify the diode forward current and reverse voltage.

3-8.2 A ±8 V square wave is applied to a device that cannot accept inputs greater than ±3.5 V. The device input current is ±600 μA. Design a suitable biased shunt clipping circuit.

3-8.3 A Zener diode shunt clipping circuit is to be used to clip a ±18 V square wave off at approximately ±7 V. The output current is to be ±1.2 mA. Design the circuit.

3-9 CLAMPING CIRCUITS

Negative and Positive Voltage Clamping Circuits

A *clamping circuit,* also known as a dc *restorer,* changes the dc voltage level of a waveform but does not affect its shape. Consider the clamping circuit shown in Fig. 3-39. When the square wave input is positive, diode D_1 is forward-biased and the output voltage equals the diode forward voltage drop V_F.

$$V_o = V_F \qquad \text{(3-44)}$$

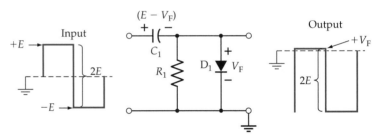

Figure 3-39 A negative voltage clamping circuit passes the complete input waveform to the output but clamps the positive peak of the output close to ground level.

During the positive half-cycle of the input, the voltage on the right side of the capacitor is $+V_F$, while that on the left side is $+E$. Thus, C_1 is charged with the polarity shown to a voltage:

$$V_C = (E - V_F) \qquad \text{(3-45)}$$

When the input goes negative, the diode is reverse-biased and has no further effect on the capacitor voltage. Also, R_1 has a very high resistance, so that it cannot discharge C_1 significantly during the negative portion of the input waveform. While the input is negative, the output voltage is the sum of the input and capacitor voltages. Since the polarity of the capacitor voltage is the same as the (negative) input, the output is

$$V_o = -(E + V_C)$$
$$= -[E + (E - V_F)]$$

or $$V_o = -(2E - V_F) \qquad \text{(3-46)}$$

The peak-to-peak output voltage ($V_{o(pp)}$) is the difference between the positive output peak (V_F) and negative output peak $-(2E - V_F)$:

$$V_{o(pp)} = V_F - [-(2E - V_F)]$$

or $$V_{o(pp)} = 2E \qquad \text{(3-47)}$$

It is seen that the peak-to-peak amplitude of the output waveform from the clamping circuit is exactly the same as the peak-to-peak input. Instead of the waveform being symmetrical above and below ground, however, the output positive peak is clamped at a level of $+V_F$.

The circuit in Fig. 3-39 is known as a *negative voltage clamping circuit,* because the output waveform is almost completely negative. Reversing the polarity of the diode and capacitor in Fig. 3-39 creates the *positive voltage clamping circuit* shown in Fig. 3-40. As illustrated, the output waveform is clamped to keep it almost completely positive.

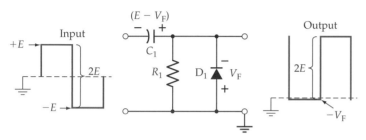

Figure 3-40   A positive voltage clamping circuit passes the complete input waveform to the output but clamps the negative peak of the output close to ground level.

Output Slope

Resistor R_1 in Figs. 3-39 and 3-40 is sometimes termed a *bleeder resistor.* Its function is to discharge the capacitor gradually over several cycles of input waveform, so that it can be charged to a new voltage level if the input changes. However, R_1 partially discharges the capacitor during the time that D_1 is reverse-biased, and this produces a *slope* or *tilt* (ΔV_c) on the output waveform, as illustrated in Fig. 3-41. The tilt voltage can be calculated from a knowledge of the waveform frequency and the circuit component values. The constant-current equation for charging or discharging a capacitor may be applied:

$$\Delta V_C = \frac{I_C \times t}{C_1}$$ (3-48)

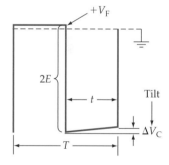

Figure 3-41   The output voltage from a clamping circuit has a slope (ΔV_C) produced by capacitor discharge. The capacitance value is determined from the acceptable slope.

Example 3-20

The diode clamping circuit in Fig. 3-42 has a ±10 V, 1 kHz square wave input. Calculate the tilt on the output waveform.

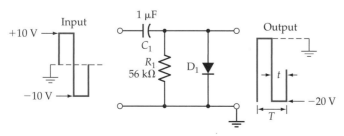

Figure 3-42   Clamping circuit for Example 3-20.

Solution

While the diode is reverse-biased,

$$V_{o(pp)} = 2E$$
$$= 2 \times 10 \text{ V}$$
$$= 20 \text{ V}$$

and
$$I_C \approx \frac{V_{o(pp)}}{R_1} = \frac{20 \text{ V}}{56 \text{ k}\Omega}$$
$$= 367 \text{ }\mu\text{A}$$

$$t = \frac{T}{2} = \frac{1}{2f} = \frac{1}{2 \times 1 \text{ kHz}}$$
$$= 500 \text{ }\mu\text{s}$$

Eq. 3-48:
$$\Delta V_C = \frac{I_C \times t}{C_1} = \frac{367 \text{ }\mu\text{A} \times 500 \text{ }\mu\text{s}}{1 \text{ }\mu\text{F}}$$
$$= 184 \text{ mV}$$

Component Determination

When a clamping circuit is designed, capacitor C_1 is selected so that it becomes completely charged in approximately five cycles of the input waveform. As already explained (and illustrated in Fig. 3-43), diode D_1 is forward-biased when C_1 is being charged. Resistor R_1 is not part of the C_1 charging circuit, because D_1 bypasses R_1. As illustrated, the C_1 charge is controlled by the signal source resistance (R_S). A capacitor is completely charged in approximately five time constants of the circuit. In the case of a resistor-capacitor circuit, the time constant is RC. So the charging time is

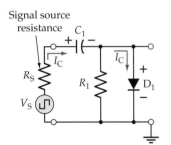

Figure 3-43 Capacitor C_1 is charged via signal source resistance R_S.

$$5 \text{ RC} = 5 \times [\text{input pulse width (PW)}]$$

or
$$\text{RC} = \text{PW}$$

This gives
$$C_1 R_S = \text{PW} \qquad \text{(3-49)}$$

When the pulse width and source resistance of the input waveform are known, C_1 can be determined from Eq. 3-49.

So long as the slope on the output waveform is very small, the capacitor discharge current is essentially a constant quantity.

$$I_C = \frac{V_o}{R_1}$$

Or
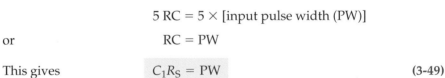
$$I_C = \frac{2E}{R_1} \qquad \text{(3-50)}$$

When the output waveform slope is specified, I_C can be calculated from Equation 3-48, and a suitable resistance for R_1 is then determined by substituting I_C into Eq. 3-50.

Example 3-21

The negative voltage clamping circuit in Fig. 3-44 has a ±8 V, 500 Hz square wave input with a 600 Ω signal source resistance. The output waveform is to have a maximum slope of 1%. Determine suitable values for R_1 and C_1.

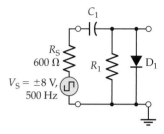

Figure 3-44 Clamping circuit for Example 3-21.

Solution

$$t = \frac{T}{2} = \frac{1}{2f} = \frac{1}{2 \times 500 \text{ Hz}}$$

$$= 1 \text{ ms}$$

$$PW = t = 1 \text{ ms}$$

From Eq. 3-49,
$$C_1 = \frac{PW}{R_S} = \frac{1 \text{ ms}}{600 \text{ Ω}}$$

$$= 1.7 \text{ μF (use 1.8 μF)}$$

$$V_{o(pp)} = 2E = 2 \times 8 \text{ V}$$

$$= 16 \text{ V}$$

$$\Delta V_C = 1\% \text{ of } V_{o(pp)} = 1\% \text{ of } 16 \text{ V}$$

$$= 0.16 \text{ V}$$

From Eq. 3-48,
$$I_C = \frac{\Delta V_C \, C_1}{t} = \frac{0.16 \text{ V} \times 1.8 \text{ μF}}{1 \text{ ms}}$$

$$= 288 \text{ μA}$$

From Eq. 3-50,
$$R_1 = \frac{2E}{I_C} = \frac{16 \text{ V}}{288 \text{ μA}}$$

$$= 55.5 \text{ kΩ (use 56 kΩ)}$$

Biased Clamping Circuit

The biased clamping circuit shown in Fig. 3-45 has a bias voltage (V_B) at the cathode of diode D_1. In this case, the output voltage cannot exceed $(V_{B1} + V_F)$, as illustrated. The capacitor charges to $(E - V_{B1} - V_F)$ when the input is positive.

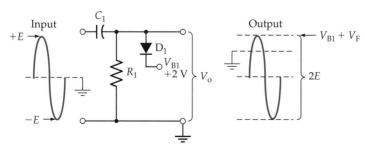

Figure 3-45 A biased clamping circuit clamps one peak of the output voltage at a selected voltage level.

When the input goes negative, the output voltage is $-(E + V_C)$, giving

$$V_o = -[E + (E - V_{B1} - V_F)]$$

$$= -2E + V_{B1} + V_F$$

The peak-to-peak output voltage is the difference between the positive and negative output peak levels.

$$V_o = (V_{B1} + V_F) - [-2E + V_{B1} + V_F]$$

$$= 2E$$

It is seen that, as for other clamping circuits, the peak-to-peak output waveform from the biased clamper is the same as the input. The positive level of the output is clamped to $(V_B + V_F)$.

Virtually any bias voltage level (positive or negative) can be used with a biased clamping circuit. The diode polarity determines whether the positive or the negative peak of the output is clamped. Care must be taken to ensure that the capacitor is connected with the correct polarity. Biased clamping circuits are designed in exactly the same way as unbiased clamping circuits.

Example 3-22

Determine the upper and lower levels of the output voltage for the circuits shown in Fig. 3-46 when the input voltage is ± 6 V. Calculate the capacitor voltage in each case.

Solution

For Fig. 3-46(a),

D_1 is forward-biased when the input goes negative.

When the input is $-E$, $V_C = V_{B1} - V_F - E$

$$= 3 \text{ V} - 0.7 \text{ V} - (-6 \text{ V})$$

$$= 8.3 \text{ V} (+ \text{ on the right})$$

$$V_o = V_{B1} - V_F$$

$$= 3 \text{ V} - 0.7 \text{ V}$$

$$= 2.3 \text{ V (low level of output)}$$

When the input is $+E$, $V_o = E + V_C$

$$= 6 \text{ V} + 8.3 \text{ V}$$

$$= 14.3 \text{ V (high level of output)}$$

For Fig. 3-46b,

D_1 is forward-biased when the input goes positive.

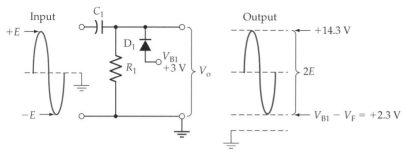

(a) Output voltage cannot go below ($V_{B1} - V_F$)

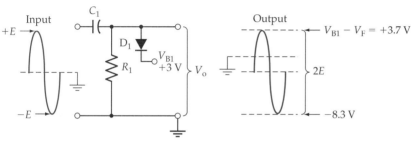

(b) Output voltage cannot go above ($V_{B1} - V_F$)

Figure 3-46 Biased clamping circuits use positive or negative bias voltages to clamp the level of one of the peaks of the output waveform above or below the bias voltage level.

When the input is $+E$, $V_C = E - V_{B1} - V_F$

$$= 6\,V - 3\,V - 0.7\,V$$

$$= 2.3\,V\ (+\ \text{on the left})$$

$$V_o = V_{B1} + V_F$$

$$= 3\,V + 0.7\,V$$

$$= 3.7\,V\ (\text{high level of output})$$

When the input is $-E$, $V_o = E + V_C$

$$= -6\,V - 2.3\,V$$

$$= -8.3\,V\ (\text{low level of output})$$

Zener Diode Clamping Circuit

Zener diode clamping circuits operate just like biased voltage clampers, with the Zener diode voltage (V_Z) replacing the bias voltage. The circuit in Fig. 3-47 has an ordinary diode (D_1) connected in series with a Zener diode (D_2). When the input voltage is at its positive peak, the output voltage is

$$V_o = V_Z + V_F$$

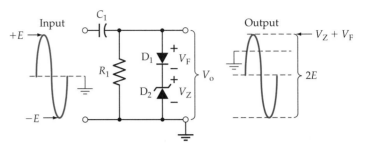

Figure 3-47   Zener diode clamping circuits clamp the output voltage at ($V_Z + V_F$). The positive peak or the negative peak of the waveform can be clamped above or below ground.

and C_1 charges (+ on the left) to

$$V_C = E - (V_Z + V_F)$$

When the input goes to its negative peak, the output voltage is

$$V_o = -(E + V_C)$$
$$V_o = -[E + E - (V_Z + V_F)]$$
$$= -(2E - V_Z - V_F)$$

The peak-to-peak output remains equal to the peak-to-peak input voltage (2E). If the polarity of the diodes (and the capacitor) are reversed, the negative output peak is clamped at $-(V_Z + V_F)$.

Practice Problems

3-9.1 A ±8 V, 300 Hz square wave is applied to the clamping circuit in Fig. 3-40. If $C_1 = 1.5\ \mu F$ and $R_1 = 33\ k\Omega$, determine the tilt on the output waveform.

3-9.2 A positive voltage clamping circuit, as in Fig. 3-40, is to have an output waveform with a maximum tilt of 0.5%. The signal is a ±7 V, 1 kHz square wave with a 500 Ω source resistance. Calculate suitable values for R_1 and C_1.

3-9.3 A clamping circuit, as in Fig. 3-47, uses a 1N754 Zener diode. Determine the upper and lower levels of the output voltage if the input is a ±12 V square wave. Calculate the capacitor voltage.

3-10 DC VOLTAGE MULTIPLIERS

Voltage Doubler

A *voltage-doubling circuit* produces an output voltage which is approximately double the peak voltage of the input waveform. Consideration of the voltage doubling circuit in Fig. 3-48 shows that it is simply a combination of two diode-capacitor clamping circuits without the discharge resistors. In fact, the circuit operation is similar to that of clamping circuits.

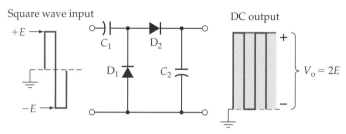

Figure 3-48 A voltage-doubling circuit produces a dc output voltage which is approximately double the peak input voltage.

When the input voltage is negative, as shown in Fig. 3-49a, diode D_1 is forward-biased and C_1 charges to $(E - V_{F1})$ with the polarity illustrated. D_2 is reverse-biased during the negative half-cycle of the input, and so the charge on C_2 is not affected at this time.

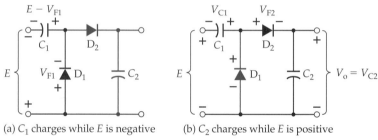

(a) C_1 charges while E is negative   (b) C_2 charges while E is positive

Figure 3-49 In a voltage-doubling circuit, capacitor C_1 charges to $(E - V_{F1})$ during the negative half-cycle of the input. Then, $+[E + (E - V_{F1}) - V_{F2}]$ is passed to C_2 during the input positive half-cycle.

Figure 3-49b shows what occurs during the input positive half-cycle. D_1 is now reverse-biased, and D_2 is forward-biased. The voltage applied to D_2 and C_2 is the sum of the input voltage and the voltage on C_1. So, as illustrated, capacitor C_2 is charged to

$$V_{C2} = E + V_{C1} - V_{F2}$$
$$= E + (E - V_{F1}) - V_{F2}$$
$$= 2(E - V_F)$$

So $\qquad\qquad V_o = 2(E - V_F)$ (3-51)

It is seen that, when the diode voltage drop is much smaller than the input voltage, the output is approximately double the peak input amplitude. The polarity of the output voltage can be reversed by reversing the polarity of the diodes and capacitors.

The output terminals of the voltage doubler are the terminals of capacitor C_2. The load current partially discharges the capacitors, producing a drop in output voltage in the same way that tilt is created on a clamping circuit output.

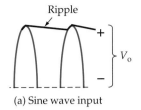

(a) Sine wave input

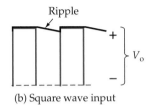

(b) Square wave input

Figure 3-50 A ripple waveform occurs at the output of a voltage doubler. The ripple amplitude depends upon the capacitance value, the input frequency, and the load current.

The repeated charge and discharge of C_1 and C_2 results in a ripple waveform on the output. A sinusoidal input waveform applied to a voltage doubler produces exactly the same type of output ripple that occurs with a half-wave rectifier (Fig. 3-50a). However, the input most often used is a dc voltage source that has been *chopped*, or converted into a square waveform (Fig. 3-50b).

Figure 3-51a shows that capacitor C_2 supplies the load current (I_L) while diode D_2 is reverse-biased (Fig. 3-51a). The discharge of C_2 accounts for half the output ripple voltage amplitude, and the discharge of C_1 produces the other half of the ripple amplitude. Equation 3-48 can be modified for calculating the capacitance of C_2.

$$C_2 = \frac{I_L t}{\Delta V_C} \qquad (3\text{-}52)$$

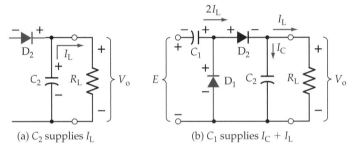

(a) C_2 supplies I_L (b) C_1 supplies $I_C + I_L$

Figure 3-51 In a voltage-doubling circuit, capacitor C_2 supplies I_L while diode D_2 is reverse biased, and C_1 supplies $2\,I_L$ when D_2 is forward-biased.

When D_2 is forward-biased (Fig. 3-51b), capacitor C_1 supplies I_L and the recharging current to C_2. The recharging current must equal I_L to maintain the full charge on C_2, so C_1 supplies $2I_L$. Applying Eq. 3-48 to calculate C_1, it is found that

$$C_1 = 2C_2 \qquad (3\text{-}53)$$

Example 3-23

Determine C_1 and C_2 for the voltage-doubling circuit in Fig. 3-52 to produce a 1% maximum output ripple. The input is a ±12 V, 5 kHz square wave.

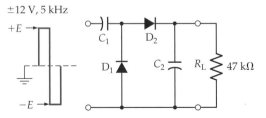

Figure 3-52 Diode voltage doubling circuit for Example 3-23.

Solution

Eq. 3-51:
$$V_o = 2(E - V_F)$$
$$= 2(12\ V - 0.7\ V)$$
$$= 22.6\ V$$
$$I_L = \frac{V_o}{R_L} = \frac{22.6\ V}{47\ k\Omega}$$
$$= 481\ \mu A$$

Capacitor discharge time $t = \dfrac{T}{2} = \dfrac{1}{2f} = \dfrac{1}{2 \times 5\ kHz}$
$$= 100\ \mu s$$

For 1% output ripple, allow 0.5% due to discharge of C_2 and 0.5% due to discharge of C_1.

$$\Delta V_C = 0.5\% \text{ of } V_{o(pp)}$$
$$= 0.5\% \text{ of } 22.6\ V$$
$$= 113\ mV$$

Eq. 3-52:
$$C_2 = \frac{I_L t}{\Delta V_C} = \frac{481\ \mu A \times 100\ \mu s}{113\ mV}$$
$$= 0.43\ \mu F \text{ (use } 0.47\ \mu F)$$

Eq. 3-53:
$$C_1 = 2C_2$$
$$= 2 \times 0.47\ \mu F$$
$$= 0.94\ \mu F \text{ (use } 1\ \mu F)$$

Multi-Stage Voltage Multipliers

A four-stage dc voltage multiplier is shown in Fig. 3-53. Compared with the voltage doubler in Fig. 3-48, this circuit consists of two cascade-connected voltage-doubling circuits. To simplify the explanation of the circuit operation, assume ideal diodes where $V_F = 0$.

- When $V_i = -E$, D_1 is forward-biased and C_1 charges via D_1 to E.
- When $V_i = +E$, point A is at $+2E$, D_1 is reverse-biased, D_2 is forward-biased, and C_2 charges via D_2 to $2E$.
- When $V_i = -E$, point A is close to ground level, point B is at $2E$ (because of V_{C2}), D_2 is reverse-biased, D_3 is forward-biased, and C_3 charges via D_3 to $2E$ volts.
- When $V_i = +E$, point A is at $+2E$, point B is at $+2E$, point C is at $+4E$ ($V_{C1} + V_{C3}$), D_4 is forward-biased, and C_4 charges via D_4 to $2E$ volts.

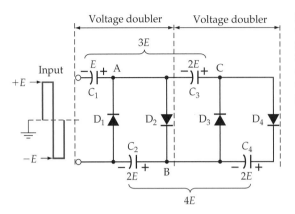

Figure 3-53 Four-stage dc voltage-multiplying circuit. Disregarding the diode voltage drops, C_1 is charged to E volts, and all other capacitors are charged to approximately $2E$. The final output is $4E$ volts.

The resulting output voltage, taken across C_2 and C_4, is $4E$ volts as illustrated. Additional stages may be added to the circuit to produce higher levels of dc output voltage. Figure 3-54 shows another way that dc voltage-multiplier circuit diagrams are often drawn. Examination of the circuit shows that it is exactly the same as Fig. 3-53.

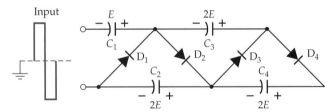

Figure 3-54 Another way of drawing the circuit of a four-stage dc voltage multiplier. The component connections are exactly as in Fig. 3-53.

Practice Problems

3-10.1 For the circuit designed in Example 3-23, determine the effect on the output ripple: (a) when R_L is changed to 27 kΩ, and (b) when the input square-wave frequency is increased to 7 kHz.

3-10.2 A voltage-doubling circuit with a 39 kΩ load resistor is to have a 20 V output. Determine the required amplitude for a 700 Hz square wave input, and calculate suitable capacitor values to produce a 0.2 V maximum output ripple. Use silicon diodes.

3-11 DIODE LOGIC CIRCUITS

AND Gate

A logic circuit produces an output voltage that is either high or low, depending upon the levels of several input voltages. The two basic logic circuits are the AND gate and the OR gate.

Figure 3-55 shows the circuit diagram of a diode AND gate. It is seen that the diode anodes are connected to a 5 V supply (V_{CC}) via resistor R_1. The circuit has a single output terminal at the diode anodes and three input terminals at the device cathodes. (An AND gate could have any number of inputs from 2 up to perhaps 50.)

If one (or more) of the input terminals is grounded, current flows from the supply through R_1 and through the forward-biased diode to ground. In this case, the output voltage is just V_F above ground (0.7 V for silicon). The output is said to be at a *low* level. When input levels of 5 V are applied to all three input terminals, none of the diodes is forward-biased, no resistor current flows, and no significant voltage drop occurs across R_1. Thus, the output voltage is equal to V_S, and is referred to as a *high* output level.

An AND gate produces a high output only when input A is high, AND input B is high, AND input C is high.

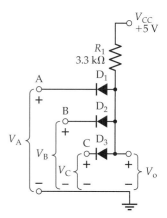

Figure 3-55 Circuit of a three-input diode AND gate. A high output occurs only when high inputs are present at input A, AND input B, AND input C.

Example 3-24

For the AND gate circuit in Fig. 3-55, determine each diode forward current level: (a) when all three inputs are *low;* (b) when only input A is *high;* (c) when inputs A and B are both *high* and C is *low.*

Solution

(a)
$$I_{R1} = \frac{V_{CC} - V_F}{R_1} = \frac{5\text{ V} - 0.7\text{ V}}{3.3\text{ k}\Omega}$$

$$= 1.3\text{ mA}$$

For each diode,

$$I_F = \frac{I_{R1}}{3} = \frac{1.3\text{ mA}}{3}$$

$$= 433\text{ }\mu\text{A}$$

(b) $I_{R1} = 1.3$ mA (as above)

$$I_{F2} = I_{F3} = \frac{I_{R1}}{2} = \frac{1.3\text{ mA}}{2}$$

$$= 650\text{ }\mu\text{A}$$

(c)
$$I_{F3} = I_{R1}$$

$$= 1.3\text{ mA}$$

OR Gate

The circuit diagram of an OR gate with three input terminals is shown in Fig. 3-56. Like the AND gate, the OR gate could have two or more inputs. It is obvious that the output voltage (V_o) of the OR gate in Fig. 3-56 is low when all three inputs are low. Now suppose that a +5 V input is applied to terminal A,

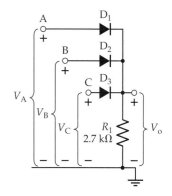

Figure 3-56 Circuit of a three-input OR gate. A high output occurs when high inputs are present at input A, OR at input B, OR at input C.

while terminals B and C remain grounded. Diode D_1 becomes forward-biased, and its cathode voltage is $V_o = (5 \text{ V} - V_F)$. The gate output voltage is high, and diodes D_2 and D_3 are reverse-biased with $+V_o$ at the cathodes and ground at the anodes. The output will be high if a high input voltage is applied to any one (or more) of the inputs.

> *An OR gate produces a high output when input A is high, OR or when input B is high, OR when input C is high.*

Diode AND and OR gates can be made with discrete components. However, *integrated-circuit* (IC) packages containing many diodes already fabricated in the form of one or more gates are available.

Practice Problems

3-11.1 The diodes in the AND gate in Fig. 3-55 are each to have a 2 mA maximum forward current. Calculate the required resistance for R_1.

3-11.2 Calculate the maximum and minimum levels of diode forward current for the OR gate in Fig. 3-56 when $V_i = 4.5$ V and $R_1 = 2.7$ kΩ.

Review Questions

Section 3-1

3-1 Draw the circuit diagram of a half-wave rectifier for producing a positive output voltage. Sketch the input and output waveforms, and explain the circuit operation.

3-2 Draw circuit diagrams to show two methods of producing a negative output voltage from a half-wave rectifier. Briefly explain.

Section 3-2

3-3 Sketch a two-diode full-wave rectifier circuit for producing a positive output voltage. Sketch the input and output waveforms and explain the circuit operation.

3-4 Draw the circuit diagram for a bridge rectifier, together with its input and output waveforms. Carefully explain the operation of the bridge rectifier circuit, identifying the forward-biased and reverse-biased diodes during each half-cycle of the input waveform.

3-5 Draw diagrams to show how a negative output voltage can be obtained from: (a) a two-diode full-wave rectifier circuit, (b) a bridge rectifier circuit. Briefly explain.

Section 3-3

3-6   Draw the circuit diagram for a dc power supply that uses a half-wave rectifier and a capacitor filter. Sketch the input and output waveform, and explain the circuit operation and the shape of the waveforms.

3-7   For a rectifier circuit with capacitor filtering, explain ripple voltage, repetitive surge current, non-repetitive surge current, and diode peak reverse voltage.

3-8   Discuss the selection of a reservoir capacitor for a dc power supply, and the importance of connecting the capacitor with the correct polarity.

3-9   List the factors involved in diode selection for a half-wave rectifier circuit. Briefly explain.

3-10   Explain the function of a transformer in a dc power supply, and discuss the factors involved in the specification of power supply transformers.

Section 3-4

3-11   Draw the circuit diagram for a dc power supply that uses a bridge rectifier and a capacitor filter circuit. Sketch the input and output waveform, and explain the circuit operation and the shape of the waveforms.

3-12   List the factors involved in diode selection for a full-wave rectifier circuit. Briefly explain.

3-13   Compare half-wave and full-wave rectifier circuits with capacitor filtering, referring to capacitor size for a given ripple amplitude, diode specification, and transformer selection.

Section 3-5

3-14   Define power supply source effect, load effect, line regulation, and load regulation. Write equations for each item.

3-15   Sketch a circuit diagram for testing a dc power supply for source effect, load effect, and ripple. List the steps in an appropriate test procedure and discuss any necessary precautions.

3-16   List possible faults that might occur in a dc power supply, and discuss the actions that should be taken.

Section 3-6

3-17   Sketch the circuit diagram for a Zener diode voltage regulator. Briefly explain the circuit operation and discuss the effects of load current.

3-18   Sketch the ac equivalent circuit for a Zener diode voltage regulator. Write equations for source effect and load effect.

Section 3-7

3-19   Draw circuit diagrams for negative and positive series clipping circuits. Show input and output waveforms and explain the operation of each circuit.

3-20   Sketch the circuit of a diode series noise clipper. Sketch input and output waveforms and explain the circuit operation.

Section 3-8

3-21 Draw circuit diagrams for negative and positive shunt clipping circuits. Sketch input and output waveforms and explain the operation of each circuit.

3-22 Sketch a shunt noise clipping circuit. Sketch input and output waveforms and explain the circuit operation.

3-23 Draw a biased shunt clipping circuit. Sketch input and output waveforms and explain the circuit operation.

3-24 A biased shunt clipper has ±9 V bias voltages. Sketch the output waveform produced by a ±12 V square wave input.

3-25 Draw a diagram for a Zener diode shunt clipping circuit together with input and output waveforms. Explain the circuit operation.

Section 3-9

3-26 Explain the difference between clipping circuits and clamping circuits. A positive voltage clamping circuit and a positive shunt clipping circuit each have a ±12 V square wave input. Sketch the output waveform from each circuit.

3-27 Draw circuit diagrams for negative and positive voltage clamping circuits. Show input and output waveforms and explain the operation of each circuit.

3-28 Draw a circuit diagram for a voltage clamping circuit using a positive bias voltage. Sketch the output waveforms and explain the circuit operation.

3-29 Draw a circuit diagram for a voltage clamping circuit using a Zener diode in place of a positive bias voltage. Sketch the output waveforms and explain the circuit operation.

Section 3-10

3-30 Draw a voltage doubling circuit. Sketch input and output waveforms and explain the circuit operation.

3-31 Draw the circuit diagram for a four-stage voltage-multiplier circuit. Briefly explain its operation, and identify the output terminals and the output voltage level relative to the input. Show how two additional diode-capacitor circuits may be added, and estimate the new output-to-input voltage ratio.

Section 3-11

3-32 Sketch the circuit diagram for a diode AND gate with five input terminals. Briefly explain the circuit operation.

3-33 Sketch the circuit diagram for a diode OR gate with five input terminals. Briefly explain the circuit operation.

Problems

Section 3-1

3-1 A half-wave rectifier circuit has a 25 V (rms) sinusoidal ac input and a 600 Ω load resistance. Calculate the peak output voltage, peak load current, and diode peak reverse voltage. Assume that $V_F = 0.7$ V.

3-2 A half-wave rectifier circuit produces a 55 mA peak current in an 820 Ω load resistor. Calculate the rms ac input voltage and the diode peak reverse voltage if $V_F = 0.7$ V.

3-3 Calculate the power dissipated in a 560 Ω load resistor connected to the output of a half-wave rectifier circuit. The ac input is 25 V (rms), and the diode voltage drop is 0.7 V.

Section 3-2

3-4 A 470 Ω load resistor is connected at the output of a bridge rectifier circuit that has a 15 V (rms) input. Calculate the peak output voltage, peak load current, and load power dissipation. Assume the diodes have 0.3 V forward voltage drop.

3-5 A bridge rectifier circuit has a 25 V (rms) sinusoidal ac input and a 600 Ω load resistance. Calculate the peak output voltage, peak load current, diode power dissipation, and diode peak reverse voltage. Assume the diodes have a V_F of 0.7 V.

3-6 A two-diode full-wave rectifier circuit (as in Fig. 3-3) dissipates 640 mW in a 400 Ω load resistor. If the diodes are silicon, calculate the peak output voltage from each half of the transformer secondary.

Section 3-3

3-7 A dc power supply consisting of a half-wave rectifier and a reservoir capacitor is required to supply 24 V with a 200 mA maximum load current. The output ripple is not to exceed ±500 mV, and the ac input frequency is 60 Hz. Determine a suitable capacitance for the reservoir capacitor.

3-8 Recalculate the capacitance approximately for the circuit in Problem 3-7, assuming that the capacitor discharge time is much greater than the charge time.

3-9 Specify the diodes required for the half-wave rectifier circuit in Problem 3-7. Select a suitable device from Appendix A, and calculate the required surge limiting resistance.

3-10 Specify a suitable transformer for the power supply in Problem 3-7.

3-11 Calculate the reservoir capacitance for a half-wave rectifier dc power supply which produces a 15 V output with a 300 mA maximum load current. The peak-to-peak output ripple voltage is to be a maximum of ±10% of V_o, and the ac input frequency is 60 Hz.

3-12 Specify the diodes required for the power supply in Problem 3-11. Select a suitable device from Appendix A, and calculate the required surge limiting resistance.

3-13 Specify a suitable transformer for the power supply in Problem 3-11.

3-14 Recalculate the capacitance for the power supply described in Problem 3-11 if the ac input frequency is to be 400 Hz.

Section 3-4

3-15 A dc power supply consisting of a bridge rectifier circuit and a reservoir capacitor is to supply 24 V with a 200 mA maximum load current. The output ripple is not to exceed ±500 mV, and the ac input frequency is 60 Hz. Determine a suitable capacitance for the reservoir capacitor.

3-16 Recalculate the capacitance for the circuit in Problem 3-15, using the approximation that the capacitor discharge time is much greater than the charge time.

3-17 Specify the diodes required for the rectifier circuit in Problem 3-15. Select a suitable device from Appendix A, and calculate the required surge limiting resistance.

3-18 Specify a suitable transformer for the power supply in Problem 3-15.

3-19 Calculate the reservoir capacitance for a dc power supply which produces an 18 V output with a 300 mA maximum load current. A bridge rectifier circuit is used, and the ac input frequency is 60 Hz. The peak-to-peak output ripple voltage is to be a maximum of $\pm10\%$ of V_o.

3-20 Specify the diodes required for the power supply in Problem 3-19. Select a suitable device from Appendix A, and calculate the required surge limiting resistance.

3-21 Recalculate the capacitance for the power supply in Problem 3-19 for a 400 Hz ac supply.

3-22 Specify a suitable transformer for the power supply in Problem 3-19.

Section 3-5

3-23 A dc power supply output voltage changes from 12 V to 11.6 V when the input drops by 10%, and from 12 V to 11.5 V when the load current increases from zero to maximum. Determine the source and load effects and the line and load regulations.

3-24 A dc power supply with a 20 V output has a 3% line regulation and a 5% load regulation. Calculate the source effect and the load effect.

3-25 Calculate the line and load regulation for the dc power supply circuit designed for Problem 3-7.

3-26 Calculate the line and load regulation for the dc power supply circuit designed for Problem 3-19.

Section 3-6

3-27 A series-connected Zener diode and resistor are to be used as a 10 V reference source with a load current less than 1 mA. The available supply voltage is 25 V. Select suitable components and calculate the effect of a ±10% change in supply voltage on the diode current.

3-28 A 5 V Zener diode voltage source is to be designed to produce maximum possible output current from a low-power Zener diode. Design the circuit to operate from a 22 V supply, and calculate the maximum load current.

3-29 Determine the source effect, load effect, and ripple rejection for the circuit designed for Problem 3-28.

3-30 Recalculate the maximum output current for the circuit designed for Problem 3-28 when the ambient temperature is raised to 100°C. The diode derating factor is 3.2 mW/°C for temperatures above 50°C.

3-31 A 1N751 Zener diode is connected in series with a 330 Ω resistor to a 25 V supply. Determine the maximum and minimum level of the Zener diode voltage, and calculate the minimum resistance that may be connected in parallel with the diode.

3-32 Calculate the source effect, load effect, and ripple rejection for the circuit in Problem 3-31.

Section 3-7

3-33 A diode-resistor negative series clipping circuit (as in Fig. 3-31a) has a ±12 V input and zero load current. Determine a suitable resistor value and specify the diode.

3-34 A device is to be protected from the negative half-cycle of a ±8 V square wave input. The device input current is 50 μA when the input is positive. Select a suitable clipping circuit, determine an appropriate resistor, and specify the diode.

3-35 Specify the diodes for the series noise clipper circuit in Fig. 3-33 if $R_1 = 270$ Ω and the input peaks are ±7 V.

Section 3-8

3-36 A positive shunt clipping circuit (as in Fig. 3-34a) with a ±10 V input is to produce a −9 V output when the load current is 500 μA. Determine a suitable resistor value and specify the diode.

3-37 A device is to be protected from the negative half-cycle of a ±6 V square wave input. The device current is 750 μA when its input is +4 V. Select a suitable shunt clipping circuit, determine an appropriate resistor, and specify the diode.

3-38 Specify the diodes for the shunt noise clipper circuit in Fig. 3-36 if $R_1 = 470$ Ω and the input peaks are ±5 V.

3-39 A ±12 V square wave is applied to a circuit that cannot accept inputs in excess of ±4 V. The circuit input current is ±100 μA. Design a suitable biased shunt clipping circuit.

3-40 Design a Zener diode shunt clipping circuit to satisfy the requirements in Problem 3-39.

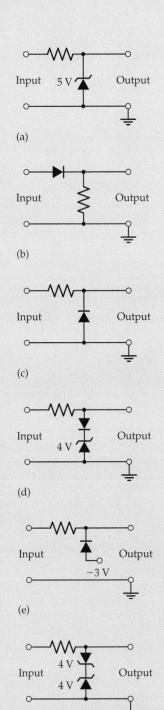

(a)

(b)

(c)

(d)

(e)

(f)

Figure 3-57 Diode clipping circuits for Problem 3-44.

3-41 A Zener diode shunt clipping circuit with a ±15 V input is to have a maximum output of ±9 V with an output current of ±750 μA. Design the circuit.

3-42 A biased shunt clipper circuit is to produce a maximum output of ±6 V to a 10 kΩ load. The input voltage peaks can be as high as ±11 V. Design the circuit.

3-43 Design a Zener diode shunt clipping circuit to satisfy the requirements in Problem 3-42.

3-44 Sketch the output waveforms from each of the circuits in Fig. 3-57 when a ±8 V square wave is applied at the input.

Section 3-9

3-45 A positive voltage-clamping circuit (as in Fig. 3-40) has $C_1 = 2$ μF and $R_1 = 27$ kΩ. Determine the tilt on the output when a ±5 V, 700 Hz square wave input is applied.

3-46 A negative voltage clamping circuit has a ±9 V, 1 kHz square wave input with a 500 Ω source resistance. The slope on the output waveform is not to exceed 0.5%. Determine suitable resistance and capacitance values.

3-47 A biased clamping circuit (as in Fig. 3-45) has a ±12 V, 330 Hz sine wave input with a 600 Ω source resistance. The positive output peak is to be clamped at 5 V, and the slope between peaks is not to exceed 1%. Determine the required bias voltage and suitable resistance and capacitance values.

3-48 Design a Zener diode clamping circuit that will perform as specified in Problem 3-47.

3-49 A Zener diode clamping circuit, as in Fig. 3-47, uses a germanium diode and a 1N746 Zener diode. The input voltage is a ±8 V, 1 kHz square wave. Determine the output voltage levels, and calculate the capacitor voltage.

3-50 Sketch the output waveforms from each of the circuits in Fig. 3-58 when a ±15 V square wave is applied at the input.

3-51 A ±20 V, 700 Hz square wave with a 500 Ω source resistance is to be clamped to a maximum level of approximately 12 V. Design a suitable Zener diode circuit that will produce a maximum output tilt of 3%.

Section 3-10

3-52 A voltage doubling circuit is to be designed to produce an output of 25 V to a 56 kΩ resistor. Determine the required amplitude for a 1 kHz square wave input; calculate suitable capacitor values if the output ripple is to be a maximum of 0.5 V.

3-53 A voltage doubling circuit with a ±10 V, 2 kHz square wave input has a 33 kΩ load resistance. Determine suitable capacitance values to give a 1% maximum output ripple.

3-54 Refer to the voltage doubling circuit designed for Problem 3-52. Determine the effect of (a) changing R_L to 33 kΩ, (b) changing the input frequency to 1.5 kHz.

Section 3-11

3-55 A three-input diode AND gate, as in Fig. 3-55, is to have a minimum diode current of 1 mA. If the circuit supply voltage is 9 V, determine a suitable resistance for R_1.

3-56 A five-input diode OR gate used diodes that require a 500 μA minimum forward current. Determine a suitable resistance value if the input voltages are +5 V.

Practice Problem Answers

3-1.1	20.51 V, 62.2 mA, 21.2 V
3-1.2	33.95 V, 48.7 V
3-2.1	32.54 V, 69.2 mA, 1.12 V
3-2.2	238 mW, 13.7 V
3-3.1	1000 μF
3-3.2	835 μF
3-3.3	25.2 V, 60 mA, 860 mA, 1N4001, 0.42 Ω,
	(115 V, 17.7 mA), (9.5 V, 216 mA)
3-4.1	500 μF
3-4.2	557 μF
3-4.3	17.5 V, 100 mA, 714 mA, 1N4001, 0.57 Ω
	(115 V, 16.8 mA), (12.1 V, 160 mA)
3-5.1	50 mV, 100 mV, 0.33%, 0.66%
3-6.1	1N756, 680 Ω, 14.7 mA
3-6.2	1N756, (220 Ω + 22 Ω), 43.8 mA
3-6.3	61.4 mV, 336 mV, 3.07×10^{-2}
3-7.1	5.6 kΩ, 7 V, 1 mA
3-7.2	4.7 kΩ, 6 V, 1 mA
3-8.1	330 Ω, 7 V, 19 mA
3-8.2	±2.8 V, 2.7 kΩ
3-8.3	1N753, 1.8 kΩ
3-9.1	550 mV
3-9.2	1 μF, 100 kΩ
3-9.3	7.5 V, −16.5 V, 4.5 V
3-10.1	1.6%, 0.65%
3-10.2	±10.7 V, 3.9 μF, 8 μF
3-11.1	680 Ω
3-11.2	1.41 mA, 469 μA

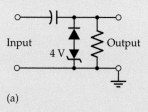

(a)

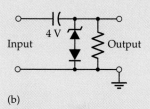

(b)

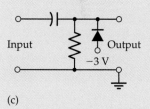

(c)

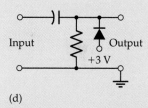

(d)

Figure 3-58 Diode clamping circuits for Problem 3-50.

CHAPTER 4
Bipolar Junction Transistors (BJTs)

CONTENTS

Objectives

You will be able to:

1   Draw block representations of *npn* and *pnp* BJTs showing depletion regions and barrier voltages. Explain the operation of each device in terms of junction bias voltages and charge carrier movement.

2   Sketch the circuit symbols for *npn* and *pnp* BJTs. Identify current directions and junction bias polarities and state typical terminal voltage levels.

3   Identify and describe the various current components in a BJT. Define BJT parameters. Write equations relating the base, emitter, and collector currents, and calculate all current levels.

4   Show how a BJT can be used for current amplification and voltage amplification. Calculate current and voltage gains.

5   Show how a BJT can be used as a switch, and explain BJT saturation.

6   Sketch typical BJT common-base, common-emitter, and common-collector characteristics. Draw circuits to show how each type of characteristic is obtained.

7   Discuss various methods of testing BJTs.

INTRODUCTION

A *bipolar junction transistor* (BJT) has three layers of semiconductor material. These are arranged either in *npn* (*n*-type—*p*-type—*n*-type) sequence or in *pnp* sequence, and each of the three layers has a terminal. A small current at the central region terminal controls the much larger total current flow through the device. This means that the transistor can be used for current amplification. As will be explained, it can also perform voltage amplification.

Because the transistor has two *pn*-junctions, its operation can be understood by applying *pn*-junction theory. The currents that flow in a transistor are similar to those that flow across a single *pn*-junction, and the transistor equivalent circuit is essentially two *pn*-junction equivalent circuits. There are three sets of current-voltage characteristics for defining the performance of a transistor.

4-1 BJT OPERATION

pnp and *npn* Transistors

A bipolar junction transistor is simply a sandwich of one type of semiconductor material (*p*-type or *n*-type) between two layers of the opposite type. A block representation of a layer of *p*-type material between two layers of *n*-type is shown in Fig. 4-1a. This is described as an *npn* transistor. Figure 4-1b shows a *pnp* transistor, consisting of a layer of *n*-type material between two layers of *p*-type. For reasons which will be understood later, the centre layer is called the *base*, one of the outer layers is termed the *emitter*, and the other outer layer is referred to as the *collector*. The emitter, base, and collector are provided with terminals, which are appropriately labelled E, B, and C. Two *pn*-junctions exist in each transistor: the *collector-base junction* and the *emitter-base junction*. Each of these junctions has the characteristics discussed in Chapters 1 and 2.

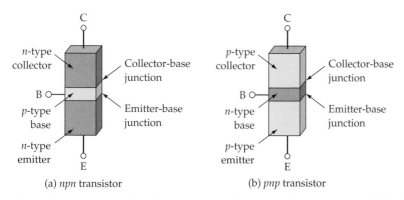

(a) *npn* transistor   (b) *pnp* transistor

Figure 4-1   Block representation of *npn* and *pnp* bipolar junction transistors (BJTs). Each device is made up of a layer of one type of semiconductor material sandwiched between two layers of the other type.

Circuit symbols for *pnp* and *npn* transistors are shown in Fig. 4-2. The arrowhead on each symbol identifies the transistor emitter terminal and indicates the conventional direction of current flow. For an *npn* transistor, the arrowhead points from the *p*-type base to the *n*-type emitter. For a *pnp* device, the arrowhead points from the *p*-type emitter to the *n*-type base. Thus, the arrowhead always points from *p* to *n*.

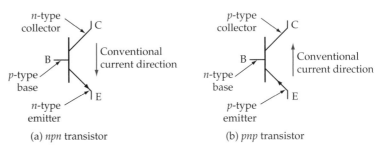

(a) *npn* transistor   (b) *pnp* transistor

Figure 4-2 BJT graphic symbols. The arrowhead identifies the emitter terminal; it always points in the conventional current direction.

Two transistor packages are illustrated in Fig. 4-3: a low-power transistor and a high-power device. Low-power transistors generally pass currents of 1 mA to 20 mA. Current levels for high-power transistors range from 100 mA to several amps. The high-power transistor is designed for mechanical connection to a heat sink in order to prevent the device from overheating. Semiconductor device packages are further investigated in Chapter 7.

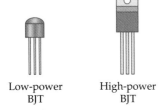

Low-power   High-power
BJT   BJT

Figure 4-3 BJTs are available in a variety of packages for various low-current and high-current applications.

npn Transistor Operation

Figure 4-4 illustrates the depletion regions and barrier voltages at the junctions of an unbiased *npn* transistor. These were explained in Chapter 1. Although it is not shown in the illustration, the centre layer of the transistor is very much

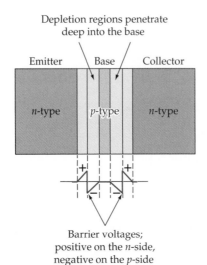

Depletion regions penetrate deep into the base

Emitter   Base   Collector

n-type   *p*-type   *n*-type

Barrier voltages;
positive on the *n*-side,
negative on the *p*-side

Figure 4-4 Depletion regions and barrier voltages at the junctions of an unbiased *npn* BJT. The barrier voltage is always positive (+) on the *n*-side and negative (−) on the *p*-side.

narrower than the two outer layers. The outer layers are also much more heavily doped than the centre layer, so that the depletion regions penetrate deep into the base, as illustrated. Because of this penetration, the distance between the two depletion regions is very short (within the base). Note that the junction barrier voltages are positive on the emitter and collector, and negative on the base of the *npn* device.

Consider Fig. 4-5, which shows an *npn* transistor with external bias voltages. For normal operation, the base-emitter (BE) junction is forward-biased and the collector-base (CB) junction is reverse-biased. Note the polarities of the external bias voltages. The forward bias at the BE junction reduces the barrier voltage and causes electrons to flow from the *n*-type emitter to the *p*-type base. The electrons are *emitted* into the base region: hence the name *emitter*. Holes also flow from the *p*-type base to the *n*-type emitter, but because the base is much more lightly doped than the collector, almost all of the current flowing across the BE junction consists of electrons entering the base from the emitter. Thus, electrons are the *majority charge carriers* in an *npn* device.

The reverse bias at the CB junction causes the CB depletion region to penetrate deeper into the base than when the junction is unbiased (see Fig. 4-5). The electrons crossing from the emitter to the base arrive quite close to the large

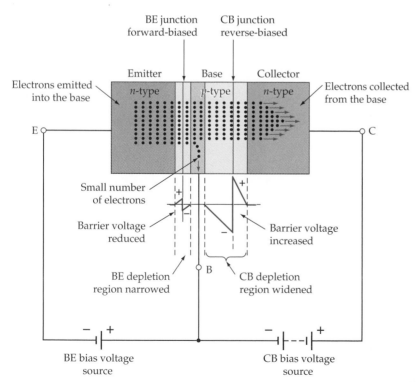

Figure 4-5 For normal BJT operation, the base-emitter junction is forward-biased and the collector-base junction is reverse-biased. Charge carriers (electrons in an *npn* device) are emitted into the base region from the emitter region, and most of them are collected by the collector region. Very few charge carriers flow through the base terminal.

negative-positive electric field (or barrier voltage) at the CB depletion region. Because electrons have a negative charge, they are drawn across the CB junction by this bias voltage. They are said to be *collected.*

Some of the charge carriers entering the base from the emitter do not reach the collector, but flow out through the base connection, as illustrated in Fig. 4-5. However, the path from the BE junction to the CB depletion region is much shorter than that to the base terminal. So only a very small percentage of the total charge carriers flow out of the base terminal. And because the base region is very lightly doped, there are few holes in the base to recombine with electrons from the emitter. The result is that about 98% of the charge carriers from the emitter are drawn across the CB junction to flow through the collector terminal and the voltage sources back to the emitter.

Another way of looking at the effect of the reverse-biased CB junction is from the point of view of minority and majority charge carriers. It has already been shown (in Section 1-6) that a reverse-biased junction opposes the flow of majority charge carriers and assists the flow of minority carriers. Majority carriers are, of course, holes from the *p*-side of a junction and electrons from the *n*-side. Minority carriers are holes from the *n*-side and electrons from the *p*-side (see Fig. 4-6). In the case of an *npn* transistor, the charge carriers arriving at the CB junction are electrons (from the emitter) travelling through the *p*-type base. Consequently, to the CB junction they appear as minority charge carriers, and the reverse bias helps them to cross the junction.

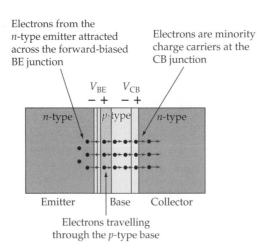

Electrons from the *n*-type emitter attracted across the forward-biased BE junction

Electrons are minority charge carriers at the CB junction

Emitter   Base   Collector

Electrons travelling through the *p*-type base

Figure 4-6 In an *npn* BJT, electrons crossing from the emitter into the base region become minority charge carriers at the collector-base junction and are pulled across the collector-base barrier voltage.

Because the BE junction is forward-biased, it has the characteristics of a forward-biased diode. Substantial current will not flow until the forward bias is about 0.7 V for a silicon device or about 0.3 V for germanium. Reducing the level of the BE bias voltage reduces the *pn*-junction forward bias and thus reduces the current that flows from the emitter through the base region to the collector. Increasing the BE bias voltage increases this current. Decreasing the BE bias voltage to zero, or reversing it, cuts the current *off* completely.

Thus, variation of the small forward-bias voltage on the BE junction controls the emitter and collector currents, and the BE controlling voltage source has to supply only the small base current. This is illustrated in Fig. 4-7.

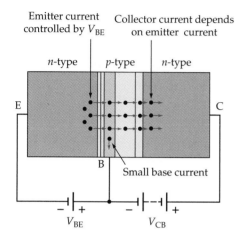

Emitter current controlled by V_{BE} Collector current depends on emitter current

n-type p-type n-type

E

C

B

Small base current

V_{BE} V_{CB}

Figure 4-7 The quantity of charge carriers crossing from the emitter to the base is controlled by the base-emitter voltage. This means that the emitter and collector current levels are controlled by the base-emitter voltage.

pnp Transistor Operation

In an unbiased *pnp* transistor, the barrier voltages are positive on the base and negative on the emitter and collector (see Fig. 4-8). As in the case of the *npn* device, the collector and emitter are heavily doped, so that the BE and CB depletion regions penetrate deep into the lightly doped base.

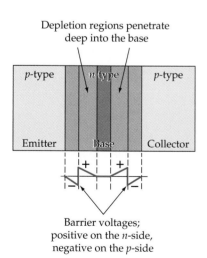

Depletion regions penetrate deep into the base

p-type n-type p-type

Emitter   Base   Collector

Barrier voltages; positive on the n-side, negative on the p-side

Figure 4-8 Depletion regions and barrier voltages at the junctions of an unbiased *pnp* BJT. The barrier voltage is always positive (+) on the *n*-side and negative (−) on the *p*-side.

A *pnp* transistor behaves exactly like an *npn* device, with the exception that the majority charge carriers are holes. As illustrated in Fig. 4-9, the BE junction is forward-biased by an external voltage source, and the CB junction is reverse-biased. Holes are emitted from the *p*-type emitter across the forward-biased BE junction into the base. In the lightly doped *n*-type base, the holes

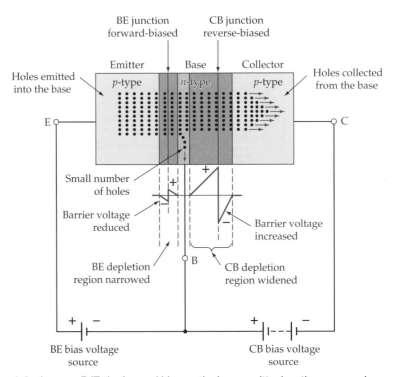

Figure 4-9　In a *pnp* BJT, the forward bias on the base-emitter junction causes charge carriers (holes) to be emitted into the base region. Most of the charge carriers are collected by being drawn across the reverse-biased collector-base junction.

find few electrons to absorb. Some of the holes flow out via the base terminal, but most are drawn across to the collector by the positive-negative electric field at the reverse-biased CB junction. Variation of the forward bias voltage at the BE junction controls the small base current and the much larger collector and emitter currents.

Bipolar Devices

Although one type of charge carrier is in the majority in a *pnp* or *npn* transistor, two types of charge carrier (holes and electrons) are involved in the current flow. Consequently, these devices are termed *bipolar junction transistors (BJT)*. This is to distinguish them from *field-effect transistors (FET)* (see Chapter 9), which are termed *unipolar* devices because they use only one type of charge carrier.

Summary

- A *bipolar junction transistor (BJT)* is an *npn* or *pnp* sandwich of semiconductor materials.
- The outer layers of the *npn* or *pnp* sandwich are called the *emitter* and the *collector*, and the centre layer is termed the *base*.

- Two *pn*-junctions are formed with *depletion regions* and *barrier voltages* at each junction.
- The barrier voltages are negative on the *p*-side and positive on the *n*-side.
- The base-emitter junction is forward-biased, so that charge carriers are *emitted* into the base.
- The collector-base junction is reverse-biased, and its depletion region penetrates deep into the base.
- The base section is made as narrow as possible so that charge carriers can easily move across from emitter to collector.
- The base is lightly doped, so that few charge carriers are available to recombine with the majority charge carriers from the emitter.
- Most charge carriers from the emitter flow to the collector; a few flow out through the base terminal.
- Variation of the base-emitter junction bias voltage alters the base, emitter, and collector currents.

Section 4-1 Review

4-1.1 Sketch a block diagram of a biased *npn* transistor. Show the depletion regions, the barrier voltages, and the movement of charge carriers across the junctions. Explain the operation of the device.

4-1.2 Repeat Question 4-1.1 for a *pnp* transistor.

4-2 BJT VOLTAGES AND CURRENTS

Terminal Voltages

The terminal voltage polarities for an *npn* transistor are shown in Fig. 4-10a. As well as conventional current direction, the direction of the arrowhead indicates the transistor bias polarities. For an *npn* transistor, the base is biased positive with respect to the emitter, and the arrowhead points from the (positive) base to the (negative) emitter. The collector is then biased to a higher positive voltage than the base.

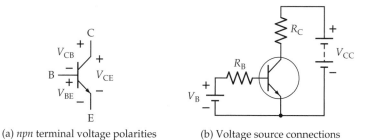

(a) *npn* terminal voltage polarities   (b) Voltage source connections

Figure 4-10 An *npn* BJT must have its base biased positive with respect to its emitter, and its collector must be biased more positive than the base.

Figure 4-10b shows that the voltage sources are usually connected to the transistor via resistors. The base bias voltage (V_B) is connected via resistor R_B, and the collector supply (V_{CC}) is connected via R_C. The negative terminals of the two voltage sources are connected at the transistor emitter terminal. V_{CC} is always much larger than V_B, and this ensures that the collector-base junction remains reverse-biased: positive on the collector (n-side) and negative on the base (p-side).

Typical transistor base-emitter voltages are similar to diode forward voltages: 0.7 V for a silicon transistor and 0.3 V for a germanium device. Typical collector voltages might be anything from 3 V to 20 V for most types of transistors.

For a *pnp* device (see Fig. 4-11a), the base is biased negative with respect to the emitter. The arrowhead points from the (positive) emitter to the (negative) base, and the collector is made more negative than the base. Figure 4-11b shows the voltage sources connected via resistors, and the source positive terminals connected at the emitter. With V_{CC} larger than V_B, the (p-type) collector is more negative than the (n-type) base, and thus the collector-base junction is kept reverse-biased.

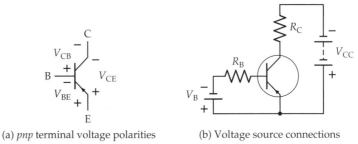

(a) *pnp* terminal voltage polarities   (b) Voltage source connections

Figure 4-11   A *pnp* BJT must have its base biased negative with respect to its emitter, and its collector must be biased more negative than the base.

All BJTs (*npn* and *pnp*) are normally operated with the collector-base junction reverse-biased and the base-emitter junction forward-biased. In the case of a switching transistor (a transistor that is not operated as an amplifier, but is either switched *on* or *off*), the collector-base junction may become forward-biased by approximately 0.5 V. Also, in transistor switching circuits (and some other circuits) the base-emitter junction can become reverse-biased.

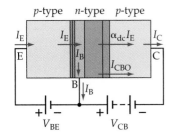

Figure 4-12   Currents in a *pnp* BJT. The collector current (I_C) approximately equals the emitter current (I_E). The base current (I_B) is very much smaller than I_E. A very small reverse leakage current (I_{CBO}) occurs at the collector-base junction.

Transistor Currents

The various current components that flow within a transistor are illustrated in Fig. 4-12. The current flowing into the emitter terminal is referred to as the *emitter current* and is identified as I_E. For the *pnp* device shown, I_E can be thought of as a flow of holes from the emitter to the base. Note that the indicated I_E direction is the conventional current direction from positive to negative. *Base current* I_B and *collector current* I_C are also shown as conventional

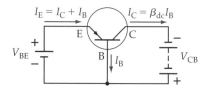

Figure 4-13 Terminal currents for a *pnp* BJT. I_E equals the sum of I_C and I_B. $I_C = \beta_{dc} I_B$ (or $I_C = h_{FE} I_B$), where β_{dc} (or h_{FE}) typically ranges from 25 to 300.

current direction. Both I_C and I_B flow *out* of the transistor while I_E flows *into* the transistor (see Figs. 4-12 and 4-13). Therefore,

$$I_E = I_C + I_B \qquad (4\text{-}1)$$

As already discussed, almost all of I_E crosses to the collector and only a small portion flows out of the base terminal. Typically 96% to 99.5% of I_E flows across the collector-base junction to become the collector current. As shown in Fig. 4-12,

$$I_C = \alpha_{dc} I_E \qquad (4\text{-}2)$$

where α_{dc} (alpha dc) is the *emitter-to-collector current gain*, or the ratio of collector current to emitter current.

From Eq. 4-2, $\alpha_{dc} = I_C/I_E$. Numerically, α_{dc} is typically 0.96 to 0.995. So, the collector current is almost equal to the emitter current, and in many circuits I_C is assumed to be equal to I_E. For reasons that will be explained later, α_{dc} is termed *the common-base dc current gain*.

Because the collector-base junction is reverse-biased, a very small *reverse saturation current* (I_{CBO}) flows across the junction (see Fig. 4-12). I_{CBO} is named the *collector-to-base leakage current*, and it is normally so small that it can be ignored.

Substituting for I_E from Equation 4-1 into Equation 4-2,

$$I_C = \alpha_{dc} (I_C + I_B) \qquad (4\text{-}3)$$

which gives

$$I_C = \frac{\alpha_{dc} I_B}{1 - \alpha_{dc}} \qquad (4\text{-}4)$$

Equation 4-4 can be rewritten as

$$I_C = \beta_{dc} I_B \qquad (4\text{-}5)$$

where

$$\beta_{dc} = \frac{\alpha_{dc}}{1 - \alpha_{dc}} \qquad (4\text{-}6)$$

β_{dc} (beta dc) is the *base-to-collector current gain*, or the ratio of collector current to base current (see Fig. 4-13). From Eq. 4-5, $\beta_{dc} = I_C/I_B$. Typically, β_{dc} ranges from 25 to 300. As will be explained, β_{dc} is also termed the *common-emitter dc current gain*. Instead of β_{dc}, another symbol for common-emitter dc current gain is h_{FE}. This originates from the *h-parameter* circuit, which is discussed in Section 6-3; h_{FE} *is the symbol used on transistor data sheets*.

Although Equations 4-1 to 4-6 were derived for a *pnp* device, they apply equally to *npn* transistors. One difference in the case of an *npn* transistor is that I_B and I_C are assumed to flow *into* the device (conventional current direction), and I_E is taken as flowing *out* (see Fig. 4-14). As explained in Section 4-1, electrons are the majority charge carriers in an *npn* transistor, and they move in a direction opposite to conventional current direction.

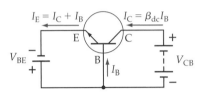

$I_E = I_C + I_B$ $I_C = \beta_{dc} I_B$

Figure 4-14 Terminal currents for an *npn* BJT. I_E equals the sum of I_C and I_B. $I_C = \beta_{dc} I_B$ (or $I_C = h_{FE} I_B$), where β_{dc} (or h_{FE}) typically ranges from 25 to 300.

Typical collector and emitter currents for low-power transistors range from 1 mA to 25 mA, and base currents are usually less than 100 μA. High-power transistors handle currents of several amps.

Example 4-1

Calculate I_C and I_E for a transistor that has $\alpha_{dc} = 0.98$ and $I_B = 100$ μA. Determine the value of β_{dc} (or h_{FE}) for the transistor.

Solution

Eq. 4-4:
$$I_C = \frac{\alpha_{dc} I_B}{1 - \alpha_{dc}} = \frac{0.98 \times 100 \text{ μA}}{1 - 0.98}$$
$$= 4.9 \text{ mA}$$

From Eq. 4-2,
$$I_E = \frac{I_C}{\alpha_{dc}} = \frac{4.9 \text{ mA}}{0.98}$$
$$= 5 \text{ mA}$$

Eq. 4-6:
$$\beta_{dc} = \frac{\alpha_{dc}}{1 - \alpha_{dc}} = \frac{0.98}{1 - 0.98}$$
$$= 49$$

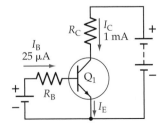

Figure 4-15 BJT circuit for Examples 4-1 and 4-2.

Example 4-2

Calculate α_{dc} and β_{dc} for the transistor (Q_1) in Fig. 4-15 if I_C is measured as 1 mA and I_B is 25 μA. Determine the new base current to give $I_C = 5$ mA.

Solution

Eq. 4-5:
$$\beta_{dc} = \frac{I_C}{I_B} = \frac{1 \text{ mA}}{25 \text{ μA}}$$
$$= 40$$

Eq. 4-1:
$$I_E = I_C + I_B = 1 \text{ mA} + 25 \text{ μA}$$
$$= 1.025 \text{ mA}$$

From Eq. 4-2,
$$\alpha_{dc} = \frac{I_C}{I_E} = \frac{1 \text{ mA}}{1.025 \text{ mA}}$$
$$= 0.976$$

Eq. 4-5:
$$I_B = \frac{I_C}{\beta_{dc}} = \frac{5 \text{ mA}}{40}$$
$$= 125 \text{ μA}$$

Practice Problems

4-2.1 Determine α_{dc} and I_B for a transistor that has I_C equal to 2.5 mA and I_E equal to 2.55 mA. Calculate β_{dc} for the transistor.

4-2.2 A transistor has measured currents of $I_C = 3$ mA and $I_E = 3.03$ mA. Calculate the new current levels when the transistor is replaced with a device that has $\beta_{dc} = 75$. Assume that I_B remains constant.

4-3 BJT AMPLIFICATION

Current Amplification

The equations derived in Section 4-2 demonstrate that a transistor can be used for current amplification. A small change in the base current (ΔI_B) produces a large change in collector current (ΔI_C) and a large emitter current change (ΔI_E) (see Fig. 4-16a and b). Rewriting Equation 4-5, the current gain from the base to collector can be stated in terms of current level changes.

$$\beta_{dc} = \frac{\Delta I_C}{\Delta I_B}$$

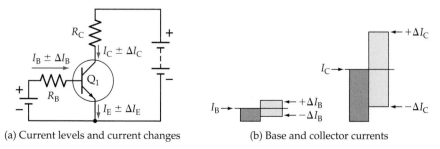

(a) Current levels and current changes   (b) Base and collector currents

Figure 4-16   BJT current amplification. Increasing and decreasing the I_B levels produces much larger changes in I_C and I_E.

The increasing and decreasing levels of input and output currents may be defined as alternating quantities. In this case, small (lower-case) letters are used for the subscripts. Thus, I_b is an ac base current, I_c is an ac collector current, and I_e is an ac emitter current. The alternating current gain from base to collector may now be stated as

$$\beta_{ac} = \frac{I_c}{I_b} \tag{4-7}$$

As in the case of dc current gain, two parameter symbols are available for common-emitter ac current gain: β_{ac} and h_{fe}. Either symbol may be used, but, once again, h_{fe} *is the symbol employed on transistor data sheets.*

Voltage Amplification

Refer to the circuit in Fig. 4-17a, and assume that the transistor (Q_1) has $\beta_{dc} = 50$. Note that the 0.7 V dc voltage source (V_B) forward biases the transistor base-emitter junction. An ac signal source (v_i) in series with V_B provides a ±20 mV input voltage. The transistor collector is connected to a 20 V dc voltage source (V_{CC}) via the 12 kΩ collector resistor (R_1).

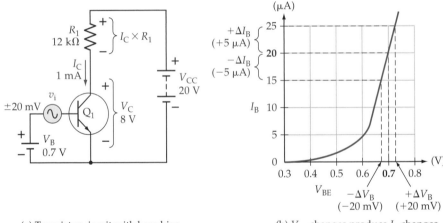

(a) Transistor circuit with base bias and signal generator

(b) V_{BE} changes produce I_B changes

Figure 4-17 BJT voltage amplification. Increasing and decreasing the (input) V_B voltage (±ΔV_B) produces I_B changes. This results in larger I_C changes and (output) V_C variations (±ΔV_C) that are larger than the input voltage changes.

If Q_1 has the I_B/V_{BE} characteristic shown in Fig. 4-17b, the 0.7 V level of V_B produces a 20 μA base current. This gives

$$I_C = \beta_{dc} I_B = 50 \times 20 \text{ μA}$$

$$= 1 \text{ mA}$$

The dc level of the transistor collector voltage can now be calculated as follows:

$$V_C = V_{CC} - (I_C R_1) = 20 \text{ V} - (1 \text{ mA} \times 12 \text{ kΩ})$$

$$= 8 \text{ V}$$

The V_C and I_C levels are shown on Fig. 4-17a.

While the ac input voltage (v_i) is zero, the transistor collector voltage remains at 8 V. When v_i causes a base voltage variation (ΔV_B) of ±20 mV, the base current changes by ±5 μA, as shown in Fig. 4-17b. The I_B change produces a change in the collector current.

$$\Delta I_C = \beta_{dc}\,\Delta I_B = 50 \times (\pm5 \ \mu A)$$

$$= \pm250 \ \mu A$$

Figure 4-18a shows that ΔI_C causes a change in the voltage drop across R_1 and thus produces a variation in the transistor collector voltage.

$$\Delta V_C = \Delta I_C\,R_1 = \pm250 \ \mu A \times 12 \ k\Omega$$

$$= \pm3 \ V$$

The circuit ac input is the base voltage change (ΔV_B), and the ac output is the collector voltage change (ΔV_C). Because the output is greater than the input, the circuit has a voltage gain: it is a *voltage amplifier*. The *voltage gain (A_v)* is the ratio of the output voltage to the input voltage.

$$A_v = \frac{\Delta V_C}{\Delta V_B} = \frac{\pm3 \ V}{\pm20 \ mV}$$

$$= 150$$

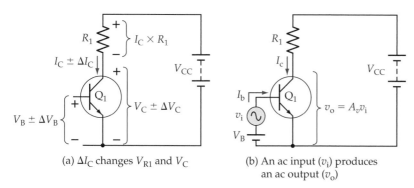

(a) ΔI_C changes V_{R1} and V_C (b) An ac input (v_i) produces an ac output (v_o)

Figure 4-18 I_C changes produce changes in the voltage drop across the collector resistor (R_1), which result in V_C changes. This means that an ac input voltage (v_i) produces an amplifier output voltage ($v_o = A_v \ v_i$).

As already discussed for current changes, the increasing and decreasing levels of voltage can be referred to as ac quantities. The ac signal voltage (v_i) produces the ac base current (I_b), and this generates the ac collector current (I_c), which produces the ac voltage change across R_1 (see Fig. 4-18b). The equation for ac voltage gain is

$$A_v = \frac{v_o}{v_i} \qquad\qquad (4\text{-}8)$$

Substituting the ac quantities for Fig. 4-17a into Eq. 4-8 gives $A_v = 150$ once again.

Example 4-3

Determine the dc collector voltage for the circuit in Fig. 4-19a if the transistor has the I_B/V_{BE} characteristics shown in Fig. 4-19b, and $\beta_{dc} = \beta_{ac} = 80$. Calculate the circuit voltage gain when v_i is ± 50 mV.

Solution

From Fig. 4-19b, $I_B = 15 \ \mu A$ for $V_B = 0.7$ V

Eq. 4-5: $I_C = \beta_{dc} I_B = 80 \times 15 \ \mu A$

$$= 1.2 \ mA$$

$$V_C = V_{CC} - (I_C R_1)$$

$$= 18 \ V - (1.2 \ mA \times 10 \ k\Omega)$$

$$= 6 \ V$$

From Fig. 4-19b, $I_b = \pm 3 \ \mu A$ for $v_i = \pm 50$ mV

$$I_c = \beta_{ac} I_b = 80 \times \pm 3 \ \mu A$$

$$= \pm 240 \ \mu A$$

$$I_c R_1 = \pm 240 \ \mu A \times 10 \ k\Omega$$

$$= \pm 2.4 \ V$$

Eq. 4-8: $A_v = \dfrac{v_o}{v_i} = \dfrac{\pm 2.4 \ V}{\pm 50 \ mV}$

$$= 48$$

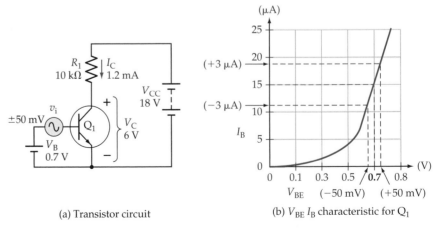

(a) Transistor circuit   (b) $V_{BE} \ I_B$ characteristic for Q_1

Figure 4-19   Basic BJT amplifier circuit and I_B/V_{BE} characteristics for Example 4-3.

4-4 BJT SWITCHING

Diode-Connected BJT

Consider the circuit in Figure 4-20a, which shows a BJT with its collector and
base terminals connected together. This is referred to as *diode-connected*, and as
will be explained, a collector current does flow as illustrated. The total current
into the base-collector terminal is $(I_C + I_B)$, which equals I_E flowing out of Q_1.
The total voltage drop across the BJT is the normal base-emitter voltage (V_{BE}),
which is typically 0.7 V for a silicon device. So, the device behaves exactly like
a diode (Fig. 4-20b).

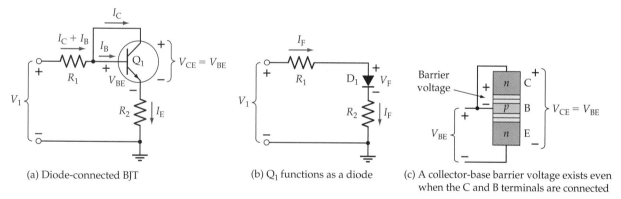

(a) Diode-connected BJT   (b) Q_1 functions as a diode   (c) A collector-base barrier voltage exists even
                                                                 when the C and B terminals are connected

Figure 4-20   A BJT with its collector and base terminals connected behaves like a diode, because the collector current (I_C) flows even when the external collector-base voltage is zero.

Now look at the circuit in Fig. 4-20c showing the transistor replaced with its
block representation. It is easy to see that charge carriers flow across the forward-
biased base-emitter junction. But since the collector-base junction has a zero voltage external bias, how do charge carriers cross it to constitute a collector current?
Recall from Section 1-5 that there is a barrier voltage at an unbiased *pn* junction,
and that its polarity is positive (+) on the *n* side and negative (−) on the *p* side.
The barrier voltage is the result of charge carriers crossing the junction to create
the depletion region, and it exists even when the collector and base terminals are
connected together. Now recall from Section 4-1 that this barrier voltage polarity
(+ on *n*, − on *p*) pulls minority charge carriers from the base into the collector.

So, a collector current will flow when the collector and base terminals are connected together, that is, when the collector-base voltage is zero.

BJT Saturation

A BJT may be used as a switch, as well as for amplification. The switching circuit in Fig. 4-21a is similar to an amplifier circuit, except that a pulse waveform input, instead of a bias voltage and ac signal source, is applied to the transistor base. When the input voltage (V_i) is at the zero level the base current (I_B) is also zero, and consequently, I_C is zero and V_{CE} equals V_{CC}.

$$V_{CE} = V_{CC} - (I_C R_2) \qquad\qquad (4\text{-}9)$$

$$\text{with } I_C = 0, V_{CE} = V_{CC}$$

When V_i is at a positive level, a base current flows as already discussed. In a switching circuit, I_B is made large enough to produce an I_C level that will cause the voltage drop across R_1 to approximately equal the supply voltage (V_{CC}).

With $I_C R_2 \approx V_{CC}$, $\qquad\qquad V_{CE} \approx V_{CC} - I_C R_2$

$$\approx 0$$

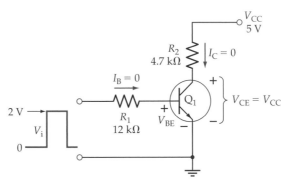

(a) BJT connected as a switch

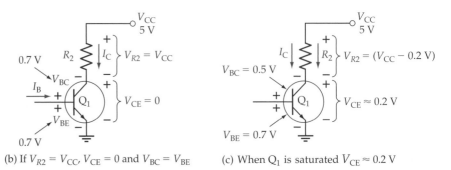

(b) If $V_{R2} = V_{CC}$, $V_{CE} = 0$ and $V_{BC} = V_{BE}$ (c) When Q_1 is saturated $V_{CE} \approx 0.2$ V

Figure 4-21 When a BJT is employed as a switch, it is either biased OFF with V_{CE} equal to the supply voltage (V_{CC}), or ON with V_{CE} equalling the collector-emitter saturation voltage ($V_{CE(sat)}$).

If V_{CE} went down to exactly zero volts, the collector-base junction would become forward-biased by 0.7 V (see Fig. 4-21b). In this case, the collector-base barrier voltage would be overcome and charge carriers from the emitter would be repelled from the collector-base junction. Consequently, there would be zero collector current. But, if I_C becomes zero, there would be no voltage drop across R_2 and the collector-emitter junction would not be forward-biased (by the $I_C R_2$ voltage drop). So, I_C does flow, but it does not become large enough to make V_{CE} equal zero.

The previous discussion about the diode-connected BJT shows that I_C flows when V_{CB} equals zero. It is found that even when the collector-base junction becomes partially forward biased, the barrier voltage is not completely overcome, and I_C continues to flow. A silicon *npn* transistor typically passes substantial collector current when $V_{BC} \approx +0.5$ V, which gives $V_{CE} \approx 0.2$ V (see Fig. 4-21c). In this situation, the collector-emitter voltage is termed the *saturation voltage* ($V_{CE(sat)}$), and $V_{CE(sat)}$ is the minimum level of collector-emitter voltage that exists when the voltage drop across the collector resistor (R_2) tends to drive V_{CE} toward zero.

The circuit in Fig. 4-21a is a BJT switching circuit. In a switching circuit (as in an amplifier) a small base current is used to control a much larger collector current, i.e., to switch the transistor between off and on.

Example 4-4

Calculate I_C, I_B, and h_{FE} for the switching circuit in Fig. 4-21a when Q_1 is switched into saturation.

Solution

From Eq. 4-9,
$$I_C = \frac{V_{CC} - V_{CE(sat)}}{R_2} = \frac{5 \text{ V} - 0.2 \text{ V}}{4.7 \text{ k}\Omega}$$

$$= 1.02 \text{ mA}$$

and
$$I_B = \frac{V_i - V_{BE}}{R_1} = \frac{2 \text{ V} - 0.7 \text{ V}}{12 \text{ k}\Omega}$$

$$= 108 \text{ }\mu\text{A}$$

Eq. 4-5:
$$h_{FE} = \frac{I_C}{I_B} = \frac{1.02 \text{ mA}}{108 \text{ }\mu\text{A}}$$

$$= 9.4$$

A question arises about the transistor current gain (h_{FE}). Suppose that Q_1 in Example 4-4 has a current gain of 50. That would give

Eq. 4-5:
$$I_C = h_{FE} I_B = 50 \times 108 \text{ }\mu\text{A}$$

$$= 5.4 \text{ mA}$$

and
$$V_{R1} = I_C R_2 = 5.4 \text{ mA} \times 4.7 \text{ k}\Omega$$

$$= 25.4 \text{ V}$$

But since V_{CC} is only 5 V, how can V_{R2} be 25.4 V? The answer is that V_{R2} cannot be larger than $(V_{CC} - V_{CE(sat)})$, and that V_{R2} determines the maximum level of I_C regardless of the I_B level and the transistor h_{FE} value.

Practice Problem

4-4.1 In a BJT switching circuit, as in Fig. 4-21a, $V_{CC} = 9$ V, $R_1 = 15$ kΩ, $R_2 = 6.8$ kΩ, and the transistor has an h_{FE} value of 25. Calculate the minimum input voltage required to switch the transistor into saturation.

4-5 COMMON-BASE CHARACTERISTICS

Common-Base Circuit

When a diode (a two-terminal device) is investigated, several levels of forward or reverse voltage are applied and the corresponding current levels are measured (see Section 2-8). The characteristics of the device are then drawn by plotting the graph of current versus voltage. Because a transistor is a three-terminal device, there are three possible connection arrangements (configurations) for investigating its characteristics. Three sets of characteristics may be constructed for each of these configurations.

Figure 4-22 shows a *pnp* transistor with its base terminal common to both the input (emitter-base) terminals and the output (collector-base) terminals. The transistor is said to be connected in *common-base* configuration. Voltmeters and ammeters are included to measure the input and output voltages and currents. Note the presence of resistor R_1 to assist in controlling the emitter current.

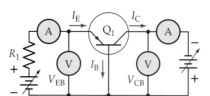

Figure 4-22 Circuit for investigating BJT common-base characteristics. The base is common to the input voltage (V_{EB}) and the output voltage (V_{CB}).

Common-Base Input Characteristics

To investigate the input characteristics, the output voltage (V_{CB}) in Fig. 4-22 is kept constant, and the input voltage (V_{EB}) is set at several convenient levels. At each input voltage, the corresponding input current (I_E) is recorded. The I_E and V_{EB} levels are then plotted to give the common-base input characteristics shown in Fig. 4-23.

Because the emitter-base junction is forward-biased, the characteristics are essentially those of a forward-biased *pn*-junction. Figure 4-23 also shows that for a given input voltage (V_{EB}), more input current flows when higher levels of collector-base (V_{CB}) voltage are used. This is because larger collector-base (reverse-bias) voltages cause the depletion region at the collector-base junction

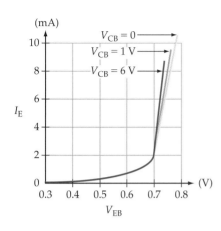

Figure 4-23 The common-base input characteristics for a BJT are (input) emitter current I_E plotted versus (input) base-emitter voltage V_{EB}. The characteristics are similar to those of a forward biased *pn*-junction.

to penetrate deeper into the base of the transistor, thus shortening the distance and reducing the resistance between the emitter-base and collector-base depletion regions.

Common-Base Output Characteristics

To prepare a table of readings for plotting the output characteristics, I_E is held constant at each of several fixed current levels, V_{CB} is adjusted in convenient steps, and the corresponding values of I_C are recorded. In Fig. 4-24 the corresponding I_C and V_{CB} levels obtained when I_E was held constant at 1 mA are plotted, and the resultant characteristic is identified as $I_E = 1$ mA. Similarly, other characteristics are plotted for I_E equal to 2 mA, 3 mA, and so on.

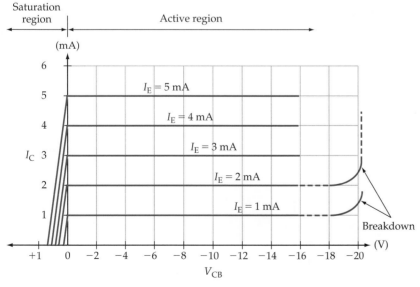

Figure 4-24 The common-base output characteristics (or collector characteristics) for a BJT are a graph of (output) collector current I_C plotted versus (output) collector-base voltage V_{CB} for various constant levels of (input) emitter current I_E. In the active region, I_C remains substantially constant for each level of I_E regardless of V_{CB}. In the saturation region, I_C is reduced to zero when V_{CB} is forward-biased.

The common-base output characteristics in Fig. 4-24 show that for each fixed level of I_E, I_C is almost equal to I_E, and I_C appears to remain constant when V_{CB} is increased. In fact, there is a very small increase in I_C with increasing V_{CB}. This is because the increase in collector-to-base bias voltage (V_{CB}) expands the collector-base depletion region and thus shortens the distance between the two depletion regions. With I_E held constant, the increase in I_C is so small that it is noticeable only for large variations in V_{CB}.

Note from Fig. 4-24 that when V_{CB} is reduced to zero, I_C still flows. This is because, even when the externally applied voltage is zero, there is still a barrier voltage existing at the collector-base junction, and this assists the flow of I_C. The charge carriers which constitute I_C are minority carriers as they cross the collector-base junction. Consequently, the reverse-bias collector-base voltage (V_{CB}) tends to pull them across the junction (see Fig. 4-12). The collector-base barrier voltage has the same (reverse-bias) polarity, and so the barrier voltage (without any external bias) also pulls the charge carriers to the collector. To stop the flow of charge carriers, the collector-base junction has to be forward-biased. As shown in Fig. 4-24, I_C is reduced to zero only when V_{CB} is increased positively.

The region of the graph for the forward-biased collector-base junction is known as the *saturation region* (see Fig. 4-24). The region in which the junction is reverse-biased is named the *active region*; this is the normal operating region for the transistor.

If an excessive reverse-bias voltage is applied to the collector-base junction, the device breakdown may occur. Breakdown, illustrated by the dashed lines in Fig. 4-24, can be caused by the same effects that make diodes break down (see Section 2-9). Breakdown can also result from the collector-base depletion region penetrating into the base (as the reverse bias is increased) until it makes contact with the emitter-base depletion region (see Fig. 4-25). This condition is known as *punch-through* or *reach-through,* and very large currents can flow when it occurs, possibly destroying the device. It is very important to maintain V_{CB} below the maximum safe limit specified by the device manufacturer. Typical maximum V_{CB} levels range from 25 V to 80 V.

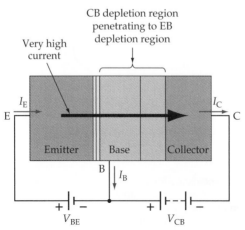

CB depletion region
penetrating to EB
depletion region

Very high
current

I_E

E

Emitter   Base   Collector

B I_B

I_C

C

V_{BE} V_{CB}

Figure 4-25 Excessive levels of (reverse bias) collector-base voltage (V_{CB}) can cause the junction depletion regions to make contact, resulting in destructive current levels.

Common-Base Current Gain Characteristics

The current gain characteristics (also termed the forward transfer characteristics) can be obtained experimentally by use of the circuit in Fig. 4-22. V_{CB} is held constant at a convenient voltage, and I_C is measured for various levels of I_E. I_C is then plotted versus I_E, and the resultant graph is identified by the V_{CB} level (see Fig. 4-26).

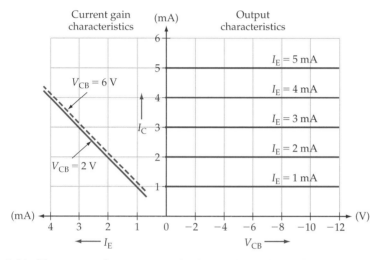

Figure 4-26 The common-base current gain characteristics for a BJT are a graph of output current (I_C) plotted versus input current (I_E) with the output voltage (V_{CB}) held constant.

The common-base current gain characteristics can be derived from the common-base output characteristics, as shown in Fig. 4-27. A vertical line is drawn through a selected V_{CB} value, and corresponding levels of I_E and I_C are read along the line. The I_C levels are then plotted versus I_E, and the characteristic is labelled with the V_{CB} used. Because almost all of I_E flows out of the collector terminal as I_C, V_{CB} has only a small effect on the current gain characteristics.

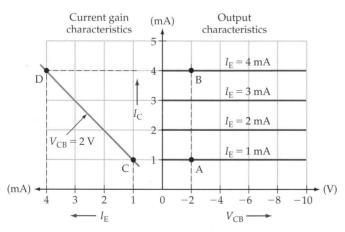

Figure 4-27 BJT common-base current gain characteristics can be derived from the output characteristics.

Example 4-5

Derive the current gain characteristic for $V_{CB} = -2$ V from the common-base output characteristics in Fig. 4-27.

Solution

Draw a vertical line at $V_{CB} = -2$ V.

Where the line intersects the characteristics at points A and B, read

$$I_C \approx 1 \text{ mA} \quad \text{for } I_E = 1 \text{ mA}$$

and $$I_C \approx 4 \text{ mA} \quad \text{for } I_E = 4 \text{ mA}$$

Plot points C and D for the output characteristic at the corresponding levels of I_C and I_E. Draw the characteristic through points C and D.

Practice Problem

4-5.1 Plot two lines of BJT common-base output characteristics from the following measured quantities:

for $I_E = 1.5$ mA: $I_C = 1.4$ mA when $V_{CB} = 0$ V
$I_C = 1.45$ mA when $V_{CB} = 10$ V

for $I_E = 3$ mA: $I_C = 2.8$ mA when $V_{CE} = 0$ V
$I_C = 2.95$ mA when $V_{CE} = 10$ V

Derive the current gain characteristics for $V_{CB} = 5$ V from the BJT output characteristics.

4-6 COMMON-EMITTER CHARACTERISTICS

Common-Emitter Circuit

Figure 4-28 shows a circuit for determining BJT common-emitter characteristics. The input voltage is applied between the base and emitter terminals, and the output is taken at the collector and emitter terminals, so that the emitter terminal is common to both input and output. Resistor R_1 is included to help

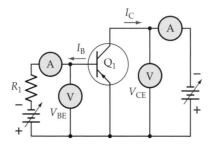

Figure 4-28 Circuit for investigating BJT common-emitter characteristics. The emitter is common to the input voltage (V_{BE}) and the output voltage (V_{CE}).

maintain the base current at a constant level. Voltage and current levels are measured as shown.

Common-Emitter Input Characteristics

To prepare a table of measured values for constructing the common-emitter input characteristics, V_{CE} is held constant, V_{BE} is set at convenient levels, and the corresponding I_B levels are recorded. I_B is then plotted versus V_{BE}, as shown in Fig. 4-29. It is seen that the common-emitter input characteristics (like the common-base input characteristics) are those of a forward-biased *pn*-junction. It should be remembered that I_B is only a small portion of the total current (I_E) that flows across the base-emitter junction.

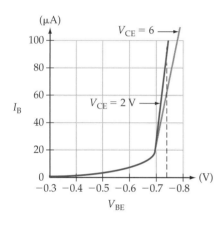

Figure 4-29 The common-emitter input characteristics for a BJT are (input) base current I_B plotted versus (input) base-emitter voltage V_{BE}. The characteristics are similar to those of a forward biased *pn*-junction.

Figure 4-29 also shows that, for a given level of V_{BE}, I_B is reduced when higher V_{CE} levels are employed. This is because the higher V_{CE} produces greater depletion region penetration into the base, reducing the distance between the collector-base and emitter-base depletion regions. Consequently, more of the charge carriers from the emitter flow across the collector-base junction, and fewer flow out through the base terminal.

Common-Emitter Output Characteristics

To obtain a table of values for plotting the common-emitter output characteristics, I_B is maintained constant at several convenient levels (in the circuit of Fig. 4-28). At each I_B level, V_{CE} is adjusted in steps and I_C is recorded at each V_{CE} step. The I_C values are plotted versus V_{CE} for each I_B level, to create the kind of output characteristics family shown in Fig. 4-30. Note that the V_{BE} and V_{CE} polarities are negative for the characteristics shown. This is because a *pnp* transistor is being investigated (see Fig. 4-28).

Because I_E is not held constant (as it is for the common-base output characteristics), the shortening of the distance between the depletion regions (when V_{CB} is increased) draws more charge carriers from the emitter to the collector. So, although I_B is constant, I_C increases to some extent with increasing V_{CE},

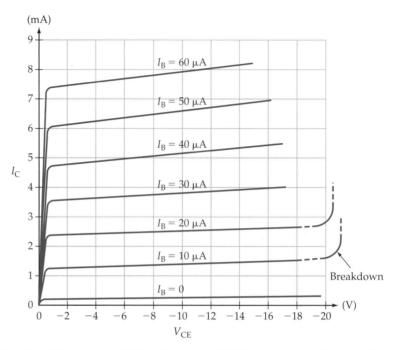

Figure 4-30   The common-emitter output characteristics (or collector characteristics) for a BJT are a graph of (output) collector current I_C plotted versus (output) collector-emitter voltage V_{CE} for various constant levels of (input) base current I_B. In the active region (above the knee of the characteristics), for each constant level of I_B, I_C increases by a small amount as V_{CE} increases. I_C is reduced to zero when V_{CE} becomes zero.

causing the slopes of the common-emitter output characteristics to be much more pronounced than those of the common-base characteristics.

On Fig. 4-30 note that I_C reduces to zero when V_{CE} becomes zero. This is because the horizontal axis voltage (V_{CE}) equals $V_{CB} + V_{BE}$ (see Fig. 4-31). At the knee of the characteristic, the collector-base junction voltage (V_{CB}) has been reduced to zero. Further reduction in V_{CE} causes the collector-base junction to be forward-biased, and this repels minority charge carriers, thus reducing I_C to zero.

The dashed lines in Fig. 4-30 show that, if V_{CE} exceeds a maximum safe voltage, I_C increases rapidly. As in the case of common-base configuration, this is due to punch-through, and it could destroy the BJT.

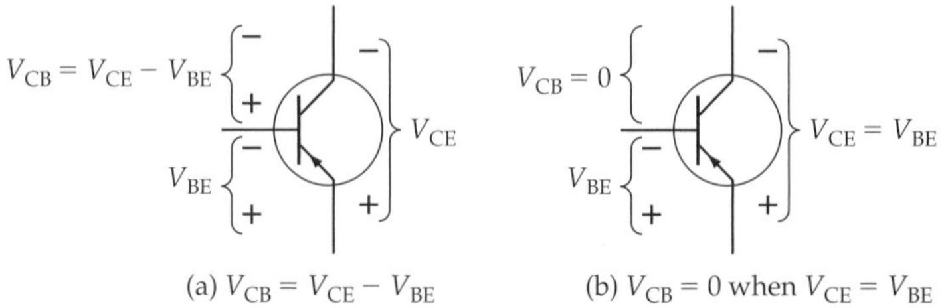

Figure 4-31   In a common-emitter circuit the collector-base junction is forward-biased when the collector-emitter voltage is reduced to zero. This cuts off the collection of charge carriers crossing the base region from the emitter, and thus reduces I_C to zero.

Example 4-6

Determine the I_B and I_C levels for a device with the characteristics in Figs. 4-29 and 4-30, when V_{BE} is 0.7 V and V_{CE} is -6 V. Calculate the value of β_{dc}.

Solution

From the input characteristics in Fig. 4-29, at $V_{BE} = 0.7$ V and $V_{CE} = -6$ V,

$$I_B \approx 20 \ \mu A$$

From the output characteristics in Fig. 4-30, at $V_{CE} = -6$ V and $I_B = 20 \ \mu A$,

$$I_C = 2.5 \ mA$$

From Eq. 4-5,

$$\beta_{dc} = \frac{I_C}{I_B} = \frac{2.5 \ mA}{20 \ \mu A}$$

$$= 125$$

Common-Emitter Current Gain Characteristics

The common-emitter current gain characteristics (see Fig. 4-32) are output current (I_C) plotted versus input current (I_B) for various fixed levels of V_{CE}. Like the common-base current gain characteristics, they can be obtained experimentally or derived from the output characteristics. To prepare the table of I_B and I_C values experimentally (using the circuit in Fig. 4-28), V_{CE} is held at a selected level, I_B is adjusted in steps, and the corresponding I_C level is recorded at each step.

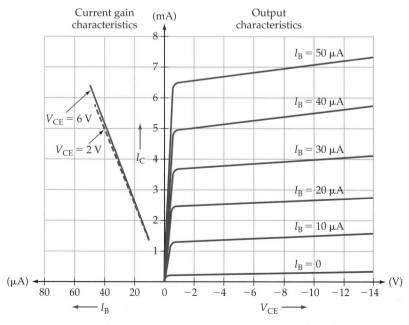

Figure 4-32   The common-emitter current gain characteristics for a BJT are a graph of output current (I_C) plotted versus input current (I_B) with the output voltage (V_{CE}) held constant.

Practice Problem

4-6.1 Plot two lines of common-emitter transistor output characteristics from the following quantities:

for $I_B = 25\ \mu A$: $I_C = 1.5$ mA when $V_{CE} = 1$ V
$I_C = 1.75$ mA when $V_{CE} = 12$ V

for $I_B = 50\ \mu A$: $I_C = 3$ mA when $V_{CE} = 1$ V
$I_C = 3.5$ mA when $V_{CE} = 12$ V

Derive the current gain characteristics for $V_{CE} = 12$ V from the device output characteristics.

4-7 COMMON-COLLECTOR CHARACTERISTICS

Output and Current Gain Characteristics

Typical representations of BJT common-collector output and current gain characteristics are shown in Figure 4-33. The output characteristics are a plot of the emitter current (I_E) versus the emitter-collector voltage (V_{EC}) for several constant levels of base current (I_B). The current gain characteristics are I_E plotted versus I_B at constant V_{EC} voltages.

It will be recalled that the common-emitter output characteristics are I_E plotted against V_{CE}, and that the common-emitter current gain characteristics are I_E plotted versus I_B. Because I_C approximately equals I_E, the common-collector

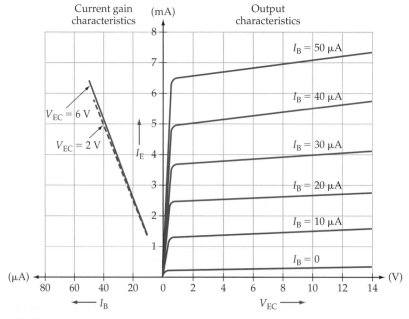

Figure 4-33 The common-collector output and current-gain characteristics are virtually identical to the common-emitter characteristics.

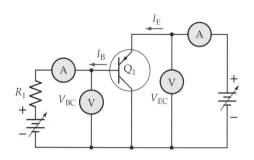

Figure 4-34 Circuit for determining BJT common-collector characteristics. This circuit is difficult to use because I_B changes each time V_{EC} is adjusted.

output and current gain characteristics are practically identical to those of the common-emitter circuit.

Common-Collector Circuit

The circuit arrangement in Fig. 4-34 shows the BJT collector terminal common to both input base-collector voltage (V_{BC}) and output emitter-collector voltage (V_{EC}), and this arrangement would seem to be suitable for determining the common-collector characteristics. However, it should be noted that V_{EC} variations affect I_B, and when a set of readings for the output characteristics is being taken, I_B has to be readjusted to its required (constant) level each time V_{EC} is adjusted.

The common-emitter circuit in Fig. 4-28 can be used for determining common-collector characteristics if the ammeter monitoring I_C is moved to the emitter circuit to indicate I_E level. This circuit allows I_B to be held constant when V_{EC} is changed, and so it is more convenient than the circuit in Fig. 4-34.

Common-Collector Input Characteristics

The common-collector input characteristics shown in Fig. 4-35 are quite different from either common-base or common-emitter input characteristics. The difference is due to the fact that the input voltage (V_{BC}) is largely determined by the output voltage (V_{EC}).

Referring to Fig. 4-34, we see that

$$V_{EC} = V_{EB} + V_{BC}$$

or

$$V_{EB} = V_{EC} - V_{BC}$$

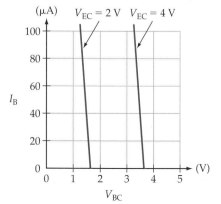

Figure 4-35 The common-collector input characteristics are quite different from common-base and common-emitter input characteristics.

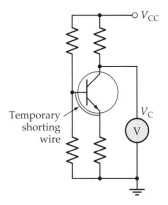

Figure 4-36 Quick in-circuit test for a BJT. Short-circuiting the base and emitter terminals should cause the collector voltage to jump to V_{CC}.

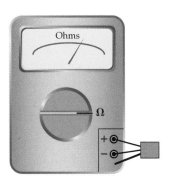

Figure 4-37 Use of an analog ohmmeter for checking the junctions of a BJT.

Figure 4-38 The diode test function on a digital multimeter can be used to check the junctions of a BJT.

Increasing the level of V_{BC} with V_{EC} held constant reduces the base-emitter voltage (V_{EB}) and thus reduces I_B. This explains the slope of the CC input characteristics.

Practice Problem

4-7.1 Referring to the common-collector input characteristics in Fig. 4-35, determine the level of V_{BC} when $I_B = 60$ μA and $V_{EC} = 2$ V, and when $I_B = 60$ μA and $V_{EC} = 4$ V.

4-8 TRANSISTOR TESTING

In-Circuit Testing

A quick test to check if a transistor is operational can be performed while the device is still connected in a circuit. Consider Fig. 4-36, which shows a voltmeter connected to measure the transistor collector voltage (V_C). The V_C measurement is noted, and then the base and emitter terminals are temporarily short-circuited, as illustrated. This should turn the transistor off, and V_C should jump approximately to the circuit supply voltage (V_{CC}). When the shorting wire is removed, V_C should return to its previous level. If the change in the V_C level does not occur, the transistor is not operational. This may be due to a faulty transistor or to some other problem in the circuit.

Ohmmeter and Digital Multimeter Tests

An ohmmeter may be used for checking the junctions of a BJT in exactly the same way a diode is tested (see Section 2-8). As in the case of a diode, a good junction (base-emitter or collector-base) indicates a low resistance when forward-biased and a high resistance when reversed (see Fig. 4-37). The resistance measured between the collector and emitter terminals of a good BJT should be high regardless of the meter terminal polarity.

The diode-testing function on a digital multimeter may be used for testing transistor junctions (see Fig. 4-38). When the junction is forward-biased, the meter displays the junction forward voltage drop, and when it is reverse-biased, either OL or an indication of the meter internal voltage is displayed. Some digital multimeters can measure the h_{FE} value of a transistor.

Characteristic Plotting

A transistor may be rapidly checked with a *curve tracer* (see Fig. 4-39). Two transistor sockets are normally provided to display the characteristics of two devices. A switch permits either socket to be selected. The characteristics of a known good transistor should be displayed first. Then the switch should be operated to display the characteristics of the transistor under investigation.

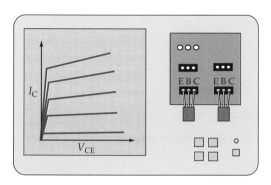

Figure 4-39 A curve tracer can be used to compare the characteristics of a BJT of unknown quality with the characteristics of an operational BJT.

Transistor characteristics may be plotted by obtaining a table of corresponding levels of current and voltage, as already discussed. A more convenient method of plotting characteristics, using an XY recorder, is illustrated in Fig. 4-40. One power supply is connected to produce base current (I_B) via resistor R_1, and the other provides I_C flow through R_2. The collector and emitter terminals are connected to the horizontal input terminals of the XY recorder, and the voltage drop across resistor R_2 provides the vertical input. The base current is set at a several convenient levels, and (at each I_B level) V_{CC} is slowly increased from zero to cause the pen to trace one line of the transistor output characteristics.

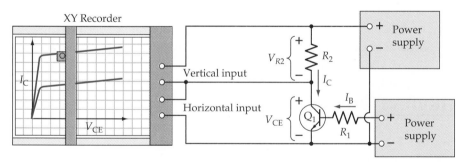

Figure 4-40 BJT characteristics can be plotted on an XY recorder. V_{CC} produces horizontal deflection, and V_{R2} causes vertical deflection.

A convenient selection for the horizontal and vertical scales of the XY recorder is 1 V/cm. If $R_2 = 1 \text{ k}\Omega$, each 1 V drop across R_2 represents 1 mA of collector current. So the I_C (vertical) scale on the characteristics is 1 mA/cm, and the V_{CE} scale is 1 V/cm.

Example 4-7

The circuit and XY recorder in Fig. 4-40 are to be used to plot transistor common-emitter output characteristics to $V_{CE(max)} = 10$ V and $I_{C(max)} = 20$ mA. Select V_{CC} and R_2, and determine appropriate V/cm scales for the XY recorder. The plotted characteristics are to be approximately 20 cm × 20 cm.

Solution

For $I_{C(max)}$ corresponding to 20 cm,

$$\text{vertical scale} = \frac{I_{C(max)}}{\text{scale length}} = \frac{20 \text{ mA}}{20 \text{ cm}}$$

$$= 1 \text{ mA/cm}$$

Select a vertical scale of 1 V/cm, so that 20 cm represents 20 V and 20 mA.

$$R_2 = \frac{20 \text{ V}}{20 \text{ mA}} = 1 \text{ k}\Omega$$

$$V_{CC} = V_{CE(max)} + (I_{C(max)} \times R_2) = 10 \text{ V} + (20 \text{ mA} \times 1 \text{ k}\Omega)$$

$$= 30 \text{ V}$$

$$\text{horizontal scale} = \frac{V_{CE(max)}}{\text{scale length}} = \frac{10 \text{ V}}{20 \text{ cm}}$$

$$= 0.5 \text{ V/cm}$$

Practice Problem

4-8.1 A BJT with a $V_{CE(max)}$ of 40 V and an $I_{C(max)}$ of 100 mA is to have its common-emitter output characteristics plotted on an XY recorder, as in Fig. 4-40. Select an appropriate supply voltage and collector resistor value, and determine suitable V/cm scales for the XY recorder to give characteristics approximately 25 cm × 25 cm.

Review Questions

Section 4-1

4-1 Draw a block diagram of an unbiased *npn* BJT. Identify each part of the device and show the depletion regions and barrier voltages. Briefly explain.

4-2 Repeat Question 4-1 for a *pnp* BJT.

4-3 Sketch a correctly biased *npn* BJT. Show the effect of the bias voltages on the depletion regions and barrier voltages. Show the movement of charge carriers through the BJT. Explain the device operation.

4-4 Repeat Question 4-3 for a correctly biased *pnp* BJT.

4-5 Draw circuit symbols for *npn* and *pnp* BJTs. Identify current directions and bias voltage polarities.

Section 4-2

4-6 Draw a sketch to show the various current components of a BJT. Briefly explain the origin of each current. Write an equation relating I_E, I_B, and I_C.

4-7 Write equations for I_C in terms of I_E, and for I_C in terms of I_B. Define α_{dc}, β_{dc}, and h_{FE} and state typical values for each.

Section 4-3

4-8   Sketch a circuit diagram to show how a BJT can be used to amplify direct current changes and alternating signal currents. Briefly explain.

4-9   Draw a circuit diagram to show how a BJT can amplify direct voltage changes and alternating signal voltages. Briefly explain.

Section 4-4

4-10   Discuss the operation of a BJT with its collector and base terminals connected together.

4-11   Draw a circuit diagram to show how a BJT can be used as a switch. Identify all currents and voltages, show typical input and output voltages, and explain BJT saturation.

Section 4-5

4-12   Explain BJT common-base configuration, and sketch a circuit for determining common-base characteristics.

4-13   Sketch typical BJT common-base input characteristics and output characteristics. Explain the shape of each set of characteristics.

4-14   Explain BJT punch-through and how it occurs.

4-15   Sketch typical BJT common-base current-gain characteristics. Explain the shape of the characteristics.

Section 4-6

4-16   Explain BJT common-emitter configuration, and draw a circuit for determining common-emitter characteristics.

4-17   Sketch typical BJT common-emitter input and output characteristics. Explain the shape of the characteristics.

4-18   Sketch typical BJT common-emitter current-gain characteristics. Explain the shape of the characteristics.

Section 4-7

4-19   Explain BJT common-collector configuration, and draw a circuit for determining common-collector characteristics.

4-20   Sketch typical BJT common-collector output and current-gain characteristics, and explain the shape of the characteristics.

4-21   Sketch typical BJT common-collector input characteristics. Discuss the shape of the characteristics.

Section 4-8

4-22   Describe how a BJT may be quickly tested while still connected in a circuit.

4-23   Discuss how a BJT may be tested by means of an ohmmeter and by use of a digital multimeter.

4-24   Sketch a diagram to show how an XY recorder may be used for plotting BJT characteristics.

Problems

Section 4-2

4-1 Calculate the values of I_C and I_E for a BJT with $\alpha_{dc} = 0.97$ and $I_B = 50\ \mu A$. Determine β_{dc} for the device.

4-2 A BJT has the following measured current levels: $I_C = 5.25$ mA and $I_B = 100\ \mu A$. Calculate I_E, α_{dc}, and β_{dc}, and determine the new I_B level to give $I_C = 15$ mA.

4-3 Calculate the collector and emitter current levels for a BJT with $\alpha_{dc} = 0.99$ and $I_B = 20\ \mu A$.

4-4 The following current measurements were made on a BJT: $I_C = 12.42$ mA and $I_B = 200\ \mu A$. Determine I_E, α_{dc}, and β_{dc}. Determine the new I_C level when I_B is 150 μA.

4-5 A BJT with measured current levels of $I_C = 16$ mA and $I_E = 16.04$ mA is replaced with another BJT that has $\beta_{dc} = 25$. Calculate the new level of I_C and I_E, assuming that the base current remains constant.

Section 4-3

4-6 The BJT in the circuit in Fig. 4-41a has the input characteristics in Fig. 4-41b. If $\beta_{dc} = 50$, calculate the collector voltage when $V_{BE} = 0.7$ V.

4-7 A ± 25 mV ac input is applied in series with the base of the BJT in Problem 4-6. Calculate the circuit voltage gain.

4-8 Determine the new V_{CE} level for the circuit in Fig. 4-41a if the V_{BE} of the BJT is adjusted to 0.75 V.

4-9 Calculate the new voltage gain for the circuit in Fig. 4-41a when R_1 is changed to 6.8 kΩ.

4-10 When the BJT in the circuit of Fig. 4-41a is changed, V_C is measured as 9 V. Assuming that I_B remains constant at 30 μA, calculate the current gain for the new BJT.

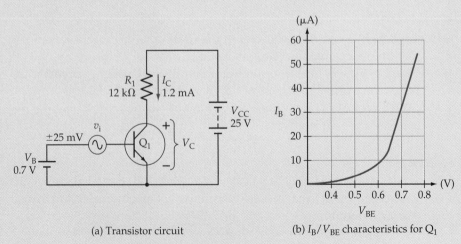

(a) Transistor circuit

(b) I_B/V_{BE} characteristics for Q_1

Figure 4-41 Circuit and device characteristics for Problems 4-6 to 4-10.

Section 4-4

4-11 A BJT connected as a switch, as in Fig. 4-21a, $V_{CC} = 12$ V, $V_i = 3$ V, $R_1 = 8.2$ kΩ, $R_2 = 3.9$ kΩ. Assuming the transistor is in saturation, calculate I_C, I_B, and the minimum h_{FE} value.

4-12 A BJT in a switching circuit has $V_{CC} = V_i = 5$ V, and a 2.5 mA collector current when saturated. If the transistor h_{FE} value is 20, calculate the maximum base resistor value.

4-13 A 2N3904 BJT connected as in Fig. 4-21a has $V_{CC} = 10$ V, $R_1 = 47$ kΩ, $R_2 = 1$ kΩ. Determine the minimum V_i level for transistor saturation.

4-14 In a switching circuit using a 2N3904 BJT, $V_{CC} = 15$ V and $V_i = 4$ V. The collector current is to be 3 mA when the transistor is saturated. Determine suitable resistances for R_1 and R_2.

Section 4-5

4-15 Derive the common-base current gain characteristics for $V_{CB} = -5$ V from the output characteristics in Fig. 4-42.

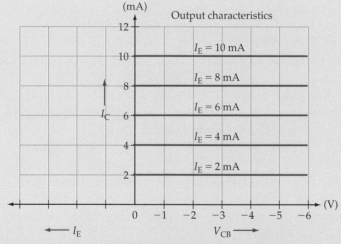

Figure 4-42 Device characteristics for Problem 4-15.

4-16 Refer to the common-base output characteristics in Fig. 4-24. Derive the current gain characteristics for $V_{CB} = -8$ V.

4-17 From the common-base input characteristics in Fig. 4-23, estimate the level of I_E when $V_{EB} = 0.72$ V and $V_{CB} = 6$ V.

Section 4-6

4-18 Derive the current gain characteristics for $V_{CB} = 9$ V from the common-emitter output characteristics in Fig. 4-43.

4-19 From the common-emitter input characteristics in Fig. 4-29, estimate the level of I_B when $V_{BE} = 0.72$ V and $V_{CE} = 2$ V.

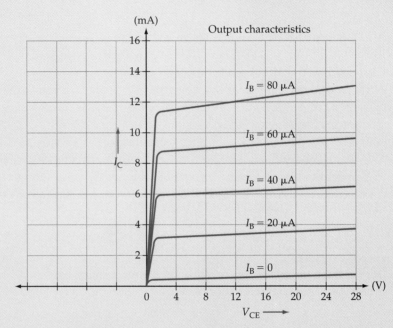

Figure 4-43   Device characteristics for Problem 4-18.

4-20   Refer to the common-emitter output characteristics in Fig. 4-30. Derive the current gain characteristics for $V_{CE} = 14$ V.

4-21   Determine the level of I_C for a BJT with the common-emitter output characteristics in Fig. 4-30 when $I_B = 60$ µA and when $I_B = 30$ µA, with V_{CE} equal to 5 V in each case. Calculate β_{dc}.

Section 4-7

4-22   Refer to the common-collector output characteristics in Fig. 4-33. Derive the current gain characteristics for $V_{EC} = 10$ V.

4-23   Calculate the input voltage for the common-collector circuit in Fig. 4-34 when $V_{EC} = 10$ V and $V_{EB} = 0.7$ V.

4-24   Determine V_{BC} from the common-collector input characteristics in Fig. 4-35 when $I_B = 20$ µA and $V_{EC} = 4$ V.

Section 4-8

4-25   An XY recorder is to be used to plot BJT common-emitter output characteristics as in Fig. 4-40. The BJT current and voltage levels are to be $V_{CE(max)} = 20$ V and $I_{C(max)} = 50$ mA. Select V_{CC} and resistor R_2, and determine appropriate V/cm scales for the XY recorder. The size of the plotted characteristics is to be approximately 25 cm × 25 cm.

4-26   Sketch a circuit diagram for using an XY recorder to determine the common-emitter input characteristics for a BJT. Select suitable circuit components and V/cm scales to plot 20 cm × 20 cm characteristics with maximum levels of $V_{BE} = 1$ V and $I_B = 100$ µA.

Practice Problem Answers

4-2.1	0.98, 50 μA, 50
4-2.2	2.25 mA, 2.28 mA
4-3.1	71.5
4-3.2	80
4-4.1	1.48 V
4-7.1	1.5 V, 3.5 V
4-8.1	1 V/cm, 250 Ω, 65 V, 1.5 V/cm

CHAPTER 5
BJT Biasing

CONTENTS

Objectives

You will be able to:

1 Draw dc load lines for transistor circuits, identify circuit Q-points, and estimate the maximum circuit output voltage variations.

2 Sketch base bias, collector-to-base bias, and voltage-divider bias circuits, and explain the operation of each circuit.

3 Analyze various types of BJT bias circuits to determine circuit voltage and current levels.

4 Trouble-shoot non-operational BJT bias circuits.

5 Design base bias, collector-to-base bias, and voltage-divider bias circuits, and select appropriate standard value components.

6 Sketch several unusual types of BJT bias circuit, and explain the operation of each circuit.

7 Analyze and design the types of bias circuit referred to in item 6.

8 Discuss the thermal stability of BJT bias circuits, and determine the effects of I_{CBO} and V_{BE} changes with temperature.

9 Sketch bias circuits for direct-coupled and capacitor-coupled switching transistors, and explain the operation of each circuit.

10 Analyze and design switching transistor bias circuits.

INTRODUCTION

Transistors used in amplifier circuits must be biased into an on state with constant (direct) levels of collector, base, and emitter current, and constant terminal voltages. The levels of I_C and V_{CE} define the transistor *dc operating point*, or *quiescent point*. The circuit that provides this state is known as a *bias circuit*. Ideally, the current and voltage levels in a bias circuit should remain absolutely constant. In practical circuits these quantities are affected by the transistor current gain (h_{FE}) and by temperature changes. The best bias circuits have the greatest stability; they hold the currents and voltages substantially constant regardless of the h_{FE} and variations in temperature. The simplest bias circuits are the least costly because they use the smallest number of components, but they are not as stable as more complex circuits.

5-1 DC LOAD LINE AND BIAS POINT

DC Load Line

The *dc load line* for a transistor circuit is a straight line drawn on the transistor output characteristics. For a common-emitter (CE) circuit, the load line is a graph of collector current (I_C) versus collector-emitter voltage (V_{CE}), for a given value of collector resistance (R_C) and a given supply voltage (V_{CC}). The load line shows all corresponding levels of I_C and V_{CE} that can exist in a particular circuit.

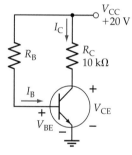

Figure 5-1 Transistor circuit with collector resistor R_C.

Consider the common-emitter circuit in Fig. 5-1. Note that the polarities of the transistor terminal voltages are such that the base-emitter junction is forward-biased and the collector-base junction is reverse-biased. These are the normal bias polarities for the transistor junctions. The dc load line for the circuit in Fig. 5-1 is drawn on the device common-emitter characteristics in Fig. 5-2.

From Fig. 5-1, the collector-emitter voltage is

$$V_{CE} = \text{(supply voltage)} - \text{(voltage drop across } R_C)$$

or

$$V_{CE} = V_{CC} - I_C R_C \qquad (5\text{-}1)$$

If the base-emitter voltage (V_{BE}) is zero, the transistor is not conducting and $I_C = 0$. Substituting the V_{CC} and R_C values from Fig. 5-1 into Eq. 5-1,

$$V_{CE} = 20 \text{ V} - (0 \times 10 \text{ k}\Omega)$$
$$= 20 \text{ V}$$

Plot *point A* on the common-emitter characteristics in Fig. 5-2 at $I_C = 0$ and $V_{CE} = 20$ V. This is one point on the dc load line.

Now assume a collector current of 2 mA, and calculate the corresponding collector-emitter voltage level.

$$V_{CE} = 20 \text{ V} - (2 \text{ mA} \times 10 \text{ k}\Omega)$$
$$= 0 \text{ V}$$

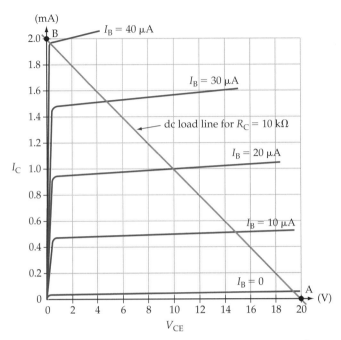

Figure 5-2 DC load line drawn upon transistor common-emitter output characteristics.

Plot *point B* on Fig. 5-2 at $V_{CE} = 0$ and $I_C = 2$ mA. The straight line drawn through point A and point B is the dc load line for $R_C = 10$ kΩ and $V_{CC} = 20$ V. If either of these two quantities is changed, a new load line must be drawn.

As already stated, the dc load line represents all corresponding I_C and V_{CE} levels that can exist in the circuit, as represented by Eq. 5-1. For example, a point plotted at $V_{CE} = 16$ V and $I_C = 1.5$ mA on Fig. 5-2 does not appear on the load line. This combination of voltage and current cannot exist in this particular circuit (Fig. 5-1). Knowing any one of I_B, I_C, or V_{CE}, it is easy to determine the other two from a dc load line drawn on the device characteristics.

It is not always necessary to have the device characteristics in order to draw the dc load line. A simple graph of I_C versus V_{CE} can be used, as demonstrated in Example 5-1.

Example 5-1

Draw the new dc load line for the circuit in Fig. 5-1 when $R_C = 12$ kΩ.

Solution

Prepare a graph for I_C versus V_{CE} (see Fig. 5-3).

When $I_C = 0$:

Eq. 5-1: $$V_{CE} = V_{CC} - I_C R_C = 20 \text{ V} - 0$$

$$= 20 \text{ V}$$

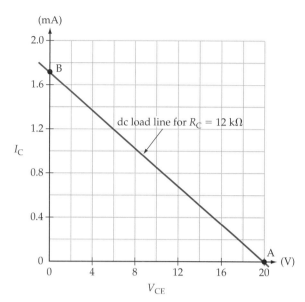

Figure 5-3 DC load line for Example 5-1.

Plot point A on Fig. 5-3 at

$$I_C = 0 \quad \text{and} \quad V_{CE} = 20 \text{ V}$$

When $V_{CE} = 0$:

From Eq. 5-1,

$$0 = V_{CC} - I_C R_C$$

giving

$$I_C = \frac{V_{CC}}{R_C} = \frac{20 \text{ V}}{12 \text{ k}\Omega}$$

$$\approx 1.7 \text{ mA}$$

Plot point B on Fig. 5-3 at

$$I_C = 1.7 \text{ mA} \quad \text{and} \quad V_{CE} = 0 \text{ V}.$$

Draw the dc load line through points A and B.

DC Bias Point (Q-Point)

The *dc bias point* or *quiescent point* (*Q-point*) (also known as the *dc operating point*) identifies the transistor collector current and collector-emitter voltage when there is no input signal at the base terminal. Thus, it defines the dc conditions in the circuit. When a signal is applied to the transistor base, I_B varies according to the instantaneous amplitude of the signal. This causes I_C to vary and consequently produces a variation in V_{CE}.

Consider the circuit in Fig. 5-4 and the 10 kΩ load line drawn for the circuit in Fig. 5-5. Assume that the bias conditions are as identified by the Q-point on the load line,

$$I_B = 20 \ \mu\text{A}, \ I_C = 1 \text{ mA}, \text{ and } V_{CE} = 10 \text{ V}$$

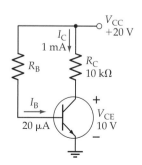

Figure 5-4 Transistor circuit with a bias point (*Q-point*) at V_{CE} = 10 V and I_C = 1 mA.

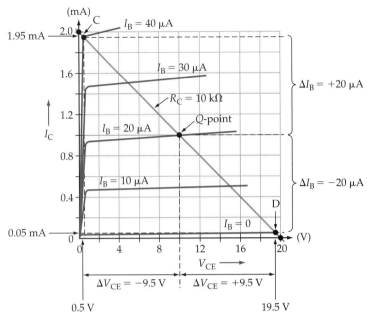

Figure 5-5 DC load line for transistor with a bias point (Q-point) at $V_{CE} = 10$ V and $I_C = 1$ mA. The transistor may be biased to any point on the dc load line.

When I_B is increased from 20 µA to 40 µA, I_C becomes approximately 1.95 mA and V_{CE} becomes 0.5 V, as illustrated at *point C* on the load line. The V_{CE} change from the Q-point is

$$\Delta V_{CE} = 10 \text{ V} - 0.5 \text{ V}$$
$$= 9.5 \text{ V}$$

So, increasing I_B by 20 µA (from 20 µA to 40 µA) caused V_{CE} to decrease by 9.5 V (from 10 V to 0.5 V).

Now look at the effect of decreasing the base current. When I_B is reduced from 20 µA to zero, I_C goes down to approximately 0.05 mA, and V_{CE} goes up to 19.5 V (*point D* on the load line in Fig. 5-5). So, the V_{CE} change is

$$\Delta V_{CE} = 19.5 - 10 \text{ V}$$
$$= 9.5 \text{ V}$$

Decreasing I_B by 20 µA (from 20 µA to zero) caused V_{CE} to increase by 9.5 V (from 10 V to 19.5 V). It is seen that with the Q-point at $I_C = 1$ mA and $V_{CE} = 10$ V, an I_B variation of ±20 µA produces a collector voltage swing of ±9.5 V.

The base current does not have to be varied by the maximum amounts discussed above; it can be increased and decreased by smaller amounts. For example, a base current change of ±10 µA (from the Q-point on Fig. 5-5) would produce a collector current change of ±0.5 mA, and a collector-emitter voltage change of ±5 V.

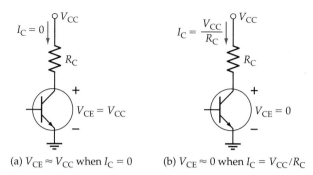

(a) $V_{CE} \approx V_{CC}$ when $I_C = 0$ (b) $V_{CE} \approx 0$ when $I_C = V_{CC}/R_C$

Figure 5-6   The collector-emitter voltage of a transistor can range from approximately zero to V_{CC}.

The maximum possible transistor collector-emitter voltage swing for a circuit can be determined without using the transistor characteristics. For convenience, it may be assumed that I_C can be driven to zero at one extreme and to V_{CC}/R_C at the other extreme (see Fig. 5-6). This changes the collector-emitter voltage from $V_{CE} = V_{CC}$ to $V_{CE} = 0$, as illustrated in Fig. 5-7. Thus, with the Q-point at the centre of the load line, the maximum possible collector voltage swing is seen to be approximately $\pm V_{CC}/2$.

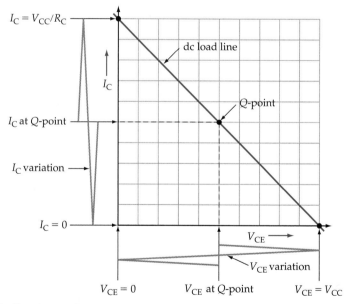

Figure 5-7   Transistor collector-emitter voltage (V_{CE}) ranges from approximately V_{CC} to zero when I_C goes from zero to V_{CC}/R_C.

Selection of *Q*-point

Suppose that, instead of being biased halfway along the load line, the transistor is biased at $I_C = 0.5$ mA and $V_{CE} = 15$ V, as shown in Fig. 5-8a. Increasing the

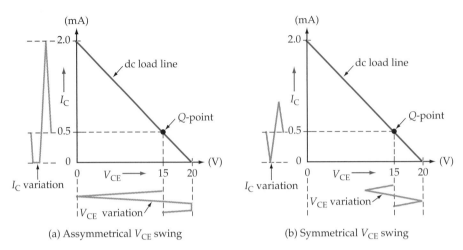

(a) Assymmetrical V_{CE} swing

(b) Symmetrical V_{CE} swing

Figure 5-8   The Q-point does not have to be at the centre of the dc load line. But its position determines the maximum symmetrical collector-emitter voltage swing.

collector current to 2 mA reduces V_{CE} to zero, giving $\Delta V_{CE} = -15$ V. Reducing I_C to zero increases V_{CE} to V_{CC}, producing $\Delta V_{CE} = +5$ V.

When used as an amplifier, the transistor output (collector-emitter) voltage must swing up and down by equal amounts; that is, the output voltage swing must be symmetrical above and below the bias point. So, the asymmetrical V_{CE} swing of -15 V and $+5$ V illustrated in Fig. 5-8a is unsuitable. If I_C is driven up and down by ± 0.5 mA (see Fig. 5-8b), a symmetrical output voltage swing of ± 5 V is obtained. This is the maximum symmetrical output voltage swing that can be achieved with the bias point shown in Fig. 5-8.

In many cases circuits are designed to have the Q-point at the centre of the load line (as in Figs. 5-5 and 5-7) to give the largest possible symmetrical output voltage swing. This is especially true for some large-signal amplifiers (see Chapter 18). Small-signal amplifiers (discussed in Chapter 12) usually require an output voltage swing not greater than ± 1 V. So, *transistors in amplifiers do not all have to be biased at the centre of the dc load line.*

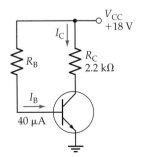

Figure 5-9   Circuit for Example 5-2.

Example 5-2

The transistor circuit in Fig. 5-9 has the collector characteristics shown in Fig. 5-10. Determine the circuit Q-point and estimate the maximum symmetrical output voltage swing. Note that $V_{CC} = 18$ V, $R_C = 2.2$ kΩ, and $I_B = 40$ μA.

Solution

Eq. 5-1:

$$V_{CE} = V_{CC} - I_C R_C$$

When $I_C = 0$:

$$V_{CE} = V_{CC} - 0$$
$$= 18 \text{ V}$$

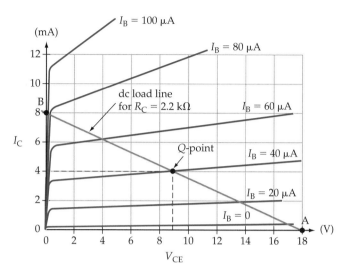

Figure 5-10 Transistor common-emitter output characteristics and dc load line for Example 5-2.

Plot point A on Fig. 5-10 at

$$I_C = 0 \quad \text{and} \quad V_{CE} = 18 \text{ V}$$

When $V_{CE} = 0$: $0 = V_{CC} - I_C R_C$

or $I_C = \dfrac{V_{CC}}{R_C} = \dfrac{18 \text{ V}}{2.2 \text{ k}\Omega}$

$$\approx 8.2 \text{ mA}$$

Plot point B on Fig. 5-10 at

$$I_C = 8.2 \text{ mA} \quad \text{and} \quad V_{CE} = 0 \text{ V}$$

Draw the dc load line through points A and B.

The Q-point is at the intersection of the load line and the $I_B = 40$ μA characteristic.

The dc bias conditions are

$$I_C \approx 4.1 \text{ mA} \quad \text{and} \quad V_{CE} \approx 9 \text{ V}$$

The maximum symmetrical output voltage swing is

$$\Delta V_{CE} \approx \pm 9 \text{ V}$$

Effect of Emitter Resistor

Figure 5-11a shows a circuit that has a resistor (R_E) in series with the transistor emitter terminal and the supply voltage connected directly to the collector terminal. In this case R_E is the dc load and Equation 5-1 is rewritten as

$$V_{CE} = V_{CC} - I_E R_E$$

The dc load line is drawn exactly as discussed, with I_E taken as equal to I_C for convenience.

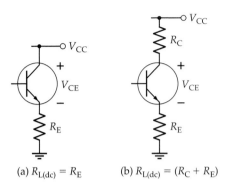

Figure 5-11 The transistor dc load is the sum of the resistors in series with the collector and emitter terminals.

(a) $R_{L(dc)} = R_E$ (b) $R_{L(dc)} = (R_C + R_E)$

In Fig. 5-11b collector and emitter resistors R_C and R_E are both present, and the total dc load in series with the transistor is $(R_C + R_E)$. For drawing the dc load line, Equation 5-1 is modified to

$$V_{CE} = V_{CC} - I_C(R_C + R_E)$$

Note that the voltage drop across the emitter resistor is actually $(I_E R_E)$, but again for convenience I_E is taken as equal to I_C.

Practice Problems

5-1.1 Draw the new dc load line and determine the Q-point for the circuit in Example 5-2 when the supply voltage is changed to 12 V.

5-1.2 A common-emitter circuit as in Fig. 5-1 has $R_C = 12$ kΩ and $V_{CC} = 15$ V. Draw the dc load line on the characteristics in Fig. 5-2. Specify the Q-point, and determine the maximum symmetrical output voltage swing if the base current is 10 μA.

5-1.3 A transister circuit as in Fig. 5-11b has $V_{CC} = 12$ V, $R_C = 820$ Ω, $R_E = 380$ Ω, and $I_B = 60$ μA. Draw the dc load line on the characteristics in Fig. 5-10, and determine the maximum symmetrical output voltage swing.

5-2 BASE BIAS

Circuit Operation and Analysis

The transistor bias arrangement shown in Fig. 5-12 is known as *base bias* and also as *fixed current bias*. The base current is a constant quantity determined by supply voltage V_{CC} and base resistor R_B. Because V_{CC} and R_B are constant quantities, I_B remains fixed at a particular level. Unlike some other bias circuits, the base current in a base bias circuit is not affected by the transistor current gain.

From Fig. 5-12, the voltage drop across R_B is $(V_{CC} - V_{BE})$, and the base current is

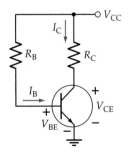

Figure 5-12 Base bias circuit, also known as fixed current bias. The base current remains constant at $I_B = (V_{CC} - V_{BE})/R_B$.

$$I_B = \frac{V_{CC} - V_{BE}}{R_B} \tag{5-2}$$

In Eq. 5-2 the base-emitter voltage (V_{BE}) is taken as 0.7 V for a silicon transistor and as 0.3 V for a germanium device.

The transistor collector current is calculated as

$$I_C = h_{FE} I_B \qquad (5\text{-}3)$$

The collector current is now used with Eq. 5-1 ($V_{CE} = V_{CC} - I_C R_C$) to calculate the collector-emitter voltage. Thus, when the supply voltage and component values are known, a base bias circuit is easily analysed to determine the circuit current and voltage levels.

Example 5-3

The base bias circuit in Fig. 5-13 has R_B = 470 kΩ, R_C = 2.2 kΩ, and V_{CC} = 18 V, and the transistor has h_{FE} = 100. Determine I_B, I_C, and V_{CE}.

Solution

Eq. 5-2: $I_B = \dfrac{V_{CC} - V_{BE}}{R_B} = \dfrac{18\text{ V} - 0.7\text{ V}}{470\text{ k}\Omega}$

$\qquad\qquad\quad = 36.8\ \mu A$

Eq. 5-3: $I_C = h_{FE}\ I_B = 100 \times 36.8\ \mu A$

$\qquad\qquad\quad = 3.68\text{ mA}$

Eq. 5-1: $V_{CE} = V_{CC} - I_C R_C = 18\text{ V} - (3.68\text{ mA} \times 2.2\text{ k}\Omega)$

$\qquad\qquad\quad = 9.9\text{ V}$

The circuit conditions are shown in Fig. 5-13.

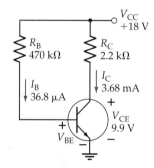

Figure 5-13   Circuit for Example 5-3.

Effect of $h_{FE(max)}$ and $h_{FE(min)}$

When the transistor dc current gain is known, it is quite easy to determine the circuit bias conditions exactly as in Example 5-3. However, in practice the precise current gain of each transistor is normally not known. The transistor is usually identified by its type number, and the maximum and minimum values of current gain can be obtained from the manufacturer's data sheet. In circuit analysis it is sometimes convenient to use a typical h_{FE} value. However, as demonstrated in Example 5-4, $h_{FE(max)}$ and $h_{FE(min)}$ must be used to calculate the range of possible levels of I_C and V_{CE}.

Example 5-4

Calculate the maximum and minimum levels of I_C and V_{CE} for the base bias circuit in Ex. 5-3 when $h_{FE(min)}$ = 50 and $h_{FE(max)}$ = 200.

Solution

The base current in this circuit is unaffected by h_{FE}, and so (as in Ex. 5-3),

$$I_B = \frac{V_{CC} - V_{BE}}{R_B} = \frac{18\text{ V} - 0.7\text{ V}}{470\text{ k}\Omega}$$

$$= 36.8\ \mu\text{A}$$

For $h_{FE(min)}$: $\quad I_C = h_{FE(min)}\ I_B = 50 \times 36.8\ \mu\text{A}$

$$= 1.84\text{ mA}$$

and $\quad V_{CE} = V_{CC} - I_C\ R_C = 18\text{ V} - (1.84\text{ mA} \times 2.2\text{ k}\Omega)$

$$= 13.95\text{ V}\quad\text{(see Fig. 5-14a)}$$

For $h_{FE(max)}$: $\quad I_C = h_{FE(max)}\ I_B = 200 \times 36.8\ \mu\text{A}$

$$= 7.36\text{ mA}$$

and $\quad V_{CE} = V_{CC} - I_C\ R_C = 18\text{ V} - (7.36\text{ mA} \times 2.2\text{ k}\Omega)$

$$= 1.8\text{ V}\quad\text{(see Fig. 5-14b)}$$

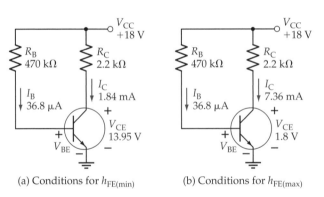

(a) Conditions for $h_{FE(min)}$　　　　(b) Conditions for $h_{FE(max)}$

Figure 5-14　Circuit current and voltage levels for $h_{FE(min)}$ and $h_{FE(max)}$ as determined in Example 5-4.

Note that the typical h_{FE} value of 100 used in Ex. 5-3 gives $I_C = 3.68$ mA and $V_{CE} = 9.9$ V. But applying the $h_{FE(min)}$ and $h_{FE(max)}$ values in Example 4-4 results in an I_C range of 1.84 mA to 7.36 mA and a V_{CE} range from 1.8 V to 13.95 V. The different I_C and V_{CE} levels determined in Examples 5-3 and 5-4 are illustrated by the three Q-points in Fig. 5-15. Transistors of a given type number always have a wide range of h_{FE} values (the h_{FE} spread), so

$h_{FE(max)}$ *and* $h_{FE(min)}$ *should always be used for practical circuit analysis.*

The base bias circuit is rarely employed because of the uncertainty of the Q-point. More predictable results can be obtained with other types of bias circuit.

Base Bias Using a *pnp* Transistor

All of the base bias circuits discussed so far use *npn* transistors. Figure 5-16 shows circuits that use *pnp* transistors. In Fig. 5-16a note that the voltage

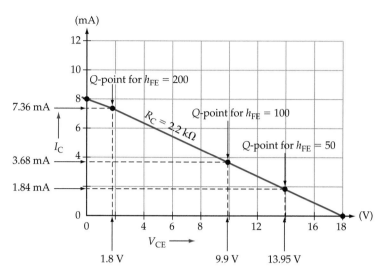

Figure 5-15   The transistor h_{FE} value has a major effect on the Q-point for a base bias circuit.

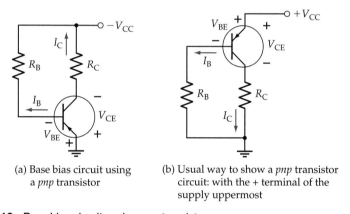

(a) Base bias circuit using a *pnp* transistor

(b) Usual way to show a *pnp* transistor circuit: with the + terminal of the supply uppermost

Figure 5-16   Base bias circuits using *pnp* transistors.

polarities and current directions are reversed compared to *npn*-transistor base bias circuits. Because it is normal to show the supply voltage with the positive terminal uppermost, the circuit of Fig. 5-16a is redrawn in Fig. 5-16b. This is the way that the circuit would usually be shown.

Equations 5-1, 5-2, and 5-3 can be applied for analyzing *pnp* transistor base bias circuits exactly as for *npn* transistor circuits.

Practice Problems

5-2.1 If the bias circuit in Example 5-3 uses a transitor with $h_{FE(min)} = 75$ and $h_{FE(max)} = 250$, determine the maximum and minimum levels of I_C and V_{CE}.

5-2.2 A base bias circuit has $V_{CC} = 24$ V, $R_B = 390$ kΩ, $R_C = 3.3$ kΩ, and $V_{CE} = 10$ V. Calculate the transistor h_{FE} value, and determine the new V_{CE} level when a new transistor is substituted with $h_{FE} = 100$.

5-3 COLLECTOR-TO-BASE BIAS

Circuit Operation and Analysis

The *collector-to-base bias* circuit shown in Fig. 5-17a has the base resistor (R_B) connected between the transistor collector and base terminals. As will be demonstrated, this circuit significantly improves bias stability for h_{FE} changes compared to base bias.

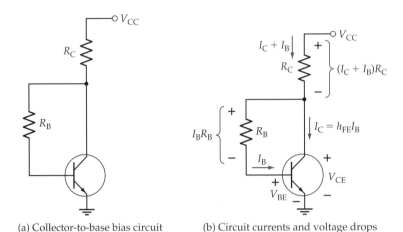

(a) Collector-to-base bias circuit   (b) Circuit currents and voltage drops

Figure 5-17   Collector-to-base bias circuits. Any change in V_{CE} changes I_B. The I_B change causes I_C to change, and this tends to return V_{CE} toward its original level.

Refer to Fig. 5-17b and note that the voltage across R_B is dependent on V_{CE}:

$$V_{CE} = V_{BE} + I_B R_B \qquad (5\text{-}4)$$

Therefore,
$$I_B = \frac{V_{CE} - V_{BE}}{R_B}$$

Also, V_{CE} is dependent on the level of I_C and I_B:

$$V_{CE} = V_{CC} - R_C(I_C + I_B) \qquad (5\text{-}5)$$

If I_C increases above the design level, there is an increased voltage drop across R_C, resulting in a reduction in V_{CE}. The reduced V_{CE} level causes I_B to be lower than its design level, and because $I_C = h_{FE} I_B$, the collector current is also reduced. Thus, an increase in I_C produces a feedback effect that tends to return I_C toward its original level. Similarly, a reduction in I_C produces an increase in V_{CE}, which increases I_B, thus tending to increase I_C back to its original level.

Analysis of this circuit is a little more complicated than base bias analysis. To simplify the process, an equation is first derived for the base current. Substituting for V_{CE} from Eq. 5-4 into Eq. 5-5,

$$V_{BE} + I_B R_B = V_{CC} - R_C(I_C + I_B)$$

Substituting $I_C = h_{FE} I_B$ (from Eq. 5-3) into the above equation,

$$V_{BE} + I_B R_B = V_{CC} - R_C (h_{FE} I_B + I_B)$$

This gives
$$I_B = \frac{V_{CC} - V_{BE}}{R_B + R_C(h_{FE} + 1)} \qquad (5\text{-}6)$$

Example 5-5

A collector-to-base bias circuit as in Fig. 5-18 has $R_B = 270$ kΩ, $R_C = 2.2$ kΩ, $V_{CC} = 18$ V, and a transistor with $h_{FE} = 100$. Analyse the circuit to determine I_B, I_C, and V_{CE}.

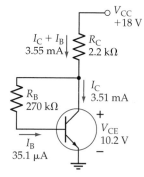

Figure 5-18   Collector-to-base bias circuit for Example 5-5.

Solution

Eq. 5-6:
$$I_B = \frac{V_{CC} - V_{BE}}{R_B + R_C(h_{FE} + 1)} = \frac{18\text{ V} - 0.7\text{ V}}{270\text{ kΩ} + 2.2\text{ kΩ}(100 + 1)}$$

$$= 35.1\ \mu A$$

Eq. 5-3:
$$I_C = h_{FE} I_B = 100 \times 35.1\ \mu A$$

$$= 3.51\text{ mA}$$

Eq. 5-5:
$$V_{CE} = V_{CC} - R_C(I_C + I_B)$$

$$= 18\text{ V} - 2.2\text{ kΩ}(3.51\text{ mA} + 35.1\ \mu A)$$

$$= 10.2\text{ V}$$

The circuit conditions are illustrated in Fig. 5-18.

Effect of $h_{FE(max)}$ and $h_{FE(min)}$

As discussed for the base bias circuit, transistors of a given type number have a wide range of h_{FE} values (h_{FE} *spread*). This affects the current and voltage levels in all bias circuits. In the collector-to-base bias circuit, the feedback from the collector to the base reduces the effects of h_{FE} spread. Thus, as demonstrated in Ex. 5-6, collector-to-base bias has greater stability than base bias for a given range of h_{FE} values.

It is important to note that, unlike the situation in a base bias circuit, *the base current in a collector-to-base bias circuit does* not *remain constant when the transistor h_{FE} value is changed.* This is also demonstrated in Example 5-6.

Example 5-6

Calculate the maximum and minimum levels of I_C and V_{CE} for the bias circuit in Example 5-5 when $h_{FE(min)} = 50$ and $h_{FE(max)} = 200$.

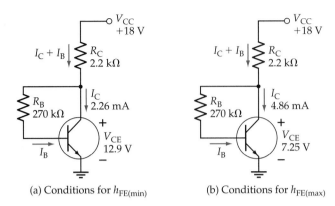

(a) Conditions for $h_{FE(min)}$ (b) Conditions for $h_{FE(max)}$

Figure 5-19 Circuit conditions for $h_{FE(min)}$ and $h_{FE(max)}$ in the collector-to-base bias circuit analyzed in Example 5-6.

Solution

For $h_{FE(min)}$:

Eq. 5-6: $I_B = \dfrac{V_{CC} - V_{BE}}{R_B + R_C\,(h_{FE} + 1)} = \dfrac{18\ V - 0.7\ V}{270\ k\Omega + 2.2\ k\Omega(50 + 1)}$

$= 42.3\ \mu A$

Eq. 5-3: $I_C = h_{FE}\,I_B = 50 \times 42.3\ \mu A$

$= 2.26\ mA$

Eq. 5-5: $V_{CE} = V_{CC} - R_C(I_C + I_B)$

$= 18\ V - 2.2\ k\Omega\ (2.26\ mA + 42.3\ \mu A)$

$= 12.9\ V$ (see Fig. 5-19a)

For $h_{FE(max)}$:

Eq. 5-6: $I_B = \dfrac{V_{CC} - V_{BE}}{R_B + R_C\,(h_{FE} + 1)} = \dfrac{18\ V - 0.7\ V}{270\ k\Omega + 2.2\ k\Omega(200 + 1)}$

$= 24.3\ \mu A$

Eq. 5-3: $I_C = h_{FE}\,I_B = 200 \times 24.3\ \mu A$

$= 4.86\ mA$

Eq. 5-5: $V_{CE} = V_{CC} - R_C(I_C + I_B)$

$= 18\ V - 2.2\ k\Omega(4.86\ mA + 24.3\ \mu A)$

$= 7.25\ V$ (see Fig. 5-19b)

The circuit Q-points are shown in Fig. 5-20.

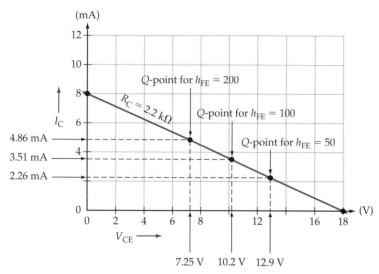

Figure 5-20   Effect of $h_{FE(min)}$ and $h_{FE(max)}$ on the Q-point for the collector-to-base bias circuit analyzed in Example 5-6.

As illustrated by the Q-points in Fig. 5-20, the typical h_{FE} value used in Ex. 5-5 gives $I_C = 3.51$ mA and $V_{CE} = 10.2$ V. Applying the $h_{FE(min)}$ and $h_{FE(max)}$ values produces an I_C range of 2.26 mA to 4.86 mA and V_{CE} levels ranging from 7.25 V to 12.9 V.

Recall that the base bias circuit analyzed in Examples 5-3 and 5-4 has the same V_{CC}, R_C, and, h_{FE} values as used in the collector-to-base bias circuit in Examples 5-5 and 5-6. Comparing the collector-to-base bias circuit Q-points in Fig. 5-20 to the base bias circuit Q-points in Fig. 5-15, it is seen that the Q-points for the collector-to-base bias circuit are much closer than those for the base bias circuit. The collector-to-base bias circuit clearly has greater stability against h_{FE} spread than the base bias circuit.

Collector-to-Base Bias Using a *pnp* Transistor

A collector-to-base bias circuit using a *pnp* transistor is illustrated in Fig. 5-21. Note that the voltage polarities and current directions are reversed compared to the *npn* transistor collector-to-base bias circuit. This circuit can be analyzed in exactly the same way as the *npn* transistor circuit.

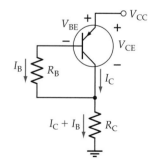

Figure 5-21   Collector-to-base bias circuit using a *pnp* transistor.

Practice Problems

5-3.1 The collector-to-base bias circuit in Example 5-5 uses a transistor with $h_{FE(min)} = 75$ and $h_{FE(max)} = 250$. Determine the new maximum and minimum levels for I_C and V_{CE}.

5-3.2 A collector-to-base bias circuit has $V_{CC} = 24$ V, $R_B = 180$ kΩ, and $R_C = 3.3$ kΩ, and V_{CE} measures as 10 V. Calculate the transistor h_{FE} value, and determine the V_{CE} level when a new transistor is substituted with $h_{FE} = 120$.

5-4 VOLTAGE-DIVIDER BIAS

Circuit Operation

Voltage divider bias is the most stable of the three basic transistor bias circuits. A voltage-divider bias circuit is shown in Fig. 5-22a, and the current and voltage conditions throughout the circuit are illustrated in Fig. 5-22b. It is seen that, as well as the collector resistor (R_C), there is an emitter resistor (R_E) connected in series with the transistor. As discussed in Section 5-1, the total dc load in series with the transistor is ($R_C + R_E$), and this total resistance must be used when drawing the dc load line for the circuit. Resistors R_1 and R_2 constitute a voltage divider that divides the supply voltage to produce the base bias voltage (V_B).

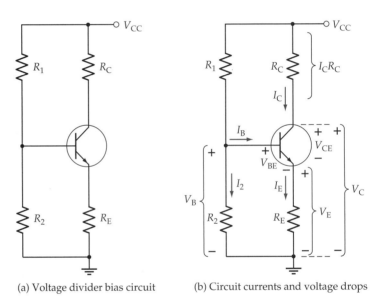

(a) Voltage divider bias circuit   (b) Circuit currents and voltage drops

Figure 5-22 Voltage-divider bias circuits. The emitter current remains constant at $I_E = (V_B - V_{BE})/R_E$.

Voltage-divider bias circuits are normally designed to have the voltage divider current (I_2) very much larger than the transistor base current (I_B). In this situation, V_B is largely unaffected by I_B, and so V_B can be assumed to remain constant.

Referring to Fig. 5-22b,

$$V_B = \frac{V_{CC} \times R_2}{R_1 + R_2}$$

(5-7)

With V_B constant, the voltage across the emitter resistor is also a constant quantity:

$$V_E = V_B - V_{BE}$$

(5-8)

This means that the emitter current is constant:

$$I_E = \frac{V_B - V_{BE}}{R_E} \qquad (5\text{-}9)$$

The collector current is approximately equal to the emitter current, so I_C is held at a constant level.

Again if we refer to Fig. 5-22b, the transistor collector voltage is

$$V_C = V_{CC} - (I_C R_C) \qquad (5\text{-}10)$$

The collector-emitter voltage is

$$V_{CE} = V_C - V_E \qquad (5\text{-}11)$$

V_{CE} can also be determined as

$$V_{CE} \approx V_{CC} - I_C(R_C + R_E) \qquad (5\text{-}12)$$

Clearly, with I_C and I_E constant, the transistor collector-emitter voltage remains constant.

It should be noted that the transistor h_{FE} value is not involved in any of the above equations.

Approximate Circuit Analysis

If the transistor base current is assumed to be much smaller than the voltage-divider current, as discussed above, the circuit currents and voltages can be readily determined by the use of Equations 5-7 to 5-12. Example 5-7 demonstrates the process.

Example 5-7

Analyze the voltage-divider bias in the circuit in Fig. 5-23 to determine the emitter voltage, collector voltage, and collector-emitter voltage.

Solution

Eq. 5-7:

$$V_B = \frac{V_{CC} \times R_2}{R_1 + R_2} = \frac{18 \text{ V} \times 12 \text{ k}\Omega}{33 \text{ k}\Omega + 12 \text{ k}\Omega}$$

$$= 4.8 \text{ V}$$

Eq. 5-8:

$$V_E = V_B - V_{BE} = 4.8 \text{ V} - 0.7 \text{ V}$$

$$= 4.1 \text{ V}$$

Eq. 5-9:

$$I_E = \frac{V_B - V_{BE}}{R_E} = \frac{4.8 \text{ V} - 0.7 \text{ V}}{1 \text{ k}\Omega}$$

$$= 4.1 \text{ mA}$$

Therefore

$$I_C \approx I_E = 4.1 \text{ mA}$$

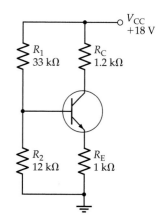

Figure 5-23 Voltage-divider bias circuit for Example 5-7.

Eq. 5-10: $V_C = V_{CC} - (I_C R_C) = 18 \text{ V} - (4.1 \text{ mA} \times 1.2 \text{ k}\Omega)$

$= 13.1 \text{ V}$

Eq. 5-11: $V_{CE} = V_C - V_E = 13.1 \text{ V} - 4.1 \text{ V}$

$= 9 \text{ V}$

The circuit conditions are illustrated in Fig. 5-24.

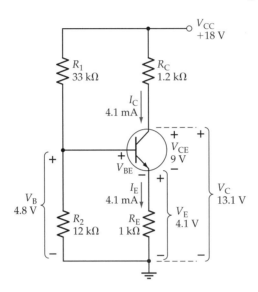

Figure 5-24 Voltage and current conditions for the circuit in Example 5-7.

Precise Circuit Analysis

To analyze a voltage-divider bias circuit precisely, the voltage divider must be replaced with its Thevenin equivalent circuit (V_T in series with R_T), as illustrated in Fig. 5-25. From Figure 5-25a,

$$V_T = \frac{V_{CC} \times R_2}{R_1 + R_2} \tag{5-13}$$

R_T is calculated as R_1 in parallel with R_2:

$$R_T = R_1 \| R_2 \tag{5-14}$$

Referring to Fig. 5-25b, we can write the following equation for the voltage drops around the base-emitter circuit:

$$V_T = I_B R_T + V_{BE} + R_E(I_B + I_C)$$

Substituting $I_C = h_{FE} I_B$,

$$V_T = I_B R_T + V_{BE} + R_E I_B(1 + h_{FE})$$

giving

$$I_B = \frac{V_T - V_{BE}}{R_T + R_E(1 + h_{FE})} \tag{5-15}$$

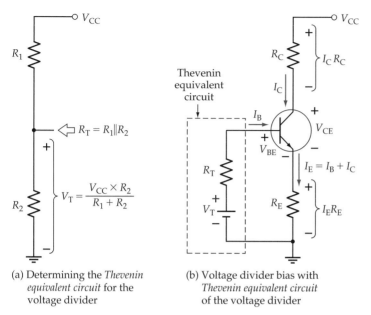

(a) Determining the *Thevenin equivalent circuit* for the voltage divider

(b) Voltage divider bias with *Thevenin equivalent circuit* of the voltage divider

Figure 5-25   For precise analysis of a voltage-divider bias circuit, the voltage divider is replaced with its Thevenin equivalent circuit.

Once I_B has been determined, I_C can be calculated from the appropriate h_{FE} value, and the transistor terminal voltages can then be calculated. The effects of the maximum and minimum h_{FE} values can also be investigated.

Example 5-8

Accurately analyze the voltage-divider bias circuit in Example 5-7 to determine I_C, V_E, V_C, and V_{CE} when the transistor $h_{FE} = 100$.

Solution

Eq. 5-13: $V_T = \dfrac{V_{CC} \times R_2}{R_1 + R_2} = \dfrac{18\text{ V} \times 12\text{ k}\Omega}{33\text{ k}\Omega + 12\text{ k}\Omega}$

$= 4.8\text{ V (see Fig. 5-26)}$

Eq. 5-14: $R_T = R_1 \| R_2 = 33\text{ k}\Omega \| 12\text{ k}\Omega$

$= 8.8\text{ k}\Omega$

Eq. 5-15: $I_B = \dfrac{V_T - V_{BE}}{R_T + R_E(1 + h_{FE})} = \dfrac{4.8\text{ V} - 0.7\text{ V}}{8.8\text{ k}\Omega + 1\text{ k}\Omega(1 + 100)}$

$= 37.3\ \mu\text{A}$

$I_C = h_{FE}\, I_B = 100 \times 37.3\ \mu\text{A}$

$= 3.73\text{ mA}$

$$I_E = I_B + I_C = 37.3\ \mu A + 3.73\ mA$$

$$= 3.77\ mA$$

$$V_E = I_E\,R_E = 3.77\ mA \times 1\ k\Omega$$

$$= 3.77\ V$$

Eq. 5-10:
$$V_C = V_{CC} - (I_C\,R_C) = 18\ V - (3.73\ mA \times 1.2\ k\Omega)$$

$$= 13.52\ V$$

Eq. 5-11:
$$V_{CE} = V_C - V_E = 13.52\ V - 3.77\ V$$

$$= 9.75\ V \quad \text{(see Fig. 5-26)}$$

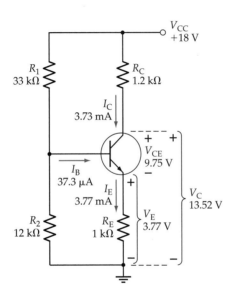

Figure 5-26   Voltage and current levels for the voltage-divider bias circuit in Example 5-8.

Example 5-9

Accurately analyze the voltage-divider bias circuit in Examples 5-7 and 5-8 for the conditions of $h_{FE(min)} = 50$.

Solution

From Ex. 5-8, $V_T = 4.8\ V$ and $R_T = 8.8\ k\Omega$

Eq. 5-15:
$$I_B = \frac{V_T - V_{BE}}{R_T + R_E(1 + h_{FE})} = \frac{4.8\ V - 0.7\ V}{8.8\ k\Omega + 1\ k\Omega(1 + 50)}$$

$$= 68.6\ \mu A$$

$$I_C = h_{FE}\,I_B = 50 \times 68.6\ \mu A$$

$$= 3.43\ mA$$

$$I_E = I_B + I_C = 68.6\ \mu A + 3.43\ mA$$

$$= 3.5\ mA$$

$$V_E = I_E R_E = 3.5\ mA \times 1\ k\Omega$$

$$= 3.5\ V$$

Eq. 5-10: $\quad V_C = V_{CC} - (I_C R_C) = 18\ V - (3.43\ mA \times 1.2\ k\Omega)$

$$= 13.9\ V$$

Eq. 5-11: $\quad V_{CE} = V_C - V_E = 13.9\ V - 3.5\ V$

$$= 10.4\ V\ (see\ Fig.\ 5\text{-}27a)$$

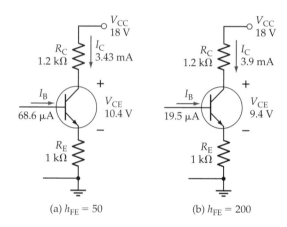

(a) $h_{FE} = 50$ (b) $h_{FE} = 200$

Figure 5-27 Voltage and current conditions produced by $h_{FE(max)}$ and $h_{FE(min)}$ in the voltage-divider bias circuit in Examples 5-9 and 5-10.

Example 5-10

Accurately analyze the voltage-divider bias circuit in Examples 5-7, 5-8, and 5-9 for the condition of $h_{FE(max)} = 200$.

Solution

From Ex. 5-8, $\quad V_T = 4.8\ V$ and $R_T = 8.8\ k\Omega$

Eq. 5-15: $\quad I_B = \dfrac{V_T - V_{BE}}{R_T + R_E(1 + h_{FE})} = \dfrac{4.8\ V - 0.7\ V}{8.8\ k\Omega + 1\ k\Omega(1 + 200)}$

$$= 19.5\ \mu A$$

$$I_C = h_{FE} I_B = 200 \times 19.5\ \mu A$$

$$= 3.9\ mA$$

$$I_E = I_B + I_C = 19.5\ \mu A + 3.9\ mA$$

$$= 3.92\ mA$$

$$V_E = I_E R_E = 3.92\ mA \times 1\ k\Omega$$

$$= 3.92\ V$$

Eq. 5-10:
$$V_C = V_{CC} - (I_C R_C)$$
$$= 18\text{ V} - (3.9\text{ mA} \times 1.2\text{ k}\Omega)$$
$$= 13.3\text{ V}$$

Eq. 5-11:
$$V_{CE} = V_C - V_E = 13.3\text{ V} - 3.92\text{ V}$$
$$= 9.4\text{ V (see Fig. 5-27b)}$$

The maximum and minimum h_{FE} values used in Examples 5-9 and 5-10 give $I_C = (3.43\text{ mA to }3.91\text{ mA})$ and $V_{CE} = (9.4\text{ V to }10.4\text{ V})$, while the typical h_{FE} value of 100 used in Example 5-8 gives $I_C = 3.73\text{ mA}$ and $V_{CE} = 9.75\text{ V}$. The I_C and V_{CE} levels determined in Examples 5-8 and 5-9 are shown in Fig. 5-27 and are illustrated by the Q-points in Fig. 5-28. Note that the analysis using the highest h_{FE} value gives results that are closest to those from the approximate analysis in Example 5-7. This is because the approximate analysis assumes that I_B is very much smaller than the voltage-divider current, and the highest h_{FE} gives the lowest I_B level.

For most practical purposes, precise analysis of voltage-divider bias circuits is unnecessary. The approximate analysis method gives quite satisfactory results.

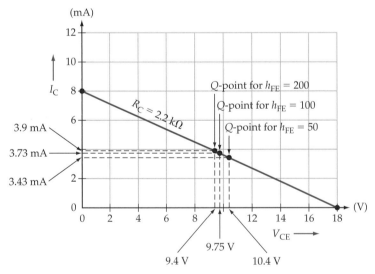

Figure 5-28 Effect of three different h_{FE} values on the Q-point for the voltage-divider bias circuit analyzed in Examples 5-8, 5-9, and 5-10.

Voltage-Divider Bias Using a *pnp* Transistor

A voltage-divider bias circuit using a *pnp* transistor is shown in Fig. 5-29. Note that the positions of the collector and emitter resistors are reversed compared to the *npn* transistor circuit. Note also that the base voltage (V_B) in Fig. 5-29 is the voltage drop across resistor R_1, not that across R_2. As in the case of other bias circuits using *pnp* transistors, the current directions and voltage polarities are the reverse of those in an *npn* circuit. Apart from these differences, a *pnp*

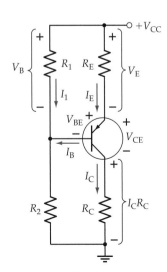

Figure 5-29 Voltage-divider bias using a *pnp* transistor.

transistor voltage-divider bias circuit is analyzed in exactly the same way as an *npn* transistor circuit.

Practice Problems

5-4.1 In a voltage-divider bias circuit (as in Fig. 5-3), $V_{CC} = 24$ V, $R_1 = 180$ kΩ, $R_2 = 56$ kΩ, $R_E = 4.7$ kΩ, and $R_C = 8.2$ kΩ. Calculate the approximate levels of $I_C, V_E, V_C,$ and V_{CE}.

5-4.2 The circuit in Practice Problem 5-4.1 uses a transistor with $h_{FE(min)} = 75$ and $h_{FE(max)} = 250$. Accurately analyze the circuit to determine the maximum and minimum levels of V_{CE}.

5-4.3 In the voltage-divider bias circuit in Fig. 5-29, $V_{CC} = 20$ V, $R_1 = 33$ kΩ, $R_2 = 100$ kΩ, $R_E = 3.9$ kΩ, and $R_C = 6.8$ kΩ. Calculate the approximate levels of V_E and V_C.

5-5 COMPARISON OF BASIC BIAS CIRCUITS

When comparing the performance of the three basic bias circuits, it must be recalled that transistor manufacturers specify maximum and minimum h_{FE} values for each transistor type number at various levels of collector current. Normally, the current gain of each individual transistor is not known, so that (as already stated) *the specified maximum and minimum values of h_{FE} must be used when analyzing (or designing) a transistor bias circuit.* It is completely impractical to use some kind of average h_{FE} value.

Consider the three basic bias circuits reproduced in Fig. 5-30. Each circuit uses an 18 V supply and has a 2.2 kΩ (total) load resistance. The circuits were analyzed in Examples 5-4, 5-6, 5-9, and 5-10 to determine the I_C and V_{CE} levels for

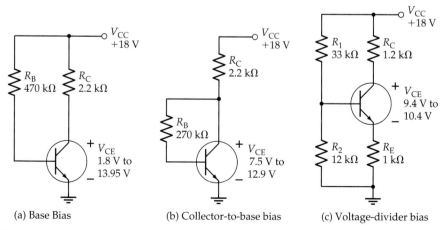

(a) Base Bias   (b) Collector-to-base bias   (c) Voltage-divider bias

Figure 5-30   Comparison of similar base bias, collector-to-base bias, and voltage-divider bias circuits.

TABLE 5-1	V_{CE} maximum and minimum levels for the bias circuits in Fig. 5-30.		
	Base bias	**Collector-to-base bias**	**Voltage-divider bias**
$V_{CE(min)}$	1.8 V	7.25 V	9.4 V
$V_{CE(max)}$	13.75 V	12.9 V	10.4 V

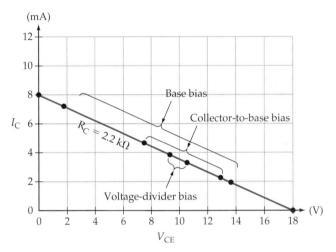

Figure 5-31 Q-point ranges for the base bias, collector-to-base bias, and voltage-divider bias circuits illustrated in Fig. 5-30.

$h_{FE(min)}$ of 50 and $h_{FE(max)}$ of 200. The V_{CE} levels calculated are indicated on each circuit and listed in Table 5-1. The Q-point ranges are also illustrated in Fig. 5-31.

To put these results in the most practical terms, suppose that the base bias circuit above is employed as part of a complex electronics system, and that its V_{CE} level is measured as 10 V. If the transistor fails and has to be replaced with another (same type number) device with an h_{FE} range of 50 to 200, the new V_{CE} level could be anywhere from 1.8 V to 13.95 V. This would normally be unacceptable. A collector-to-base bias circuit in the same circumstance would have a V_{CE} ranging from 7.25 V to 12.9 V when the transistor is replaced. For the voltage-divider bias circuit in this situation, the measurable V_{CE} range with a new transistor would be 9.4 V to 10.4 V.

Clearly, collector-to-base bias gives more predictable bias conditions than base bias. Voltage-divider bias gives the most predictable bias conditions, because it has the greatest stability against h_{FE} spread.

Because of its excellent stability, voltage-divider bias is almost always preferable. However, collector-to-base bias, or some variation of it, is used in many circuits. Its one major advantage is that the feedback from the collector prevents the transistor from going into *saturation* (see Section 5-10). Base bias is most often used in switching circuits. This is also explained in Section 5-10.

5-6 TROUBLESHOOTING BJT BIAS CIRCUITS

Voltage Measurement

When a bias circuit is constructed in a laboratory situation, the supply voltage (V_{CC}) and the voltage levels at the transistor terminals (V_C, V_B, and V_E) should be measured with respect to the ground (or negative supply terminal), as illustrated in Fig. 5-32. When the measured voltages are not as expected, the circuit must be further investigated to locate the fault.

Common Errors

The following is a list of errors that are commonly made by both experienced and inexperienced individuals:

- Power supply not switched *on*
- Power supply current limiter control incorrectly set
- Cables incorrectly connected to the power supply
- Volt-ohm-milliammeter (VOM) function incorrectly selected
- Wrong VOM terminals used
- Incorrect component connections
- Incorrect resistor values used
- Resistors in the wrong places

Obviously, it is important to connect the cables correctly to the power supply and to switch the power supply *on*. However, with many things to think about, people often neglect basic items. Similarly, it is just too easy to connect the VOM cables to the wrong terminals or to select the wrong function. With plug-in-type breadboards, the sockets are so small and close together that incorrect connections often occur. These items should all be checked before looking for other circuit faults. Care should also be taken when selecting resistors. For example, using a 100 Ω resistor instead of a 100 kΩ resistor can have serious consequences. Polarized capacitors connected with the wrong polarity have a leakage current that can affect the bias conditions. They can also explode (see Section 3-3).

Open-Circuited and Short-Circuited Components

Open-circuited and short-circuited resistors and capacitors are not likely to be sources of error in laboratory-constructed circuits. It is true that when a resistor overheats or a capacitor dielectric breaks down, either component may offer a short-circuit or an open-circuit between its terminals. However, such component breakdowns are unusual in small signal amplifiers and other low-power circuits. Incorrect connections that simulate component short-circuits and open-circuits are a more common cause of difficulties, and these should be considered first where faulty components are suspected.

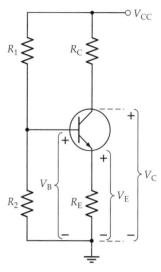

Figure 5-32 When a bias circuit is tested, all voltage levels should first be measured with respect to the ground (or negative supply terminal).

Base Bias

Figure 5-33 illustrates typical sources of error in a base bias circuit. Normally in a base bias circuit $V_{BE} \approx 0.7$ V and V_C is anywhere in the range of 2 V to $(V_{CC} - 2$ V). If $V_C = V_{CC}$, as in Fig. 5-33a, R_1 or one of its connecting leads might be open-circuited or the transistor may have an open-circuited junction. Alternatively, R_2 might be shorted. When $V_C \approx 0$, as in Fig. 5-33b, R_1 and R_2 could be in the wrong places; that is, the high-value and low-value resistors might be interchanged. Other possibilities are that R_2 is open-circuited or the transistor CE terminals are short-circuited.

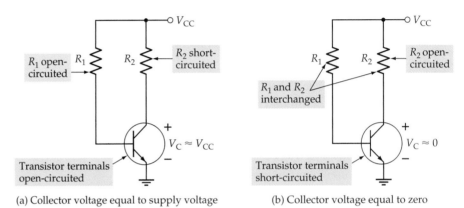

(a) Collector voltage equal to supply voltage   (b) Collector voltage equal to zero

Figure 5-33   The measured transistor terminal voltages on a base bias circuit can give an indication of possible circuit faults.

Collector-to-Base Bias

Some reasons for incorrect voltage levels in a collector-to-base bias circuit are illustrated in Fig. 5-34. These are similar to those discussed for the base bias circuit. When V_C equals V_{CC}, either resistor R_1 or one of the transistor terminals

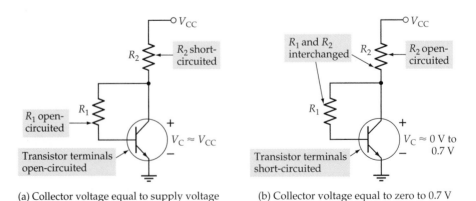

(a) Collector voltage equal to supply voltage   (b) Collector voltage equal to zero to 0.7 V

Figure 5-34   Typical incorrect terminal voltages in a collector-to-base bias circuit and probable circuit faults.

is likely to be open-circuited (see Fig. 5-34a). If V_C is in the range of 0 V to 0.7 V, R_1 and R_2 might be interchanged, the device terminals might be short-circuited, or R_2 might be open-circuited (see Fig. 5-34b).

Voltage-Divider Bias

Figure 5-35 shows unsuitable measured voltages and probable errors in a voltage-divider bias circuit. If V_C equals V_{CC}, either R_1 or R_4 might be open-circuited. Alternatively, R_2 or R_3 might be short-circuited (see Fig. 5-35a), or the transistor terminals might be open-circuited. When V_C approximately equals V_E, R_2 might be open-circuited, or R_1 and R_2 might be interchanged (see Fig. 5-35b). Alternatively, the transistor terminals might be short-circuited.

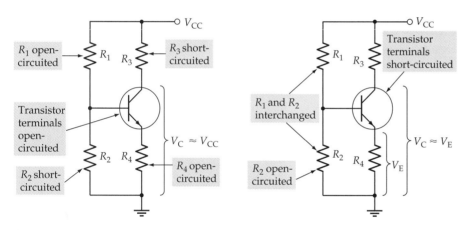

(a) Collector voltage equal to supply voltage (b) Collector voltage equal to emitter voltage

Figure 5-35 Unsatisfactory measured voltages on a voltage-divider bias circuit, and probable circuit faults.

5-7 BIAS-CIRCUIT DESIGN

Designing bias circuits can be amazingly simple. Usually, it is just a matter of determining the required voltage across each resistor and the appropriate current levels. Then the resistor values are calculated by application of Ohm's law.

Designs often begin with specification of the supply voltage and the required levels of I_C and V_{CE}. The resistor values are calculated to meet these requirements, and standard value resistors are selected. (See Appendix B.) Usually, resistors with a tolerance of $\pm 10\%$ are selected wherever possible. These are less expensive than $\pm 5\%$ and $\pm 1\%$ components.

Base Bias Circuit Design

A base bias circuit is very easily designed by application of Equations 5-1 and 5-2. This is illustrated in Fig. 5-36 and demonstrated in Example 5-11.

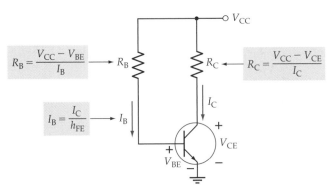

Figure 5-36   Base bias circuit design.

Example 5-11

Design a base bias circuit (as in Fig. 5-36) to have $V_{CE} = 5$ V and $I_C = 5$ mA. The supply voltage is 15 V, and the h_{FE} of the transistor $= 100$.

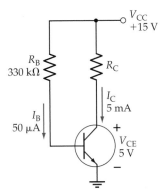

Figure 5-37   Base bias circuit designed in Example 5-11.

Solution

From Eq. 5-1,

$$R_C = \frac{V_{CC} - V_{CE}}{I_C} = \frac{15\text{ V} - 5\text{ V}}{5\text{ mA}}$$

$$= 2\text{ k}\Omega \text{ (use 1.8 k}\Omega \text{ or 2.2 k}\Omega \text{ standard value)}$$

$$I_B = \frac{I_C}{h_{FE}} = \frac{5\text{ mA}}{100}$$

$$= 50\ \mu\text{A (see Fig. 5-37)}$$

From Eq. 5-2,

$$R_B = \frac{V_{CC} - V_{BE}}{I_B} = \frac{15\text{ V} - 0.7\text{ V}}{50\ \mu\text{A}}$$

$$= 286\text{ k}\Omega \text{ (use 270 k}\Omega \text{ or 330 k}\Omega \text{ standard value)}$$

When the standard value resistors are selected, a decision must be made whether to use the next-smaller resistance value or the next-larger value. In general, it is best to select the resistance value that tends to increase the transistor collector-emitter voltage, thus keeping V_{CE} from approaching zero. In Example 5-11, selecting $R_C = 1.8$ kΩ (instead of 2.2 kΩ) gives the smallest voltage drop across R_C and results in a larger V_{CE} than the (5 V) design value. Selecting $R_B = 330$ kΩ (instead of 270 kΩ) gives the lowest I_B and consequently keeps I_C at a minimum, to produce the highest V_{CE}.

After the circuit is designed, it should be analyzed using the selected standard-value components and the maximum and minimum h_{FE} values for the transistor (see Section 5-2).

Collector-to-Base Bias Circuit Design

The design procedure for a collector-to-base bias circuit is similar to that for base bias, with the exception that the voltage and current levels are different for calculating R_B and R_C. In collector-to-base bias, the voltage across R_B is $(V_C - V_{BE})$ and the current through R_C is $(I_B + I_C)$. The design equations are shown in Fig. 5-38, and the design procedure is demonstrated in Example 5-12.

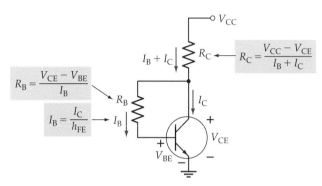

Figure 5-38 Collector-to-base bias circuit design.

Example 5-12

Design a collector-to-base bias circuit (as in Fig. 5-38) to have $V_{CE} = 5$ V and $I_C = 5$ mA when the supply voltage is 15 V and the transistor h_{FE} is 100.

Solution

$$I_B = \frac{I_C}{h_{FE}} = \frac{5\ \text{mA}}{100}$$

$$= 50\ \mu\text{A (see Fig. 5-39)}$$

From Eq. 5-5,

$$R_C = \frac{V_{CC} - V_{CE}}{I_C + I_B} = \frac{15\ \text{V} - 5\ \text{V}}{5\ \text{mA} + 50\ \mu\text{A}}$$

$$= 1.98\ \text{k}\Omega\ (\text{use } 1.8\ \text{k}\Omega \text{ or } 2.2\ \text{k}\Omega \text{ standard value})$$

From Eq. 5-4,

$$R_B = \frac{V_{CE} - V_{BE}}{I_B} = \frac{5\ \text{V} - 0.7\ \text{V}}{50\ \mu\text{A}}$$

$$= 86\ \text{k}\Omega\ (\text{use } 82\ \text{k}\Omega \text{ or } 100\ \text{k}\Omega \text{ standard value})$$

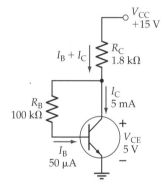

Figure 5-39 Collector-to-base bias circuit designed in Example 5-12.

Once again decisions must be made (in Example 5-12) about selecting the next-larger or the next-smaller standard value resistors. As in the case of the base bias circuit in Example 5-11, choosing the smaller value for R_C and the larger value for R_B tends to produce a larger V_{CE} than the specified level. The design should be analyzed using the selected standard-value components and the transistor $h_{FE(max)}$ and $h_{FE(min)}$ values.

Voltage-Divider Bias Circuit Design

When one is designing a voltage-divider bias circuit, the voltage-divider current (I_2 in Fig. 5-40) should be selected to be much larger than the transistor base current (I_B). This makes the base voltage (V_B) a stable quantity largely unaffected by the h_{FE} value of the transistor. However, a high level of I_2 results in small resistance values for R_1 and R_2, and (as explained in Chapter 12) this gives the circuit an undesirably low input impedance.

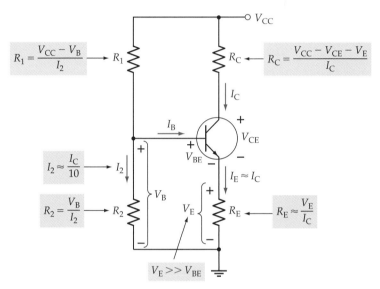

Figure 5-40 Voltage-divider bias circuit design.

A rule-of-thumb for the selection of I_2 is to use a voltage-divider current approximately equal to one-tenth of the transistor collector current.

$$I_2 = \frac{I_C}{10} \qquad (5\text{-}16)$$

This gives reasonably large values for R_1 and R_2 while still keeping I_2 much larger than I_B.

As already stated, if V_E is not specified, it should be selected to be much larger than the transistor V_{BE}:

$$V_E \gg V_{BE}$$

This is because V_{BE} can vary from transistor to transistor, and it can also change with changes in temperature. Making V_E very much larger than V_{BE} minimizes the effect of V_{BE} changes on the circuit bias conditions. As another rule-of-thumb, V_E is selected as 5 V regardless of the supply voltage. When V_{CC} is low (for example, 9 V), V_E can be as low as 3 V without changes in V_{BE} having any significant effect on the bias conditions.

Figure 5-40 shows the equations used for calculating each resistor value, and Example 5-13 demonstrates the design procedure.

Example 5-13

Design the voltage divider bias circuit (as in Fig. 5-40) to have $V_{CE} = V_E = 5$ V and $I_C = 5$ mA when the supply voltage is 15 V. Assume the transistor h_{FE} is 100.

Solution

$$R_E = \frac{V_E}{I_E} \approx \frac{V_E}{I_C} = \frac{5\text{ V}}{5\text{ mA}}$$

$$= 1\text{ k}\Omega \text{ (standard value)}$$

$$R_C = \frac{V_{CC} - V_{CE} - V_E}{I_C} = \frac{15\text{ V} - 5\text{ V} - 5\text{ V}}{5\text{ mA}}$$

$$= 1\text{ k}\Omega \text{ (standard value) (see Fig. 5-41)}$$

Eq. 5-16:
$$I_2 = \frac{I_C}{10} = \frac{5\text{ mA}}{10}$$

$$= 500\ \mu\text{A}$$

From Eq. 5-8,
$$V_B = V_E + V_{BE} = 5\text{ V} + 0.7\text{ V}$$

$$= 5.7\text{ V}$$

$$R_2 = \frac{V_B}{I_2} = \frac{5.7\text{ V}}{500\ \mu\text{A}}$$

$$= 11.4\text{ k}\Omega \text{ (use 12 k}\Omega \text{ standard value)}$$

$$R_1 = \frac{V_{CC} - V_B}{I_2} = \frac{15\text{ V} - 5.7\text{ V}}{500\ \mu\text{A}}$$

$$= 18.6\text{ k}\Omega \text{ (use 18 k}\Omega \text{ standard value)}$$

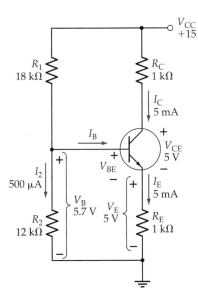

Figure 5-41 Voltage divider bias circuit designed in Example 5-13.

The resistor values calculated in Example 5-13 are conveniently close to standard values. When this is not the case, some thought must be given to deciding whether to select the higher or lower standard value. As explained for the other bias circuits, it is best to select the resistance value that tends to increase the transistor collector-emitter voltage, in order to keep V_{CE} from approaching zero. As with all bias circuit designs, the designed circuit should be analyzed using the standard value components and the transistor $h_{FE(max)}$ and $h_{FE(min)}$ values.

Designing with Standard Resistor Values

In all circuit designs a suitable standard resistor should normally be selected when each resistor value is calculated rather than after the design is completed. Then, the new resistor voltage drop or current level should be determined before the next component value is calculated. This is demonstrated in Example 5-14.

Example 5-14

Design the voltage-divider bias circuit to operate from a 12 V supply. The bias conditions are to be $V_{CE} = 3$ V, $V_E = 5$ V, and $I_C = 1$ mA.

Solution

$$R_4 = \frac{V_E}{I_E} \approx \frac{V_E}{I_C} = \frac{5 \text{ V}}{1 \text{ mA}}$$

$$= 5 \text{ k}\Omega \text{ (use a 4.7 k}\Omega \text{ standard value)}$$

With $I_C = 1$ mA and $R_4 = 4.7$ kΩ, V_E becomes

$$V_E = I_C R_4 = 1 \text{ mA} \times 4.7 \text{ k}\Omega$$

$$= 4.7 \text{ V}$$

and

$$V_C = V_E + V_{CE} = 4.7 \text{ V} + 3 \text{ V}$$

$$= 7.7 \text{ V}$$

$$V_{R3} = V_{CC} - V_C = 12 \text{ V} - 7.7 \text{ V}$$

$$= 4.3 \text{ V}$$

$$R_3 = \frac{V_{R3}}{I_C} = \frac{4.3 \text{ V}}{1 \text{ mA}}$$

$$= 4.3 \text{ k}\Omega \text{ (use a 3.9 k}\Omega \text{ standard value to reduce } V_{R3} \text{ and increase } V_{CE}.)$$

$$V_B = V_E + V_{BE} = 4.7 \text{ V} + 0.7 \text{ V}$$

$$= 5.4 \text{ V (see Fig. 5-42)}$$

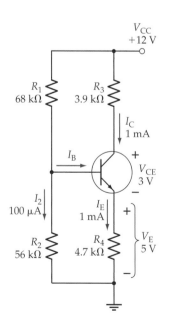

Figure 5-42 Voltage-divider bias circuit for Example 5-14.

Eq. 5-16:
$$I_2 = \frac{I_C}{10} = \frac{1\ \text{mA}}{10}$$

$$= 100\ \mu\text{A}$$

$$R_2 = \frac{V_B}{I_2} = \frac{5.4\ \text{V}}{100\ \mu\text{A}}$$

$$= 54\ \text{k}\Omega\ (\text{use } 56\ \text{k}\Omega\ \text{standard value})$$

With $R_2 = 56\ \text{k}\Omega$ and $V_B = 5.4\ \text{V}$, I_2 becomes

$$I_2 = \frac{V_B}{R_2} = \frac{5.4\ \text{V}}{56\ \text{k}\Omega}$$

$$= 96.4\ \mu\text{A}$$

$$R_1 = \frac{V_{CC} - V_B}{I_2} = \frac{12\ \text{V} - 5.4\ \text{V}}{96.4\ \mu\text{A}}$$

$$= 68.5\ \text{k}\Omega\ (\text{use } 68\ \text{k}\Omega\ \text{standard value})$$

Practice Problems

5-7.1 Design a base bias circuit so that $V_{CE} = 4$ V and $I_C = 3$ mA. The supply voltage is 12 V, and the transistor h_{FE} is 125.

5-7.2 Design a collector-to-base bias circuit in which $V_{CE} = 4$ V and $I_C = 2$ mA. The supply voltage is 18 V and the transistor h_{FE} is 90.

5-7.3 Design a voltage-divider bias circuit in which $V_{CE} = V_E = 6$ V and $I_C = 1.5$ mA. The supply voltage is 24 V, and the transistor h_{FE} is 80.

5-7.4 Design a voltage-divider bias circuit in which $V_{CE} = 7$ V, $V_E = 6$ V, and $I_C = 2$ mA. The supply voltage is 20 V.

5-8 MORE BIAS CIRCUITS

Base Bias and Collector-to-Base Bias with Emitter Resistors

The circuit in Fig. 5-43(a) is a base bias circuit with the addition of emitter resistor R_3. This can be shown to improve the circuit stability, although the improvement is not as great as can be achieved with collector-to-base bias. Figure 5-43b shows an emitter resistor added to a collector-to-base bias circuit. Analyzing this circuit and comparing it to a similar collector-to-base bias circuit, we find that there is no improvement in circuit stability. An advantage to the use of an emitter resistor is that it improves the circuit input resistance (see Section 6-5).

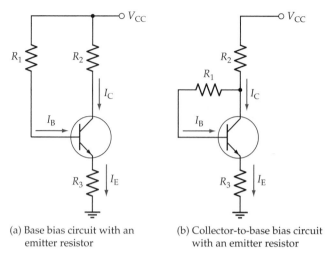

(a) Base bias circuit with an
emitter resistor

(b) Collector-to-base bias circuit
with an emitter resistor

Figure 5-43 An emitter resistor improves the stability of base bias, but has no effect on collector-to-base bias.

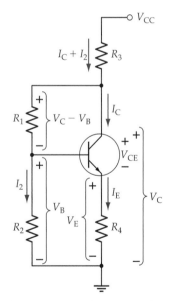

Figure 5-44 A combination of voltage-divider bias and collector-to-base bias provides excellent stability.

Voltage-Divider and Collector-to-Base Combination

Figure 5-44 shows a voltage-divider bias circuit with resistor R_1 connected to the transistor collector instead of to the supply. Thus, the circuit combines collector-to-base bias with voltage-divider bias. An analysis of the circuit shows that this combination produces even greater bias stability than voltage-divider bias alone.

The design procedure for this circuit is similar to voltage-divider bias design, except that for calculating the resistances of R_1 and R_3, the voltage across R_1 is $(V_C - V_B)$ instead of $(V_{CC} - V_B)$, and the current through R_3 is $(I_C + I_2)$.

Example 5-15

Design the bias circuit in Fig. 5-44 to operate from an 18 V supply. The bias conditions are to be $V_{CE} = 9$ V, $V_E = 4$ V, and $I_C = 4$ mA.

Solution

$$R_4 = \frac{V_E}{I_E} \approx \frac{V_E}{I_C} = \frac{4\ V}{4\ mA}$$

$$= 1\ k\Omega \text{ (standard value)}$$

$$V_B = V_E + V_{BE} = 4\ V + 0.7\ V$$

$$= 4.7\ V \text{ (see Fig. 5-45)}$$

Eq. 5-16:
$$I_2 = \frac{I_C}{10} = \frac{4\ mA}{10}$$

$$= 400\ \mu A$$

$$R_2 = \frac{V_B}{I_2} = \frac{4.7\ V}{400\ \mu A}$$

$$= 11.8\ k\Omega \text{ (use 12 k}\Omega \text{ standard value)}$$

With $R_2 = 12\ k\Omega$, I_2 becomes,

$$I_2 = \frac{V_B}{R_2} = \frac{4.7\ V}{12\ k\Omega}$$

$$= 392\ \mu A$$

$$R_1 = \frac{V_C - V_B}{I_2} = \frac{(9\ V + 4\ V) - 4.7\ V}{392\ \mu A}$$

$$= 21.2\ k\Omega \text{ (use 22 k}\Omega \text{ standard value)}$$

$$V_{R3} = V_{CC} - V_C = 18\ V - (9\ V + 4\ V)$$

$$= 5\ V$$

$$R_3 = \frac{V_{R3}}{I_C + I_2} = \frac{5\ V}{4\ mA + 392\ \mu A}$$

$$= 1.14\ k\Omega \text{ (use a 1.2 k}\Omega \text{ standard value)}$$

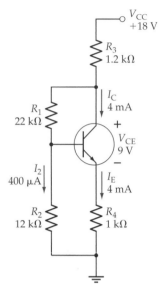

Figure 5-45 Circuit designed in Example 5-15.

Emitter-Current Bias

The *emitter-current bias* circuit in Fig. 5-46a uses a *plus and minus voltage supply* ($+V_{CC}$ and $-V_{EE}$) and has the transistor base grounded via resistor R_1. This is similar to voltage-divider bias, in fact, as illustrated in Fig. 5-46b, a voltage

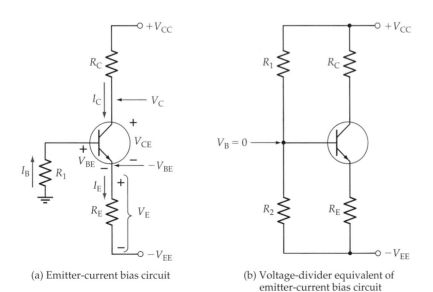

(a) Emitter-current bias circuit

(b) Voltage-divider equivalent of emitter-current bias circuit

Figure 5-46 Emitter-current bias using a plus-minus supply voltage is similar to voltage-divider bias.

divider (R_1 and R_2) could be used to provide V_B instead of grounding the base via R_1.

So long as there is very little voltage drop across base resistor R_1, the circuit in Fig. 5-46a (like voltage-divider bias) has excellent bias stability. Normally, R_1 is selected to have a voltage drop much smaller than the transistor base-emitter voltage:

$$I_{B(\text{max})} R_1 \ll V_{BE}$$

A reasonable rule of thumb to use is

$$I_{B(\text{max})} R_1 \approx \frac{V_{BE}}{10} \qquad (5\text{-}17)$$

This means that the transistor base voltage can be treated as ground level, and the voltage at the emitter terminal is always V_{BE} below ground (see Fig. 5-46a). So the voltage across R_E is a constant quantity:

$$V_E = V_{EE} - V_{BE}$$

The transistor collector voltage is

$$V_C = V_{CC} - V_{RC}$$

and, as illustrated, the collector-emitter voltage is

$$V_{CE} = V_C + V_{BE}$$

Example 5-16

Design the bias circuit in Fig. 5-47 to operate from a ± 9 V supply. The bias conditions are to be $V_C = 5$ V and $I_C = 1$ mA, and the transistor has $h_{FE} = 70$.

Solution

$$V_E = V_{EE} - V_{BE} = 9\text{ V} - 0.7\text{ V}$$
$$= 8.3\text{ V}$$

$$R_3 = \frac{V_E}{I_E} \approx \frac{V_E}{I_C} = \frac{8.3\text{ V}}{1\text{ mA}}$$
$$= 8.3\text{ k}\Omega \text{ (use 8.2 k}\Omega \text{ standard value)}$$

New level of I_E:

$$I_E = \frac{V_E}{R_3} = \frac{8.3\text{ V}}{8.2\text{ k}\Omega}$$
$$= 1.01\text{ mA}$$

$$V_{R2} = V_{CC} - V_C = 9\text{ V} - 5\text{ V}$$
$$= 4\text{ V}$$

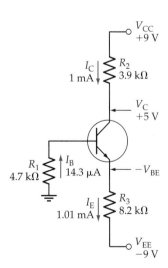

Figure 5-47 Emitter current bias circuit designed in Example 5-16.

$$R_2 = \frac{V_{R2}}{I_C} \approx \frac{V_{R2}}{I_E} = \frac{4\,V}{1.01\,mA}$$

$$= 3.96\,k\Omega \text{ (use 3.9 k}\Omega \text{ standard value)}$$

$$I_B = \frac{I_C}{h_{FE}} = \frac{1\,mA}{70}$$

$$= 14.3\,\mu A$$

$$V_{R1} \approx \frac{V_{BE}}{10} = \frac{0.7\,V}{10}$$

$$= 70\,mV$$

$$R_1 = \frac{V_{R1}}{I_B} \approx \frac{70\,mV}{14.3\,\mu A}$$

$$= 4.9\,k\Omega \text{ (use 4.7 k}\Omega \text{ standard value)}$$

The circuit in Fig. 5-47 can be quickly analyzed almost by just looking at it. The transistor emitter terminal is 0.7 V below ground, and the voltage across R_E is

$$V_E = 9\,V - 0.7\,V = 8.3\,V$$

This gives

$$I_E \approx \frac{V_E}{R_E} = \frac{8.3\,V}{8.2\,k\Omega}$$

$$\approx 1\,mA$$

and

$$V_C = V_{CC} - V_{RC} = 9\,V - (1\,mA \times 3.9\,k\Omega)$$

$$= 5.1\,V$$

Note that, as in the case of voltage divider bias, the transistor h_{FE} value is not used in the approximate circuit analysis. For accurate analysis, the same approach as used for voltage divider bias must be employed: the voltage source at the transistor base must be replaced with its Thevenin equivalent circuit.

Practice Problems

5-8.1 Design the bias circuit in Fig. 5-43a to have $V_{CE} = 4$ V, $V_E = 5$ V, and $I_C = 1.5$ mA. The supply voltage is 18 V and the transistor h_{FE} is 70.

5-8.2 A bias circuit that combines voltage divider and collector-to-base bias (as in Fig. 5-44) is to have $V_{CE} = 5$ V and $I_C = 2$ mA. Design the circuit using $V_{CC} = 15$ V.

5-8.3 Design an emitter-current bias circuit (as in Figure 5-46a) to have $V_C = 6$ V and $I_C = 3$ mA. The supply voltage is ±10 V and the transistor h_{FE} is 120.

5-9 THERMAL STABILITY OF BIAS CIRCUITS

V_{BE} and I_{CBO} Variations

Many transistor circuits are required to operate over a wide temperature range. So another aspect of bias circuit stability is *thermal stability*, or how stable I_C and V_{CE} remain when the circuit temperature changes.

Measures to deal with the effects of h_{FE} variations have already been discussed. These apply whether the different h_{FE} values are due to temperature changes or to h_{FE} differences from one transistor to another.

The base-emitter voltage (V_{BE}) and the collector-base reverse saturation current (I_{CBO}) are the two temperature-sensitive quantities that largely determine the thermal stability of a transistor circuit (see Fig. 5-48a). The base-emitter and collector-base *pn*-junctions have the temperature characteristics discussed in Section 1-6. For a silicon transistor, V_{BE} changes by approximately -1.8 mV/°C, and I_{CBO} approximately doubles for every 10°C rise in temperature. These effects are illustrated by the characteristics in Fig. 5-48b and c.

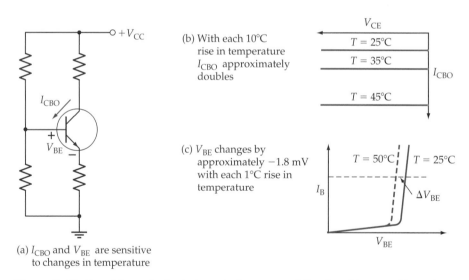

(b) With each 10°C rise in temperature I_{CBO} approximately doubles

(c) V_{BE} changes by approximately -1.8 mV with each 1°C rise in temperature

(a) I_{CBO} and V_{BE} are sensitive to changes in temperature

Figure 5-48 Transistor collector-to-base leakage current (I_{CBO}) and base-emitter voltage (V_{BE}) can be seriously affected by temperature change.

An increase in I_{CBO} causes I_C to be larger, and the increase in I_C raises the collector-base junction temperature. This, in turn, results in a further increase in I_{CBO}. The effect is cumulative, so that the end result might be a substantial increase in collector current. This could produce a significant shift in the circuit Q-point, or in the worst case, I_C *might keep on increasing until the transistor collector-base junction overheats and burns out.* This effect is known as *thermal runaway.* Measures taken to avoid thermal runaway are similar to those required for good bias stability against h_{FE} spread.

Changes in V_{BE} may also produce significant changes in I_C and consequently in the circuit Q-point. However, because of the possibility of thermal runaway, changes in I_{CBO} are the most important. The thermal stability of a bias circuit is assessed by calculating a stability factor.

Stability Factor

The *stability factor (S)* of a circuit is the ratio of the change in collector current to the change in collector-base leakage current.

$$S = \frac{\Delta I_C}{\Delta I_{CBO}}$$

or
$$\Delta I_C = S \times \Delta I_{CBO} \qquad\qquad (5\text{-}18)$$

The value of S depends on the circuit configuration and on the resistor values. The minimum value of S is 1. This means that if I_{CBO} increases by 1 μA, I_C will increase by 1 μA. If a circuit has an S of 50, then $\Delta I_C = 50 \times \Delta I_{CBO}$. A stability factor of 50 (or larger) is considered poor, while a factor of 10 or less is considered good.

An equation for the stability factor of a bias circuit can be derived by writing an equation for the circuit I_C and investigating the effect of changes in I_{CBO}. The stability factors for the three basic bias circuit types (reproduced in Fig. 5-49) can be shown to be as follows:

For base bias:

$$S = 1 + h_{FE} \qquad\qquad (5\text{-}19)$$

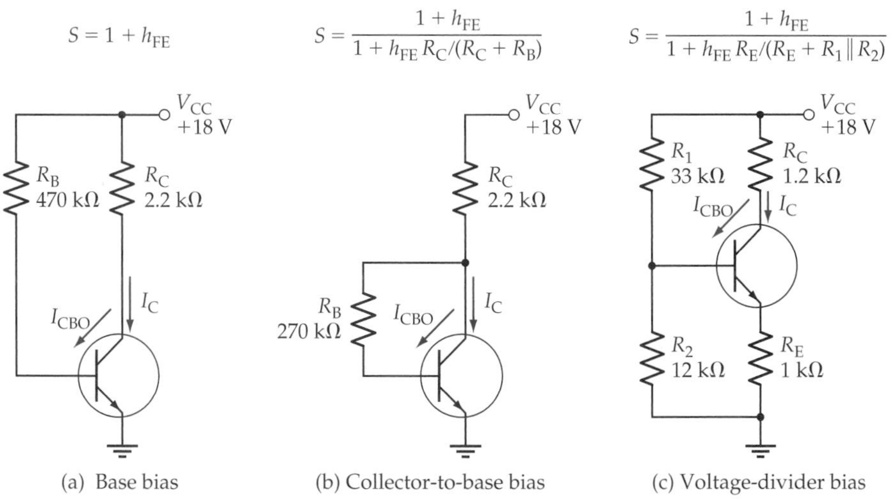

Figure 5-49   Stability factor equations for the three basic bias circuits.

For collector-to-base bias:

$$S = \frac{1 + h_{FE}}{1 + h_{FE}\, R_C/(R_C + R_B)} \tag{5-20}$$

For voltage-divider bias:

$$S = \frac{1 + h_{FE}}{1 + h_{FE}\, R_E/(R_E + R_1 \| R_2)} \tag{5-21}$$

The change in I_{CBO} over a given temperature range can be calculated by recalling that I_{CBO} doubles for every 10°C increase in temperature. The temperature change (ΔT) is divided by 10 to give the number of 10°C changes (n). If the starting level of collector-base leakage current is $I_{CBO(1)}$, the new level is

$$I_{CBO(2)} = I_{CBO(1)} \times 2^n \tag{5-22}$$

The change in I_{CBO} and the circuit stability factor can be used to determine the change in I_C (Eq. 5-18). Then the resulting change in V_{CE} can be investigated.

Example 5-17

Calculate the stability factors for the three bias circuits shown in Fig. 5-49 if each circuit uses a transistor with $h_{FE} = 100$. Note that these are the bias circuits analyzed in Examples 5-4, 5-6, and 5-8.

Solution

For base bias:

Eq. 5-19: $S = 1 + h_{FE} = 1 + 100$

$\qquad\qquad = 101$

For collector-to-base bias:

Eq. 5-20: $S = \dfrac{1 + h_{FE}}{1 + h_{FE}\, R_C/(R_C + R_B)}$

$\qquad\qquad = \dfrac{1 + 100}{1 + [100 \times 2.2\ k\Omega/(2.2\ k\Omega + 270\ k\Omega)]}$

$\qquad\qquad \approx 56$

For voltage-divider bias:

Eq. 5-21: $S = \dfrac{1 + h_{FE}}{1 + h_{FE}\, R_E/(R_E + R_1 \| R_2)}$

$\qquad\qquad = \dfrac{1 + 100}{1 + [100 \times 1\ k\Omega/(1\ k\Omega + 33\ k\Omega \| 12\ k\Omega)]}$

$\qquad\qquad = 8.2$

Example 5-18

Determine the I_C change produced in each bias circuit referred to in Example 5-17 when the circuit temperature increases from 25°C to 105°C and $I_{CBO} = 15$ nA at 25°C. The circuits are reproduced in Fig. 5-50.

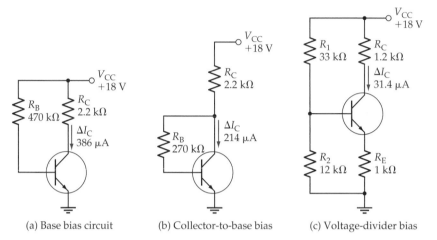

(a) Base bias circuit   (b) Collector-to-base bias   (c) Voltage-divider bias

Figure 5-50 Effect of temperature change from 25°C to 105°C on bias circuits, as determined in Example 5-18.

Solution

$$\Delta T = 105°C - 25°C$$

$$= 80°C$$

or $$\Delta T \text{ in } 10°C \text{ steps} = \frac{80°C}{10°C}$$

$$= 8$$

Eq. 5-22: $$I_{CBO(2)} = I_{CBO(1)} \times 2^n = 15\,\text{nA} \times 2^8$$

$$= 3.84\,\mu\text{A}$$

$$\Delta I_{CBO} = I_{CBO(2)} - I_{CBO(1)} = 3.84\,\mu\text{A} - 15\,\text{nA}$$

$$\approx 3.83\,\mu\text{A}$$

For base bias:

Eq. 5-18: $$\Delta I_C = S \times \Delta I_{CBO} = 101 \times 3.83\,\mu\text{A}$$

$$\approx 386\,\mu\text{A}$$

For collector-to-base bias:

Eq. 5-18: $$\Delta I_C = S \times \Delta I_{CBO} = 56 \times 3.83\,\mu\text{A}$$

$$= 214\,\mu\text{A}$$

For voltage-divider bias:

Eq. 5-18: $$\Delta I_C = S \times \Delta I_{CBO} = 8.2 \times 3.83\,\mu\text{A}$$

$$= 31.4\,\mu\text{A}$$

The changes in the collector current produced by the increase in I_{CBO} (as calculated in Examples 5-17 and 5-18) are small compared to the effects of transistor h_{FE} spread. However, many transistors have higher levels of I_{CBO} (at 25°C) than the 15 nA used in Example 5-18 (some have lower levels). It is important to note that the bias circuit that offers the best thermal stability is also the one that gives the best stability against h_{FE} spread—voltage-divider bias.

Effect of V_{BE} Changes

Consider the voltage-divider bias circuits in Fig. 5-51a and b, which each have $V_{CC} = 12$ V and $I_C = 1$ mA. From Eq. 5-9,

$$I_C \approx I_E = \frac{V_B - V_{BE}}{R_E}$$

Assuming that V_B remains substantially constant, an equation for I_C change with V_{BE} change can be written

$$\Delta I_C \approx \Delta I_E = \frac{\Delta V_{BE}}{R_E} \tag{5-23}$$

As discussed in Section 5-7, the emitter resistor voltage should typically be chosen to be 5 V ($V_E \gg V_{BE}$). This is to ensure that I_C is not significantly affected by changes in V_{BE}. Thus, the circuit in Fig. 5-51a (with $V_E \approx 5$ V) has greater stability against V_{BE} changes than the one in Fig. 5-51b, which was designed for $V_E = V_{CC}/10$ ($V_E = 1.2$ V). This is demonstrated in Example 5-19.

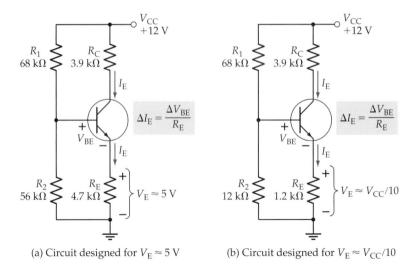

(a) Circuit designed for $V_E \approx 5$ V   (b) Circuit designed for $V_E \approx V_{CC}/10$

Figure 5-51   A voltage-divider bias circuit with $V_E \approx 5$ V is affected less by temperature changes than one with $V_E \approx 1$ V.

Example 5-19

Determine the change in I_C produced in each of the two circuits in Fig. 5-51 by the effect of changes in V_{BE} over a temperature range of 25°C to 125°C.

Solution

$$\Delta T = 125°C - 25°C$$

$$= 100°C$$

$$\Delta V_{BE} = \Delta T \times (1.8\ mV/°C) = 100°C \times 1.8\ mV$$

$$= 180\ mV$$

For Fig. 5-51a:

$$\Delta I_C \approx \Delta I_E = \frac{\Delta V_{BE}}{R_E} = \frac{180\ mV}{4.7\ k\Omega}$$

$$= 38\ \mu A$$

For Fig. 5-51b:

$$\Delta I_C \approx \Delta I_E = \frac{\Delta V_{BE}}{R_E} = \frac{180\ mV}{1.2\ k\Omega}$$

$$= 150\ \mu A$$

Diode Compensation

The use of a diode to compensate for changes in V_{BE} is illustrated in Fig. 5-52. In this case,

$$V_B = V_{R2} + V_{D1}$$

and

$$I_C \approx I_E = \frac{V_{R2} + V_{D1} - V_{BE}}{R_E} \qquad (5\text{-}24)$$

When V_{BE} changes by ΔV_{BE}, the diode voltage changes by an approximately equal amount (ΔV_{D1}). ΔV_{BE} and ΔV_{D1} tend to cancel each other, leaving I_C largely constant at

$$I_C \approx I_E = \frac{V_{R2}}{R_E}$$

Base bias and collector-to-base bias are affected less by changes in V_{BE} than is voltage-divider bias. This can easily be demonstrated by considering the equations for the base current in each case.

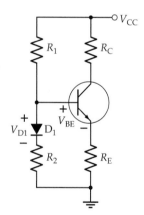

Figure 5-52 A diode can be used to compensate for changes in V_{BE} in a voltage-divider bias circuit.

Practice Problems

5-9.1 Calculate the stability factor for the bias circuits designed in Practice Problems 5-7.1, 5-7.2, and 5-7.3.

5-9.2 The circuit temperature of each of the bias circuits in Practice Problem 5-9.1 increases from 25°C to 125°C. Calculate the collector current change in each circuit if the transistor $I_{CBO} = 10\ nA$ at 25°C.

5-9.3 Calculate the I_C change produced by ΔV_{BE} in the circuit designed in Example 5-14 when the temperature changes from −35°C to +100°C.

5-10 BIASING BJT SWITCHING CIRCUITS

Direct-Coupled Switching Circuit

When a transistor is used as a switch (see Section 4-4), it is either biased off to $I_C = 0$ or biased on to its maximum collector current level. Figure 5-53 illustrates the two conditions. Note that the circuit in Fig. 5-53 is termed a *direct-coupled switching circuit* because the signal source is directly connected to the circuit. In Fig. 5-53a the negative polarity of the base input voltage (V_S) biases the transistor (Q_1) off. In this case, the only current flowing is the collector base leakage current (I_{CBO}), which is normally so small that it can be ignored. The transistor collector-emitter voltage is

$$V_{CE} = V_{CC} - (I_C R_C)$$

With Q_1 off (see Fig. 5-53a),

$$V_{CE} \approx V_{CC}$$

In Fig. 5-53b V_S is positive, and it biases Q_1 on to the maximum possible I_C level. The collector current is limited only by the collector supply voltage (V_{CC}) and the collector resistor (R_C). So,

$$I_C R_C \approx V_{CC} \tag{5-25}$$

and

$$V_{CE} = V_{CC} - (I_C R_C) \approx 0$$

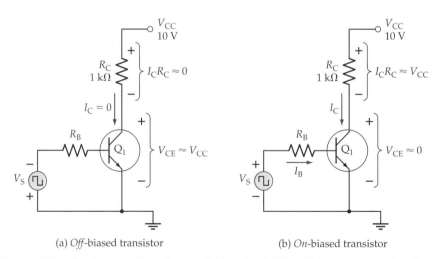

(a) *Off*-biased transistor   (b) *On*-biased transistor

Figure 5-53   Direct-coupled transistor switching circuit. When V_B is zero or negative, $I_C = 0$ and $V_{CE} \approx V_{CC}$. When V_B is positive, I_B flows, driving the transistor into saturation.

Now consider Fig. 5-54, which shows the output characteristics and dc load line for the switching circuit in Fig. 5-53. The load line is drawn by the usual process of plotting point A at ($I_C = 0$ and $V_{CE} = V_{CC}$) and point B at ($V_{CE} = 0$ and $I_C = V_{CC}/R_C$) (see Section 5-1).

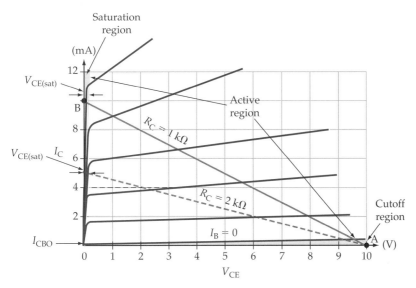

Figure 5-54   DC load line for a transistor switching circuit. When *on*, the transistor is operating in the saturation region. When *off*, it is in the cutoff region.

The 1 kΩ load line shows that, when $I_B = 0$, I_C is close to zero at

$$I_C = I_{CBO}$$

At this point the transistor is said to be *cut off*. The region of the characteristics below $I_B = 0$ is termed the *cutoff region*. When I_C is at its maximum level, the transistor is said to be *saturated*, and the collector-emitter voltage is the *saturation voltage* ($V_{CE(sat)}$).

$$V_{CE} = V_{CE(sat)}$$

The region of the transistor characteristics at $V_{CE(sat)}$ is termed the *saturation region*. The region between saturation and cutoff is the *active region*, which is where a transistor is normally biased for amplification. From the load line, it is seen that $V_{CE(sat)}$ is dependent upon the I_C level. For the 2 kΩ load line shown as a dashed line in Fig. 5-54, $V_{CE(sat)}$ is smaller than for $R_C = 1$ kΩ.

Returning to Fig. 5-53b, we can seen that I_B is a constant quantity:

$$I_B = \frac{V_S - V_{BE}}{R_B} \qquad \text{(5-26)}$$

Also, the level of I_C depends upon I_B:

$$I_C = h_{FE} \times I_B$$

The collector current can be determined from Eq. 5-25 (which assumes that $V_{CE(sat)} = 0$), and the base current can be calculated from Eq. 5-26. Then the minimum required current gain for the transistor is

$$h_{FE(1)} = \frac{I_C}{I_B} \qquad \text{(5-27)}$$

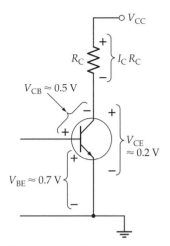

Figure 5-55 When a transistor is in saturation, the CB junction is forward-biased.

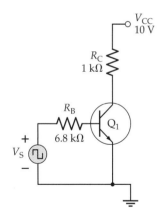

Figure 5-56 Direct-coupled switching circuit for Example 5-20.

If the transistor h_{FE} value is less than the calculated $h_{FE(1)}$ value, I_C will be lower than the level required for transistor saturation. If the actual h_{FE} of the transistor is greater than the calculated $h_{FE(1)}$, I_C tends to be greater than the current level required to saturate the transistor. However, I_C cannot exceed V_{CC}/R_C. Consequently, an h_{FE} value larger than $h_{FE(1)}$ will adjust down to $h_{FE(1)}$. An h_{FE} value lower than $h_{FE(1)}$ cannot adjust up. So, to ensure that the transistor in a switching circuit saturates, it must have an h_{FE} value equal to or greater than the $h_{FE(1)}$ calculated for the circuit.

$V_{CE(sat)}$ is typically 0.2 V for a low-current silicon transistor, while V_{BE} is normally 0.7 V. Consider the circuit in Fig. 5-53b once again. If $V_{BE} = 0.7$ V and $V_{CE(sat)} = 0.2$ V, the transistor base is 0.5 V more positive than the collector (see Fig. 5-55). This means that the collector-base junction, which is usually reverse-biased, is forward-biased when the transistor is in saturation. With the collector-base junction forward-biased, fewer charge carriers from the emitter are drawn across to the collector, and the device current gain is lower than normal.

Example 5-20

Calculate the minimum h_{FE} for the transistor in the circuit in Fig. 5-56 to be in saturation when $V_{CC} = 10$ V, $R_C = 1$ kΩ, $R_B = 6.8$ kΩ, and $V_S = 5$ V. Determine the transistor V_{CE} level when $h_{FE} = 10$.

Solution

$h_{FE(1)}$ calculation:

From Eq. 5-25,
$$I_C \approx \frac{V_{CC}}{R_C} = \frac{10 \text{ V}}{1 \text{ k}\Omega}$$
$$= 10 \text{ mA}$$

Eq. 5-26:
$$I_B = \frac{V_S - V_{BE}}{R_B} = \frac{5 \text{ V} - 0.7 \text{ V}}{6.8 \text{ k}\Omega}$$
$$= 632 \text{ }\mu\text{A}$$

Eq. 5-27:
$$h_{FE(1)} = \frac{I_C}{I_B} = \frac{10 \text{ mA}}{632 \text{ }\mu\text{A}}$$
$$= 15.8$$

When $h_{FE} = 10$,
$$I_C = h_{FE} I_B = 10 \times 632 \text{ }\mu\text{A}$$
$$= 6.32 \text{ mA}$$

$$V_{CE} = V_{CC} - (I_C R_C) = 10 \text{ V} - (6.32 \text{ mA} \times 1 \text{ k}\Omega)$$
$$= 3.68 \text{ V}$$

Capacitor-Coupled Switching Circuit

The base bias circuit in Fig. 5-57 is similar to circuits of that type that have already been considered, with the exception that the transistor is biased into saturation. Although too unpredictable for biasing amplifier circuits, base bias is quite satisfactory for switching circuits. The transistor in Fig. 5-57 is in a *normally-on* state with $V_{CE} = V_{CE(sat)}$. The capacitor-coupled (pulse waveform) input turns the device off, giving $V_{CE} = V_{CC}$.

The transistor base current is

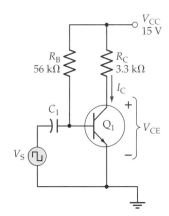

$$I_B = \frac{V_{CC} - V_{BE}}{R_B} \qquad (5\text{-}28)$$

The collector current can be determined from Eq. 5-25, and then the minimum required h_{FE} value can be calculated from Eq. 5-27.

Figure 5-57 Capacitor-coupled switching circuit with the transistor biased *on* into saturation.

Example 5-21

A new transistor is to be substituted for Q_1 in the circuit in Fig. 5-57. Calculate the minimum h_{FE} required for the transistor.

Solution

From Eq. 5-25, $$I_C \approx \frac{V_{CC}}{R_C} = \frac{15\text{ V}}{3.3\text{ k}\Omega}$$

$$= 4.55\text{ mA}$$

Eq. 5-28: $$I_B = \frac{V_{CC} - V_{BE}}{R_B} = \frac{15\text{ V} - 0.7\text{ V}}{56\text{ k}\Omega}$$

$$= 255\ \mu\text{A}$$

Eq. 5-27: $$h_{FE(1)} = \frac{I_C}{I_B} = \frac{4.55\text{ mA}}{255\ \mu\text{A}}$$

$$= 17.8$$

Another type of capacitor-coupled switching circuit is illustrated in Fig. 5-58. In this case, resistor R_B keeps the transistor base-emitter voltage at zero, to ensure that the device is in a *normally-off* state. The capacitor-coupled input voltage switches the transistor *on* into saturation.

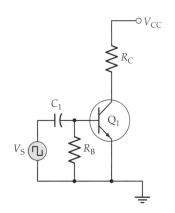

Switching Circuit Design

The resistance of R_C for any one of the switching circuits discussed can be calculated by using the specified V_{CC} and I_C levels with Eq. 5-25. The transistor $h_{FE(min)}$ value can be used with I_C to determine the minimum I_B level required for transistor saturation. However, uncertainty about the actual transistor current gain can be avoided by using an h_{FE} value of 10. It is very unlikely that the

Figure 5-58 Capacitor-coupled switching circuit with the transistor biased *off*.

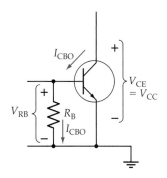

Figure 5-59 A voltage drop is produced across R_B by the collector-base leakage current (I_{CBO}).

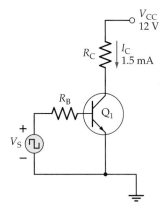

Figure 5-60 Direct-coupled switching circuit for Example 5-22.

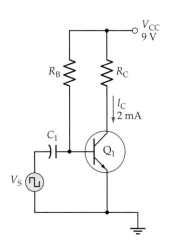

Figure 5-61 Capacitor-coupled switching circuit for Example 5-23.

transistor will have an $h_{FE(min)}$ less than 10, so a circuit designed in this way will work with virtually any low-current transistor.

The resistance of R_B for the direct-coupled circuit in Fig. 5-53 is calculated from Eq. 5-26. For the normally-on capacitor-coupled circuit in Fig. 5-57, Eq. 5-28 is used for determining the R_B value.

A different approach must be taken for calculating a suitable resistance for R_B in the normally-off capacitor-coupled circuit in Fig. 5-58. The current that flows through R_B when the transistor is off and the allowable voltage drop across R_B must be considered (see Fig. 5-59). The resistor current is the collector-base leakage current (I_{CBO}) and the maximum voltage drop produced by I_{CBO} must be much smaller than the normal transistor V_{BE} when the device is on. I_{CBO} at maximum transistor temperature is unlikely to exceed 5 μA, and a V_{RB} of 0.1 V will normally keep the transistor biased off. So a maximum resistance value for R_B is calculated as follows:

$$R_B = \frac{0.1 \text{ V}}{5 \text{ μA}} = 20 \text{ k}\Omega$$

An R_B value of 22 kΩ or lower is usually suitable for keeping the transistor biased off.

Example 5-22

The circuit in Fig. 5-60 uses a V_{CC} of 12 V, the collector current is to be approximately 1.5 mA, and V_S is +5 V. Determine suitable resistances for R_B and R_C.

Solution

From Eq. 5-25, $R_C \approx \dfrac{V_{CC}}{I_C} = \dfrac{12 \text{ V}}{1.5 \text{ mA}}$

$= 8 \text{ k}\Omega$ (use 8.2 kΩ standard value)

From Eq. 5-27, $I_B = \dfrac{I_C}{h_{FE(min)}} = \dfrac{1.5 \text{ mA}}{10}$

$= 150 \text{ μA}$

From Eq. 5-26, $R_B = \dfrac{V_S - V_{BE}}{I_B} = \dfrac{5 \text{ V} - 0.7 \text{ V}}{150 \text{ μA}}$

$= 28.6 \text{ k}\Omega$ (use 27 kΩ to ensure that I_B is larger than required for saturation)

Example 5-23

Determine suitable resistor values for the capacitor-coupled switching circuit in Fig. 5-61.

Solution

From Eq. 5-25, $R_C \approx \dfrac{V_{CC}}{I_C} = \dfrac{9\text{ V}}{2\text{ mA}}$

$= 4.5\text{ k}\Omega$ (use 4.7 kΩ standard value)

From Eq. 5-27, $I_B = \dfrac{I_C}{h_{FE(min)}} = \dfrac{2\text{ mA}}{10}$

$= 200\text{ }\mu\text{A}$

From Eq. 5-28, $R_B = \dfrac{V_{CC} - V_{BE}}{I_B} = \dfrac{9\text{ V} - 0.7\text{ V}}{200\text{ }\mu\text{A}}$

$= 41.5\text{ k}\Omega$ (use 39 kΩ)

Practice Problems

5-10.1 A direct-coupled transistor switching circuit as in Fig. 5-56 has $V_{CC} =$ 15 V, $V_S = 9$ V, $R_C = 3.3$ kΩ, and $R_B = 22$ kΩ. Calculate the minimum h_{FE} for the transistor.

5-10.2 A normally-on capacitor-coupled switching circuit (as in Fig. 5-57) uses a transistor with $h_{FE(min)} = 25$. The supply voltage is 12 V, and $R_C = 4.7$ kΩ. Determine a suitable resistance for R_B.

5-10.3 A direct-coupled transistor switching circuit has $V_{CC} = 5$ V and $V_S = 3$ V. Calculate suitable resistances for R_C and R_B to give $I_C = 2.5$ mA.

Review Questions

Section 5-1

5-1 Identify the components that constitute the dc load in a BJT bias circuit. Explain the procedure for drawing the dc load line on the transistor CE output characteristics.

5-2 Explain the selection of a Q-point for a transistor bias circuit, and discuss the limitations on the output voltage swing.

Section 5-2

5-3 Sketch a base bias circuit (a) using an *npn* transistor, and (b) using a *pnp* transistor. In each case show the polarity of V_{CC}, V_{BE}, and V_{CE}. Show the direction of I_C and I_B.

5-4 Explain the operation of the base bias circuits drawn for Question 5-3, and write equations for I_B, I_C, and V_{CE}.

Section 5-3

5-5 Sketch a collector-to-base bias circuit using an *npn* transistor. Show the polarity of V_{CC}, V_{BE}, and V_{CE}. Show the direction of I_C and I_B.

5-6 Explain the operation of the collector-to-base bias circuit drawn for Question 5-5. Write equations for I_B, I_C, and V_{CE}.

5-7 Sketch a collector-to-base bias circuit that uses a *pnp* transistor. Show all voltage polarities and current directions, and discuss the circuit operation.

Section 5-4

5-8 Sketch a voltage-divider bias circuit using an *npn* transistor. Show all voltage polarities and current directions.

5-9 Explain the operation of the voltage-divider bias circuit drawn for Question 5-8, and write approximate equations for V_B, I_E, I_C, and V_{CE}.

5-10 Explain the procedure for precise analysis of a voltage-divider bias circuit.

5-11 Sketch a voltage-divider bias circuit using an *pnp* transistor. Show all voltage polarities and current directions, and discuss the circuit operation.

Section 5-5

5-12 Compare base bias, collector-to-base bias, and voltage-divider bias circuit with regard to stability of the transistor collector voltage with spread in h_{FE} value. Discuss the advantages and disadvantages of the three types of bias circuit.

Section 5-6

5-13 List common errors involved in constructing and testing a transistor bias circuit.

5-14 Suggest possible errors in a base bias circuit (a) when $V_C \approx V_{CC}$ and (b) when $V_C \approx 0$.

5-15 Suggest possible errors in a collector-to-base bias circuit (a) when $V_C \approx V_{CC}$ and (b) when $V_C \approx 0$.

5-16 Suggest possible errors in a voltage-divider bias circuit (a) when $V_C \approx V_{CC}$ and (b) when $V_C \approx V_E$.

Section 5-7

5-17 Write the equations for calculating R_B and R_C for a base bias circuit.

5-18 Write equations for calculating R_B and R_C for a collector-to-base bias circuit.

5-19 For a voltage-divider bias circuit, write equations for calculating R_E, R_C, and the voltage-divider resistors.

Section 5-8

5-20 Sketch a base bias circuit that uses an emitter resistor with an *npn* transistor. Briefly discuss the circuit operation.

5-21 Sketch an *npn* transistor bias circuit that uses a combination of voltage-divider bias and collector-to-base bias. Show all voltage polarities and current directions, and explain the circuit operation.

5-22 Repeat Question 5-21 for a circuit that uses a *pnp* transistor.

5-23 Sketch an emitter-current bias circuit using a plus and minus supply. Explain the circuit operation and compare it to voltage-divider bias.

Section 5-9

5-24 Discuss the thermal stability of a transistor bias circuit with regard to I_{CBO} and V_{BE}. State approximations for the variation in V_{BE} and I_{CBO} with temperature changes.

5-25 Define the stability factor (S) for a transistor bias circuit. Compare the three basic bias circuits with regard to thermal stability.

5-26 Show how a voltage-divider bias circuit may be compensated for V_{BE} changes with temperature. Derive an equation for I_C.

Section 5-10

5-27 Draw a direct-coupled transistor switching circuit that may be turned on or off by application of an input voltage. Explain the circuit operation.

5-28 Sketch the typical output characteristics and dc load line for a transistor used as a switch. Identify the regions of the characteristics and briefly explain.

5-29 Sketch the circuit of a capacitor-coupled transistor switch biased in (a) a normally-on condition, and (b) a normally-off condition. Explain the operation of each circuit.

Problems

Section 5-1

5-1 Plot the dc load line for the transistor circuit in Fig. 5-62 on the output characteristics shown in Fig. 5-63 (a) when $V_{CC} = 15$ V and $R_C = 7.5$ kΩ and (b) when $V_{CC} = 12$ V and $R_C = 8$ kΩ. Specify the Q-point in each case if $I_B = 20$ μA.

5-2 A circuit with the configuration and characteristics shown in Figs. 5-62 and 5-63 has $V_{CC} = 10$ V, $R_C = 4.7$ kΩ, and $I_B = 10$ μA. Draw the dc load line and specify the circuit Q-point.

5-3 Determine the maximum symmetrical output voltage swing for each of the two circuits in Problem 5-1.

5-4 Determine the maximum symmetrical output voltage swing for the circuit in Problem 5-2.

5-5 Using the transistor characteristics in Fig. 5-63, draw the dc load line for the circuit in Fig. 5-64. Determine I_C for the circuit when $I_B = 10$ μA and calculate V_{RC}, V_{RE}, and V_{CE}.

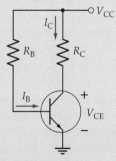

Figure 5-62 Circuit for Problem 5-1.

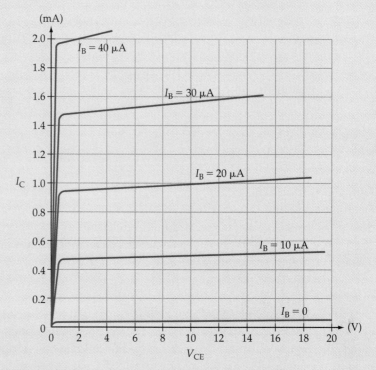

Figure 5-63   Transistor characteristics for Problems 5-1 to 5-5 and 5-25.

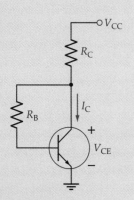

Figure 5-64   Circuit for Problem 5-5.

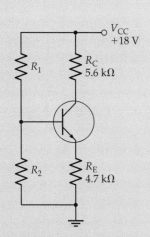

Figure 5-65   Circuit for Problem 5-11.

5-6   A circuit with the characteristics in Fig. 5-10 has $V_{CC} = 7.5$ V, $R_C = 1.2$ kΩ, and $I_B = 40\ \mu$A. Draw the dc load line, specify the circuit Q-point, and determine the upper and lower limits of V_{CE}.

Section 5-2

5-7   A base bias circuit (as in Fig. 5-62) has $V_{CC} = 15$ V, $R_C = 1.8$ kΩ, and $R_B = 120$ kΩ. Assuming $V_{BE} = 0.7$ V and $h_{FE} = 50$, determine I_C and V_{CE}.

5-8   Determine the maximum and minimum I_C and V_{CE} levels for the circuit in Problem 5-7 when the transistor used has $h_{FE(max)} = 60$ and $h_{FE(min)} = 20$.

5-9   A base bias circuit has $V_{CC} = 20$ V, $R_C = 5.6$ kΩ, $R_B = 270$ kΩ, and $V_{CE} = 10$ V. Determine the transistor h_{FE} value, and calculate the new V_{CE} when a transistor with $h_{FE} = 40$ is substituted.

5-10   A base bias circuit with $V_{CC} = 12$ V, $R_C = 3.9$ kΩ, and $R_B = 330$ kΩ has a measured V_{CE} of 3 V. Determine the transistor h_{FE} value.

Section 5-3

5-11   A collector-to-base bias circuit (as in Fig. 5-65) has $V_{CC} = 15$ V, $R_C = 1.8$ kΩ, $R_B = 39$ kΩ, and $h_{FE} = 50$. Determine the I_C and V_{CE} levels.

5-12   Calculate the maximum and minimum levels of I_C and V_{CE} for the circuit in Problem 5-11 when the transistor used has $h_{FE(max)} = 60$ and $h_{FE(min)} = 20$.

5-13 A collector-to-base bias circuit has $V_{CC} = 15$ V, $R_C = 5.6$ kΩ, $R_B = 82$ kΩ, and $V_{CE} = 5$ V. Determine the transistor h_{FE} value, and calculate the new V_{CE} level when a transistor with $h_{FE} = 50$ is substituted.

5-14 Determine the new I_C and V_{CE} levels for the circuit in Problem 5-11 when the supply voltage is changed to 20 V.

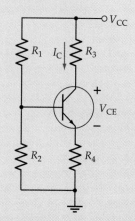

Figure 5-66 Circuit for Problem 5-15.

Section 5-4

5-15 The voltage-divider bias circuit in Fig. 5-66 has $V_{CC} = 15$ V, $R_1 = 6.8$ kΩ, $R_2 = 3.3$ kΩ, $R_3 = 900$ Ω, $R_4 = 900$ Ω, and $h_{FE} = 50$. Analyze the circuit approximately to determine the levels of I_C and V_{CE}.

5-16 Precisely analyze the circuit in Problem 5-15 to determine the maximum and minimum levels of I_C and V_{CE} when $h_{FE(max)} = 60$ and $h_{FE(min)} = 20$.

5-17 A voltage-divider bias circuit with a 25 V supply has $R_C = 4.7$ kΩ, $R_E = 3.3$ kΩ, $R_1 = 33$ kΩ, $R_2 = 12$ kΩ, and $h_{FE} = 50$. Use the approximate analysis method to calculate the V_{CE} level.

5-18 Precisely analyze the circuit in Problem 5-17 to determine I_C and V_{CE} when the transistor has $h_{FE} = 60$.

5-19 A transistor circuit using voltage-divider bias has the following components: $R_C = 2.2$ kΩ, $R_E = 3.3$ kΩ, $R_1 = 6.8$ kΩ, $R_2 = 4.7$ kΩ. The supply voltage is $V_{CC} = 15$ V. Analyze the circuit approximately to determine V_{CE}.

5-20 A voltage-divider bias circuit has $V_{CC} = 15$ V, $R_C = 2.7$ kΩ, $R_E = 2.2$ kΩ, $R_1 = 22$ kΩ, $R_2 = 12$ kΩ. Calculate V_{CE} level.

Section 5-5

5-21 Using the transistor characteristics in Fig. 5-10, plot the dc load line and maximum and minimum Q-points for the circuits in Problems 5-8, 5-12, and 5-16.

Section 5-6

5-22 Analyze the circuit in Problem 5-7 to investigate the effect of interchanging the base and collector resistors.

5-23 Analyze the circuit in Problem 5-11 to investigate the effect of interchanging the base and collector resistors.

5-24 Analyze the circuit in Problem 5-15 to investigate the effect of interchanging the voltage-divider resistors.

Section 5-7

5-25 A base bias circuit with $V_{CC} = 18$ V uses a transistor with the characteristics in Fig. 5-63. The circuit is to have $V_{CE} = 9$ V and $I_C = 2$ mA. Plot the Q-point, draw the dc load line, and determine the required resistance for R_C.

5-26 A base bias circuit with a 12 V supply uses a transistor with $h_{FE} = 70$. Design the circuit so that $I_C = 2$ mA and $V_{CE} = 9$ V.

5-27 Analyze the circuit designed for Problem 5-26 to determine the maximum and minimum levels of V_{CE} when the transistor type used has $h_{FE(min)} = 30$ and $h_{FE(max)}$ of 200.

5-28 A base bias circuit has $V_{CC} = 20$ V, $R_C = 6.8$ kΩ, and the transistor $h_{FE} = 120$. Calculate the base resistance value required to give a V_{CE} of 5 V.

5-29 Design a collector-to-base bias circuit to have $I_C = 3.5$ mA and $V_{CE} = 12$ V. The supply voltage is $V_{CC} = 20$ V, and the transistor $h_{FE} = 80$.

5-30 Analyze the circuit designed for Problem 5-29 to determine the maximum and minimum levels of V_{CE} when the transistor type used has $h_{FE(min)} = 20$ and $h_{FE(max)} = 100$.

5-31 A collector-to-base bias circuit has $I_C = 3$ mA and $V_{CE} = 10$ V when $V_{CC} = 25$ V, and the transistor $h_{FE} = 80$. Calculate suitable resistor values.

5-32 A collector-to-base bias circuit has $V_{CC} = 30$ V, $R_C = 8.2$ kΩ, and the transistor $h_{FE} = 100$. Calculate the base resistance value required to give $V_{CE} = 7$ V.

5-33 Design a voltage-divider bias circuit to have $V_{CE} = 10$ V, $I_C = 1$ mA, and $V_{CC} = 30$ V.

5-34 Precisely analyze the circuit designed for Problem 5-33 to determine the maximum and minimum levels of V_{CE} when the transistor type used has $h_{FE(min)} = 40$ and $h_{FE(max)} = 80$.

5-35 A voltage-divider bias circuit with $V_{CC} = 20$ V and $R_C = 6$ kΩ uses a transistor with $h_{FE} = 80$. Calculate suitable resistor values to give $V_{CE} = 8$ V.

5-36 Design a voltage-divider bias circuit using a 2N3906 transistor. The supply voltage is 22 V, the collector current is to be 10 mA, and the collector-emitter voltage is to be 4 V.

5-37 Precisely analyze the circuit in Problem 5-35 to determine V_{CE} and I_C.

5-38 Design a voltage-divider bias circuit to have $V_{CE} = 4$ V, $V_E = 5$ V, and $I_C = 1.3$ mA. The supply is 18 V.

Section 5-8

5-39 Design an emitter-current bias circuit (as in Fig. 5-67) to have $V_{CE} = 6$ V and $I_C = 10$ mA. The supply voltage is $V_{CC} = \pm 18$ V and the transistor has $h_{FE} = 100$.

5-40 Precisely analyze the circuit in Problem 5-39 to determine V_{CE} and I_C.

5-41 An emitter-current bias circuit has a 10 kΩ emitter resistor and a V_{CC} of ± 20 V. Calculate suitable base and collector resistor values to give $V_{CE} = 8$ V when $h_{FE} = 75$.

5-42 Determine suitable resistor values for the circuit in Fig. 5-68.

5-43 A circuit employing collector-to-base bias and voltage-divider bias as in Fig. 5-68 uses a 30 V supply. Design the circuit to have $V_{CE} = 10$ V and $I_C = 10$ mA.

Section 5-9

5-44 Determine the stability factors for the circuits in Problems 5-7, 5-11, and 5-15.

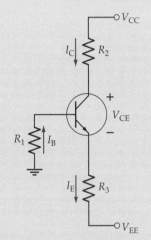

Figure 5-67 Emitter current bias circuit for Problem 5-39.

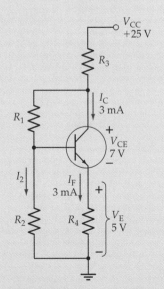

Figure 5-68 Circuit for Problem 5-42.

5-45 Calculate the changes in I_C produced in each of the circuits referred to in Problem 5-44 when the circuit temperature changes from 25°C to 75°C. The transistor collector-base leakage current is specified as 25 nA at 25°C.

5-46 The voltage-divider bias circuit in Problems 5-33 and 5-34 is to have a maximum I_C change of 15 μA due to change in I_{CBO} when the circuit temperature increases from 25°C to 35°C. Determine the maximum acceptable I_{CBO} for the transistor at 25°C.

5-47 Determine the I_C change produced in the circuits designed in Problems 5-33, 5-35, and 5-39 by the effect of transistor V_{BE} changes over a temperature range of 25°C to 75°C.

5-48 A voltage-divider bias circuit is to have a maximum I_C change of 50 μA due to change in V_{BE} when the circuit temperature increases from 25°C to 75°C. Determine the minimum acceptable emitter resistance.

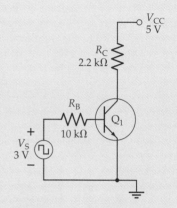

Figure 5-69 Circuit for Problem 5-49.

Section 5-10

5-49 Calculate $h_{FE(min)}$ for the transistor in the direct-coupled switching circuit in Fig. 5-69.

5-50 The normally-on capacitor-coupled switching circuit in Fig. 5-70 uses a transistor with $h_{FE(min)} = 40$. Determine a suitable resistance for R_B.

5-51 A direct-coupled transistor switching circuit (as in Fig. 5-69) with $V_{CC} = 9$ V and $V_S = 6$ V is to have $I_C = 5$ mA. Calculate suitable resistances for R_C and R_B.

5-52 A direct-coupled transistor switching circuit has $V_{CC} = 9$ V and $V_S = 1.6$ V. The resistor values are $R_C = 2.7$ kΩ, and $R_B = 4.7$ kΩ. Calculate the minimum transistor h_{FE} value for saturation. Will the transistor be saturated if R_C is changed to 1 kΩ and $h_{FE} = 40$?

5-53 The normally-off capacitor-coupled switching circuit in Fig. 5-71 uses a transistor with $h_{FE(min)} = 40$. Determine suitable resistances for R_B and R_C.

5-54 A normally-on capacitor-coupled switching circuit has $V_{CC} = 12$ V and $R_C = 3.9$ kΩ. Determine a suitable resistance for R_B.

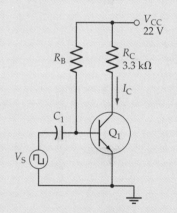

Figure 5-70 Circuit for Problem 5-50.

Practice Problem Answers

5-1.1   Q-point (40 μA, 4 V, 3.6 mA)
5-1.2   Q-point (10 μA, 0.5 mA, 9 V), $\Delta V_o = \pm 6$ V
5-1.3   Q-point (60 μA, 6.25 mA, 4.5 V), $\Delta V_o \approx \pm 4.5$ V
5-2.1   (9.2 mA, 2.76 mA), (11.93 V, 0 V)
5-2.2   71, 4.3 V
5-3.1   (5.26 mA, 2.97 mA), (11.38 V, 6.38 V)
5-3.2   81, 7.93 V
5-4.1   1.06 mA, 15.3 V, 10.3 V
5-4.2   11.8 V, 10.7 V
5-4.3   4.26 V, 12.7 V

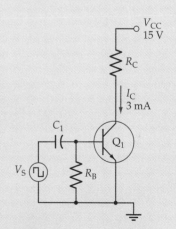

Figure 5-71 Circuit for Problem 5-53.

5-7.1   470 kΩ, 2.7 kΩ
5-7.2   150 kΩ, 6.8 kΩ
5-7.3   100 kΩ, 39 kΩ, 8.2 kΩ, 3.9 kΩ
5-7.4   (56 kΩ + 5.6 kΩ), 27 kΩ, 3.9 kΩ, 2.7 kΩ
5-8.1   (560 kΩ + 15 kΩ), 5.6 kΩ, 3.3 kΩ
5-8.2   22 kΩ, 27 kΩ, 2.7 kΩ, 2.2 kΩ
5-8.3   2.7 kΩ, 1.2 kΩ, 3.3 kΩ
5-9.1   126, 18.6, 7.5
5-9.2   1.29 mA, 190 μA, 76.5 μA
5-9.3   51.7 μA
5-10.1   12.1
5-10.2   100 kΩ
5-10.3   2.2 kΩ, 8.2 kΩ

CHAPTER 6
AC Analysis of BJT Circuits

CONTENTS

Objectives

You will be able to:

1 Explain the need for coupling and bypass capacitors in transistor circuits, and draw ac equivalents for circuits containing capacitors.

2 Draw ac load lines for basic transistor circuits.

3 Sketch and explain transistor *r*-parameter and *h*-parameter models.

4 Show how *h*-parameters can be derived from device characteristics.

5 Convert between CE, CC, and CB *h*-parameters, and between *h*-parameters and *r*-parameters.

6 Sketch *h*-parameter circuits for various CE, CC, and CB transistor circuits.

7 Analyze various CE, CC, and CB transistor circuits to determine input resistance, output resistance, voltage gain, current gain, and power gain.

8 Compare the performance of CE, CC, and CB circuits.

INTRODUCTION

A transistor circuit used as an ac amplifier normally has the signal source capacitor-coupled to the circuit input terminal. A load resistor at the output is also usually capacitor-coupled to the circuit. Capacitor-coupling ensures that the source and load do not affect the dc bias conditions in the circuit. An ac load line can be drawn on the transistor output characteristics to study the ac performance of a circuit. All transistor circuits, however complicated, are based on one of the three basic circuit configurations—common-emitter, common-collector, or common-base. The three basic circuits have different input impedances, output impedances, and voltage gains, and these are most easily determined by the use of *h*-parameters. Multi-stage transistor circuits can also be analyzed by application of *h*-parameters.

6-1 COUPLING AND BYPASSING CAPACITORS

Coupling Capacitors

To use a transistor circuit to amplify or otherwise process an ac signal, the signal source must be connected to the circuit input. If the source is directly connected to the input, as illustrated in Fig. 6-1a, the circuit bias conditions will be altered. Figure 6-1b shows that the signal source resistance (r_S) is in parallel with voltage divider resistor R_2 when the signal source is directly

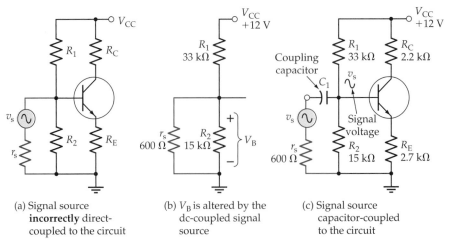

(a) Signal source **incorrectly** direct-coupled to the circuit

(b) V_B is altered by the dc-coupled signal source

(c) Signal source capacitor-coupled to the circuit

Figure 6-1 AC input signals must be capacitor-coupled to the input terminal of a transistor circuit. Direct coupling would alter the circuit bias conditions.

connected to the circuit. Thus, the bias voltage is altered to

$$V_B = \frac{V_{CC} \times (r_S \| R_2)}{R_1 + (r_S \| R_2)}$$

instead of

$$V_B = \frac{V_{CC} \times R_2}{R_1 + R_2}$$

Figure 6-1c shows the use of capacitor C_1 to couple the signal source to the circuit input. Because C_1 is an open-circuit to direct currents, r_S does not affect the level of V_B. Capacitor C_1 behaves as a short-circuit for the ac signals, so that the signal voltage (v_S) appears at the transistor base, as illustrated. In this case, the signal is said to be *ac-coupled* to the circuit input, and C_1 is referred to as a *coupling capacitor.* Input coupling capacitors are normally used with all types of bias circuits, otherwise the circuit bias conditions would be altered.

A coupling capacitor is usually required at the output of a transistor circuit (as well as at the input) to couple to a load resistor or to another amplification stage. Figures 6-2a and b show the effect of directly coupling a load (R_L) to the circuit output. The collector supply voltage is reduced from V_{CC} to

$$V = \frac{V_{CC} \times R_L}{R_C + R_L}$$

and the collector resistance becomes

$$R = R_C \| R_L$$

This has the effect of altering the circuit dc load line and Q-point.

The use of an output coupling capacitor (C_2) is illustrated in Fig. 6-2c. Like the input coupling capacitor, C_2 offers a dc open circuit and behaves as an ac short circuit. Thus, it passes the ac output waveform to the load without affecting the circuit bias conditions.

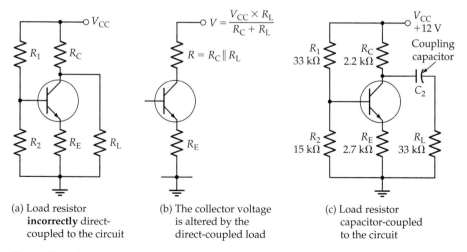

(a) Load resistor **incorrectly** direct-coupled to the circuit

(b) The collector voltage is altered by the direct-coupled load

(c) Load resistor capacitor-coupled to the circuit

Figure 6-2 Output loads must be capacitor-coupled to the output terminal of a transistor circuit. Direct coupling would alter the circuit bias conditions.

Example 6-1

Calculate the base bias voltage for the circuit in Fig. 6-1c when no signal source is present, and when the signal source is directly connected as in Fig. 6-1a.

Solution

With no signal source,

$$V_B = \frac{V_{CC} \times R_2}{R_1 + R_2} = \frac{12\ V \times 15\ k\Omega}{33\ k\Omega + 15\ k\Omega}$$

$$= 3.75\ V$$

With the signal source directly connected,

$$V_B = \frac{V_{CC} \times (r_S \| R_2)}{R_1 + (r_S \| R_2)} = \frac{12\ V \times (600\ \Omega \| 15\ k\Omega)}{33\ k\Omega + (600\ \Omega \| 15\ k\Omega)}$$

$$= 0.21\ V$$

AC Degeneration

In the discussion on collector-to-base bias, it was explained that an increase in I_B would produce a decrease in V_{CE}, and this would tend to cancel the original increase in I_B. This, of course, is an effect that produces good bias stability; however, this same reaction occurs when an ac signal is applied to the circuit input for amplification. The voltage change at the transistor collector (produced by the ac input) is fed back to the base, where it tends to partially cancel the signal. The effect is termed *ac degeneration*, and it can result in a very low voltage gain.

Figure 6-3a shows how ac degeneration is eliminated in collector-to-base bias circuits. R_B is replaced with two approximately equal resistors (R_{B1} and R_{B2}), which add up to R_B. A *bypass capacitor* (C_B) is connected from the junction of R_{B1} and R_{B2} to the ground terminal, as illustrated. Capacitor C_B offers a short circuit to ac signals, so that there is no feedback from the transistor collector to the base. Figure 6-3b shows that, when C_B is replaced with a short circuit,

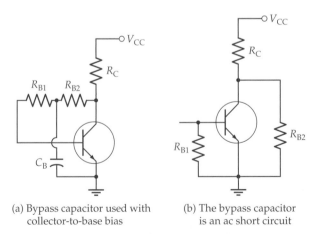

(a) Bypass capacitor used with collector-to-base bias

(b) The bypass capacitor is an ac short circuit

Figure 6-3 A collector-to-base bias circuit must use two resistors for R_B with the junction of the two ac-grounded via a capacitor. This is necessary to eliminate ac degeneration.

R_{B1} and R_{B2} appear in parallel with the circuit input and output respectively. Because they are large-value resistors, they do not affect the circuit voltage gain.

Emitter Bypassing

Voltage-divider bias circuits also suffer ac degeneration because of the presence of the emitter resistor (R_E). In this case, the ac signal voltage is developed across R_E and the transistor BE junction in series, instead of just across the BE junction. The problem is eliminated by the *emitter bypass* capacitor (C_E) illustrated in Fig. 6-4, which provides an ac short circuit across R_E.

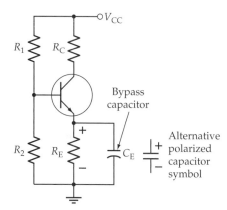

Figure 6-4 A resistor connected in series with the transistor emitter must be ac short-circuited by means of a bypass capacitor to ensure maximum amplification.

Capacitor Polarity

Bypass capacitors are usually polarized (electrolytic-type), and coupling capacitors can also be polarized, so it is very important that capacitors be correctly connected (see Section 3-3). Electrolytic capacitors tend to have a high leakage current when incorrectly connected, so even if they do not explode, they can affect circuit bias conditions. The capacitor positive terminal must be connected to the more positive of the two circuit points where the capacitor is installed.

The capacitor positive terminal is represented by the straight line on the capacitor graphic symbol, or is identified by the plus sign on the alternative capacitor symbol (see Fig. 6-4). The transistor emitter terminal in Fig. 6-4 is more positive than ground level, as illustrated. Consequently, the positive terminal of the emitter bypass capacitor is connected at the transistor emitter, and the negative terminal goes to ground.

Figure 6-5 shows a circuit with correctly connected coupling and bypass capacitors. The dc voltage level at the right side of C_1 is +0.7 V, and the left side is grounded via the signal source. So, the polarity is *plus* on the right, *minus* on the left. The voltage at the junction of resistors R_{B1} and R_{B2} is +3.2 V; consequently, the positive terminal of C_2 must be connected at that point, and C_2 negative terminal is grounded. Output coupling capacitor C_3 has its left side at +5.7 V, and the right side is grounded via R_L, requiring the capacitor to be connected as illustrated.

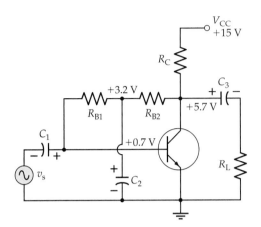

Figure 6-5 Polarized capacitors must be correctly connected to avoid high leakage current.

Practice Problem

6-1.1 Calculate the transistor collector voltage for the circuit in Fig. 6-2c with C_2 present and with R_L directly connected.

6-2 AC LOAD LINES

AC Equivalent Circuits

Capacitors behave as short-circuits to ac signals, so in the *ac equivalent circuit* for a transistor circuit all capacitors must be replaced with short-circuits. Power supplies also behave as ac short-circuits, because the dc supply voltage is not affected by ac signals. Also, all power supplies have large-value capacitors at the output terminals, and these will offer short-circuits to ac signals. Substituting short-circuits for the power supply and all capacitors in the circuit in Fig. 6-6a gives the ac equivalent circuit in Fig. 6-6b. If R_L is present, as shown, it appears in parallel with R_C in the ac equivalent circuit.

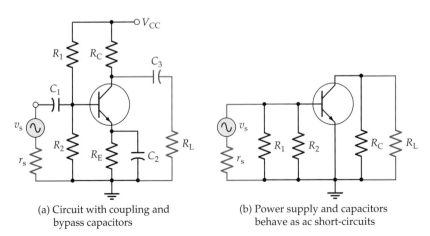

(a) Circuit with coupling and bypass capacitors

(b) Power supply and capacitors behave as ac short-circuits

Figure 6-6 Voltage-divider bias circuit with coupling and bypass capacitors, and the ac equivalent circuit.

Once the ac equivalent circuit is drawn, the circuit ac performance can be investigated by drawing an ac load line and by substituting a transistor model for the device.

AC Load Lines

The dc load for the circuit in Fig. 6-6a is $(R_C + R_E)$; consequently, the dc load line is drawn (as discussed in Section 5-1) for a total resistance of $(R_C + R_E)$. Because the emitter resistor is bypassed by the capacitor in Fig. 6-6a, resistor R_E is not part of the circuit ac load. If external load R_L were not present, the circuit ac load would simply be R_C. With R_L capacitor-coupled to the circuit output, the ac load is $R_C \| R_L$. An ac *load line* may now be drawn to represent the ac performance of the circuit.

When there is no input signal, the transistor voltage and current conditions are exactly as indicated by the Q-point on the dc load line. An ac signal causes the transistor voltage and current levels to vary above and below the Q-point. Therefore, the Q-point is common to both the ac and dc load lines. Starting from the Q-point, another point is found on the ac load line by taking a convenient collector current change (usually $\Delta I_C = I_{CQ}$) and calculating the corresponding collector-emitter voltage change (ΔV_{CE}) (see Fig. 6-8). As demonstrated in Ex. 6-2, *point C* is plotted by measuring ΔI_C and ΔV_{CE} from the Q-point. The ac load line is then drawn through points C and Q, as illustrated.

Example 6-2

Draw the dc and ac load lines for the transistor circuit in Fig. 6-7, using the transistor common-emitter characteristics in Fig. 6-8.

Solution

Drawing the dc load line:

$$R_{L(dc)} = R_C + R_E = 2.2 \text{ k}\Omega + 2.7 \text{ k}\Omega$$
$$= 4.9 \text{ k}\Omega$$

The equation for V_{CE} is

$$V_{CE} = V_{CC} - I_C (R_C + R_E)$$

When $I_C = 0$, $V_{CE} = V_{CC} = 20 \text{ V}$

Plot point A at $I_C = 0$ and $V_{CE} = 20 \text{ V}$

When $V_{CE} = 0$, $I_C = \dfrac{V_{CC}}{R_C + R_E} = \dfrac{20 \text{ V}}{4.9 \text{ k}\Omega}$
$$= 4.08 \text{ mA}$$

Plot point B at $I_C = 4.08 \text{ mA}$ and $V_{CE} = 0$

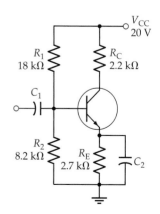

Figure 6-7 Voltage-divider bias circuit with a bypassed emitter resistor. The dc load is $(R_C + R_E)$, and the ac load is R_C when there is no capacitor-coupled load.

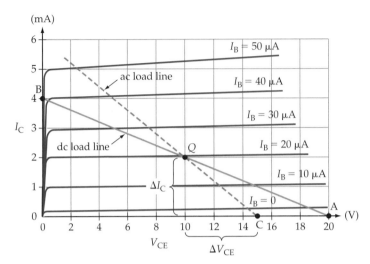

Figure 6-8 The ac load line for a transistor circuit is drawn through the Q-point.

Draw the dc load line through points A and B.

$$V_B = \frac{V_{CC} \times R_2}{R_1 + R_2} = \frac{20\ V \times 8.2\ k\Omega}{18\ k\Omega + 8.2\ k\Omega}$$

$$= 6.3\ V$$

$$V_E = V_B - V_{BE} = 6.3\ V - 0.7\ V$$

$$= 5.6\ V$$

$$I_C \approx I_E = \frac{V_E}{R_E} = \frac{5.6\ V}{2.7\ k\Omega}$$

$$= 2.07\ mA$$

Mark the Q-point on the dc load line at

$$I_C = 2.07\ mA$$

Drawing the ac load line:
When there is no external R_L,

$$R_{L(ac)} = R_C = 2.2\ k\Omega$$

When I_C changes by $\Delta I_C = 2.07\ mA$,

$$\Delta V_{CE} = \Delta I_C \times R_C$$

$$= 2.07\ mA \times 2.2\ k\Omega$$

$$= 4.55\ V$$

Plot point C at $\Delta I_C = 2.07\ mA$ and $\Delta V_{CE} = 4.55\ V$ from the Q-point.
Draw the ac load line through points C and Q.

The Output Voltage Swing

In Section 5-1 it was explained that the maximum symmetrical output voltage swing ($\pm V_{o(max)}$) from a common-emitter circuit depends upon the Q-point position on the dc load line. In the case of a circuit with a bypassed emitter resistor or a capacitor-coupled load, the maximum symmetrical output swing depends upon the Q-point position on the ac load line. For the ac load line shown in Fig. 6-8,

$$V_{o(max)} \approx \pm 4.5 \text{ V}$$

Practice Problem

6-2.1 Draw the new ac load line for the transistor circuit in Example 6-2 when a 5.6 kΩ external load is capacitor-coupled to the circuit output. Determine the new maximum possible symmetrical output voltage swing.

6-3 TRANSISTOR MODELS AND PARAMETERS

T-Equivalent Circuit

Because a transistor consists of two *pn*-junctions with a common centre block, it should be possible to use two *pn*-junction ac equivalent circuits as the transistor model. Figure 6-9 shows the ac equivalent circuit for a transistor connected in common-base configuration. Resistor r_e represents the BE junction resistance, r_c represents the CB junction resistance, and r_b represents the resistance of the base region which is common to both junctions. Junction capacitances C_{BE} and C_{BC} are also included.

If the transistor equivalent circuit is simply left as a combination of resistances and capacitances, it could not account for the fact that most of the emitter current flows out of the collector terminal as collector current. To represent this, a current generator is included in parallel with r_c and C_{BC}. The current generator is given the value αI_e, where $\alpha = I_c/I_e$.

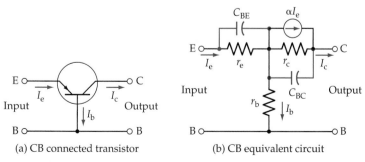

(a) CB connected transistor   (b) CB equivalent circuit

Figure 6-9   The *T*-equivalent circuit, or *r*-parameter equivalent circuit, for a transistor is essentially a combination of two *pn*-junction equivalent circuits.

The complete circuit is known as the *T-equivalent circuit*, or the *r-parameter equivalent circuit*. The equivalent circuit can be rearranged in common-emitter or common-collector configuration.

The currents in Fig. 6-9 are designated I_b, I_c, and I_e (instead of I_B, I_C, and I_E) to indicate that they are ac quantities rather than dc. The circuit parameters r_c, r_b, r_e, and α are also ac quantities.

r-Parameters

In Fig. 6-9, r_e represents the ac resistance of the forward-biased BE junction; it has a low resistance value (typically 25 Ω). The resistance of the reverse-biased CB junction (r_c) is high (typically 100 kΩ to 1 MΩ). The base region resistance (r_b) depends upon the doping density of the base material. Usually, r_b ranges from 100 Ω to 300 Ω.

C_{BE} is the capacitance of a forward-biased *pn*-junction, and C_{BC} is that of a reverse-biased junction (see Section 2-6). At medium and low frequencies the junction capacitances may be neglected. Instead of the current generator (αI_e) in parallel with r_c, a voltage generator ($\alpha I_e r_c$) may be used in series with r_c. Two transistor *r*-parameter models are shown in Fig. 6-10, one using a current generator and the other using a voltage generator.

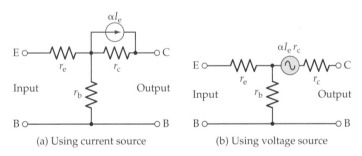

(a) Using current source   (b) Using voltage source

Figure 6-10 Transistor low-frequency *r*-parameter ac equivalent circuits.

Determination of *r*e

Because r_e is the ac resistance of the BJT forward-biased base-emitter junction, it can be determined from a plot of I_E versus V_{BE}. As illustrated in Fig. 6-11,

$$r_e = \frac{\Delta V_{BE}}{\Delta I_E} \tag{6-1}$$

This is similar to the determination of the dynamic resistance (r_d) for a forward-biased diode (see Section 2-2). As in the case of a diode, the ac resistance for the transistor BE junction can be calculated in terms of the current crossing the junction:

$$r'_e = \frac{26 \text{ mV}}{I_E} \tag{6-2}$$

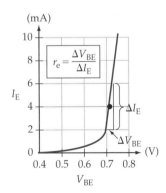

Figure 6-11 Determination of r_e from the transistor base-emitter junction forward characteristic.

Like r_d' for a diode, r_e' does not include the resistance of the device semiconductor material. Consequently, r_e' is slightly smaller than the actual measured value of r_e for a given transistor.

Equation 6-2 applies only to transistors at a temperature of 25°C. For determination of r_e' at higher or lower temperatures, the equation must be modified.

$$r_e' = \frac{26 \text{ mV}}{I_E} \left[\frac{T + 273°C}{298°C} \right] \tag{6-3}$$

h-Parameters

It has been shown that transistor circuits can be represented by an *r*-parameter or *T*-equivalent circuit. In circuits involving more than a single transistor, analysis by *r*-parameters can be virtually impossible. The *hybrid parameters* (or *h-parameters*) are much more convenient for circuit analysis. These are used only for ac circuit analysis, although dc current gain factors are also expressed as *h*-parameters. Transistor *h*-parameter models simplify transistor circuit analysis by separating the input and output stages of a circuit to be analyzed.

In Fig. 6-12 a common-emitter *h-parameter equivalent circuit* is compared with a common-emitter *r*-parameter circuit. In each case an external collector resistor (R_C) is included, as well as a signal source voltage (v_s) and source resistance (r_s). Note that the output current generator in the *r*-parameter circuit has a value of αI_e, which equals βI_b.

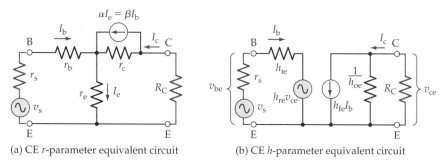

(a) CE *r*-parameter equivalent circuit   (b) CE *h*-parameter equivalent circuit

Figure 6-12 Comparison of common-emitter *r*-parameter and *h*-parameter equivalent circuits.

The input to the *h*-parameter circuit is represented as an input resistance (h_{ie}) in series with a voltage source ($h_{re}v_{ce}$). Looking at the *r*-parameter circuit, it is seen that a change in output current I_c causes a voltage variation across r_e. This means that a voltage is *fed back* from the output to the input. In the *h*-parameter circuit, this feedback voltage is represented as the portion h_{re} of the output voltage v_{ce}. The parameter h_{re} is appropriately termed the *reverse voltage transfer ratio*.

The output of the *h*-parameter circuit is represented as an output resistance ($1/h_{oe}$) in parallel with a current generator ($h_{fe}I_b$), where I_b is the (input) base current. So, $h_{fe}I_b$ is produced by the input current I_b, and it divides between the device output resistance $1/h_{oe}$ and the collector resistor R_C. I_c is the current passed to R_C. This compares with the r-parameter equivalent circuit, where some of the generator current (βI_b) flows through r_c. The current generator parameter (h_{fe}) is termed the *forward transfer current ratio*. The *output conductance* is h_{oe}, so that $1/h_{oe}$ is a resistance.

r_π Equivalent Circuit

An approximate h-parameter model for a transistor CE circuit is shown in Fig. 6-13a. In this case, the feedback generator ($h_{re}v_{ce}$ in Fig. 6-12b) is omitted. The effect of $h_{re}v_{ce}$ is normally so small that it can be neglected for most practical purposes.

The approximate h-parameter circuit is reproduced in Fig. 6-13b with the components labelled as r-parameters: $r_\pi = h_{ie}$, $\beta I_b = h_{fe}I_b$, and $r_c = 1/h_{oe}$. This circuit, known as a *hybrid-π* model, is sometimes used instead of the h-parameter circuit.

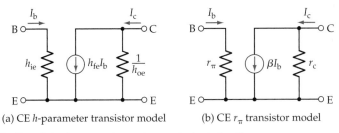

(a) CE *h*-parameter transistor model (b) CE r_π transistor model

Figure 6-13 Transistor *h*-parameter and r_π equivalent circuits.

Definition of *h*-Parameters

The e in the subscript of h_{ie} identifies the parameter as a *common-emitter* quantity, and the i signifies that it is an *input* resistance. Common-base and common-collector input resistances are designated h_{ib} and h_{ic}, respectively.

As an ac input resistance, h_{ie} can be defined as the ac input voltage divided by the ac input current.

$$h_{ie} = \frac{V_{be}}{I_b}$$

This is usually stated as

$$h_{ie} = \frac{V_{be}}{I_b}\bigg|_{V_{CE}} \tag{6-4}$$

This means that the collector-emitter voltage (V_{CE}) must remain constant when h_{ie} is measured.

The input resistance can also be defined in terms of changes in dc levels:

$$h_{ie} = \frac{\Delta V_{BE}}{\Delta I_B} \bigg|_{V_{CE}} \qquad (6\text{-}5)$$

Equation 6-5 can be used for determining h_{ie} from the transistor common-emitter input characteristics. As illustrated in Fig. 6-14, ΔV_{BE} and ΔI_B are measured at one point on the characteristics, and h_{ie} is calculated.

The reverse transfer ratio h_{re} can also be defined in terms of ac quantities, or as the ratio of dc current changes from a plot of V_{BE} versus V_{CE}. In both cases, the input current (I_B) must be held constant.

$$h_{re} = \frac{\Delta V_{BE}}{\Delta V_{CE}} \bigg|_{I_B} \qquad (6\text{-}6)$$

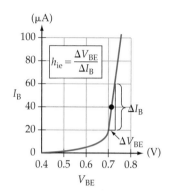

Figure 6-14 Derivation of h_{ie} from the CE input characteristics.

The forward current transfer ratio h_{fe} can similarly be defined in terms of ac quantities, or as the ratio of dc current changes. In both cases, the output voltage (V_{CE}) must be held constant.

$$h_{fe} = \frac{I_c}{I_b} \bigg|_{V_{CE}} \qquad (6\text{-}7)$$

or

$$h_{fe} = \frac{\Delta I_C}{\Delta I_B} \bigg|_{V_{CE}} \qquad (6\text{-}8)$$

Equation 6-8 can be used to determine h_{fe} from the CE current gain characteristics. Figure 6-15 shows the measurement of ΔI_C and ΔI_B at one point on the characteristics for the h_{fe} calculation.

The output conductance, h_{oe}, is the ratio of ac collector current to ac collector-emitter voltage, and its value can be determined from the common-emitter output characteristics (see Fig. 6-16).

$$h_{oe} = \frac{I_c}{V_{ce}} \bigg|_{I_B} \qquad (6\text{-}9)$$

or

$$h_{oe} = \frac{\Delta I_C}{\Delta V_{CE}} \bigg|_{I_B} \qquad (6\text{-}10)$$

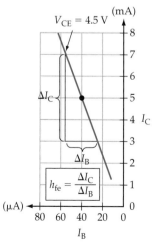

Figure 6-15 Derivation of h_{fe} from the CE current gain characteristics.

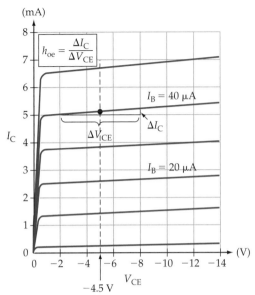

Figure 6-16 Derivation of h_{oe} from the CE output characteristics.

Example 6-3

Determine h_{fe} and h_{oe} for $V_{CE} = 4.5$ V and $I_B = 40$ μA from the transistor characteristics in Figs. 6-15 and 6-16.

Solution

From the current gain characteristics at $V_{CE} = 4.5$ V and $I_B = 40$ μA,

$$\Delta I_C = 4\,\text{mA} \quad \text{and} \quad \Delta I_B \approx 30\,\mu\text{A}$$

Eq. 6-8:
$$h_{fe} = \frac{\Delta I_C}{\Delta I_B} = \frac{4\,\text{mA}}{30\,\mu\text{A}}$$
$$= 133$$

From the output characteristics at $V_{CE} = 4.5$ V and $I_B = 40$ μA,

$$\Delta I_C \approx 400\,\mu\text{A} \quad \text{and} \quad \Delta V_{CE} = 6\,\text{V}$$

Eq. 6-9:
$$h_{oe} = \frac{\Delta I_C}{\Delta V_{CE}} = \frac{0.2\,\text{mA}}{6\,\text{V}}$$
$$= 33.3\,\mu\text{S (micro Siemens)}$$

$$1/h_{oe} = \frac{1}{33.3\,\mu\text{S}}$$
$$= 30\,\text{k}\Omega$$

Common-Base and Common-Collector *h*-Parameters

The common-base and common-collector *h*-parameters are defined similarly to common-emitter *h*-parameters. They may also be derived from the CB and CC characteristics. Common-base parameters are identified as h_{ib}, h_{fb}, etc., and common-collector parameters are designated h_{ic}, h_{fc}, and so on. Common-emitter, common-base, and common-collector h-parameter equivalent circuits are further investigated in Sections 6-4 to 6-8.

Parameter Relationships

Device manufacturers do not list the values of all parameters on transistor data sheets. Usually, only the CE h-parameters are stated. However, CB and CC h-parameters can be determined from the CE h-parameters. r-Parameters can also be calculated from the CE h-parameters. Table 6-1 shows the parameter relationships.

TABLE 6-1   Conversion from CE h-parameters to CB and CC h-parameters and to r-parameters

CE to CB h-parameters	CE to CC h-parameters	CE h-parameters to r-parameters
$h_{ib} \approx \dfrac{h_{ie}}{1 + h_{fe}}$	$h_{ic} = h_{ie}$	$\alpha \approx \dfrac{h_{fe}}{1 + h_{fe}}$
$h_{rb} \approx \dfrac{h_{ie} h_{oe}}{1 + h_{fe}} - h_{re}$	$h_{rc} = 1 - h_{re}$	$r_c = \dfrac{1 + h_{fe}}{h_{oe}}$
$h_{fb} \approx \dfrac{h_{fe}}{1 + h_{fe}}$	$h_{fc} = 1 + h_{fe}$	$r_e = \dfrac{h_{ie}}{1 + h_{fe}} \approx r_e'$
$h_{ob} \approx \dfrac{h_{oe}}{1 + h_{fe}}$	$h_{oc} = h_{oe}$	$r_b = h_{ie} - \dfrac{h_{re}(1 + h_{fe})}{h_{oe}}$
		$r_\pi = h_{ie}$
		$\beta = h_{fe}$

Example 6-4

Calculate h_{fc}, h_{ob}, and α from the parameters determined in Example 6-3.

Solution

From Table 6-1:
$$h_{fc} = 1 + h_{fe} = 1 + 133$$
$$= 134$$

$$h_{ob} \approx \frac{h_{oe}}{1 + h_{fe}} \approx \frac{33.3\ \mu S}{1 + 133}$$
$$= 249\ mS$$

$$\alpha \approx \frac{h_{fe}}{1 + h_{fe}} \approx \frac{133}{1 + 133}$$
$$= 0.993$$

Example 6-5

A transistor in a circuit has its current levels measured as follows: $I_B = 20 \ \mu A$ and $I_C = 1$ mA. Estimate the CE input resistance. Determine r_π and β.

Solution

Eq. 6-2:
$$r'_e = \frac{26 \ mV}{I_E} \approx \frac{26 \ mV}{1 \ mA}$$
$$= 26 \ \Omega$$

$$h_{fe} \approx \frac{I_C}{I_B} \approx \frac{1 \ mA}{20 \ \mu A}$$
$$= 50$$

$$h_{ie} = (1 + h_{fe}) \ r'_e = (1 + 50) \times 26 \ \Omega$$
$$\approx 1.33 \ k\Omega$$

$$r_\pi = h_{ie} = 1.33 \ k\Omega$$

$$\beta = h_{fe} = 50$$

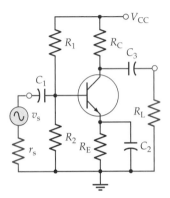

Figure 6-17 Transistor common-emitter amplifier with coupling and bypass capacitors.

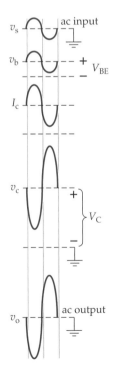

Figure 6-18 Voltage and current waveforms in a common-emitter amplifier.

Practice Problems

6-3.1 Determine h_{fe} and h_{oe} at $V_{CE} = 6$ and $I_B = 20 \ \mu A$ from the common-emitter output and current gain characteristics in Fig. 4-30.

6-3.2 Calculate α and r_c from the h-parameter values determined in Problem 6-3.1.

6-3.3 Calculate h_{ie} for a transistor at a temperature of 50°C when the emitter current is 1.3 mA and $h_{fe} = 80$.

6-4 COMMON-EMITTER CIRCUIT ANALYSIS

Common-Emitter Circuit

Consider the transistor amplifier circuit shown in Fig. 6-17. When the capacitors are regarded as ac short circuits, it is seen that the circuit input terminals are the transistor base and emitter, and the output terminals are the collector and the emitter. So, the emitter terminal is *common* to both input and output, and the circuit configuration is termed *common-emitter* (CE).

The current and voltage waveforms for the CE circuit in Fig. 6-17 are illustrated in Fig. 6-18. It is seen that there is a 180° *phase shift* between the input and output waveforms. This can be understood by considering the effect of a positive-going input signal. When v_s increases in a positive direction, it increases the transistor base-emitter voltage (V_{BE}). The increase in V_{BE} raises the level of I_C, thereby increasing the voltage drop across R_C, and thus reducing the level of the collector voltage (V_C). The changing level of V_C is capacitor-coupled to the circuit output to produce the ac output voltage (v_o). As v_s increases in a positive direction, v_o goes in a negative direction, as illustrated. Similarly, when v_s changes in a negative direction, the resultant decrease in V_{BE} reduces the I_C level, thereby reducing V_{RC} and producing a positive-going output.

The circuit in Fig. 6-17 has an *input impedance* (Z_i), and an *output impedance* (Z_o). These can cause voltage division of the circuit input and output voltages, as illustrated in Fig. 6-19. So, for most transistor circuits, Z_i and Z_o are important parameters. The circuit voltage amplification (A_v), or *voltage gain*, depends on the transistor parameters and on resistors R_C and R_L.

h-Parameter Equivalent Circuit

The first step in ac analysis of a transistor circuit is to draw the ac equivalent circuit, by substituting short-circuits in place of the power supply and capacitors. When this is done for the circuit in Fig. 6-17, it gives the ac equivalent circuit shown in Fig. 6-20a (reproduced from Fig. 6-6b). The *h*-parameter circuit is now drawn simply by replacing the transistor in the ac equivalent circuit with its *h*-parameter model (from Fig. 6-12b). This gives the CE *h*-parameter equivalent circuit in Fig. 6-20b. Note that the feedback voltage generator ($h_{re}v_c$) in Fig. 6-12b is omitted in Fig. 6-20b. As discussed, the effect of $h_{re}v_c$ in a CE circuit is unimportant for most practical purposes.

The current directions and voltage polarities in Fig. 6-20b are those that occur when the instantaneous level of the input voltage is moving in a positive direction.

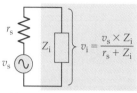

(a) Division of signal voltage

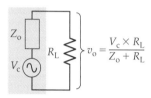

(b) Division of output voltage

Figure 6-19 Circuit input and output voltages are divided by the effects of Z_i and Z_o.

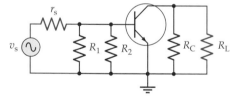

(a) ac equivalent circuit for CE transistor circuit

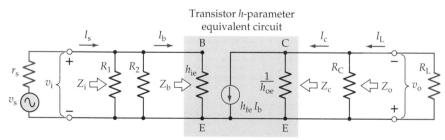

(b) *h*-parameter equivalent circuit for CE circuit

Figure 6-20 A common-emitter *h*-parameter circuit is drawn by first replacing the power supply and all capacitors with short-circuits; then the *h*-parameter model is substituted for the transistor.

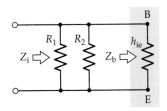

Figure 6-21 The transistor input impedance in a CE circuit is Z_b, and the circuit input impedance is $Z_b \| R_1 \| R_2$.

Input Impedance

The input section of the h-parameter circuit in Fig. 6-20b, reproduced in Fig. 6-21, shows that the input impedance at the transistor terminals is

$$Z_b \approx h_{ie} \tag{6-11}$$

A typical value of h_{ie} for a low-current transistor is 1.5 kΩ. If h_{ie} is not known, it can be estimated by first calculating r_e' from Eq. 6-2 or Eq. 6-3, as appropriate. Then $r_e' = h_{ib}$, and $h_{ie} = (1 + h_{fe})h_{ib}$ Z_b is the input impedance at the device base terminal. At the circuit input terminals, resistors R_1 and R_2 are seen to be in parallel with Z_b. So the circuit input impedance is

$$Z_i = R_1 \| R_2 \| Z_b \tag{6-12}$$

Output Impedance

'Looking into' the output of the CE h-parameter circuit reproduced in Fig. 6-22, that the output impedance at the transistor collector is

$$Z_c \approx 1/h_{oe} \tag{6-13}$$

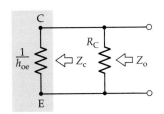

Figure 6-22 The output impedance for a CE circuit is $R_C \| (1/h_{oe})$.

At the circuit output terminals, resistor R_C is in parallel with Z_c. So the circuit output impedance is

$$Z_o = R_C \| (1/h_{oe}) \tag{6-14}$$

Because $1/h_{oe}$ is typically 50 kΩ and R_C is usually much less than 50 kΩ, the circuit output impedance is approximately equal to R_C. Using this information, it is possible to tell the output impedance of a CE circuit simply by reading the resistance of R_C.

Voltage Gain

The circuit voltage gain is given by the equation

$$A_v = \frac{v_o}{v_i}$$

Figures 6-23a and b, reproduced from Fig. 6-20, show that

$$v_i = I_b h_{ie} \quad \text{and} \quad v_o = -I_c (R_C \| R_L)$$

and so

$$A_v = \frac{-I_c(R_C \| R_L)}{I_b h_{ie}}$$

I_c/I_b can be replaced with h_{fe}. The voltage gain equation now becomes

$$A_v = \frac{-h_{fe}(R_C \| R_L)}{h_{ie}} \tag{6-15}$$

The minus sign in Equation 6-15 indicates that v_o is 180° out of phase with v_i. (When v_i increases, v_o decreases, and vice versa.) If the appropriate h-parameters

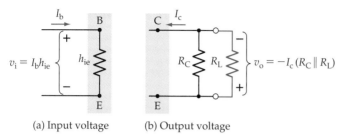

(a) Input voltage   (b) Output voltage

Figure 6-23   Derivation of equations for v_i and v_o in a CE circuit.

as well as R_C and R_L are known, the voltage gain of a CE circuit can be quickly estimated. If the values $R_C = 4.7$ kΩ, $h_{fe} = 70$, and $h_{ie} = 1.5$ kΩ are used, and $R_L \gg R_C$ is assumed, a typical CE voltage gain is -220.

Substituting from Table 6-1, the CE voltage gain equation can be rewritten as

$$A_v \approx \frac{-(R_C \| R_L)}{h_{ib}} \qquad (6\text{-}16)$$

Since h_{ib} equals r'_e, Eq. 6-16 can be restated:

$$A_v \approx \frac{-(R_C \| R_L)}{r'_e}$$

Current Gain

The transistor ac current gain is

$$h_{fe} = \frac{I_c}{I_b} \qquad (6\text{-}17)$$

This is the device current gain, *not* the circuit current gain. Examination of the circuit in Fig. 6-20 shows that the signal current (I_s) is divided between h_{ie} and the bias resistors [$R_B = (R_1 \| R_2)$]. Also, the output current from the transistor collector (I_c) is divided between R_C and R_L. The overall circuit current gain can be shown to be

$$A_i = \frac{h_{fe} R_C R_B}{(R_C + R_L)(R_C + h_{ie})} \qquad (6\text{-}18)$$

In most practical applications, the current gain of a transistor amplifier is normally *not* an important quantity.

Power Gain

The power gain of an amplifier can be shown to be

$$A_p = A_v \times A_i \qquad (6\text{-}19)$$

Like current gain, the power gain of a transistor amplifier is largely unimportant for most practical applications.

Summary of Typical CE Circuit Performance

Device input impedance	$Z_b \approx h_{ie}$
Circuit input impedance	$Z_i = R_1 \| R_2 \| Z_b$
Device output impedance	$Z_c \approx 1/h_{oe}$
Circuit output impedance	$Z_o \approx R_C \| (1/h_{oe}) \approx R_C$
Circuit voltage gain	$A_v = \dfrac{-h_{fe}(R_C \| R_L)}{h_{ie}} \approx \dfrac{-(R_C \| R_L)}{r'_e}$

The common-emitter circuit has good voltage gain, with 180° phase shift, medium input impedance, and relatively high output impedance. As a voltage amplifier, the CE circuit is by far the most often used of the three basic circuit configurations.

Example 6-6

The transistor in the CE circuit in Fig. 6-24 has the following parameters: $h_{ie} = 2.1$ kΩ, $h_{fe} = 75$, and $h_{oe} = 1$ µS. Calculate the circuit input impedance, output impedance, and voltage gain. This circuit is designed in Example 5-14.

Solution

From Eq. 6-11 and Eq. 6-12,

$$Z_i = R_1 \| R_2 \| h_{ie} = 68 \text{ k}\Omega \| 56 \text{ k}\Omega \| 2.1 \text{ k}\Omega$$
$$= 1.97 \text{ k}\Omega$$

Eq. 6-14:
$$Z_o = R_C \| (1/h_{oe}) = 3.9 \text{ k}\Omega \| [1/(1 \text{ µS})]$$
$$\approx 3.9 \text{ k}\Omega$$

Eq. 6-15:
$$A_v = \frac{-h_{fe}(R_C \| R_L)}{h_{ie}} = \frac{-75 \times (3.9 \text{ k}\Omega \| 82 \text{ k}\Omega)}{2.1 \text{ k}\Omega}$$

$$= -133$$

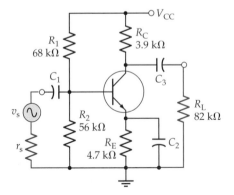

Figure 6-24 CE circuit for Example 6-6 (designed in Example 5-14).

Example 6-7

In a CE circuit (as in Fig. 6-24), $I_C = 1.5$ mA, $R_C = 4.7$ kΩ, and $R_L = 56$ kΩ. Estimate r'_e, and calculate the circuit voltage gain.

Solution

Eq. 6-2:
$$r'_e = \frac{26 \text{ mV}}{I_C} \approx \frac{26 \text{ mV}}{1.5 \text{ mA}}$$

$$= 17.3 \ \Omega$$

Eq. 6-16:
$$A_v \approx \frac{-(R_C \| R_L)}{r'_e} = \frac{-(4.7 \text{ k}\Omega \| 56 \text{ k}\Omega)}{17.3 \ \Omega}$$

$$= -250$$

Practice Problems

6-4.1 Calculate Z_i, Z_o, and A_v for a CE circuit (as in Fig. 6-24) with the following components and parameters: $R_1 = 18$ kΩ, $R_2 = 8.2$ kΩ, $R_C = 5.6$ kΩ, $R_E = 2.7$ kΩ, $R_L = 68$ kΩ, $h_{ie} = 1$ kΩ, $h_{fe} = 100$, and $h_{oe} = 1.67$ μS.

6-4.2 A CE circuit has a 0.8 mA emitter current, a 6.8 kΩ collector resistor, and a 68 kΩ capacitor-coupled load. Calculate the circuit voltage gain at 25°C and at 80°C.

6-5 CE CIRCUIT WITH UNBYPASSED EMITTER RESISTOR

h-Parameter Equivalent Circuit

When an unbypassed emitter resistor (R_E) is present in a CE circuit, as in Fig. 6-25a, it is also present in the ac equivalent circuit (Fig. 6-25b). R_E must also be shown in the *h*-parameter circuit between the transistor emitter terminal

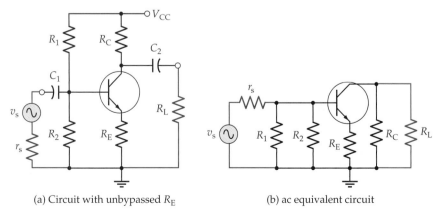

(a) Circuit with unbypassed R_E (b) ac equivalent circuit

Figure 6-25   CE circuit with unbypassed emitter resistor, and ac equivalent circuit.

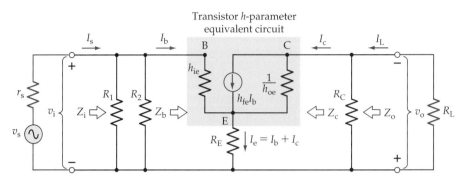

Figure 6-26 *h*-Parameter equivalent circuit for CE circuit with unbypassed emitter resistor.

and the circuit common input-output terminal (Fig. 6-26). As with the previous *h*-parameter circuit, the current directions and voltage polarities shown are those that occur when the instantaneous input voltage is positive-going. The presence of R_E without a bypass capacitor significantly affects the circuit input impedance and voltage gain.

Input Impedance

An equation for the input impedance at the transistor base can be determined from v_i and I_b. From Fig. 6-26,

$$v_i = I_b h_{ie} + I_e R_E$$
$$= I_b h_{ie} + R_E(I_b + I_c)$$
$$= I_b h_{ie} + R_E(I_b + h_{fe} I_b)$$
$$= I_b[h_{ie} + R_E(1 + h_{fe})]$$

and
$$\frac{v_i}{I_b} = h_{ie} + R_E(1 + h_{fe})$$

or
$$Z_b = h_{ie} + R_E(1 + h_{fe}) \qquad (6\text{-}20)$$

Equation 6-20 gives the input impedance at the transistor base terminal. The circuit input impedance is again given by Eq. 6-12:

$$Z_i = R_1 \| R_2 \| Z_b$$

Examination of Eq. 6-20 shows that it is possible to estimate very quickly the input impedance at the transistor base in a CE circuit with an unbypassed emitter resistor. For example, a circuit with $R_E = 1\,\text{k}\Omega$ and $h_{fe} = 100$ has $Z_b \approx 100\,\text{k}\Omega$.

Output Impedance

The output impedance at the transistor collector can be shown to be substantially increased (above $1/h_{oe}$) by the presence of the unbypassed emitter resistor. Consequently, the circuit output impedance should be taken as

$$Z_o \approx R_C$$

Voltage Gain

From the derivation of the input impedance equation,

$$v_i = I_b[h_{ie} + R_E(1 + h_{fe})]$$

and

$$v_o = -I_c(R_C \| R_L)$$

So

$$A_v = \frac{v_o}{v_i} = \frac{-I_c(R_C \| R_L)}{I_b[h_{ie} + R_E(1 + h_{fe})]}$$

giving

$$A_v = \frac{-h_{fe}(R_C \| R_L)}{h_{ie} + R_E(1 + h_{fe})} \qquad (6\text{-}21)$$

This reduces to

$$A_v = \frac{-(R_C \| R_L)}{r'_e + R_E}$$

Usually,

$$R_E \gg r'_e$$

so

$$A_v \approx \frac{-(R_C \| R_L)}{R_E} \qquad (6\text{-}22)$$

The voltage gain of a CE circuit with an unbypassed emitter resistor can be quickly estimated from Eq. 6-22. For the circuit in Fig. 6-25a, with $R_C = 4.7$ kΩ, $R_E = 1$ kΩ, and $R_L \gg R_C$, the gain is $A_v \approx -4.7$.

Performance Summary for CE Circuit with Unbypassed R_E

Device input impedance $Z_b = h_{ie} + R_E(1 + h_{fe}) \approx h_{fe} R_E$

Circuit input impedance $Z_i \approx R_B \| Z_b$

$$= R_1 \| R_2 \| Z_b$$

Circuit output impedance $Z_o \approx R_C$

Circuit voltage gain $A_v = \frac{-h_{fe}(R_C \| R_L)}{h_{ie} + R_E(1 + h_{fe})}$

Circuit voltage gain $A_v \approx \frac{-(R_C \| R_L)}{R_E}$

The most significant feature of the performance of a CE circuit with an unbypassed emitter resistor is that its voltage gain is much lower than it would be normally. Its input impedance is also much higher than Z_i for a CE circuit that has R_E bypassed.

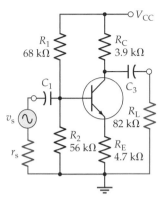

Figure 6-27 CE circuit for Example 6-8.

Example 6-8

The CE circuit in Fig. 6-27 is the same as in Fig. 6-24, but with bypass capacitor C_2 omitted. The transistor parameters are $h_{ie} = 2.1$ kΩ, $h_{fe} = 75$, and $h_{oe} = 1$ μS (as in Example 6-6). Calculate the circuit input and output impedance and voltage gain.

Solution

Eq. 6-20:
$$Z_b = h_{ie} + R_E(1 + h_{fe}) = 2.1 \text{ k}\Omega + 4.7 \text{ k}\Omega(1 + 75)$$
$$= 359 \text{ k}\Omega$$

Eq. 6-12:
$$Z_i = R_1 \| R_2 \| Z_b = 68 \text{ k}\Omega \| 56 \text{ k}\Omega \| 359 \text{ k}\Omega$$
$$= 28.3 \text{ k}\Omega$$

$$Z_o \approx R_c = 3.9 \text{ k}\Omega$$

Eq. 6-21:
$$A_v = \frac{-h_{fe}(R_C \| R_L)}{h_{ie} + R_E(1 + h_{fe})} = \frac{-75 \times (3.9 \text{ k}\Omega \| 82 \text{ k}\Omega)}{2.1 \text{ k}\Omega + 4.7 \text{ k}\Omega(1 + 75)}$$
$$= -0.78$$

Practice Problem

6-5.1 Calculate Z_i, Z_o, and A_v for a CE circuit with an unbypassed emitter resistor (as in Fig. 6-27) with the following component values: $R_1 = 18$ kΩ, $R_2 = 8.2$ kΩ, $R_C = 5.6$ kΩ, $R_E = 2.7$ kΩ, $R_L = 68$ kΩ, $h_{ie} = 1$ kΩ, $h_{fe} = 100$, and $h_{oe} = 1.66$ μS.

6-6 COMMON-COLLECTOR CIRCUIT ANALYSIS

Common-Collector Circuit

In the *common-collector* (CC) circuit shown in Fig. 6-28, the external load (R_L) is capacitor-coupled to the transistor emitter terminal. The circuit uses

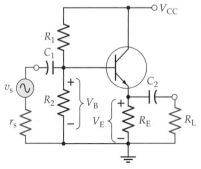

Figure 6-28 Transistor common-collector amplifier circuit, also called an emitter follower.

voltage-divider bias to derive the transistor base voltage (V_B) from the supply. The transistor collector terminal is directly connected to V_{CC}; no collector resistor is used. The circuit output voltage is developed across the emitter resistor (R_E), and there is no bypass capacitor.

To understand the operation of a CC circuit, note that V_B is a constant quantity and that $V_E = V_B - V_{BE}$. When a signal is applied via C_1 to the transistor base, V_B increases and decreases as the signal goes positive and negative. If the signal voltage (v_s) increases to $+0.5$ V, V_B is increased by 0.5 V. V_E also increases by 0.5 V, because V_{BE} remains substantially constant, and $V_E = V_B - V_{BE}$. The change in V_E is coupled via C_2 to give an ac output voltage ($v_o = 0.5$ V) (see the waveforms in Fig. 6-29). Similarly, when v_s decreases to -0.5 V, both V_B and V_E decrease by 0.5 V, giving $v_o = -0.5$ V.

It is seen that the ac output voltage from a CC circuit is essentially the same as the input voltage; there is no voltage gain or phase shift. Thus, the CC circuit can be said to have a voltage gain of 1. (Actually, the output voltage can be shown to be slightly smaller than the input because of a very small change in V_{BE}.)

The fact that the CC output voltage follows the changes in signal voltage gives the circuit its other name: *emitter follower*.

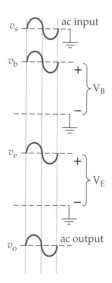

Figure 6-29 Voltage waveforms in a common-collector amplifier circuit.

h-Parameter Equivalent Circuit

As for a CE circuit, the power supply and capacitors must be replaced with short circuits to study the CC ac performance. This gives the CC ac equivalent circuit in Fig. 6-30a. The input terminals of the ac equivalent circuit are seen to be the transistor base and collector, and the output terminals are the emitter and collector. Because the collector terminal is common to both input and output, the circuit configuration is named *common-collector*.

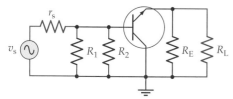

(a) ac equivalent circuit for CC transistor circuit

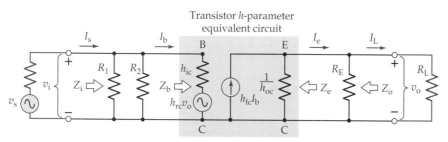

(b) *h*-parameter equivalent circuit for CC circuit

Figure 6-30 Common-collector amplifier ac equivalent circuit and *h*-parameter circuit.

The CC h-parameter circuit is now drawn by substituting the transistor h-parameter model into the ac equivalent circuit, to give the circuit in Fig. 6-30b. The current directions and voltage polarities indicated in Fig. 6-30b are, once again, those that are produced by a positive-going signal voltage. It should be noted that $h_{rc} = 1$ for a CC circuit; all of v_o is fed back to the input. So, unlike the case of a CE circuit, the feedback generator cannot be omitted in the equivalent circuit of a CC amplifier.

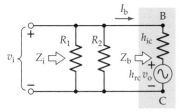

Figure 6-31 Input impedance determination for a CC circuit.

Input Impedance

The input impedance for the CC circuit is determined by first writing an equation for the input voltage. Referring to Fig. 6-30 and Fig. 6-31,

$$v_i = I_b h_{ic} + h_{rc} v_o$$
$$= I_b h_{ic} + I_e(R_E \| R_L)$$
$$= I_b h_{ic} + h_{fc} I_b(R_E \| R_L)$$
$$= I_b[h_{ic} + h_{fc}(R_E \| R_L)]$$

which gives

$$\frac{v_i}{I_b} = h_{ic} + h_{fc}(R_E \| R_L)$$

or

$$Z_b = h_{ic} + h_{fc}(R_E \| R_L) \qquad (6\text{-}23)$$

Equation 6-23 is similar to the equation for the transistor input impedance in a CE circuit with an unbypassed emitter resistor (see Eq. 6-20), except that R_L is now in parallel with R_E. The circuit input impedance is again given by Eq. 6-12,

$$Z_i = R_1 \| R_2 \| Z_b$$

Using Eq. 6-23, we can quickly estimate the input impedance at the transistor base in a CC circuit. A circuit with $R_E = 1\text{ k}\Omega$ and $h_{fe} = 100$ has $Z_b \approx 100\text{ k}\Omega$ if $R_L \gg R_E$.

Output Impedance

As already explained, the ac voltage at the output of a CC circuit is all fed back to the input. This fact is used in determining the output impedance (Z_e) at the emitter terminal. The signal voltage is assumed to be zero, and v_o is used to calculate I_e. With $v_s = 0$, I_b is produced by the fed-back voltage ($h_{rc} v_o = v_o$) (see Fig. 6-32a):

$$I_b = \frac{v_o}{h_{ic} + (R_1 \| R_2 \| r_s)}$$

and

$$I_e = h_{fc} I_b = \frac{h_{fc} v_o}{h_{ic} + (R_1 \| R_2 \| r_s)}$$

So

$$\frac{v_o}{I_e} = \frac{[h_{ic} + (R_1 \| R_2 \| r_s)]}{h_{fc}} \qquad \text{(see Fig. 6-32b)}$$

giving

$$Z_e = \frac{h_{ic} + (R_1 \| R_2 \| r_s)}{h_{fc}} \qquad (6\text{-}24)$$

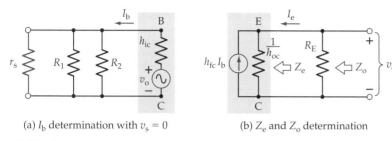

(a) I_b determination with $v_s = 0$ (b) Z_e and Z_o determination

Figure 6-32 Determination of output impedance for a CC circuit.

Note that the output impedance at the emitter terminal is

$$\frac{h_{ic} + (Z \text{ in series with the base terminal})}{h_{fc}}$$

It is interesting to compare this to the base input impedance (Eq. 6-23), which is

$$h_{ic} + h_{fc} (Z \text{ in series with the emitter terminal})$$

Equation 6-24 gives the device output impedance. (Actually, $1/h_{oc}$ should be included, but it has a negligible effect on Z_e.) The circuit output impedance also involves R_E (see Fig. 6-32b).

$$Z_o = Z_e \| R_E \qquad (6\text{-}25)$$

R_E is usually much larger than Z_e, so that $Z_o \approx Z_e$.

Voltage Gain

As already explained, the ac output voltage from a CC circuit is almost exactly equal to the input voltage. However, the precise equation for A_v is easily derived by considering the ac output and input voltages as determined from Fig. 6-30b:

$$v_i = I_b[h_{ic} + h_{fc}(R_E \| R_L)]$$

and
$$v_o = I_e(R_E \| R_L)$$

So
$$A_v = \frac{v_o}{v_i} = \frac{I_e(R_E \| R_L)}{I_b[h_{ic} + h_{fc}(R_E \| R_L)]}$$

This reduces to
$$A_v = \frac{(R_E \| R_L)}{h_{ib} + (R_E \| R_L)} \qquad (6\text{-}26)$$

Recall that $r'_e \approx h_{ib}$. Usually, $R_E \| R_L$ is so much larger than h_{ib} (or r'_e) that the CC circuit voltage gain is simply taken as

$$A_v \approx 1 \qquad (6\text{-}27)$$

Summary of CC Circuit Performance

Device input impedance $Z_b = h_{ic} + h_{fc}(R_E \| R_L)$

Circuit input impedance $Z_i \approx R_B \| Z_b = R_1 \| R_2 \| Z_b$

Device output impedance $Z_e = \dfrac{h_{ic} + (R_1 \| R_2 \| r_s)}{h_{fc}}$

Circuit output impedance $Z_o = Z_e \| R_E$

Circuit voltage gain $A_v = \dfrac{(R_E \| R_L)}{h_{ib} + (R_E \| R_L)} \approx 1$

A CC circuit has a voltage gain of 1, no phase shift between input and output, high input impedance, and low output impedance. Because of its high Z_i, low Z_o, and unity gain, the CC circuit is normally used as a *buffer amplifier*, placed between a high impedance signal source and a low impedance load.

Example 6-9

The transistor in the CC circuit in Fig. 6-33 has the following parameters: $h_{ie} = 2.1$ kΩ and $h_{fe} = 75$. (a) Calculate the circuit input and output impedance with R_L not connected. (b) Calculate Z_i and A_v with R_L connected.

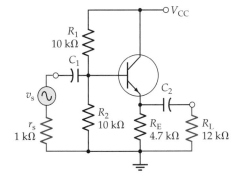

Figure 6-33 CC circuit for Example 6-9.

Solution

(a) R_L *not connected*

From Table 6-1, $h_{ic} = h_{ie} = 2.1$ kΩ

and $h_{fc} = 1 + h_{fe} = 76$

Eq. 6-23: $Z_b = h_{ic} + h_{fc}(R_E \| R_L) = 2.1$ k$\Omega + 76(4.7$ k$\Omega)$

            $= 359.3$ kΩ

Eq. 6-12: $Z_i = R_1 \| R_2 \| Z_b = 10$ k$\Omega \| 10$ k$\Omega \| 359.3$ kΩ

            $= 4.93$ kΩ

Eq. 6-24: $Z_e = \dfrac{h_{ic} + (R_1 \| R_2 \| r_s)}{h_{fc}} = \dfrac{2.1\text{ k}\Omega + (10\text{ k}\Omega \| 10\text{ k}\Omega \| 1\text{ k}\Omega)}{76}$

            $= 38.6$ Ω

Eq. 6-25: $Z_o = Z_e \| R_E = 38.6\ \Omega \| 4.7\ k\Omega$

$= 38.3\ \Omega$

(b) R_L connected

Eq. 6-23: $Z_b = h_{ic} + h_{fc}(R_E \| R_L) = 2.1\ k\Omega + 76(4.7\ k\Omega \| 12\ k\Omega)$

$= 258.8\ k\Omega$

Eq. 6-12: $Z_i = R_1 \| R_2 \| Z_b = 10\ k\Omega \| 10\ k\Omega \| 258.8\ k\Omega$

$= 4.91\ k\Omega$

From Table 6-1, $h_{ib} = \dfrac{h_{ie}}{1 + h_{fe}} = \dfrac{2.1\ k\Omega}{1 + 75}$

$= 27.6\ \Omega$

Eq. 6-26: $A_v = \dfrac{(R_E \| R_L)}{h_{ib} + (R_E \| R_L)} = \dfrac{(4.7\ k\Omega \| 12\ k\Omega)}{27.6\ \Omega + (4.7\ k\Omega \| 12\ k\Omega)}$

$= 0.99$

Practice Problem

6-6.1 Calculate Z_i and Z_o for a CC circuit (as in Fig. 6-33) with the following component values and parameters: $R_1 = 56\ k\Omega$, $R_2 = 39\ k\Omega$, $R_E = 5.6\ k\Omega$, $R_L = 22\ k\Omega$, $r_s = 400\ \Omega$, $h_{ie} = 1\ k\Omega$, and $h_{fe} = 100$.

6-7 COMMON-BASE CIRCUIT ANALYSIS

Common-Base Circuit

The common-base (CB) circuit in Fig. 6-34 is very similar to a CE circuit, except that the input signal is applied to the transistor emitter terminal (via C_2) instead of to the base. Also, there is no bypass capacitor across the R_E, but the

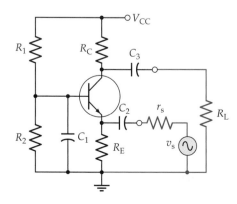

Figure 6-34 In a common-base circuit, the signal is capacitor-coupled to the transistor emitter terminal, and the output is taken from the collector.

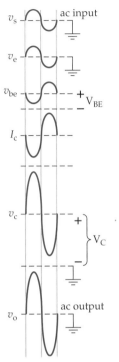

Figure 6-35 Voltage and current waveforms for a common-base amplifier circuit.

base terminal is ac grounded via capacitor C_1. Because the base is ac grounded, all of the signal voltage is developed across the transistor base-emitter junction. As for a CE circuit, the output voltage is developed across the collector resistor (R_C).

A positive-going signal voltage at the input of a CB circuit pushes the transistor emitter in a positive direction while the base voltage remains fixed. Thus, a positive-going signal reduces the base-emitter voltage (see the waveforms in Fig. 6-35). The reduction in V_{BE} reduces the transistor collector current, and the I_C reduction reduces V_{RC} and, consequently, causes the transistor collector voltage to rise. The rise in V_C is effectively a rise in the circuit output voltage (v_o). It is seen that a positive-going input voltage produces a positive-going output. It can also be demonstrated that a negative-going input produces a negative-going output. There is no phase shift from input to output in a CB circuit.

h-Parameter Equivalent Circuit

The CB ac equivalent circuit is drawn, as always, by replacing the supply voltage and capacitors with short circuits. This gives the circuit in Fig. 6-36a, which shows that the transistor base terminal (grounded via capacitor C_1) is common to both input and output. Hence the name *common base*.

The CB *h*-parameter circuit is now drawn by substituting the transistor *h*-parameter model into the ac equivalent circuit, giving the circuit in Fig. 6-36b. Once again, the current directions and voltage polarities indicated in the *h*-parameter circuit are those produced by a positive-going signal voltage. Note that the feedback voltage generator ($h_{rb} v_o$) is not included in the CB *h*-parameter circuit. This is because the feedback voltage effect is so small that it can be neglected when we are deriving practical approximate equations for the

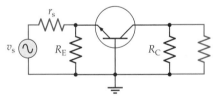

(a) ac equivalent circuit for CB transistor circuit

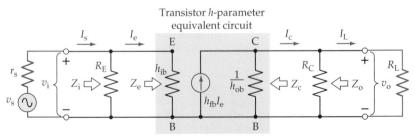

(b) *h*-parameter equivalent circuit for CB circuit

Figure 6-36 Common-base amplifier, ac equivalent circuit, and *h*-parameter circuit.

circuit performance. This corresponds with the CE h-parameter circuit, but *not* with the CC h-parameter circuit, where the feedback voltage is very important.

Input Impedance

Fig. 6-36b shows that the input impedance to the transistor emitter is simply h_{ib}.

$$Z_e = h_{ib} \tag{6-28}$$

Typically h_{ib} is around 21 Ω for a low-current transistor ($h_{ib} \approx r'_e$, see Section 6-3). Resistor R_E is in parallel with Z_e, so the circuit input impedance is

$$Z_i = Z_e \| R_E \tag{6-29}$$

R_E is usually much larger than Z_e, so $Z_i \approx Z_e$.

Output Impedance

'Looking into' the collector-base terminals of the CB h-parameter circuit in Fig. 6-36b, we can see a very large resistance ($1/h_{ob}$).

$$Z_c = 1/h_{ob} \tag{6-30}$$

Once again, Z_c is the device output impedance, and the circuit output impedance is R_C in parallel with Z_c.

$$Z_o = Z_c \| R_c \tag{6-31}$$

Because Z_c is normally much larger than R_C, $Z_o \approx R_C$.

As in the case of a CE circuit, the output impedance of a CB circuit can be determined by just reading the resistance of R_C.

Voltage Gain

The input voltage equation for the circuit in Fig. 6-36b is

$$v_i = I_e h_{ib}$$

and the output voltage equation is

$$v_o = I_c(R_C \| R_L)$$
$$= h_{fb} I_e (R_C \| R_L)$$

So

$$\frac{v_o}{v_i} = \frac{h_{fb} I_e (R_C \| R_L)}{I_e h_{ib}}$$

giving

$$A_v = \frac{h_{fb}(R_C \| R_L)}{h_{ib}} \tag{6-32}$$

Referring to Table 6-1, we can show Eq. 6-32 to be exactly the same as the common-emitter voltage gain,

$$A_v = \frac{h_{fe}(R_C \| R_L)}{h_{ie}}$$

So, similar common-base and common-emitter circuits have equal voltage gains.

Effect of Unbypassed Base Resistors

If capacitor C_1 is not present in the CB circuit in Fig. 6-34, the transistor base is *not* ac short-circuited to ground. So, a resistance ($R_B = R_1 \| R_2$) must be shown in the ac equivalent circuit (Fig. 6-37a) and in the *h*-parameter circuit (Fig. 6-37b). The presence of the unbypassed base resistors can substantially affect the transistor input impedance and the circuit voltage gain. Analysis of the *h*-parameter circuit shows that

$$Z_e = h_{ib} + R_B(1 - h_{fb}) \tag{6-33}$$

and

$$A_v = \frac{h_{fb}(R_C \| R_L)}{h_{ib} + R_B(1 - h_{fb})} \tag{6-34}$$

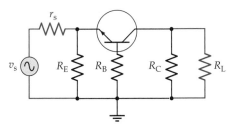

(a) ac equivalent for CB circuit with unbypassed base

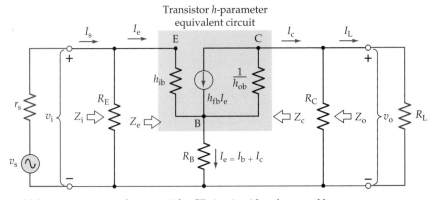

(b) *h*-parameter equivalent circuit for CB circuit with unbypassed base

Figure 6-37 AC equivalent circuit and *h*-parameter circuit for a CB amplifier with an unbypassed base resistor.

Comparing Eq. 6-33 to Eq. 6-28 (Z_e with C_1 present), it is seen that $R_B(1 - h_{fb})$ is added to h_{ib} to give Z_e without C_1 present. Eq. 6-34 is similar to Eq. 6-32 (A_v with C_1 present), except that $R_B(1 - h_{fb})$ is again added to h_{ib} to give the CB circuit voltage gain without C_1 (Eq. 6-34). Typically $h_{fb} = 0.99$ and $(1 - h_{fb}) = 0.01$.

Summary of CB Circuit Performance

With the base bypassed to ground:

Device input impedance	$Z_e = h_{ib}$
Circuit input impedance	$Z_i = Z_e \| R_E$
Device output impedance	$Z_c = \dfrac{1}{h_{ob}}$
Circuit output impedance	$Z_o \approx R_c$
Circuit voltage gain	$A_v = \dfrac{h_{fb}(R_C \| R_L)}{h_{ib}}$

With the base unbypassed:

Device input impedance	$Z_e = h_{ib} + R_B(1 - h_{fb})$
Circuit voltage gain	$A_v = \dfrac{h_{fb}(R_C \| R_L)}{h_{ib} + R_B(1 - h_{fb})}$

A CB circuit has good voltage gain and relatively high output impedance, like a CE circuit. But unlike a CE circuit, a CB circuit has a very low input impedance, and this makes it unsuitable for most voltage amplifier applications. It is normally employed only as a high-frequency amplifier. The transistor base terminal in a CB circuit should always be bypassed to ground. When this is not done, the input impedance is increased and the voltage amplification is substantially reduced.

Example 6-10

The transistor in the CB circuit in Fig. 6-38 has the following parameters: $h_{ie} = 2.1$ kΩ and $h_{fe} = 75$. Calculate the circuit input and output impedances and voltage gain.

Solution

From Table 6-1,
$$h_{ib} = \frac{h_{ie}}{1 + h_{fe}} = \frac{2.1 \text{ k}\Omega}{1 + 75}$$
$$= 27.6 \ \Omega$$

and
$$h_{fb} = \frac{h_{fe}}{1 + h_{fe}} = \frac{75}{1 + 75}$$
$$= 0.987$$

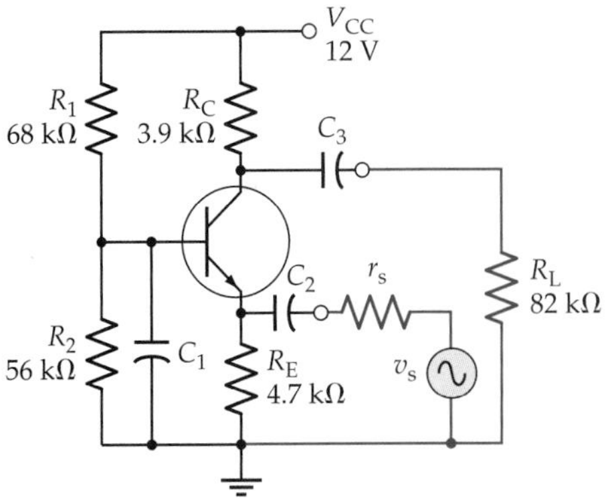

Figure 6-38   CB circuit for Example 6-10.

Eq. 6-29: $Z_i = h_{ib} \| R_E = 27.6\ \Omega \| 4.7\ \text{k}\Omega$

                   $= 27.4\ \Omega$

Eq. 6-31: $Z_o = (1/h_{ob}) \| R_C \approx R_C$

                   $= 3.9\ \text{k}\Omega$

Eq. 6-32: $A_v = \dfrac{h_{fb}(R_C \| R_L)}{h_{ib}} = \dfrac{0.987\ (3.9\ \text{k}\Omega \| 82\ \text{k}\Omega)}{27.6\ \Omega}$

                   $= 133$

Example 6-11

Calculate the input impedance and voltage gain for the CB circuit in Ex. 6-10 when capacitor C_1 is disconnected.

Solution

Eq. 6-33: $Z_e = h_{ib} + R_B(1 - h_{fb}) = 27.6\ \Omega + (68\ \text{k}\Omega \| 56\ \text{k}\Omega)(1 - 0.987)$

            $= 426.8\ \Omega$

Eq. 6-29: $Z_i = Z_e \| R_E = 426.8\ \Omega \| 4.7\ \text{k}\Omega$

            $= 391\ \Omega$

Eq. 6-34: $A_v = \dfrac{h_{fb}(R_C \| R_L)}{h_{ib} + R_B(1 - h_{fb})} = \dfrac{0.987\ (3.9\ \text{k}\Omega \| 82\ \text{k}\Omega)}{27.6\ \Omega + (68\ \text{k}\Omega \| 56\ \text{k}\Omega)(1 - 0.987)}$

            $= 8.6$

Practice Problems

6-7.1 Calculate Z_i, Z_o, and A_v for a CB circuit (as in Fig. 6-38) with the following component values and parameters: $R_1 = 56\ \text{k}\Omega$, $R_2 = 39\ \text{k}\Omega$, $R_C = 5.6$ kΩ, $R_E = 3.3\ \text{k}\Omega$, $R_L = 47\ \text{k}\Omega$, $h_{ie} = 1\ \text{k}\Omega$, and $h_{fe} = 100$.

6-7.2 Determine Z_i and A_v for the circuit in Problem 6-7.1 when capacitor C_1 is disconnected.

6-8 COMPARISON OF CE, CC, AND CB CIRCUITS

Table 6-2 compares Z_i, Z_o, and A_v for CE, CC, and CB circuits. As already discussed, the CE circuit has high voltage gain, medium input impedance, high output impedance, and a 180° phase shift from input to output. The CC circuit has high input impedance, low output impedance, a voltage gain of 1, and no phase shift. The CB circuit offers low input impedance, high output impedance, high voltage gain, and no phase shift.

TABLE 6-2	Comparison of common-emitter, common-collector, and common-base circuits			
Circuit Configuration	Z_i	Z_o	A_v	**Phase Shift**
CE	medium	$\approx R_C$	high	180°
CC	high	low	≈ 1	0
CB	low	$\approx R_C$	high	0

Device manufacturers normally only list the CE h-parameters on a transistor data sheet. Although these can be converted to CC and CB parameters, it is convenient to use CE parameters for all three types of circuits. Table 6-3 gives the circuit impedance equations in terms of CE h-parameters. In the case of input and output impedances, it is helpful to think in terms of the terminal being *looked into*.

TABLE 6-3	Common-emitter, common-collector, and common-base circuit equations using CE h-parameters		
Circuit Configuration	Z_b	Z_e	Z_c
CE	h_{ie}	—	$1/h_{oe}$
CE with unbypassed R_E	$h_{ie} + R_E(1 + h_{fe})$	—	$1/h_{oe}$
CC	$h_{ie} + (R_E \| R_L)(1 + h_{fe})$	$\dfrac{h_{ie} + R_B}{1 + h_{fe}}$	—
CB	—	$\dfrac{h_{ie}}{1 + h_{fe}}$	$\dfrac{1 + h_{fe}}{h_{oe}}$
CB with unbypassed R_B	—	$\dfrac{h_{ie} + R_B}{1 + h_{fe}}$	$\dfrac{1 + h_{fe}}{h_{oe}}$

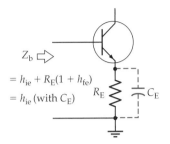

$Z_b \Rightarrow$

$= h_{ie} + R_E(1 + h_{fe})$

$= h_{ie}$ (with C_E)

Figure 6-39 Impedance at the transistor base terminal.

Impedance at the Transistor Base

Consideration of each type of circuit shows that the input impedance (Z_i) depends upon which transistor terminal is involved. In both the CE and CC circuits, the input signal is applied to the transistor base terminal. So, Z_i is the impedance 'looking into' the base. Figure 6-39 shows that, for a CE circuit with an unbypassed emitter resistor,

$$Z_b = h_{ie} + R_E(1 + h_{fe})$$

When R_E is bypassed, the $R_E(1 + h_{fe})$ portion can be treated as zero, so that

$$Z_b = h_{ie}$$

The CC equations for Z_b are almost identical to the CE equations, except that they use h_{ic} and h_{fc}, which are essentially equal to h_{ie} and h_{fe} (see Table 6-3). A rough approximation for the base input impedance of any transistor circuit with an unbypassed emitter resistor is

$$Z_b \approx h_{fe}(\textit{impedance in series with the emitter})$$

The circuit input impedance for both CE and CC configurations is Z_b in parallel with the bias resistors:

$$Z_i = Z_b \| R_1 \| R_2$$

Impedance at the Transistor Emitter

The transistor emitter is the output terminal for a CC circuit and the input terminal for a CB circuit. So, the device impedance in both cases is the impedance 'looking into' the transistor emitter terminal (Z_e). Although the Z_e equations for the two circuits look different, they can be shown to give exactly the same result in similar situations. For a CC circuit or a CB circuit with an unbypassed base (using CE parameters),

$$Z_e = \frac{h_{ie} + R_B}{1 + h_{fe}}$$

where R_B is the impedance when 'looking back' from the transistor base. As shown in Fig. 6-40, R_B includes the signal source resistance (r_s) for a CC circuit. A CB circuit normally has its base bypassed as shown, so that the impedance at the emitter is

$$Z_e = \frac{h_{ie}}{1 + h_{fe}}$$

For all circuit arrangements, the impedance at the transistor emitter terminal is

$$Z_e = \frac{\textit{Impedance in series with the base}}{1 + h_{fe}}$$

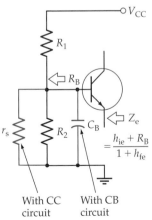

With CC circuit   With CB circuit

Figure 6-40 Impedance at the transistor emitter terminal.

Impedance at the Transistor Collector

The output for CE and CB circuits is taken from the transistor collector terminal. So, the impedance 'looking into' the collector is the device output impedance. This is normally a very large quantity at low and medium signal frequencies. As illustrated in Fig. 6-41, the circuit output impedance at the collector is essentially

$$Z_o \approx R_C$$

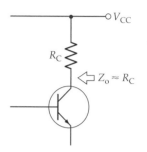

Figure 6-41 Impedance at the transistor collector terminal.

Voltage gain

In the case of a circuit with an unbypassed emitter resistor, the ac voltage at the emitter follows the ac input at the transistor base. So, a CC circuit (an emitter follower) has a voltage gain of 1.

With both CE and CB circuits, the ac input (v_i) is developed across the base-emitter junction, and the ac output (v_o) is produced at the transistor collector terminal (see Fig. 6-42). Thus (as discussed in Section 6-7), the magnitude of the voltage gain is the same for CB and CE circuits with similar component values and transistor parameters. The CE voltage gain equation can be used for the CB circuit, with the omission of the minus sign that indicates CE phase inversion. Although the voltage gains are equal for similar CB and CE circuits, the low input impedance of the CB circuit can substantially attenuate the signal voltage, and result in a low amplitude output. This is demonstrated in Example 6-12.

Figure 6-42 CE and CB voltage gain.

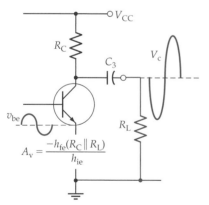

$$A_v = \frac{-h_{fe}(R_C \| R_L)}{h_{ie}}$$

Example 6-12

A 50 mV signal with a 600 Ω source resistance is applied to the circuit in Fig. 6-43. Calculate v_o for (a) CE circuit operation with v_s at the transistor base and R_E bypassed, and (b) CB circuit operation with v_s at the emitter and the base resistors bypassed. The transistor parameters are $h_{ie} = 1.5$ kΩ and $h_{fe} = 100$.

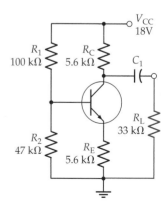

Figure 6-43 Circuit for Example 6-12.

Solution

(a) CE *circuit*

$$A_v = \frac{h_{fe}(R_C \| R_L)}{h_{ie}} = \frac{100\,(5.6\text{ k}\Omega \| 33\text{ k}\Omega)}{1.5\text{ k}\Omega}$$

$$= 319$$

$$Z_b = h_{ie} = 1\text{ k}\Omega$$

$$Z_i = Z_b \| R_1 \| R_2 = 1\text{ k}\Omega \| 100\text{ k}\Omega \| 47\text{ k}\Omega$$

$$= 970\ \Omega$$

$$v_i = \frac{v_s \times Z_i}{r_s + Z_i} = \frac{50\text{ mV} \times 970\ \Omega}{600\ \Omega + 970\ \Omega}$$

$$= 30.9\text{ mV}$$

$$v_o = A_v \times v_i = 319 \times 30.9\text{ mV}$$

$$= 9.9\text{ V}$$

(b) CB *circuit*

$$A_v = \frac{h_{fe}(R_C \| R_L)}{h_{ie}} = 319$$

From Table 6-3,
$$Z_e = \frac{h_{ie}}{1 + h_{fe}} = \frac{1.5\text{ k}\Omega}{1 + 100}$$

$$= 14.85\ \Omega$$

$$Z_i = Z_e \| R_E = 14.85\ \Omega \| 5.6\text{ k}\Omega$$

$$= 14.81\ \Omega$$

$$v_i = \frac{v_s \times Z_i}{r_s + Z_i} = \frac{50\text{ mV} \times 14.81\ \Omega}{600\ \Omega + 14.81\ \Omega}$$

$$= 1.2\text{ mV}$$

$$v_o = A_v \times v_i = 319 \times 1.2\text{ mV}$$

$$= 383\text{ mV}$$

Practice Problems

6-8.1 Using the appropriate CE equations from Table 6-3, calculate Z_i, Z_o, and A_v for a CB circuit (as in Fig. 6-38) with the following component values and parameters: $R_1 = 56$ kΩ, $R_2 = 39$ kΩ, $R_C = 5.6$ kΩ, $R_E = 3.3$ kΩ, $R_L = 47$ kΩ, $h_{ie} = 1$ kΩ, and $h_{fe} = 100$.

6-8.2 Determine v_o for the circuit in Problem 6-8.1 when a 30 mV ac signal with $r_s = 400\ \Omega$ is capacitor-coupled to the transitor emitter.

Review Questions

Section 6-1

6-1   Sketch the circuit of a CE amplifier with a capacitor-coupled signal source and capacitor-coupled load. Explain the need for coupling capacitors, and discuss the correct polarity for connecting the capacitors.

6-2   Explain what is meant by ac degeneration in a transistor circuit. Show how ac degeneration is eliminated in voltage-divider bias and collector-to-base bias circuits.

Section 6-2

6-3   Sketch the ac equivalent circuit for a CE circuit with voltage-divider bias, a bypassed emitter resistor, a capacitor-coupled signal source, and a capacitor-coupled load.

6-4   Sketch the ac equivalent circuit for a CE circuit with collector-to-base bias, a capacitor-coupled signal source, and a capacitor-coupled load.

6-5   Discuss the purpose of an ac load line for a transistor circuit. State how ac load lines differ from dc load lines for circuits with bypassed emitter resistors, and for circuits with capacitor-coupled loads.

Section 6-3

6-6   Sketch the T-equivalent (r-parameter) circuit for a transistor connected in CB configuration. Identify each component of the circuit and discuss its origin. Sketch a simplified form of the transistor low-frequency equivalent circuit.

6-7   Sketch the h-parameter equivalent circuit for a transistor connected in CE configuration. Name each component of the circuit and discuss its origin.

6-8   Define h_{ie}, h_{fe}, h_{oe}, and h_{FE}, and show how each parameter may be derived from the transistor characteristics. State typical h-parameter values for a low-current transistor.

6-9   Sketch an approximate h-parameter CE transistor model that may be used for most purposes. Briefly explain.

6-10   Draw a CE transistor hybrid-π model. Briefly explain.

6-11   Define r'_e, and write equations for calculating r'_e at 25°C and at other temperatures.

Section 6-4

6-12   Briefly explain the operation of the CE circuit described in Question 6-3.

6-13   Draw an h-parameter equivalent circuit for the CE circuit described in Question 6-3. Identify all components.

6-14 Write equations for Z_i, Z_o, and A_v for the h-parameter circuit in Question 6-13.

6-15 Draw an h-parameter equivalent circuit for the CE circuit described in Question 6-4. Identify all components.

6-16 Write equations for Z_i, Z_o, and A_v for the h-parameter circuit in Question 6-15.

6-17 Sketch a base-biased *pnp* transistor CE circuit. Include a capacitor-coupled signal source and load resistor. Draw the h-parameter equivalent circuit identifying all components.

6-18 Write equations for Z_i, Z_o, and A_v for the h-parameter circuit in Question 6-17.

Section 6-5

6-19 Referring to the CE circuit described in Question 6-3, explain the effect of removing the bypass capacitor from R_E.

6-20 Draw an h-parameter equivalent circuit for the CE circuit in Question 6-19.

6-21 Write equations for Z_i, Z_o, and A_v for the h-parameter circuit in Question 6-20.

6-22 Sketch a collector-to-base biased *pnp* transistor CE circuit with two centre-bypassed base resistors and an unbypassed emitter resistor. Include a capacitor-coupled signal source and load resistor. Draw the h-parameter equivalent circuit and identify all components.

6-23 Write equations for Z_i, Z_o, and A_v for the h-parameter circuit in Question 6-22.

Section 6-6

6-24 Sketch a practical *npn* transistor CC circuit that uses base bias. Include a capacitor-coupled signal source and load resistor. Explain the circuit operation.

6-25 Draw the h-parameter equivalent circuit for the circuit in Question 6-24. Identify all components.

6-26 Write equations for Z_i, Z_o, and A_v for the h-parameter circuit in Question 6-25.

6-27 Sketch the h-parameter equivalent for a CC circuit with voltage divider bias, a capacitor-coupled signal source, and a capacitor-coupled load.

6-28 Write the Z_i and Z_o equations for the h-parameter circuit in Question 6-27.

6-29 Sketch a base-biased *pnp* transistor CC circuit. Include a capacitor-coupled signal source and load resistor.

6-30 Draw the h-parameter equivalent circuit for the circuit in Question 6-29.

Section 6-7

6-31 Sketch a practical *npn* transistor CB circuit that uses collector-to-base bias. Include a capacitor-coupled signal source and load resistor. Explain the circuit operation.

6-32 Draw the *h*-parameter equivalent circuit for the circuit in Question 6-31. Identify all components.

6-33 Write equations for Z_i, Z_o, and A_v for the CB *h*-parameter circuit in Question 6-32.

6-34 Sketch a practical *pnp* transistor CB circuit that uses voltage-divider bias. Include a capacitor-coupled signal source and load resistor.

6-35 Sketch the *h*-parameter equivalent for a CB circuit with voltage-divider bias, a capacitor-coupled signal source, and a capacitor-coupled load.

6-36 Write the Z_i, Z_o, and A_v equations for the CB *h*-parameter circuit in Question 6-35.

Section 6-8

6-37 Compare the performance of CE, CC, and CB circuits, and discuss typical applications of each type of circuit.

6-38 Write equations for the impedances seen 'looking into' the base, emitter, and collector terminals of a transistor in a circuit. Briefly explain each equation.

Problems

Section 6-1

6-1 Calculate V_{CE} for the circuit in Fig. 6-43 when the load is capacitor-coupled, as shown, and when the load is directly connected to the collector terminal.

6-2 Determine V_B for the circuit in Fig. 6-43 when a 600 Ω signal source is capacitor-coupled to the input, and when the signal source is directly connected.

Section 6-2

6-3 The transistor characteristics for the circuit in Fig. 6-43 are shown in Fig. 6-44. Draw the dc and ac load lines for the circuit when it is operated in CE configuration with R_E bypassed. Determine the maximum possible symmetrical output voltage swing.

6-4 Draw the new ac load line for the circuit in Problem 6-3 when R_L is changed to 12 kΩ.

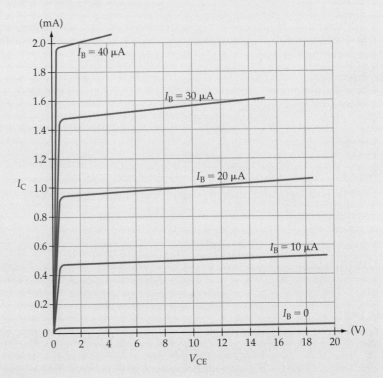

Figure 6-44

6-5 Using the device characteristics in Fig. 6-44, draw the dc and ac load lines for the CE circuit shown in Fig. 6-45.

6-6 The CE circuit in Fig. 6-46 uses collector-to-base bias with a bypass capacitor. Using the device characteristics in Fig. 6-44, draw the dc and ac load lines for the circuit. The transistor has $h_{FE} = 100$.

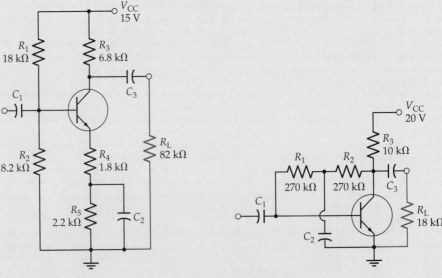

Figure 6-45

Figure 6-46

Section 6-3

6-7   Referring to the CE input and output characteristics in Figs. 4-29 and 4-30, determine h_{ie}, h_{oe}, and h_{fe} for $V_{CE} = 6$ V and $I_B = 40$ μA.

6-8   A transistor in a CE circuit has a constant I_B level of 40 μA. When $V_{CE} = 5$ V, I_C is measured as 4.9 mA, and when V_{CE} is adjusted to 10 V, I_C becomes 5 mA. Calculate h_{oe} and h_{FE}.

6-9   A transistor has V_{CE} maintained constant at 7.5 V and I_C measured at several levels of I_B as follows:

I_B	25	50	75	100	(μA)
I_C	3.06	6.12	9.2	12.3	(mA)

Determine the h_{fe} value for the transistor.

6-10   The transistor in Problem 6-9 has V_{CE} maintained constant at 7.5 V and I_B measured at several levels of V_{BE} as follows:

V_{BE}	0.68	0.72	0.76	0.8	(V)
I_B	20	40	60	80	(μA)

Determine the h_{ie} value for the transistor.

6-11   The transistor in Problems 6-9 and 6-10 has I_B maintained constant at 40 μA, and corresponding levels of I_C and V_{CE} measured as follows:

V_{CE}	2	4	6	8	(V)
I_C	4.7	4.8	4.9	5	(mA)

Determine the h_{oe} value for the transistor.

6-12   Calculate α and r_c for the transistor referred to in Problems 6-9 to 6-11.

6-13   Determine the value of r'_e for a transistor with a 2 mA emitter current if its junction temperature is 80°C.

6-14   The collector and base currents of a transistor are measured as 1.28 mA and 20 μA, respectively. Calculate r'_e, h_{ie}, and r_π for the transistor.

Section 6-4

6-15   The CE circuit in Fig. 6-47 has the following transistor parameters: $h_{ie} = 1$ kΩ, $h_{fe} = 85$, and $h_{oe} = 2$ μS. Calculate Z_i, Z_o, and A_v.

6-16   Recalculate Z_i, Z_o, and A_v for the circuit in Problem 6-15 when the transistor is replaced with one having $h_{ie} = 1.4$ kΩ and $h_{fe} = 55$.

6-17   The base-biased CE circuit in Fig. 6-48 has the following transistor parameters: $h_{ie} = 1.3$ kΩ, $h_{fe} = 40$, and $h_{oe} = 1.5$ μS. Calculate Z_i, Z_o, and A_v.

6-18   The collector-to-base biased CE circuit in Fig. 6-49 has the following transistor parameters: $h_{ie} = 1.3$ kΩ, $h_{fe} = 40$, and $h_{oe} = 1.5$ μS. Calculate Z_i, Z_o, and A_v.

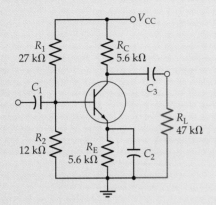

Figure 6-47

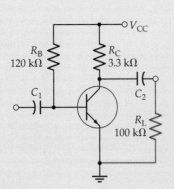

Figure 6-48

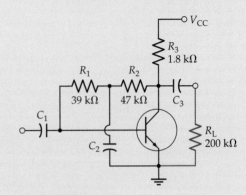

Figure 6-49

6-19 Recalculate Z_i, Z_o, and A_v for the circuit in Problem 6-17 when the transistor is replaced with one having $h_{ie} = 1.5\,\text{k}\Omega$, $h_{oe} = 1\,\mu\text{S}$, and $h_{fe} = 100$.

6-20 The transistor in the circuit in Problem 6-18 is replaced with one having $h_{ie} = 1.5\,\text{k}\Omega$, $h_{oe} = 1\,\mu\text{S}$, and $h_{fe} = 100$. Recalculate Z_i, Z_o, and A_v.

Section 6-5

6-21 For the circuit in Problem 6-15, recalculate Z_i and A_v when the bypass capacitor is removed from R_E.

6-22 The circuit in Problem 6-15 has R_E replaced with two series-connected resistors: $R_{E1} = 1.5\,\text{k}\Omega$ and $R_{E2} = 3.9\,\text{k}\Omega$. Calculate Z_i and A_v when only R_{E1} is capacitor-bypassed, and when only R_{E2} is capacitor-bypassed.

Section 6-6

6-23 The CC circuit in Fig. 6-50 has the following transistor parameters: $h_{ie} = 1\,\text{k}\Omega$, $h_{fe} = 85$, and $h_{oe} = 2\,\mu\text{S}$. Calculate Z_i and Z_o.

6-24 Recalculate Z_i and Z_o for Problem 6-23 when the transistor is replaced with one having $h_{ie} = 1.4\,\text{k}\Omega$ and $h_{fe} = 55$.

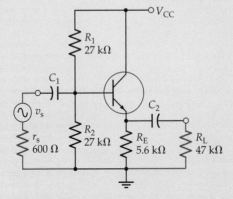

Figure 6-50

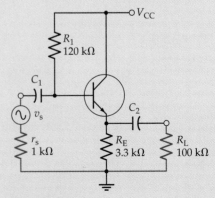

Figure 6-51

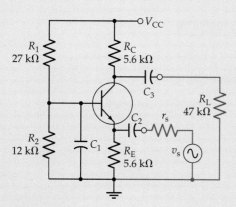

Figure 6-52

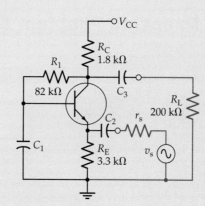

Figure 6-53

6-25 The CC circuit in Fig. 6-51 has the following transistor parameters: $h_{ie} = 1.3$ kΩ and $h_{fe} = 40$. Calculate Z_i and Z_o.

6-26 A CC circuit using voltage-divider bias (as in Fig. 6-50) has the following component values: $R_1 = 33$ kΩ, $R_2 = 47$ kΩ, $R_E = 15$ kΩ, and $R_L = 47$ kΩ. The transistor parameters are $h_{ie} = 1.2$ kΩ and $h_{fe} = 120$. Calculate Z_i and Z_o.

Section 6-7

6-27 The CB circuit in Fig. 6-52 has a transistor with the following parameters: $h_{ie} = 1$ kΩ, $h_{fe} = 85$, and $h_{oe} = 2$ μS. Calculate Z_i, Z_o, and A_v.

6-28 Recalculate Z_i and A_v for the CB circuit in Problem 6-27 when the transistor is replaced with one having $h_{ie} = 1.4$ kΩ and $h_{fe} = 55$.

6-29 Calculate Z_i, Z_o, and A_v for the CB circuit in Fig. 6-53. The transistor parameters are: $h_{ie} = 1.3$ kΩ, $h_{fe} = 40$, and $h_{oe} = 1.5$ μS.

6-30 Recalculate Z_i, Z_o, and A_v for the circuit in Problem 6-29 when the transistor is replaced with one having $h_{ie} = 1.5$ kΩ, $h_{oe} = 1$ μS, and $h_{fe} = 100$.

6-31 A CB circuit (as in Fig. 6-52) has the following component values and transistor parameters: $R_1 = 33$ kΩ, $R_2 = 22$ kΩ, $R_C = 3.3$ kΩ, $R_E = 3.9$ kΩ, $R_L = 39$ kΩ, $h_{ie} = 1.6$ kΩ, $h_{fe} = 200$, and $h_{oe} = 1.5$ μS. Calculate Z_i, Z_o, and A_v.

Section 6-8

6-32 The transistor in the CE circuit in Problem 6-15 is replaced with a 2N3904 transistor (see data sheet A-5 in Appendix A). Calculate the maximum and minimum values of Z_i, Z_o, and A_v.

6-33 The CC circuit in Problem 6-23 is reconstructed to use a 2N3903 transistor. Calculate the maximum and minimum values of Z_i and Z_o.

6-34 The transistor in the CB circuit in Problem 6-31 is replaced with a 2N3903 transistor. Calculate the maximum and minimum values of Z_i and Z_o.

Practice Problem Answers

6-1.1	9.5 V, 8.9 V
6-2.1	point C at 2.07 mA and 3.3 V from Q
6-3.1	133, 20 μS
6-3.2	0.993, 6.7 MΩ
6-3.3	1.76 kΩ
6-4.1	849 Ω, 5.6 kΩ, -517
6-4.2	190, 161
6-5.1	5.52 kΩ, 5.6 kΩ, -1.87
6-6.1	21.9 kΩ, 14 Ω
6-7.1	9.87 Ω, 5.6 kΩ, 500
6-7.2	224 Ω, 20.6
6-8.1	9.87 Ω, 5.6 kΩ, 500
6-8.2	361 mV

CHAPTER 7
Fabrication of Semiconductor Devices and ICs

CONTENTS

Objectives

You will be able to:

1   Describe the process of preparing semiconductor material for device manufacture, and explain diffusion and epitaxial growth.
2   Draw sketches to show the fabrication of alloy and diffused diodes.
3   Explain the manufacturing requirements for transistors to produce satisfactory current gain, frequency response, power dissipation, etc.

4   Describe the various transistor fabrication methods and their performance characteristics.
5   Discuss manufacturing methods for the production of integrated circuits.
6   Draw sketches to show the construction of monolithic IC components, and discuss problems that occur with the use of IC components.
7   Sketch various diode, transistor, and IC packages, and show the terminal identification systems.

INTRODUCTION

The method employed to manufacture a semiconductor device determines its electrical characteristics and, thus, dictates its applications. For example, low-current fast-switching diodes and transistors must be fabricated differently from high-power devices. The process that produces a transistor with high current gain may result in undesirable large junction capacitances. A technique that gives a small junction capacitance may produce devices with low breakdown voltages. An *integrated circuit* (*IC*) consists of many transistors, diodes, and other components in one package. Usually, all the components are fabricated on one small silicon chip.

7-1 PROCESSING OF SEMICONDUCTOR MATERIALS

Preparation of Silicon and Germanium

Silicon is one of the commonest elements on earth. It occurs as silicon dioxide (SiO_2) and as silicates, compounds of silicon and other materials. Germanium is derived from zinc or copper ores. When converted to bulk metal, silicon and germanium contain large quantities of impurities. Both metals must be carefully refined before they can be used for the manufacture of semiconductor devices.

Semiconductor material is normally *polycrystalline* after it is refined. This means that it is made up of many individual formations of atoms with no overall fixed pattern. For use in semiconductors, the metal must be converted into *single-crystal* material; that is, all of its atoms must be arranged into a single pattern.

In its final condition for device manufacture, silicon and germanium are in the form of single-crystal bars about 2.5 cm in diameter and perhaps 30 cm long. The bars are sliced into disc-shaped wafers about 0.4 cm thick, and the wafers are polished to a mirror surface. Several thousand devices are usually fabricated on the surface of each wafer; then the wafers are scribed and cut like glass (see Fig. 7-1).

Diffusion

For processing, semiconductor wafers contained in an enclosure are heated by means of radio frequency (RF) heating coils. This is illustrated in Fig. 7-2. When wafers of *n*-type material are raised to a high temperature in an atmosphere containing *p*-type impurity atoms, some of the impurities *diffuse* into each wafer. This converts the outer layer of the *n*-type material into *p*-type (Fig. 7-3). The diffusion process can be continued by further heating the wafers in an atmosphere containing *n*-type impurity atoms, so that *npn* layers are produced, as illustrated. Because the diffusion process is very slow (about 2.5 μm per hour), very narrow diffused regions can be created by careful timing.

(a) Semiconductor bar sliced into wafers

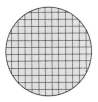

(b) Scribed wafer

Figure 7-1 Bars of single-crystal semiconductor material are sliced into wafers for the manufacture of semiconductor devices.

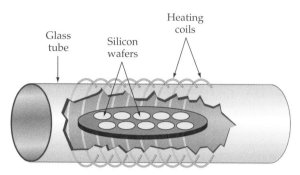

Figure 7-2   Semiconductor wafers are heated in an atmosphere containing *n*-type or *p*-type impurities that diffuse into the surface of the wafers.

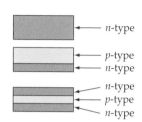

Figure 7-3 The diffusion process converts *n*-type semiconductor into *p*-type, and vice versa.

Epitaxial Growth

The *epitaxial* process is very similar to the diffusion process, except that silicon or germanium atoms are contained in the gas surrounding the semiconductor wafers. The semiconductor atoms in the gas grow (accumulate) on the wafer in the form of a thin layer (Fig. 7-4). This layer is single-crystal material and it may be *p*-type or *n*-type, according to the impurity content of the gas. The epitaxial layer may be 'doped' by the diffusion process.

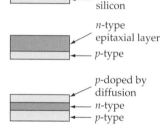

Figure 7-4 Epitaxial growth adds layers of *n*-type or *p*-type material to the surface of a wafer.

Section 7-1 Review

7-1.1 Briefly explain polycrystalline material, single-crystal material, diffusion, and epitaxial growth.

7-2 DIODE FABRICATION AND PACKAGING

Alloy and Diffused Diodes

Two commonly used techniques for diode manufacture are the *alloy* method and the *diffusion* method. To construct an alloy diode, a *pn*-junction is formed by melting a tiny pellet of aluminum (or some other *p*-type impurity) on the surface of an *n*-type crystal. Alternatively, an *n*-type impurity may be melted on the surface of a *p*-type crystal. The process is illustrated in Fig. 7-5a.

Figure 7-5b shows the diffusion technique for diode construction. When an *n*-type semiconductor is heated in a chamber containing an acceptor impurity in vapour form, some of the acceptor atoms are diffused (or absorbed) into the *n*-type crystal. This produces a *p*-region in the *n*-type material and thus creates a *pn*-junction. The size of the *p*-region can be precisely defined by uncovering only part of the *n*-type material during the diffusion process (the rest of the surface has a thin coating of silicon dioxide). Metal contacts are electroplated onto each region for connecting leads.

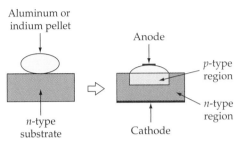

(a) Alloy diode fabrication

Figure 7-5 Diode fabrication methods. For the alloy diode, a pellet containing *p*-type impurities is melted into *n*-type material. A diffused diode is created by heating a wafer of *n*-type material in an atmosphere containing *p*-type impurities.

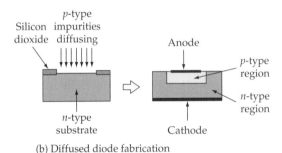

(b) Diffused diode fabrication

The diffusion technique lends itself to the simultaneous fabrication of many hundreds of diodes on one small disc of semiconductor material. This process is also used in the production of transistors and integrated circuits.

Diode Packaging

As explained in Section 2-1, semiconductor diodes vary widely in size, depending on their application. Figure 7-6 shows typical single diode packages for low-, medium-, and high-current devices. In Fig. 7-7 multiple diode packages are illustrated. Two diodes in a surface mount package (see Section 7-7) are shown in Fig. 7-7a, and two in a power-transistor-type package are illustrated in Fig. 7-7b. Note from the circuits that the cathode terminals are internally commoned in each case.

A rectifier assembly is shown in Fig. 7-7c. This consists of four diodes connected to function as a bridge rectifier and is contained in one package.

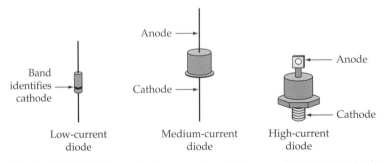

Figure 7-6 Individual diodes vary in size according to the current they are required to pass.

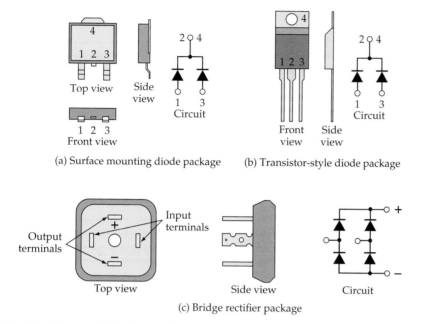

(a) Surface mounting diode package   (b) Transistor-style diode package

(c) Bridge rectifier package

Figure 7-7 Some multiple diode packages.

One side of the package has a metal plate for mounting on a heat sink (see Section 8-8). The two output terminals are identified as + and −, and the other two terminals are for the ac input.

Section 7-2 Review

7-2.1 Briefly describe the construction of alloy diodes and diffused diodes.

7-3 TRANSISTOR CONSTRUCTION AND PERFORMANCE

Current Gain

Good current gain requires that most charge carriers from the emitter pass rapidly to the collector. Thus there should be little outflow of charge carriers through the base terminal and few carrier recombinations within the base. These conditions dictate a very narrow, lightly doped base (see Fig. 7-8).

High Power

High-power transistors must have large emitter-base surfaces to provide the required quantity of charge carriers. Large collector-base surfaces are also needed to dissipate power without overheating the collector-base junctions (Fig. 7-8).

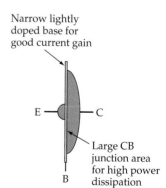

Narrow lightly doped base for good current gain

Large CB junction area for high power dissipation

Figure 7-8 High current gain requires a narrow lightly doped transistor base, and high power dissipation demands a large CB junction area.

Frequency Response

For the highest possible frequency response, the transistor base region must be very narrow to ensure a short transit time of charge carriers from emitter to collector (see Fig. 7-9). Input capacitance must also be kept to a minimum, and this requires a small-area emitter-base junction as well as a highly resistive (lightly doped) base region.

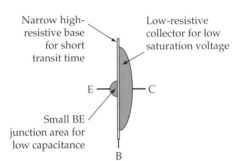

Narrow high-resistive base for short transit time

Low-resistive collector for low saturation voltage

Small BE junction area for low capacitance

E — C

B

Figure 7-9   A small BE junction area and narrow base are required for high frequency performance and fast switching. Low saturation voltage is promoted by a low-resistive collector region.

Because power transistors require large junction surfaces, and high-frequency performance demands small junction areas, there is a conflict in high-frequency power transistors. To keep the junction area to a minimum and still provide adequate charge-carrier emission, the emitter-base junction is usually in the form of a long thin zigzag strip.

Switching Transistors

Fast switching demands the same low junction capacitance required for good high-frequency performance. A switching transistor should have a low *saturation voltage* and a short storage time (see Section 8-5). For low saturation voltages, the collector region must have low resistivity (Fig. 7-9). Short storage time requires fast recombination of charge carriers left in the depletion region at the collector-base junction. Fast recombination is helped by additional doping of the collector with gold atoms.

Breakdown and Punch-Through

The transistor collector-base junction is normally reverse-biased, so the maximum collector voltage is limited by the junction reverse breakdown voltage. To achieve high breakdown voltages, either the collector or the base must be very lightly doped. If the base is more lightly doped than the collector, the collector-base depletion region penetrates deeply into the base. This will cause transistor breakdown by *punch-through* when it links with the emitter-base depletion region. To avoid punch-through, the collector region close to the base is usually more lightly doped than the base. The collector-base depletion region then spreads into the collector rather than into the base.

Section 7-3 Review

7-3.1 Briefly discuss the requirements for constructing transistors with (a) good current gain, and (b) high-frequency response.

7-3.2 Explain the design requirements for transistors with (a) high power dissipation, and (b) high breakdown voltage.

7-3.3 List the requirements for switching transistor construction.

7-4 TRANSISTOR FABRICATION

Alloy Transistors

For manufacture of *alloy transistors*, single-crystal *n*-type wafers are scribed into many small sections, or *dice*, each of which forms the substrate for one transistor. A small pellet of *p*-type material is melted on one surface of each section until it partially penetrates and forms an alloy with the substrate (Fig. 7-10a), thus creating a *pn*-junction. The process is repeated on the other side of the wafer to constitute a *pnp* transistor.

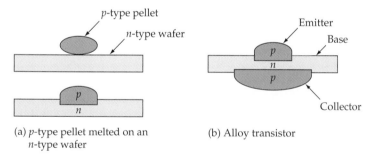

(a) *p*-type pellet melted on an *n*-type wafer

(b) Alloy transistor

Figure 7-10 A *pnp* alloy transistor is constructed by melting *p*-type pellets on the surface of an *n*-type wafer.

One of the junctions has a large area and one has a small area, as illustrated in Fig. 7-10b. The small-area junction becomes the emitter-base junction, and the larger one becomes the collector-base junction. One reason for this is that the large-area junction most easily collects all the charge carriers emitted from the small-area junction. Another more important reason is that most of the power is dissipated at the transistor collector-base junction.

Suppose that a silicon transistor has $I_C = 10$ mA and $V_{CE} = 10$ V (see Fig. 7-11). The total power dissipated in the transistor is

$$P = V_{CE} \times I_C = 10\text{ V} \times 10\text{ mA}$$
$$= 100\text{ mW}$$

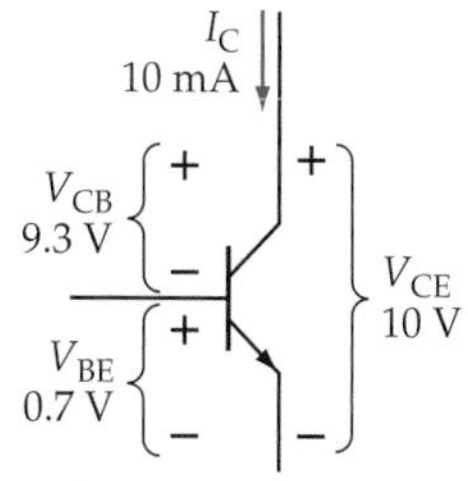

Figure 7-11 Most of the power dissipation in a transistor occurs at the collector-base junction.

The base-emitter voltage $V_{BE} = 0.7$ V, and the collector-base voltage is

$$V_{CB} = V_{CE} - V_{BE} = 10 \text{ V} - 0.7 \text{ V}$$

$$= 9.3 \text{ V}$$

The power dissipated at the base-emitter junction is

$$V_{BE} \times I_C = 0.7 \text{ V} \times 10 \text{ mA}$$

$$= 7 \text{ mW}$$

and the power dissipated at the collector-base junction is

$$V_{CB} \times I_C = 9.3 \text{ V} \times 10 \text{ mA}$$

$$= 93 \text{ mW}$$

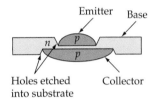

Emitter   Base

Holes etched   Collector
into substrate

Figure 7-12 A very narrow base region gives microalloy transistors good high-frequency performance.

Microalloy Transistors

Because very narrow base widths are difficult to obtain with alloy transistors, they are not suitable for high-frequency applications. To improve the high-frequency performance, holes are first etched into the wafer from each side, leaving a very thin layer (Fig 7-12). By a plating process, surfaces of impurity material are formed on each side of the thin n-type portion. Heat is then applied to alloy the impurities into the base region. This process produces very thin base regions and good high-frequency performance. The device is termed a *microalloy* transistor.

Microalloy-Diffused Transistors

In microalloy transistors, the collector-base depletion region penetrates deeply into the very thin base. Thus, a major disadvantage is that punch-through can occur at very low collector voltages. *Microalloy-diffused* transistors use a substrate that is initially undoped. After the holes are etched into each side of the wafer to produce the thin base region, the base is doped by diffusion from the collector side. The diffusion can be carefully controlled so that the base region is heavily doped at the collector side, with the doping becoming progressively less until the material is almost intrinsic (undoped) at the emitter. With this kind of doping, the collector-base depletion region penetrates only a short distance into the base, so that high punch-through voltages are achieved.

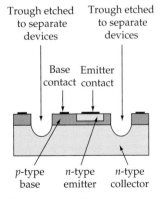

Trough etched   Trough etched
to separate   to separate
devices   devices

Base   Emitter
contact contact

p-type n-type n-type
base   emitter   collector

Figure 7-13 Mesa transistors are created by etching troughs between individual devices to give low collector-base leakage currents.

Diffused Mesa

In the production of *mesa transistors*, several thousand transistors are simultaneously formed on the wafer by the diffusion process. As shown in Fig. 7-13, the main body of the wafer becomes the n-type collectors, the diffused p-regions become the bases, and the final n-regions are the emitters. Metal films are deposited on the base and emitter surfaces for contacts.

The transistors could be separated by the usual process of scribing lines on the surface of the wafer and breaking it into individual units. However, this produces very rough edges that result in relatively high leakage currents between collector and base. So, before cutting the wafer, the transistors are isolated by etching away the unwanted portions of the diffused area to form troughs between devices. As illustrated, the base and emitter regions now project above the main wafer, which forms the collector region. This is the mesa structure. The narrow base widths that can be achieved by the diffusion process make the mesa transistor useful at very high frequencies.

Epitaxial Mesa

One of the disadvantages of the process just described is that, because the collector region is highly resistive, diffused mesa transistors have a high saturation voltage (see Section 4-4). Such devices are unsuitable for saturated switching applications. This same characteristic (high collector resistance) is desirable for high punch-through voltage. One method of achieving both high punch-through voltage and low saturation level uses the epitaxial process to produce an *epitaxial mesa* transistor.

Starting with a low-resistive (highly doped) wafer, a thin, highly resistive epitaxial layer is grown. This layer becomes the collector, and the base and emitter are diffused as already discussed. The arrangement is illustrated in Fig. 7-14. The punch-through voltage is high because the collector-base depletion region spreads deepest into the lightly doped collector. Saturation voltage is low because the collector region is very narrow and the main body of the wafer through which collector current must flow has a very low resistance.

Diffused Planar Transistor

In all the transistors described previously, the collector-base junction is exposed (within the transistor package), and substantial charge carrier leakage can occur at the junction surface. In the *diffused planar* transistor illustrated in Fig. 7-15, the collector-base junction is covered with a layer of silicon dioxide. This construction gives a very low collector-base leakage current (I_{CBO}), typically 0.1 nA.

Annular Transistor

A problem that occurs particularly with *pnp* planar transistors is the *induced channel*. This results when a relatively high voltage is applied to the silicon dioxide surface (at one of the terminals). Consider the *pnp* structure shown in Fig. 7-16a. If the surface of the silicon dioxide has a positive voltage, minority charge carriers within the lightly doped *p*-type substrate are attracted by the positive potential. The minority charge carriers concentrate at the upper edge of the substrate and form an *n*-type channel from the base to the edge of the device. This becomes an extension of the *n*-type base region and results in charge carrier leakage at the exposed edge of the collector-base junction.

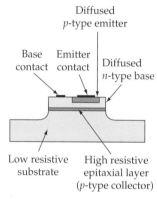

Figure 7-14 Epitaxial mesa transistor. The epitaxial process, combined with the mesa structure, gives high breakdown voltages and low saturation levels.

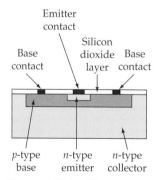

Figure 7-15 A diffused planar transistor has the collector-base junction covered, so that I_{CBO} is reduced to very low levels.

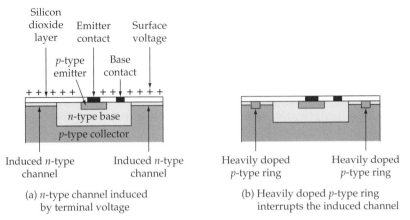

Figure 7-16 Annular transistors have a heavily doped *p*-type ring included to interrupt unwanted *n*-type channels that are induced by high voltages.

The problem arises because the *p*-type substrate is highly resistive. If it were heavily doped with *p*-type charge carriers, the concentration of *n*-type carriers would be absorbed (the electrons would be swallowed by holes). The introduction of a heavily doped *p*-type ring around the base, as in Fig. 7-16b, interrupts the induced channel and isolates the collector-base junction from the device surface. The *annular transistor*, therefore, is a high-voltage device with the low collector-base leakage of the planar transistor.

Section 7-4 Review

7-4.1 Briefly explain the construction of alloy, microalloy, and microalloy-diffused transistors.

7-4.2 Describe diffused mesa and epitaxial mesa transistor construction.

7-4.3 Discuss the construction of diffused planar and annular transistors.

7-5 INTEGRATED CIRCUITS

IC Types

An *integrated circuit* (IC) consists of several interconnected transistors, resistors, and so on, all contained in one small package with external connecting terminals. The circuit may be entirely self-contained, requiring only input and output connections and a supply voltage. Alternatively, a few external components may have to be connected to make the circuit operate.

Integrated circuits may be classified in terms of their function as *analog* or *digital*. Analog ICs (also termed *linear* ICs) may be amplifiers, voltage regulators, and so on. Digital ICs (logic gates and counting circuits) contain transistors that are normally either in a switched-*off* or switched-*on* state.

The techniques used in manufacturing integrated circuits provide another method of classification. The major IC manufacturing technique is termed *monolithic*. Other fabrication techniques are *thin-film*, *thick-film*, and *hybrid*.

Monolithic Integrated Circuits

In a *monolithic* integrated circuit all components are fabricated by the diffusion process on a single chip of silicon. Interconnections between components are provided on the surface of the structure, and external connecting wires are taken out to terminals, as illustrated in Fig. 7-17. The vast majority of integrated circuits are monolithic because the diffusion process is the most economical for mass production.

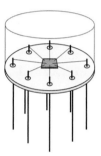

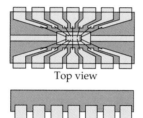

Top view

Side view

Figure 7-17 Monolithic integrated circuits consist of a single-chip circuit contained in a metal or plastic enclosure with connecting pins.

(a) Can-type IC enclosure   (b) Plastic IC enclosure

Thin Film

Thin-film integrated circuits are constructed by depositing films of conducting material on the surface of a glass or ceramic base. By controlling the width and thickness of the films, and by using different materials selected for their resistivity, resistors and conductors are fabricated. Capacitors are produced by sandwiching a film of insulating oxide between two conducting films. Inductors are made by depositing metal film in a spiral formation. Transistors and diodes are usually tiny discrete components connected into the circuit.

Thick Film

Thick-film integrated circuits are sometimes referred to as *printed thin-film circuits*. In this process, silk-screen printing techniques are employed to create the desired circuit pattern on a ceramic substrate. The inks used are pastes which have conductive, resistive, or dielectric properties. After printing, the circuits are fired at a high temperature in a furnace to fuse the films to the substrate. Thick-film passive components are fabricated in the same way as those in thin-film circuits. As with thin-film circuits, active components are added as separate devices. A portion of a thick-film circuit is shown in Fig. 7-18.

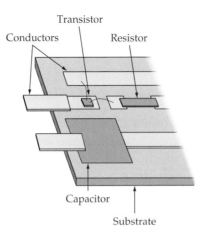

Transistor

Conductors   Resistor

Figure 7-18 Enlarged portion of a thick film integrated circuit.

Capacitor

Substrate

Integrated circuits produced by thin- or thick-film techniques usually have better component tolerances and give better high-frequency performance than monolithic integrated circuits.

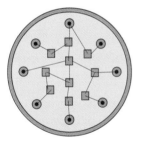

Figure 7-19 Uncovered top view of hybrid integrated circuit. Individual components are interconnected to form a single circuit.

Hybrid

Figure 7-19 illustrates the structure of a *hybrid* or *multichip* integrated circuit. As the name implies, the circuit is constructed by interconnecting a number of individual chips. The active components are diffused transistors or diodes. The passive components may be groups of diffused resistors or capacitors on a single chip, or they may be thin-film components. Connections between chips are provided by fine wire or metal film.

Like thin-film and thick-film ICs, multichip circuits usually have better performance than monolithic circuits. Although the process is too expensive for mass production, multichip techniques are quite economical for small quantities and are frequently used as prototypes for monolithic integrated circuits.

Section 7-5 Review

7-5.1 Briefly define monolithic, thin-film, thick-film, and hybrid integrated circuits.

7-6 IC COMPONENTS AND CIRCUITS

Transistors and Diodes

The epitaxial planar diffusion process described in Section 7-4 is normally employed for the manufacture of IC transistors and diodes. Collector, base, and emitter regions are diffused into a silicon substrate, and surface terminals are provided for connection, as illustrated in Fig. 7-20a.

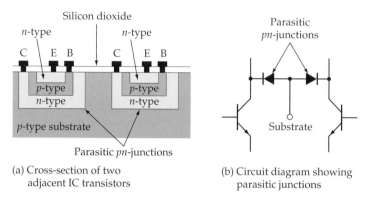

(a) Cross-section of two
    adjacent IC transistors

(b) Circuit diagram showing
    parasitic junctions

Figure 7-20 Integrated circuit components have unwanted (parasitic) interconnecting *pn*-junctions that must be kept reverse-biased.

In discrete transistors the substrate is normally used as a collector. If this were done with transistors in a monolithic integrated circuit, all transistors fabricated on one substrate would have their collectors connected together. For this reason, separate collector regions must be diffused into the substrate.

Even though separate collector regions are formed, they are not completely isolated from the substrate. Figure 7-20a shows that an unwanted (*parasitic*) *pn*-junction is formed by the substrate and the transistor collector region. If the circuit is to function correctly, these junctions must never become forward-biased. Thus, in the case of a *p*-type substrate, the substrate must always be kept negative with respect to the transistor collectors (Fig. 7-20b). This requires that the substrate be connected to the most negative terminal of the circuit supply.

The parasitic junctions can affect the circuit performance even when they are reverse-biased. The junction reverse leakage current can be a serious problem in circuits that operate at very low current levels. The capacitance of the reverse-biased junction can affect the circuit high-frequency performance, and the junction breakdown voltage imposes limits on the usable level of supply voltage. All these factors can be minimized by using highly resistive (lightly doped) material for the substrate.

Integrated circuit diodes are fabricated by diffusion exactly like transistors. Two of the regions are used to form one *pn*-junction. Alternatively, the collector region of a transistor may be connected directly to the base base (Fig. 7-21), so that the device behaves as a diode while operating similarly to a saturated transistor (see Section 4-4).

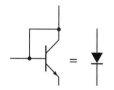

Figure 7-21 A transistor can be made to behave as a diode by connecting the collector and base.

Circuits

Some IC internal circuitry differs substantially from discrete component circuits. This is because of similarities between devices that are all fabricated at the same time on the same substrate. One such circuit, known as a *current mirror* is illustrated in Fig. 7-22. As explained above, BJTs are frequently used as diodes in integrated circuits. The current mirror uses such a diode to control

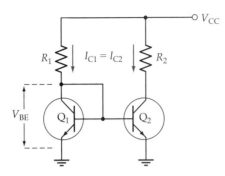

Figure 7-22 Current mirror. Q_1 is diode-connected, and because $V_{BE2} = V_{BE1}$, $I_{C2} = I_{C1}$.

the current level in an adjacent BJT. Transistor Q_1 in Fig. 7-22 is connected to function as a diode, with its base and emitter terminals commoned with Q_2, which functions as a transistor. The voltage drop across Q_1 (collector and base to emitter) is identified as V_{BE}, and so the current through Q_1 is

$$ I_{C1} = \frac{V_{CC} - V_{BE}}{R_1} $$

V_{BE} is usually 0.7 V, but the precise value of V_{BE1} depends upon the level of I_{C1}. Because the base and emitter terminals of Q_1 and Q_2 are commoned, V_{BE1} and V_{BE2} are the same voltage. Also, because Q_1 and Q_2 are identical devices, each transistor will have the same level of collector current when they have the same base-emitter voltage. Consequently, I_{C2} (the collector current in Q_2) *mirrors* the current in Q_1. If I_{C1} is altered by adjustment of R_1, I_{C2} changes to match the change in I_{C1}. In this situation Q_2 is biased to any desired level of I_C without the use of a voltage divider or an emitter resistor.

Resistors

The resistivity of semiconductor material is a function of doping density, and so resistors can be produced by doping strips of material and providing terminals (Fig. 7-23). The range of resistor values that may be produced by the diffusion process varies from ohms to hundreds of kilohms. The typical tolerance, however, may be no better than ±5% and may even be as high as ±20%. However, if all resistors are diffused at the same time, then the *tolerance ratio* can be good. For example, several resistors with the same nominal value may all have a ±20% tolerance but have actual resistance values within a few percent of each other.

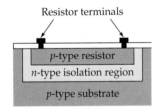

Figure 7-23 A monolithic IC resistor is created by doping a strip of semiconductor material to give the desired resistance.

Another method of producing resistors for integrated circuits uses the thin-film technique. In this process a metal film is deposited on a glass or silicon dioxide surface. The thickness, width, and length of the film are regulated to give a desired resistance value. Since diffused resistors can be processed while diffusing transistors, the diffusion technique is the least expensive and the most frequently used.

Capacitors

Because all *pn*-junctions have capacitance, capacitors may be produced by fabricating suitable junctions. As in the case of other diffused components, parasitic junctions are unavoidable. Both the parasitic and the main junction must be kept reverse-biased to avoid direct current flow. The depletion region width and junction capacitance vary with changes in reverse bias; so for capacitance value stability, a dc reverse bias greater than maximum signal voltage must be maintained across the junction.

Integrated circuit capacitors may also be fabricated by using the silicon dioxide surface layer as a dielectric. A heavily doped *n*-region is diffused to form one plate of the capacitor. The other plate is created by depositing a film of aluminum on the silicon dioxide (see Fig. 7-24). Voltages of any polarity may be employed with this type of capacitor, and the breakdown voltage is very much larger than that for diffused capacitors. The junction areas available for creation of IC capacitors are very small indeed, so that normally only picofarad capacitance values are possible.

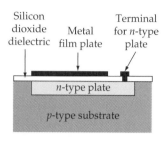

Figure 7-24 IC capacitor using silicon dioxide as the dielectric. The lower plate is a heavily doped *n*-type semiconductor, and the upper plate is a metal film.

Section 7-6 Review

7-6.1 Discuss the fabrication of integrated circuit transistors and diodes.

7-6.2 Explain how IC resistors and capacitors are constructed.

7-7 TRANSISTOR AND IC PACKAGING

Discrete Transistor Packaging

Many low-power transistors are encapsulated in resin with protruding metal connecting leads, as illustrated in Fig. 7-25. This is known as a TO-92 package. Note the emitter, base, and collector terminal connections. These are in the sequence E, B, and C, from left to right, if we look at the bottom of the transistor with the flat side uppermost. This kind of package is economical, but it has a limited operating temperature range.

Figure 7-26 shows another method of low-power transistor packaging where the device is hermetically scaled in a metal can. Depending on the size of the can, this is identified as a TO-5 to TO-18 package. The transistor is first mounted with its collector in mechanical and electrical contact with a heat-conducting metal plate. Wires (insulated from the plate) pass through for the emitter and base connections, and the covering metal can is welded to the plate. Looking at the bottom of the transistor, and moving clockwise from the tab, the terminals are identified as E, B, and C. The metal can enclosure gives a greater temperature range and greater power dissipation than resin encapsulation.

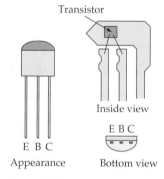

Figure 7-25 Low-power resin-encapsulated transistor; TO-92 package.

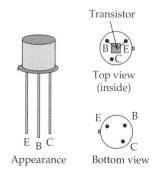

Figure 7-26 Low-power transistor, enclosed in a metal can; TO-18 package.

For high-power transistor packaging, a sealed can (TO-3) is often used (Fig. 7-27a). In this case the heat-conducting plate is much larger than in the TO-5 package, and it is designed for mounting directly on a heat sink. Connecting pins are provided for the base and the emitter, and the collector connection is made by means of the metal plate. Note the terminal identifications, again looking at the bottom of the device.

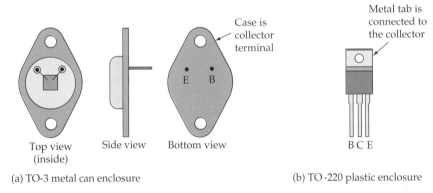

(a) TO-3 metal can enclosure

(b) TO-220 plastic enclosure

Figure 7-27 Metal can-type and plastic-encapsulated high-power transistors.

Figure 7-27b shows a plastic-encapsulated power transistor with a metal tab for fastening to a heat sink. Once again, note the transistor terminal identifications, and note that the the metal tab is connected to the collector. Several other types of package are available for low-power and high-power devices.

IC Packages

Like semiconductor devices, integrated circuits must be packaged to provide mechanical protection and terminals for electrical connection. Several standard packages in general use are illustrated in Fig. 7-28.

The metal-can-type of container affords electromagnetic shielding for the IC chip which cannot be obtained with the plastic packages (Fig. 7-28a). Note that the terminals are numbered clockwise from the tab, looking at the bottom of the can.

Plastic *dual-in-line* (*DIP*) packages (Fig. 7-28b) are more economical than metal cans and are widely used for industrial and consumer applications. The terminal numbering system again follows the bottom-view clockwise rule, starting from a notch in the plastic. A dimple, or other marker, is sometimes used instead of a notch. This is located on top of the package close to pin 1. DIP-type packages are much more convenient for circuit board use than cans, because the lead arrangement and flat package allow greater circuit densities. The surface mounting package shown in Fig. 7-28c is most economical for printed circuit boards because the terminals are soldered directly onto the board without the need for drilling holes.

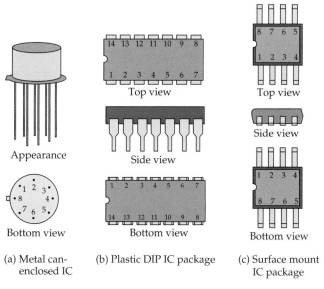

Appearance | Side view | Side view

Bottom view | Bottom view | Bottom view

(a) Metal can-enclosed IC | (b) Plastic DIP IC package | (c) Surface mount IC package

Figure 7-28   Integrated circuits are usually packaged in metal cans, dual-in-line plastic (DIP), or surface-mount plastic enclosures.

Section 7-7 Review

7-7.1 Sketch the terminal arrangements for TO-92, TO-5, and TO-3 transistor packages, and identify the emitter, base, and collector terminals on each package.

7-7.2 Sketch metal can and DIP integrated circuit packages, and show the terminal numbering systems.

Review Questions

Section 7-1

7-1   Describe the process of preparing semiconductor material for device manufacture.

7-2   Explain diffusion and epitaxial growth, and discuss the application of each process to transistor manufacture.

Section 7-2

7-3   Draw sketches to show the construction of alloy diodes and diffused diodes. Briefly explain in each case.

Section 7-3

7-4 Explain the various transistor manufacturing requirements for maximum performance with respect to (a) current gain, (b) power dissipation, (c) frequency response, (d) switching response, and (e) breakdown voltage.

Section 7-4

7-5 Describe the microalloy and microalloy-diffusion techniques for transistor manufacture. Explain the advantages and disadvantages of transistors made by these techniques.

7-6 Using sketches, explain diffused mesa and epitaxial mesa transistors. Discuss the advantages and disadvantages of mesa transistors.

7-7 Explain the manufacturing process for diffused-planar and annular transistors, and discuss the advantages and disadvantages of each.

7-8 Show that most of the power dissipation in a transistor occurs at the collector-base junction.

Section 7-5

7-9 Briefly explain the thin-film and thick-film methods of integrated circuit manufacture, and discuss their advantages and disadvantages.

7-10 Using illustrations, explain the fabrication process for monolithic integrated circuits. Discuss the advantages and disadvantages of monolithic ICs.

7-11 Describe hybrid integrated circuits.

Section 7-6

7-12 Sketch the construction and circuit diagram for two monolithic integrated circuit transistors showing the parasitic components. State any precautions necessary in the use of the circuit.

7-13 Briefly explain how diodes, resistors, and capacitors are fabricated in monolithic integrated circuits.

7-14 Sketch the circuit of a current mirror, and explain how it functions.

Section 7-7

7-15 Sketch TO-92, TO-18, and TO-3 transistor packages, and identify the emitter, base, and collector terminals.

7-16 Sketch metal-can and DIP integrated circuit packages, and show the terminal numbering system.

CHAPTER 8
BJT Specifications and Performance

CONTENTS

Objectives

You will be able to:

1 Identify important parameters on transistor data sheets for small-signal, high-power, high-frequency, and switching transistors, and determine the parameter values.

2 Calculate circuit power gains and power output changes in decibels.

3 Define Miller effect. Using the data sheet specifications, determine the input capacitance for transistor circuits.

4 Explain the effects of input and stray capacitances on BJT circuit frequency response, and calculate cutoff frequencies for CE, CB, and CC circuits.

5 Determine transistor switching times from their data sheets, and calculate device turn-on and turn-off times.

6 Define noise figure and noise factor, and estimate circuit output noise levels originating from resistors and transistors.

7 Using power derating factors and derating graphs, determine transistor maximum power dissipations at any temperature.

8 Sketch a thermal resistance circuit for a transistor with a heat sink. Determine the required thermal resistance for transistor heat sinks, and select suitable heat sinks for given circuit conditions.

INTRODUCTION

The performance characteristics and parameters for each type of transistor are specified on a data sheet published by the device manufacturer. The specifications must be correctly interpreted to achieve optimum transistor performance and to avoid breakdowns. Transistors must be selected to have a specified maximum collector-emitter voltage greater than the voltage applied to the CE terminals. The specified maximum collector current must be greater than the highest I_C level that may flow in the circuit. The maximum power that may be dissipated in a device is normally listed for a case temperature of 25°C. This value must be derated for operation at higher temperatures, and heat sinks must be employed where appropriate. Transistor cutoff frequency, switching times, noise figure, and input capacitance can all be very important in various applications.

8-1 TRANSISTOR DATA SHEETS

When selecting a transistor for a particular application, the data sheets provided by device manufacturers must be consulted. Typical transistor data sheets are shown in data sheets A-5 to A-9 in Appendix A, and portions are reproduced in this chapter.

Most data sheets start with the device type number at the top of the page, a descriptive title, and a list of major applications for the device. This information is usually followed by mechanical data in the form of an illustration showing the appearance of the package, as well as indicating which leads are collector, base, and emitter (see Fig. 8-1).

The absolute maximum ratings of the transistor at a temperature of 25°C are listed next. These are the maximum voltages, currents, and power dissipation

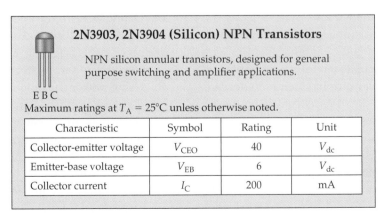

2N3903, 2N3904 (Silicon) NPN Transistors

NPN silicon annular transistors, designed for general purpose switching and amplifier applications.

E B C
Maximum ratings at T_A = 25°C unless otherwise noted.

Characteristic	Symbol	Rating	Unit
Collector-emitter voltage	V_{CEO}	40	V_{dc}
Emitter-base voltage	V_{EB}	6	V_{dc}
Collector current	I_C	200	mA

Figure 8-1 Portion of 2N3903 and 2N3904 transistor data sheet listing maximum terminal voltages and collector current.

that the device can survive without breaking down. It is very important that these ratings never be exceeded; otherwise, failure of the device is quite possible. *For reliability, the maximum ratings should not even be approached.* The maximum transistor ratings must also be adjusted downward for operation at temperatures greater than 25°C.

After the absolute maximum ratings, the data sheet normally shows a complete list of electrical characteristics for the device. (See data sheet A-5 in Appendix A.) Again, these are specified at 25°C, and allowances are necessary for temperature variations.

A complete understanding of all the quantities specified on a data sheet will not be achieved until circuit analysis and design are studied. Some of the most important quantities are considered below. It is important to note that the ratings for transistors are stated for specified circuit conditions. The ratings are no longer valid if these conditions change.

V_{CBO}	*Collector-base voltage*—maximum dc voltage for reverse-biased collector-base junction.
V_{CEO}	*Collector-emitter voltage*—maximum collector-emitter dc voltage with base open-circuited.
V_{EBO}	*Emitter-base voltage*—maximum emitter-base reverse-bias dc voltage.
I_C	*Collector current*—maximum dc collector current.
I_{CBO} or I_{CO}	*Collector cutoff current*—dc collector current with the collector-base junction reverse-biased and the emitter open-circuited.
$V_{CE(sat)}$	*Collector-emitter saturation voltage*—collector-emitter voltage with the device in saturation.
h_{FE}	*Static forward current transfer ratio*—common emitter ratio of dc collector current and base current.
NF	*Noise figure*—ratio of total noise output to total noise input expressed as a decibel (dB) ratio. Defines the amount of noise added by the device.
f_{hfe} or $f_{\alpha e}$	*Common-emitter cutoff frequency*—common-emitter operating frequency at which the device current gain falls to 0.707 of its normal (mid-frequency) value.
f_{hfb} or $f_{\alpha b}$	*Common-base cutoff frequency*—as above, for common base.

From the section of the 2N3903 and 2N3904 data sheet reproduced in Fig. 8-1, the maximum collector-emitter voltage is specified as $V_{CEO} = 40$ V. Obviously, this device should not be used in any circuit with a supply exceeding 40 V; otherwise the transistor might break down. Preferably, the circuit supply voltage should always be less than the specified maximum V_{CEO}.

The maximum V_{EB} is listed as 6 V. This is the highest voltage that may be applied *in reverse* across the base-emitter junction of the transistor. If a greater voltage is used, the base-emitter junction of the transistor is likely to break

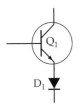

Figure 8-2 A diode connected in series with a transistor emitter protects the BE junction against excessive reverse voltage.

down. The *maximum (reverse) V_{EB} for most transistors is 5 V*. In circumstances where the applied voltage might exceed this level, a clipping circuit may be used (see Sections 3-7 and 3-8). Alternatively, a diode can be connected in series with the transistor emitter to increase the breakdown voltage, as shown in Fig. 8-2. The normal forward-biased base-emitter voltage for a transistor is, of course, approximately 0.7 V for a silicon transistor and approximately 0.3 V for a germanium transistor.

Figure 8-3 shows sections of a transistor data sheet which specify the *dc current gain*, h_{FE}, and the *small signal* (or ac) *current gain*, h_{fe}. For a 2N3904 device with $I_C = 10$ mA, h_{FE} is listed as 100 minimum and 300 maximum (see the solid underlined quantities). If the transistor collector current is to be about 1 mA, then the minimum h_{FE} is 70 (dashed underline). The maximum h_{FE} for $I_C = 1$ mA is not specified. It is unlikely that $h_{FE(max)}$ would exceed 300 for this condition. However, because the manufacturer does not list a maximum value, it remains uncertain. This situation illustrates the importance of reliable bias circuits that make I_C and V_{CE} largely independent of h_{FE} (see Chapter 5).

The *small-signal current gain* (h_{fe}) for a 2N3904 is listed in Fig. 8-3 as 100 minimum and 400 maximum. Once again, this parameter is specified for a particular set of bias conditions. The signal frequency at which h_{fe} was measured is specified as 1 kHz. Section 8-3 discusses the fact that h_{fe} decreases at high frequencies.

2N3903, 2N3904

Characteristic		Symbol	Min	Max	Unit
DC current gain		h_{FE}			–
($I_C = 0.1$ mA dc, $V_{CE} = 1$ V dc)	2N3903		20	–	
	2N3904		40	–	
($I_C = 1.0$ mA dc, $V_{CE} = 1$ V dc)	2N3903		35	–	
	2N3904		70	–	
($I_C = 10$ mA dc, $V_{CE} = 1$ V dc)	2N3903		50	150	
	2N3904		100	300	
($I_C = 50$ mA dc, $V_{CE} = 1$ V dc)	2N3903		30	–	
	2N3904		60	–	
($I_C = 100$ mA dc, $V_{CE} = 1$ V dc)	2N3903		15	–	
	2N3904		30	–	
Small signal current gain		h_{fe}			–
($I_C = 1$ mA dc, $V_{CE} = 10$ V, f = 1 kHz)					
	2N3903		50	200	
	2N3904		100	400	

Figure 8-3 Portion of 2N3903 and 2N3904 transistor data sheet listing dc current gain and small-signal ac current gain.

Example 8-1

Referring to the 2N3905 specification in Appendix A-6, determine the h_{FE} range for $I_C = 10$ mA, and the h_{fe} range for $f = 1$ kHz. Also determine the emitter-base breakdown voltage.

Solution

At $I_C = 10$ mA, $h_{FE} = 50$ to 150

At $f = 1$ kHz, $h_{fe} = 50$ to 200

$$V_{(BR)EBO} = 5 \text{ V}$$

Practice Problem

8-1.1 From the data sheet A-7 in Appendix A, determine the following quantities for a 2N3251 transistor: $h_{FE(min)}$ at $I_C = 1$ mA, $h_{fe(min)}$ at $f = 1$ kHz, $V_{CE(max)}$, $V_{EB(max)}$, and noise figure.

8-2 DECIBELS AND HALF-POWER POINTS

Power Measurement in Decibels

The *power gain* (A_p) of an amplifier may be expressed in terms of the log of the ratio of output power (p_o) to input power (p_i). This is illustrated in Fig. 8-4.

$$A_p = \log_{10}\left(\frac{p_o}{p_i}\right) \text{ bels}$$

or $$A_p = 10 \log_{10}\left(\frac{p_o}{p_i}\right) \text{ decibels (dB)} \qquad (8\text{-}1)$$

Thus, the *decibel* is the unit of *power gain*, or *power level change*.

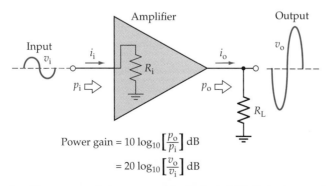

$$\text{Power gain} = 10 \log_{10}\left[\frac{p_o}{p_i}\right] \text{ dB}$$
$$= 20 \log_{10}\left[\frac{v_o}{v_i}\right] \text{ dB}$$

Figure 8-4 Amplifer power gain is measured in decibels (dB). Output power level changes are also measured in decibels.

The power dissipated in a resistance is (v^2/R).

So
$$A_p = 10 \log_{10} \left(\frac{(v_o)^2/R_L}{(v_i)^2/R_i} \right) \text{ dB}$$

When $R_i = R_L$,
$$A_p = 10 \log_{10} \left(\frac{v_o}{v_i} \right)^2 \text{ dB}$$

or
$$A_p = 20 \log_{10} \left(\frac{v_o}{v_i} \right) \text{ dB} \qquad (8\text{-}2)$$

Also, $P_o = i_o^2 R_L$ and $P_i = i_i^2 R_i$

So
$$A_p = 20 \log_{10} \left(\frac{i_o}{i_i} \right) \text{ dB} \qquad (8\text{-}3)$$

It is important to note that *the above equations are correct only when the load resistance connected to the circuit output is equal to the circuit input resistance.* However, in the electronics industry, these equations are sometimes applied regardless of the relationship between input and load resistances!

Amplifier *output power level changes* (ΔP_o) are also measured in decibels, and in this case the equations become

$$\Delta P_o = 10 \log_{10} \left(\frac{p_2}{p_1} \right) \text{ dB} \qquad (8\text{-}4)$$

where p_1 is the initial output power level and p_2 is the new level.

Also,
$$\Delta P_o = 20 \log_{10} \left(\frac{v_2}{v_1} \right) \text{ dB} \qquad (8\text{-}5)$$

and
$$\Delta P_o = 20 \log_{10} \left(\frac{i_2}{i_1} \right) \text{ dB} \qquad (8\text{-}6)$$

where v_1 and i_1 are the initial output voltage and current levels, and v_2 and i_2 are the new levels. Equations 8-4 to 8-6 are correct so long as the amplifier load resistance remains constant. Examples 8-2 and 8-3 demonstrate that the output power of an amplifier is reduced by 3 dB when the measured power falls to half its normal level or when the measured voltage falls to 0.707 of its normal level.

Example 8-2

The output power from an amplifier is 50 mW when the signal frequency is 5 kHz. The power output falls to 25 mW when the frequency is increased to 20 kHz. Calculate the output power change in decibels.

Solution

Eq. 8-4:

$$\Delta P_\text{o} = 10 \log_{10}\left(\frac{p_2}{p_1}\right) = 10 \log_{10}\left(\frac{25 \text{ mW}}{50 \text{ mW}}\right)$$

$$= -3 \text{ dB}$$

Half-Power Points

Figure 8-5 shows a typical graph of amplifier output voltage or power plotted versus frequency. Note that the frequency scale is logarithmic. The output normally remains constant over a middle range of frequencies and falls off at low and high frequencies owing to effects explained in Sections 8-3 and 8-4. The gain over this middle range is termed the *mid-frequency gain*. The low frequency and high frequency at which the gain falls by 3 dB are designated f_1 and f_2 respectively. This range (from f_1 to f_2) is normally considered the useful range of operating frequency for the amplifier, and the frequency difference ($f_2 - f_1$) is termed the *amplifier bandwidth* (*BW*).

Frequencies f_1 and f_2 are termed the *half-power points,* or the *3 dB points*. This is because, as shown in Example 8-2, the power output is −3 dB from its normal level when p_2 is half p_1. When the amplifier output is expressed as a voltage on the graph of frequency response, the 3 dB points (f_1 and f_2) occur when v_2 is 0.707 v_1. This relationship is demonstrated in Example 8-3.

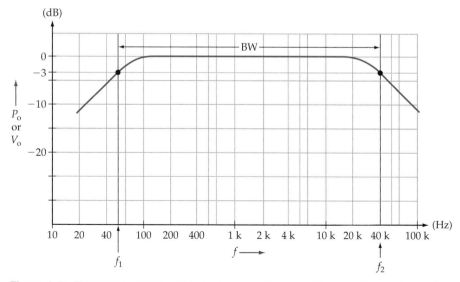

Figure 8-5   The output voltage and output power of an amplifier normally remain constant over a middle band of signal frequencies and fall off at high and low frequencies.

Example 8-3

The output voltage of an amplifier is measured as 1 V at 5 kHz and as 0.707 V at 20 kHz. Calculate the output power change.

Solution

Eq. 8-5:
$$\Delta P_{\text{o}} = 20 \log_{10}\left(\frac{v_2}{v_1}\right) = 20 \log_{10}\left(\frac{0.707 \text{ V}}{1 \text{ V}}\right)$$

$$= -3 \text{ dB}$$

Practice Problem

8-2.1 The output voltage of an amplifier is 2 V when the signal frequency is 3 kHz, and 0.5 V when the signal frequency is increased to 50 kHz. Calculate the decibel change in output power.

8-3 BJT CUTOFF FREQUENCY AND CAPACITANCES

Device Cutoff Frequency

All transistors have junction capacitances (see Sections 2-6 and 6-3). The junction capacitances and the transit time of charge carriers through the semiconductor material limit the high-frequency performance of the device. This limitation is expressed as a *cutoff frequency* (f_α), which is the frequency at which the transistor current gain falls to 0.707 of its gain at low and medium frequencies. The cutoff frequency can be expressed in two ways, the *common emitter cutoff frequency* $f_{\alpha e}$, and the *common base cutoff frequency* ($f_{\alpha b}$). $f_{\alpha e}$ is the frequency at which the common emitter current gain (h_{fe}) falls to 0.707 × (mid-frequency h_{fe}). $f_{\alpha b}$ is the frequency at which the common base current gain (h_{fb}) falls to 0.707 × (mid-frequency h_{fb}). It can be shown that

$$f_{\alpha b} = h_{\text{fe}}\, f_{\alpha e} \tag{8-7}$$

Equation 8-7 demonstrates that $f_{\alpha b}$ can be identified as the *current gain-bandwidth product*, or *total bandwidth* (f_{T}). f_{T} is the designation normally used on transistor data sheets. The data sheet portion in Fig. 8-6 shows that f_{T} is 250 MHz for a 2N3903 transistor and 300 MHz for a 2N3904.

Junction Capacitances

As already discussed, all BJTs have junction capacitances. These are the base-collector capacitance (C_{bc}) and the base-emitter capacitance (C_{be}) (see Fig. 8-7). The device data sheet portion in Fig. 8-8 shows that the capacitances are usually listed as an *output capacitance* (C_{obo}), which is the same as C_{bc}, and as an

Characteristic	Symbol	Min	Max	Unit
Small-Signal Characteristics				
Current-gain-bandwidth product (I_C = 10 mA dc, V_{CE} = 20 V, f = 100 MHz) 2N3903 2N3904	f_T	 250 300	 — —	MHz

Figure 8-6 Specification of transistor cutoff frequency on the 2N3903 and 2N3904 data sheet.

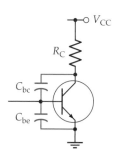

Figure 8-7 All transistors have junction capacitances.

Characteristic	Symbol	Min	Max	Unit
Output capacitance (V_{CB} = 5 Vdc, I_E = 0, f = 1.0 MHz)	C_{obo}	—	4.0	pF
Input capacitance (V_{BE} = 0.5 Vdc, I_C = 0, f = 1.0 MHz)	C_{ibo}	—	8.0	pF

Figure 8-8 Junction capacitance specifications for 2N3903 and 2N3904 transistors.

input capacitance (C_{ibo}), which corresponds to C_{be}. It should be noted that in both cases, the capacitance values are specified for reverse-biased junctions with zero current levels. C_{obo} varies when V_{CB} is altered; however, the changes are usually a maximum of approximately ±3 pF. C_{ibo} changes substantially from the specified capacitance when the base-emitter junction is forward-biased (as it always is for linear BJT operation).

When an emitter current flows across the BE junction, there is a diffusion capacitance (see Section 2-6) at the junction. This is directly proportional to the current level. It can be shown that

$$C_{be} \approx \frac{6.1\, I_E}{f_T} \tag{8-8}$$

Miller Effect

Figure 8-9 shows an input signal ($+\Delta V_i$) applied to the base of a transistor connected in CE configuration. If the circuit voltage amplification is $-A_v$, then the collector voltage change is

$$\Delta V_o = -A_v \times \Delta V_i$$

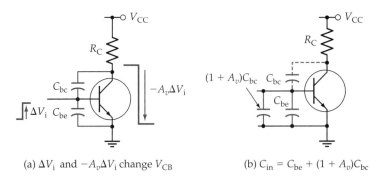

(a) ΔV_i and $-A_v\Delta V_i$ change V_{CB} (b) $C_{in} = C_{be} + (1 + A_v)C_{bc}$

Figure 8-9   Because of the Miller effect, the input capacitance for a common-emitter amplifier is $C_{in} = C_{be} + (1 + A_v)C_{bc}$.

Note that, because of the phase reversal between input and output, the col-
lector voltage is *reduced* by $(A_v\,\Delta V_i)$ when the base voltage is *increased* by ΔV_i.
The increase in V_B and decrease in V_C result in a total collector-base voltage
reduction of

$$\Delta V_{CB} = \Delta V_i + A_v\Delta V_i$$
$$= \Delta V_i(1 + A_v)$$

This voltage change appears across the collector-base capacitance.

From the equation $Q = C \times \Delta V$, it is found that the charge supplied to the
input of the circuit is

$$Q = C_{bc} \times \Delta V_i(1 + A_v)$$

or
$$Q = (1 + A_v)C_{bc} \times \Delta V_i$$

Therefore, *'looking into'* the base, the collector-base capacitance appears to be
$(1 + A_v)C_{bc}$. So the capacitance is *amplified* by a factor of $(1 + A_v)$. This is known
as the *Miller effect*.

The total input capacitance (C_{in}) to the transistor is $(1 + A_v)C_{bc}$ in parallel
with the base-emitter capacitance (C_{be}):

$$C_{in} = C_{be} + (1 + A_v)C_{bc} \qquad (8\text{-}9)$$

It should be noted that the Miller effect occurs only with amplifiers that
have a 180° phase shift between input and output (an inverting amplifier).
Consequently, it occurs with CE circuits but not with CB and emitter-follower
circuits.

Example 8-4

The transistor in the circuit in Fig. 8-10 has $I_C = 1$ mA, $h_{fe} = 50$, $h_{ie} = 1.3$ kΩ,
$f_T = 250$ MHz, and $C_{bc} = 5$ pF. Calculate the input capacitance when the
circuit is operated as a CE amplifier with R_E bypassed.

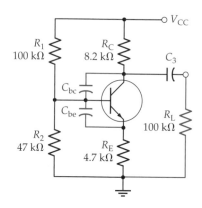

Figure 8-10 The upper cutoff frequency for a transistor circuit can be limited by the input capacitance.

Solution

From Eq. 6-12, $|A_v| = \dfrac{h_{fe}(R_C\|R_L)}{h_{ie}} = \dfrac{50 \times (8.2\ \text{k}\Omega\|100\ \text{k}\Omega)}{1.3\ \text{k}\Omega}$

$= 291$

Eq. 8-8: $C_{be} \approx \dfrac{6.1\ I_E}{f_T} \approx \dfrac{6.1 \times 1\ \text{mA}}{250\ \text{MHz}}$

$= 24.4\ \text{pF}$

Eq. 8-9: $C_{in} = C_{be} + (1 + A_v)C_{bc} = 24.4\ \text{pF} + [(1 + 291) \times 5\ \text{pF}]$

$= 1.48\ \text{nF}$

Practice Problem

8-3.1 A CE circuit with a voltage gain of 100 has $I_C = 0.75$ mA, $C_{bc} = 3$ pF, and $f_T = 300$ MHz. Calculate C_{in} when a 45 pF capacitor is connected across the collector-base terminals.

8-4 BJT CIRCUIT FREQUENCY RESPONSE

Coupling and Bypass Capacitor Effects

Consider the typical transistor amplifier frequency response illustrated in Fig. 8-5. As explained in Section 8-2, the amplifier voltage gain is constant over a middle range of signal frequencies, and it falls at the low and high ends of the frequency range.

The gain fall-off at low signal frequencies is due to the effect of coupling and bypass capacitors. Recall that the reactance of a capacitor is $X_C = 1/(2\pi f C)$. At medium and high frequencies, the factor f makes X_C very small,

Fundamentals of Electronic Devices and Circuits

so that all coupling and bypass capacitors behave as ac short circuits. At low frequencies, X_C is large enough to divide the voltages across the capacitors and series resistances (see Fig. 8-11). As the signal frequency gets lower, the capacitive reactance increases, more of the signal is lost across the capacitors, and the circuit gain continues to fall. Coupling and bypass capacitors are further investigated in Section 12-1.

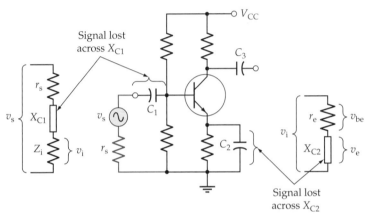

Figure 8-11 The low-frequency fall-off in voltage gain in a transistor amplifier is due to signal loss across coupling and bypass capacitors.

Input-Capacitance Effect on CE and CB Circuits

The input capacitance of an amplifier (discussed in Section 8-3) reduces the circuit gain by 3 dB when the capacitive impedance equals the resistance in parallel with the input (see Fig. 8-12). That is when

$$X_{Ci} = Z_i \| r_s \qquad (8\text{-}10)$$

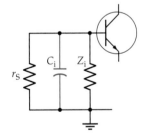

Figure 8-12 The input resistance (Z_i) and the signal source resistance (r_s) are in parallel with the circuit input capacitance (C_{in}).

Thus all circuits have an *input-capacitance-limited upper cutoff frequency* ($f_{2(i)}$). As already explained, the input capacitance of a CE circuit is amplified by the Miller effect, but Miller effect does not occur in a CB circuit. Consequently, a CB circuit operates to a much higher signal frequency than a similar CE circuit.

Example 8-5

Calculate the input-capacitance-limited upper cutoff frequency for the circuit in Fig. 8-13 (reproduced from Fig. 8-10): (a) when the circuit is used in CE configuration with R_E bypassed, and (b) when operating as a CB circuit with the base bypassed to ground. As in Example 8-4, $C_{bc} = 5$ pF and $C_{be} = 24.4$ pF. Also, $h_{fe} = 50$, $h_{ie} = 1.3$ kΩ, $h_{ib} = 24.5$ Ω ($= r'_e$), and $r_s = 600$ Ω.

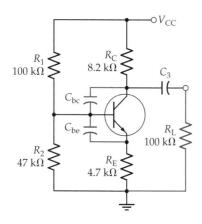

Figure 8-13 Circuit that may be connected to function in either CE or CB configuration.

Solution

(a) Common emitter circuit:

$$Z_i = R_1 \| R_2 \| h_{ie} = 100 \text{ k}\Omega \| 47 \text{ k}\Omega \| 1.3 \text{ k}\Omega$$

$$= 1.25 \text{ k}\Omega$$

$$r_s \| Z_i = 600 \text{ }\Omega \| 1.25 \text{ k}\Omega$$

$$= 405 \text{ }\Omega$$

From Ex. 8-4, $\qquad C_{in} = 1.48 \text{ nF}$ (due to Miller effect)

From Eq. 8-10, $\qquad f_{2(i)} = \dfrac{1}{2\pi C_{in}(r_s \| Z_i)} = \dfrac{1}{2\pi \times 1.48 \text{ nF} \times 405 \text{ }\Omega}$

$$= 266 \text{ kHz}$$

(b) Common base circuit:

$$Z_i = R_E \| h_{ib} = 4.7 \text{ k}\Omega \| 24.5 \text{ }\Omega$$

$$= 24.4 \text{ }\Omega$$

$$r_s \| Z_i = 600 \text{ }\Omega \| 24.4 \text{ }\Omega$$

$$= 23.4 \text{ }\Omega$$

$$C_{in} = C_{be} + C_{bc} = 24.4 \text{ pF} + 5 \text{ pF}$$

$$= 29.4 \text{ pF}$$

From Eq. 8-10, $\qquad f_{2(i)} = \dfrac{1}{2\pi C_{in}(r_s \| Z_i)} = \dfrac{1}{2\pi \times 29.4 \text{ pF} \times 23.4 \text{ }\Omega}$

$$= 231 \text{ MHz}$$

Input-Capacitance Effects on Emitter Follower

When a transistor is used as an emitter follower, its BE junction voltage is not significantly altered by the ac input signal, because virtually all of v_i appears at the emitter as v_o (see Fig. 8-14). Since there is no Miller effect to amplify C_{bc},

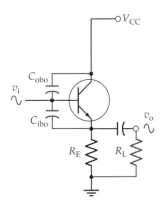

Figure 8-14 There is no Miller effect input-capacitance amplification with an emitter follower circuit.

the input capacitance is $C_{ibo} \| C_{obo}$. So the input-capacitance-limited cutoff frequency for an emitter follower is very much higher than that for a CE circuit.

Stray Capacitance

Figure 8-15 illustrates the fact that *stray capacitances* (C_{si} and C_{so}) exist in all transistor circuits. This is capacitance between connecting wires and ground, and normally it is extremely small. The stray capacitance at the device base is usually much smaller than the input capacitance at the base, so that it can usually be neglected. When this is not the case, the stray capacitance must be included with the input capacitance for CE circuit cutoff-frequency calculations.

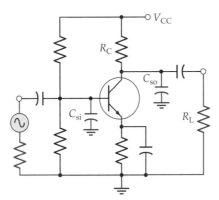

Figure 8-15 Stray capacitance can cause a reduction in the transistor amplifier gain at high frequencies.

At the circuit output, the impedance of the stray capacitance (C_{so}) is very high at low and medium signal frequencies, so that it has no effect on the gain. At high frequencies, the stray capacitive impedance becomes small enough to shunt away some of the output current, and thus it reduces the circuit gain. In Fig. 8-15, it can be seen that the circuit ac load consists of the output stray capacitance (C_{so}) in parallel with R_C and R_L. Rewriting the voltage gain equation for CE and CB circuits gives

$$|A_v| = \frac{h_{fe}(R_C \| R_L \| X_{cso})}{h_{ie}}$$

If the transistor cutoff frequency has not caused the gain to fall off at a lower frequency, then the gain falls by 3 dB when $X_{cso} = R_C \| R_L$. This is an *output-capacitance-limited cutoff frequency* ($f_{2(o)}$).

If the voltage gain falls by 3 dB at $f_{\alpha e}$ because of the transistor, and by 3 dB at the same frequency because of the stray capacitance, then the gain is down by 6 dB (see point A in Fig. 8-16). Consequently, the amplifier upper cutoff frequency is lower than $f_{\alpha e}$. This is illustrated by point B in Fig. 8-16.

As discussed, there is an additional 3 dB attenuation when $X_{cso} = R_C \| R_L$. If $X_{cso} = 2(R_C \| R_L)$ at $f_{\alpha e}$, the additional attenuation can be shown to be 1 dB. With $X_{cso} = 5(R_C \| R_L)$ at $f_{\alpha e}$, there is only a 0.2 dB additional attenuation (point C).

In some circumstances it is desirable to set the upper cutoff frequency of a circuit well below the transistor cutoff frequency (see point D in Fig. 8-16).

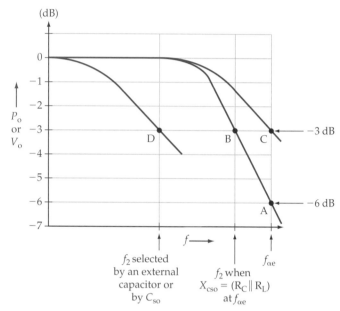

Figure 8-16   The BJT cutoff frequency (f_T) determines the upper cutoff frequency of a circuit so long as the impedance of the output stray capacitance (C_{so}) is much larger than $R_C \| R_L$. When X_{CSO} is much smaller than $R_C \| R_L$, the circuit upper cutoff frequency is determined by C_{so}.

This is done simply by connecting a capacitor from the collector terminal to ground exactly as C_{so} is shown in Fig. 8-15. The capacitance value is calculated at the desired cutoff frequency to give

$$X_c = R_C \| R_L \tag{8-11}$$

Example 8-6

A transistor with $f_T = 50$ MHz and $h_{fe} = 50$ is employed in the CE amplifier in Fig. 8-17. Determine the upper 3 dB frequency for the device. Calculate the capacitance required for C_4 to give a 60 kHz upper cutoff frequency. Assume that $R_L \gg R_C$.

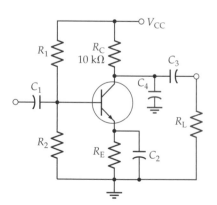

Figure 8-17   Circuit for Example 8-6.

Solution

From Eq. 8-7,

$$f_{\alpha e} = \frac{f_T}{h_{fe}} = \frac{50 \text{ MHz}}{50}$$

$$= 1 \text{ MHz}$$

From Eq. 8-11,

$$C_4 = \frac{1}{2\pi f_{2(o)} R_C} = \frac{1}{2 \times \pi \times 60 \text{ kHz} \times 10 \text{ k}\Omega}$$

$$= 265 \text{ pF}$$

Practice Problems

8-4.1 In a CE amplifier circuit $R_C = 5.6$ kΩ and $R_L = 56$ kΩ. The transistor used has $h_{fe} = 60$, $h_{ie} = 1.5$ kΩ, $C_{be} = 192$ pF, and $C_{bc} = 6$ pF. The signal source resistance is 1 kΩ, and the voltage divider bias resistors are $R_1 = 82$ kΩ and $R_2 = 39$ kΩ. Determine the input-capacitance-limited upper cutoff frequency.

8-4.2 A transistor in a single-stage CE amplifier has $h_{fe} = 75$ and $f_T = 12$ MHz. The load resistance is $R_C \| R_L = 20$ kΩ. Determine the upper 3 dB frequency (f_2) for the circuit. Also, calculate the output stray capacitance that will produce an additional 3 dB attenuation at f_2.

8-5 TRANSISTOR SWITCHING TIMES

Turn-On Times

For transistor switching circuits (see Sections 4-4 and 5-10), the switching speed of the device can be an important quantity. Consider the circuit in Fig. 8-18a. When the base input current is applied, the transistor does not switch *on* immediately. Like frequency response, the switching time is affected by junction capacitance and the transit time of electrons across the junctions. The time between the application of the input pulse and the beginning of collector current flow is termed the *delay time* (t_d) (see Fig. 8-18b). Even when the transistor begins to switch *on*, a finite time elapses before I_C reaches its maximum level. This is known as the *rise time* (t_r). The rise time is specified as the time required for I_C to go from 10% to 90% of its maximum level. As illustrated, the *turn-on time* (t_{on}) is the sum of t_d and t_r.

Turn-Off Times

When the input current is switched *off*, I_C does not go to zero until after a *turn-off time*, t_{off}, made up of a *storage time* (t_s) and a *fall time* (t_f), as illustrated. The fall time is specified as the time required for I_C to go from 90% to 10% of its

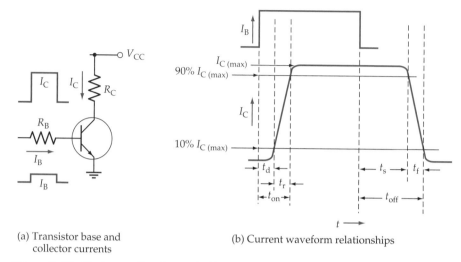

(a) Transistor base and
    collector currents

(b) Current waveform relationships

Figure 8-18   The turn-on time for a switching transistor is the sum of the delay time (t_d) and the rise time (t_r). The transistor turn-off time is the sum of the storage time (t_s) and the fall time (t_f).

maximum level. The storage time is the result of charge carriers being trapped in the depletion region when a junction polarity is reversed. When a transistor is in a saturated *on* condition (see Section 5-10), both the collector-base and emitter-base junctions are forward-biased. At switch-*off*, both junctions are reverse-biased. Before I_C begins to fall, the stored charge carriers must be withdrawn or made to recombine with opposite-type charge carriers.

For a fast-switching transistor, t_{on} and t_{off} must be of the order of nanoseconds. The excerpt from a 2N3904 transistor data sheet in Fig. 8-19 specifies the following switching times: $t_d = 35$ ns, $t_r = 35$ ns, $t_s = 200$ ns, and $t_f = 50$ ns.

Characteristic		Symbol	Min	Max	Unit
Switching Characteristics (2N3904)					
Delay time	$V_{CC} = 3$ V, $V_{BE} = 0.5$ V,	t_d	—	35	ns
Rise time	$I_C = 10$ mA, $I_{B1} = 1$ mA	t_r	—	35	ns
Storage time	$V_{CC} = 3$ V, $I_C = 10$ mA,	t_s	—	200	ns
Fall time	$I_{B1} = I_{B2} = 1$ mA	t_f	—	50	ns

Figure 8-19   Transistor switching times, as specified on a data sheet.

Example 8-7

The circuit shown in Fig. 8-18a uses a 2N3904 transistor and has an input pulse with a 5 μs pulse width (PW). Determine the time from the beginning of I_C until the transistor turns *off*.

Solution

I_C begins at t_d (at 10% of $I_{C(max)}$) after the start of the input pulse, and ceases at $(t_s + t_f)$ (at 10% of $I_{C(max)}$) after the end of the input pulse.

$$t = PW - t_d + t_s + t_f = 5 \text{ μs} - 35 \text{ ns} + 200 \text{ ns} + 50 \text{ ns}$$

$$= 5.215 \text{ μs}$$

Rise Time and Cutoff Frequency

Just how fast a circuit can switch from off to on is related to its upper cutoff frequency (f_H). It can be shown that

$$f_H = \frac{0.35}{t_r} \qquad (8\text{-}12)$$

So, for example, if the output pulse from the circuit in Fig. 8-18a has a rise time of 100 ns and the input pulse has a very much smaller rise time,

$$f_H = \frac{0.35}{100 \text{ ns}} = 3.5 \text{ MHz}$$

It should be noted that in this case f_H is the same as f_{ae}, because a common-emitter circuit is involved (see Section 8-3). Pulse testing can be used for rapidly determining the upper cutoff frequency of a circuit or device.

Switching Time Improvement

The turn-on time can be shortened by overdriving the transistor, that is, applying a larger base current than required for transistor saturation. However, a larger I_B level produces an increased storage time and thus lengthens the device turn-off time. Turn-off time can be reduced by the application of a large negative input voltage to rapidly discharge the forward-biased junctions at switch-off. But the effect of this would be to make the turn-on time longer, because the transistor base would have to be raised from its negative level before switch-on can begin.

Ideally for fast switching, the transistor base-emitter voltage should start at zero, and I_B should be large at switch-on but should rapidly settle down to the minimum level required for transistor saturation. Also, a large negative voltage should be applied for switch-off to rapidly discharge the forward-biased junctions, but this voltage should quickly return to zero. These desirable conditions are achieved by connecting a capacitor (C_1) in parallel with the base resistor, as in Fig. 8-20a.

Capacitor C_1, known as a *speed-up capacitor*, short-circuits R_B when a positive input ($+V_S$) is applied, and thus the initial level of base current is

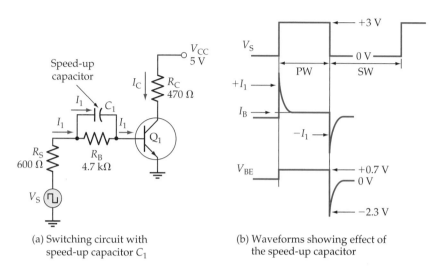

(a) Switching circuit with
speed-up capacitor C_1

(b) Waveforms showing effect of
the speed-up capacitor

Figure 8-20   Transistor switching times can be improved by the use of a speed-up capacitor.

increased to shorten the switch-on time. Without C_1 in the circuit, the base current is calculated from

Eq. 5-26:
$$I_B = \frac{V_S - V_{BE}}{R_B + R_S}$$

With C_1,
$$I_1 = \frac{V_S - V_{BE}}{R_S} \qquad (8\text{-}13)$$

The capacitor-charging current spike flows into the base terminal (see Fig. 8-20b).

When the capacitor is completely charged, I_1 returns to the normal I_B level for Q_1 saturation, and the capacitor voltage is then

$$V_{C1} = I_B R_B$$

At switch-off, V_S goes to zero and $-V_{C1}$ is applied to the transistor base, thus improving Q_1 switch-off time. After switch-off, V_{C1} returns to zero (discharged by R_B), so that there is no negative base-emitter voltage to affect the turn-on time.

A suitable speed-up capacitance value is calculated by allowing C_1 to charge by 60% during the specified transistor turn-on time. This will keep I_1 close to its maximum level during a reduced turn-on time, so as to turn Q_1 on as quickly as possible. C_1 charges via the signal source resistance (R_S), and it can be shown that for a 60% charge,

$$t_{on} = R_S C_1$$

giving
$$C_1 = \frac{t_{on}}{R_S} \qquad (8\text{-}14)$$

Capacitor C_1 has the undesirable effect of limiting the maximum signal frequency that may be used with the circuit. After Q_1 switches on, C_1 has to be completely charged before switch-off begins. Also, after Q_1 switches off, C_1 has to be completely discharged to near zero volts before another switch-on pulse can be applied. A capacitor will charge or discharge by more than 99% in a time period of $5CR$. Charge to C_1 is via R_S and discharge is via R_B. The time between switch-on and switch-off is the pulse width (PW) (see Fig. 8-20b), and this should not be less than

$$PW_{(min)} = 5R_SC_1 \qquad (8\text{-}15)$$

The time between switch-off and switch-on is the space width (SW) (Fig. 8-20b), which should be a minimum of

$$SW_{(min)} = 5R_BC_1 \qquad (8\text{-}16)$$

The maximum signal frequency that may be used with a given switching circuit is dictated by the sum of the minimum pulse and space widths:

$$T = PW + SW$$

giving
$$f_{(max)} = \frac{1}{PW_{(min)} + SW_{(min)}} \qquad (8\text{-}17)$$

Example 8-8

The circuit in Fig. 8-20a uses a transistor with a 100 ns turn-on time. Calculate a suitable speed-up capacitor value, and determine the maximum signal frequency that may be used with the circuit.

Solution

Eq. 8-14:
$$C_1 = \frac{t_{on}}{R_S} = \frac{100 \text{ ns}}{600 \text{ }\Omega}$$

$$= 167 \text{ pF (use 160 pF standard value)}$$

Eq. 8-15:
$$PW_{(min)} = 5\,R_S\,C_1 = 5 \times 600\text{ }\Omega \times 160\text{ pF}$$
$$= 0.48 \text{ }\mu s$$

Eq. 8-16:
$$SW_{(min)} = 5\,R_B\,C_1 = 5 \times 4.7\text{ k}\Omega \times 160\text{ pF}$$
$$= 3.76 \text{ }\mu s$$

Eq. 8-17:
$$f_{(max)} = \frac{1}{PW_{(min)} + SW_{(min)}} = \frac{1}{0.48 \text{ }\mu s + 3.76 \text{ }\mu s}$$
$$= 236 \text{ kHz}$$

Practice Problems

8-5.1 Calculate the turn-on time and turn-off time for a transistor with $t_d = 10\,\text{ns}$, $t_r = 12$ ns, $t_s = 15$ ns, and $t_f = 12$ ns. Also determine the time from the beginning of a 100 ns input pulse to the end of the BJT *on* time.

8-5.2 The upper cutoff frequency for a switching circuit (as in Fig. 8-18) is measured as 1.7 MHz. Calculate the rise time of the output when a pulse with negligible rise time is applied as input.

8-5.3 A direct-coupled switching circuit using a 2N3904 BJT has a 12 kΩ base resistance and a 350 Ω source resistance. Determine a suitable speed-up capacitor value. Calculate the minimum input pulse width and space width that should be used with the circuit.

8-6 TRANSISTOR CIRCUIT NOISE

Unwanted signals at the output of an electronics system are termed *noise*. The noise amplitude may be large enough to severely distort or completely swamp the wanted signals. Consequently, the noise level dictates the minimum signal amplitude that can be handled. Noise originates as atmospheric noise from outside the system and as circuit noise generated within resistors and devices.

Consider a conductive material at room temperature (Fig. 8-20). The motion of free electrons drifting around in the material constitutes a flow of many tiny random electric currents. These currents cause minute voltage drops, which appear across the ends (or terminals) of the material. Because the number of free electrons available and the random motion of the electrons both increase as temperature rises, the generated voltage amplitude is proportional to temperature. This unwanted, randomly varying voltage is termed *thermal noise*.

Thermal noise is generated in resistors, and when the resistors are at the input stage of an amplifier, the noise is amplified and produced as an output. Noise from other resistors is not amplified as much as that from the resistors right at the input; consequently, only the input stage resistors need be considered in noise calculations. Noise is also generated within a transistor, and like resistors, the input stage transistor of an amplifier is the most important because its noise is amplified more than that from any other stage.

Because thermal noise is an alternating quantity, its rms output level from any amplifier is dependent upon the bandwidth of the amplifier. It can be shown that the rms noise voltage generated in a resistance is

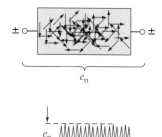

Figure 8-21 Noise voltages are generated within a resistor by random movements of electrons.

$$e_n = \sqrt{4kTBR} \qquad (8\text{-}18)$$

where k = Boltzmann's constant = 1.374×10^{-23} J/K
          (i.e., joules per degree Kelvin)
       T = absolute temperature (Kelvin)
       R = resistance in ohms
       B = circuit bandwidth

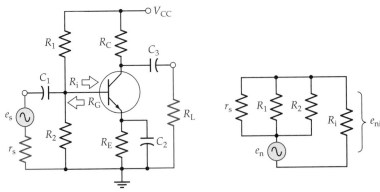

(a) Common-emitter amplifier circuit   (b) Equivalent noise input circuit

Figure 8-22 Thermal noise at the input of an amplifier depends upon the equivalent resistance of the bias and signal source resistances.

In the circuit shown in Fig. 8-22a, R_1 and R_2 are bias resistors and e_s is a signal voltage with source resistance r_s. The total noise-generating resistance (R_G) in parallel with the amplifier input terminal is

$$R_G = r_s \| R_1 \| R_2 \qquad (8\text{-}19)$$

In the noise equivalent circuit (Fig. 8-22b), e_n is the noise voltage generated by R_G. If the device input resistance is R_i, then the noise voltage is divided:

$$e_{ni} = e_n \times \frac{R_i}{R_i + R_G} \qquad (8\text{-}20)$$

If the amplifier voltage gain is A_v, the output noise due to R_G is

$$e_{no} = A_v e_{ni} \qquad (8\text{-}21)$$

With a circuit collector resistance R_C and load resistance R_L, the noise output power produced by R_G is

$$P_{nG} = \frac{e_{no}^2}{R_C \| R_L} \qquad (8\text{-}22)$$

To specify the amount of noise generated by a transistor, manufacturers usually quote a *noise figure*. To arrive at this figure, the transistor noise output is measured under specified bias conditions and with a specified source resistance, temperature, and noise bandwidth. The noise figure defines the amount of noise added by the transistor to the noise generated by the specified resistance (R_G) at the input. Recall that R_G is the combined bias and signal source resistances, as seen from the amplifier input.

Noise Characteristics (2N4104)

Parameter	Test conditions	Min	Max
NF Spot noise figure	$V_{CE} = 5$ V, $I_C = 30$ μA, $R_G = 10$ kΩ, $f = 10$ Hz		15 dB
	$V_{CE} = 5$ V, $I_C = 30$ μA, $R_G = 10$ kΩ, $f = 100$ Hz		4 dB
	$V_{CE} = 5$ V, $I_C = 5$ μA, $R_G = 50$ kΩ, $f = 1$ kHz		1 dB
	$V_{CE} = 5$ V, $I_C = 5$ μA, $R_G = 50$ kΩ, $f = 10$ kHz		1 dB

Figure 8-23   Data sheet specification of noise characteristics for a 2N4104 transistor.

The section of data sheet in Fig. 8-23 specifies *spot noise figures* for a 2N4104 transistor. This means that the noise has been measured for a bandwidth of 1 Hz. The bias conditions are listed because the transistor noise can be affected by V_{CE} and I_C. Note that the specified I_C levels are very low (5 μA to 30 μA), because transistor noise increases with increasing current levels.

The *noise factor* (F) is the total circuit noise power output divided by noise output power from the source resistor.

$$F = \frac{P_{no}}{P_{nG}} \qquad (8\text{-}23)$$

The *noise figure* (NF) is the decibel value of F:

$$NF = 10 \log_{10} F \qquad (8\text{-}24)$$

Ideally, a transistor would add no noise to the circuit, and its noise figure would be zero. Obviously, the smallest possible transistor noise figure is the most desirable.

If the circuit in which the transistor is employed does not have the source resistance and the bias conditions specified, the specified noise figure does not apply. In this case the noise figures can still be used to compare transistors, but for accurate estimations of noise, a new measurement of the noise figure must be made. From Eq. 8-23, the total noise output power due to R_G and the input transistor is

$$P_{no} = F \times P_{nG} \qquad (8\text{-}25)$$

Example 8-9

Calculate the noise output voltage for the amplifier in Fig. 8-24 if the transistor is completely noiseless. The circuit voltage gain is 600, the base input resistance is 3 kΩ, and the cutoff frequencies are $f_1 = 100$ Hz and $f_2 = 40$ kHz. The circuit temperature is 25°C.

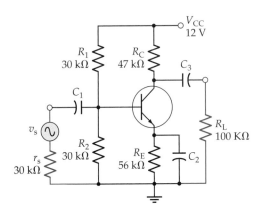

Figure 8-24 Circuit for Examples 8-9 and 8-10.

Solution

Eq. 8-19: $R_G = r_s \| R_1 \| R_2 = 30 \text{ k}\Omega \| 30 \text{ k}\Omega \| 30 \text{ k}\Omega$

$$= 10 \text{ k}\Omega$$

$$T = 25°C = (273 + 25) \text{ K}$$

$$= 298 \text{ K (degrees Kelvin)}$$

$$B = f_2 - f_1 = 40 \text{ kHz} - 100 \text{ Hz}$$

$$= 39.9 \text{ kHz}$$

Eq. 8-18: $e_n = \sqrt{4\,k\,T\,B\,R}$

$$= \sqrt{4 \times 1.37 \times 10^{-23} \times 298 \times 39.9 \text{ kHz} \times 10 \text{ k}\Omega}$$

$$= 2.6 \text{ μV}$$

Eq. 8-20: $e_{ni} = e_n \times \dfrac{R_i}{R_i + R_G} = 2.6 \text{ μV} \times \dfrac{3 \text{ k}\Omega}{3 \text{ k}\Omega + 10 \text{ k}\Omega}$

$$= 0.59 \text{ μV}$$

Eq. 8-21: $e_{no} = A_v e_{ni} = 600 \times 0.59 \text{ μV}$

$$= 354 \text{ μV}$$

Example 8-10

The circuit in Example 8-9 uses a 2N4104 transistor with the following bias conditions: $I_C = 30$ μA and $V_{CE} = 5$ V. Calculate the total noise output voltage for the amplifier.

Solution

From Fig. 8-23, $NF = 4$ dB

From Eq. 8-24, $F = \text{antilog}\left(\dfrac{NF}{10}\right) = \text{antilog}\left(\dfrac{4\text{ dB}}{10}\right)$

$\qquad\qquad = 2.51$

Eq. 8-25: $P_{no} = F \times P_{nG}$

So $\dfrac{v_n^2}{R_C \| R_L} = F \times \dfrac{e_{no}^2}{R_C \| R_L}$

and $v_n = \sqrt{F} \times e_{no} = \sqrt{2.51} \times 354\ \mu V$

$\qquad\qquad = 560\ \mu V$

Practice Problems

8-6.1 A transistor circuit has $A_V = 800$, $B = 20$ kHz, $T = 40°C$, and $R_i = 2$ kΩ. Determine the noise output voltage produced by bias resistors with an equivalent resistance of 15 kΩ.

8-6.2 Calculate the total noise output voltage for the circuit shown in Problem 8-6.1 if the transistor has a noise figure of 6 dB.

8-7 TRANSISTOR POWER DISSIPATION

Maximum Power Dissipation

Consider the portion of the data sheet for the 2N3903 and 2N3904 BJT reproduced in Fig. 8-25. The *total device dissipation* (P_D) is specified as 625 mW at a maximum ambient temperature (T_A) of 25°C. This means that ($V_{CE} \times I_C$) must not exceed 625 mW when the surrounding air temperature is 25°C. Figure 8-1 lists the maximum I_C and V_{CE} for the 2N3903 and 2N3904 as 200 mA and 40 V respectively. If these two quantities exist simultaneously at the transistor, the power dissipation is

$$P_D = V_{CE} \times I_C = 40\text{ V} \times 200\text{ mA}$$

$$= 8\text{ W}$$

which is much greater than the 625 mW specified maximum.

2N3903, 2N3904			
Total device dissipation @ $T_A = 25°C$	P_D	625	mW
Derate above = 25°C		5	mW/°C

Figure 8-25 Power dissipation specification for low-power transistors.

If the maximum V_{CE} is used with a 2N3904 transistor, the maximum collector current must not exceed

$$I_C = \frac{P_{D(max)}}{V_{CE}} = \frac{625 \text{ mW}}{40 \text{ V}}$$

$$= 15.6 \text{ mA}$$

When the transistor has to pass the maximum I_C, the V_{CE} should be limited to

$$V_{CE} = \frac{P_{D(max)}}{I_C} = \frac{625 \text{ mW}}{200 \text{ mA}}$$

$$= 3.1 \text{ V}$$

If the air temperature is greater than 25°C, the transistor maximum power dissipation must be derated. From Fig. 8-25, the *power derating factor* for 2N3903 and 2N3904 transistors is 5 mW/°C. This may be used to calculate the device maximum power dissipation at any air temperature.

$$P_{D(T2)} = P_{D(25°C)} - D(T_2 - 25°C) \tag{8-26}$$

where D is the power dissipation derating factor.

Example 8-11

Calculate the maximum I_C level that may be used with a 2N3904 transistor at 55°C when V_{CE} is 10 V.

Solution

Eq. 8-26: $P_{D(T2)} = P_{D(25°C)} - D(T_2 - 25°C)$

$$= 625 \text{ mW} - (5 \text{ mW/°C})(55°C - 25°C)$$

$$= 475 \text{ mW}$$

$$I_C = \frac{P_D}{V_{CE}} = \frac{475 \text{ mW}}{10 \text{ V}}$$

$$= 47.5 \text{ mA}$$

The data sheet portion in Fig. 8-26 shows the voltage, current, and power ratings for a 2N3055 high-power transistor. The maximum dissipation is specified as 115 W at a *case temperature* (T_C) of 25°C. Note that for a power transistor the *case* temperature is specified, instead of the air temperature. The derating factor is 0.667 W/°C.

The maximum power dissipation at any temperature may be calculated for the 2N3055 by use of Eq. 8-26. Alternatively, a *power derating graph* can be constructed for reading the device maximum power dissipation at any temperature. This is demonstrated in Example 8-12.

2N3055 Silicon NPN transistor

Characteristic	Symbol	Value	Unit
Collector-emitter voltage	V_{CEO}	60	Vdc
Collector-emitter voltage	V_{CER}	70	Vdc
Collector-base voltage	V_{CB}	100	Vdc
Emitter-base voltage	V_{EB}	7	Vdc
Collector current — continuous	I_C	15	Adc
Base current	I_B	7	Adc
Total power dissipation @ $T_C = 25°C$	P_D	115	Watts
Derate above 25°C		0.667	W/°C

Figure 8-26   Part of data sheet for 2N3055 power transistor.

Example 8-12

Draw a power derating graph for a 2N3055 transistor, and determine the device maximum power dissipation at 100°C.

Solution

Prepare a horizontal temperature scale to 200°C, and a vertical power dissipation scale to 115 W, as in Fig. 8-27.

Plot point A at $P_D = 115$ W and $T = 25°C$.

When $P_D = 0$,
$$T = 25°C + \frac{P_D}{D} = 25°C + \frac{15\ W}{0.667\ W/°C}$$
$$= 197°C$$

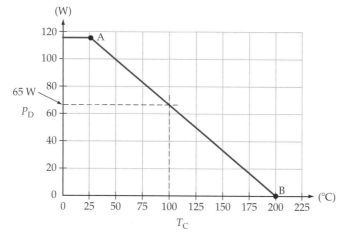

Figure 8-27   Power derating graph for 2N3055 transistor.

Plot point B at $P_D = 0$ and $T = 197°C$.

Draw the power derating graph through points A and B.

From the graph, at $T = 100°C$,

$$P_D = 65 \text{ W}$$

Maximum Power Dissipation Curve

When the maximum power that may be dissipated in a transistor is determined, the maximum I_C level may be calculated for any given V_{CE}, or vice versa. The corresponding voltage and current levels may be more easily determined by drawing a *maximum power dissipation curve* on the transistor output characteristics. To draw this curve, the greatest power that may be dissipated at the operating temperature is first calculated. Then, the corresponding collector current levels for the maximum power dissipation are calculated, using convenient collector-emitter voltages. The curve on the device characteristics is plotted, using these current and voltage levels. Example 8-13 demonstrates the process.

The transistor voltage and current conditions must at all times be maintained in the portion of the characteristics below the maximum power-dissipation curve. This means, for example, that all point on load lines must be below the curve.

Example 8-13

Draw a maximum power dissipation curve for $P_D = 80$ W on the I_C/V_{CE} characteristics in Fig. 8-28.

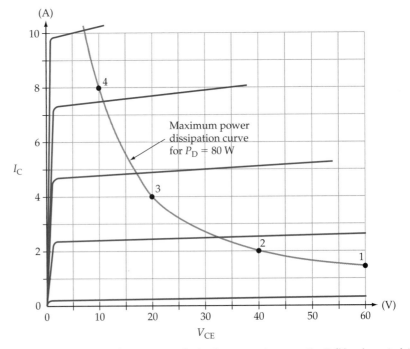

Figure 8-28   Transistor maximum power dissipation curve drawn on the I_C/V_{CE} characteristics.

Solution

When V_{CE} = 60 V,
$$I_C = \frac{P_D}{V_{CE}} = \frac{80\ W}{60\ V}$$
$$= 1.3\ A$$

Plot point 1 on the characteristics at
$$V_{CE} = 60\ V \quad \text{and} \quad I_C = 1.3\ A$$

When V_{CE} = 40 V,
$$I_C = \frac{80\ W}{40\ V} = 2\ A \quad (\text{point 2})$$

When V_{CE} = 20 V,
$$I_C = \frac{80\ W}{20\ V} = 4\ A \quad (\text{point 3})$$

When V_{CE} = 10 V,
$$I_C = \frac{80\ W}{10\ V} = 8\ A \quad (\text{point 4})$$

Draw the maximum power dissipation curve through the points on the graph.

Practice Problems

8-7.1 Draw a power derating graph for a transistor with $P_{D(25°C)}$ = 310 mW and D = 2.81 mV/°C. Read the device maximum power dissipation at 80°C from the graph.

8-7.2 Using $V_{CE(max)}$ = 40 V, determine four suitable points for drawing the maximum power dissipation curve for the transistor in Practice Problem 8-7.1 when T_C = 75°C.

8-8 HEAT SINKING

When power is dissipated in a transistor, the heat generated must flow from the collector-base junction to the case and then to the surrounding atmosphere. When only a very small amount of power is involved, as in a small-signal transistor, the surface area of the transistor case is normally large enough to allow all of the heat to escape. For the large power dissipations that can occur in high-power transistors, the transistor surface area is not large enough. *Heat sinks* must be used to increase the area in contact with the atmosphere. For small transistors, the clip-on *star-type* heat sinks illustrated in Fig. 8-29a may be used. For higher-power transistors, sheet-metal and aluminum-extrusion heat sinks are available, as illustrated in Fig. 8-29b and c.

Figure 8-30a shows the cross-section of a high-power transistor fastened to a heat sink. The heat generated at the collector-base junction must flow from the junction to the transistor case, then from the case to the heat sink, and

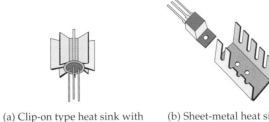

Figure 8-29 Heat sinks are used to conduct heat from a transistor case so that the device will not overheat when power is dissipated.

(a) Clip-on type heat sink with TO-18 transistor enclosure

(b) Sheet-metal heat sink with TO-220 transistor

(c) Aluminum extrusion heat sink with TO-3 transistor

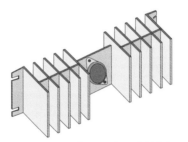

Collector-base junction

Mica gasket

Transistor case

Heat sink

(a) Transistor and heat sink

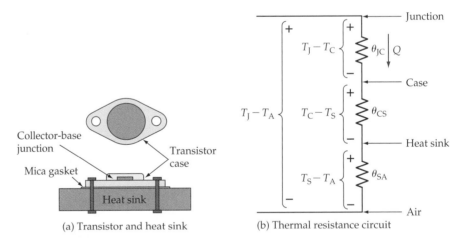

(b) Thermal resistance circuit

Figure 8-30 The thermal resistances between the collector-base junction of a transistor and the air surrounding the transistor heat sink constitute a series thermal circuit. The thermal equivalent of Ohm's Law may be applied for circuit analysis.

finally from the heat sink to the surrounding air. In many cases, a mica gasket is inserted between the transistor case and the heat sink for electrical insulation (see Fig. 8-30a). Each part of the path that the heat must pass through has a *thermal resistance*. These are:

θ_{JC}—*junction-to-case* thermal resistance
θ_{CS}—*case-to-sink* thermal resistance
θ_{SA}—*sink-to-air* thermal resistance

Figure 8-30b shows the *thermal equivalent circuit* for the transistor and heat sink. This consists of the three thermal resistances connected in series. The size of heat sink required for a transistor with a given power dissipation may be determined from the thermal equivalent circuit.

The temperature difference between the transistor collector-base junction and the air surrounding the heat sink ($T_J - T_A$) causes the dissipated power (Q) to flow through each of the thermal resistances in turn. The thermal resistance series circuit is analogous to an electrical series resistive circuit. The power flow in the thermal circuit is similar to the current flow in the electrical circuit. Also, the temperature drop across each thermal resistance is analogous to the voltage drop across each electrical resistance (see Fig. 8-30b). Ohm's Law may be applied to a thermal series circuit exactly as in the case of an electrical series circuit.

For an electrical series resistive circuit, $I = E/R$ or

$$\text{current flow} = \frac{\text{voltage difference}}{\text{total resistance}}$$

For a series thermal resistance circuit,

$$\text{power flow} = \frac{\text{temperature difference}}{\text{total thermal resistance}}$$

or
$$Q = \frac{T_J - T_A}{\theta_{JC} + \theta_{CS} + \theta_{SA}} \qquad (8\text{-}27)$$

When the temperatures are in degrees Celsius (°C) and the thermal resistances are in *degrees Celsius per watt* (°C/W), Q is the power dissipated in watts (W).

The value of θ_{JC} depends upon the transistor case style, and it is usually specified on the data sheet. See the excerpt from a 2N3055 data sheet in Fig. 8-31. θ_{CS} is dependent on the transistor case and on the mechanical contact between the case and the heat sink. The contact may be *dry*, have a layer of heat-conducting compound, or have an insulating gasket. Table 8-1 shows typical θ_{CS} values for three case styles and three contact conditions. θ_{SA} is determined by the size and style of the heat sink (see Table 8-2 and Appendix A-10).

Thermal Characteristic (2N3055)			
Characteristic		Max	Unit
Junction-to-case thermal resistance	θ_{JC}	1.52	°C/W

Figure 8-31 Section of 2N3055 data sheet showing the transistor junction-to-case thermal resistance.

TABLE 8-1 Typical case-to-sink thermal resistances for various mechanical connections

Transistor Case	θ_{CS} (°C/W) Metal-to-Metal		θ_{CS} (°C/W) With Mica Insulator and Compound
	Dry	With Compound	
TO-220	1.2	0.7	1.0
TO-3	0.6	0.4	0.5
TO-66	1.5	0.5	2.3

TABLE 8-2 Typical sink-to-air thermal resistances for various heat sinks

Heat Sink Style	Transistor Case	Wakefield Heat Sink Type	Natural Convection θ_{SA}	θ_{SA}
Star-type clip-on	TO-18	201	65°C/W	65°C @ 1 W
Sheet metal with fins	TO-220	289	25°C/W	50°C @ 2 W
Aluminum extrusion	TO-3	401	2.7°C/W	80°C @ 30 W
" "	"	403	1.8°C/W	55°C @ 30 W
" "	"	421	1.2°C/W	58°C @ 50 W
" "	"	423	0.94°C/W	47°C @ 50 W
" "	"	441	0.59°C/W	47°C @ 80 W
" "	"	621	5°C/W	75°C @ 15 W

Equation 8-27 may be used to calculate θ_{SA} when all other quantities are known. The smallest heat sink with a thermal resistance equal to or lower than the calculated value is then selected. Alternatively, when a given heat sink is used, Equation 8-27 may be applied to calculate the transistor junction temperature.

Another method used to specify heat sink thermal resistances is shown in Table 8-2. For the Wakefield 621 heat sink, θ_{SA} is listed as 75°C at 15 W for natural convection. This means that the maximum temperature difference between the case and the surrounding air will be 75°C when the heat sink is dissipating 15 W. This can be redefined as

$$\theta_{SA} = \frac{75°C}{15\ W} = 5°C/W$$

Example 8-14

A transistor with $V_{CE} = 20$ V and $I_C = 1$ A has a 1°C/W junction-to-case thermal resistance. The TO-3 case is fastened directly to the heat sink, and conducting compound is used. Calculate the thermal resistance for a heat sink that will keep the maximum junction temperature at 90°C when the ambient temperature is 25°C. Select a suitable heat sink from Appendix A-10.

Solution

$$Q = V_{CE}\,I_C = 20\text{ V} \times 1\text{ A}$$

$$= 20\text{ W}$$

From Table 8-1, $\theta_{CS} = 0.4°C/W$ (TO-3 case, direct contact with compound)

From Eq. 8-27,

$$\theta_{SA} = \frac{T_J - T_A}{Q} - (\theta_{JC} + \theta_{CS})$$

$$= \frac{90°C - 25°C}{20\text{ W}} - (1°C/W + 0.4°C/W)$$

$$= 1.85°C/W$$

From Table 8-2 (and Appendix A-10), the 403, 421, 423, and 441 all have θ_{SA} less than 1.85°C/W for natural convection. Select the smallest and least expensive of the three—the 403.

Practice Problems

8-8.1 A 2N6121 transistor (see Appendix A-9) is fastened to a heat sink with the use of a mica insulator and compound. Calculate the thermal resistance of a suitable heat sink if the device is to dissipate 10 W and the maximum junction temperature is to be 100°C. The ambient temperature is 25°C.

8-8.2 A TO-3 style transistor with $T_{J(max)} = 140°C$ and $\theta_{JC} = 0.6°C/W$ is fastened dry to a heat sink with a $\theta_{CA} = 1.3\ °C/W$. Determine the maximum power that may be dissipated when $T_A = 40\ °C$.

Review Questions

Section 8-1

8-1 Define the following BJT quantities listed on a device data sheet: V_{CBO}, V_{CEO}, V_{EBO}, and $V_{CE(sat)}$.

8-2 Define the following BJT quantities: I_C, I_{CBO}, I_{CO}, h_{FE}, and h_{fe}.

Section 8-2

8-3 Write the equations for determining power gains and power level changes in decibels (a) when two power levels are measured and (b) when two voltage levels are measured.

8-4   Sketch a typical frequency response graph for an amplifier, and identify the upper and lower cutoff frequencies and the bandwidth. Briefly explain.

Section 8-3

8-5   Define $f_{\alpha e}$, $f_{\alpha b}$, and f_T for a BJT. Briefly explain.

8-6   Explain CB and CE junction capacitances, and identify the quantities that determine junction capacitance values.

8-7   Describe the Miller effect, and derive an equation for the input capacitance of an inverting amplifier.

Section 8-4

8-8   Identify the quantities that cause the gain of an amplifier to fall off at low frequencies. Briefly explain.

8-9   Discuss the quantities that cause the gain of an amplifier to fall off at high frequencies.

Section 8-5

8-10   Sketch the waveforms of input and output currents for a switching transistor. Show the various switching times involved and explain the origin of each.

8-11   Draw the diagram of a direct-coupled BJT switching circuit using a speed-up capacitor. Explain how the capacitor affects the device switching times, and discuss the limits of input pulse width and space width that can be used with the circuit.

Section 8-6

8-12   Explain thermal noise, and discuss the effect of resistor noise at the input of a transistor. Define noise figure and noise factor for a transistor.

Section 8-7

8-13   Sketch a typical power-temperature derating graph for a transistor. Briefly explain.

8-14   Describe how a transistor maximum power dissipation graph may be drawn on the I_C/V_{CE} characteristics.

Section 8-8

8-15   Sketch the cross-section of a power transistor and heat sink. Label the thermal resistances in the power dissipation path.

8-16   Draw the thermal resistance equivalent circuit for a transistor and heat sink. Write the equation relating power dissipation, temperature difference, and thermal resistance.

Problems

Section 8-1

8-1   From Appendix A-9, determine the following quantities for a 2N6125 transistor: maximum V_{CE}, maximum collector current, maximum V_{BE}, $h_{FE(max)}$, and $h_{FE(min)}$ at $I_C = 1.5$ A.

8-2   From the 2N3055 data sheet A-8 in Appendix A, determine the following quantities: maximum collector-base voltage, maximum collector current, maximum emitter-base voltage, h_{FE} minimum, and h_{FE} maximum at $I_C = 4$ A.

Section 8-2

8-3   The output power from an amplifier is 100 mW when the signal frequency is 1 kHz. When the signal frequency is increased to 25 kHz, the output power falls to 75 mW. Calculate the decibel change in output power.

8-4   Calculate the power gain for an amplifier with equal input and load resistances when $v_i = 100$ mV and $v_o = 3$ V.

8-5   The output voltage of an amplifier is 2 V when the signal frequency is 1 kHz. Calculate the new output voltage when it has fallen by 4 dB.

Section 8-3

8-6   The input capacitance of a CE circuit is measured as 800 pF. The circuit has $R_C = 7$ kΩ, and the device has the following parameters: $h_{fe} = 60$ and $h_{ie} = 1.5$ kΩ. If the base-emitter capacitance is 15 pF, calculate the collector-base capacitance.

8-7   A transistor with $h_{fe} = 100$, $h_{ie} = 2.2$ kΩ, $C_{cb} = 3$ pF, $I_C = 1.2$ mA, and $f_T = 4$ MHz is connected as a CE amplifier with $R_C \| R_L = 6.8$ kΩ. Calculate the amplifier input capacitance.

8-8   Calculate the new value of C_{in} for the amplifier in Problem 8-7 when a 100 pF capacitor is connected (a) between emitter and base and (b) between collector and base.

8-9   For a common emitter amplifier, $A_v = 50$, $C_{in} = 414$ pF, and $C_{bc} = 8$ pF. Calculate the base-emitter capacitance.

Section 8-4

8-10   The circuit in Problem 8-7 uses voltage-divider bias with $R_1 = 56$ kΩ and $R_2 = 15$ kΩ. The signal source resistance is 3.3 kΩ. Calculate the input-capacitance-limited upper cutoff frequency for the circuit.

8-11   Repeat Problem 8-10 for the circuit connected to function as a CB amplifier.

8-12   A voltage-divider bias circuit (as in Fig. 8-24) has the following components: $R_1 = 68$ kΩ, $R_2 = 47$ kΩ, $R_C = 5.6$ kΩ, $R_E = 4.7$ kΩ. The supply voltage is 12 V, and the transistor has $f_T = 35$ MHz and $C_{bc} = 3.5$ pF.

Calculate the input-capacitance-limited upper cutoff frequency when the circuit is connected as a CE amplifier with R_E bypassed. The signal source resistance is 1.5 kΩ.

8-13 Repeat Problem 8-12 for the circuit employed as a CB amplifier with the transistor base bypassed to ground.

8-14 The circuit in Problem 8-12 is modified to function as an emitter follower with R_C shorted and the output taken from the emitter. Determine the input-capacitance-limited upper cutoff frequency for the circuit.

8-15 A transistor employed in an amplifier has $h_{fe} = 75$ and $f_{\alpha b} = 12$ MHz. The collector load is $R_C \| R_L = 20$ kΩ, and there is 100 pF stray capacitance at the output. Determine the stray-capacitance-limited upper 3 dB frequency.

8-16 A transistor amplifier with $R_C \| R_L$ equal to 15 kΩ has a 75 kHz upper 3 dB frequency. Assuming that f_2 is determined by the stray capacitance at the transistor collector terminal, calculate the value of the stray capacitance.

Section 8-5

8-17 Determine the turn-on and turn-off times for a switching circuit with a 2N3251 transistor (see Appendix A-7).

8-18 If the circuit in Problem 8-17 has a 2 kHz square wave input at the transistor base, calculate the time from the instant that the collector current first reaches 90% of its maximum level until it falls to 10% of maximum.

8-19 A switching circuit is to be turned *on* and *off* by a 1 MHz square wave input. The output rise and fall times are not to exceed 10% of the transistor *on* time. Determine the maximum value for t_r and t_f.

8-20 A direct-coupled switching circuit in which $R_B = 4.7$ kΩ uses a BJT with $t_{on} = 150$ ns. The signal source resistance is $R_S = 400$ Ω. Determine a suitable value of speed-up capacitor, and calculate the maximum pulse input frequency.

8-21 Determine a suitable value of speed-up capacitor for the BJT switching circuit in Fig. 5-56. Assume that the signal source resistance is 500 Ω and that the transistor has $t_{on} = 200$ ns. Calculate the minimum input pulse width that can be used with the circuit.

Section 8-6

8-22 An amplifier with a 2N4104 input transistor (see Fig. 8-23) has 3 dB points at 2 kHz and 10 kHz respectively. The transistor bias conditions are: $V_{CE} = 5$ V and $I_C = 5$ μA; the amplifier voltage gain is 40. Calculate the noise output voltage at 25°C if $R_G = 50$ kΩ and $R_i = 10$ kΩ.

8-23 A transistor amplifier where $A_v = 100$, $B = 15$ kHz, and $R_i = 12$ kΩ has input bias resistors equivalent to $R_G = 33$ kΩ. If the maximum noise voltage at the output is not to exceed 100 μV, determine the largest tolerable noise figure for the input transistor.

8-24   An amplifier with $B = 10$ kHz, $R_G = 5$ kΩ, and $R_i = 3.3$ kΩ has a maximum output noise of 150 μV. Assuming that the transistor is noiseless, determine the amplifier voltage gain.

Section 8-7

8-25   Calculate the maximum ambient temperature for a transistor with a 200 mW power dissipation if $P_{D(25°C)} = 310$ mW and $D = 2.81$ mW/$°$C.

8-26   The device in Problem 8-25 is to be operated at 80°C maximum ambient with a 5 mA collector current. Determine the maximum V_{CE} level that may be used.

8-27   Referring to Appendix A-9, draw a power derating graph for a 2N6121 transistor and determine the maximum power dissipation at a case temperature of 75°C.

8-28   Calculate the maximum case temperature for a 2N3251 transistor when its power dissipation is 0.25 W.

8-29   A 2N3055 transistor is to be operated at a maximum case temperature of 125°C. Construct a suitable I_C/V_{CE} graph, and draw the maximum power dissipation curve for the device at this temperature.

8-30   For the $P_D = 80$ W maximum power dissipation curve in Fig. 8-28, draw the dc load line for the smallest possible value of R_C when the circuit supply voltage is 50 V. Determine the value of $R_{C(min)}$.

Section 8-8

8-31   A 2N3055 transistor has a V_{CE} of 10 V and an I_C of 500 mA. The case temperature is not to exceed 30°C when the ambient temperature is 25°C. The device is fastened to a heat sink by means of a mica gasket and compound. Calculate the maximum sink-to-air thermal resistance for a suitable heat sink.

8-32   A 2N3055 transistor dissipating 9 W is directly fastened to a NC403 heat sink by means of compound (see Appendix A-10). If the ambient temperature is 25°C, calculate the transistor case temperature and junction temperature.

8-33   A 2N4900 transistor has a specified maximum power dissipation of 25 W at a case temperature of 25°C, and a derating factor of 0.143 W/$°$C. If the transistor is to dissipate 20 W, determine its maximum case temperature and calculate the maximum sink-to-air thermal resistance for a suitable heat sink. Assume that the TO-66 case is directly connected to the heat sink with compound.

8-34   A TO-220 case style transistor with a V_{CE} of 30 V and an I_C of 500 mA is fastened to a heat sink with a mica gasket and compound. Calculate the maximum sink-to-air thermal resistance for a suitable heat sink if the transistor maximum junction temperature is 150°C and the junction-to-case thermal resistance is 2.7°C/W.

Practice Problem Answers

8-1.1	90, 100, 40 V, 5 V, 6 dB
8-2.1	−12 dB
8-3.1	4.86 nF
8-4.1	191 kHz
8-4.2	49.7 pF
8-5.1	22 ns, 27 ns, 127 ns
8-5.2	206 ns
8-5.3	200 pF, 0.35 μs, 12 μs
8-6.1	2.14 mV
8-6.2	4.28 mV
8-7.1	(310 mW, 25°C), (0, 135°C)
8-7.2	(40 V, 4.24 mA), (30 V, 5.65 mA), (20 V, 8.48 mA), (10 V, 16.95 mA)
8-8.1	3.38°C/W
8-8.2	40 W

CHAPTER 9
Field Effect Transistors

CONTENTS

Objectives

You will be able to:

1   Explain the operation of *n*-channel and *p*-channel junction field effect transistors (JFETs).

2   Draw typical JFET characteristics. Identify the regions of the characteristics and all important current and voltage levels.

3   For JFETs, define saturation current, pinch-off voltage, forward transfer admittance, output admittance, and drain-source ON resistance.

4   Determine JFET parameter values from manufacturers' data sheets.

5   From the data sheet information, draw the maximum and minimum transfer characteristics for any given JFET type number.

6   Solve problems involving JFET characteristics and parameters.

7   Show how a JFET can be used for voltage amplification and for switching.

8   Explain the operation of enhancement-mode and depletion-enhancement-mode MOSFETs.

9   Draw typical drain and transfer characteristics for MOSFETs, and discuss the differences between MOSFETs and JFETs.

10 Solve problems involving MOSFET characteristics and parameters.

11 Explain the operation of VMOS-FETs, sketch typical characteristics, and discuss the advantages of VMOS.

12 Sketch circuit symbols for JFETs, MOSFETs, and VFETs. Identify all terminals, current directions, and voltage polarities.

INTRODUCTION

A *field effect transistor* (*FET*) is a voltage-operated device that can be used in amplifiers and switching circuits, similarly to a bipolar transistor. Unlike a BJT, a FET requires virtually no input current. This gives it an extremely high input resistance, which is its most important advantage over a BJT. There are two major categories of field effect transistors: *junction* FETs and MOSFETS. These are further subdivided into *p*-channel and *n*-channel devices.

9-1 JUNCTION FIELD EFFECT TRANSISTORS

n-Channel JFET

The operating principle of an *n-channel junction field effect transistor* (*JFET*) is illustrated by the block representation in Fig. 9-1a. A piece of *n*-type semiconductor material, referred to as the *channel*, is sandwiched between two smaller pieces of *p*-type (the *gates*). The ends of the channel are designated the *drain* (*D*) and the *source* (*S*), and the two pieces of *p*-type material are connected together and their terminal is named the *gate* (*G*).

With the gate left unconnected and a drain-source voltage (V_D) applied (positive at the drain, negative at the source), a *drain current* (I_D) flows, as shown in Fig. 9-1a. When a gate-source voltage (V_{GS}) is applied with the gate negative with respect to the source (Fig. 9-1b), the gate-channel *pn*-junctions are reverse-biased. The channel is more lightly doped than the gate material, so the depletion regions penetrate deep into the channel. Because the depletion regions are regions depleted of charge carriers, they behave as insulators. The result is that the channel is narrowed, its resistance is increased, and I_D is reduced. When the negative gate-source bias voltage is further increased, the depletion regions meet as the centre of the channel (Fig. 9-1c), and I_D is cut off.

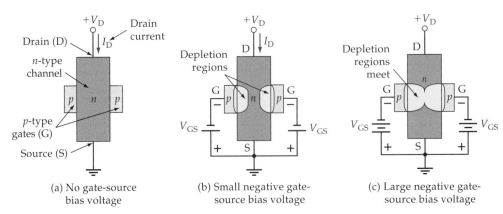

(a) No gate-source bias voltage

(b) Small negative gate-source bias voltage

(c) Large negative gate-source bias voltage

Figure 9-1 An *n*-channel JFET consists of an *n*-type channel with *p*-type gate regions on each side.

An ac signal applied to the gate causes the reverse gate-source voltage to increase as the instantaneous level of the signal goes negative, and to decrease when the signal is positive-going. This causes the gate-channel depletion regions to widen and decrease successively. When the signal goes negative, the depletions widen, the channel resistance is increased, and the drain current decreases. As the signal goes positive, the depletion regions recede, the channel resistance is reduced, and the drain current increases. It is seen that the FET gate-source voltage controls the drain current. The gate-channel *pn*-junctions are maintained in reverse bias, so the gate current is normally extremely low, much lower than the base current for a BJT.

The name *field effect transistor* comes from the fact that the depletion regions in the channel are produced by the electric field at the reverse-biased gate-channel junctions. The term *unipolar device* is sometimes applied to a *FET*, because, unlike a bipolar transistor, the current consists of only one type of charge carrier: electrons in the case of an *n*-channel device.

Circuit symbols for an *n*-channel JFET are shown in Fig. 9-2. As in the case of all semiconductor device symbols, the arrowhead points from *p*-type to *n*-type. For an *n*-channel device, the arrowhead points from the *p*-type gate to the *n*-type channel. This is the direction of conventional current flow if the junctions become forward-biased. Some device manufacturers use the symbol in Fig. 9-2a with the gate directly opposite the source terminal. Others show the gate centralized between the drain and source (Fig. 9-2b). This can sometimes make circuit diagrams confusing unless the drain and source terminals are clearly identified. The symbol in Fig. 9-2c is used when the terminals of the two gate regions are provided with separate connecting leads. In this case, the device is referred to as a *tetrode-connected FET*.

p-Channel JFET

In a *p*-channel JFET, shown in block form in Fig. 9-3a, the channel is a *p*-type semiconductor, and the gates are *n*-type. The drain-source voltage (V_D) is applied negative to the drain and positive to the source, as illustrated, and the drain current flows (in the conventional direction) from source to drain. To reverse-bias the gate-channel junctions, the *n*-type gate regions must be made positive with respect to the *p*-type channel. So the bias voltage is applied positive on the gate terminals and negative on the source.

A positive-going signal at the gate terminal of a *p*-channel JFET increases in the gate-channel junction reverse bias, causing the depletion regions to penetrate further into the channel. This increases the channel resistance and decreases the drain current. Conversely, a negative-going signal narrows the depletion regions, reduces the channel resistance, and increases the drain current.

Circuit symbols for a *p*-channel JFET are shown in Fig. 9-3b. The arrowheads again point from the *p*-type material to the *n*-type, in this case, from the *p*-type channel to the *n*-type gate.

(a) Circuit symbol for *n*-channel JFET

(b) Alternative circuit symbol

(c) Tetrode-connected JFET

Figure 9-2 Circuit symbols for an *n*-channel JFET. Like all semiconductor symbols, the arrowheads point from the *p*-type to the *n*-type material.

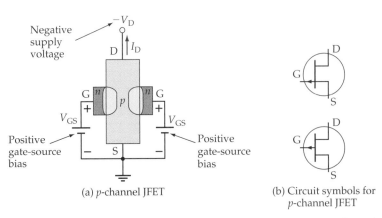

(a) p-channel JFET

(b) Circuit symbols for p-channel JFET

Figure 9-3 A p-channel JFET operates in exactly the same way as an n-channel device, except that the current directions and voltage polarities are reversed.

JFET Fabrication and Packaging

Junction field-effect transistors are normally manufactured by the diffusion process (see Chapter 7). This type of construction is illustrated in Fig. 9-4. Starting with a p-type substrate, an n-channel is diffused; then, p-type impurities are diffused into the channel. Finally, metal terminal connections are deposited through holes in the silicon dioxide surface, as illustrated.

The n-type region is the FET channel, and the two p-type regions constitute the gates. With this symmetrical construction, the drain and source terminals are interchangeable. Other fabrication techniques produce device geometry that is not symmetrical. In these cases, interchanging the drain and source terminals would radically affect the performance of the device.

Figure 9-5 shows several FET packages which are similar to BJT enclosures. Note the device terminal identifications in each case.

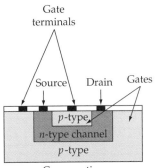

Figure 9-4 Cross-section and top view of n-channel JFET. With this type of construction, the drain and source are interchangeable.

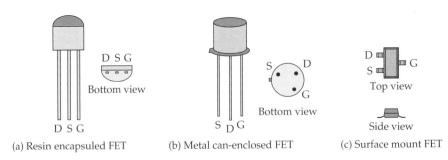

(a) Resin encapsuled FET

(b) Metal can-enclosed FET

(c) Surface mount FET

Figure 9-5 Various JFET enclosures.

Section 9-1 Review

9-1.1 Sketch a block representation of an n-channel JFET. Indicate voltages and current direction, and explain the device operation.

9-1.2 Repeat Question 9-1.1 for a p-channel JFET.

9-1.3 Sketch circuit symbols for n-channel and p-channel JFETs. Identify the terminals and briefly explain.

9-2 JFET CHARACTERISTICS

Depletion regions

An *n*-channel JFET block representation is shown in some detail in Fig. 9-6. With a drain-source voltage applied as illustrated, I_D flows in the direction shown, producing voltage drops along the channel. Consider the voltage drops from the source terminal (*S*) to points A, B, and C within the channel. Point A is positive with respect to the source; alternatively, it can be stated that *S* is negative with respect to *A*. Because the gate blocks are connected to *S*, the gates are negative with respect to point *A* by a voltage V_A. This causes the depletion regions to penetrate into the channel by an amount proportional to V_A.

The voltage drop between point B and the source is V_B, which is less than V_A. Consequently, at point B on the channel the gates are at $-V_B$ with respect to the channel, and the depletion region penetration is less than at point A. From point C to the source terminal, the voltage drop (V_C) is less than V_B. Thus, the gate-channel reverse bias (at point C) is V_C volts, and the depletion region penetration is less than at points A or B. The differing voltage drops along the channel, and the resulting variation in gate-channel reverse bias, account for the shape of the depletion region penetration of the channel.

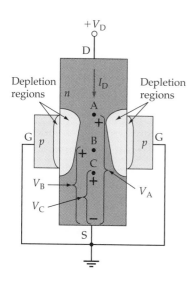

Figure 9-6 Drain current in an *n*-channel JFET causes voltage drops along the channel which reverse-bias the gate channel junctions. This produces different levels of depletion region penetration into the channel.

Drain Characteristics with $V_{GS} = 0$

Figure 9-7 shows a circuit for determining the drain-current versus drain-source voltage characteristic for an *n*-channel JFET with $V_{GS} = 0$. V_{DS} is increased in convenient steps from zero, and I_D is measured at each V_{DS} level. This produces a table of I_D/V_{DS} values for plotting the characteristic shown in Fig. 9-8.

Referring to the characteristic, it is seen that when $V_{DS} = 0$, $I_D = 0$. There is no channel voltage drop, and so the voltage between the gate and all points on the channel is zero, and there is no depletion region penetration. When V_{DS} is

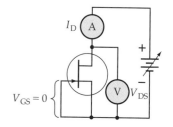

Figure 9-7 Circuit for obtaining the I_D/V_{DS} characteristic for an *n*-channel junction field effect transistor with $V_{GS} = 0$.

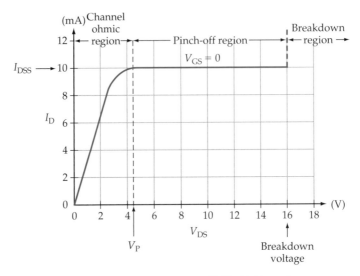

Figure 9-8 I_D/V_{DS} characteristic for an n-channel JFET with $V_{GS} = 0$.

increased by a small amount (less than 1 V), a small drain current flows, producing some voltage drop along the channel. This results in some depletion penetration of the channel (as explained for Fig. 9-6), but it is so small that it has no significant effect on the channel resistance. With further small increases in V_{DS}, the drain current increase is approximately linear and the channel behaves as an almost constant-value resistance (see Fig. 9-8).

The channel continues to behave as a fixed-value resistance until the voltage drops along it become large enough to produce considerable depletion region penetration. At this stage the channel resistance begins to be affected by the depletion regions. Further increases in V_{DS} now produce smaller I_D increases, as shown by the curved part of the characteristic. The increased I_D levels, in turn, cause more depletion region penetration and greater channel resistance. Eventually, a *saturation level* of I_D is reached where further V_{DS} increase seems to have no effect on I_D.

At the point on the characteristic where I_D levels off, the drain current is referred to as the *drain-source saturation current* (I_{DSS}) (10 mA in Fig. 9-8). The shape of the depletion regions in the channel at the I_{DSS} level is such that they appear to *pinch off* the channel (see Fig. 9-6). Thus, the drain source voltage at this point is termed the *pinch-off voltage* (V_P) (4.5 V in Fig. 9-8.) The region of the characteristic where I_D is constant is called the *pinch-off region*, as illustrated. The channel mostly behaves like a resistance between the points where $V_{DS} = 0$ and $V_{DS} = V_P$; so this part of the characteristic is referred to as the *channel ohmic region*.

If V_{DS} is continuously increased (in the pinch-off region), a voltage is reached at which the (reverse-biased) gate-channel junctions break down (see Fig. 9-8). When this happens, I_D increases rapidly and the device may be destroyed. The pinch-off region of the characteristic is the normal operating region for the FET.

Drain Characteristics with External Bias

A circuit for obtaining the I_D/V_{DS} characteristics for an *n*-channel JFET when an external gate-source bias (V_{GS}) is applied is shown in Fig. 9-9. In this case, V_{GS} is set to a convenient (negative) level (such as -1 V). V_{DS} is increased in steps, and the corresponding level of I_D is noted at each V_{DS} step. The I_D/V_{DS} characteristic for a V_{GS} of -1 V is then plotted, as illustrated in Fig. 9-10.

When a -1 V external gate-source bias voltage is applied, the gate-channel junctions are reverse-biased even when $I_D = 0$. So when $V_{DS} = 0$ the depletion regions are already penetrating to some depth into the channel. Because of this, a smaller voltage drop along the channel (smaller than when $V_{GS} = 0$) will increase the depletion regions to the point at which they produce channel pinch-off. Consequently, when $V_{GS} = -1$ V the pinch-off voltage is reached at a lower I_D level than when $V_{GS} = 0$. The $V_{GS} = -1$ V characteristic in Fig. 9-10 has $V_P = 3.5$ V.

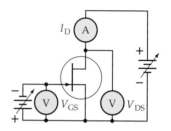

Figure 9-9 Circuit for obtaining the I_D/V_{DS} characteristic for an *n*-channel junction field effect transistor with various gate-source bias voltages.

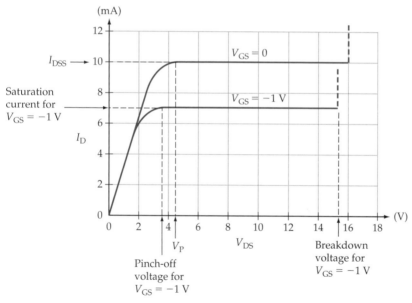

Figure 9-10 I_D/V_{DS} characteristics for $V_{GS} = 0$ and $V_{GS} = -1$ for an *n*-channel JFET.

A family of *drain characteristics* can be obtained by using several levels of negative gate-source bias voltage (see Fig. 9-11). If a positive V_{GS} is used, a higher level of I_D can be produced, as shown by the characteristic for $V_{GS} = +0.5$ V. However, V_{GS} is normally kept negative to avoid the possibility of forward-biasing the gate-channel junctions.

The dashed line from the zero point on the characteristics in Fig. 9-11 is drawn through the points at which I_D saturates for each level of gate-source bias voltage. When $V_{GS} = 0$, I_D saturates at I_{DSS}, and the characteristic shows $V_P = 4.5$ V. When a -1 V external bias is applied, the gate-channel junctions still require -4.5 V to achieve pinch-off. This means that a drop of 3.5 V instead of 4.5 V is now required along the channel, and the lower voltage drop

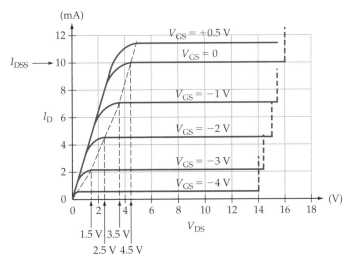

Figure 9-11   Family of I_D/V_{DS} characteristics for an *n*-channel JFET with various levels of V_{GS}.

is achieved with a lower I_D. Similarly, when $V_{GS} = -2$ V and -3 V, pinch-off is achieved with 2.5 V and 1.5 V, respectively, along the channel. The 2.5 V and 1.5 V drops are, of course, obtained with further reduced I_D levels.

Suppose a -4.5 V gate-source bias is applied to a device with the characteristics shown in Fig. 9-11. This is a V_{GS} level equal to the pinch-off voltage V_P. Without any additional channel voltage drop produced by I_D, the depletion regions penetrate so deep into the channel that they meet in the middle, completely cutting I_D *off*. So, a gate-source bias equal to the pinch-off voltage reduces I_D to zero. The bias voltage required to do this is termed the *gate cut-off voltage* ($V_{GS(off)}$), and, as explained, $V_{GS(off)} = V_P$. Note in Fig. 9-11 that the drain-source voltage at which breakdown occurs is reduced as the negative gate-source bias voltage is increased. This is because $-V_{GS}$ adds to the reverse bias at the junctions.

Example 9-1

Plot the I_D/V_{DS} characteristic for a JFET from the following table of values obtained with $V_{GS} = 0$. Determine I_{DSS} and V_P from the characteristics.

V_{DS} (V)	0	1	2	2.5	3	3.5	3.75	4	6	9
I_D (mA)	0	3	6	7	7.5	7.8	8	8	8	8

Solution

On Fig. 9-12, plot *point 1* where $V_{DS} = 0$ and $I_D = 0$.

Plot *point 2* at $V_{DS} = 1$ V and $I_D = 3$ mA,

and so on through $V_{DS} = 9$ V and $I_D = 8$ mA.

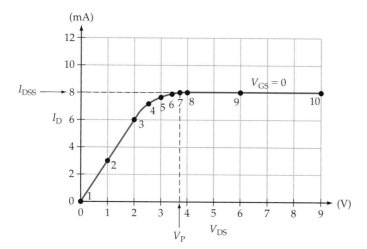

Figure 9-12   FET characteristics for Example 9-1.

Draw the drain characteristic for $V_{GS} = 0$ through *points 1 to 10*.

From the characteristic, $I_{DSS} = 8$ mA and $V_P = 3.75$ V.

Transfer Characteristics

The *transfer characteristics* for an *n*-channel JFET are a plot of I_D versus V_{GS}. The gate-source voltage of an FET controls the level of the drain current; so the transfer characteristic shows how I_D is controlled by V_{GS}. As illustrated in Fig. 9-13a, the transfer characteristic extends from $I_D = I_{DSS}$ at $V_{GS} = 0$, to $I_D = 0$ at $V_{GS} = -V_{GS(off)}$.

Figure 9-13b shows a circuit for determining experimentally a table of quantities for plotting the transfer characteristic of a given FET. The drain-source

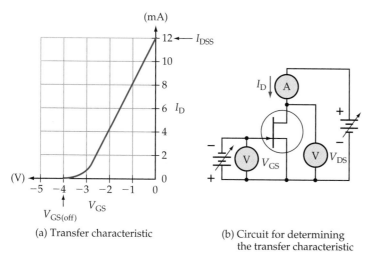

(a) Transfer characteristic

(b) Circuit for determining the transfer characteristic

Figure 9-13   The transfer characteristics for a FET are a plot of I_D versus V_{GS}.

voltage is maintained constant, V_{GS} is adjusted in convenient steps, and the corresponding levels of V_{GS} and I_D are recorded.

The transfer characteristic for a FET can be derived from the drain characteristics. A line is drawn vertically on the drain characteristics to represent a constant V_{DS} level. The corresponding I_D and V_{GS} values along this line are noted and then used to plot the transfer characteristic. The process is demonstrated in Example 9-2.

Example 9-2

Derive the transfer characteristic for $V_{DS} = 8$ V from the FET drain characteristics in Fig. 9-11.

Solution

On Fig. 9-11 draw a vertical line at $V_{DS} = 8$ V.

From the intersection of the line and the characteristics read the following quantities:

V_{GS} (V)	0	−1	−2	−3	−4
I_D (mA)	10	7	4.5	2.2	0.5

On Fig. 9-14, plot *point 1* at $V_{GS} = 0$ and $I_D = 10$ mA.

Plot *point 2* at $V_{GS} = -1$ V and $I_D = 7$ mA,

and so on through $V_{GS} = -4$ V and $I_D = 0.5$ mA.

Draw the transfer characteristic through the points.

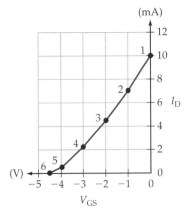

Figure 9-14   Transfer characteristics for Example 9-2.

p-Channel JFET Characteristics

Figure 9-15 shows a circuit for obtaining the characteristics of a *p*-channel JFET. Note the direction of the arrowhead on the FET symbol, and the drain current direction. Note also the supply voltage polarity and the polarity of the gate-source bias voltage. The drain terminal is negative with respect to the source, and the gate terminal is positive with respect to the source. To obtain a table of quantities for plotting a drain characteristic, V_{GS} is maintained constant at the desired (positive) level, $-V_{DS}$ is increased in steps from zero, and the I_D levels are noted at each step.

Typical p-channel JFET drain characteristics and transfer characteristics are shown in Fig. 9-16. It is seen that these are similar to the characteristics for an n-channel JFET, except for the voltage polarities. In Fig. 9-16, when $V_{GS} = 0$ and $I_{DSS} = 15$ mA, progressively more positive levels of V_{GS} reduce I_D toward cutoff $V_{GS(off)} = +6$ V. Using V_{GS} of -0.5 V produces a higher I_D than when $V_{GS} = 0$. As in the case of the n-channel JFET, forward bias at the gate-channel junctions should be avoided; consequently, negative V_{GS} levels are normally not used with a p-channel JFET.

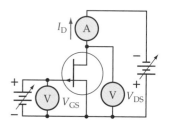

Figure 9-15 Circuit for determining the characteristics of a p-channel JFET.

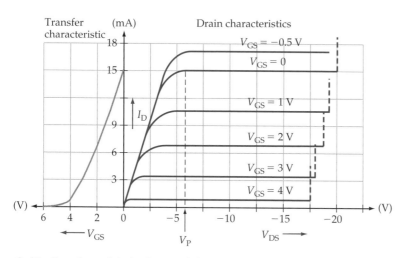

Figure 9-16 Transfer and drain characteristics for a p-channel JFET.

The transfer characteristic for a p-channel device can be obtained experimentally or can be derived from the drain characteristics, just as for an n-channel FET.

Practice Problem

9-2.1 The following table of I_D/V_{DS} values for a FET was obtained with a V_{GS} of 0. Plot the drain characteristic and determine I_{DSS} and V_P.

V_{DS} (V)	0	1	2	2.5	3	4	6	8	10	12
I_D (mA)	0	2	4.5	5.3	5.5	5.5	5.5	5.5	5.5	5.5

9-3 JFET DATA SHEETS AND PARAMETERS

Maximum Ratings

Typical FET data sheets A-11 and A-12 are shown in Appendix A, and part of a FET data sheet is reproduced in Fig. 9-17. As with other device data sheets, a device type number and brief description are usually given at the top. Maximum ratings follow, and then the electrical characteristics are stated for specific bias conditions. From Fig. 9-17, the maximum drain-source voltage (V_{DS}) for the 2N5457 to 2N5459 devices is 25 V, and the maximum drain-gate voltage (V_{DG}) is also 25 V. This means, for example, that if a -5 V gate-source bias is used, the drain-source voltage should not exceed

$$V_{DS} = V_{DG(max)} - V_{GS} = 25 \text{ V} - 5 \text{ V}$$

$$= 20 \text{ V}$$

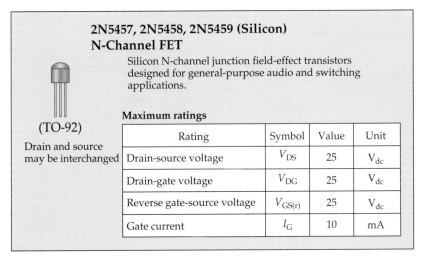

2N5457, 2N5458, 2N5459 (Silicon)
N-Channel FET

Silicon N-channel junction field-effect transistors designed for general-purpose audio and switching applications.

(TO-92)

Drain and source may be interchanged

Maximum ratings

Rating	Symbol	Value	Unit
Drain-source voltage	V_{DS}	25	V_{dc}
Drain-gate voltage	V_{DG}	25	V_{dc}
Reverse gate-source voltage	$V_{GS(r)}$	25	V_{dc}
Gate current	I_G	10	mA

Figure 9-17 Part of data sheet for *n*-channel JFET.

Note in Fig. 9-17 that the maximum reverse gate-source voltage ($V_{GS(r)}$) is specified as 25 V. This is considerably greater than the (typically 5 V) maximum base-emitter reverse voltage for a BJT.

No maximum drain current is specified in Fig. 9-17, but this can be calculated from the maximum power dissipation and the V_{DS} level. The specified gate current (I_G) is the maximum gate current if the gate-channel junctions become forward-biased.

Saturation Current and Pinch-off Voltage

The *drain-source saturation current* (I_{DSS}) and the *pinch-off voltage* (V_P or $V_{GS(off)}$) have already been discussed in Section 9-2. Values for these are listed in the excerpt from a JFET data sheet showing the 'off characteristics' and 'on characteristics' in Fig. 9-18. It can be seen that $V_{GS(off)}$ for a 2N5457 (underline) ranges

2N5457, 2N5458, 2N5459
Off Characteristics

Characteristic	Symbol	Min	Typ	Max	Unit
Gate-source cutoff voltage (V_{DS} = 15 V dc, I_D = 10 nA dc)	$V_{GS(off)}$				V dc
2N5457		0.5	—	6.0	
2N5458		1.0	—	7.0	
2N5459		2.0	—	8.0	

On Characteristics

Characteristic	Symbol	Min	Typ	Max	Unit
Zero gate voltage drain current (V_{DS} = 15 V dc, V_{GS} = 0)	I_{DSS}				mA dc
2N5457		1.0	3.0	5.0	
2N5458		2.0	6.0	9.0	
2N5459		4.0	9.0	16.0	

Figure 9-18   Gate-source cutoff voltage and drain-source saturation current specifications for *n*-channel JFETs.

from a minimum of −0.5 V to a maximum of −6 V. Also, the I_{DSS} level for a 2N5457 (dashed underline) is a minimum of 1 mA and a maximum of 5 mA.

The FET transfer characteristic approximately follows the equation

$$I_D = I_{DSS}\left[1 - \frac{V_{GS}}{V_{GS(off)}}\right]^2 \qquad (9\text{-}1)$$

When I_{DSS} and $V_{GS(off)}$ are known, a table of corresponding values of I_D and V_{GS} can be determined from Eq. 9-1. These may be used to construct the FET transfer characteristic. Because of the wide range of specified values for I_{DSS} and $V_{GS(off)}$, the transfer characteristic can differ substantially from one device to another one with the same type number. This creates a problem in FET bias circuits (see Chapter 10).

Example 9-3

Using the information provided on the data sheet (Fig. 9-18), construct the minimum and maximum transfer characteristics for a 2N5459 JFET.

Solution

From Fig. 9-18, $V_{GS(off)} = -2$ V (min), −8 V (max)

and $I_{DSS} = 4$ mA (min), 16 mA (max)

To construct the minimum transfer characteristic, substitute $V_{GS(off)(min)}$ and $I_{DSS(min)}$ into Eq. 9-1, together with convenient levels of V_{GS}.

Eq. 9-1:
$$I_D = I_{DSS}\left[1 - \frac{V_{GS}}{V_{GS(off)}}\right]^2$$

For $V_{GS} = 0$, $\quad I_D = 4\,\text{mA}[1 - (0/2\,\text{V})]^2 = 4\,\text{mA}$

Plot point 1 for the minimum transfer characteristic on Fig. 9-19 at $V_{GS} = 0$ and $I_D = 4\,\text{mA}$.

For $V_{GS} = 0.25\,V_{GS(off)}$, $\quad I_D = 2.25\,\text{mA}$: $\quad$ point 2

For $V_{GS} = 0.5\,V_{GS(off)}$, $\quad I_D = 1\,\text{mA}$: $\quad$ point 3

For $V_{GS} = 0.75\,V_{GS(off)}$, $\quad I_D = 0.25\,\text{mA}$: $\quad$ point 4

For $V_{GS} = V_{GS(off)}$, $\quad I_D = 0$: $\quad$ point 5

The minimum transfer characteristic is now drawn through points 1 to 5.

For the maximum transfer characteristic, the above process is repeated using $V_{GS(off)} = -8\,\text{V}$ and $I_{DSS} = 16\,\text{mA}$.

For $V_{GS} = 0$, $\quad I_D = 16\,\text{mA}[1 - (0/8\,\text{V})]^2 = 16\,\text{mA}$

Plot point 6 for the maximum transfer characteristic on Fig. 9-19 at $V_{GS} = 0$ and $I_D = 16\,\text{mA}$.

For $V_{GS} = 0.25\,V_{GS(off)}$, $\quad I_D = 9\,\text{mA}$ $\quad$ point 7

For $V_{GS} = 0.5\,V_{GS(off)}$, $\quad I_D = 4\,\text{mA}$ $\quad$ point 8

For $V_{GS} = 0.75\,V_{GS(off)}$, $\quad I_D = 1\,\text{mA}$ $\quad$ point 9

For $V_{GS} = V_{GS(off)}$, $\quad I_D = 0$ $\quad$ point 10

Draw the maximum transfer characteristic through points 6 to 10.

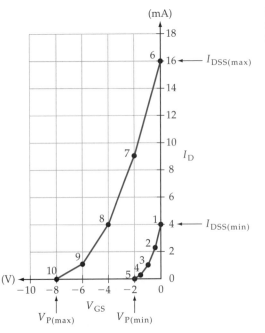

Figure 9-19 Maximum and minimum transfer characteristic for a 2N5459 JFET constructed for Example 9-3.

Forward Transfer Admittance

The *forward transfer admittance* (Y_{fs}), which is also known as the *transconductance* (g_m or g_{FS}), for a FET defines how the drain current is controlled by the gate-source voltage.

$$Y_{fs} = \frac{\text{variation in } I_D}{\text{variation in } V_{GS}} \quad \text{(When } V_{DS} \text{ remains constant)}$$

$$Y_{fs} = \frac{\Delta I_D}{\Delta V_{GS}}\bigg|_{V_{DS}} \tag{9-2}$$

The units used for Y_{fs} are *microSiemens* (μS), which can be restated as *microamps per volt* ($\mu A/V$). *MilliSiemens* (mS), or *milliamps/volt* (*mA/V*) may also be used. For a FET with $Y_{fs} = 2$ mA/V, I_D changes by 2 mA when V_{GS} is altered by 1 V. In the portion of a FET data sheet in Fig. 9-20, the Y_{fs} units are μmhos. The *mho* (ohm written backwards) is another (older) name for the *Siemen*, the unit of conductance. For the 2N5457 (underlined) the value of Y_{fs} is specified as a minimum of 1000 μmhos, and a maximum of 5000 μmhos. An inverted ohm symbol is also sometimes used instead of the Siemens symbol.

2N5457, 2N5458, 2N5459
Dynamic Characteristics

Characteristic	Symbol	Min	Typ	Max	Unit
Forward Transfer Admittance ($V_{DS} = 15$ V dc, $V_{GS} = 0$, $f = 1$ kHz)	Y_{fs}				μmhos
2N5457		1000	3000	5000	
2N5458		1500	4000	5500	
2N5459		2000	4500	6000	

Figure 9-20 JFET forward transfer admittance specification.

Because Y_{fs} defines the relationship between I_D and V_{GS}, it can be determined from the slope of the FET transfer characteristic. This is demonstrated in Example 9-4 and illustrated in Fig. 9-21.

Example 9-4

Determine Y_{fs} at $V_{GS} = -1$ V and $V_{GS} = -4$ V from the 2N5459 FET transfer characteristic in Fig. 9-21.

Solution

At $V_{GS} = -1$ V:

Eq. 9-2:

$$Y_{fs} = \frac{\Delta I_D}{\Delta V_{GS}} = \frac{4.3 \text{ mA}}{1.25 \text{ V}}$$

$$= 3.4 \text{ mA/V} = 3400 \text{ } \mu S$$

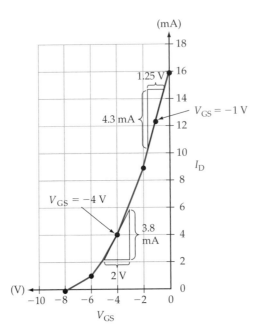

Figure 9-21 The value of the forward transfer admittance for a FET can be determined from the slope of the transfer characteristic.

At $V_{GS} = -4$ V:

$$Y_{fs} = \frac{\Delta I_D}{\Delta V_{GS}} = \frac{3.8 \text{ mA}}{2 \text{ V}}$$

$$= 1.9 \text{ mA/V} = 1900 \ \mu S$$

The transfer characteristic for a FET is defined by Eq. 9-1, and Y_{fs} is determined from the slope of the transfer characteristic. So an equation for Y_{fs} can be derived by differentiating Eq. 9-1.

$$Y_{fs} = \frac{2 I_{DSS}}{V_{GS(off)}} \left[1 - \frac{V_{GS}}{V_{GS(off)}} \right] \qquad (9\text{-}3)$$

Equation 9-3 may be used to calculate the value of Y_{fs} for any V_{GS} level. As in Eq. 9-1, since the negative sign for V_{GS} is already included, only the numerical value should be entered.

Example 9-5

Using Eq. 9-3, determine Y_{fs} at $V_{GS} = -1$ V and at $V_{GS} = -4$ V for a 2N5458 JFET with a maximum transfer characteristic (see Fig. 9-18).

Solution
From the data sheet (Fig. 9-18), the 2N5458 maximum transfer characteristic is defined by

$$V_{GS(off)(max)} = -7 \text{ V}$$

and

$$I_{DSS(max)} = 9 \text{ mA}$$

At $V_{GS} = -1$ V:

Eq. 9-3: $Y_{fs} = \dfrac{2\,I_{DSS}}{V_{GS(off)}}\left[1 - \dfrac{V_{GS}}{V_{GS(off)}}\right] = \dfrac{2 \times 9\text{ mA}}{7\text{ V}}\left[1 - \dfrac{1\text{ V}}{7\text{ V}}\right]$

$= 2200\ \mu S$

At $V_{GS} = -4$ V:

Eq. 9-3: $Y_{fs} = \dfrac{2\,I_{DSS}}{V_P}\left[1 - \dfrac{V_{GS}}{V_P}\right] = \dfrac{2 \times 9\text{ mA}}{7\text{ V}}\left[1 - \dfrac{4\text{ V}}{7\text{ V}}\right]$

$= 1100\ \mu S$

Output Admittance

The *drain resistance* (r_d) of a FET is the ac resistance between drain and source terminals when the device is operating in the pinch-off region of its drain characteristics. It is also the slope of the drain characteristics in the pinch-off region (see Fig. 9-22). Since the drain characteristics are almost flat, r_d is not easily determined from the characteristic. Because r_d is usually the output resistance of the FET at the drain terminal, it may also be expressed as an *output admittance* ($Y_{os} = 1/r_d$).

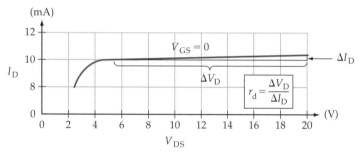

Figure 9-22   The ac drain resistance of a FET (r_d) is the slope of the drain characteristic. The output admittance (Y_{fs}) is the inverse of r_d.

The output admittance is defined as

$$Y_{os} = \frac{\text{variation in } I_D}{\text{variation in } V_{DS}} \quad \text{(When } V_{GS} \text{ remains constant)}$$

$$Y_{os} = \frac{\Delta I_D}{\Delta V_{DS}}\bigg|_{V_{GS}} \qquad (9\text{-}4)$$

From Appendix A-11, Y_{os} for the 2N5457-5459 JFETs is 10 μS typically, and 50 μS maximum. This corresponds to a drain resistance of 100 kΩ typically, and 20 kΩ minimum.

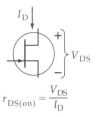

$$r_{DS(on)} = \frac{V_{DS}}{I_D}$$

Figure 9-23 The drain-source ON resistance ($r_{DS(on)}$) of a FET is the channel resistance.

(a) Gate reverse current

(b) Gate input resistance

Figure 9-24 Gate cutoff current and input resistance.

Drain-Source ON Resistance

The *drain-source ON resistance* ($r_{DS(on)}$) (also designated $R_{D(on)}$) is a dc quantity, not to be confused with the ac drain-source resistance (r_{ds}). $r_{DS(on)}$ is the resistance of the FET channel when the depletion regions are absent; when the device is biased ON in the channel ohmic region of the drain characteristics (see Fig. 9-23). In this condition, the voltage drop along the channel from drain to source is ($I_D \times r_{DS(on)}$). This is the *drain-source ON voltage* ($V_{DS(on)}$), which is similar to the $V_{CE(sat)}$ of a BJT.

The drain-source ON resistance might typically be 60 Ω or lower (see Appendix A-12). It can be an important quantity for FETs used in switching circuits. $V_{DS(on)}$ can be much smaller than $V_{CE(sat)}$, making FET gates superior to BJT gates for some applications.

Gate Cutoff Current and Input Resistance

The gate-channel junctions in a JFET are *pn*-junctions, and because they are normally reverse-biased, a minority charge carrier current flows. This is the *gate-source cutoff current* (I_{GSS}), or *gate reverse current* (see Fig. 9-24a). For a 2N5457 FET, I_{GSS} is specified as 1 nA at 25°C, and 200 nA at 100°C (see Appendix A-11).

The *gate input resistance* (R_{GS}) is the resistance of the reverse-biased gate-channel junctions (Fig. 9-24b), and it is inversely proportional to I_{GSS}. Typical values of R_{GS} for a JFET are 10^9 Ω at 25°C and 10^7 Ω at 100°C.

Breakdown Voltage

There are several ways of expressing JFET breakdown voltage. BV_{DGO} is the *drain-gate breakdown voltage* with the source terminal open-circuited. BV_{GSS} is the *gate-source breakdown voltage* with the drain terminal shorted to the source. Both are a specification of the voltage at which the gate-channel junctions might break down. Appendix A-11 shows a BV_{GSS} of 25 V for a 2N5457 (this is identified as $V_{(BR)GSS}$ on the data sheet).

Noise Figure

A FET usually has a much lower level of thermal noise than a BJT. This is because there are very few charge carriers crossing junctions in a FET. As in the case of a BJT, the *FET noise figure* (NF) is specified as a *spot noise figure* at a particular frequency and bias conditions and for a given resistance at the input. The figure varies if any of these conditions are altered. Noise calculations for a FET circuit are performed exactly as for a BJT circuit.

Capacitances

Terminal capacitances for FETs may be specified as *gate-drain capacitance* (C_{gd}), *gate-source capacitance* (C_{gs}), and *drain-source capacitance* (C_{ds}). The input capacitance is sometimes expressed as the *common-source input capacitance* (C_{iss} or C_{gss}).

This is the gate-source capacitance measured with the drain terminal shorted to the source (Fig. 9-25). In this case, a *reverse transfer capacitance* (C_{rss}) is also specified, C_{rss} being another term for C_{gd}. These quantities are very important for FET high-frequency and switching circuits. From the 2N5457 specification in data sheet A-11 in Appendix A, the maximum capacitance values are $C_{iss} = 7$ pF and $C_{rss} = 3$ pF.

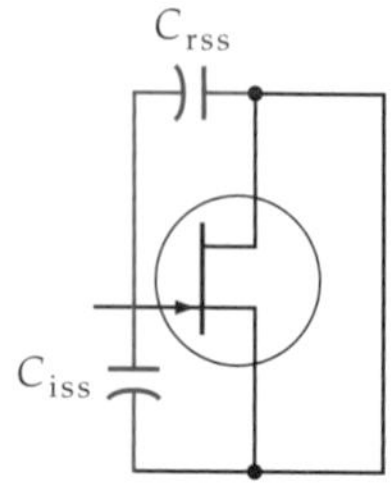

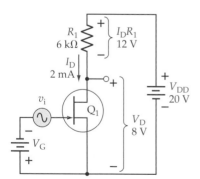

Figure 9-25 FET common-source input capacitances and reverse transfer capacitance.

Practice Problems

9-3.1 Construct the maximum transfer characteristic for a 2N5458 JFET (see data sheet A-11 in Appendix A). From the transfer characteristic, determine the value of Y_{fs} at $V_{GS} = -4$ V.

9-3.2 Use Eq. 9-3 to calculate the value of Y_{fs} for a 2N5459 JFET at $V_{GS} = -1$ V and at $V_{GS} = -4$ V.

9-3.3 From the data sheet A-12 in Appendix A, determine the following quantities for a 2N4860 FET: maximum pinch-off voltage, maximum drain-source saturation current, and drain-source ON resistance.

9-4 FET AMPLIFICATION AND SWITCHING

Amplification

Consider the *n*-channel JFET circuit in Fig. 9-26. Note that drain-source terminals are provided with a dc supply (V_{DD}), connected via the drain resistor (R_1). The gate-source junctions are reverse-biased by the gate voltage (V_G). An ac signal generator (with voltage v_i) is connected in series with the gate terminal. As already discussed, field effect transistors are voltage-operated devices. The drain current is controlled by the gate-source voltage. A change in gate-source voltage (ΔV_{GS}) produces a change in drain current (ΔI_D), and this causes a variation in the voltage drop across the resistor connected in series with the drain terminal.

Figure 9-26 FET voltage amplifer circuit. An ac input (v_i) produces a variation in I_D which varies the voltage drop across resistor R_1. This results in an (amplified) ac output voltage.

Suppose that, for the circuit in Fig. 9-26, $V_{DD} = 20$ V, $R_1 = 6$ kΩ, and V_G is adjusted to give $I_D = 2$ mA. Also, assume that the FET has a forward transfer admittance of $Y_{fs} = 4000$ μS. The dc level of the drain voltage is

$$V_D = V_{DD} - (I_D R_1) = 20 \text{ V} - (2 \text{ mA} \times 6 \text{ k}\Omega)$$

$$= 8 \text{ V}$$

Now assume that the instantaneous level of v_i is +50 mV. This produces an increase:

$$\Delta I_D = Y_{fs} \times v_i = 4000 \text{ μS} \times 50 \text{ mV}$$

$$= 0.2 \text{ mA}$$

The new level of drain voltage is

$$V_D = V_{DD} - R_1 (I_D + \Delta I_D) = 20 \text{ V} - 6 \text{ k}\Omega(2 \text{ mA} + 0.2 \text{ mA})$$

$$= 6.8 \text{ V}$$

So V_D changed from 8 V to 6.8 V when the ac input increased from zero to 50 mV.

Or $\qquad \Delta V_D = 6.8 \text{ V} - 8 \text{ V}$

$$= -1.2 \text{ V}$$

When the instantaneous level of v_i is -50 mV,

$$\Delta I_D = Y_{fs} \times v_i = 4000 \text{ μS} \times (-50 \text{ mV})$$

$$= -0.2 \text{ mA}$$

This gives a new drain voltage:

$$V_D = V_{DD} - R_1 (I_D + \Delta I_D) = 20 \text{ V} - 6 \text{ k}\Omega(2 \text{ mA} - 0.2 \text{ mA})$$

$$= 9.2 \text{ V}$$

So $\qquad \Delta V_D = 9.2 \text{ V} - 8 \text{ V}$

$$= 1.2 \text{ V}$$

It is seen that an ac input of $v_i = \pm 50$ mV produces an output voltage change at the FET drain terminal (ΔV_D) of ± 1.2 V. This can be stated as an ac output voltage v_o of ± 1.2 V. The circuit voltage amplification is

$$A_v = \frac{v_o}{v_i} = \frac{1.2 \text{ V}}{50 \text{ mV}}$$

$$= 24$$

Example 9-6

A 2N5457 JFET is used in the amplifier circuit in Fig. 9-27. Calculate the maximum and minimum output voltage produced by a ± 100 mV ac input. Calculate the circuit voltage gain in each case.

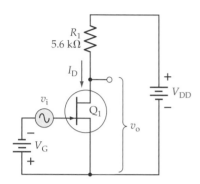

Figure 9-27 FET amplifier circuit for Ex. 9-6.

Solution

From Appendix A-11,

$$Y_{fs(max)} = 5000 \ \mu S \quad \text{and} \quad Y_{fs(min)} = 1000 \ \mu S$$

For $Y_{fs(max)}$,

$$\Delta I_D = Y_{fs(max)} \times v_i = 5000 \ \mu S \times (\pm 100 \ mV)$$

$$= \pm 0.5 \ mA$$

and

$$v_o = \Delta I_D \times R_i = \pm 0.5 \ mA \times 5.6 \ k\Omega$$

$$= \pm 2.8 \ V$$

$$A_v = \frac{v_o}{v_i} = \frac{\pm 2.8 \ V}{\pm 100 \ mV}$$

$$= 28$$

For $Y_{fs(min)}$,

$$\Delta I_D = Y_{fs(min)} \times v_i = 1000 \ \mu S \times (\pm 100 \ mV)$$

$$= \pm 0.1 \ mA$$

and

$$v_o = \Delta I_D \times R_1 = \pm 0.1 \ mA \times 5.6 \ k\Omega$$

$$= \pm 0.56 \ V$$

$$A_v = \frac{v_o}{v_i} = \frac{\pm 0.56 \ V}{\pm 100 \ mV}$$

$$= 5.6$$

FET Switching

The direct-coupled JFET switching circuit in Fig. 9-28a is similar to the BJT switching circuit in Fig. 8-18a, but there are important differences in the operation and performance of the two circuits.

In Section 9-3 it is explained that a FET can be biased *on* to produce the lowest possible *drain-source ON voltage* ($V_{DS(on)}$), which corresponds to $V_{CE(sat)}$ for a BJT. $V_{DS(on)}$ occurs when the device has a minimum channel resistance known as the *drain-source ON resistance* ($r_{DS(on)}$). As illustrated in Fig. 9-28b, the output voltage is

$$V_{DS(on)} = I_D \times r_{DS(on)}$$

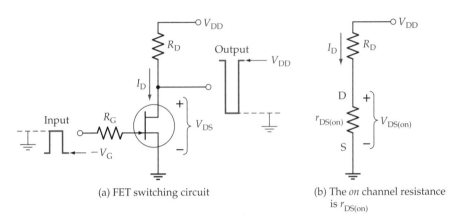

(a) FET switching circuit

(b) The *on* channel resistance is $r_{DS(on)}$

Figure 9-28   Direct-coupled *n*-channel JFET switching circuit. The output switches from V_{DD} to $I_D \times r_{DS(on)}$ when the input changes from $-V_G$ to zero.

Like a BJT switch, a FET switching circuit has turn-on and turn-off times that are made up of delay time, rise time, storage time, and fall time. FET switching times tend to be shorter than BJT times because a FET does not have a forward-biased junction with a diffusion capacitance (like the BJT base-emitter junction).

FET switching circuits are covered further in Section 10-11.

Practice Problem

9-4.1 A FET amplifier circuit, as in Fig 9-27, has $V_{DD} = 15$ V, $R_1 = 4.7$ kΩ, and $v_i = \pm 30$ mV. If V_G is adjusted so that $I_D = 1.5$ mA, calculate V_D. Also, calculate the typical voltage gain for the circuit if Q_1 is a 2N5459 with $Y_{fs} = 4500$ μS.

9-5 MOSFETs

Enhancement MOSFET

Figure 9-29 shows the construction of a *metal oxide semiconductor FET (MOSFET)*, also known as an *insulated gate FET*. Starting with a high-resistive *p*-type substrate, two blocks of heavily-doped *n*-type material are diffused into the substrate, and then the surface is coated with a layer of silicon dioxide. Holes are cut through the silicon dioxide to make contact with the *n*-type blocks. Metal is deposited through the holes for source and drain terminals, as illustrated, and a metal plate is deposited on the surface area between drain and source. As will be explained, this plate functions as a gate.

Consider the situation illustrated in Fig. 9-30a. The drain terminal of the MOSFET is positive with respect to the source, and the gate is open-circuited. The two *n*-type blocks and the *p*-type substrate form back-to-back *pn*-junctions

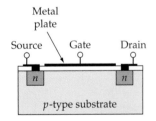

Figure 9-29   Metal oxide semiconductor FET (MOSFET) construction.

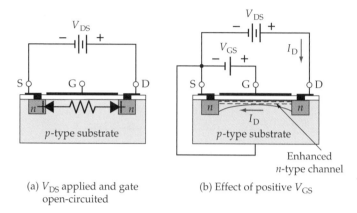

(a) V_{DS} applied and gate
open-circuited

(b) Effect of positive V_{GS}

Figure 9-30   Effect of $+V_{DS}$ on the MOSFET with the gate terminal open-circuited and with $+V_{GS}$ applied to the gate.

connected by the resistance of the p-type material, as illustrated. The pn-junction close to the drain terminal is reverse-biased, so that only a very small (reverse-leakage) current flows from D to S.

Now assume that the source terminal is connected to the substrate and that a positive gate voltage is applied, as shown in Fig. 9-30b. Negative (minority) charge carriers within the substrate are attracted to the (positive) plate that constitutes the gate. Since these charge carriers (electrons) cannot cross the silicon dioxide to the gate, they accumulate close to the surface of the substrate, as shown. The minority charge carriers constitute an n-type channel between drain and source, and as the gate-source voltage is made more positive, more electrons are attracted into the channel, causing the channel resistance to decrease. A drain current flows along the channel between the D and S terminals, and because the channel resistance is controlled by the gate-source voltage (V_{GS}), the drain current is also controlled by V_{GS}. The channel conductivity is said to be *enhanced* by the positive gate-source voltage, and so the device is known as an *enhancement-mode MOSFET (EMOSFET or EMOS transistor)*.

Typical drain and transfer characteristics for an n-channel EMOS device are shown in Fig. 9-31. Note on both characteristics that the drain current increases as the positive gate-source bias voltage is increased. Because the gate of the MOSFET is insulated from the channel, there is no gate-source leakage current and the device has an extremely high (gate) input resistance: typically 10^{15} Ω or greater. Typical forward transfer admittance values for this type of (low-power) MOSFET range from 1 mS (1 mA/V) to a maximum of perhaps 6 mS, which is similar to JFET Y_{fs} values.

Two graphic symbols for the n-channel EMOS transistor are shown in Fig. 9-32. One symbol shows the source and substrate connected internally, while the other has a separate substrate terminal. The line representing the device channel is broken into three sections to indicate that the channel does not exist until an appropriate gate voltage is applied. To show that the device has an insulated gate, the gate symbol does not make direct contact with the

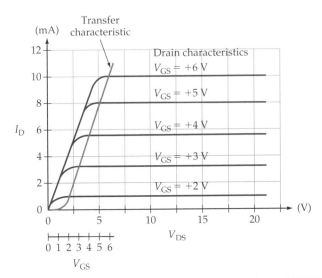

Figure 9-31 Typical drain and transfer characteristics for an *n*-channel EMOS transistor.

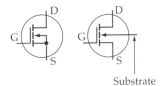

Figure 9-32 Circuit symbols for an *n*-channel EMOSFET.

channel. The arrowhead points from the *p*-type substrate to the *n*-type channel.

A *p*-channel EMOS transistor is constructed by starting with an *n*-type substrate and diffusing *p*-type drain and source blocks, as illustrated in Fig. 9-33a. The device characteristics are similar to those in Fig. 9-31, except that all voltage polarities and current directions are reversed. The drain-source voltage is negative, and a negative gate-source voltage is required to create the *p*-type channel. The arrowheads in the circuit symbols are also reversed (see Fig. 9-33b).

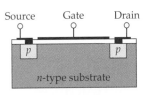

(a) *p*-channel EMOSFET   (b) Circuit symbol

Figure 9-33 Construction and circuit symbol for a *p*-channel EMOSFET.

Depletion-Enhancement MOSFET

The device cross-section shown in Fig. 9-34a is similar to that for an EMOS transistor, except that a lightly doped *n*-type channel is included between the drain and source blocks. When a positive drain-source voltage (V_{DS}) is applied, a drain current (I_D) flows even when the gate-source voltage (V_{GS}) is zero. If a negative V_{GS} is applied, as shown in Fig. 9-34b, some of the negative charge carriers are repelled from the gate and driven out of the *n*-type channel. This creates a depletion region in the channel, as illustrated, causing an increase in channel resistance and a decrease in drain current. The effect is similar to that in an *n*-channel JFET. Because of the channel depletion regions, the device can be termed a *depletion-mode MOSFET*.

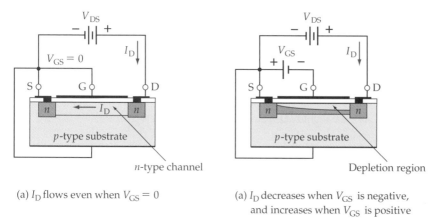

(a) I_D flows even when $V_{GS} = 0$

(a) I_D decreases when V_{GS} is negative, and increases when V_{GS} is positive

Figure 9-34   *n*-Channel depletion-enhancement MOSFET.

Now consider what happens when a positive gate-source voltage is applied. Additional *n*-type charge carriers are attracted from the substrate into the channel, decreasing its resistance and increasing the drain current. So the depletion-mode MOSFET can also be operated as an enhancement-mode device. Thus, these devices can be referred to as *depletion-enhancement MOS-FETs* or *DE-MOSFETs*; however, the terms DMOSFET is normally applied.

Typical drain and transfer characteristics for a DMOSFET are shown in Fig. 9-35. The device operates in the depletion mode when V_{GS} is negative, and in the enhancement mode when positive levels of V_{GS} are used.

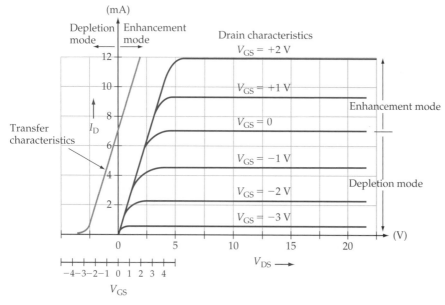

Figure 9-35   Drain and transfer characteristics for an *n*-channel depletion-enhancement MOSFET.

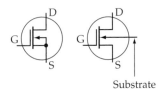

Figure 9-36 Circuit symbols for an *n*-channel depletion-enhancement MOSFET.

The circuit symbols used for DMOSFETs are similar to those already discussed for EMOSFETs, except that the line representing the channel is made solid to show that a channel is present when $V_{GS} = 0$ (see Fig. 9-36).

Example 9-7

From the DMOSFET characteristics in Fig. 9-35, determine the forward transfer admittance at $V_{GS} = 0$.

Solution

At $V_{GS} = -2$ V, $\qquad\qquad I_D = 2.2$ mA

and at $V_{GS} = +2$ V, $\qquad\quad I_D = 11.8$ mA

$$\Delta I_D = 11.8 \text{ mA} - 2.2 \text{ mA} = 9.6 \text{ mA}$$

and $\qquad\qquad\qquad\qquad \Delta V_{GS} = +2 \text{ V} - (-2 \text{ V}) = 4 \text{ V}$

Eq. 9-2: $\qquad\qquad\qquad Y_{fs} = \dfrac{\Delta I_D}{\Delta V_{GS}} = \dfrac{9.6 \text{ mA}}{4 \text{ V}}$

$$= 2.4 \text{ mA/V} = 2400 \text{ μS}$$

VMOSFET

A disadvantage of the MOSFET types already discussed is that the minimum channel length depends upon the dimensions of the photographic masks used in the manufacturing process. Shorter channel lengths can be produced by changing the geometry of the MOSFET to create a vertical channel (instead of horizontal). The channel length is then easily determined by the diffusion process. The shorter channel results in lower resistance, greater power dissipation, higher frequency response, and larger forward transfer admittance values than are possible with other MOSFETs.

Figure 9-37a shows a device referred to as a VMOSFET because of its V-shaped configuration, and because it uses a vertical channel between drain

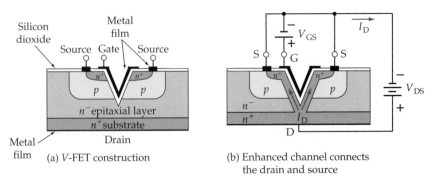

(a) *V*-FET construction

(b) Enhanced channel connects the drain and source

Figure 9-37 VMOSFET construction and operation. The vertical channel gives an improved frequency response, lower channel resistance, and greater power dissipation than other FETs.

and source. The V-cut penetrates from the surface of the device through n^+, p, and n^- layers almost to the n^+ substrate. The n^+ layers are low-resistive, and the n^- is a high-resistive region. The silicon dioxide layer covers both the horizontal surface and the V-cut surface. The gate is a metal film deposited on the silicon dioxide surface in the V-cut. The drain terminal is at the bottom of the n^+ substrate, and the source connection is made to the top n^+ region and to the p region.

The VMOSFET operates in the enhancement mode; no channel exists until a positive gate source voltage is applied. An n-type channel is created as shown in Fig. 9-37b when V_{GS} is made positive and a current flows vertically from drain to source.

Some versions of VMOSFETs use a vertical channel but do not use the V-cut. Motorola currently manufactures devices referred to as TMOS, because the drain current flow tends to be T-shaped. International Rectifier uses the term Hexfet, because of the hexagonal configuration employed in the device construction.

Typical VMOSFET drain and transfer characteristics are shown in Fig. 9-38. It can be seen that they are similar to the enhancement-mode device characteristics in Fig. 9-31, although much higher drain current levels are usually involved with a VMOSFET than with other MOSFETs. The characteristics shown are for a device capable of dissipating 40 W.

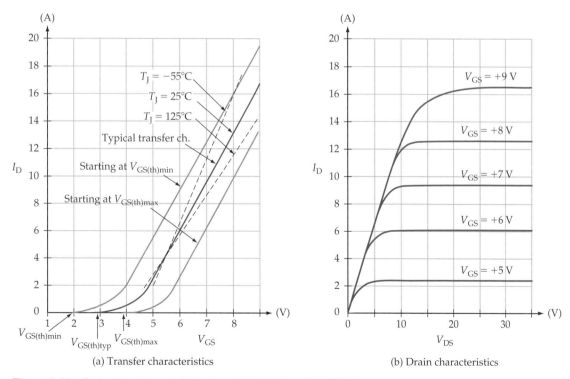

(a) Transfer characteristics (b) Drain characteristics

Figure 9-38 Typical transfer and drain characteristics for a VMOSFET.

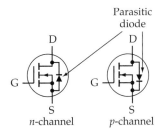

Figure 9-39 Circuit symbols for VMOSFETs.

The VMOSFET transfer characteristic is seen to be approximately linear over most of its length, and it is curved (non-linear) only at the low current levels. For the typical characteristic shown, I_D commences at the typical gate threshold voltage ($V_{GS(th)typ}$), and the slope of the linear portion of the graph is set by the typical g_{FS} value (same as Y_{FS}). It should be noted that for a device with a given type number, I_D might begin at $V_{GS(th)min}$ or $V_{GS(th)max}$, and that the slope of the characteristic depends upon the actual g_{FS} for the device. Temperature changes can also have a serious effect on the characteristics, as shown by the dashed lines.

The graphic symbols used for VMOS devices are the same as for other enhancement-mode MOSFETs. However, a *parasitic pn*-junction is present between the p and n^- layers in a VMOSFET, and this must not be allowed to become forward-biased. So the diode is usually shown on the device symbol (see Fig. 9-39).

VMOS devices can be described as high-voltage, EMOSFETs. They are very suitable for high-frequency or fast switching applications. Normally, they have much larger forward transfer admittance (g_{FS}) values and lower drain-source ON resistances than other FETs. Figure 9-40 lists some performance data for two representative (low-power and high-power) VMOSFETs. Note that the 80 mS minimum g_{FS} for the 2N7002 is substantially larger than the 4 mS typical for other FET types. The IRF520 has a typical g_{FS} of 2.9 S, which is greater again than that for the 2N7002. FET voltage amplification is directly proportional to the forward transfer admittance; so VMOSFET circuits can have much greater voltage gains than circuits using other FETs.

2N7002

$V_{DS(max)}$	$I_{D(max)}$	$P_{D(max)}$	$r_{d(on)}$	g_{FS}	$V_{GS(th)}$
60 V	115 mA	200 mW	7.5 Ω	80 mS (min)	1 V (min) 2.5 V (max)

IRF520

$V_{DS(max)}$	$I_{D(max)}$	$P_{D(max)}$	$r_{d(on)}$	g_{FS}	$V_{GS(th)}$
100 V	8 A	40 W	0.3 Ω	1.5 S (min) 2.9 S (typ)	2 V (min) 4 V (max)

Figure 9-40   Performance data for low-power and high-power VMOSFETs.

MOSFET specifications normally list *gate threshold voltage* ($V_{GS(th)}$) values. The drain current remains at zero with gate-source voltage levels below $V_{GS(th)}$ and begins to flow when V_{GS} is increased above $V_{GS(th)}$. So $V_{GS(th)}$ is similar to the pinch-off voltage for a JFET. Like the JFET pinch-off voltage, $V_{GS(th)}$ has maximum and minimum values for each device type number.

For a JFET, I_{DSS} is the drain current when $V_{GS} = 0$. The EMOSFET parameter identified as I_{DSS} is still the drain current at zero gate-source voltage.

However, because the device is *off* when $V_{GS} = 0$, I_{DSS} is a drain-source leakage current with a level measured in microamps. For EMOSFET devices, I_D is usually specified as $I_{D(on)}$ at $V_{GS} = 10$ V.

EMOSFETs are often applied in situations where they are biased *on* to a low drain current level, and then I_D is increased to a desired higher level by the application of a signal voltage. The precise levels of the required gate-source bias and signal voltages can be determined only from the transfer characteristic for each individual device. However, for a given I_D level beyond the curved part of the characteristic, a rough approximation for the required V_{GS} can be calculated from the maximum $V_{GS(th)}$ and the typical g_{FS} values, as follows:

$$V_{GS} \approx V_{GS(th)max} + \frac{I_D}{g_{FS(typ)}} \qquad (9\text{-}5)$$

Example 9-8

Using the information in Fig. 9-40, calculate the approximate gate-source voltage required to produce a 7 A drain current in an IRF520. Determine the drain-source ON voltage and the device power dissipation at $I_D = 7$ A.

Solution

Eq. 9-5:
$$V_{GS} \approx V_{GS(th)(max)} + \frac{I_D}{g_{FS(typ)}} = 4 \text{ V} + \frac{7 \text{ A}}{2.9 \text{ S}}$$

$$= 6.4 \text{ V}$$

$$V_{DS(on)} = I_D \times r_{D(on)} = 7 \text{ A} \times 0.3 \ \Omega$$

$$= 2.1 \text{ V}$$

$$P_D = I_D \times V_{DS(on)} = 7 \text{ A} \times 2.1 \text{ V}$$

$$= 14.7 \text{ W}$$

Handling MOSFETs

Figure 9-41 shows the kind of label normally found on packages containing MOSFETs. This means that the devices can be very easily destroyed by *electrostatic discharge (ESD)*, that is, the discharge of static electricity that accumulates on individuals or objects. Usually the very thin layer of silicon dioxide at the gate breaks down when an excessive gate-source voltage is applied. Special precautions are necessary to protect MOSFETs from ESD:

1. Always store MOSFETs in closed conductive containers. They are usually packaged with the terminals inserted into conducting foam rubber.
2. Use a work station with a grounded anti-static bench-top pad and a grounded anti-static floor pad.
3. Wear anti-static clothing and grounded wrist bands.

Figure 9-41 Packaging symbol used to indicate that the enclosed devices can be destroyed by static electricity.

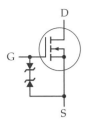

Figure 9-42 MOSFET with built-in Zener diodes for protection against static electricity.

4. Use a soldering iron with a grounded tip.
5. Avoid touching the device terminals.

Some MOSFETs have built-in protection against static electricity. Back-to-back series-connected Zener diodes are located between the gate and source terminals, as illustrated in Fig. 9-42. When the gate-source voltage is high enough to reverse-bias one of the diodes into breakdown, the other diode is forward-biased. Thus the gate-source voltage cannot exceed $\pm(V_Z + V_F)$.

Section 9-5 Review

9-5.1 Using illustrations, explain the operation of n-channel EMOSFET.
9-5.2 Sketch circuit symbols for n-channel and p-channel EMOS and DMOS field effect transistors.
9-5.3 Sketch typical transfer characteristics for DMOS devices.
9-5.4 Briefly explain the operation of VMOSFETs.

Practice Problems

9-5.1 Determine the forward transfer admittance at $I_D = 8$ mA for the EMOS device with the characteristics in Fig. 9-31.
9-5.2 Calculate the approximate gate-source voltage required to produce a 7.5 A drain current in an EMOSFET that has a maximum $V_{GS(th)} = 2.5$ V and typical $g_{FS} = 1.5$ S.

Review Questions

Section 9-1

9-1 Sketch a block representation for an n-channel JFET, showing bias voltages, depletion regions, and current directions. Label the device terminals and explain its operation. Explain the effect of increasing levels of negative gate-source voltage.
9-2 Repeat Question 9-1 for a p-channel JFET, explaining the effect of increasing levels of positive gate-source voltage.
9-3 Sketch circuit symbols for n-channel and p-channel JFETs. Label the terminals in each case, show supply and bias voltage polarities, and indicate the drain current directions.
9-4 Sketch the cross-section of a diffused JFET. Label all parts of the device and briefly explain.

Section 9-2

9-5 Draw a circuit diagram for obtaining the drain and transfer characteristics for an n-channel JFET. Briefly explain how a table of values should be determined for each characteristic.

9-6 Sketch a typical drain characteristic for $V_{GS} = 0$ for an n-channel JFET. Explain the shape of the characteristic, identify the regions, and indicate the important current and voltage levels.

9-7 Draw a typical drain characteristics family for an n-channel JFET. Identify the V_{GS} for each characteristic. Briefly explain.

9-8 Draw a typical transfer characteristic for an n-channel JFET. Identify the drain-source saturation current and pinch-off voltage.

9-9 Sketch typical drain and transfer characteristics for a p-channel JFET. Show typical current and voltage scales and gate-source voltage levels.

Section 9-3

9-10 Define the saturation current and pinch-off voltage for a JFET, and state typical values for each quantity.

9-11 Define forward-transfer admittance, transconductance, output admittance, and drain resistance for a JFET. State typical values for each quantity.

9-12 Define drain-source ON resistance, drain-source ON voltage, and gate cutoff current for a JFET. State typical values.

Section 9-4

9-13 Sketch a basic voltage amplifier circuit using a JFET, and explain its operation.

9-14 Sketch a typical JFET switching circuit and explain its operation. Define $r_{DS(on)}$ and $V_{DS(on)}$.

Section 9-5

9-15 Draw a cross-section diagram for an enhancement-mode MOSFET. Label all parts of the device and explain its operation.

9-16 Sketch typical drain and transfer characteristics for an n-channel EMOS transistor. Show typical current and voltage scales and gate-source voltage levels.

9-17 Sketch graphic symbols for n-channel and p-channel EMOS devices. Label the terminals in each case, show the supply and bias voltage polarities, and indicate the drain and source current directions.

9-18 Draw a diagram of the cross-section of a depletion-enhancement-mode MOSFET. Identify all parts of the device and explain its operation.

9-19 Sketch typical drain and transfer characteristics for an n-channel DMOSFET. Show typical current and voltage scales and gate-source voltage levels.

9-20 Sketch graphic symbols for n-channel and p-channel DMOS transistors. Label the terminals in each case, show the supply and bias voltage polarities, and indicate the drain and source current directions.

9-21 Draw a diagram of the cross-section of a VMOSFET. Identify all parts of the device, explain its operation, and discuss its advantages compared to other MOSFETs.

9-22 Sketch typical drain and transfer characteristics for an *n*-channel VMOSFET. Briefly explain the characteristics.

9-23 Sketch graphic symbols for *n*-channel and *p*-channel VMOSFETs. Label the terminals in each case, show the supply and bias voltage polarities, and indicate the drain and source current directions.

Problems

Section 9-2

9-1 Plot the I_D/V_{DS} characteristic for an *n*-channel JFET from the following table of current and voltage levels. Determine I_{DSS} and V_P from the characteristics.

$V_{GS} = 0$	V_{DS} (V)	0	1.25	2	2.5	5	10	15	25
	I_D (mA)	0	2	3	4	4.8	4.9	4.95	5

$V_{GS} = -1$ V	V_{DS} (V)	0	1	2	2.5	5	10	15	25
	I_D (mA)	0	1	2	2.3	2.4	2.5	2.6	2.7

$V_{GS} = -2$ V	V_{DS} (V)	0	1	1.5	5	10	15	25
	I_D (mA)	0	0.5	0.8	0.9	0.95	0.99	1

9-2 Derive the device transfer characteristic from the drain characteristics plotted for Problem 9-1.

9-3 Plot the I_D/V_{DS} characteristic for a *p*-channel JFET from the following table of current and voltage. Determine I_{DSS} and V_P from the characteristics.

$V_{GS} = 0$	$-V_{DS}$ (V)	0	1	2.2	3	5	7	9.5	15	25
	I_D (mA)	0	2	4	6	8	8.6	9	9.2	9.3

$V_{GS} = +1$ V	$-V_{DS}$ (V)	0	1	2.5	3	5	6	10	25
	I_D (mA)	0	2	4	5	5.8	6	6.2	6.3

$V_{GS} = +2$ V	$-V_{DS}$ (V)	0	1.2	2.4	3	5	6	10	25
	I_D (mA)	0	2	3	3.5	3.9	4	4.2	4.3

9-4 Derive the device transfer characteristic from the drain characteristics plotted for Problem 9-3.

Section 9-3

9-5 Determine the value of Y_{fs} at $V_{GS} = -2$ V from the transfer characteristics constructed for Problem 9-2.

9-6 Determine the value of Y_{fs} at $V_{GS} = 1$ V from the transfer characteristics constructed for Problem 9-4.

9-7 From the information in Appendix A-12, determine the following quantities for a 2N4857 JFET: maximum drain-source voltage, maximum power dissipation, maximum pinch-off voltage, maximum drain-source saturation current, and drain-source ON resistance.

9-8 Using the information in Appendix A-11, construct the maximum transfer characteristics for a 2N5457 JFET.

9-9 Determine the value of Y_{fs} at $V_{GS} = -4$ V and at $V_{GS} = -2$ V from the transfer characteristic constructed for Problem 9-8.

9-10 Using the information in Appendix A-12, construct the maximum transfer characteristics for a 2N4858 JFET.

9-11 Determine the value of Y_{fs} at $V_{GS} = -2$ V and at $V_{GS} = -1$ V from the characteristic constructed for Problem 9-10.

9-12 Using the maximum values of I_{DSS} and $V_{GS(off)}$ for a 2N5457 JFET, calculate Y_{fs} at $V_{GS} = -4$ V and at $V_{GS} = -2$ V.

9-13 Using the maximum values of I_{DSS} and $V_{GS(off)}$ for a 2N4858 JFET, calculate Y_{fs} at $V_{GS} = -2$ V and at $V_{GS} = -1$ V.

9-14 Calculate Y_{os} and r_d from the following measurements made on a JFET: $I_D = 9$ mA when $V_{GS} = 0$ and $V_{DS} = 5$ V, and $I_D = 9.25$ mA when $V_{GS} = 0$ and $V_{DS} = 15$ V.

9-15 Referring to the specification in Appendix A-11, determine the following quantities for a 2N5457 JFET: maximum drain-source voltage, maximum power dissipation, maximum pinch-off voltage, and maximum drain-source saturation current.

Section 9-4

9-16 The circuit in Fig. 9-43 uses a 2N5459 JFET with the maximum transfer characteristics shown in Fig. 9-19. Calculate V_{DS} when $V_{GS} = -1$ V and when $V_{GS} = -2$ V. Calculate the circuit voltage gain.

9-17 The 2N5459 in the circuit in Fig. 9-43 is replaced with a 2N5457. Calculate the maximum and minimum variations in V_{DS} produced by a ± 0.1 V change at the gate. Calculate the circuit voltage gain.

9-18 Determine a suitable JFET forward transfer admittance for the amplifier circuit in Fig. 9-43 to have a minimum voltage gain of 15 when R_1 is changed to 2.2 kΩ.

9-19 A JFET amplifier circuit, as in Fig. 9-43, is to have a minimum voltage gain of 5. If the FET used is a 2N5458, determine a suitable resistance for R_1.

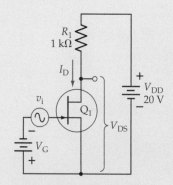

Figure 9-43 Circuit for Problems 9-16 to 9-19.

Section 9-5

9-20 Plot the transfer characteristics for a DMOSFET from the following table of measured current and voltage levels. From the characteristics, determine the value of Y_{fs} at $V_{GS} = 0$.

V_{GS} (V)	−3	−2	−1	0	+1	+2	+3
I_D (mA)	0.5	2.7	4.8	7	9	11.3	13.4

9-21 A VMOSFET has $r_{d(on)} = 0.5\ \Omega$, $g_{FS} = 2.5$ S, and $V_{GS(th)} = 2$ V. Calculate the approximate V_{GS} required to produce a 5 A drain current. Calculate $V_{DS(on)}$ and the device power dissipation when $I_D = 5$ A.

9-22 From the 2N7002 data sheet information in Fig. 9-40, determine $V_{DS(on)}$ and the power dissipation when $I_D = 50$ mA.

9-23 Calculate the approximate gate-source voltage, drain-source ON voltage, and power dissipation in an IRF520 with a 5 A drain current. Refer to the partial specification in Fig. 9-40.

Practice Problem Answers

9-2.1	5.5 mA, 3 V
9-3.1	1127 μS
9-3.2	3160 μS, 1975 μS
9-3.3	−6 V, 100 mA, 40 Ω
9-4.1	7.95 V, 21.15
9-5.1	5 mS
9-5.2	7.5 V

CHAPTER 10
FET Biasing

CONTENTS

Objectives

You will be able to:

1   Draw dc load lines for FET circuits, identify circuit Q-points, and estimate the maximum voltage variations in the circuit output.
2   Sketch gate bias, self-bias, and voltage-divider bias circuits, and explain the operation of each circuit.
3   Analyze various types of FET bias circuits by using the device maximum and minimum transfer characteristics to determine voltage and current levels.
4   Troubleshoot non-operational FET bias circuits.
5   Design gate-bias, self-bias, and voltage-divider bias circuits, and select appropriate standard value components.
6   Sketch constant-current and other unusual types of bias circuit, and explain the operation of each circuit.
7   Analyze and design each type of bias circuit referred to in Item 6.
8   Construct a universal transfer characteristic, and use it in the analysis and design of FET bias circuits.
9   Sketch various bias circuits for MOSFET devices and explain the operation of each circuit.
10 Analyze and design MOSFET bias circuits.
11 Analyze and design bias circuits used with FET switches.

INTRODUCTION

As in bipolar transistor circuits, field effect transistors used in amplifier (or other) circuits must be biased into an *on* state with constant current and terminal voltage levels. The widely differing maximum and minimum transfer characteristics for a FET can result in a wide range of drain current levels. This is similar to the situation caused by $h_{FE(max)}$ and $h_{FE(min)}$ in a BJT bias circuit. For reasonable upper and lower drain current limits in a FET circuit, source resistors and voltage-divider networks must be employed. A graphical approach is most convenient for the analysis or design of FET bias circuits.

10-1 DC LOAD LINE AND BIAS POINT

DC Load Line

The *dc load line* for a FET circuit is drawn on the device output characteristics (or drain characteristics) in exactly the same way as for a BJT circuit (see Section 5-1). Refer to the *n*-channel FET circuit in Fig. 10-1. The source terminal is grounded, the gate is biased via resistor R_G to a negative voltage $(-V_G)$, and the drain terminal is supplied from $+V_{DD}$ via resistor R_D. The dc load line for this circuit is a graph of drain current (I_D) versus drain-source voltage (V_{DS}) for a given value of drain resistance and a given supply voltage.

From Fig. 10-1, the drain-source voltage is

$$V_{DS} = \text{(supply voltage)} - \text{(voltage drop across } R_D)$$

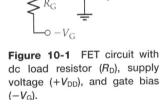

Figure 10-1 FET circuit with dc load resistor (R_D), supply voltage $(+V_{DD})$, and gate bias $(-V_G)$.

or
$$V_{DS} = V_{DD} - I_D R_D \qquad (10\text{-}1)$$

Substituting V_{DD}, R_D, and convenient levels of I_D into Eq. 10-1 produces corresponding V_{DS} and I_D values. These values are then plotted on the device characteristics, and the dc load line is drawn through them.

Example 10-1

Draw the dc load line for the circuit in Fig. 10-1 using the FET characteristics in Fig. 10-2.

Solution

When $I_D = 0$:

Eq. 10-1: $\qquad\qquad V_{DS} = V_{DD} - I_D R_D = 22\text{ V} - 0$
$$= 22\text{ V}$$

Plot point A on Fig. 10-2 at

$$I_D = 0 \quad \text{and} \quad V_{DS} = 22\text{ V}$$

When $V_{DS} = 0$:

from Eq. 10-1,

$$0 = V_{DD} - I_D R_D$$

giving

$$I_D = \frac{V_{DD}}{R_D} = \frac{22\text{ V}}{2\text{ k}\Omega}$$

$$= 11\text{ mA}$$

Plot point B on Fig. 10-2 at

$$I_D = 11\text{ mA} \quad \text{and} \quad V_{DS} = 0\text{ V}$$

Draw the dc load line through points A and B.

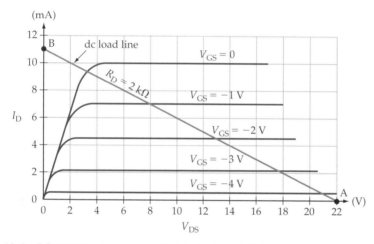

Figure 10-2 DC load line drawn upon FET drain characteristics.

As already explained, the dc load line represents all corresponding I_D and V_{DS} levels that can exist in a FET circuit, as represented by Equation 10-1. A point plotted at $V_{DS} = 16$ V and $I_D = 10$ mA on Fig. 10-2 does not appear on the load line, and so this combination of voltage and current cannot exist in this particular circuit (Fig. 10-1). The straight line drawn in Example 10-1 is the dc load line for $R_D = 2$ kΩ and $V_{DD} = 22$ V. If either of these two quantities is changed, a new load line must be drawn.

Bias Point (*Q*-Point)

Just as in the case of a BJT circuit, the dc *bias point*, or *quiescent point* (*Q-point*), identifies the device current and terminal voltages when there is no ac input signal. When a signal is applied to the gate, I_D varies according to the instantaneous amplitude of the signal, producing a variation in V_{DS}.

For a FET amplifier circuit, V_{DS} must remain in the pinch-off region of the characteristics. This means that it must not be allowed to fall below the gate-source cutoff voltage ($V_{GS(off)}$ in Fig. 10-3), to avoid going into the channel

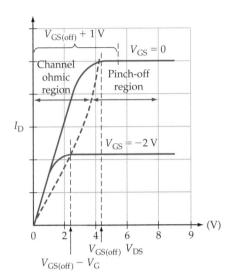

Figure 10-3 In a FET amplifier circuit, the drain-source voltage must be a minimum of $V_{GS(off)}$ + 1 V.

ohmic region. Therefore, in a FET bias circuit the drain-source voltage should always be a minimum of ($V_{GS(off)}$ + 1 V), as illustrated.

$$V_{DS(min)} = V_{GS(off)} + 1 \text{ V} \qquad\qquad (10\text{-}2)$$

Where an external bias voltage (V_G) is included (as in Fig. 10-1), the pinch-off voltage for that bias level on the device characteristics is $V_{GS(off)} - V_G$ (see Fig. 10-3). Consequently, the minimum drain-source voltage may be reduced to

$$V_{DS(min)} = V_{GS(off)} - V_G + 1 \text{ V} \qquad\qquad (10\text{-}3)$$

The bias point for a BJT can be selected halfway along the load line, to give the maximum possible symmetrical output voltage variation (see Section 5-1). To achieve the same result with a FET, the bias point must be halfway between $V_{DS} = V_{DD}$ and $V_{DS} = V_{GS(off)}$. This is illustrated at Q_1 in Fig. 10-4. When maximum V_{DS} variations are not required, the FET bias point can be at any convenient point on the dc load line (Q_2 in Fig. 10-4), just as in the case of a BJT amplifier.

Example 10-2

Estimate the maximum symmetrical output voltage swing for the 2 kΩ load line in Fig. 10-4 when the FET is biased to give a Q-point at V_{DS} = 11 V, and when it is biased to give V_{DS} = 20 V.

Solution

For a Q-point at V_{DS} = 11 V:

When I_D is reduced to zero, $\Delta V_{DS} = 22 \text{ V} - 11 \text{ V}$

$$= 11 \text{ V}$$

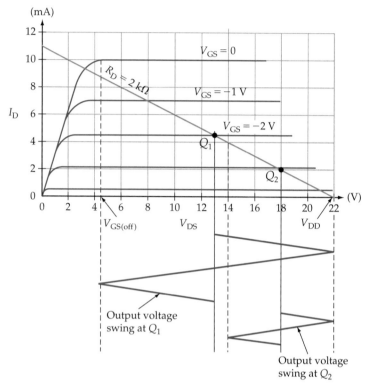

Figure 10-4 The maximum output voltage swing from a FET amplifier circuit depends upon the Q-point.

When I_D is increased to 9 mA, $\quad \Delta V_{DS} = 11\ V - 4\ V$
$$= 7\ V$$

So the maximum symmetrical output voltage swing is
$$V_{o(max)} = \pm 7\ V$$

For a Q-point at $V_{DS} = 20\ V$:

When I_D is reduced to zero, $\quad \Delta V_{DS} = 22\ V - 20\ V$
$$= 2\ V$$

When I_D is increased to 9 mA, $\quad \Delta V_{DS} = 20\ V - 4\ V$
$$= 16\ V$$

So the maximum symmetrical output voltage swing is
$$V_{o(max)} = \pm 2\ V$$

Effect of Source Resistor

Figure 10-5a shows a circuit that has a resistor (R_S) in series with the FET source terminal, and the supply voltage connected directly to the drain. In this case R_S is the dc load, and Eq. 10-1 is rewritten as
$$V_{DS} = V_{DD} - (I_D R_S)$$
The dc load line is drawn exactly as already described.

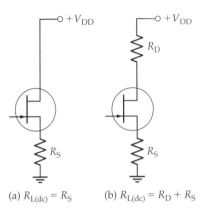

Figure 10-5 When a source resistor (R_S) is included in a circuit, R_S is the dc load or part of the total load.

(a) $R_{L(dc)} = R_S$ (b) $R_{L(dc)} = R_D + R_S$

In Fig. 10-5b, drain and source resistors R_D and R_S are both present, and the total dc load in series with the FET is ($R_D + R_S$). For drawing the dc load line, Eq. 10-1 is modified to

$$V_{DS} = V_{DD} - I_D (R_D + R_S)$$

Practice Problems

10-1.1 A 1 kΩ source resistor (R_s) is added to the circuit of Fig. 10-1, as in Fig. 10-5b. Draw the dc load line on the characteristic in Fig. 10-2, and determine V_{DS} when $V_{GS} = -2$ V.

10-1.2 Determine the maximum symmetrical output voltage swing for the circuit in Problem 10-1.1 when the Q-point is at $V_{DS} = 17$ V.

10-2 GATE BIAS

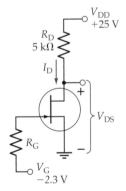

Figure 10-6 FET circuit with gate bias. The gate-source voltage $V_{GS} = -V_G$.

Circuit Operation

Consider the *gate bias* circuit shown in Fig. 10-6. The FET gate terminal is connected via resistor R_G to a bias voltage V_G. If the gate is connected directly to the bias source (instead of using R_G), any ac signal applied to the gate would be short-circuited to V_G.

Since there is no gate current to produce a voltage drop across R_G, the gate-source voltage remains constant at V_G. Depending on the device transfer characteristics, V_G sets the level of I_D in the FET. The I_D level can be readily determined if the transfer characteristic is available for the particular FET used in the circuit. If the device is replaced, I_D is determined by the transfer characteristic of the new device. Like the base current in a BJT base bias circuit, the FET gate voltage does not change when the device characteristics change.

Circuit Analysis

As discussed in Section 9-3, each FET of a given type number has maximum and minimum transfer characteristics, and these must be used to determine the maximum and minimum possible levels of I_D in the circuit. A gate bias circuit is most easily analyzed by drawing a *bias line* vertically on the transfer characteristics from $V_{GS} = -V_G$. The levels of $I_{D(min)}$ and $I_{D(max)}$ are read from the intersections of the bias line with the transfer characteristics, and the maximum and minimum drain-source voltages can then be calculated from these I_D levels.

Example 10-3

The gate bias circuit in Fig. 10-6 uses a FET with the transfer characteristics shown in Fig. 10-7. Determine the maximum and minimum levels of I_D and V_{DS} for the circuit.

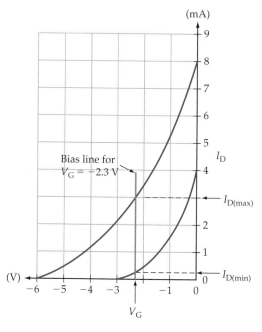

Figure 10-7 The possible drain current levels in a gate bias circuit are determined simply by drawing a bias line vertically on the transfer characteristics from the point where $V_{GS} = V_G$.

Solution

Draw the bias line on the transfer characteristics at

$$V_{GS} = V_G = -2.3 \text{ V}$$

From the bias line, $I_{D(min)} = 0.25$ mA and $I_{D(max)} = 3$ mA

$$V_{DS(min)} = V_{DD} - (I_{D(max)} R_D) = 25 \text{ V} - (3 \text{ mA} \times 5 \text{ k}\Omega)$$
$$= 10 \text{ V}$$

$$V_{DS(max)} = V_{DD} - (I_{D(min)} R_D) = 25 \text{ V} - (0.25 \text{ mA} \times 5 \text{ k}\Omega)$$
$$= 23.75 \text{ V}$$

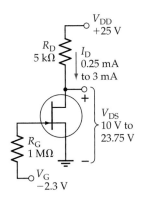

Figure 10-8 Maximum and minimum voltage and current levels for the gate bias circuit in Example 10-3.

The maximum and minimum I_D and V_{DS} levels determined in Ex. 10-3 are illustrated by the circuit current and voltage levels in Fig. 10-8, and by the Q-points plotted in Fig. 10-9. As already explained, JFETs with a given type number have a wide range of transfer characteristics; therefore FET maximum and minimum transfer characteristics should always be used for practical circuit analysis.

The gate bias technique produces a wide range of I_D and V_{DS} levels for any FET with a particular type number. It is possible to adjust V_G to set V_{DS} to a desired voltage. However, this is normally done only in an experimental situation. Slightly more complicated bias circuits must be employed for more predictable voltage and current conditions.

A gate bias circuit can be analyzed without drawing the FET transfer characteristics. Since Equation 9-1 is used for plotting the transfer characteristics, it can be applied directly to determine the drain current levels.

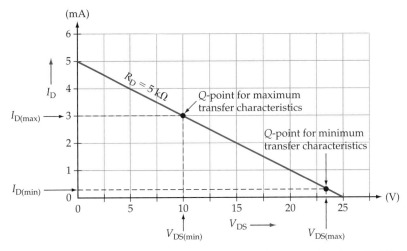

Figure 10-9 Q-points produced by the maximum and minimum transfer characteristics with the gate bias circuit in Example 10-3.

Example 10-4

Use Eq. 9-1 to determine $I_{D(max)}$ and $I_{D(min)}$ for the bias circuit in Example 10-3.

Solution

From Fig. 10-7, $I_{DSS(max)} = 8$ mA and $V_{P(max)} = 6$ V

Eq. 9-1: $\quad I_{D(max)} = I_{DSS(max)}\left[1 - \dfrac{V_{GS}}{V_{GS(off)(max)}}\right]^2 = 8\text{ mA}\left[1 - \dfrac{2.3\text{ V}}{6\text{ V}}\right]^2$

$= 3.04$ mA

From Fig. 10-7, $I_{DSS(min)} = 4$ mA and $V_{P(min)} = 3$ V

Eq. 9-1: $\quad I_{D(min)} = I_{DSS(min)}\left[1 - \dfrac{V_{GS}}{V_{GS(off)(min)}}\right]^2 = 4\text{ mA}\left[1 - \dfrac{2.3\text{ V}}{3\text{ V}}\right]^2$

$= 0.22$ mA

Gate Bias for a *p*-Channel JFET

A gate bias circuit using a *p*-channel JFET is shown in Fig. 10-10a. This is similar to *n*-channel JFET gate bias circuits except that V_{DD} is a negative voltage and V_G is positive. Figure 10-10b shows the same *p*-channel JFET circuit with a positive supply voltage. In this case, the FET source terminal is connected to V_{DD} and R_D is connected to ground. To reverse-bias the gate-channel junctions, the gate terminal must be positive with respect to the source. So gate bias voltage V_G must be positive with respect to V_{DD}, as illustrated. The *p*-channel gate bias circuit can be analyzed in exactly the same way as an *n*-channel circuit.

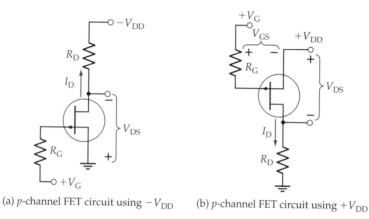

(a) *p*-channel FET circuit using $-V_{DD}$ (b) *p*-channel FET circuit using $+V_{DD}$

Figure 10-10   JFET bias circuits using *p*-channel devices.

Practice Problems

10-2.1 A gate bias circuit has $V_{DD} = 20$ V, $R_D = 2.7$ kΩ, $R_G = 1$ MΩ, and $V_G = -1.5$ V. If the FET has the characteristics shown in Fig. 10-7, determine the maximum and minimum V_{DS} levels for the circuit.

10-2.2 The FET in the circuit in Problem 10-2.1 is changed to one that has $I_{DSS} = 6$ mA and $V_{GS(off)} = 4.5$ V. Calculate the circuit V_{DS}.

10-3 SELF-BIAS

Circuit Operation

In a *self-bias* JFET circuit, gate-source bias is provided by the voltage drop across a resistor in series with the device source terminal. Consider the *n*-channel JFET self-bias circuit illustrated in Fig. 10-11. The voltage drop across resistor R_S is

$$V_S = I_D R_S$$

If $I_D = 1$ mA and $R_S = 1$ kΩ, then $V_S = 1$ V. This means that the source terminal in Fig. 10-11 is 1 V positive with respect to ground. Alternatively, ground

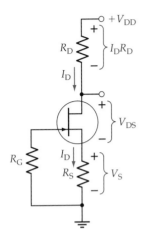

Figure 10-11 JFET self-bias circuit using an *n*-channel device. The gate-source bias voltage is $V_{GS} = -I_D R_S$.

can be said to be 1 V negative with respect to the source terminal. Because the FET gate terminal is grounded via R_G (and there is no voltage drop across R_G), the gate is 1 V negative with respect to the source. That is, the gate-source bias voltage is -1 V. So, for a self-bias circuit,

$$V_{GS} = -I_D R_S \qquad (10\text{-}4)$$

The fact that I_D determines V_{GS} and that V_{GS} sets the I_D level means that there is a feedback effect tending to control I_D. Thus, if I_D increases when the device is changed, the increased voltage across R_S results in an increased (negative) gate-source voltage that tends to lower I_D back toward its original level. Similarly, a fall in I_D produces a reduced V_{GS} which tends to raise I_D toward its original level.

Circuit Analysis

As for all FET bias circuits, a graphical technique is the most convenient method for analysis of a self-bias circuit. In this case, the bias line drawn on the transfer characteristics is quite different from that for the gate bias circuit. The value of R_S and convenient I_D levels are inserted into Eq. 10-4 to determine corresponding V_{GS} levels. The I_D and V_{GS} values are plotted on the transfer characteristics, and the bias line is drawn through them (see Fig. 10-12). The intersection points of the bias line and the transfer characteristics gives the maximum and minimum I_D levels.

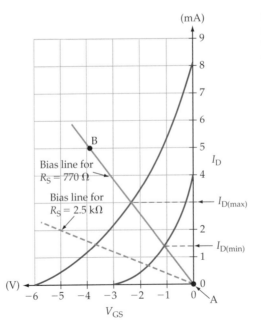

Figure 10-12 The bias line for a self-bias circuit is drawn from the point where $I_D = 0$ and $V_{GS} = 0$ at a slope determined by R_S.

Summing the voltage drops across R_D, R_S, and the FET drain-source terminals in Fig. 10-11 gives

$$V_{DD} = I_D R_D + V_{DS} + I_D R_S \qquad (10\text{-}5)$$

Equation 10-5 may be used to calculate the maximum and minimum levels of V_{DS} when $I_{D(max)}$ and $I_{D(min)}$ have been determined.

Example 10-5

Analyze the self-bias circuit in Fig. 10-13 to determine $I_{D(max)}$, $I_{D(min)}$, $V_{DS(max)}$, and $V_{DS(min)}$. The transfer characteristics for the FET are given in Fig. 10-12.

Solution

When $I_D = 0$:

Eq. 10-4: $\qquad\qquad V_{GS} = -I_D R_S = 0$

Plot point A on Fig. 10-12 at

$$I_D = 0 \quad \text{and} \quad V_{GS} = 0$$

When $I_D = 5$ mA:

Eq. 10-4: $\qquad\qquad V_{GS} = -I_D R_S = -(5 \text{ mA} \times 770 \text{ }\Omega)$

$$= -3.85 \text{ V}$$

Plot point B on Fig. 10-12 where

$$I_D = 5 \text{ mA} \quad \text{and} \quad V_{GS} = -3.85 \text{ V}$$

Draw the bias line for $R_S = 770 \text{ }\Omega$ through points A and B.
It can be seen that where the bias line intersects the transfer characteristics,

$$I_{D(max)} = 3 \text{ mA} \quad \text{and} \quad I_{D(min)} = 1.4 \text{ mA}$$

From Eq. 10-5, $\qquad V_{DS(min)} = V_{DD} - I_{D(max)}(R_D + R_S)$

$$= 25 \text{ V} - 3 \text{ mA}(4.23 \text{ k}\Omega + 770 \text{ }\Omega)$$

$$= 10 \text{ V}$$

and $\qquad\qquad V_{DS(max)} = V_{DD} - I_{D(min)}(R_D + R_S)$

$$= 25 \text{ V} - 1.4 \text{ mA}(4.23 \text{ k}\Omega + 770 \text{ }\Omega)$$

$$= 18 \text{ V}$$

The maximum and minimum I_D and V_{DS} levels determined in Ex. 10-5 are shown in the circuit in Fig. 10-14 and are illustrated by the Q-points plotted in Fig. 10-15. Consideration of Example 10-6 shows that closer levels of $I_{D(max)}$ and $I_{D(min)}$ could be obtained by using a larger resistance for R_S. However, as shown by the (dashed) bias line for $R_S = 2.5 \text{ k}\Omega$ in Fig. 10-12, I_D is substantially reduced when larger values of R_S are used.

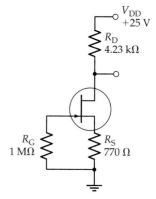

Figure 10-13 FET self-bias circuit for Example 10-5.

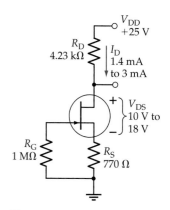

Figure 10-14 Maximum and minimum voltage and current levels for the self-bias circuit in Example 10-5.

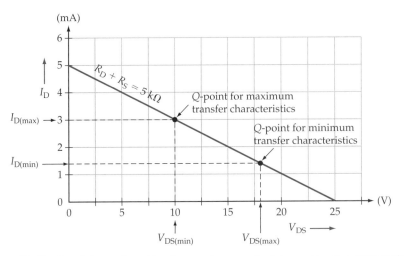

Figure 10-15 *Q*-points produced by the maximum and minimum transfer characteristics with the self-bias circuit in Example 10-5.

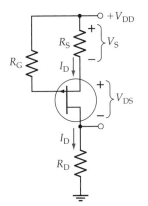

Figure 10-16 JFET self-bias circuit using a *p*-channel device.

Self-Bias for a *p*-Channel JFET

A self-biased circuit using a *p*-channel JFET is shown in Fig. 10-16. The gate-source bias voltage is once again provided by the voltage drop across source resistor R_S. In this case, the source is negative with respect to the supply voltage ($+V_{DD}$). Because the gate is connected to $+V_{DD}$ via R_G, the gate is positive with respect to the source terminal. The positive level of V_{GS} provides the reverse bias at the gate-channel junctions of the *p*-channel FET. A *p*-channel self-biased circuit can be analyzed in exactly the same way as an *n*-channel circuit.

Practice Problem

10-3.1 A self-bias circuit has $V_{DD} = 22$ V, $R_D = 3.3$ kΩ, $R_G = 1$ MΩ, and $R_S = 3.3$ kΩ. Using the FET characteristics in Fig. 10-12, analyze the circuit to determine the maximum and minimum V_{DS}.

10-4 VOLTAGE-DIVIDER BIAS

Circuit Operation

For the self-bias circuit, it was seen that increasing the resistance of R_S brings $I_{D(max)}$ and $I_{D(min)}$ closer together, but that increased R_S values result in lower I_D levels. As will be demonstrated, *voltage-divider bias* allows R_S to be increased without making I_D very small.

Figure 10-17 shows that a voltage-divider bias circuit combines the use of a source resistor (R_S) with a gate bias voltage (V_G). The gate bias voltage is derived from the supply by means of the voltage divider (R_1 and R_2), and the source voltage (V_S) depends upon R_S and I_D. The gate terminal and the source terminal are both positive with respect to ground, and the gate-source voltage

(V_{GS}) is the difference between V_G and V_S. Because V_S is larger than V_G, the FET source terminal is at a higher level than the gate. So the gate is negative with respect to the source, and the gate-channel junctions are reverse-biased.

As in the case of a self-biased circuit, the voltage drop across R_S increases if I_D increases. This produces an increase in $-V_{GS}$, which drives I_D back toward its original level. A decrease in I_D causes a reduction in $-V_{GS}$, which tends to increase I_D.

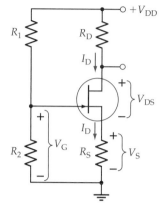

Figure 10-17 JFET voltage-divider bias circuit. The gate-source voltage is $V_{GS} = V_G - V_S$.

Circuit Analysis

Once again, the graphical technique of drawing a bias line on the FET transfer characteristics is the most convenient way of analyzing a voltage-divider bias circuit. An equation relating V_{GS} and I_D must be derived, so that corresponding values of each quantity may be calculated for plotting the bias line.

As discussed,

$$V_{GS} = V_G - I_D R_S \qquad (10\text{-}6)$$

Because there is no gate current,

$$V_G = \frac{V_{DD} \times R_2}{R_1 + R_2} \qquad (10\text{-}7)$$

Referring to Eq. 10-6, it is seen that

when $I_D = 0$, $\qquad\qquad V_{GS} = V_G$

Since this is a positive quantity, the bias line is drawn from $+V_G$, as shown in Fig. 10-18. It is seen that this gives higher drain current levels than those

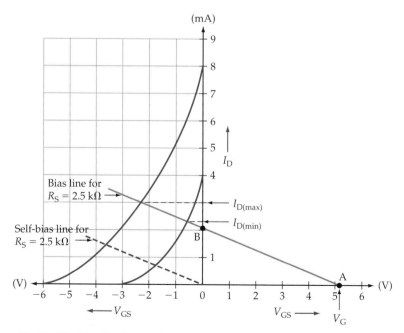

Figure 10-18 The bias line for a voltage-divider bias circuit is drawn from the point where $I_D = 0$ and $V_{GS} = +V_G$ at a slope determined by R_S.

obtained when the bias line is drawn from $I_D = 0$ and $V_{GS} = 0$ (dashed line in Fig. 10-18). The analysis of a voltage-divider bias circuit is demonstrated in Example 10-6.

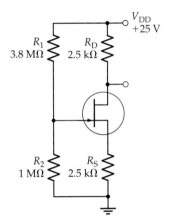

Figure 10-19 FET voltage-divider bias circuit for Example 10-6.

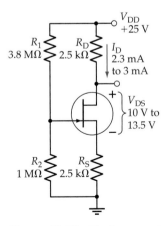

Figure 10-20 Maximum and minimum voltage and current levels for the voltage-divider bias circuit in Example 10-6.

Example 10-6

Analyze the bias circuit in Fig. 10-19 to determine $I_{D(max)}$, $I_{D(min)}$, $V_{DS(max)}$, and $V_{DS(min)}$. Figure 10-18 shows the transfer characteristics for the FET.

Solution

Eq. 10-7:
$$V_G = \frac{V_{DD} \times R_2}{R_1 + R_2} = \frac{25\text{ V} \times 1\text{ M}\Omega}{3.8\text{ M}\Omega + 1\text{ M}\Omega}$$
$$= 5.2\text{ V}$$

When $I_D = 0$:

Eq. 10-6:
$$V_{GS} = V_G - I_D R_S = 5.2\text{ V} - 0$$
$$= 5.2\text{ V}$$

Plot point A on Fig. 10-18 at
$$I_D = 0 \quad \text{and} \quad V_{GS} = +5.2\text{ V}$$

When $V_{GS} = 0$:

From Eq. 10-6,
$$I_D = \frac{V_G}{R_S} = \frac{5.2\text{ V}}{2.5\text{ k}\Omega}$$
$$= 2.08\text{ mA}$$

Plot point B on Fig. 10-18 at
$$I_D = 2.08\text{ mA} \quad \text{and} \quad V_{GS} = 0$$

Draw the bias line for $R_S = 2.5\text{ k}\Omega$ through points A and B. Where the bias line intersects the transfer characteristics, read
$$I_{D(max)} = 3\text{ mA} \quad \text{and} \quad I_{D(min)} = 2.3\text{ mA}$$

From Eq. 10-5,
$$V_{DS(min)} = V_{DD} - I_{D(max)}(R_D + R_S)$$
$$= 25\text{ V} - 3\text{ mA}(2.5\text{ k}\Omega + 2.5\text{ k}\Omega)$$
$$= 10\text{ V}$$

and
$$V_{DS(max)} = V_{DD} - I_{D(min)}(R_D + R_S)$$
$$= 25\text{ V} - 2.3\text{ mA}(2.5\text{ k}\Omega + 2.5\text{ k}\Omega)$$
$$= 13.5\text{ V}$$

The maximum and minimum I_D and V_{DS} levels determined for the voltage-divider bias circuit in Example 10-6 are shown on the circuit diagram in Fig. 10-20 and are illustrated by the Q-points plotted in Fig. 10-21.

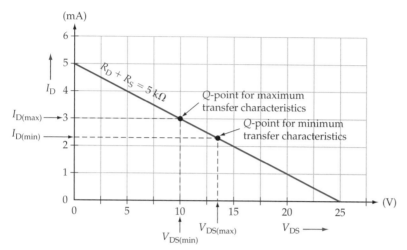

Figure 10-21   Q-points produced by the maximum and minimum transfer characteristics with the voltage-divider bias circuit in Example 10-6.

Voltage-Divider Bias for a *p*-Channel JFET

A voltage-divider bias circuit using a *p*-channel JFET is shown in Fig. 10-22. As for the *n*-channel circuit, the gate-source bias voltage is the difference between V_G and V_S. In this case, these are both negative with respect to the positive supply voltage (V_{DD}). Because V_S is larger than V_G, the gate terminal is positive with respect to the source, and this reverse-biases the gate-channel junctions. A *p*-channel voltage-divider bias circuit is analyzed in exactly the same way as an *n*-channel circuit.

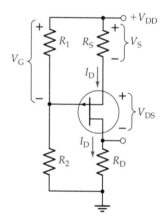

Figure 10-22   JFET voltage-divider bias circuit using a *p*-channel device.

Practice Problem

10-4.1   A voltage-divider bias circuit has $V_{DD} = 30$ V, $R_D = R_S = 4.7$ kΩ, $R_1 = 1$ MΩ, and $R_2 = 150$ kΩ. Using the FET characteristics in Fig. 10-18, analyze the circuit to determine the maximum and minimum V_{DS} levels.

10-5 COMPARISON OF BASIC JFET BIAS CIRCUITS

The three basic FET bias circuits (gate bias, self-bias, and voltage-divider bias) are similar in performance to the three basic BJT bias circuits (base bias, collector-to-base bias, and voltage-divider bias). When comparing the performance of FET bias circuits, it should be recalled that each FET type number has maximum and minimum transfer characteristics. The particular transfer characteristic for each individual FET is normally not known. Therefore, as already stated, *the specified maximum and minimum transfer characteristics must be used when analysing (or designing) a FET bias circuit.*

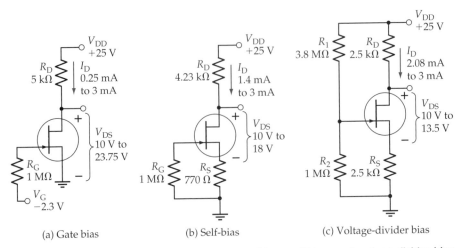

Figure 10-23 Comparison of similar JFET gate bias, self-bias, and voltage-divider bias circuits.

Consider the basic JFET bias circuits reproduced in Fig. 10-23. Each circuit uses a 25 V supply and a 5 kΩ (total) series resistance, and each uses a FET with the same maximum and minimum transfer characteristics. The circuits are analysed in Examples 10-3, 10-5, and 10-6 to determine the maximum and minimum I_D and V_{DS} levels. The values calculated are indicated on each circuit in Fig. 10-23, and the Q-point ranges are illustrated in Fig. 10-24.

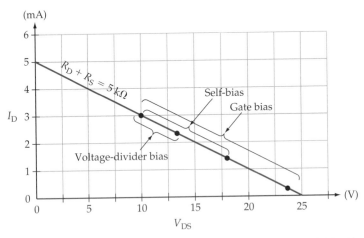

Figure 10-24 Q-point ranges for the gate bias, self-bias, and voltage-divider bias circuits shown in Fig. 10-23.

If the FET in the gate bias circuit fails and has to be replaced with another device (of the same type number), the new V_{DS} could be anywhere from 10 V to 23.75 V. The self-bias circuit in the same circumstance (with the same FET type number) would have a V_{DS} ranging from 10 V to 18 V when the FET is replaced. For the voltage-divider bias circuit in this situation, the measurable

V_{DS} range with a new device would be 10 V to 13.5 V. Clearly, self-bias gives more predictable bias conditions than gate bias, and voltage-divider bias gives the most predictable bias conditions. Because of its better performance, voltage-divider bias is usually preferred.

Section 10-5 Review

10-5.1 Briefly discuss the three basic FET bias circuits, comparing the effect of the bias lines on the maximum and minimum I_D levels.

10-6 TROUBLESHOOTING JFET BIAS CIRCUITS

Voltage Measurement

Troubleshooting procedures for FET bias circuits are similar to those for BJT bias circuits. The major difference is that there is only one junction in the FET (the gate-channel junction) that might become short-circuited or open-circuited. To determine if the circuit is functioning correctly, the FET terminal voltages should all be measured with respect to ground, as illustrated in Fig. 10-25. This is exactly as done for a BJT circuit, and, here again, if the voltage levels are not satisfactory, the circuit must be further investigated to locate the fault.

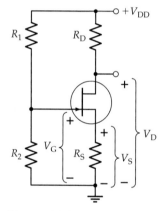

Figure 10-25 When testing a FET bias circuit, all voltage levels should first be measured with respect to the ground or negative supply terminal.

Common Errors

Before proceeding further, the following list of common errors (reproduced from Section 5-6) should be considered. These apply equally to FET and BJT circuits.

- Power supply not switched *on*.
- Power supply current limiter control set incorrectly.
- Cables incorrectly connected to the power supply.
- Volt-ohm-milliammeter (VOM) function incorrectly selected.
- Wrong VOM terminals used.
- Incorrect component connections.
- Incorrect resistor values.
- Resistors in the wrong places.

Gate Bias

Figure 10-26 illustrates typical sources of error in a gate bias circuit. Normally, the gate should be negative with respect to ground and the drain voltage should be anywhere in the range of perhaps 3 V to $(V_{DD} - 2 \text{ V})$. If $V_D \approx V_{DD}$,

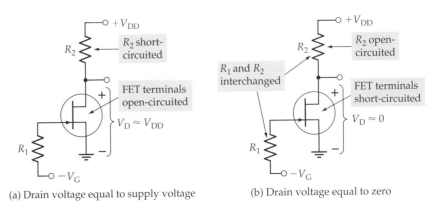

(a) Drain voltage equal to supply voltage   (b) Drain voltage equal to zero

Figure 10-26 The measured FET terminal voltages in a gate bias circuit can indicate possible circuit errors.

as in Fig. 10-26a, the FET terminals might be open-circuited, or R_2 might be shorted. When $V_D \approx 0$, as in Fig. 10-26b, R_1 and R_2 might be interchanged (the high-value and low-value resistors in the wrong places). Other possibilities are that R_2 is open-circuited or the FET terminals are short-circuited.

Self-Bias

Possible reasons for incorrect voltage levels in a self-bias circuit are illustrated in Fig. 10-27. These are similar to those discussed for gate bias. If V_D is approximately equal to V_{DD} (Fig. 10-27a), resistor R_3 or one of the transistor terminals could to be open-circuited, or R_1 and R_3 might have been interchanged. Alternatively, R_2 might be short-circuited. If V_D equals V_S (Fig. 10-27b), the FET transistor terminals might be short-circuited, or R_1 and R_2 might be in the wrong places (interchanged).

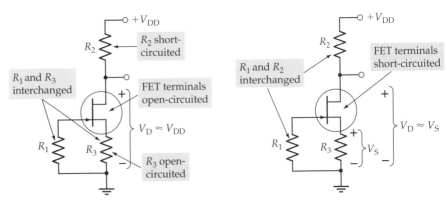

(a) Drain voltage equal to supply voltage   (b) Drain voltage equal to source voltage

Figure 10-27 Typical incorrect terminal voltages, and probably circuit faults, in a FET self-bias circuit.

Voltage-Divider Bias

Unsatisfactory voltage measurements and probable errors in a voltage-divider bias circuit are shown in Fig. 10-28. If V_D equals V_{DD} (Fig. 10-28a), R_1 or R_4 might be open-circuited. Alternatively, R_2 or R_3 might be short-circuited, or the FET terminals may be open-circuited. When V_D approximately equals V_S (Fig. 10-28b), R_2 might be open-circuited, or R_1 and R_2 might be interchanged. Another possibility is that the FET terminals are short-circuited.

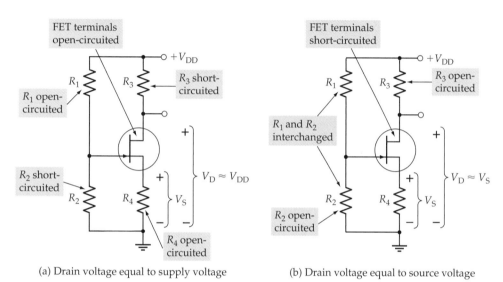

(a) Drain voltage equal to supply voltage   (b) Drain voltage equal to source voltage

Figure 10-28 Unsatisfactory terminal voltages, and probably circuit faults, in a FET voltage-divider bias circuit.

Section 10-6 Review

10-6.1 Briefly discuss the procedure for testing FET bias circuits. List possible faults for each bias circuit type when (a) $V_{DS} \approx 0$ and (b) $V_D \approx V_{DD}$.

10-7 JFET BIAS CIRCUIT DESIGN

Design Approach

The design of FET bias circuits is just as simple as the design of BJT bias circuits. One major difference is that FET circuit design normally uses a graphical approach involving the drawing of a bias line on the device transfer characteristics, as in the case of FET circuit analysis. If the type number of the FET to be used in the circuit is known, the maximum and minimum transfer characteristics are available or can be plotted. A maximum drain current ($I_{D(\max)}$) is selected at a point on the maximum transfer characteristic (see Fig. 10-29), and the bias line is drawn through this point. As will be explained, this procedure

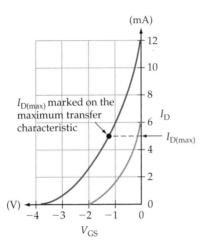

Figure 10-29 The first step in designing a FET bias circuit is the selection of I_D on the maximum transfer characteristic.

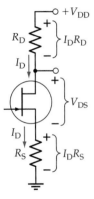

Figure 10-30 When the maximum transfer characteristic is not used in FET circuit design calculations, the I_D level can be greater than the design value and the FET can be forced into the channel ohmic region of its characteristic.

applies to all types of bias circuits. The selected level of $I_{D(max)}$ is used to calculate the values of resistors connected to the drain and source terminals.

Instead of using the maximum transfer characteristics, suppose that the minimum characteristics, or some characteristic between the maximum and minimum, is used in the circuit design. Now, if the FET happens to have the maximum transfer characteristic for the device type number, I_D will be larger than the design value. Consequently, the resistor voltage drops ($I_D R_D$ and $I_D R_S$) will be greater than intended (see Fig. 10-30). In this case, V_{DS} will be smaller than its design level, the FET may be forced into the channel ohmic region of its characteristics, and the circuit is unlikely to function correctly. Therefore, *always use the device maximum transfer characteristic when designing a FET bias circuit.*

As already explained, a FET has a very high input resistance, so high-value bias resistors can be used at the gate terminal. However, there are disadvantages to using extremely high resistance values. A charge accumulated at the gate can take a long time to *leak off* through a very high resistance. In this case, the gate voltage might not be a stable quantity, and consequently, the drain current could be unpredictable. High-value bias resistors also generate unwanted thermal noise (see Section 8-6), and they readily pick-up stray radiofrequency signals. For these reasons, 1 MΩ is normally a reasonable upper limit for bias resistors.

In all situations where resistor values are calculated, the closest standard-value components are selected (see Appendix B). Normally, ±10% tolerance resistors are used, and the more expensive ±5% and ±1% components are employed only where greater precision is required.

Gate Bias Circuit Design

The design of a gate bias circuit is very simple. As explained, the desired maximum I_D level is marked on the maximum transfer characteristics for the FET. The bias line is drawn vertically through this point, and the required gate bias voltage ($-V_G$) is indicated where the bias line intersects the V_{GS} scale. Alternatively, V_{GS} can be calculated by rewriting Equation 9-1:

$$V_{GS} = V_{GS(off)}\left[1 - \sqrt{I_D/I_{DSS}}\right] \qquad (10\text{-}8)$$

Because the circuit design uses $I_{D(max)}$, the voltage drop across R_D is a maximum, and consequently, the drain-source voltage (V_{DS}) is a minimum. So the drain resistor is calculated as

$$R_D = \frac{V_{DD} - V_{DS(min)}}{I_{D(max)}} \qquad (10\text{-}9)$$

As discussed, the bias resistor (R_G) is usually selected as 1 MΩ, although a lower resistance may be used.

Figure 10-31 shows the equations involved in designing a gate bias circuit, and Example 10-7 demonstrates the design procedure.

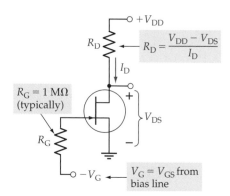

Figure 10-31   FET gate bias circuit design.

Example 10-7

Design a gate bias circuit (as in Fig. 10-31) to have $I_{D(max)} = 3$ mA and $V_{DS(min)} = 10$ V. The FET to be used has the transfer characteristics in Fig. 10-32, and the supply voltage is $V_{DD} = 25$ V.

Solution

On the maximum transfer characteristic, mark

$$I_{D(max)} = 3 \text{ mA (point X on Fig. 10-32)}$$

Draw the bias line vertically through point X.
Read the required bias voltage where the bias line intersects the V_{GS} scale:

$$V_{GS} = -2.3 \text{ V}$$

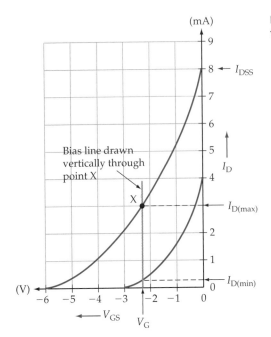

Figure 10-32 FET transfer characteristic and bias line for Example 10-7.

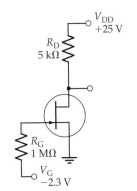

Figure 10-33 FET gate bias circuit designed in Example 10-7.

Alternatively,

Eq. 10-8:
$$V_{GS} = V_{GS(off)}\left[1 - \sqrt{(I_D/I_{DSS(max)})}\right]$$
$$= -6\,V\left[1 - \sqrt{(3\,mA/8\,mA)}\right]$$
$$= -2.3\,V$$

Select
$$R_G = 1\,M\Omega$$

Eq. 10-9:
$$R_D = \frac{V_{DD} - V_{DS(min)}}{I_{D(max)}} = \frac{25\,V - 10\,V}{3\,mA}$$
$$= 5\,k\Omega \text{ (use 4.7 k}\Omega \text{ standard value)}$$

From Fig. 10-32, note that $I_{D(min)} = 0.25\,mA$

The calculated resistor values are shown in Fig. 10-33, and the circuit is analyzed in Example 10-3.

Self-Bias Circuit Design

As for gate bias, the procedure for designing a self-bias circuit (Fig. 10-34) begins with $I_{D(max)}$ being marked on the FET maximum transfer characteristic (Fig. 10-35). The bias line is then drawn through this point and the point where $I_{DS} = 0$ and $V_{GS} = 0$. The value of the source resistor (R_S) is determined from the slope of the bias line:

$$R_S = \frac{\Delta V_{GS}}{\Delta I_D} \tag{10-10}$$

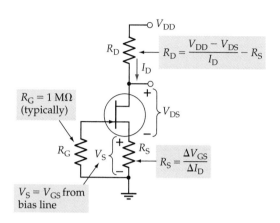

Figure 10-34 FET self-bias circuit design.

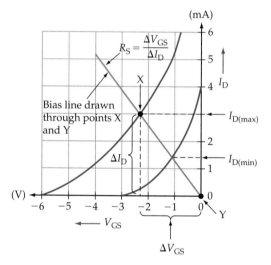

Figure 10-35 FET transfer characteristic and bias line for Example 10-8.

The total resistance in series with the FET drain-source terminals (Fig. 10-34) is ($R_D + R_S$). So Equation 10-9 can be modified to calculate this quantity:

$$R_D + R_S = \frac{V_{DD} - V_{DS(min)}}{I_{D(max)}} \qquad (10\text{-}11)$$

With R_S already calculated, R_D is easily determined. Gate resistor R_G is normally selected as 1 MΩ or lower, as discussed.

Figure 10-34 shows the equations for designing a self-bias circuit, and Example 10-8 demonstrates the design procedure.

Example 10-8

Design a self-bias circuit (as in Fig. 10-34) to have $I_{D(max)} = 3\,\text{mA}$ and $V_{DS(min)} = 10\,\text{V}$. The FET transfer characteristics are shown in Fig. 10-35, and the supply voltage is $V_{DD} = 25\,\text{V}$.

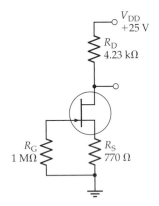

Figure 10-36 FET self-bias circuit designed in Example 10-8.

Solution

On the maximum transfer characteristics, mark

$$I_{D(max)} = 3 \text{ mA (point X on Fig. 10-35)}$$

Mark point Y at $I_D = 0$ and $V_{DS} = 0$

Draw the bias line through points X and Y.

From the bias line, $\Delta V_{GS} = 2.3$ V and $\Delta I_D = 3$ mA

Eq. 10-10: $$R_S = \frac{\Delta V_{GS}}{\Delta I_D} = \frac{2.3 \text{ V}}{3 \text{ mA}}$$

$$\approx 770 \ \Omega \text{ (use 750 } \Omega, \pm 5\% \text{ standard value)}$$

Eq. 10-11: $$R_D + R_S = \frac{V_{DD} - V_{DS(min)}}{I_{D(max)}} = \frac{25 \text{ V} - 10 \text{ V}}{3 \text{ mA}}$$

$$= 5 \text{ k}\Omega$$

$$R_D = (R_D + R_S) - R_S = 5 \text{ k}\Omega - 770 \ \Omega$$

$$= 4.23 \text{ k}\Omega \text{ (use 3.9 k}\Omega \text{ standard value)}$$

Select $$R_G = 1 \text{ M}\Omega$$

From Fig. 10-35, note that $I_{D(min)} = 1.4$ mA

The calculated resistor values are shown in Fig. 10-36, and the circuit is analyzed in Example 10-5.

Voltage-Divider Bias Circuit Design

As always, circuit design starts with the specification of V_{DD}, $I_{D(max)}$, and $V_{DS(min)}$. The maximum drain current is plotted on the device maximum transfer characteristic; then either the resistance of R_S or the level of V_G must be determined before the bias line can be drawn for a voltage-divider bias circuit.

With $V_{DS(min)}$ specified, the rest of the supply voltage ($V_{DD} - V_{DS(min)}$) is split between V_S and V_{RD}. It is shown in Chapter 11 that the voltage gain of a FET amplifier stage is directly proportional to R_D. Maximum gain is achieved by selecting R_D as large as possible, and for a given level of I_D this requires the greatest possible voltage V_{RD}. However, the best bias stability is achieved by selecting R_S as large as possible, and this dictates maximum V_S. A reasonable compromise between these two conflicting requirements (maximum V_{RD} and maximum V_S) is simply to make the two voltage drops (and the two resistors) equal. Equation 10-11 is used to calculate ($R_D + R_S$), and this is divided to give two equal resistors.

Once R_S is determined, the bias line can be drawn through $I_{D(max)}$ with the slope of the line set by R_S. This requires that the V_{GS} scale be extended positively, as shown in Fig. 10-37.

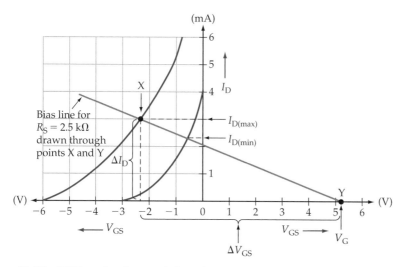

Figure 10-37 FET transfer characteristic and bias line for voltage-divider bias circuit.

From Equation 10-10,

$$\Delta V_{GS} = \Delta I_D \times R_S \qquad \text{(10-12)}$$

The gate bias voltage (V_G) is read from the intersection of the bias line and the (extended) V_{GS} scale (Fig. 10-37). R_2 is normally the smallest resistor in the voltage divider. So R_2 is usually selected as 1 MΩ, for the reasons already discussed for R_G in the gate bias and self-bias circuits. R_1 is then calculated from R_2 and the resistor voltage drops. From Eq. 10-7,

$$R_1 = \frac{(V_{DD} - V_G) \times R_2}{V_G} \qquad \text{(10-13)}$$

Another approach that can be used to determine R_S is to specify a minimum drain current and mark $I_{D(min)}$ on the minimum drain characteristic (after marking $I_{D(max)}$). The bias line is then drawn through the $I_{D(max)}$ and $I_{D(min)}$ points. The slope of the bias line and its intersection with the V_{GS} scale dictate R_S and V_G, respectively.

The design equations for a FET voltage-divider bias circuit are shown on Fig. 10-38, and Example 10-9 demonstrates the design procedure.

Example 10-9

Design a voltage-divider bias circuit (as in Fig. 10-38) to have $I_{D(max)} = 3$ mA and $V_{DS(min)} = 10$ V. Use the FET transfer characteristics in Fig. 10-37 and a supply voltage of $V_{DD} = 25$ V.

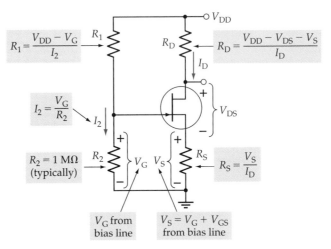

Figure 10-38   FET voltage-divider bias circuit design.

Solution

Eq. 10-11:
$$R_D + R_S = \frac{V_{DD} - V_{DS(min)}}{I_{D(max)}} = \frac{25 \text{ V} - 10 \text{ V}}{3 \text{ mA}}$$

$$= 5 \text{ k}\Omega$$

Select
$$R_D = R_S = \frac{R_D + R_S}{2} = \frac{5 \text{ k}\Omega}{2}$$

$$= 2.5 \text{ k}\Omega \text{ (use 2.2 k}\Omega \text{ standard value)}$$

On the maximum transfer characteristics, mark

$$I_{D(max)} = 3 \text{ mA (point X on Fig. 10-37)}$$

For $\Delta I_D = 3$ mA,

Eq. 10-12:
$$\Delta V_{GS} = \Delta I_D \times R_S = 3 \text{ mA} \times 2.5 \text{ k}\Omega$$

$$= 7.5 \text{ V}$$

Mark point Y at $\Delta I_D = 3$ mA and $\Delta V_{DS} = 7.5$ V from point X.

Draw the bias line through points X and Y.

From the bias line, $V_G = 5.2$ V

Alternatively,

Eq. 10-8:
$$V_{GS} = V_{GS(off)}\left[1 - \sqrt{(I_D/I_{DSS(max)})}\right] = -6 \text{ V}\left[1 - \sqrt{(3 \text{ mA}/8 \text{ mA})}\right]$$

$$= -2.3 \text{ V}$$

$$V_S = I_D \times R_S = 3 \text{ mA} \times 2.5 \text{ k}\Omega$$

$$= 7.5 \text{ V}$$

$$V_G = V_S - V_{GS} = 7.5 \text{ V} - 2.3 \text{ V}$$

$$= 5.2 \text{ V}$$

Select $R_2 = 1 \text{ M}\Omega$

Eq. 10-13: $R_1 = \dfrac{(V_{DD} - V_G) \times R_2}{V_G} = \dfrac{(25 \text{ V} - 5.2 \text{ V}) \times 1 \text{ M}\Omega}{5.2 \text{ V}}$

$$= 3.8 \text{ M}\Omega \text{ (use 3.9 M}\Omega \text{ standard value)}$$

From Fig. 10-37, note that $I_{D(min)} = 2.3 \text{ mA}$

The calculated resistor values are shown in Fig. 10-39, and the circuit is analyzed in Example 10-6.

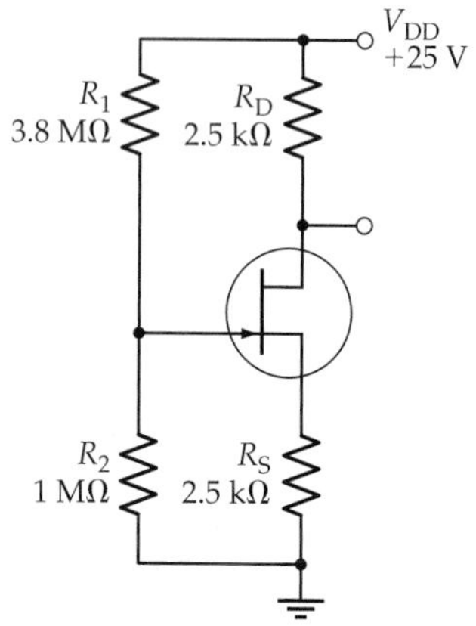

Figure 10-39 FET voltage-divider bias circuit designed in Example 10-9.

Designing with Standard-Value Resistors

Each time a resistor value is calculated, the nearest (higher or lower) standard value should be selected, as shown in the preceding examples. Sometimes it will be appropriate to choose a resistance higher than the calculated value, and sometimes a lower value should be used. For example, if a minimum current level is used to calculate resistance, it is usually best to choose a lower resistance value so that the current is higher than the specified minimum. Regardless of how the standard values are selected, the new current level or voltage drop produced by the standard-value component should be calculated before proceeding with the design. This approach is employed in design examples in the next section.

Practice Problems

10-7.1 A JFET gate bias circuit using $V_{DD} = 30$ V is to have $I_{D(max)} = 4$ mA and $V_{DS(min)} = 11$ V. Using the transfer characteristics in Fig. 10-32, design the circuit.

10-7.2 Design a JFET self-bias circuit to have $I_{D(max)} = 2$ mA and $V_{DS(min)} = 8$ V. The supply voltage is $V_{DD} = 20$ V, and the FET transfer characteristics are shown in Fig. 10-35.

10-7.3 Design a JFET voltage-divider bias circuit to have $I_{D(max)} = 1.5$ mA and $V_{DS(min)} = 10$ V. The supply voltage $V_{DD} = 27$ V, and the FET transfer characteristics are shown in Fig. 10-37.

10-8 MORE JFET BIAS CIRCUITS

Use of Plus/Minus Supplies

When plus/minus supply voltages are to be used with a FET bias circuit, the gate terminal is usually grounded via R_G, as illustrated in Fig. 10-40a. In this case, the circuit is essentially a voltage-divider bias circuit with the gate bias voltage equal to the level of the negative supply voltage ($V_G = V_{SS}$). The (reverse) gate-source bias voltage ($-V_{GS}$) makes the source terminal $+V_{GS}$ above ground. So the voltage across R_S is

$$V_S = V_{SS} + V_{GS} \tag{10-14}$$

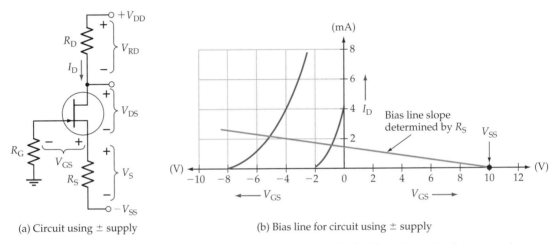

(a) Circuit using ± supply   (b) Bias line for circuit using ± supply

Figure 10-40   Circuit and bias line for voltage-divider bias using a plus/minus supply.

When the voltage levels are understood, the type of circuit in Fig. 10-40a can be designed by a procedure similar to that for a voltage-divider bias circuit. $I_{D(max)}$ is plotted on the maximum transfer characteristics, and V_{GS} is read from the horizontal axis. V_{GS} can also be determined by substituting $I_{D(max)}$, $I_{DSS(max)}$, and $V_{GS(off)(max)}$ into Eq. 10-8. Then, with the desired maximum voltage drops known, R_D and R_S are readily calculated. R_G is selected as 1 MΩ or less for the reasons already discussed.

The analysis procedure for this circuit is also like that for a voltage-divider bias circuit. One point on the bias line is plotted on the transfer characteristics where $V_G = V_{SS}$, as shown on Fig. 10-40b, and the bias line is drawn at the slope determined by R_S. As always, the bias line intersections with the maximum and minimum characteristics give the $I_{D(max)}$ and $I_{D(min)}$ levels.

Drain Feedback

The *drain feedback* bias circuit in Fig. 10-41 is essentially a voltage-divider bias circuit. However, instead of V_G being derived from the supply voltage, the

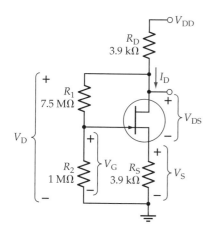

Figure 10-41   FET drain feedback bias circuit.

FET drain voltage (V_D) is divided by R_1 and R_2 to produce V_G. Modifying Eq. 10-7 gives

$$V_G = \frac{V_D \times R_2}{R_1 + R_2} \tag{10-15}$$

The circuit is designed for a particular level of V_D and I_D, and the voltage drop across R_S helps to stabilize I_D, as in other voltage-divider bias circuits. The voltage drop across R_D also provides feedback that helps to correct changes in I_D. When I_D is greater than the design value, V_{RD} is increased, V_D is reduced, and thus V_G is lowered to drive I_D back toward its original level. The two feedback effects (V_{RD} changes and V_{RS} changes) tend to keep $I_{D(min)}$ and $I_{D(max)}$ closer than in an ordinary voltage-divider bias circuit.

The design procedure for this circuit is the same as for voltage-divider bias with the exception that the voltage drop across R_1 is ($V_D - V_G$) instead of ($V_{DD} - V_G$). For circuit analysis a bias line should be drawn on the transfer characteristics, as always. This requires the equation relating V_{GS} and I_D:

$$V_{GS} = \frac{R_2(V_{DD} - I_D R_D)}{R_1 + R_2} - I_D R_S \tag{10-16}$$

Example 10-10

Design the drain feedback circuit in Fig. 10-41 to use a 2N5458 FET with a 20 V supply and to have an $I_{D(max)}$ of 1.5 mA and a $V_{DS(min)}$ of 8 V.

Solution

Eq. 10-11: $R_D + R_S = \dfrac{V_{DD} - V_{DS(min)}}{I_{D(max)}} = \dfrac{20\ V - 8\ V}{1.5\ mA}$

$= 8\ k\Omega$

Select $$R_D = R_S = \frac{R_D + R_S}{2} = \frac{8\ k\Omega}{2}$$

$$= 4\ k\Omega\ (\text{use } 3.9\ k\Omega\ \text{standard value})$$

From data sheet A-11 in Appendix A, $I_{DSS(max)} = 9$ mA and $V_{GS(off)(max)} = 7$ V (for a 2N5458)

Eq. 10-8:

$$V_{GS(max)} = V_{GS(off)(max)}\left[1 - \sqrt{(I_D/I_{DSS(max)})}\right] = -7\ \text{V}\left[1 - \sqrt{(1.5\ \text{mA}/9\ \text{mA})}\right]$$

$$= -4.14\ \text{V}$$

$$V_S = I_D R_S = 1.5\ \text{mA} \times 3.9\ k\Omega$$

$$= 5.85\ \text{V}$$

$$V_G = V_S + V_{GS(max)} = 5.85\ \text{V} - 4.14\ \text{V}$$

$$= 1.7\ \text{V}$$

$$V_D = V_{DD} - I_D R_D = 20\ \text{V} - (1.5\ \text{mA} \times 3.9\ k\Omega)$$

$$= 14.2\ \text{V}$$

Select $$R_2 = 1\ M\Omega$$

From Eq. 10-15, $$R_1 = \frac{(V_D - V_G) \times R_2}{V_G} = \frac{(14.2\ \text{V} - 1.7\ \text{V}) \times 1\ M\Omega}{1.7\ \text{V}}$$

$$= 7.4\ M\Omega\ (\text{use } 7.5\ M\Omega \pm 5\%\ \text{standard value})$$

Constant-Current Bias

As already discussed, the bias lines with the smallest slope give the closest levels of $I_{D(min)}$ and $I_{D(max)}$. In the two circuits in Fig. 10-42 the bipolar transistor (Q_2) keeps the FET source current constant. Consequently, as illustrated in Fig. 10-43, the bias line is drawn horizontally on the transfer characteristics, showing that $I_{D(min)} = I_{D(max)}$ with different V_{GS} levels.

Both circuits provide bias voltage V_B to the base of Q_2, to give

$$I_D = I_C \approx I_E = \frac{V_B - V_{BE}}{R_E} \tag{10-17}$$

The circuit in Fig. 10-42a uses voltage-divider bias to derive V_B from V_{DD}, and in Fig. 10-42b $V_B = V_{EE}$. In both cases, the FET gate is connected to the base of Q_2 via resistor R_G, and $V_{GS} = -V_{CB}$.

Designing each of the circuits in Fig. 10-42 simply involves selecting appropriate voltage and current levels and calculating the resistors. The emitter resistor voltage drop (V_E) should typically be a minimum of 5 V, but satisfactory

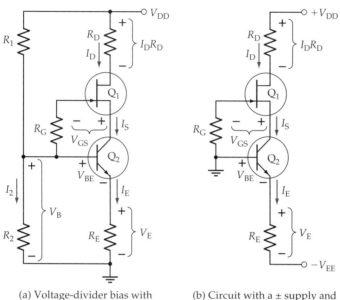

(a) Voltage-divider bias with
constant current source

(b) Circuit with a ± supply and
a constant current source

Figure 10-42 FET circuits with BJT constant-current bias circuits. In each case, I_S is held constant at the I_C level.

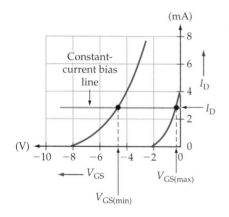

Figure 10-43 Bias line for FET constant-current bias circuits.

results can be obtained with 3 V. It can, of course, be much larger than 5 V, as would normally be the case in Fig. 10-42b. In both circuits,

$$V_B = V_E + V_{BE}$$

The V_B calculated in this way can be used to determine the ratio of R_1 and R_2 in Fig. 10-42a. For the circuit in Fig. 10-42b, V_{EE} dictates the V_B and V_E levels.

The maximum V_{GS} is determined by plotting I_D on the maximum transfer characteristic for the FET and then reading $V_{GS(max)}$ at I_D. Alternatively, V_{GS} can be calculated from Equation 10-8. The source terminal voltage is more positive than the gate. The equation for V_S is

$$V_S = V_B - V_{GS} \tag{10-18}$$

The analysis of each of the circuits in Fig. 10-42 is very simple. Once V_B is determined, I_D is calculated from Eq. 10-17. $V_{GS(max)}$ and $V_{GS(min)}$ are found by drawing the bias line for I_D on the transfer characteristics or by substituting I_D, $V_{GS(off)}$, and I_{DSS} into Eq. 10-8.

Example 10-11

Design the constant-current bias circuit in Fig. 10-42b to use a 2N5459 FET with a ± 20 V supply, and to have $I_{D(max)} = 3$ mA and $V_{DS(min)} = 9$ V.

Solution

$$V_E = V_{EE} - V_{BE} = 20\text{ V} - 0.7\text{ V}$$

$$= 19.3\text{ V}$$

$$R_E = \frac{V_E}{I_D} = \frac{19.3\text{ V}}{3\text{ mA}}$$

$$= 6.4\text{ k}\Omega \text{ (use 6.8 k}\Omega \text{ standard value)}$$

I_D becomes

$$\frac{V_E}{R_E} = \frac{19.3\text{ V}}{6.8\text{ k}\Omega}$$

$$= 2.83\text{ mA}$$

From data sheet A-11 in Appendix A,

$$I_{DSS(max)} = 16\text{ mA and } V_{GS(off)(max)} = 8\text{ V (for a 2N5459)}$$

Eq. 10-8:

$$V_{GS(max)} = V_{GS(off)}[1 - \sqrt{(I_D/I_{DSS})}]$$

$$= -8\text{ V}[1 - \sqrt{(2.83\text{ mA}/16\text{ mA})}]$$

$$= -4.6\text{ V}$$

Eq. 10-18:

$$V_S = V_B - V_{GS} = 0 - (-4.6\text{ V})$$

$$= +4.6\text{ V}$$

$$V_{RD} = V_{DD} - V_{DS(min)} - V_S = 20\text{ V} - 9\text{ V} - 4.6\text{ V}$$

$$= 6.4\text{ V}$$

$$R_D = \frac{V_{RD}}{I_D} = \frac{6.4\text{ V}}{2.83\text{ mA}}$$

$$= 2.26\text{ k}\Omega \text{ (use 2.2 k}\Omega \text{ standard value)}$$

Select

$$R_G = 1\text{ M}\Omega$$

Figure 10-44 shows the circuit voltage and current levels when the FET has $I_{DSS(max)}$ and $V_{GS(off)(max)}$, as used in the design. When the FET does not have the maximum parameters, only V_{GS} and V_{DS} should change.

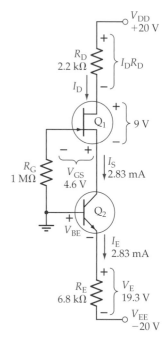

Figure 10-44 Constant-current bias circuit designed in Example 10-11.

Practice Problems

10-8.1 Design the voltage-divider bias circuit in Fig. 10-40a to use $V_{DD} = \pm 18$ V. The circuit is to have $I_{D(max)} = 2.5$ mA and $V_{DS(min)} = 7$ V, and the FET has the transfer characteristics in Fig. 10-37.

10-8.2 The constant-current bias circuit in Fig. 10-42b is to have a 2N5457 FET, $V_{DD} = 30$ V, and $V_{EE} = -12$ V. The bias conditions are to be $I_{D(max)} = 2$ mA and $V_{DS(min)} = 8$ V. Design the circuit.

10-8.3 Design a drain feedback bias circuit, as in Fig. 10-41, to have $I_{D(max)} = 2.5$ mA and $V_{DS(min)} = 9$ V. A 2N5457 FET is to be used, and the supply is $V_{DD} = 24$ V.

10-9 USE OF UNIVERSAL TRANSFER CHARACTERISTIC

Universal Transfer Characteristic

A *universal transfer characteristic* for a FET is simply a transfer characteristic plotted with $I_{DSS} = 1$ and $V_{GS(off)} = 1$. Then, instead of the scales being calibrated in milliamps and volts, they are marked as the ratios I_D/I_{DSS} and $V_{GS}/V_{GS(off)}$. To construct the universal transfer characteristic, Eq. 9-1 is rewritten:

$$\frac{I_D}{I_{DSS}} = \left[1 - \frac{V_{GS}}{V_{GS(off)}} \right]^2 \qquad (10\text{-}19)$$

Now, by substituting convenient values for $V_{GS}/V_{GS(off)}$ into Eq. 10-19, we can calculate corresponding values of I_D/I_{DSS}. These are used to plot the characteristic shown in Fig. 10-45.

The universal transfer characteristic can be used to analyze or design a circuit with virtually any JFET. This saves the necessity of plotting the transfer characteristics for each device. The specified values of I_{DSS} and $V_{GS(off)}$ must be known for the device. Then, the characteristic is used in the same way as an actual transfer characteristic, except that V_{GS} levels must be converted to $V_{GS}/V_{GS(off)}$ ratios, and I_D levels must be converted to I_D/I_{DSS} before they can be plotted on the graph. Similarly, readings taken from the graph are converted back to V_{GS} and I_D levels by multiplying them by $V_{GS(off)}$ and I_{DSS} respectively.

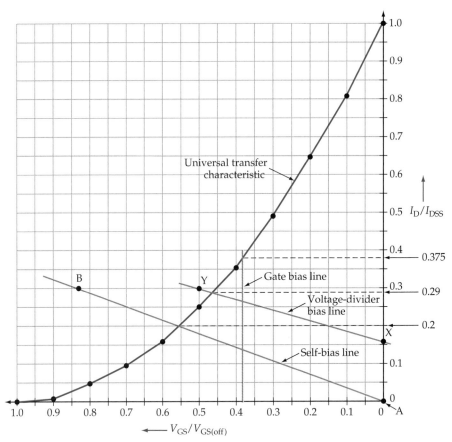

$$\longleftarrow V_{GS}/V_{GS(off)}$$

Figure 10-45 A universal transfer characteristic for a FET is constructed using Eq. 9-1 to determine corresponding levels of I_D and V_{GS} with $I_{DSS} = 1$ and $V = 1$.

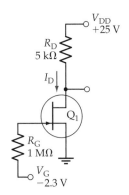

Figure 10-46 The gate bias circuit in Example 10-3 can be analyzed by use of the universal transfer characteristic.

Circuit Analysis

Consider the gate bias circuit analyzed in Ex. 10-3 and reproduced in Fig. 10-46. The gate voltage is given as -2.3 V, and the device used has $I_{DSS(max)} = 8$ mA and $V_{P(max)} = -6$ V. The bias line is drawn vertically on the universal transfer characteristic at

$$\frac{V_{GS}}{V_{GS(off)}} = \frac{2.3 \text{ V}}{6 \text{ V}} = 0.38$$

The gate bias line drawn on Fig. 10-45 (at $V_{GS}/V_{GS(off)} = 0.38$) intersects the universal characteristic at the point where $I_D/I_{DSS} = 0.375$, giving

$$I_{D(max)} = 0.375 \, I_{DSS} = 0.375 \times 8 \text{ mA}$$

$$= 3 \text{ mA}$$

This is the $I_{D(max)}$ level determined in Ex. 10-3.

The minimum I_D for a gate bias circuit can be determined from the universal transfer characteristic by using the $I_{DSS(min)}$ and $V_{GS(off)(min)}$ values. Bias lines can also be drawn for analysis of self-bias circuits and voltage-divider bias circuits. Examples 10-12 and 10-13 demonstrate the analysis procedures.

Example 10-12

Use the FET universal transfer characteristic to determine $I_{D(max)}$ and $V_{DS(min)}$ in the self-bias circuit shown in Fig. 10-47.

Solution

From data sheet A-11 in Appendix A, the 2N5457 has $I_{DSS(max)} = 5$ mA and $V_{GS(off)(max)} = 6$ V

When $I_D = 0$, $\qquad\qquad V_{GS} = V_S = 0$

Therefore $\qquad\qquad I_D/I_{DSS} = 0$ and $V_{GS}/V_{GS(off)} = 0$

Plot point A on the universal transfer characteristic in Fig. 10-45 at

$$I_D/I_{DSS} = 0 \quad \text{and} \quad V_{GS}/V_{GS(off)} = 0$$

When $I_D = 1.5$ mA, $\qquad V_{GS} = I_D \times R_S = 1.5 \text{ mA} \times 3.3 \text{ k}\Omega$

$$= 4.95 \text{ V}$$

Now, $\qquad\qquad \dfrac{I_D}{I_{DSS}} = \dfrac{1.5 \text{ mA}}{5 \text{ mA}} = 0.3$

and $\qquad\qquad \dfrac{V_{GS}}{V_{GS(off)}} = \dfrac{4.95 \text{ V}}{6 \text{ V}} = 0.825$

Plot point B on the universal transfer characteristic at

$$I_D/I_{DSS} = 0.3 \quad \text{and} \quad V_{GS}/V_{GS(off)} = 0.825$$

Draw the self-bias line through points A and B, and where the bias line intersects the universal transfer characteristic, read

$$I_D/I_{DSS} = 0.2$$

giving $\qquad\qquad I_{D(max)} = 0.2\, I_{DSS} = 0.2 \times 5 \text{ mA}$

$$= 1 \text{ mA}$$

$$V_{DS(min)} = V_{DD} - I_D(R_D + R_S)$$

$$= 20 \text{ V} - 1 \text{ mA}(3.3 \text{ k}\Omega + 3.3 \text{ k}\Omega)$$

$$= 13.4 \text{ V}$$

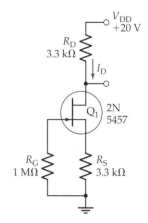

Figure 10-47 Self-bias circuit for Example 10-12.

Example 10-13

Using the FET universal transfer characteristic, determine $I_{D(max)}$ and $V_{DS(min)}$ for the voltage-divider bias circuit in Fig. 10-48.

Solution

From data sheet A-11 in Appendix A, the 2N5458 has $I_{DSS(max)} = 9$ mA and $V_{GS(off)(max)} = 7$ V

Eq. 10-7: $\qquad V_G = \dfrac{V_{DD} \times R_2}{R_1 + R_2} = \dfrac{22 \text{ V} \times 1 \text{ M}\Omega}{4.7 \text{ M}\Omega + 1 \text{ M}\Omega}$

$$= 3.9 \text{ V}$$

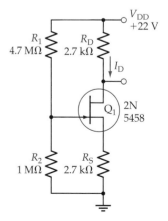

Figure 10-48 Voltage-divider bias circuit for Example 10-13.

When $V_{GS} = 0$, $\quad V_{GS}/V_{GS(off)} = 0$

and $\qquad\qquad\qquad\qquad I_D R_S = V_G$

so $\qquad\qquad\qquad I_D = \dfrac{V_G}{R_S} = \dfrac{3.9\ \text{V}}{2.7\ \text{k}\Omega} = 1.4\ \text{mA}$

and $\qquad\qquad\qquad \dfrac{I_D}{I_{DSS}} = \dfrac{1.4\ \text{mA}}{9\ \text{mA}} = 0.16$

Plot point X on the universal transfer characteristic in Fig. 10-45 at

$$I_D/I_{DSS} = 0.16 \quad \text{and} \quad V_{GS}/V_{GS(off)} = 0$$

When $V_{GS}/V_{GS(off)} = 0.5$, $\quad V_{GS} = 0.5\ V_{GS(off)} = 0.5 \times (-7\ \text{V})$

$$= -3.5\ \text{V}$$

From Eq. 10-6, $\qquad I_D R_S = V_G - V_{GS} = 3.9\ \text{V} - (-3.5\ \text{V})$

$$= 7.4\ \text{V}$$

so $\qquad\qquad\qquad I_D = \dfrac{I_D R_S}{R_S} = \dfrac{7.4\ \text{V}}{2.7\ \text{k}\Omega} = 2.7\ \text{mA}$

$$\dfrac{I_D}{I_{DSS}} = \dfrac{2.7\ \text{mA}}{9\ \text{mA}} = 0.3$$

Plot point Y on the universal transfer characteristic where

$$I_D/I_{DSS} = 0.3 \quad \text{and} \quad V_{GS}/V_{GS(off)} = 0.5$$

Draw the voltage-divider bias line through points X and Y, and where the bias line intersects the universal transfer characteristic, read

$$I_D/I_{DSS} = 0.29$$

giving $\qquad\qquad\qquad I_D = 0.29\ I_{DSS} = 0.29 \times 9\ \text{mA}$

$$= 2.61\ \text{mA}$$

$$V_{DS} = V_{DD} - I_D(R_D + R_S)$$

$$= 22\ \text{V} - 2.61\ \text{mA}(2.7\ \text{k}\Omega + 2.7\ \text{k}\Omega)$$

$$= 7.9\ \text{V}$$

Practice Problems

10-9.1 The self-bias circuit in Fig. 10-47 has its FET changed to a 2N5448. Use the FET universal transfer characteristic to determine $I_{D(max)}$ and $V_{DS(min)}$ for the circuit.

10-9.2 Using the FET universal transfer characteristic, analyze the circuit in Example 10-6 to determine $V_{DS(min)}$.

10-9.3 A gate bias circuit, as in Fig. 10-46, has a 2N5458. The circuit has $V_{DD} = 20$ V, $V_G = -4.5$ V, and $R_D = 4.7$ kΩ. Use the FET universal transfer characteristic to determine $I_{D(max)}$ and $V_{DS(min)}$.

10-10 MOSFET BIASING

DMOS Bias Circuits

DMOS bias circuits are similar to JFET bias circuits. Any of the FET bias circuits already discussed can be used to produce a negative V_{GS} level for an n-channel MOSFET or a positive V_{GS} for a p-channel device. In this case, both devices would be operating in the depletion mode, just like JFETs. To operate an n-channel DMOS in the enhancement mode, the gate terminal must be made positive with respect to the source. A p-channel DMOS in the enhancement mode requires the gate to be negative with respect to the source.

Consider the four MOSFET bias circuits shown in Fig. 10-49, and assume that each device has the transfer characteristics in Fig. 10-50. In Fig. 10-49a the gate-source bias voltage is zero, so the bias line is drawn vertically on the transfer characteristics at $V_{GS} = 0$, as shown in Fig 10-50. The FET in Fig. 10-49b has a positive gate-source bias voltage; consequently, its bias line must be drawn vertically at $V_{GS} = +V_G$. The self-bias circuit in Fig. 10-49c has its bias line drawn from $I_D = 0$ and $V_{GS} = 0$ at a slope determined by R_S, exactly as for a JFET self-bias circuit. Similarly, the bias line for the voltage divider bias circuit in Fig. 10-49d is drawn from the point where $I_D = 0$ and $V_{GS} = +V_G$ with a slope set by R_S.

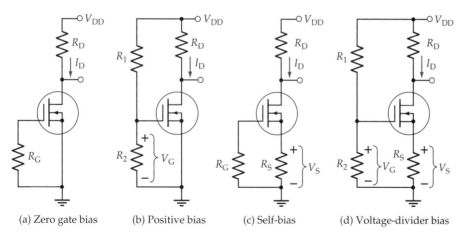

(a) Zero gate bias   (b) Positive bias   (c) Self-bias   (d) Voltage-divider bias

Figure 10-49 Various DMOS bias circuits.

As always, $I_{D(min)}$ and $I_{D(max)}$ for any circuit are indicated by the bias line intersections with the maximum and minimum transfer characteristics. Fig. 10-50 shows that in some cases the FET is operating in depletion mode, and in other cases it is operating in enhancement mode. Analysis and design procedures for these circuits are essentially the same as for similar JFET bias circuits.

MOSFETs can also be used with a plus/minus supply voltage, as discussed in Section 10-8.

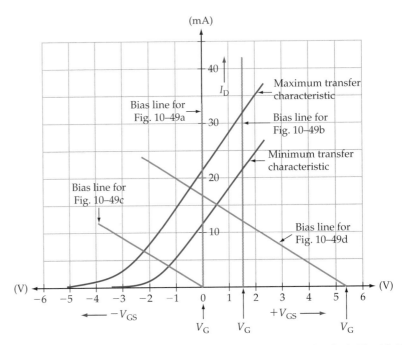

Figure 10-50   DMOS transfer characteristics and bias lines for the circuits in Fig. 10-49.

EMOS Bias Circuits

Enhancement MOSFETs (such as the VMOS and TMOS devices discussed in Section 9-5) must have positive gate-source bias voltages in the case of n-channel devices, and negative V_{GS} levels for a p-channel FET. Thus, the gate bias circuit in Fig. 10-49b and the voltage-divider bias circuit in Fig. 10-49d are suitable for EMOS devices. In each case, the bias line is drawn exactly as already discussed.

Example 10-14

Determine I_D and V_{DS} for the bias circuit in Fig. 10-51, assuming that the device has the transfer characteristic in Fig. 10-52.

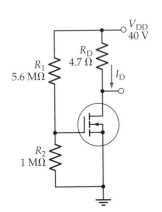

Figure 10-51   EMOS gate bias circuit for Ex. 10-14.

Solution

$$V_G = \frac{V_{DD} \times R_2}{R_1 + R_2} = \frac{40\ \text{V} \times 1\ \text{M}\Omega}{5.6\ \text{M}\Omega + 1\ \text{M}\Omega}$$

$$= 6.06\ \text{V}$$

Draw the bias line vertically on the transfer characteristic at $V_{GS} = 6.06$ V (see Fig. 10-52).

From the point where the bias line intersects the transfer characteristic,

$$I_D = 6.2\ \text{A}$$

$$V_{DS} = V_{DD} - I_D R_D = 40\ \text{V} - (6.2\ \text{mA} \times 4.7\ \Omega)$$

$$= 10.9\ \text{V}$$

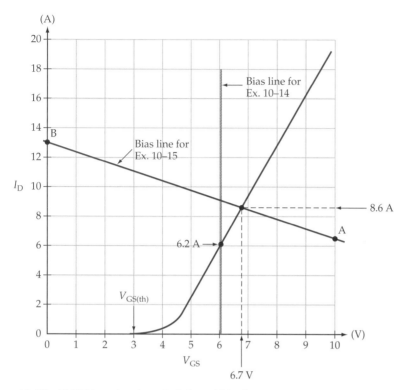

Figure 10-52   EMOS transfer characteristic and bias lines.

The *drain-to-gate* bias circuit shown in Fig. 10-53 is uniquely suitable for an EMOSFET. The FET gate is directly connected to the drain terminal via resistor R_G, so that $V_{GS} = V_{DS}$. If the drain current is higher than the design level, a higher voltage drop is produced across R_D. This tends to reduce V_{GS} and thus reduce I_D back toward the design level. Similarly, a lower I_D than intended produces a lower $I_D R_D$ voltage drop. This results in a higher V_{GS}, which drives I_D back up toward the desired current. Corresponding levels of I_D and V_{GS} can be calculated and plotted on the device transfer characteristics to draw the circuit bias line.

Example 10-15

Analyze the bias circuit in Fig. 10-53 to determine the typical levels of I_D and V_{DS}. Figure 10-52 shows the typical transfer characteristic for the device.

Solution

When $V_{GS} = V_{DS} = 10$ V:

$$I_D = \frac{V_{DD} - V_{GS}}{R_D} = \frac{20\text{ V} - 10\text{ V}}{1.5\ \Omega}$$

$$= 6.7\text{ A}$$

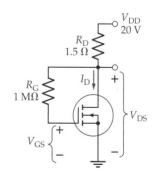

Figure 10-53   EMOS drain-to-gate bias circuit.

Plot point A on Fig. 10-52 at $V_{GS} = 10$ V and $I_D = 6.7$ A

When $V_{GS} = 0$:

$$I_D = \frac{V_{DD} - V_{GS}}{R_D} = \frac{20\text{ V} - 0}{1.5\ \Omega}$$

$$= 13\text{ A}$$

Plot point B on Fig. 10-52 at $V_{GS} = 0$ and $I_D = 13$ A

Draw the bias line for $R_D = 1.5\ \Omega$ through points A and B.

Where the bias line intersects the transfer characteristics,

$$I_D = 8.6\text{ A} \quad \text{and} \quad V_{GS} = 6.7\text{ V}$$

$$V_{DS} = V_{GS} = 6.7\text{ V}$$

Practice Problems

10-10.1 The MOSFET voltage-divider bias circuit in Fig. 10-49d has $V_{DD} = 30$ V, $R_1 = 330$ kΩ, $R_2 = 150$ kΩ, and $R_D = R_S = 680\ \Omega$. The FET has the transfer characteristics in Fig. 10-50. Analyze the circuit to determine $V_{DS(min)}$ and $V_{DS(max)}$.

10-10.2 A drain-to-gate bias circuit, as in Fig. 10-53, has $V_{DD} = 35$ V, and a MOSFET with the transfer characteristics in Fig. 10-52. Calculate an approximate resistance for R_D to give $V_{DS} = 5$ V.

10-11 BIASING FET SWITCHING CIRCUITS

JFET Switching

A field-effect transistor in a switching circuit is normally in an *off* state with zero drain current, or in an *on* state with a very small drain-source voltage. When the FET is *off*, there is a drain-source leakage current so small that it can almost always be neglected. When the device is *on*, the drain-source voltage drop depends on the channel resistance ($r_{DS(on)}$) and the drain current (I_D)

$$V_{DS(on)} = I_D r_{DS(on)} \tag{10-20}$$

Field-effect transistors designed specifically for switching applications have very low channel resistances. For example, the 2N4856 has $r_{DS(on)} = 25\ \Omega$ (see data sheet A-12 in Appendix A). With low I_D levels, $V_{DS(on)}$ can be much smaller than the 0.2 V $V_{CE(sat)}$ typical for a BJT. This is an important advantage of a FET switch over a BJT switch.

Direct-Coupled JFET Switching Circuit

A direct-coupled JFET switching circuit is shown in Fig. 10-54a, and the circuit waveforms are illustrated in Fig. 10-54b. When $V_i = 0$, the FET gate and source voltages are equal, and there is no depletion region penetration into the channel. The output voltage is now $V_o = V_{DS(on)}$, as expressed by Equation 10-20. When V_i exceeds the FET gate-source cutoff voltage, the device is switched *off*, and the output voltage goes to V_{DD}, as illustrated.

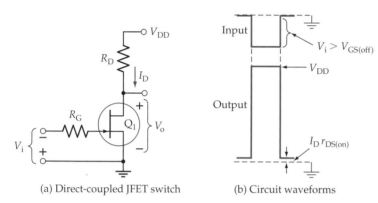

(a) Direct-coupled JFET switch   (b) Circuit waveforms

Figure 10-54 Direct-coupled JFET switching circuit and the waveform of the circuit input and output voltages.

Assuming that $V_{DS(on)}$ is very small, the drain current level is determined from the equation

$$V_{DD} \approx I_D R_D \qquad (10\text{-}21)$$

Equation 10-21 can be used to calculate R_D when V_{DD} and I_D are specified, or to calculate I_D when R_D is known. The I_D level can then be employed to determine $V_{DS(on)}$. The lowest drain current that can be used must be very much greater than the drain-source leakage current specified for the device.

To switch the FET *off*, V_i should exceed the maximum gate-source cutoff voltage. However, V_i must not be so large that the drain-gate voltage ($V_{DG} = V_{DD} + V_i$) approaches the breakdown voltage. A rule of thumb is to select the *off* input voltage 1 V larger than $V_{GS(off)(max)}$.

$$V_i = -(V_{GS(off)(max)} + 1\text{ V}) \qquad (10\text{-}22)$$

The gate resistor (R_G) in the circuit in Fig. 10-54 is provided solely to limit any gate current in the event that the gate-source junctions become forward-biased. The circuit might operate satisfactorily with an R_G of 1 MΩ; however, high-value resistors can slow the switching speed of the circuit. Thus, much smaller resistance values are often used for R_G.

Example 10-16

Design the JFET switching circuit in Fig. 10-54 to have $V_{DS(on)}$ not greater than 200 mV. A 2N4856 FET is to be used with a 12 V supply.

Solution

From data sheet A-12 in Appendix A, $r_{DS(on)} = 25\ \Omega$ and $V_{GS(off)} = 10$ V (max)

From Eq. 10-20,
$$I_D = \frac{V_{DS(on)}}{r_{DS(on)}} = \frac{200\ \text{mV}}{25\ \Omega}$$
$$= 8\ \text{mA}$$

From Eq. 10-21,
$$R_D \approx \frac{V_{DD}}{I_D} = \frac{12\ \text{V}}{8\ \text{mA}}$$
$$= 1.5\ \text{k}\Omega\ \text{(standard value)}$$

Eq. 10-22:
$$V_i = -(V_{GS(off)(max)} + 1\ \text{V}) = -(10\ \text{V} + 1\ \text{V})$$
$$= -11\ \text{V}$$

Capacitor-Coupled JFET Switching Circuits

Two capacitor-coupled JFET switching circuits are shown in Fig. 10-55. The FET in Fig. 10-55a is normally-*on* because it has $V_{GS} = 0$, and the device in Fig. 10-55b is normally-*off* with $-V_{GS}$ greater than the pinch-off voltage. In both circuits, the FET is switched *on* or *off* by a capacitor-coupled input pulse. The design procedure for these circuits uses the equations already discussed for the direct-coupled JFET switching circuit.

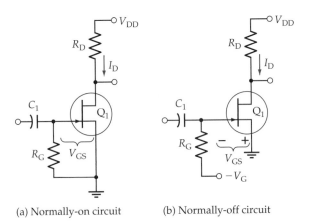

(a) Normally-on circuit   (b) Normally-off circuit

Figure 10-55 Normally-on and normally-off capacitor-coupled JFET switching circuits.

MOSFET Switching

Figure 10-56 shows two capacitor-coupled MOSFET switching circuits. In Fig. 10-56a, the FET is biased *off* because $V_{GS} = 0$. A positive-going input signal is required to turn the device *on*. The FET in Fig. 10-56b is biased *on* by the positive V_{GS} provided by the voltage divider (R_1 and R_2). In this case, a

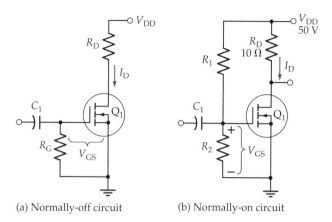

(a) Normally-off circuit   (b) Normally-on circuit

Figure 10-56   MOSFET normally-*on* and normally-*off* capacitor-coupled switching circuits.

negative-going input voltage must be applied to turn the FET *off*. Equations 10-20 and 10-21 can be applied to these circuits to calculate I_D and $V_{DS(on)}$. To set the device *on* to a desired level of drain current, the transfer characteristics can be employed for determining V_{GS} if they are available. Alternatively, Equation 9-5 can be used to estimate the required gate-source bias voltage, as explained in Section 9-5. To ensure that the FET is *off*, the gate-source voltage must be driven below the minimum threshold voltage for the device.

Example 10-17

The MOSFET in the switching circuit in Fig. 10-56b is to be biased *on* to have the smallest possible $V_{DS(on)}$. Determine suitable resistances for R_1 and R_2, and calculate $V_{DS(on)}$. The device has the transfer characteristics shown in Fig. 10-52, and its drain-source *on* resistance is 0.25 Ω.

Solution

From Eq. 10-21,
$$I_D \approx \frac{V_{DD}}{R_D} = \frac{50\ \text{V}}{10\ \Omega}$$
$$= 5\ \text{A}$$

From the transfer characteristics in Fig. 10-52,

at $I_D = 5$ A, $V_{GS} \approx 5.7$ V

Select $R_2 = 1$ MΩ

From Eq. 10-13, $R_1 = \dfrac{(V_D - V_{GS}) \times R_2}{V_G} = \dfrac{(50\ \text{V} - 5.7\ \text{V}) \times 1\ \text{M}\Omega}{5.7\ \text{V}}$

$= 7.7$ MΩ (use 6.8 MΩ to make $V_{GS} > 5.7$ V to ensure that the FET is biased *on*)

Eq. 10-20: $V_{DS(on)} = I_D r_{DS(on)} = 5\ \text{A} \times 0.25\ \Omega$

$= 1.25$ V

Practice Problems

10-11.1 A normally-off JFET switching circuit, as in Fig. 10-55b, is to use a 2N4861 FET with a 30 V supply, and is to have an $I_{D(max)}$ of 10 mA. Calculate R_D, $V_{DS(on)}$, and the required V_G level.

10-11.2 The normally-off MOSFET switching circuit in Fig. 10-56a is to be employed to pass 15 A through a 3 Ω resistor. The device used has the transfer characteristics shown in Fig. 10-52, and $r_{DS(on)} = 0.2\ \Omega$. Calculate approximate levels for the supply and input voltages.

Review Questions

Section 10-1

10-1 Identify the components that constitute the dc load in a FET bias circuit. Explain the procedure for drawing the dc load line on the FET drain characteristics.

10-2 Explain the selection of a Q-point on a FET dc load line, and discuss the limitations on the output voltage swing.

Section 10-2

10-3 Sketch a gate bias circuit using an *n*-channel JFET. Identify the polarities of V_{DD}, V_{DS}, V_G, and V_{GS}. Show the I_D direction. Briefly explain the circuit operation.

10-4 Repeat Question 10-3 for a gate bias circuit using a *p*-channel JFET.

10-5 Write equations for $V_{DS(max)}$ and $V_{DS(min)}$ for a JFET gate bias circuit.

10-6 State a typical maximum resistance for gate resistor R_G in a gate bias circuit. Briefly explain.

10-7 Explain why the device maximum and minimum transfer characteristics should be used in FET bias circuit analysis.

Section 10-3

10-8 Sketch a self-bias circuit using an *n*-channel JFET. Identify the polarities of V_{DD}, V_{DS}, V_S, V_G, and V_{GS}. Briefly explain the circuit operation.

10-9 Repeat Question 10-8 for a self-bias circuit using a *p*-channel JFET.

10-10 Write equations for $V_{DS(max)}$ and $V_{DS(min)}$ for a FET self-bias circuit. Write the equation used for drawing the circuit bias line on the transfer characteristics.

Section 10-4

10-11 Sketch a voltage-divider bias circuit using an *n*-channel JFET. Identify the polarities of V_{DD}, V_{DS}, V_S, V_G, and V_{GS}. Briefly explain the circuit operation.

10-12 Repeat Question 10-11 for a voltage-divider bias circuit using a p-channel JFET.

10-13 Write equations for $V_{DS(max)}$ and $V_{DS(min)}$ for a FET voltage-divider bias circuit. Write the equation used for drawing the circuit bias line on the transfer characteristics.

Section 10-5

10-14 Compare the performance of the three basic JFET bias circuits in terms of the differences between $I_{D(max)}$ and $I_{D(min)}$ in each circuit and the predictability of circuit Q-points.

Section 10-6

10-15 Briefly explain the procedure for testing a FET bias circuit. List common errors made when bias circuits are tested.

10-16 List possible errors in a JFET bias circuit with $V_D \approx V_{DD}$ in the case of (a) gate bias, (b) self-bias, and (c) voltage-divider bias.

10-17 List possible errors in a JFET bias circuit with $V_{DS} \approx 0$ in the case of (a) gate bias, (b) self-bias, and (c) voltage-divider bias.

Section 10-7

10-18 Write the equation for calculating R_D in a gate bias circuit.

10-19 Explain why the FET maximum drain characteristics should be used when designing a JFET bias circuit.

10-20 Write the equation for calculating R_D and R_S in a JFET self-bias circuit.

10-21 Write the equation for determining R_D, R_S, R_1, and R_2 in a JFET voltage-divider bias circuit.

Section 10-8

10-22 Sketch an n-channel JFET voltage-divider bias circuit using a plus/minus supply. Identify the polarities of V_{DD}, V_{DS}, V_S, V_G, and V_{GS}. Briefly explain the circuit operation.

10-23 Repeat Question 10-22 for a similar circuit using a p-channel JFET.

10-24 Sketch an n-channel JFET drain feedback bias circuit. Identify all voltage polarities, and explain the circuit operation.

10-25 Repeat Question 10-24 for a drain feedback bias circuit using a p-channel JFET.

10-26 Sketch an n-channel JFET constant-current bias circuit using voltage-divider bias at the gate. Identify all voltage polarities and current directions, and explain the circuit operation.

10-27 Repeat Question 10-26 for a constant-current bias circuit using a p-channel JFET.

10-28 Sketch an n-channel JFET constant-current bias circuit using a plus/minus supply. Label all voltage polarities and current directions, and explain the circuit operation.

10-29 Repeat Question 10-28 for a constant-current bias circuit using a p-channel JFET.

Section 10-9

10-30 Write the equation used for constructing a JFET universal transfer characteristic. Briefly explain the universal transfer characteristic and its application.

Section 10-10

10-31 Sketch n-channel DMOS gate bias circuits using (a) $V_{GS} = 0$, (b) $V_{GS} = +V_G$. Identify all voltage polarities and current directions, and explain the operation of each circuit.

10-32 Draw the diagram of a self-bias circuit using an n-channel DMOS transistor. Show all voltage polarities and current directions, and explain the circuit operation.

10-33 Draw a voltage-divider bias circuit using an n-channel DMOSFET. Show all voltage polarities and current directions, and explain the circuit operation.

10-34 Sketch approximate maximum and minimum transfer characteristics for a DMOSFET. On the transfer characteristics, draw typical bias lines for the circuits in Questions 10-31 to 10-33. Briefly explain.

10-35 Draw a circuit diagram for a drain-to-gate bias circuit using an n-channel EMOSFET. Explain the circuit operation.

10-36 Draw a circuit diagram for a drain-to-gate bias circuit using a p-channel EMOSFET. Explain.

Section 10-11

10-37 Sketch a circuit diagram for a direct-coupled switching circuit using an n-channel JFET. Sketch input and output waveforms, and explain the circuit operation.

10-38 Draw circuit diagrams for normally-*on* and normally-*off* capacitor-coupled JFET switching circuits. Explain the operation of each circuit.

10-39 Sketch circuit diagrams for normally-*on* and normally-*off* capacitor-coupled switching circuits using EMOSFETs. Explain the operation of each circuit.

Problems

Section 10-1

10-1 The circuit in Fig. 10-57 has a JFET with the characteristics in Fig. 10-58. Draw the dc load line for the circuit, and select a suitable gate voltage to give the maximum possible symmetrical output voltage swing.

10-2 Estimate the maximum possible symmetrical output voltage swing for the circuit in Problem 10-1 when $V_G = -2$ V.

10-3 A JFET circuit like the one in Fig. 10-57 is to have $V_{DS} = 10$ V and $I_D = 2$ mA. If $V_{DD} = 18$ V, draw the dc load line on the drain characteristics in Fig. 10-58, and determine the required resistance for R_D.

10-4 Using the drain characteristics in Fig. 10-58, draw the dc load line for the circuit in Fig. 10-59.

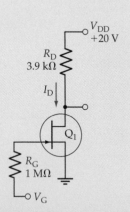

Figure 10-57

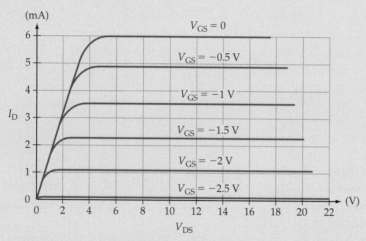

Figure 10-58

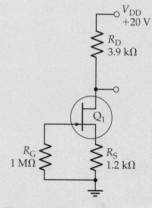

Figure 10-59

10-5 Determine the I_D and V_{DS} levels for the circuit in Problem 10-4 if the gate-source voltage is -1.5 V.

10-6 A JFET circuit as in Fig. 10-57 has $R_D = 4.7$ kΩ. The bias conditions are to be $V_{DS} = 8$ V and $I_D = 1.5$ mA. Draw the dc load line on the drain characteristics in Fig. 10-58, and determine a suitable supply voltage.

Section 10-2

10-7 The gate bias circuit in Fig. 10-60 has a JFET with the transfer characteristics in Fig. 10-61. Analyze the circuit to calculate $V_{DS(max)}$ and $V_{DS(min)}$.

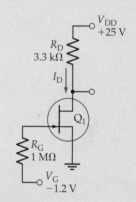

Figure 10-60

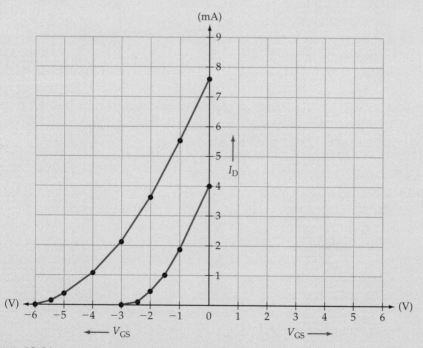

Figure 10-61

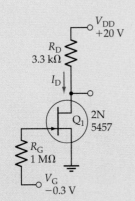

Figure 10-62

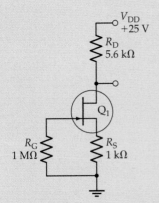

Figure 10-63

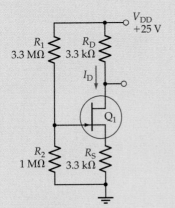

Figure 10-64

10-8 Repeat Problem 10-7 using Equation 9-1 instead of the transfer characteristics.

10-9 The JFET used in a gate bias circuit has the transfer characteristics in Fig. 10-61. If $R_D = 2.2$ kΩ, $V_{DD} = 20$ V, and $V_G = -2$ V, draw the bias line and calculate $V_{DS(max)}$ and $V_{DS(min)}$.

10-10 Repeat Problem 10-9 using Equation 10-8 instead of the transfer characteristics.

10-11 Analyze the circuit in Fig. 10-62 to determine $V_{DS(max)}$ and $V_{DS(min)}$.

Section 10-3

10-12 Determine the maximum and minimum levels of V_{DS} for the JFET self-bias circuit in Fig. 10-63. Assume that the device has the transfer characteristics in Fig. 10-61.

10-13 A JFET self-bias circuit (as in Fig. 10-63) has $V_{DD} = 30$ V, $R_D = 4.7$ kΩ, $R_S = 820$ Ω, and $R_G = 1$ MΩ. If the FET is a 2N5457, calculate the minimum level of V_{DS}.

10-14 Recalculate $V_{DS(min)}$ for the circuit in Problem 10-13 when the device is changed to a 2N5458.

Section 10-4

10-15 Assuming that the JFET in the voltage-divider bias circuit in Fig. 10-64 has the transfer characteristics in Fig. 10-61, determine the maximum and minimum V_{DS} levels.

10-16 A JFET voltage-divider bias circuit (as in Fig. 10-64) has $V_{DD} = 20$ V, $R_D = R_S = 2.7$ kΩ, $R_1 = 7.7$ MΩ, and $R_2 = 1$ MΩ. Using the JFET transfer characteristics in Fig. 10-61, determine the maximum and minimum V_{DS} levels.

10-17 A JFET with the transfer characteristics in Fig. 10-61 is connected in a voltage-divider bias circuit with $V_{DD} = 22$ V. The circuit components are: $R_D = R_S = 3.9$ kΩ, $R_1 = 7.8$ MΩ, and $R_2 = 1$ MΩ. Calculate $V_{DS(max)}$ and $V_{DS(min)}$.

10-18 A 2N5457 JFET is used in a voltage-divider bias circuit with $V_{DD} = 20$ V. The circuit components are: $R_D = 4.7$ kΩ, $R_S = 3.9$ kΩ, $R_1 = 1.2$ MΩ, and $R_2 = 300$ kΩ. Determine $V_{DS(min)}$.

Section 10-7

10-19 A JFET gate bias circuit is to have $I_{D(max)} = 5.5$ mA and $V_{DS(min)} = 7$ V. The supply voltage is 25 V, and the FET transfer characteristics are shown in Fig. 10-61. Determine the required bias voltage and suitable resistor values.

10-20 Design a gate bias circuit using a 2N5457 JFET. The circuit is to have $I_{D(max)} = 4.5$ mA and $V_{DS(min)} = 7.5$ V, and the supply voltage is to be 20 V.

10-21 A JFET self-bias circuit is to have $V_{DD} = 25$ V, $I_{D(max)} = 2.5$ mA, and $V_{DS(min)} = 7$ V. Using the FET transfer characteristics shown in Fig. 10-61, design the circuit.

10-22 Design a self-bias circuit using a 2N5457 JFET with $V_{DD} = 20$ V, $I_{D(max)} = 2$ mA, and $V_{DS(min)} = 7.5$ V.

10-23 A JFET voltage-divider bias circuit is to have $V_{DD} = 25$ V, $I_{D(max)} = 2.5$ mA, and $V_{DS(min)} = 7$ V. Using the device transfer characteristics shown in Fig. 10-61, design the circuit.

10-24 Design a JFET voltage-divider bias circuit to have $V_{DD} = 20$ V, $I_{D(max)} = 2$ mA, and $V_{DS(min)} = 7.5$ V. Use a 2N5457 JFET.

10-25 A JFET voltage-divider bias circuit is to have $V_{DD} = 22$ V, $I_{D(max)} = 1.5$ mA, and $V_{DS(min)} = 9$ V. Using the device transfer characteristics in Fig. 10-61, design the circuit.

10-26 A JFET with the device transfer characteristics in Fig. 10-61 is used in a voltage-divider bias circuit with $V_{DD} = 20$ V and $R_D = 4.7$ kΩ. Determine suitable values for R_1, R_2, and R_S to maintain I_D within the limits of 1 mA to 1.3 mA.

Section 10-8

10-27 Using the transfer characteristics in Fig. 10-61, design the circuit shown in Fig. 10-65.

10-28 Analyze the circuit designed for Problem 10-27 to determine the levels of $V_{DS(max)}$ and $V_{DS(min)}$.

10-29 A JFET circuit like the one in Fig. 10-65 is to have the following voltage and current levels: $V_{DD} = 20$ V, $V_{SS} = -10$ V, $I_{D(max)} = 1.5$ mA, and $V_{DS(min)} = 8$ V. Design the circuit to use a 2N5458 JFET. Select suitable standard-value resistors.

10-30 Analyze the circuit designed for Problem 10-29 to determine $V_{DS(min)}$.

10-31 Design the circuit shown in Fig. 10-66 using the JFET transfer characteristics in Fig. 10-37. Select suitable standard-value resistors.

10-32 Analyze the circuit designed for Problem 10-31 to determine $V_{DS(max)}$ and $V_{DS(min)}$.

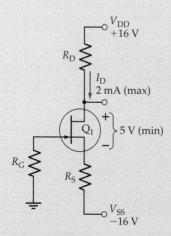

Figure 10-65

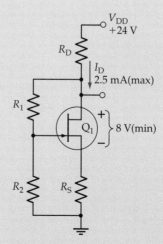

Figure 10-66

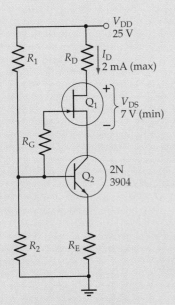

Figure 10-67

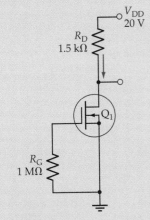

Figure 10-68

10-33 A JFET circuit like the one in Fig. 10-66 is to have the following voltage and current levels: $V_{DD} = 22$ V, $I_{D(max)} = 1.5$ mA, and $V_{DS(min)} = 9$ V. Design the circuit to use a JFET with the transfer characteristics in Fig. 10-61. Select suitable standard-value resistors.

10-34 Analyze the circuit designed for Problem 10-33 to determine the maximum and minimum levels of V_{DS}.

10-35 Design the JFET circuit shown in Fig. 10-67 using the transfer characteristics in Fig. 10-61. Select suitable standard-value resistors.

10-36 Analyze the circuit designed for Problem 10-35 to determine all voltage and current levels.

Section 10-9

10-37 Use the FET universal transfer characteristic in Fig. 10-45 to determine $V_{DS(min)}$ for the gate bias circuit in Fig. 10-62.

10-38 Use the FET universal transfer characteristic in Fig. 10-45 to determine $V_{DS(min)}$ for the circuit in Fig. 10-63. The actual device transfer characteristics are shown in Fig. 10-61.

10-39 Assuming that a 2N5458 JFET is connected in the voltage-divider bias circuit in Fig. 10-64, determine $V_{DS(min)}$ using the FET universal transfer characteristic in Fig. 10-45.

Section 10-10

10-40 The MOSFET in the gate bias circuit in Fig. 10-68 has the transfer characteristics in Fig. 10-69. Draw the bias line and calculate $V_{DS(max)}$ and $V_{DS(min)}$.

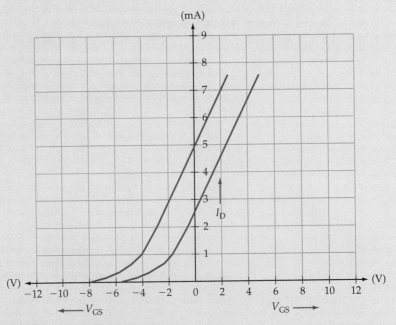

Figure 10-69

10-41 Using the transfer characteristics in Fig. 10-69, design the MOSFET gate bias circuit shown in Fig. 10-70. Select suitable standard-value resistors.

10-42 Analyze the circuit designed for Problem 10-41 to determine all voltage and current maximum and minimum levels.

10-43 Design the MOSFET voltage-divider bias circuit shown in Fig. 10-71 using the transfer characteristics in Fig. 10-69. Select suitable standard-value resistors.

10-44 Analyze the circuit designed for Problem 10-43 to determine all voltage and current maximum and minimum levels.

10-45 Using the transfer characteristics in Fig. 10-52, determine suitable resistors for the MOSFET drain-to-gate bias circuit in Fig. 10-72. Calculate the drain-source voltage.

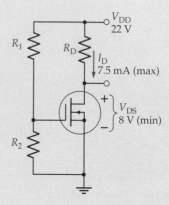

Figure 10-70

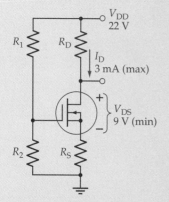

Figure 10-71

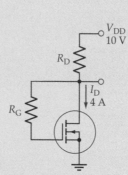

Figure 10-72

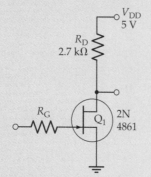

Figure 10-73

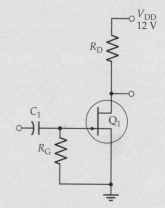

Figure 10-74

Section 10-11

10-46 Calculate V_{DS} for the direct-coupled JFET switching circuit in Fig. 10-73 when the FET is *off* and when it is *on*. Data sheet A-12 in Appendix A shows the data sheet for the 2N4861.

10-47 A direct-coupled JFET switching circuit (as in Fig. 10-73) is to be designed to use a 2N4861, and $V_{DD} = 18$ V. V_{DS} is not to exceed 100 mV when the device is *on*. Determine suitable resistor values and a suitable input voltage amplitude.

10-48 The JFET switching circuit in Fig. 10-74 is to have $V_{DS} \approx 50$ mV when the device is *on*. Select a FET from the 2N4856 to 2N4861 range, calculate appropriate resistor values, and determine a suitable input voltage amplitude.

10-49 The switching circuit in Fig. 10-75 is required to pass 5 A through a 5 Ω resistance. Calculate a suitable $r_{DS(on)}$ for the device. Determine a suitable input voltage amplitude from the transfer characteristic in Fig. 10-52.

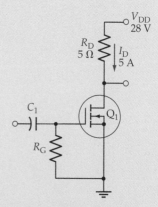

Figure 10-75

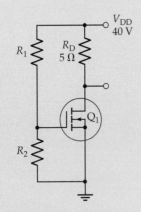

V_{DD}
40 V

R_1

R_D
5 Ω

Q_1

R_2

Figure 10-76

10-50 The MOSFET in the switching circuit in Fig. 10-76 has the transfer characteristic in Fig. 10-52 and $r_{DS(on)} = 0.2\ \Omega$. Calculate the required drain current and suitable values for resistors R_1 and R_2.

Practice Problem Answers

10-1.1   8 V
10-1.2   ±5 V
10-2.1   8.4 V, 17.6
10-2.2   12.7 V
10-3.1   8.8 V, 16.1 V
10-4.1   14 V, 18.7 V
10-7.1   −1.7 V, 1 MΩ, 4.7 kΩ
10-7.2   1 MΩ, 3.9 kΩ, 1.5 kΩ
10-7.3   4.7 MΩ, 1 MΩ, 5.6 kΩ, 5.6 kΩ
10-8.1   1 MΩ, 3.3 kΩ, 8.3 kΩ
10-8.2   1 MΩ, 10 kΩ, 5.6 kΩ
10-8.3   (2.2 MΩ + 220 kΩ), 1 MΩ, 2.7 kΩ, 2.7 kΩ
10-9.1   12.4 V
10-9.2   10 V
10-9.3   14.1 V
10-10.1 9.6 V, 12.3 V
10-10.2 12 Ω
10-11.1 3.3 kΩ, 0.6 V, −5 V
10-11.2 48 V, 8.6 V

CHAPTER 11
AC Analysis of FET Circuits

CONTENTS

Objectives

You will be able to:

1   Explain the need for coupling and bypass capacitors in FET circuits, and draw ac equivalent circuits for FET circuits containing capacitors.

2   Draw ac load lines for basic FET circuits.

3   Sketch ac equivalent circuits for various FET common-source, common-drain, and common-gate circuits.

4   Analyze various CS, CD, and CG FET circuits to determine input resistance, output resistance, and voltage gain.

5   Compare the performance of CS, CD, and CG circuits, and compare FET and BJT circuits.

6   Calculate the cutoff frequencies for FET circuits.

INTRODUCTION

Like BJT circuits, FET amplifiers have ac signals capacitor-coupled to the circuit input, and loads capacitor-coupled to the output terminals. Bypass capacitors are also employed both in FET and BJT circuits to ensure maximum ac voltage gains. Because of the ac-coupled load and the ac-bypassed components, the ac load for a given circuit is different from the dc load; consequently, the ac load line is different from the dc load line.

There are three basic FET circuit configurations: common-source, common-drain, and common-gate. These are similar to the BJT common-emitter, common-collector, and common-base circuits respectively. The common-source circuit is capable of voltage amplification, and it has a high input impedance. The common-drain circuit is a buffer amplifier, with a voltage gain of approximately one, high input impedance, and low output impedance. The common-gate circuit has voltage gain, low input impedance, and good high-frequency performance.

11-1 COUPLING, BYPASSING, AND AC LOAD LINES

Coupling Capacitors

This subject is treated for BJT circuits in Section 6-1, and the discussion there applies equally to FET circuits. As explained, coupling capacitors are required at a circuit input to couple a signal source to the circuit without affecting the bias conditions. Similarly, loads are capacitor-coupled to the circuit output to avoid the change in bias conditions produced by direct coupling. Input and output coupling capacitors (C_1 and C_3) are shown in the FET circuit in Fig. 11-1.

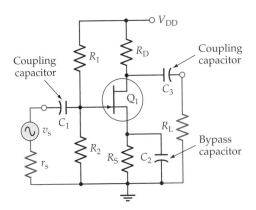

Figure 11-1 FET circuit with coupling and bypass capacitors.

Bypassing Capacitors

Bypass capacitors are also just as necessary in FET circuits as in BJT circuits. Bypass capacitor C_2 in Fig. 11-1 provides an ac short-circuit across resistor R_S.

As will be shown, if C_2 is not present, R_S substantially reduces the ac voltage gain of the circuit.

Figure 11-2 illustrates another situation where a bypassing capacitor is required. The MOSFET drain-to-gate bias circuit shown would have its voltage gain reduced by feedback from the drain to the gate via R_G (ac degeneration) if capacitor C_2 were not present. The feedback is eliminated by splitting R_G into two equal resistors and ac shorting the junction to ground via C_2.

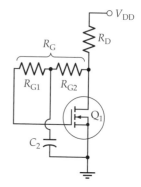

Figure 11-2 MOSFET circuit with bypass capacitor to eliminate ac degeneration.

AC Load Lines

Once again, this is a subject that applies equally to BJT and FET circuits (see Section 6-2). The dc load for the FET circuit in Fig. 11-1 is $(R_D + R_S)$. With R_S ac-bypassed and R_L absent, the ac load is R_D. With the capacitor-coupled load present in Fig. 11-1, the ac load is $R_D \| R_L$. The dc load line is drawn in the usual way, and the Q-point is marked; then the ac load line is drawn through the Q-point, as shown in Fig. 11-3.

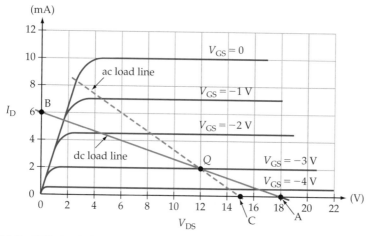

Figure 11-3 FET circuit dc and ac load lines.

Example 11-1

Draw the dc and ac load lines for the transistor circuit in Fig. 11-4 using the transistor common-emitter characteristics in Fig. 11-3, and identifying the Q-point current as $I_D = 2$ mA.

Solution

DC load line:

$$R_{L(dc)} = R_D + R_S = 1.5 \text{ k}\Omega + 1.5 \text{ k}\Omega$$

$$= 3 \text{ k}\Omega$$

The equation for V_{DS} is $\quad V_{DS} = V_{DD} - I_D(R_D + R_S)$

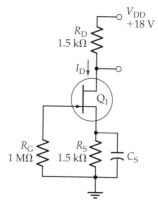

Figure 11-4 Circuit for Example 11-1.

When $I_D = 0$, $V_{DS} = V_{DD} = 18\ \text{V}$

Plot point A at $I_D = 0$ and $V_{DS} = 18\ \text{V}$

When $V_{DS} = 0$, $I_D = \dfrac{V_{DD}}{R_D + R_S} = \dfrac{18\ \text{V}}{3\ \text{k}\Omega}$

$$= 6\ \text{mA}$$

Plot point B at $I_D = 6\ \text{mA}$ and $V_{DS} = 0$

Draw the dc load line through points A and B.

Mark the Q-point on the dc load line at $I_D = 2\ \text{mA}$.

AC load line:

$$R_{L(ac)} = R_D = 1.5\ \text{k}\Omega\ \text{(when } R_S \text{ is bypassed)}$$

When I_D changes 2 mA,

$$\Delta V_{DS} = \Delta I_D \times R_D = 2\ \text{mA} \times 1.5\ \text{k}\Omega$$

$$= 3\ \text{V}$$

Plot point C at $\Delta I_D = 2\ \text{mA}$ and $\Delta V_{DS} = 3\ \text{V}$ from the Q-point.

Draw the ac load line through points C and Q.

Practice Problem

11-1.1 In the circuit in Fig. 11-1, $R_D = 2.2\ \text{k}\Omega$, $R_S = 1.8\ \text{k}\Omega$, $R_L = 15\ \text{k}\Omega$, and $V_{DD} = 20\ \text{V}$. Determine the dc and ac load resistances, and draw both load lines on the characteristic in Fig 11-3. Assume that the Q-point drain current is 2.5 mA.

11-2 FET MODELS AND PARAMETERS

FET Equivalent Circuit

The complete equivalent circuit for a field effect transistor is shown in Fig. 11-5a. It is seen that the source terminal is common to both input and output. Therefore this is a *common-source* equivalent circuit. Resistor R_{GS} between the gate and source terminals is the resistance of the reverse-biased gate-source junction, and C_{gs} is the junction capacitance. So a signal applied to the input 'sees' R_{GS} in parallel with C_{gs}.

The output stage of the equivalent circuit is represented as a current source ($Y_{fs}\ v_{gs}$) supplying current to the drain resistance (r_d). Since Y_{fs} is the forward transfer admittance for the FET, and v_{gs} is the ac signal voltage developed across

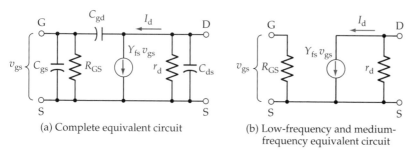

(a) Complete equivalent circuit

(b) Low-frequency and medium-
frequency equivalent circuit

Figure 11-5 The complete equivalent circuit for a FET includes capacitances between terminals. For low- and medium-frequency operations, the capacitances are omitted.

the gate-source terminals, the ac drain current is $(Y_{fs}\, v_{gs})$. The drain-source capacitance (C_{ds}) appears in parallel with r_d, and the gate-drain capacitance (C_{gd}) is shown connected between the input and output stages.

For low- and medium-frequency operations the capacitances can be neglected, and the equivalent circuit is then as shown in Fig. 11-5b. This is the FET *model* (or equivalent circuit) normally used in ac circuit analysis.

Equivalent Circuit Parameters

The parameters used in the equivalent circuit are discussed in Section 9-3. As explained above, R_{GS} is a junction reverse resistance with a $10^9\ \Omega$ typical value for a JFET. Because its resistance is so high, R_{GS} is often regarded as an open circuit. Instead of R_{GS} being listed on a device data sheet, the gate-source reverse current is usually specified, and a value for R_{GS} can be calculated from this quantity.

The forward transfer admittance $(Y_{fs}$ or $g_{fs})$ varies widely for different types of FET. For a small-signal or switching JFET, Y_{fs} typically ranges from 1000 μS to 5000 μS. For an EMOSFET, Y_{fs} might be 2.9 S.

The drain resistance (r_d) shown in the equivalent circuit is the ac resistance offered between the drain and source terminals of the FET when operating in the pinch-off region of the drain characteristics. This quantity is usually defined in terms of an output admittance (Y_{os}) which equals $1/r_d$. Typical values of r_d range from 20 kΩ to 100 kΩ for a JFET.

Section 11-2 Review

11-2.1 Sketch the low- and medium-frequency equivalent circuit for a FET. Identify all components, list typical component values, and briefly explain how the equivalent circuit represents the device.

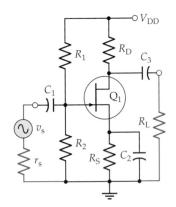

Figure 11-6 FET common-source amplifier circuit.

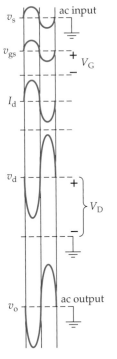

Figure 11-7 Voltage and current waveforms in a common-source circuit.

11-3 COMMON-SOURCE CIRCUIT ANALYSIS

Common-Source Circuit

The circuit of a FET common-source amplifier is shown in Fig. 11-6. With the capacitors treated as ac short-circuits, the circuit input terminals are the gate and source, and the output terminals are the drain and the source. So the source terminal is *common* to both input and output, and the circuit configuration is known as common-source (CS).

The current and voltage waveforms for the CS circuit in Fig. 11-6 are illustrated in Fig. 11-7. The 180° *phase shift* between the input and output waveforms can be understood by considering the effect of a positive-going input signal. An increase in v_s increases the FET gate-source voltage (V_{GS}), thus raising the level of I_D and increasing the voltage drop across R_D. This produces a decrease in the level of the drain voltage (V_D), which is capacitor-coupled to the circuit output as a negative-going ac output voltage (v_o). Consequently, as v_s increases in a positive direction, v_o changes in a negative direction, as illustrated. Conversely, when v_s changes in a negative direction, the resultant decrease in V_{GS} reduces I_D and produces a positive-going output.

The circuit in Fig. 11-6 has an *input impedance* (Z_i), an *output impedance* (Z_o), and a *voltage gain* (A_v). Equations for these quantities can be determined by ac analysis of the circuit.

Common-Source Equivalent Circuit

The first step in ac analysis of a FET (or BJT) circuit is to draw the ac equivalent circuit, by substituting ac short-circuits in place of the power supply and capacitors. When this is done for the circuit in Fig. 11-6, the ac equivalent circuit shown in Fig. 11-8a is created. For ac analysis, the FET equivalent circuit (from Fig. 11-5b) is now substituted in place of the device. This results in the CS ac equivalent circuit in Fig. 11-8b.

The current directions and voltage polarities in Fig. 11-8b are those that occur when the instantaneous level of the input voltage is moving in a positive direction.

Input Impedance

Considering the input section of the equivalent circuit in Fig. 11-8b we can see that the input impedance at the FET terminals is

$$Z_g = R_{gs} \tag{11-1}$$

Recall that a typical value of R_{gs} for a low-current JFET is 10^9 Ω. At the circuit input terminals, resistors R_1 and R_2 are seen to be in parallel with Z_g. So the circuit input impedance is

$$Z_i = R_1 \| R_2 \| Z_g \tag{11-2}$$

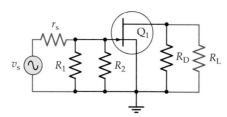

(a) ac equivalent circuit for FET CS circuit

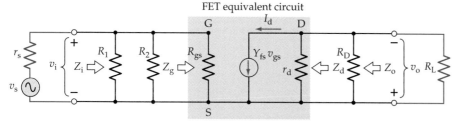

(b) Substitution of the FET equivalent circuit into the CS ac equivalent circuit

Figure 11-8   A common-source ac equivalent circuit is drawn by first replacing the power sup-ply and all capacitors with short-circuits. Then the FET model is substituted in place of the device.

Because Z_g is so large, the circuit input impedance is almost always deter-mined by the bias resistors.

$$Z_i = R_1 \| R_2 \qquad\qquad (11\text{-}3)$$

So the input impedance of a FET circuit can usually be determined simply by calculating the equivalent resistance of the bias resistors. However, as already pointed out, very high input resistance is one of the most important properties of a FET. When it is necessary to achieve the highest possible input resistance, the signal source may be directly connected in series with the gate terminal, and in this case $Z_i \approx Z_g$.

In the case of the self-biased circuit in Fig. 11-4, the circuit input impedance is equal to the gate bias resistor R_G.

Output Impedance

'Looking into' the output of the CS equivalent circuit in Fig. 11-8b, the output impedance at the FET drain terminal is

$$Z_d = r_d \qquad\qquad (11\text{-}4)$$

At the circuit output terminals, resistor R_D is in parallel with Z_d. So the circuit output impedance is

$$Z_o = R_D \| r_d \qquad\qquad (11\text{-}5)$$

Because r_d is usually 100 kΩ, and R_D is usually much lower than 50 kΩ, the circuit output impedance is approximately equal to R_D. With this information, the output impedance of a CS circuit can be discovered simply by reading the resistance of R_D.

$$Z_o \approx R_D \qquad (11\text{-}6)$$

Voltage Gain
Amplifier voltage gain is given by the equation

$$A_v = \frac{v_o}{v_i}$$

From Fig. 11-8b, $\qquad v_o = I_d\,(r_d\|R_D\|R_L)$

and $\qquad I_d = -Y_{fs}\,v_i$

So $\qquad A_v = \dfrac{-Y_{fs}\,v_i(r_d\|R_D\|R_L)}{v_i}$

or $\qquad A_v = -Y_{fs}(r_d\|R_D\|R_L) \qquad (11\text{-}7)$

The minus sign in Eq. 11-7 indicates that v_o is 180° out of phase with v_i. (When v_i increases, v_o decreases, and vice versa.) It is seen that the voltage gain of a common source amplifier is directly proportional to the Y_{fs} of the FET. If the appropriate Y_{fs} value and the resistance of R_D and R_L are known, the voltage gain of a CS circuit can be quickly estimated. For $Y_{fs} = 5000\ \mu S$ and $R_D = 4.7\ k\Omega$, and with $R_L \gg R_D$ and $r_d \gg R_D$, the voltage gain is −23.5.

The voltage gain of a FET common-source amplifier is typically about one-tenth of the gain of a BJT common-emitter circuit. Low voltage gain, caused by the low Y_{fs} value, is the major disadvantage of JFET circuits compared to BJT circuits. The same is generally true of MOSFET circuits, except for EMOSFET devices (VMOS and TMOS), which have much larger Y_{fs} values. However, EMOSFETs use relatively large drain currents, and they are normally unsuitable for small-signal applications.

Summary of Typical CS Circuit Performance

Device input impedance $\qquad Z_g = R_{gs}$

Circuit input impedance $\qquad Z_i = R_1\|R_2\|Z_g \approx R_1\|R_2$

Device output impedance $\qquad Z_d = r_d$

Circuit output impedance $\qquad Z_o = R_D\|r_d \approx R_D$

Circuit voltage gain $\qquad A_v = -Y_{fs}(r_d\|R_D\|R_L)$

The common-source circuit has voltage gain, 180° phase shift, high input impedance, and relatively high output impedance.

Example 11-2

Using the FET typical parameters, calculate the input impedance, output impedance, and voltage gain for the common-source circuit in Fig. 11-9.

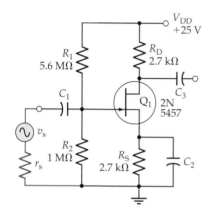

Figure 11-9   Circuit for Example 11-2.

Solution

From the 2N5457 data sheet,

$$r_d = \frac{1}{Y_{os}} = \frac{1}{10\ \mu S}$$

$$= 100\ k\Omega$$

and $Y_{fs} = 3000\ \mu S$

Eq. 11-3: $Z_i = R_1 \| R_2 = 1\ M\Omega \| 5.6\ M\Omega$

$$= 848\ k\Omega$$

Eq. 11-5: $Z_o = R_D \| r_d = 2.7\ k\Omega \| 100\ k\Omega$

$$= 2.63\ k\Omega$$

Eq. 11-7: $A_v = -Y_{fs}(r_d \| R_D \| R_L) = -3000\ \mu S\ (100\ k\Omega \| 2.7\ k\Omega)$

$$= -7.9$$

Practice Problems

11-3.1 Calculate the typical input impedance, output impedance, and voltage gain for the circuit in Fig. 11-4 if the FET is a 2N5458.

11-3.2 A circuit similar to the one in Fig. 11-4 has $R_G = 470\ k\Omega$, $R_D = 3.9\ k\Omega$, and a 2N5457 FET. Determine the typical input impedance, output impedance, and voltage gain for the circuit.

11-4 CS CIRCUIT WITH UNBYPASSED SOURCE RESISTOR

Equivalent Circuit

When an unbypassed source resistor (R_S) is present in a FET common-source circuit, as shown in Fig. 11-10a, it also appears in the ac equivalent circuit (Fig. 11-10b). In the complete equivalent circuit, R_S must be shown connected between the FET source terminal and the circuit common input-output terminal (Fig. 11-11). As with the previous equivalent circuit, the current directions and voltage polarities shown in Fig. 11-11 are those that occur when the instantaneous input voltage is positive-going. The presence of R_S without a bypass capacitor significantly affects the circuit voltage gain.

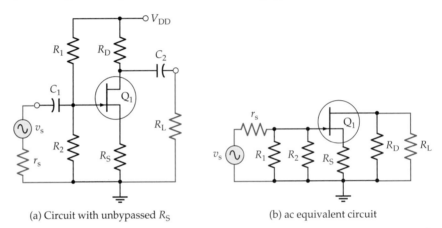

(a) Circuit with unbypassed R_S (b) ac equivalent circuit

Figure 11-10 FET common-source circuit with unbypassed source resistor and ac equivalent circuit.

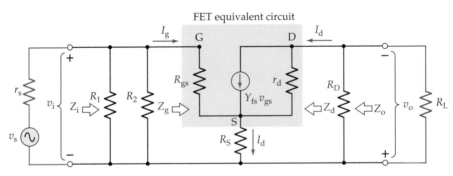

Figure 11-11 An unbypassed source resistor in a FET circuit must be included in the complete ac equivalent circuit.

Input Impedance

An equation for the input impedance at the FET gate can be determined from v_i and I_g. From Fig. 11-11,

$$v_i = v_{gs} + I_d R_S = v_{gs} + Y_{fs}v_{gs}R_S$$

$$= v_{gs}(1 + Y_{fs}R_S)$$

and
$$I_g = \frac{v_{gs}}{R_{gs}}$$

so
$$Z_g = \frac{v_i}{I_g} = \frac{v_{gs}(1 + Y_{fs}R_S)}{v_{gs}/R_{gs}}$$

or
$$Z_g = R_{gs}(1 + Y_{fs}R_S) \qquad\qquad (11\text{-}8)$$

Equation 11-8 gives the input impedance at the FET gate terminal. The circuit input impedance is again given by Eq. 11-2:

$$Z_i = R_1 \| R_2 \| Z_g$$

In this case, since Z_g is much larger than $R_1 \| R_2$, the circuit input impedance is determined by the gate bias resistors.

Eq. 11-3:
$$Z_i \approx R_1 \| R_2$$

Output Impedance

To calculate the circuit output impedance, the ac signal voltage (v_s) is assumed to be zero, and an ac voltage (v_o) is applied at the output (see Fig. 11-12a). The ac output current (I_d) is calculated in terms of v_o; then Z_o is determined as v_o divided by I_d. Figure 11-12a shows that when v_s is zero, the ac voltage across source resistor R_S is applied as a gate-source voltage:

$$v_{gs} = I_d R_S$$

Actually, $I_d R_S$ is divided across $r_s \| R_G$ and R_{gs}. However, $R_{gs} \gg r_s \| R_G$, so that all of $I_d R_S$ is effectively applied as a gate-source voltage. The v_{gs} produced in this way generates an ac drain current which opposes the drain current produced by v_o,

$$I = -Y_{fs}v_{gs} = -Y_{fs}(I_d R_S)$$

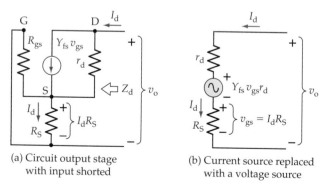

(a) Circuit output stage
with input shorted

(b) Current source replaced
with a voltage source

Figure 11-12 Analysis of the output impedance of a common-source circuit with an unbypassed source resistor.

Converting the current generator ($Y_{fs}v_{gs}$ in parallel with r_d) to a voltage generator ($Y_{fs}v_{gs}r_d$ in series with r_d) gives the equivalent circuit in Fig. 11-12b. The equation for the voltage drops around this circuit is

$$I_d(r_d + R_S) = v_o - (Y_{fs}\,I_d\,R_S\,r_d)$$

giving

$$v_o = I_d(r_d + R_S + Y_{fs}\,R_S\,r_d)$$

and

$$Z_d = \frac{v_o}{I_d} = r_d + R_S + Y_{fs}R_S r_d$$

or

$$Z_d = r_d + R_S + Y_{fs}R_S r_d \tag{11-9}$$

The circuit output impedance is

Eq. 11-5: $$Z_o = R_D \| Z_d$$

Because the output impedance at the FET drain terminal is much larger than the drain resistor (R_D), the output impedance of the circuit with the unbypassed source resistor is still

Eq. 11-6: $$Z_o \approx R_D$$

Voltage Gain

From the derivation of the input impedance equation,

$$v_i = v_{gs}(1 + Y_{fs}R_S)$$

Neglecting r_d,

$$v_o = I_d(R_D \| R_L) = -Y_{fs}v_{gs}(R_D \| R_L)$$

So

$$A_v = \frac{v_o}{v_i} = \frac{-Y_{fs}v_{gs}(R_D \| R_L)}{v_{gs}(1 + Y_{fs}R_S)}$$

giving

$$A_v = \frac{-Y_{fs}(R_D \| R_L)}{1 + Y_{fs}R_S} \tag{11-10}$$

Usually,

$$Y_{fs}R_S \gg 1$$

So

$$A_v = \frac{-(R_D \| R_L)}{R_S} \tag{11-11}$$

The voltage gain of a FET common-source circuit with an unbypassed source resistor can be quickly estimated from Equation 11-11. For the circuit in Fig. 11-10a with $R_D = 4.7 \text{ k}\Omega$, $R_S = 2.2 \text{ k}\Omega$, and $R_L \gg R_D$, $A_v \approx -2.1$.

Summary of Performance of CS Circuit with Unbypassed R_S

Device input impedance	$Z_g = R_{gs}(1 + Y_{fs}R_S)$
Circuit input impedance	$Z_i = R_G \| Z_g \approx R_1 \| R_2$
Circuit output impedance	$Z_o \approx R_D$
Circuit voltage gain	$A_v = \dfrac{-Y_{fs}(R_D \| R_L)}{1 + Y_{fs}R_S}$
Circuit voltage gain	$A_v \approx \dfrac{-(R_D \| R_L)}{R_S}$

The most significant feature of the performance of a CS circuit with an unbypassed source resistor is that its voltage gain is much lower than that for a CS circuit with R_S bypassed.

Example 11-3

For the common-source circuit in Fig. 11-13, calculate the gate input impedance, the drain output impedance, the circuit input and output impedances, and the voltage gain. Use the typical parameters for the FET.

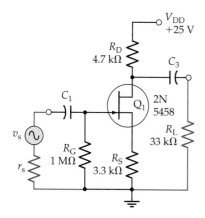

Figure 11-13   Circuit for Example 11-3.

Solution

From the 2N5458 data sheet,

$$r_d = \frac{1}{Y_{os}} = 100 \text{ k}\Omega$$

$$Y_{fs} = 4000 \text{ }\mu S$$

and

$$I_G = 1 \text{ nA} \quad \text{when } V_{GS} = 15 \text{ V}$$

So

$$R_{gs} = \frac{V_{GS}}{I_G} = \frac{15 \text{ V}}{1 \text{ nA}}$$

$$= 15 \times 10^9 \text{ }\Omega$$

Eq. 11-8: $\qquad Z_g = R_{gs}(1 + Y_{fs}R_S) = 15 \times 10^9 \ \Omega[1 + (4 \ mS \times 3.3 \ k\Omega)]$

$$= 2.13 \times 10^{11} \ \Omega$$

From Eq. 11-3, $\qquad Z_i = R_G = 1 \ M\Omega$

Eq. 11-9: $\qquad Z_d = r_d + R_S + (Y_{fs}R_S r_d)$

$$= 100 \ k\Omega + 3.3 \ k\Omega + (4 \ mS \times 3.3 \ k\Omega \times 100 \ k\Omega)$$

$$\approx 1.4 \ M\Omega$$

Eq. 11-5: $\qquad Z_o = R_D \| Z_d = 4.7 \ k\Omega \| 1.4 \ M\Omega$

$$= 4.68 \ k\Omega$$

Eq. 11-10: $\qquad A_v = \dfrac{-Y_{fs}(R_D\|R_L)}{1 + Y_{fs}R_S} = \dfrac{-4 \ mS \times (4.7 \ k\Omega \| 33 \ k\Omega)}{1 + (4 \ mS \times 3.3 \ k\Omega)}$

$$= -1.16$$

or Eq. 11-11: $\qquad A_v \approx \dfrac{-(R_D\|R_L)}{R_S} = \dfrac{4.7 \ k\Omega \| 33 \ k\Omega}{3.3 \ k\Omega}$

$$= -1.24$$

Practice Problems

11-4.1 Calculate the gate input impedance, drain output impedance, circuit input and output impedance, and voltage gain for the circuit in Fig. 11-4 when the source bypass capacitor is removed. Assume that the FET is a 2N5458.

11-4.2 Determine the items listed in Problem 11-4.1 for the circuit in Fig. 11-9 when the source bypass capacitor is removed.

11-5 COMMON-DRAIN CIRCUIT ANALYSIS

Common-Drain Circuit

The FET common-drain circuit shown in Fig. 11-14 has the output voltage developed across the source resistor (R_S). The external load (R_L) is capacitor-coupled to the source terminal of the FET, and the gate bias voltage (V_G) is derived from V_{DD} by means of voltage-divider resistors R_1 and R_2. No resistor is connected in series with the drain terminal, and no source bypass capacitor is employed.

To understand the operation of the circuit in Fig. 11-14, note that the dc gate voltage (V_G) is a constant quantity and that the source voltage is

$$V_S = V_G + V_{GS}$$

When an ac signal is applied to the gate via capacitor C_1, the gate voltage is increased and decreased as the instantaneous level of the signal voltage rises and falls. Also, V_{GS} remains substantially constant, so the source voltage increases

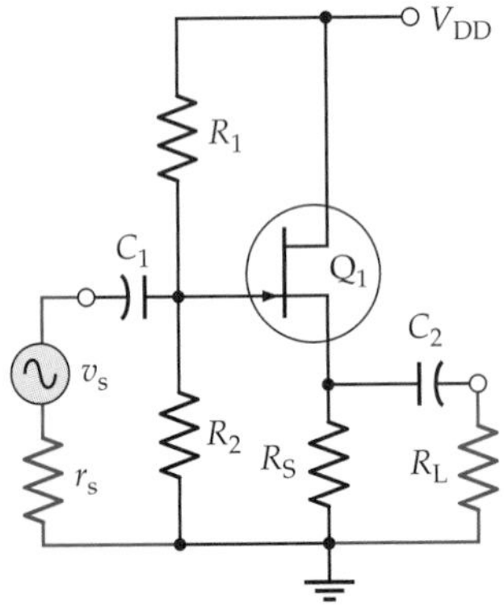

Figure 11-14 Common-drain circuit, also called a source follower. The ac input is applied to the FET gate, and the output is taken from the source terminal.

and decreases with the gate voltage. (See the waveforms in Fig. 11-15.) Thus, the ac output voltage is closely equal to the ac input voltage, and the circuit can be said to have unity gain. AC analysis of the circuit shows that v_o is slightly smaller than v_i. Because the output voltage at the source terminal follows the signal voltage at the gate, the common-drain circuit is also known as a source follower.

Common-Drain Equivalent Circuit

As in the case of other circuits, the supply voltage and coupling capacitors in Fig. 11-14 must be replaced with short-circuits in order to study the circuit ac performance. This gives the common-drain ac equivalent circuit in Fig. 11-16a. The input terminals of the ac equivalent circuit are seen to be the FET gate and drain, and the output terminals are the source and drain. Because the drain terminal is common to both input and output, the circuit configuration is named common-drain (CD).

The complete CD ac equivalent circuit is drawn by substituting the FET model into the ac equivalent circuit (Fig. 11-16b). The indicated current directions and voltage polarities are, once again, those that are produced by a positive-going signal voltage.

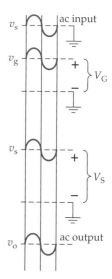

Figure 11-15 Voltage waveforms in a common-drain circuit.

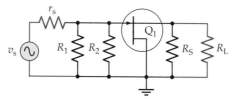

(a) ac equivalent circuit for FET CD circuit

Figure 11-16 A common-drain ac equivalent circuit is drawn by first replacing the power supply and all capacitors with short-circuits. Then the FET model is substituted for the device.

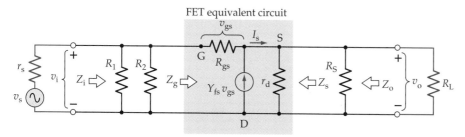

(b) Substitution of the FET equivalent circuit into the CD ac equivalent circuit

Input Impedance

'Looking into' the gate and drain terminals in Fig. 11-16b is similar to 'looking into' the input of a common-source amplifier with an unbypassed source resistor. Therefore, the common-drain Z_g equation is derived in the same way as Eq. 11-8. One important difference is that the common-drain circuit has the external load resistor (R_L) in parallel with R_S. Modifying Equation 11-8 gives

$$Z_g = R_{gs}[1 + Y_{fs}(R_S \| R_L)] \tag{11-12}$$

As discussed in Section 11-4, the equation for Z_g gives a very large value. This is important only when the signal source is connected directly in series with the gate terminal. With the signal capacitor coupled to the input, the circuit input impedance is given by Eq. 11-2:

$$Z_i = R_1 \| R_2 \| Z_g$$

This reduces to Eq. 11-3:

$$Z_i \approx R_1 \| R_2$$

Output Impedance

In a common-drain circuit, any variation in the output voltage (v_o) has an effect on the FET gate-source voltage. To determine the impedance when 'looking into' the source terminal, the signal voltage (in Fig. 11-16b) is assumed to be zero, and the output current (I_s) is calculated in terms of v_o. The circuit is redrawn in Fig. 11-17 with resistor R_G representing the gate bias resistors ($R_G = R_1 \| R_2$ in Fig. 11-16a). If $v_s = 0$, v_o is divided across $R_G \| r_s$ and R_{gs} (see Fig. 11-17). Because $R_{gs} \gg (R_G \| r_s)$, effectively all of v_o is applied as an ac gate-source voltage. Therefore,

$$I_s = Y_{fs} v_o$$

and
$$Z_{ss} = \frac{v_o}{I_s} = \frac{v_o}{Y_{fs} v_o} \qquad \text{(see Fig. 11-17)}$$

or
$$Z_{ss} = 1/Y_{fs}$$

and
$$Z_s = Z_{ss} \| r_d$$

Therefore
$$Z_s = (1/Y_{fs}) \| r_d \qquad\qquad (11\text{-}13)$$

Equation 11-13 gives the device output impedance. The circuit output impedance also involves R_S.

$$Z_o = (1/Y_{fs}) \| R_S \| r_d \qquad\qquad (11\text{-}14)$$

The drain resistance is usually much larger than R_S, and R_S is usually much larger than $1/Y_{fs}$, so that $Z_o \approx 1/Y_{fs}$. If $Y_{fs} = 5000 \ \mu S$, $Z_o = 200 \ \Omega$.

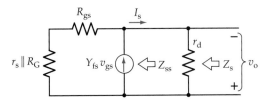

Figure 11-17 Analysis of the output impedance of a common-drain circuit.

Voltage Gain

From Fig. 11-16b,

$$v_o = I_s(R_S||R_L) = Y_{fs}v_{gs}(R_S||R_L)$$

and

$$v_i = v_{gs} + v_o = v_{gs} + Y_{fs}v_{gs}(R_S||R_L)$$

$$= v_{gs}[1 + Y_{fs}(R_S||R_L)]$$

So

$$A_v = \frac{v_o}{v_i} = \frac{Y_{fs}v_{gs}(R_S||R_L)}{v_{gs}[1 + Y_{fs}(R_S||R_L)]}$$

or

$$A_v = \frac{Y_{fs}(R_S||R_L)}{1 + Y_{fs}(R_S||R_L)} \qquad (11\text{-}15)$$

As already discussed, the ac output voltage from a CD circuit is usually closely equal to the input voltage, the voltage gain is normally taken as

$$A_v \approx 1 \qquad (11\text{-}16)$$

Summary of CD Circuit Performance

Device input impedance $Z_g = R_{gs}[1 + Y_{fs}(R_S||R_L)]$

Circuit input impedance $Z_i \approx R_1||R_2$

Device output impedance $Z_s = (1/Y_{fs})||r_d$

Circuit output impedance $Z_o = (1/Y_{fs})||R_S||r_d \approx (1/Y_{fs})||R_S$

Circuit voltage gain $A_v = \dfrac{Y_{fs}(R_S||R_L)}{1 + Y_{fs}(R_S||R_L)}$

Circuit voltage gain $A_v \approx 1$

A common-drain circuit has a voltage gain approximately equal to 1, no phase shift between input and output, very high input impedance, and low output impedance. Because of its high Z_i, low Z_o, and unity gain, the CD circuit is usually used as a *buffer amplifier* between a high-impedance signal source and a low-impedance load.

Example 11-4

The common-drain circuit in Fig. 11-18 has a FET with $Y_{fs} = 3000$ μS, $R_{gs} = 100$ MΩ, and $r_d = 50$ kΩ. Determine the input and output impedances at the device terminals, and the circuit input and output impedances. Calculate the circuit voltage gain.

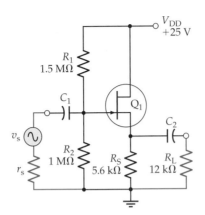

Figure 11-18 Circuit for Example 11-4.

Solution

Eq. 11-12:

$$Z_g = R_{gs}[1 + Y_{fs}(R_S \| R_L)]$$
$$= 100 \text{ M}\Omega[1 + 3000 \text{ }\mu\text{S}(5.6 \text{ k}\Omega \| 12 \text{ k}\Omega)]$$
$$= 1.25 \times 10^9 \text{ }\Omega$$

Eq. 11-2:

$$Z_i \approx R_1 \| R_2 = 1.5 \text{ M}\Omega \| 1 \text{ M}\Omega$$
$$= 600 \text{ k}\Omega$$

Eq. 11-13:

$$Z_s = (1/Y_{fs}) \| r_d = (1/3000 \text{ }\mu\text{S}) \| 50 \text{ k}\Omega$$
$$= 331 \text{ }\Omega$$

Eq. 11-14:

$$Z_o = (1/Y_{fs}) \| R_S \| r_d = 333 \text{ }\Omega \| 5.6 \text{ k}\Omega \| 50 \text{ k}\Omega$$
$$= 312 \text{ }\Omega$$

Eq. 11-15:

$$A_v = \frac{-Y_{fs}(R_S \| R_L)}{1 + Y_{fs}(R_S \| R_L)} = \frac{3000 \text{ }\mu\text{S}(5.6 \text{ k}\Omega \| 50 \text{ k}\Omega)}{1 + [3000 \text{ }\mu\text{S}(5.6 \text{ k}\Omega \| 50 \text{ k}\Omega)]}$$
$$= 0.94$$

Practice Problem

11-5.1 Calculate the device and circuit input and output impedances for a common-drain circuit (as in Fig. 11-18) with the following components: $R_1 = 470 \text{ k}\Omega$, $R_2 = 330 \text{ k}\Omega$, $R_S = 3.3 \text{ k}\Omega$, and $R_L = 47 \text{ k}\Omega$. Assume that the FET is a 2N5458.

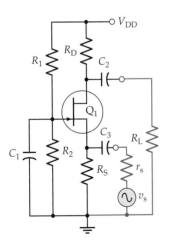

Figure 11-19 In a common-gate circuit the signal is capacitor-coupled to the FET source terminal, and the output is taken from the drain.

11-6 COMMON-GATE CIRCUIT ANALYSIS

Common-Gate Circuit

The FET common-gate (CG) circuit shown in Fig. 11-19 uses voltage-divider bias. The ac output is taken from the drain terminal, and an external load (R_L) is capacitor-coupled to the drain, exactly as in the case of a common-source circuit. Unlike a CS circuit, the ac input for the CG circuit is applied to the FET source terminal (via C_3), and a capacitor (C_1) is included to ac short-circuit the gate to ground. Because the gate is grounded, all of the ac input voltage appears across the gate-source terminals.

The current and voltage waveforms for the CG circuit shown in Fig. 11-20 illustrate the circuit operation. When the instantaneous level of the ac signal voltage increases, the source voltage increases while the gate voltage remains constant. Consequently, the gate-source voltage decreases, the drain current is reduced, and the output voltage at the drain terminal is increased. Similarly, when v_i decreases, v_{gs} increases, I_d increases, and v_d decreases. Thus, an increase in v_i causes an increase in v_o, and a decrease in v_i produces a decrease in v_o; so there is no phase shift between the input and the output. (See the voltage and current waveforms in Fig. 11-20.)

Common-Gate Equivalent Circuit

As in the case of other circuits, the supply voltage and capacitors in Fig. 11-19 are replaced with short-circuits in order to study the circuit ac performance. This gives the common-gate ac equivalent circuit in Fig. 11-21a. The input terminals of the ac equivalent circuit are seen to be the FET source and gate, and the output terminals are the drain and gate. Since the gate terminal is common to both input and output, the circuit is named common-gate.

The complete CG ac equivalent circuit is drawn by substituting the FET model into the ac equivalent circuit (Fig. 11-21b). As always, the current directions and voltage polarities indicated are those produced by a positive-going signal voltage.

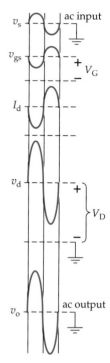

Figure 11-20 Voltage and current waveforms in a common-gate circuit.

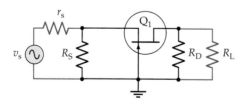

(a) ac equivalent circuit for FET CG circuit

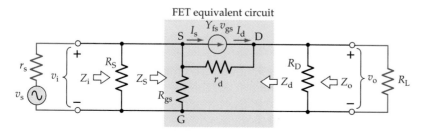

(b) Complete common-gate ac equivalent circuit

Figure 11-21 A common-gate ac equivalent circuit is drawn by first replacing the power supply and all capacitors with short-circuits; then the FET model is substituted for the device.

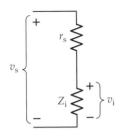

Figure 11-22 Because a common-gate circuit has a low input impedance, some of the signal voltage is lost when it is divided across r_s and Z_i.

Input Impedance

'Looking into' the source and gate terminals of the CG circuit in Fig. 11-21b is similar to 'looking into' the output of a common-drain circuit. Therefore, the equation for the common-gate device input impedance (Z_s) is derived in the same way as the equation for Z_s in the common-drain circuit.

$$Z_s = 1/Y_{fs} \tag{11-17}$$

The circuit input impedance is R_S in parallel with the device input impedance.

So
$$Z_i = (1/Y_{fs}) \| R_S \tag{11-18}$$

With a typical Y_{fs} of 5000 μS, the input impedance of a common-gate circuit is around 200 Ω. Low input impedance is the major disadvantage of the common-gate circuit, because the signal voltage is divided across r_s and Z_i (see Fig. 11-22).

$$v_i = \frac{v_s \times Z_i}{r_s + Z_i} \tag{11-19}$$

Output Impedance

The output of a common-gate circuit is taken from the drain terminal, as in the case of a common-source circuit. So the common-gate output impedance is the same as the common-source output impedance. At the drain terminal:

Eq. 11-4:　　　　　　　　　　　$Z_d = r_d$

and the circuit output impedance is given by

Eq. 11-5:　　　　　　　　　　　$Z_o = R_D \| r_d$

Voltage Gain

From Fig. 11-21b,

$$v_o = I_d(R_D \| R_L) = Y_{fs}v_i(R_D \| R_L)$$

and
$$A_v = \frac{v_o}{v_i} = \frac{Y_{fs}v_i(R_D \| R_L)}{v_i}$$

or
$$A_v = Y_{fs}(R_D \| R_L) \tag{11-20}$$

Equation 11-20 is similar to Equation 11-7 for the voltage gain of a common-source circuit, except that there is no minus sign in Equation 11-20. So the voltage gain for a common-gate circuit is the same as the voltage gain for a common-source circuit with the same component values. As already discussed, there is no phase shift between the input and output of a common-gate circuit; hence the absence of the minus sign in Equation 11-20.

Effect of Unbypassed Gate Resistors

If capacitor C_1 is not present in the CG circuit in Fig. 11-19, the FET gate is *not* ac-short-circuited to ground. So a resistance ($R_G = R_1 \| R_2$) must be included in the ac equivalent circuit (see Fig. 11-23) and in the complete CG ac equivalent circuit. The presence of the unbypassed gate resistors affects the circuit input impedance and voltage gain. Analysis of the equivalent circuit shows that

$$Z_i = (1/Y_{fs}) \frac{R_G + R_{gs}}{R_{gs}} \qquad (11\text{-}21)$$

and that

$$A_v = Y_{fs}(R_D \| R_L) \frac{R_{gs}}{R_G + R_{gs}} \qquad (11\text{-}22)$$

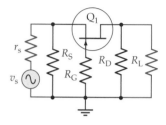

Figure 11-23   When the gate bypass capacitor is omitted in a CG circuit, the equivalent resistance of the gate bias resistors (R_G) must be included for circuit analysis.

When $R_{gs} \gg R_G$ (which is usually the case), Eq. 11-21 gives $Z_i = 1/Y_{fs}$, and Eq. 11-22 gives $A_v = Y_{fs}(R_D \| R_L)$, which are the equations for a CG circuit with the gate bypassed to ground. The Z_i and A_v effects occur only in situations where very high-value bias resistors are used in a CG circuit. Figure 11-24 shows a CG circuit that uses self-bias. In this case, R_G is omitted because there is no signal applied at the gate input; the gate is directly grounded and no gate bypass capacitor is required.

Summary of CG Circuit Performance

With the gate bypassed to ground:

Device input impedance $Z_s = 1/Y_{fs}$

Circuit input impedance $Z_i = (1/Y_{fs}) \| R_S$

Device output impedance $Z_d = r_d$

Circuit output impedance $Z_o = R_D \| r_d$

Circuit voltage gain $A_v = Y_{fs}(R_D \| R_L)$

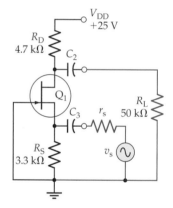

Figure 11-24   A common-gate circuit that uses self-bias can have its gate directly grounded.

With the gate unbypassed:

Circuit input impedance $Z_i = (1/Y_{fs}) \dfrac{R_G + R_{gs}}{R_{gs}}$

Circuit voltage gain $A_v = Y_{fs}(R_D \| R_L) \dfrac{R_{gs}}{R_G + R_{gs}}$

A common-gate circuit has a voltage gain, no phase shift between input and output, low input impedance, and relatively high output impedance.

Example 11-5

The common-gate circuit in Fig. 11-24 has a FET with $Y_{fs} = 3000$ μS and $r_d = 50$ kΩ. Determine the device and circuit input and output impedances, and the circuit voltage gain. Calculate the overall voltage gain if the signal source impedance is $r_s = 600$ Ω.

Solution

Eq. 11-17: $\qquad Z_s = 1/Y_{fs} = 1/3000$ μS

$\qquad\qquad\qquad = 333$ Ω

Eq. 11-18: $\qquad Z_i = (1/Y_{fs})\|R_S = 333\ \Omega\|3.3\ \text{k}\Omega$

$\qquad\qquad\qquad = 302$ Ω

Eq. 11-4: $\qquad Z_d = r_d = 50$ kΩ

Eq. 11-5: $\qquad Z_o = R_D\|r_d = 4.7\ \text{k}\Omega\|50\ \text{k}\Omega$

$\qquad\qquad\qquad = 4.3$ kΩ

Eq. 11-20: $\qquad A_v = Y_{fs}(R_D\|R_L) = 3000\ \mu\text{S} \times (4.7\ \text{k}\Omega\|50\ \text{k}\Omega)$

$\qquad\qquad\qquad = 12.9$

If Eq. 11-19 and 11-20 are combined, the overall voltage gain is

$$A_v = \frac{Y_{fs}(R_D\|R_L)Z_i}{r_s + Z_i} = \frac{12.9 \times 302\ \Omega}{600\ \Omega + 302\ \Omega}$$

$$= 4.3$$

Practice Problems

11-6.1 The common-gate circuit in Fig. 11-19 has the following components: $R_1 = 3.9$ MΩ, $R_2 = 2.2$ MΩ, $R_D = 3.3$ kΩ, $R_S = 1.5$ kΩ, and $R_L = 27$ kΩ. The FET has $Y_{fs} = 3.5$ mS and $r_d = 70$ kΩ, and the signal source has $r_s = 1$ kΩ. Calculate the device and circuit input and output impedances, the circuit voltage gain, and the overall voltage gain.

11-6.2 If the FET in Problem 11-6.1 has $R_{gs} = 10$ MΩ, determine the circuit input impedance and voltage gain when C_1 is removed.

11-7 COMPARISON OF FET AND BJT CIRCUITS

CS, CD, and CG Circuit Comparison

Table 11-1 compares Z_i, Z_o, and A_v for CS, CD, and CG circuits. The CS circuit has voltage gain, high input impedance, high output impedance, and a 180° phase shift from input to output. The CD circuit has high input impedance,

TABLE 11-1   Comparison of common-source, common-drain, and common-gate circuits

Circuit Configuration	Z_i	Z_o	A_v	Phase Shift
CS	$\approx R_G$	$\approx R_D$	$-Y_{fs}(R_D\|R_L)$	180°
CD	$\approx R_G$	$\approx 1/Y_{fs}$	≈ 1	0
CG	$\approx 1/Y_{fs}$	$\approx R_D$	$Y_{fs}(R_D\|R_L)$	0

low output impedance, a voltage gain of approximately 1, and no phase shift. The CG circuit offers low input impedance, high output impedance, voltage gain, and no phase shift.

Impedance at the FET Gate

Consideration of each type of circuit shows that the input or output impedance depends upon which device terminal is involved. In both the CS and CD circuits, the input signal is applied to the FET gate terminal, and so Z_i is the impedance when 'looking into' the gate. Figure 11-25 shows that for CS and CD circuits the gate input impedance is

$$Z_g = R_{gs}$$

and the circuit input impedance is

$$Z_i = R_G \| R_{gs}$$

where R_G is the equivalent resistance of the bias resistors. Because the reverse-biased gate-source resistance (R_{gs}) is so large, an unbypassed source resistor has no significant effect on Z_i at the gate. The circuit input impedance is largely determined by R_G in both the CS and CD circuits.

$$Z_i \approx R_G$$
$$= R_1 \| R_2 \text{ (in the case of voltage-divider bias)}$$

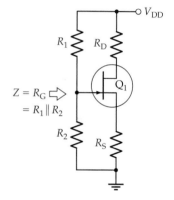

Figure 11-25 FET circuit impedance at the gate terminal.

Impedance at the FET Source

The FET source is the output terminal for a CD circuit and the input terminal for a CG circuit. The device impedance in each case is the impedance seen on looking into the source (see Fig. 11-26):

$$Z_s = 1/Y_{fs}$$

The circuit impedance at the source terminal must include the source resistor:

$$Z_i \text{ or } Z_o = (1/Y_{fs}) \| R_S$$

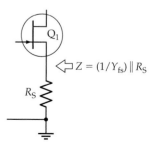

Figure 11-26 FET circuit impedance at the source terminal.

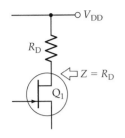

Figure 11-27 FET circuit impedance at the drain terminal.

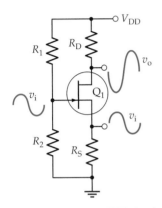

Figure 11-28 In a FET circuit that has an unbypassed R_S, v_i applied to the gate also appears at the source terminal.

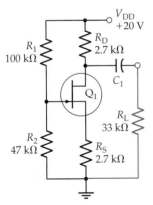

Figure 11-29 Circuit for Example 11-6.

Impedance at the FET Drain

The output for CS and CG circuits is produced at the FET drain terminal. So the impedance seen when looking into the drain is the device output impedance (r_d). As illustrated in Fig. 11-27, the circuit output impedance for any circuit with the output taken from the FET drain terminal is

$$Z_o = R_D \| r_d \approx R_D$$

Voltage Gain

In a circuit with an unbypassed source resistor, the ac voltage at the source terminal follows the ac input at the gate (see Fig. 11-28). A CD circuit (a source follower) has a voltage gain of approximately 1. A CS circuit with an unbypassed source resistor has v_i developed across R_S, and so

$$v_o \approx v_i \times \frac{R_D}{R_S}$$

In a CG circuit and in a CS circuit with R_S bypassed, the ac input (v_i) is developed across the gate-source terminals, and the ac output (v_o) is produced at the drain terminal. Thus, the voltage gain is the same for CS and CG circuits with similar component values and FET parameters. The CS voltage gain equation can be used for the CG circuit, with the omission of the minus sign that indicates CS phase inversion.

Although the voltage gains are equal for similar CS and CG circuits, the low input impedance of the CG circuit can substantially attenuate the signal voltage and result in a low amplitude output. This is demonstrated in Example 11-6.

Example 11-6

A 50 mV signal with a 600 Ω source resistance is applied to the circuit in Fig. 11-29. Calculate the ac output voltage when the conditions are: (a) CS circuit operation with the signal coupled to the gate and R_S bypassed, and (b) CG circuit operation with the signal applied to the source and the gate bias resistors bypassed. The FET's forward transfer admittance $Y_{fs} = 6000\ \mu S$.

Solution

(a) CS circuit:

$$A_v = -Y_{fs}(R_D \| R_L) = -6000\ \mu S\ (2.7\ k\Omega \| 33\ k\Omega)$$

$$\approx -15$$

$$Z_i = R_1 \| R_2 = 100\ k\Omega \| 47\ k\Omega$$

$$= 32\ k\Omega$$

$$v_i = \frac{v_s \times Z_i}{r_s \times Z_i} = \frac{50\ mV \times 32\ k\Omega}{600\ \Omega + 32\ k\Omega}$$

$$= 49.1\ mV$$

$$v_o = A_v \times v_i = -15 \times 49.1 \text{ mV}$$

$$= 736 \text{ mV}$$

(b) CG circuit:

$$A_v = Y_{fs}(R_D \| R_L) = 6000 \text{ μS} \, (2.7 \text{ k}\Omega \| 33 \text{ k}\Omega)$$

$$= 15$$

$$Z_i = (1/Y_{fs}) \| R_S = (1/6000 \text{ μS}) \| 2.7 \text{ k}\Omega$$

$$= 157 \text{ }\Omega$$

$$v_i = \frac{v_s \times Z_i}{r_s \times Z_i} = \frac{50 \text{ mV} \times 157 \text{ }\Omega}{600 \text{ }\Omega + 157 \text{ }\Omega}$$

$$= 10.4 \text{ mV}$$

$$v_o = A_v \times v_i = 15 \times 10.6 \text{ mV}$$

$$= 156 \text{ mV}$$

FET-BJT Circuit Comparison

Table 11-2 compares Z_i, Z_o, and A_v for the basic FET and BJT circuits. The BJT CE and CB circuits have much higher voltage gains than the corresponding FET CS and CG circuits, whereas the FET CS and CD circuits have much higher input impedances than the BJT CE and CC circuits. Apart from these differences, BJTs and FETs can generally perform similar functions. High-frequency, fast-switching, and high-power devices of both types are available. In some switching applications, FETs have the advantage of a smaller voltage drop ($V_{DS(on)}$) than that across BJTs ($V_{CE(sat)}$). This is discussed in Section 10-11.

TABLE 11-2   Typical quantities for basic FET and BJT circuits

FET Circuit	A_v	Z_i	Z_o	BJT Circuit	A_v	Z_i	Z_o
CS	-25	500 kΩ	5 kΩ	CE	-250	5 kΩ	5 kΩ
CD	≈ 1	500 kΩ	200 Ω	CC	≈ 1	5 kΩ	20 Ω
CG	25	200 Ω	5 kΩ	CB	250	20 Ω	5 kΩ

Practice Problem

11-7.1 A FET circuit with voltage-divider bias has the following component values: $R_1 = 560$ kΩ, $R_2 = 390$ kΩ, $R_D = 3.9$ kΩ, $R_S = 3.3$ kΩ, $R_L = 50$ kΩ. Assuming that the FET is a 2N5457 with typical parameters, calculate the impedances looking into the drain, source, and gate terminals.

11-8 FREQUENCY RESPONSE OF FET CIRCUITS

Low-Frequency Response

The low-frequency response of FET circuits is determined by exactly the same considerations as for BJT circuits. The lower cutoff frequency is normally set by a source bypass capacitor, and it can be affected by coupling capacitors. This is considered in detail in Chapter 12.

High-Frequency Response

Unlike BJTs, a device cutoff frequency is not normally specified for a FET. Instead, FETs intended for high-frequency operation have the parameters listed as measured at a specified high frequency. The low-frequency parameters are also normally listed. For example, a 2N5484 JFET has the following parameters:

At f = 1 kHz	min	max
Output conductance (Y_{os})		50 μS
Forward transconductance (Y_{fs})	3000 μS	6000 μS

At f = 100 MHz		
Input admittance (Y_{is})		100 μS
Output conductance (Y_{os})		75 μS
Forward transconductance (Y_{fs})	2500 μS	

The device inter-terminal capacitances (C_{gs}, C_{gd}, and C_{ds}) are important quantities in determining the performance of a FET circuit, and these are used in the same way as the BJT capacitances. An input capacitance limited cutoff frequency ($f_{2(i)}$) and an output capacitance limited cutoff frequency ($f_{2(o)}$) can be calculated exactly as explained for BJT circuits in Section 8-4. The input capacitance is amplified by the Miller effect in the case of a CS circuit (an inverting amplifier). Equation 8-9 can be rewritten for the FET circuit as follows:

$$C_{in} = C_{gs} + (1 + A_v)\, C_{gd} \tag{11-23}$$

The FET capacitances are specified as the input capacitance (C_{iss}), the reverse transfer capacitance (C_{rss}), and the output capacitance (C_{oss}). Reverse transfer capacitance is another name for gate-drain capacitance. The input capacitance is the sum of the gate-source capacitance and the gate-drain capacitance. The output capacitance is simply the drain-source capacitance.

$$C_{rss} = C_{gd} \tag{11-24}$$

$$C_{iss} = C_{gs} + C_{gd} \tag{11-25}$$

$$C_{oss} = C_{ds} \tag{11-26}$$

Typical input capacitance values for a 2N5484 high-frequency JFET are: $C_{rss} = 1$ pF, $C_{iss} = 5$ pF, and $C_{oss} = 2$ pF. Because these are extremely small, the circuit performance can be easily affected by stray capacitance.

Example 11-7

Calculate the input capacitance limited cutoff frequency for the circuit in Fig. 11-30 when operated as a CS circuit with R_S bypassed. Assume that there is no additional stray capacitance at the input terminals and that the FET has the following parameters: $C_{rss} = 1$ pF, $C_{iss} = 5$ pF, $Y_{fs} = 2500$ μS, and $Y_{os} = 75$ μS.

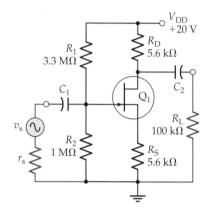

Figure 11-30   Circuit for Example 11-17.

Solution

Eq. 11-24: $\qquad\qquad C_{gd} = C_{rss} = 1$ pF

From Eq. 11-25, $\qquad C_{gs} = C_{iss} - C_{rss}$

$$= 5\,\text{pF} - 1\,\text{pF}$$

$$= 4\,\text{pF}$$

$$|A_v| = Y_{fs}[(1/Y_{os})\|R_D\|R_L]$$

$$= 2500\ \mu\text{S}\,[(1/75\ \mu\text{S})\|5.6\ \text{k}\Omega\|100\ \text{k}\Omega]$$

$$= 9.5$$

Eq. 11-23: $\qquad C_{in} = C_{gs} + (1 + A_v)C_{gd} = 4\,\text{pF} + (1 + 9.5)1\,\text{pF}$

$$= 14.5\,\text{pF}$$

Eq. 11-2: $\qquad Z_i = R_1\|R_2 = 3.3\ \text{M}\Omega\|1\ \text{M}\Omega$

$$= 767\ \text{k}\Omega$$

From Eq. 8-10, $\qquad f_{2(i)} = \dfrac{1}{2\pi\,C_{in}(r_s\|Z_i)} = \dfrac{1}{2\pi \times 14.5\,\text{pF} \times (600\ \Omega\,\|\,767\ \text{k}\Omega)}$

$$= 18.3\,\text{MHz}$$

Practice Problems

11-8.1 The FET circuit in Fig. 11-30 is reconnected to function as a common-gate circuit with the gate terminal bypassed to ground and no source bypass capacitor. Calculate the input capacitance limited cutoff frequency.

11-8.2 The circuit in Ex. 11-7 has 5 pF of stray capacitance at the output terminal. Calculate the output capacitance limited cutoff frequency.

Review Questions

Section 11-1

11-1 Sketch a practical common-source amplifier circuit using voltage-divider bias. Show a capacitor-coupled signal source and a capacitor-coupled load. Explain the need for coupling capacitors, and discuss the correct polarity for connecting the capacitors.

11-2 Explain what is meant by ac degeneration in a FET circuit. Show how ac degeneration is eliminated in voltage-divider bias and self-bias circuits.

11-3 Discuss the purpose of an ac load line for a FET circuit. State how ac load lines differ from dc load lines for circuits with bypassed source resistors and for circuits with capacitor-coupled loads.

Section 11-2

11-4 Sketch the complete equivalent circuit for a field effect transistor connected in common-source configuration. Identify each component of the circuit and discuss its origin. Sketch a simplified form of the FET equivalent circuit for low- and medium-frequency applications.

11-5 Define each parameter in the FET low-frequency equivalent circuit, and state typical values.

Section 11-3

11-6 Sketch a practical common-source amplifier circuit using voltage-divider bias. Draw ac voltage and current waveforms for the circuit, and explain its operation.

11-7 Show what happens to a CS circuit when the capacitors and dc supply are replaced with ac short-circuits. Explain.

11-8 Draw the complete ac equivalent for a CS circuit with voltage divider bias, a bypassed source resistor, a capacitor-coupled signal source, and a capacitor-coupled load.

11-9 Draw the complete ac equivalent for a CS circuit with self-bias, a bypassed source resistor, a capacitor-coupled signal source, and a capacitor-coupled load. Briefly explain.

11-10 Write equations for Z_i, Z_o, and A_v for the FET circuit in Question 11-8.

11-11 Draw the complete ac equivalent for a CS circuit with gate bias, a bypassed source resistor, a capacitor-coupled signal source, and a capacitor-coupled load. Briefly explain.

11-12 List the performance characteristics of a CS circuit, and briefly discuss its usual applications.

Section 11-4

11-13 Draw the complete ac equivalent for a CS circuit with voltage-divider bias, a capacitor-coupled signal source, a capacitor-coupled load, and *no* bypass capacitor across the source resistor. Explain the circuit.

11-14 Write equations for Z_i, Z_o, and A_v for the circuit described in Question 11-13.

11-15 Draw the complete ac equivalent for the circuit in Question 11-11 when the source bypass capacitor is removed.

Section 11-5

11-16 Sketch a practical common-drain amplifier circuit using voltage-divider bias. Draw ac voltage and current waveforms for the circuit, and explain its operation.

11-17 Show what happens to a CD circuit when the capacitors and dc supply are replaced with ac short-circuits. Explain.

11-18 Draw the complete ac equivalent for a CD circuit with voltage-divider bias, a capacitor-coupled signal source, and a capacitor-coupled load. Explain the circuit.

11-19 Draw the complete ac equivalent for a CD circuit with self-bias, a capacitor-coupled signal source, and a capacitor-coupled load. Explain.

11-20 Write equations for Z_i, Z_o, and A_v for the FET circuits in Questions 11-18 and 11-19.

11-21 List the performance characteristics of a CD circuit, and briefly discuss its usual applications.

Section 11-6

11-22 Sketch a practical common-gate amplifier circuit using voltage-divider bias. Draw ac voltage and current waveforms for the circuit, and explain its operation.

11-23 Show what happens to a CG circuit when the capacitors and dc supply are replaced with ac short-circuits. Explain.

11-24 Draw the complete ac equivalent for a CG circuit with voltage-divider bias, a capacitor-coupled signal source, and a capacitor-coupled load. Explain the circuit.

11-25 Sketch a practical common-gate amplifier circuit using self-bias. Draw the complete ac equivalent for the circuit.

11-26 Write equations for Z_i, Z_o, and A_v for the FET circuits in Questions 11-24 and 11-25.

11-27 List the performance characteristics of a CG circuit, and briefly discuss its usual applications.

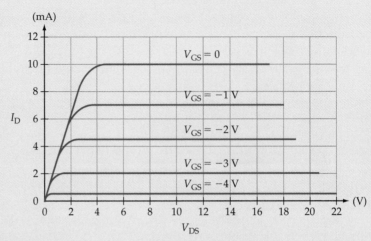

Figure 11-31

Section 11-7

11-28 Compare the performance of CS, CD, and CG circuits, and discuss typical applications of each type of circuit.

11-29 Write equations for the impedances *seen on looking into* the gate, source, and drain terminals of a FET in a circuit. Briefly explain each equation.

11-30 Construct a table to compare A_v, Z_i, and Z_o for basic FET and BJT circuits.

Problems

Section 11-1

11-1 The drain characteristics for the circuit in Fig. 11-29 are shown in Fig. 11-31. Draw the dc and ac load lines for the circuit when it is operated in CS configuration with R_S bypassed. Assume that the Q-point drain current is 1.8 mA.

11-2 Draw the new ac load line for the circuit in Problem 11-1 when a 12 kΩ external load resistor (R_L) is capacitor-coupled to the FET drain terminal.

11-3 Using the device characteristics in Fig. 11-31, draw the dc and ac load lines for the CS circuit shown in Fig. 11-32. Take the Q-point drain current as 2 mA.

Section 11-3

11-4 The FET in the CS circuit in Fig. 11-33 has $Y_{fs} = 6000$ μS and $r_d = 70$ kΩ. Calculate the circuit input impedance, output impedance, and voltage gain.

11-5 Recalculate Z_i, Z_o, and A_v for the circuit in Problem 11-4 when the FET is replaced with one having $Y_{fs} = 3500$ μS and $r_d = 95$ kΩ.

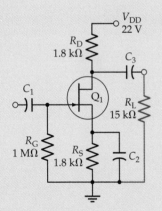

Figure 11-32

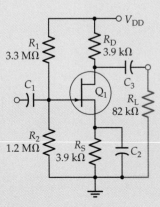

Figure 11-33

11-6 Determine the circuit input impedance, output impedance, and voltage gain for the CS circuit in Fig. 11-34.

11-7 Determine the circuit input impedance, output impedance, and voltage gain for the CS circuit in Fig. 11-35.

11-8 Determine the new values of input impedance, output impedance, and voltage gain for the CS circuit in Fig. 11-35 when a 2N5457 FET is used and R_L is changed to 18 kΩ.

11-9 A CS amplifier is to have a minimum voltage gain of 12. If $R_L = 150$ kΩ and a 2N5459 FET is used, determine a suitable resistance for R_D. Calculate the typical gain.

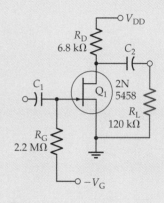

Figure 11-34

Section 11-4

11-10 Calculate the new values of input impedance, output impedance, and voltage gain for the circuit in Problem 11-4 when the source bypass capacitor is removed.

11-11 Calculate the new values of input impedance, output impedance, and voltage gain for the circuit in Problem 11-7 when the source bypass capacitor is removed.

11-12 The source resistor in the circuit in Problem 11-4 is replaced with two series-connected resistors (R_{S1} and R_{S2}). R_{S2} is bypassed and R_{S1} is left unbypassed. Calculate a suitable resistance for R_{S1} to give a voltage gain of 5.

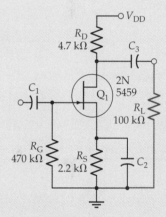

Figure 11-35

Section 11-5

11-13 The FET in the CD circuit in Fig. 11-36 has $Y_{fs} = 6000$ μS and $r_d = 70$ kΩ. Calculate the circuit input impedance, output impedance, and voltage gain.

11-14 Recalculate Z_i, Z_o, and A_v for the circuit in Problem 11-13 when the FET has $Y_{fs} = 3500$ μS and $r_d = 95$ kΩ.

11-15 Determine the input impedance, output impedance, and voltage gain for the CD circuit in Fig. 11-37.

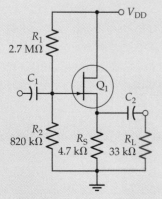

Figure 11-36

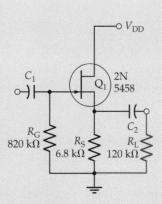

Figure 11-37

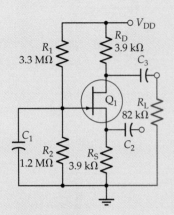

Figure 11-38

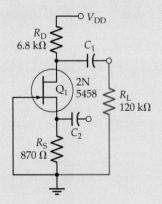

Figure 11-39

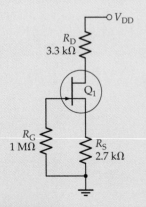

Figure 11-40

11-16 A CD circuit as in Fig. 11-36 has the following components: $R_1 = 2.2$ MΩ, $R_2 = 1$ MΩ, $R_S = 6.8$ kΩ, and $R_L = 150$ kΩ. The FET parameters are $Y_{fs} = 5000$ μS and $r_d = 120$ kΩ. Calculate the input impedance, output impedance, and voltage gain.

11-17 Determine the typical input impedance, output impedance, and voltage gain for the CD circuit in Problem 11-16 when a 2N5457 FET is used and R_L is changed to 39 kΩ.

11-18 A CD amplifier is to have an output impedance of approximately 200 Ω. Select a suitable FET from data sheet A-11 in Appendix A. If $R_S = 1$ kΩ, calculate the typical, minimum, and maximum values of Z_o. Calculate the typical voltage gain for the circuit.

Section 11-6

11-19 The FET in the CG circuit in Fig. 11-38 has $Y_{fs} = 6000$ μS and $r_d = 70$ kΩ. Calculate the circuit input impedance, output impedance, and voltage gain.

11-20 A 120 mV ac signal with $r_s = 300$ Ω is applied as input to the CG circuit in Problem 11-19. Calculate the ac output voltage of the circuit.

11-21 Recalculate Z_i, Z_o, and A_v for the circuit in Problem 11-19 when the FET is replaced with one having $Y_{fs} = 3500$ μS and $r_d = 95$ kΩ.

11-22 Determine the input impedance, output impedance, and voltage gain for the CG circuit in Fig. 11-39.

11-23 Recalculate the values of input impedance, output impedance, and voltage gain for the circuit in Problem 11-22 when a 2N5457 FET is used and R_L is changed to 18 kΩ.

11-24 Determine the input impedance, output impedance, and voltage gain for a CG circuit, as in Fig. 11-39, when a 2N5457 FET is used and the resistors are as follows: $R_S = 2.2$ kΩ, $R_D = 3.3$ kΩ, and $R_L = 100$ kΩ.

11-25 An ac signal with $r_s = 600$ Ω is applied to the source input of the CG circuit in Problem 11-23. Calculate the signal voltage required if the ac output is to be 4 V.

Section 11-7

11-26 The circuit in Fig. 11-40 uses a FET that has $Y_{fs} = 5000$ μS and $r_d = 100$ kΩ. Calculate the circuit impedances when 'looking into' each terminal.

11-27 A 30 mV ac signal with $r_s = 300$ Ω is applied to the circuit in Fig. 11-40. Determine the ac output voltage when the circuit is operated as (a) a CS circuit with R_S bypassed and (b) a CG circuit with the gate bypassed to ground. The FET used has $Y_{fs} = 7000$ μS.

11-28 A 2N5458 FET is used in a circuit with voltage-divider bias and the following components: $R_1 = 560$ kΩ, $R_2 = 390$ kΩ, $R_D = 3.3$ kΩ, $R_S = 2.7$ kΩ, and $R_L = 68$ kΩ. Calculate the impedances seen on 'looking into' the drain, source, and gate terminals. Use the FET typical parameters.

Section 11-8

11-29 The circuit in Fig. 11-40 has a FET with the following high-frequency parameters: $Y_{fs} = 5000\ \mu S$, $Y_{is} = 67\ \mu S$, $Y_{os} = 50\ \mu S$, $C_{iss} = 7\ pF$, $C_{rss} = 3\ pF$, and $C_{oss} = 3\ pF$. Calculate the input capacitance limited cutoff frequency when the circuit is used as a CS amplifier with the source terminal bypassed to ground. The signal source has a 1 kΩ resistance.

11-30 Repeat Problem 11-29 for the circuit reconnected as a CG amplifier with the gate bypassed to ground.

11-31 The circuit in Problems 11-29 and 11-30 has 8 pF stray capacitance at the device drain terminal. Determine the output capacitance limited cutoff frequency.

11-32 The CS amplifier in Fig. 11-9 uses a 2N5484 JFET. Calculate the input capacitance limited cutoff frequency for the circuit. Assume that $r_s = 4.7\ k\Omega$.

11-33 Recalculate the input capacitance limited cutoff frequency for the circuit in Problem 11-32 when it is connected to function as a CG circuit.

Practice Problem Answers

11-1.1   4 kΩ, 1.92 kΩ
11-3.1   1 MΩ, 1.48 kΩ, −5.5
11-3.2   470 kΩ, 3.75 kΩ, −11.25
11-4.1   1 MΩ, 1.5 kΩ, −1.3
11-4.2   848 kΩ, 2.7 kΩ, −0.89
11-5.1   194 kΩ, 232 Ω, 0.93
11-6.1   286 Ω, 240 Ω, 70 kΩ, 3.15 kΩ, 10.3, 2
11-6.2   274 Ω, 9.1
11-7.1   3.75 kΩ, 303 Ω, 230 kΩ
11-8.1   115 MHz, 6 MHz

CHAPTER 12
Small-Signal Amplifiers

CONTENTS

Objectives

You will be able to:
Sketch the following small-signal amplifier circuits, explain the operation of each circuit, design each circuit for a given specification, and analyze each circuit to determine its performance:

1   Single-stage BJT CE amplifier
2   Single-stage FET CS amplifier
3   Two-stage BJT CE capacitor-coupled amplifier
4   Direct-coupled two-stage BJT amplifier
5   Direct-coupled two-stage BJT amplifier circuits using complementary transistors

6   Direct-coupled two-stage BJT amplifier with emitter-follower output stage
7   DC feedback pair with CE output
8   DC feedback pair with CC output
9   Two-stage capacitor-coupled BIFET
10 Two-stage direct-coupled BIFET
11 Differential amplifier
12 Common-base amplifier
13 Cascode amplifier

INTRODUCTION

A *small-signal amplifier* accepts low-voltage ac inputs and produces amplified outputs. The ac outputs are also relatively low-amplitude voltages, usually no larger than ±1 V. BJT and FET bias circuit design and ac analysis of single-stage circuits are treated in earlier chapters. This chapter covers the design of small-signal amplifier circuits to meet given specifications for voltage gain, load resistance, supply voltage, frequency response, and so on.

A single-stage BJT circuit may be employed as a small-signal amplifier, but two cascaded stages give much greater amplification. For very high input impedance, a FET may be used as an input stage with a BJT as the second stage.

The calculation of circuit resistor values merely involves the application of Ohm's law after suitable voltage and current levels have been chosen throughout the circuit. Each capacitor value is determined in terms of the circuit cutoff frequency and the resistance in series with the capacitor.

12-1 SINGLE-STAGE COMMON-EMITTER AMPLIFIER

Specification

Bias circuit design for the CE amplifier in Fig. 12-1 is explained in Chapter 5, and ac analysis of the circuit is covered in Chapter 6. Design of this circuit (or any other circuit) normally begins with a specification that might list the supply voltage, minimum voltage gain, frequency response, signal source impedance, and load impedance.

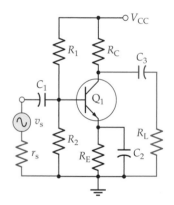

Figure 12-1 Common-emitter circuit design begins with a specification for the circuit performance: voltage gain, load impedance, frequency response, etc.

Selection of I_C, R_C, and R_E

Designing for a particular voltage gain requires the use of *ac negative feedback* to stabilize the gain (see Chapter 13). The circuit shown in Fig. 12-1 has no provision for negative feedback; consequently, it is designed to achieve the largest possible voltage gain. From Eq. 6-15, the voltage gain of a CE circuit is

$$A_v = \frac{-h_{fe}(R_C \| R_L)}{h_{ie}} \tag{12-1}$$

Because A_v is directly proportional to $R_C \| R_L$, designing for the greatest voltage gain would seem to require that the largest possible collector resistance be chosen. However, a very large value of R_C might make the collector current too small for satisfactory transistor operation. For most small-signal transistors, I_C should not be less than 500 μA. A good minimum I_C to aim for is 1 mA. Special low-noise transistors operate with much lower collector current levels.

The transistor h_{fe} value is related to I_C; so a high I_C might be chosen to give the largest h_{fe}, again to achieve the greatest A_v. But a high I_C level results in a small R_C value for a given voltage drop (V_{RC}). So a high I_C might actually result in a low A_v value, although h_{fe} might be relatively large.

For a given level of I_C, the largest possible voltage drop across R_C gives the greatest R_C value ($R_C = V_{RC}/I_C$). To make V_{RC} as large as possible, V_{CE} and V_E should be held to a minimum (see Fig. 12-2a). The collector-emitter voltage should usually be around 3 V. This is large enough to ensure that the transistor operates linearly, and it also allows for a collector voltage swing of ±1 V, which is usually adequate for a small-signal amplifier. Another consideration in selecting R_C is that there is nothing to be gained by making R_C larger than R_L. In fact, R_C should normally be very much smaller than R_L, so that R_L has little effect on the circuit voltage gain.

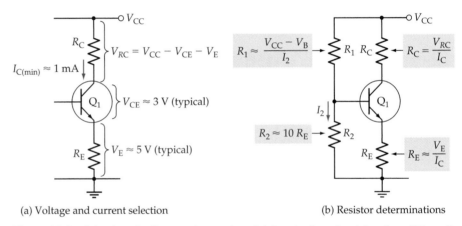

(a) Voltage and current selection   (b) Resistor determinations

Figure 12-2   Selection of voltage and current, and determination of resistors for a CE amplifier circuit.

For good bias stability, the emitter-resistor voltage drop (V_E) should be much larger than the base-emitter voltage (V_{BE}) (see Section 5-7). This is because $V_E = V_B - V_{BE}$, and when $V_E \gg V_{BE}$, V_E will be only slightly affected by any variation in V_{BE} (due to temperature change or other effects). Consequently, I_E and I_C remain fairly stable at $I_C \approx I_E = V_E/R_E$. A minimum V_E of 5 V gives good bias stability in most circumstances (see Fig. 12-2a). With supply voltages less than 10 V, V_E might have to be reduced to 3 V to allow for reasonable levels of V_{CE} and V_{RC}. Normally, an emitter-resistor voltage drop less than 3 V is likely to produce poor bias stability.

Once V_E, V_{CE}, and I_C are selected, V_{RC} is determined:

$$V_{RC} = V_{CC} - V_{CE} - V_E$$

Then, R_C and R_E are calculated:

$$R_C = \frac{V_{RC}}{I_C} \quad \text{and} \quad R_E \approx \frac{V_E}{I_C} \quad \text{(see Fig. 12-2b)}$$

Bias Resistors

As explained in Section 5-7, selection of the voltage divider current (I_2) as $I_C/10$ gives good bias stability and reasonably high input resistance. Where the input resistance is not important, I_2 may be made equal to I_C for excellent bias stability. The bias resistors are calculated as follows:

$$R_2 = \frac{V_B}{I_2} \quad \text{and} \quad R_1 \approx \frac{V_{CC} - V_B}{I_2}$$

Selecting $R_2 = 10\,R_E$ gives $I_2 \approx I_C/10$ (Fig. 12-2b). The precise level of I_2 can be calculated as $I_2 = V_B/R_2$, and this can be used in the equation for R_1.

Bypass Capacitor

All capacitors should be selected to have the smallest possible capacitance value, both to minimize the physical size of the circuit and for economy (large capacitors are the most expensive). Because each capacitor has its highest impedance at the lowest operating frequency, the capacitor values are calculated at the lowest signal frequency that the circuit is required to amplify. This frequency is the circuit *lower cutoff frequency*, or *low 3 dB frequency* (f_1) (see Fig. 12-3).

Bypass capacitor C_2 in Fig. 12-1 is normally the largest capacitor in the circuit because it has the smallest resistance in series with it (r'_e) (see Section 8-4). So C_2 is selected to set f_1 at the desired frequency. Equation 6-21 (for voltage gain) was developed for a *CE* circuit with an unbypassed emitter resistor (R_E). Rewriting the equation to include X_{C2} in parallel with R_E gives

$$A_v = \frac{-h_{fe}(R_C \| R_L)}{h_{ie} + (1 + h_{fe})(R_E \| X_{C2})}$$

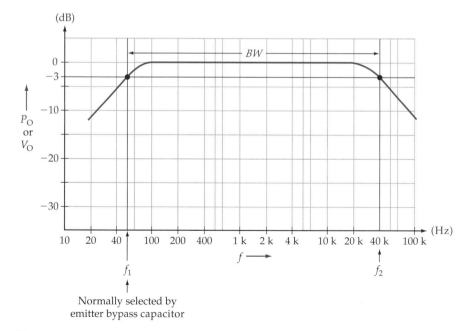

Figure 12-3   Typical frequency response for a transistor amplifier.

Since normally $R_E \gg X_{C2}$, R_E can be omitted. Also, X_{C2} is capacitive; therefore

$$A_v = \frac{-h_{fe}(R_C \| R_L)}{\sqrt{\{h_{ie}^2 + [(1 + h_{fe})(X_{C2})]^2\}}} \tag{12-2}$$

When $h_{ie} = (1 + h_{fe})\, X_{C2}$,

$$|A_v| = \frac{-h_{fe}(R_C \| R_L)}{h_{ie}\sqrt{(1 + 1)}} = \frac{\text{mid-frequency gain}}{\sqrt{2}}$$

$$= (\text{mid-frequency gain}) -3\ \text{dB}$$

Therefore, at f_1,

$$h_{ie} = (1 + h_{fe})\, X_{C2}$$

or

$$X_{C2} = \frac{h_{ie}}{1 + h_{fe}} \text{ at } f_1$$

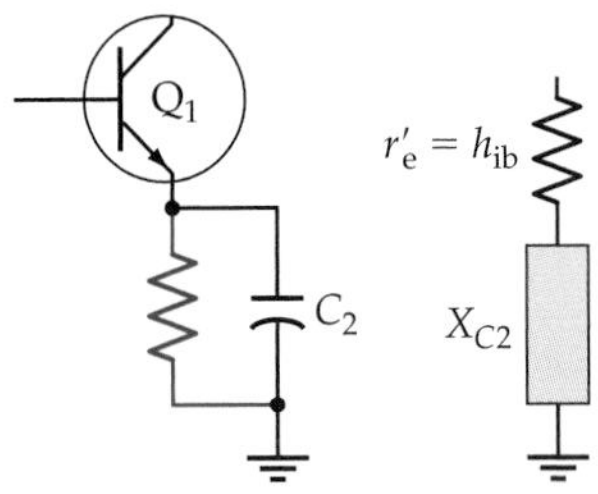

From Table 6-1 (in Section 6-3),

$$h_{ib} = \frac{h_{ie}}{1 + h_{fe}} = (\text{impedance seen on 'looking into' the emitter})$$

So

$$X_{C2} = h_{ib} \text{ at } f_1 \tag{12-3}$$

or

$$X_{C2} = r_e' \text{ at } f_1$$

At f_1,

$$X_{C2} = \text{impedance 'looking into' the emitter}$$
$$= \text{impedance in } series \text{ with } X_{C2} \text{ (see Fig. 12-4)}$$

Figure 12-4 In a CE amplifier circuit, the signal voltage is divided across the emitter bypass capacitor and the emitter resistance (h_{ib} or r_e').

Equation 12-3 give the smallest value for the bypass capacitor. When selecting a standard value for the capacitor, the next larger value should be chosen. This will give a cutoff frequency slightly lower than the f_1 value used in the calculations.

It is important to note that *the emitter bypass capacitor* is calculated in terms of the *resistance 'seen looking into' the transistor emitter terminal* (the resistance in series with C_2). The capacitance is not determined in relation to the resistance of the emitter resistor (R_E). Example 12-1 demonstrates the results of correct and incorrect choice of bypass capacitor.

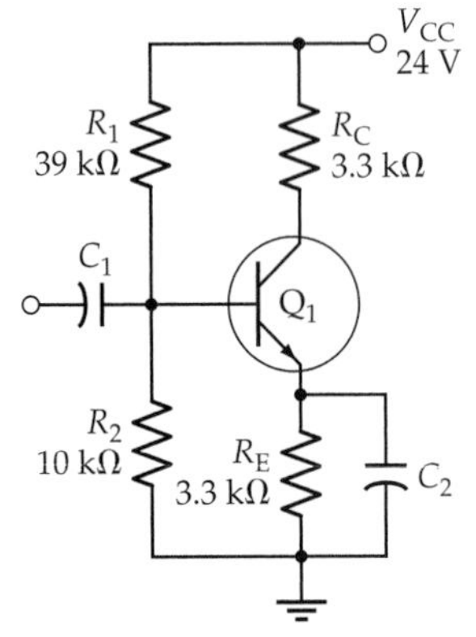

Example 12-1

The circuit shown in Fig. 12-5 uses a transistor with $h_{fe} = 50$, $h_{ie} = 1\ \text{k}\Omega$, and $h_{ib} = 20\ \Omega$. The circuit lower cutoff frequency (f_1) is to be 100 Hz. Calculate the following quantities:

(a) The required capacitance for C_2.
(b) A_v with the emitter terminal completely bypassed to ground.
(c) A_v when $f = 100$ Hz, using the C_2 value selected.
(d) A_v when $f = 100$ Hz and when C_2 is (incorrectly) selected as $X_{C2} = R_E/10$.

Figure 12-5 Common-emitter circuit for Example 12-1.

Solution

(a) C_2 determination:

Eq. 12-3: $X_{C2} = h_{ib} = 20\ \Omega$

$$C_2 = \frac{1}{2\pi f_1 X_{C2}} = \frac{1}{2 \times 100\ \text{Hz} \times 20\ \Omega}$$

$$= 79.6\ \mu\text{F (use 80 }\mu\text{F standard value)}$$

(b) Voltage gain:

Eq. 6-15: $A_v = \dfrac{-h_{fe}(R_C\|R_L)}{h_{ie}} = \dfrac{-50 \times 3.3\ \text{k}\Omega}{1\ \text{k}\Omega}$

$$= -165$$

(c) A_v at 100 Hz:

at $f = 100$ Hz, $X_{C2} = 20\ \Omega$:

Eq. 12-2: $A_{v(1)} = \dfrac{-h_{fe}(R_C\|R_L)}{\sqrt{\{h_{ie}^2 + [(1 + h_{fe})X_{C2}]^2\}}}$

$$= \frac{-50 \times 3.3\ \text{k}\Omega}{\sqrt{\{(1\ \text{k}\Omega)^2 + [(1 + 50) \times 20\ \Omega]^2\}}}$$

$$= -116 = A_v/\sqrt{2}$$

(d) A_v at 100 Hz when C_2 is *incorrectly* determined from $X_{C2} = R_E/10$:

From Eq. 12-1, $A_{v(x)} = \dfrac{-h_{fe}(R_C\|R_L)}{\sqrt{\{h_{ie}^2 + [(1 + h_{fe})(R_E/10)]^2\}}}$

$$= \frac{-50 \times 3.3\ \text{k}\Omega}{\sqrt{\{(1\ \text{k}\Omega)^2 + [(1 + 50) \times (3.3\ \text{k}\Omega/10)]^2\}}}$$

$$= -9.8$$

In Ex. 12-1 the circuit voltage gain is -165 when the emitter terminal is completely bypassed to ground. This is the mid-frequency gain (A_v) for the amplifier. With a correctly selected bypass capacitor, the voltage gain ($A_{v(1)}$) at a signal frequency of 100 Hz is -116. This is $1/\sqrt{2}$ times the mid-frequency gain, or 3 dB below the mid-frequency gain, which is to be expected because the bypass capacitor is selected to give a 3 dB gain reduction at 100 Hz. When the impedance of the bypass capacitor is *incorrectly* selected as $R_E/10$, the voltage gain ($A_{v(x)}$) at 100 Hz is much smaller than the required $A_v/\sqrt{2}$.

Coupling Capacitors

Because the emitter bypass capacitor sets the circuit low 3 dB frequency, the coupling capacitors (C_1 and C_3 in Fig. 12-1) should have only a negligible effect on the frequency response of the circuit. Figure 12-6a illustrates the fact that

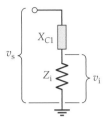

(a) The signal voltage is divided across X_{C1} and Z_i

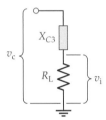

(b) The collector ac voltage is divided across X_{C3} and R_L

Figure 12-6 Input and output voltages can be divided across coupling capacitors and input and load resistances.

X_{C1} and Z_i constitute a voltage divider. If X_{C1} is too large, the circuit ac input voltage (v_i) will be significantly smaller than the signal voltage (v_s). Similarly, X_{C3} and R_L attenuate the ac voltage at the transistor collector, so that the ac output voltage (v_o) can be smaller than the ac collector voltage (v_c) (Fig. 12-6b). To minimize the effects of C_1 and C_3, the reactance of each coupling capacitor is selected to be approximately equal to one-tenth of the impedance in series with it at the lowest operating frequency for the circuit (f_1):

$$X_{C1} = \frac{Z_i + r_s}{10} \text{ at } f_1 \qquad \text{(12-4)}$$

$$X_{C3} = \frac{Z_o + R_L}{10} \text{ at } f_1 \qquad \text{(12-5)}$$

Usually, $R_L \gg Z_o$, and often $Z_i \gg r_s$, so that Z_o and r_s can be omitted in Equations 12-4 and 12-5. Once again, the equations give minimum capacitance values, so that the next larger standard values should always be chosen for C_1 and C_3.

Equations 12-4 and 12-5 give an impression that approximately 10% of the signal and output voltages are lost across C_1 and C_3. This would be true if the quantities were resistive. However, X_{C1} and X_{C3} are capacitive while Z_i and R_L are usually resistive. So, when the actual loss is calculated, it is found to be only about 0.5% for each capacitor.

Another approach to the selection of coupling capacitors is to make X_{C1} equal to Z_i at two octaves below f_1. Thus,

$$X_{C1} = Z_i \text{ at } f_1/4 \qquad \text{(12-6)}$$

The output coupling capacitor is then determined by making the impedance of C_3 equal to R_L at two octaves below $f_1/4$:

$$X_{C3} = R_L \text{ at } f_1/16 \qquad \text{(12-7)}$$

When Equations 12-6 and 12-7 are used to determine the values of the coupling capacitors, it can be shown that the capacitor attenuations are less than 5% of A_v.

Shunting Capacitors

Sometimes an amplifier is required to have a particular *upper cutoff frequency* (f_2 in Fig. 12-3). The transistor must be selected to have a much higher cutoff frequency than f_2. The upper cutoff frequency for the circuit can then be set by connecting a small capacitor from the transistor collector terminal to ground (C_4 in Fig. 12-7). The effect of the shunt capacitance at the transistor collector is discussed in Section 8-4.

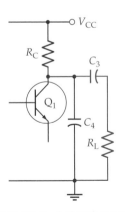

Figure 12-7 A shunting capacitor can be used at the output of a common-emitter circuit to select the upper 3 dB frequency.

Design Calculations

Example 12-2

Calculate suitable resistor values for the common-emitter amplifier in Fig. 12-8.

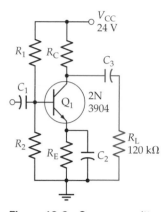

Figure 12-8   Common-emitter amplifier for Example 12-2.

Solution

$$R_C \ll R_L$$

Select
$$R_C = \frac{R_L}{10} = \frac{120 \text{ k}\Omega}{10}$$

$$= 12 \text{ k}\Omega \text{ (standard value)}$$

Select
$$V_E = 5 \text{ V and } V_{CE} = 3 \text{ V}$$

$$V_{RC} = V_{CC} - V_{CE} - V_E = 24 \text{ V} - 3 \text{ V} - 5 \text{ V}$$

$$= 16 \text{ V}$$

$$I_C = \frac{V_{RC}}{R_C} = \frac{16 \text{ V}}{12 \text{ k}\Omega}$$

$$= 1.3 \text{ mA}$$

$$R_E = \frac{V_E}{I_C} = \frac{5 \text{ V}}{1.3 \text{ mA}}$$

$$= 3.75 \text{ kV} \quad \text{(Use a 3.9 k}\Omega \text{ standard value to}$$
$$\text{make } I_C \text{ a little less than the design}$$
$$\text{level, thus ensuring that } V_{CE} \text{ is not}$$
$$\text{less than 3 V.)}$$

$$R_2 = 10 \, R_E = 10 \times 3.9 \text{ k}\Omega$$

$$= 39 \text{ k}\Omega \text{ (standard value)}$$

$$I_2 = \frac{V_E + V_{BE}}{R_2} = \frac{5 \text{ V} + 0.7 \text{ V}}{39 \text{ k}\Omega}$$

$$\approx 146 \text{ }\mu\text{A}$$

$$R_1 = \frac{V_{CC} - V_B}{I_2} = \frac{24 \text{ V} - 5.7 \text{ V}}{146 \text{ }\mu\text{A}}$$

$$\approx 125 \text{ k}\Omega \text{ (use 120 k}\Omega \text{ standard value)}$$

Example 12-3

Determine suitable capacitor values for the CE amplifier in Ex. 12-2. The signal source resistance is 600 Ω, and the lower cutoff frequency is to be 100 Hz.

Solution

From data sheet A-5 in Appendix A for the 2N3904,

$$h_{fe} = 100$$

Eq. 6-2:
$$r'_e = \frac{26 \text{ mV}}{I_E} = \frac{26 \text{ mV}}{1.3 \text{ mA}}$$

$$= 20 \ \Omega$$

Eq. 12-3:
$$X_{C2} = r'_e \text{ at } f_1$$

Therefore,
$$C_2 = \frac{1}{2\pi f_1 X_{C2}} = \frac{1}{2\pi \times 100 \times 20 \ \Omega}$$

$$= 79.5 \ \mu\text{F (use 80 } \mu\text{F standard value)}$$

$$h_{ie} = (1 + h_{fe}) \, r'_e = (1 + 100) \times 20 \ \Omega$$

$$\approx 2 \ \text{k}\Omega$$

$$Z_i = R_1 \| R_2 \| h_{ie} = 120 \ \text{k}\Omega \| 39 \ \text{k}\Omega \| 2 \ \Omega$$

$$= 1.87 \ \text{k}\Omega$$

From Eq. 12-4,
$$C_1 = \frac{1}{2\pi f_1 (Z_i + r_s)/10}$$

$$= \frac{1}{2\pi \times 100 \text{ Hz} \times (1.87 \ \text{k}\Omega + 600 \ \Omega)/10}$$

$$= 6.4 \ \mu\text{F (use 6.8 } \mu\text{F standard value)}$$

From Eq. 12-5,
$$C_3 = \frac{1}{2\pi f_1 (R_C + R_L)/10}$$

$$= \frac{1}{2\pi \times 100 \text{ Hz} \times (12 \ \text{k}\Omega + 120 \ \text{k}\Omega)/10}$$

$$= 0.12 \ \mu\text{F (standard value)}$$

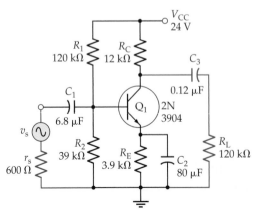

Figure 12-9 CE amplifier circuit designed in Examples 12-2 and 12-3.

The complete common-source circuit designed in Examples 12-2 and 12-3 is shown in Fig. 12-9.

12-2 SINGLE-STAGE COMMON-SOURCE AMPLIFIER

Specification

Bias circuit design for the common-source amplifier in Fig. 12-10 is treated in Chapter 10, and ac analysis of FET circuits is covered in Chapter 11. As with the common-emitter BJT circuit in Section 12-1, the design begins with specification of the supply voltage, amplification, frequency response, load impedance, and so on.

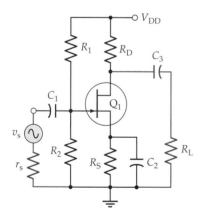

Figure 12-10 Design of a common-source circuit design begins with a specification for the circuit performance.

Selection of I_D, R_D, and R_S

Since the circuit shown in Fig. 12-10 has no provision for negative feedback, it should be designed to achieve the largest possible gain. From Equation 11-7, the voltage gain of a CS circuit is

$$A_v = -Y_{fs}\,(R_D\|R_L)$$

Because A_v is directly proportional to $R_D\|R_L$, a design for the greatest voltage gain normally requires that the largest possible drain resistance be chosen. However, a very large value of R_D might make the drain current too small for satisfactory FET operation. Also, low I_D levels give small Y_{fs} values, which result in lower ac voltage gain. Furthermore, R_D should normally be much smaller than R_L, so that R_L will have little effect on the circuit voltage gain.

For a given I_D, the largest possible voltage drop across R_D gives the greatest R_D value ($R_D = V_{RD}/I_D$). To make V_{RD} as large as possible, V_{DS} and V_S should be held to a minimum (see Fig. 12-11a). The drain-source voltage should usually be $V_{DS(min)} = (V_{GS(off)(max)} + 1 \text{ V})$. This is large enough to ensure that the FET operates in the pinch-off region of its characteristics. It also allows for a drain voltage swing of ± 1 V, which is usually adequate for a small-signal amplifier. If the gate-source bias voltage (V_{GS}) is other than zero, then, as illustrated in Fig. 12-11a, the minimum V_{DS} should be calculated as follows:

$$V_{DS(min)} = V_{GS(off)(max)} + 1 \text{ V} - V_{GS} \qquad (12\text{-}8)$$

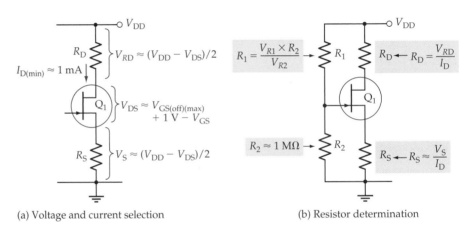

(a) Voltage and current selection   (b) Resistor determination

Figure 12-11 Voltage and current selection and resistor determination for a CS amplifier circuit.

Recall from Chapter 10 that for good bias stability, the source resistor voltage drop (V_S) should be as large as possible. Where the supply voltage is small, V_S may be reduced to a minimum to allow for the minimum level of V_{DS}. A reasonable approach for most FET circuits is to calculate the the sum of V_S and V_{RD} from the equation

$$V_S + V_{RD} = V_{DD} - V_{DS(min)}$$

and then make

$$V_S = V_{RD} = \frac{V_{DD} - V_{DS(min)}}{2} \qquad (12\text{-}9)$$

Once V_S, V_{RD}, and I_D are selected, R_D and R_S are calculated:

$$R_D = \frac{V_{RD}}{I_D} \quad \text{and} \quad R_S \approx \frac{V_S}{I_D} \quad \text{(see Fig. 12-11b)}$$

As already discussed in Section 10-4, a bias line should be drawn upon the FET transfer characteristics to determine a suitable gate bias voltage (V_G). The maximum drain current selected ($I_{D(max)}$) is plotted on the maximum transfer characteristics for the FET used. The bias line is then drawn through this point with a slope of $1/R_S$. The gate bias voltage is read from the intersection of the bias line and the V_G scale. As an alternative to drawing the bias line, V_{GS} can be read from the transfer characteristic when $I_{D(max)}$ is plotted. Then,

$$V_G = V_S - V_{GS}$$

Instead of using the FET transfer characteristics, $V_{GS(off)(max)}$ and $I_{DSS(max)}$ can be substituted into Eq. 10-8 to calculate the V_{GS} level.

Bias Resistors

With a voltage-divider bias circuit (as in Fig. 12-11b), R_2 is usually selected as 1 MΩ or less. Smaller resistance values may be used where a lower input impedance is acceptable. Larger resistance values may also be used; however, as explained in Section 10-7, there are distinct disadvantages to using resistances higher than 1 MΩ. With R_2 determined, R_1 is calculated from R_2 and the ratio of V_{R1} to V_{R2}

$$R_1 = \frac{V_{R1} \times R_2}{V_{R2}} \qquad (12\text{-}10)$$

Capacitors

As for a BJT capacitor-coupled circuit, coupling and bypass capacitors should be selected to have the smallest possible capacitance values. The largest capacitor in the circuit (source bypass capacitor C_2 in Fig. 12-10) sets the circuit low 3 dB frequency (f_1). Equation 11-10 was developed for the voltage gain of a common-source circuit with an unbypassed source resistor (R_S). Rewriting the equation to include X_{C2} in parallel with R_S gives

$$A_v = \frac{-Y_{fs}(R_D\|R_L)}{1 + Y_{fs}(R_S\|X_{C2})}$$

Normally $R_S \gg X_{C2}$, so R_S can be omitted. And since X_{C2} is capacitive,

$$A_v = \frac{-Y_{fs}(R_D\|R_L)}{\sqrt{[1 + (Y_{fs}X_{C2})^2]}} \qquad (12\text{-}11)$$

When $Y_{fs}X_{C2} = 1$,

$$|A_v| = \frac{-Y_{fs}(R_D\|R_L)}{\sqrt{(1 + 1)}} = \frac{\text{mid-frequency gain}}{\sqrt{2}}$$

$$= (\text{mid-frequency gain}) - 3 \text{ dB}$$

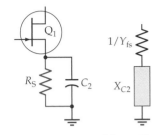

Figure 12-12 In a CS amplifier circuit, the signal voltage can be divided across the source bypass capacitor and the source resistance ($1/Y_{fs}$).

Therefore, at f_1,

$$Y_{fs}\,X_{C2} = 1$$

or

$$X_{C2} = \frac{1}{Y_{fs}} \text{ at } f_1 \qquad\qquad\qquad (12\text{-}12)$$

From Section 11-7,

$$Z_s = \frac{1}{Y_{fs}} = \text{(impedance seen on 'looking into' the FET source)}$$

So, at f_1, $X_{C2} = \text{(impedance 'looking into' the source terminal)}$
$= \text{(impedance in \textit{series} with } X_{C2}) \text{ (see Fig. 12-12)}$

Equation 12-12 gives the smallest value for the source bypass capacitor. When a standard value is selected, the next larger capacitance value should be chosen. This will give a cutoff frequency slightly lower than the f_1 value used in the calculations.

As explained in Section 12-1, it is important to note that the bypass capacitor is calculated in terms of the resistance seen *when looking into* the device terminal (the resistance in series with C_2). *The capacitance value is not determined in terms of the parallel resistor* (R_S).

The input and output coupling capacitors should have almost zero effect on the circuit frequency response. In Section 12-1 it is explained that X_{C1} in series with Z_i and X_{C3} in series with R_L constitute voltage dividers that can attenuate the ac input and output voltages. To minimize the attenuation, the reactance of each coupling capacitor is selected to be approximately one-tenth of the impedance in series with it at the lowest operating frequency for the circuit (f_1). Equations 12-4 and 12-5 apply once again for the calculation of the minimum C_1 and C_2 values.

Eq. 12-4: $$X_{C1} = \frac{Z_i + r_s}{10} \text{ at } f_1$$

Eq. 12-5: $$X_{C2} = \frac{Z_o + R_L}{10} \text{ at } f_1$$

As in BJT circuits, R_L is usually much larger than Z_o, and Z_i is often much larger than r_s; therefore Z_o and r_s can often be omitted in Equations 12-4 and 12-5. As always, since the equations give minimum capacitance values, the next larger standard values should always be chosen for C_1 and C_3.

Alternatively, as explained in Section 12-1, Equations 12-6 and 12-7 may be used for determining coupling capacitor values.

Design Calculations

Example 12-4

Calculate suitable resistor values for the common-source circuit in Fig. 12-13.

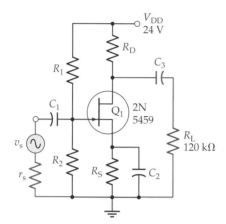

Figure 12-13   Common-source amplifier circuit for Example 12-4.

Solution

Refer to the 2N5459 transfer characteristics in Fig. 9-19. A reasonable level of $I_{D(max)}$ might be 1.5 mA. Y_{fs} becomes smaller at lower I_D levels.

Eq. 10-8:
$$V_{GS} = V_{GS(off)} [1 - \sqrt{(I_D/I_{DSS})}]$$

Using $V_{GS(off)(max)}$ and $I_{DSS(max)}$ from data sheet A-11 in Appendix A (and from Fig. 9-19) gives

$$V_{GS} = 8 \text{ V}[1 - \sqrt{(1.5 \text{ mA}/16 \text{ mA})}]$$

$$= 5.6 \text{ V}$$

Eq. 12-8:
$$V_{DS(min)} = V_{GS(off)(max)} + 1 \text{ V} - V_{GS} = 8 \text{ V} + 1 \text{ V} - 5.6 \text{ V}$$

$$= 3.4 \text{ V}$$

Eq. 12-9:
$$V_S = V_{RD} = \frac{V_{DD} - V_{DS(min)}}{2} = \frac{24 \text{ V} - 3.4 \text{ V}}{2}$$

$$= 10.3 \text{ V}$$

$$R_S = R_D = \frac{V_{RD}}{I_D} = \frac{10.3 \text{ V}}{1.5 \text{ mA}}$$

$$= 6.87 \text{ k}\Omega \text{ (use 6.8 k}\Omega \text{ standard value)}$$

R_D should be $\ll R_L$
6.8 k$\Omega \ll$ 120 kΩ

$$V_{R2} = V_G = V_S - V_{GS} = 10.3 \text{ V} - 5.5 \text{ V}$$

$$= 4.8 \text{ V}$$

$$V_{R1} = V_{DD} - V_G = 24 \text{ V} - 4.8 \text{ V}$$

$$= 19.2 \text{ V}$$

Select
$$R_2 = 1 \text{ M}\Omega$$

Eq. 12-10: $\qquad R_1 = \dfrac{V_{R1} \times R_2}{V_{R2}} = \dfrac{19.2\ \text{V} \times 1\ \text{M}\Omega}{4.8\ \text{V}}$

$\qquad\qquad\qquad = 4\ \text{M}\Omega$ (Use 4.7 MΩ standard value to ensure that V_G is slightly lower than the design level. This will give a lower I_D and a higher V_{DS} than the minimum design level.)

Example 12-5

Determine suitable capacitor values for the common-source amplifier in Ex. 12-4. The signal source resistance is 600 Ω, and the lower cutoff frequency is to be 100 Hz.

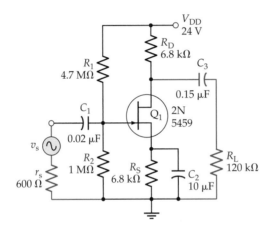

Figure 12-14 Common-source circuit designed in Examples 12-4 and 12-5.

Solution

From data sheet A-11 in Appendix A, for the 2N5459:

$$Y_{\text{fs(max)}} = 6000\ \mu\text{S}$$

Eq. 12-12: $\qquad X_{C2} = \dfrac{1}{Y_{\text{fs}}} = \dfrac{1}{6000\ \mu\text{S}}$

$\qquad\qquad\qquad \approx 167\ \Omega$ (see Fig. 12-14)

$$C_2 = \dfrac{1}{2\pi f_1 X_{C2}} = \dfrac{1}{2\pi \times 100\ \text{Hz} \times 167\ \Omega}$$

$\qquad\qquad = 9.5\ \mu\text{F}$ (use 10 μF standard value)

$$Z_i = R_1 \| R_2 = 4.7\ \text{M}\Omega \| 1\ \text{M}\Omega$$

$\qquad\qquad = 825\ \text{k}\Omega$

From Eq. 12-4, $\qquad C_1 = \dfrac{1}{2\pi f_1 (Z_i + r_s)/10}$

$$= \dfrac{1}{2\pi \times 100\ \text{Hz} \times (825\ \text{k}\Omega + 600\ \Omega)/10}$$

$\qquad\qquad = 0.019\ \mu\text{F}$ (use 0.02 μF standard value)

From Eq. 12-5,

$$C_3 = \frac{1}{2\pi f_1(R_D + R_L)/10}$$

$$= \frac{1}{2\pi \times 100 \text{ Hz} \times (6.8 \text{ k}\Omega + 120 \text{ k}\Omega)/10}$$

$$= 0.13 \text{ }\mu\text{F (use 0.15 }\mu\text{F standard value)}$$

Practice Problems

12-2.1 A common-source amplifier (see Fig. 12-13) is to use a 2N5458 FET and a 25 V supply. The external load is 100 kΩ. Determine suitable resistor values for the circuit.

12-2.2 Determine suitable capacitor values for the circuit in Problem 12-2.1 to give a 30 Hz lower cutoff frequency.

12-3 CAPACITOR-COUPLED TWO-STAGE CE AMPLIFIER

Circuit

A capacitor-coupled two-stage amplifier circuit is shown in Fig. 12-15. Each stage is similar to the single-stage circuit in Fig. 12-9. Stage 1 output is capacitor-coupled (via C_3) to the input of stage 2. The signal is applied to the input of stage 1, and the load is coupled to the output of stage 2. The signal is amplified by stage 1, and the output of stage 1 is amplified by stage 2, so that the overall voltage gain is much greater than the gain of a single stage. Two amplifier stages used in this way are referred to as *cascaded*. As illustrated by the circuit waveforms in Fig. 12-16, the signal voltage is phase-shifted through 180° by stage 1, and through a further 180° by stage 2. Consequently, the overall phase shift from input to output is zero (or 360°).

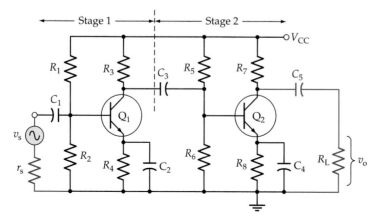

Figure 12-15 A two-stage capacitor-coupled CE amplifier consisting of two similar CE circuits with capacitor coupling between stages.

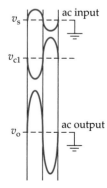

Figure 12-16 Voltage waveforms for a two-stage CE circuit.

Circuit Design

The simplest approach to design of this circuit is to make each stage identical. Then, when stage 2 has been designed, the components for stage 1 are chosen as $R_1 = R_5$, $R_2 = R_6$, $R_3 = R_7$, $R_4 = R_8$, $C_1 = C_3$, and $C_2 = C_4$. Note that C_3 is calculated in terms of the input resistance to stage 2, which should be identical to that for stage 1 if the stages are otherwise identical. C_5 is calculated in the usual way for an output coupling capacitor.

One way the design of a two-stage circuit differs from that of a single-stage circuit is in the determination of emitter bypass capacitors C_2 and C_4. Consider Equation 12-3, which selects the lower cutoff frequency (f_1) for a single stage:

$$X_C = h_{ib} = r_e'$$

If this equation is used to calculate the capacitance of C_2 and C_4 in the two-stage circuit, it is found that the gain of stage 1 is down by 3 dB at f_1, and the gain of stage 2 is also down by 3 dB at f_1 (see Fig. 12-17). This means that the overall gain of the circuit is down by a total of 6 dB at frequency f_1. For a 3 dB decrease in overall voltage gain at f_1, the bypass capacitors must be calculated to give a 1.5 dB reduction in the gain of each stage. A 1.5 dB reduction in stage gain at f_1 is achieved by making the reactance of the emitter bypass capacitor 0.65 times the impedance in series with the capacitor. Thus, the equation for the bypass capacitor impedance now becomes

$$X_{C2} = X_{C4} = 0.65\, r_e' \text{ at } f_1 \qquad (12\text{-}13)$$

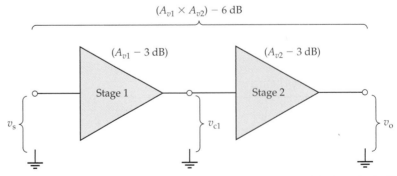

Figure 12-17   If the voltage gain of each stage of a two-stage circuit is down by 3 dB at f_1, the overall voltage gain is down by 6 dB.

Example 12-6

Design a two-stage capacitor-coupled amplifier, as shown in Fig. 12-15, to fulfil the specification used for Examples 12-2 and 12-3: 2N3904 transistors, $V_{CC} = 24$ V, $R_L = 120$ kΩ, $f_1 = 100$ Hz.

Solution

Except for the emitter bypass capacitors, each stage is designed exactly as in Examples 12-2 and 12-3. Therefore, $R_3 = R_7 = 12$ kΩ, $R_4 = R_8 = 3.9$ kΩ, $R_1 = R_5 = 120$ kΩ, $R_2 = R_6 = 39$ kΩ, $C_1 = C_3 = 6.8$ µF, $C_5 = 0.12$ µF.

From data sheet A-5 in Appendix A and Example 12-3, we find that $h_{fe} = 100$, and $r_e' = 20$ Ω. See Fig. 12-18.

Eq. 12-13:

$$X_{C4} = 0.65\, r_e' = 0.65 \times 20\ \Omega$$

$$\approx 13\ \Omega$$

$$C_2 = C_4 = \frac{1}{2\pi f_1 X_{C4}} = \frac{1}{2\pi \times 100\ \text{Hz} \times 13\ \Omega}$$

$$= 123\ \mu F\ (\text{use } 150\ \mu F \text{ standard value})$$

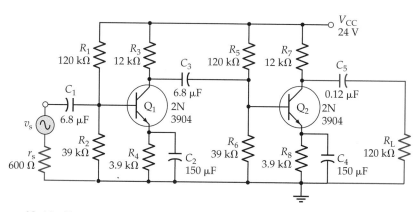

Figure 12-18   Two-stage capacitor-coupled CE circuit designed in Example 12-6.

Circuit Analysis

AC analysis of a two-stage, capacitor-coupled circuit is similar to analysis of single-stage circuits, as covered in Chapter 6. The *h*-parameter equivalent circuit for the amplifier circuit in Fig. 12-15 is drawn in Fig. 12-19. Once again, this equivalent circuit is produced by replacing the supply voltage and all capacitors with short-circuits, and substituting device h-parameter models for each transistor. The resulting equivalent circuit simply consists of two single-stage *h*-parameter equivalent circuits. As always, voltage polarities and current

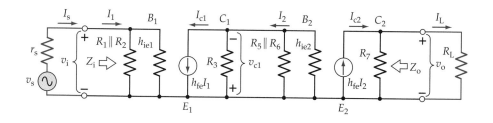

Figure 12-19 The *h*-parameter equivalent circuit for a two-stage CE amplifier is drawn by cascading the equivalent circuits for each stage.

directions are indicated for an instantaneous positive-going signal voltage. The voltage polarities show a $180°$ phase shift between v_i and v_{c1}, and a further $180°$ phase shift between v_{c1} and v_o.

The performance equations for the two-stage circuit are readily derived from the equations for a single-stage circuit. Input and output impedance are exactly the same as for the single-stage circuit:

$$Z_i = R_1 \| R_2 \| h_{ie} \tag{12-14}$$

$$Z_o \approx R_7 \tag{12-15}$$

The voltage gain equation for the second stage is exactly the same as the single-stage gain equation. The capacitor-coupled load at the collector terminal of the first stage is the input impedance of the second stage. So the voltage gain equations are

$$A_{v1} = \frac{-h_{fe1}(R_3 \| Z_{i2})}{h_{ie1}} \tag{12-16}$$

and

$$A_{v2} = \frac{-h_{fe2}(R_7 \| R_L)}{h_{ie2}} \tag{12-17}$$

The overall voltage gain is found by multiplying the two individual stage gains:

$$A_v = A_{v1} \times A_{v2} \tag{12-18}$$

Equations for current gain and power gain can also be derived from the single-stage equations, but these quantities are normally not important.

Example 12-7

Analyze the two-stage amplifier circuit designed in Example 12-6 to determine its input impedance, output impedance, and minimum overall voltage gain.

Solution

Eq. 12-14: $Z_i = R_1 \| R_2 \| h_{ie} = 120 \text{ k}\Omega \| 39 \text{ k}\Omega \| 2 \text{ k}\Omega$

$= 1.87 \text{ k}\Omega$

Eq. 12-15: $Z_o \approx R_7 = 12 \text{ k}\Omega$

From Eq. 12-14: $Z_{i2} = R_5 \| R_6 \| h_{ie2} = 120 \text{ k}\Omega \| 39 \text{ k}\Omega \| 2 \text{ k}\Omega$

$= 1.87 \text{ k}\Omega$

Eq. 12-16: $A_{v1} = \dfrac{-h_{fe1}(R_3 \| Z_{i2})}{h_{ie1}} = \dfrac{-100 \times (12 \text{ k}\Omega \| 1.87 \text{ k}\Omega)}{2 \text{ k}\Omega}$

≈ -81

Eq. 12-17:
$$A_{v2} = \frac{-h_{fe2}(R_7\|Z_L)}{h_{ie2}} = \frac{-100 \times (12\ k\Omega\|120\ k\Omega)}{2\ k\Omega}$$

$$= -545$$

Eq. 12-18:
$$A_v = A_{v1} \times A_{v2} = (-81) \times (-545)$$

$$= 44\ 145$$

Practice Problems

12-3.1 Determine suitable resistor values for a capacitor-coupled two-stage CE amplifier with a 15 V supply and an 82 kΩ external load.

12-3.2 Determine suitable capacitor values for the circuit in Problem 12-3.1. The circuit uses 2N3903 BJTs, the lower cutoff frequency is 70 Hz, and the signal source resistance is 1 kΩ.

12-3.3 Calculate the overall voltage gain for the circuit in Problems 12-3.1 and 12-3.2.

12-4 DIRECT-COUPLED TWO-STAGE CIRCUITS

Direct-Coupled Circuit

For economy, the number of components used in any circuit should be kept to a minimum. The use of direct coupling between stages is one way of eliminating components. Figure 12-20 shows a two-stage circuit that has the base of transistor Q_2 directly coupled to the collector of Q_1. Comparing this to the capacitor-coupled circuit in Fig. 12-15, it is seen that the two bias resistors for stage 2 and the interstage coupling capacitor have been eliminated. This is a saving of only three components; however, the total savings can be considerable when many similar circuits are to be manufactured.

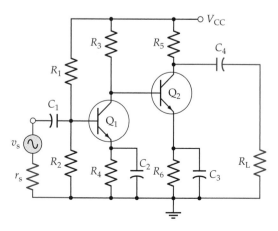

Figure 12-20 Two-stage direct-coupled CE amplifier. Because Q_2 base is directly coupled to Q_1 collector, the inter-stage coupling capacitor and the bias resistors for Q_2 are eliminated.

Circuit Design

The first step in the design of the circuit in Fig. 12-20 involves determining a suitable level of base bias voltage for Q_2. This is done by estimating satisfactory levels of V_{E1} and V_{CE1} for transistor Q_1. Then,

$$V_{B2} = V_{C1} = V_{E1} + V_{CE1}$$

$$V_{E2} = V_{B2} - V_{BE}$$

and

$$V_{R5} = V_{CC} - V_{E2} - V_{CE2}$$

As explained in Section 12-1, appropriate dc voltage levels for small-signal amplifiers are $V_{CE} = 3$ V and $V_E = 5$ V, except in the case of a very low supply voltage.

As for all amplifier circuits, the resistor at the collector of the output transistor (R_5 in Fig. 12-20) should be much smaller than the external load resistor. Once R_5 is selected, I_{C2} can be calculated from the voltage V_{R5} already determined. If I_{C2} looks too small for satisfactory operation of the transistor, a suitable current level should be selected and a new value of R_5 calculated. The collector current of Q_1 is determined by making I_{C1} very much greater than the base current for Q_2. This is done to ensure that I_{B2} has a negligible effect on the bias conditions of Q_1. Normally, just making I_{C1} equal to I_{C2} is the simplest way to achieve the desired effect.

Example 12-8

Calculate suitable resistor values for a two-stage, direct-coupled amplifier, as in Fig. 12-20. The circuit is to use 2N3903 transistors, a 14 V supply, a 40 kΩ load resistance, and a signal source with a 600 Ω resistance.

Solution

Select

$$V_{E1} = 5 \text{ V}, V_{CE1} = V_{CE2} = 3 \text{ V}$$

$$V_{B2} = V_{C1} = V_{E1} + V_{CE1} = 5 \text{ V} + 3 \text{ V}$$

$$= 8 \text{ V}$$

$$V_{E2} = V_{B2} - V_{BE} = 8 \text{ V} - 0.7 \text{ V}$$

$$= 7.3 \text{ V}$$

$$V_{R5} = V_{CC} - V_{E2} - V_{CE2} = 14 \text{ V} - 7.3 \text{ V} - 3 \text{ V}$$

$$= 3.7 \text{ V}$$

Select

$$R_5 = \frac{R_L}{10} = \frac{40 \text{ k}\Omega}{10}$$

$$= 4 \text{ k}\Omega \text{ (use 3.9 k}\Omega \text{ standard value)}$$

$$I_{C2} = \frac{V_{R5}}{R_5} = \frac{3.7 \text{ V}}{3.9 \text{ k}\Omega}$$

$$= 949 \text{ } \mu\text{A}$$

$$R_6 = \frac{V_{E2}}{I_{C2}} = \frac{7.3 \text{ V}}{949 \text{ μA}}$$

$$= 7.7 \text{ kΩ (use 8.2 kΩ and recalculate } I_{C2})$$

$$I_{C2} = \frac{V_{E2}}{R_6} = \frac{7.3 \text{ V}}{8.2 \text{ kΩ}}$$

$$= 890 \text{ μA}$$

$I_{C1} \gg I_{B2}$, select $\quad I_{C1} = 1 \text{ mA}$

$$V_{R3} = V_{CC} - V_{C1} = 14 \text{ V} - 8 \text{ V}$$

$$= 6 \text{ V}$$

$$R_3 = \frac{V_{R3}}{I_{C1}} = \frac{6 \text{ V}}{1 \text{ mA}}$$

$$= 6 \text{ kΩ (use 5.6 kΩ and recalculate } I_{C1}$$
$$\text{in order to keep } V_{B2} \approx 8 \text{ V)}$$

$$I_{C1} = \frac{V_{R3}}{R_3} = \frac{6 \text{ V}}{5.6 \text{ kΩ}}$$

$$= 1.07 \text{ mA}$$

$$R_4 = \frac{V_{E1}}{I_{C1}} = \frac{5 \text{ V}}{1.07 \text{ mA}}$$

$$\approx 4.7 \text{ kΩ (standard value)}$$

$$V_{R2} = V_{B1} = V_{E1} + V_{BE} = 5 \text{ V} + 0.7 \text{ V}$$

$$= 5.7 \text{ V}$$

$$V_{R1} = V_{CC} - V_{B1} = 14 \text{ V} - 5.7 \text{ V}$$

$$= 8.3 \text{ V}$$

$$R_2 = 10 \, R_4 = 10 \times 4.7 \text{ kΩ}$$

$$= 47 \text{ kΩ}$$

$$I_2 = \frac{V_{B1}}{R_2} = \frac{5.7 \text{ V}}{47 \text{ kΩ}}$$

$$= 121 \text{ μA}$$

Eq. 12-10: $\qquad R_1 = \dfrac{V_{R1} \times R_2}{V_{R2}} = \dfrac{8.3 \text{ V} \times 47 \text{ kΩ}}{5.7 \text{ V}}$

$$= 68.4 \text{ kΩ (use 68 kΩ standard value)}$$

Example 12-9

Determine suitable capacitor values for the two-stage direct-coupled amplifier in Example 12-8. The lower cutoff frequency is to be 75 Hz. Assume $r_e' = 26\ \Omega$.

Solution

From data sheet A-5 in Appendix A for the 2N3903,

$$h_{fe(min)} = 50$$

$$h_{ie} = (1 + h_{fe})r_e' = 51 \times 26\ \Omega$$

$$\approx 1.3\ k\Omega$$

$$Z_i = R_1 \| R_2 \| h_{ie} = 68\ k\Omega \| 47\ k\Omega \| 1.3\ k\Omega$$

$$= 1.24\ k\Omega$$

Eq. 12-4:
$$X_{C1} = \frac{Z_i + r_s}{10} = \frac{1.24\ k\Omega\ +\ 600\ \Omega}{10}$$

$$= 184\ \Omega$$

$$C_1 = \frac{1}{2\pi f_1 X_{C1}} = \frac{1}{2\pi \times 75\ Hz \times 184\ \Omega}$$

$$= 11.5\ \mu F\ (\text{use } 15\ \mu F \text{ standard value})$$

Eq. 12-13:
$$X_{C2} = X_{C3} = 0.65\ r_e' = 0.65 \times 26\ \Omega$$

$$\approx 16.9\ \Omega$$

$$C_2 = C_3 = \frac{1}{2\pi f_1 X_{C2}} = \frac{1}{2\pi \times 75\ Hz \times 16.9\ \Omega}$$

$$= 125\ \mu F\ (\text{use } 150\ \mu F \text{ standard value})$$

From Eq. 12-5,
$$X_{C4} = \frac{R_5 + R_L}{10} = \frac{3.9\ k\Omega\ +\ 40\ k\Omega}{10}$$

$$= 4.39\ k\Omega$$

$$C_4 = \frac{1}{2\pi f_1 X_{C4}} = \frac{1}{2\pi \times 75\ Hz \times 4.39\ k\Omega}$$

$$= 0.48\ \mu F\ (\text{use } 0.47\ \mu F \text{ standard value})$$

Circuit Analysis

Analysis procedure for a two-stage direct-coupled circuit is similar to that for analysis of a two-stage capacitor-coupled circuit. The *h*-parameter equivalent for the circuit of Fig. 12-20 is exactly like that in Fig. 12-19, except that the bias resistors for the second stage are omitted. The voltage gain and impedance equations are as determined for the capacitor-coupled circuit. Because the component numbers differ slightly for the two circuits, care must be taken in substituting components into the equations. For example, from Equation 12-17, the voltage gain of the second stage of the circuit in Fig. 12-20 is

$$A_{v2} = \frac{-h_{fe2}(R_5 \| R_L)}{h_{ie2}}$$

Example 12-10

Calculate the minimum overall voltage gain for the circuit in Fig. 12-21.

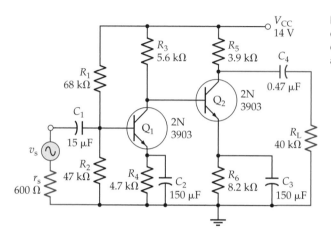

Figure 12-21   Two-stage direct-coupled CE amplifer designed in Examples 12-8 and 12-9.

Solution

Eq. 12-16:

$$A_{v1} = \frac{-h_{fe1}(R_3 \| Z_{i2})}{h_{ie1}} = \frac{-h_{fe1}(R_3 \| h_{ie2})}{h_{ie1}}$$

$$= \frac{-50 \times (5.6\ \text{k}\Omega \| 1.3\ \text{k}\Omega)}{1.3\ \text{k}\Omega}$$

$$\approx -40.6$$

Eq. 12-17:

$$A_{v2} = \frac{-h_{fe2}(R_5 \| R_L)}{h_{ie2}} = \frac{-50 \times (3.9\ \text{k}\Omega \| 40\ \text{k}\Omega)}{1.3\ \text{k}\Omega}$$

$$= -137$$

Eq. 12-18:

$$A_v = A_{v1} \times A_{v2} = (-40.6) \times (-137)$$

$$= 5562$$

Use of Complementary Transistors

The direct-coupled, two-stage circuit illustrated in Fig. 12-22 is similar to that in Fig. 12-20, except that transistor Q_2 is a *pnp* device. The transistors are selected to have similar characteristics and parameters although one is *npn* and the other is *pnp*. This means that they are *complementary transistors*. Examination of Appendices A-5 and A-6 reveals that the 2N3903 and the 2N3905 are complementary, as are the 2N3904 and the 2N3906.

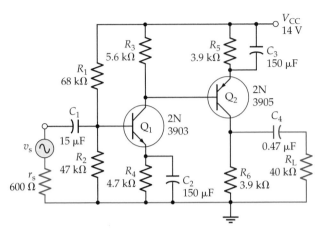

Figure 12-22 Direct-coupled two-stage CE amplifier using complementary transistors. Q_1 and Q_2 have similar parameters, although one is an *npn* transistor and the other is a *pnp* device.

Suppose the circuit in Fig. 12-22 is to be designed to use a 14 V supply, as in Example 12-8. Allowing $V_{E1} = 5$ V and $V_{CE1} = 3$ V, the base voltage for Q_2 (measured from $+V_{CC}$) is

$$V_{B2} = V_{R3} = 6 \text{ V}$$

and

$$V_{R5} = V_{B2} - V_{BE} = 6 \text{ V} - 0.7 \text{ V}$$

$$= 5.3 \text{ V}$$

Therefore,

$$V_{R6} = V_{CC} - V_{E2} - V_{CE2} = 14 \text{ V} - 5.3 \text{ V} - 3 \text{ V}$$

$$= 5.7 \text{ V}$$

The use of complementary transistors in Fig. 12-22 reduces the voltage drop across the emitter resistor of Q_2 compared to the situation in Fig. 12-20. This makes more voltage available for dropping across the Q_2 collector resistor.

The design procedure for the circuit using complementary transistors is very similar to the procedure for designing the non-complementary circuit in Fig. 12-20.

Practice Problems

12-4.1 A direct-coupled two-stage CE amplifier using complementary transistors (see Fig. 12-22) is to be designed to operate from a 25 V supply and to use a 2N3904 and a 2N3906 transistor. The capacitor-coupled load is 56 kΩ. Determine suitable resistor values.

12-4.2 Determine suitable capacitor values for the circuit in Problem 12-4.1 to give a 30 Hz lower cutoff frequency. Assume that $r_s = 300$ Ω.

12-4.3 Calculate the minimum overall voltage gain for the circuit in Problems 12-4.1 and 12-4.2.

12-5 TWO-STAGE CIRCUIT
WITH EMITTER FOLLOWER OUTPUT

Circuit

Figure 12-23 shows another direct-coupled circuit. This time the second stage is a common collector circuit, or emitter follower. Stage 2 gives the circuit a very low output impedance but has unity voltage gain. Stage 1 still has substantial voltage gain.

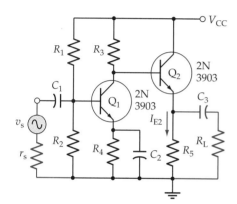

Figure 12-23   Two-stage direct-coupled CE amplifier consisting of a CE input stage and a CC (emitter follower) output stage for low output impedance.

The design procedure for this circuit is similar to the procedures already discussed. As with the circuit in Fig. 12-20, the Q_2 base bias voltage is the collector voltage of Q_1. A suitable level of emitter current I_{E2} is selected, and R_E is calculated to be $(V_{B2} - V_{BE})/I_{E2}$. Coupling capacitor C_1 is once again calculated from the equation $X_{C1} = Z_i/10$. The load resistor (R_L) may be relatively small, which means that coupling capacitor C_3 must be large. Therefore, both C_2 and C_3 are used to determine the low 3 dB frequency of the circuit by using the 0.65 factor employed in Equation 12-14.

It is important to realize that the emitter follower in Fig. 12-23 is a small-signal circuit. In fact, this type of circuit functions best when the amplitude of the ac output voltage is much smaller than the dc base-emitter voltage of the output transistor (V_{BE}). With ac output voltages that approach or exceed V_{B2}, the base-emitter junction of Q_2 may become reverse-biased when the output voltage moves rapidly in a negative direction. In this case, the circuit will not function correctly. Where large output voltages and low output impedance are required, a complementary emitter follower is used (see Section 19-4).

With the second stage operating as a small-signal emitter follower, resistor R_5 in Fig. 12-23 does *not* have to be very much smaller than external load R_L. However, for satisfactory operation, Q_2 emitter current (I_{E2}) should be greater than the peak ac load current (i_p). The peak output current is calculated by dividing the desired peak output voltage (v_p) by the load resistance: $i_p = v_p/R_L$.

Example 12-11

Determine suitable resistor values for the two-stage, amplifier circuit in Fig. 12-23. The circuit is to use 2N3903 transistors, the supply voltage is 20 V, the load resistance is 100 Ω, and the output voltage is to be ±100 mV.

Solution

$$i_p = \frac{V_P}{R_L} = \frac{100 \text{ mV}}{100 \text{ }\Omega}$$

$$= 1 \text{ mA}$$

$I_{E2} > i_p$, select $I_{E2} = 2 \text{ mA}$

Select $V_{E1} = 5 \text{ V}$ and $V_{CE1} = 3 \text{ V}$

$$V_{B2} = V_{C1} = V_{E1} + V_{CE1} = 5 \text{ V} + 3 \text{ V}$$

$$= 8 \text{ V}$$

$$V_{E2} = V_{B2} - V_{BE} = 8 \text{ V} - 0.7 \text{ V}$$

$$= 7.3 \text{ V}$$

$$R_5 = \frac{V_{E2}}{I_{E2}} = \frac{7.3 \text{ V}}{2 \text{ mA}}$$

$$= 3.65 \text{ k}\Omega \text{ (use 3.3 k}\Omega \text{ standard value)}$$

$I_{C1} \gg I_{B2}$, select $I_{C1} = 1 \text{ mA}$

$$V_{R3} = V_{CC} - V_{B2} = 20 \text{ V} - 8 \text{ V}$$

$$= 12 \text{ V}$$

$$R_3 = \frac{V_{R3}}{I_{C1}} = \frac{12 \text{ V}}{1 \text{ mA}}$$

$$= 12 \text{ k}\Omega \text{ (standard value)}$$

$$R_4 = \frac{V_{E1}}{I_{C1}} = \frac{5 \text{ V}}{1 \text{ mA}}$$

$$= 5 \text{ k}\Omega \text{ (use 4.7 k}\Omega \text{ standard value)}$$

$$V_{B1} = (I_{C1} \times R_4) + V_{BE} = (1 \text{ mA} \times 4.7 \text{ k}\Omega) + 0.7 \text{ V}$$

$$= 5.4 \text{ V}$$

Select $R_2 = 10 \, R_4 = 10 \times 4.7 \text{ k}\Omega$

$$= 47 \text{ k}\Omega$$

Eq. 12-10: $$R_1 = \frac{V_{R1} \times R_2}{V_{R2}} = \frac{(20 \text{ V} - 5.4 \text{ V}) \times 47 \text{ k}\Omega}{5.4 \text{ V}}$$

$$= 127 \text{ k}\Omega \text{ (use 120 k}\Omega \text{ standard value)}$$

Example 12-12

Calculate suitable capacitor values for the circuit designed in Example 12-11. The signal source resistance is 600 Ω, and the low 3-dB frequency is 150 Hz.

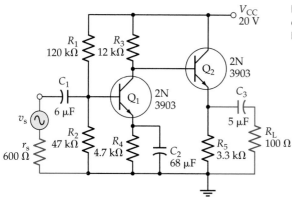

Figure 12-24   Two-stage direct-coupled amplifier designed in Examples 12-11 and 12-12.

Solution

See Fig. 12-24.

Eq. 6-2:
$$r'_{e1} = \frac{26 \text{ mV}}{I_{E1}} = \frac{26 \text{ mV}}{1 \text{ mA}}$$

$$= 26 \text{ }\Omega$$

$$h_{ie1} = (1 + h_{fe}) \, r'_{e1} = (1 + 50) \times 26 \text{ }\Omega$$

$$\approx 1.3 \text{ k}\Omega$$

$$Z_i = R_1 \| R_2 \| h_{ie1} = 120 \text{ k}\Omega \| 47 \text{ k}\Omega \| 1.3 \text{ k}\Omega$$

$$= 1.25 \text{ k}\Omega$$

$$X_{C1} = \frac{Z_i + r_s}{10} = \frac{1.25 \text{ k}\Omega + 600 \text{ }\Omega}{10}$$

$$= 185 \text{ }\Omega$$

$$C_1 = \frac{1}{2\pi f_1 X_{C1}} = \frac{1}{2\pi \times 150 \times 185 \text{ }\Omega}$$

$$= 5.8 \text{ }\mu\text{F (use 6 }\mu\text{F standard value)}$$

$$X_{C2} = 0.65 \, r'_{e1} \text{ at } f_1 = 0.65 \times 26 \text{ }\Omega$$

$$= 16.9 \text{ }\Omega$$

$$C_2 = \frac{1}{2\pi f_1 X_{C2}} = \frac{1}{2\pi \times 150 \times 16.9 \text{ }\Omega}$$

$$= 63 \text{ }\mu\text{F (use 68 }\mu\text{F standard value)}$$

$$r'_{e2} = \frac{26 \text{ mV}}{I_{E2}} = \frac{26 \text{ mV}}{2 \text{ mA}}$$

$$= 13 \text{ }\Omega$$

$$Z_o = R_5 \| (r'_{e2} + R_3/h_{fe2}) = 3.3\ k\Omega \| (13\ \Omega + 12\ k\Omega/50)$$

$$= 231\ \Omega$$

$$X_{C3} \approx 0.65\ (R_L + Z_o) = 0.65\ (100\ \Omega + 231\ \Omega)$$

$$= 215\ \Omega$$

$$C_3 = \frac{1}{2\pi f_1 X_{C3}} = \frac{1}{2\pi \times 150\ Hz \times 215\ \Omega}$$

$$= 4.85\ \mu F\ (\text{use } 5\ \mu F \text{ standard value})$$

Example 12-13

Analyze the two-stage transistor amplifier circuit in Fig. 12-24 to determine the minimum overall voltage gain.

Solution

$$r'_{e2} = \frac{26\ mV}{I_{E2}} = \frac{26\ mV}{2\ mA}$$

$$= 13\ \Omega$$

$$h_{ie2} = h_{ic2} = h_{fe} \times r'_{e2}$$

$$= 50 \times 13\ \Omega$$

$$= 650\ \Omega$$

From Eq. 6-23, $\quad Z_{i2} = h_{ic2} + h_{fc2}(R_5\|R_L) = 650\ \Omega + 51\ (3.3\ k\Omega\|100\ \Omega)$

$$= 5.6\ k\Omega$$

Eq. 12-16: $\quad A_{v1} = \dfrac{-h_{fe1}(R_3\|Z_{i2})}{h_{ie1}} = \dfrac{-50 \times (12\ k\Omega\|5.6\ k\Omega)}{1.3\ k\Omega}$

$$= -147$$

From Eq. 6-27, $\quad A_{v2} \approx 1$

From Eq. 12-18, $\quad A_v = A_{v1} \times A_{v2} = -147 \times 1$

$$= -147$$

Practice Problems

12-5.1 A two-stage direct-coupled BJT amplifier with an emitter-follower output is to be designed to operate from a 15 V supply. The capacitor-coupled load is 60 Ω. Using 2N3906 (*pnp*) transistors, determine suitable resistor values for the circuit.

12-5.2 Determine suitable capacitor values for the circuit in Problem 12-5.1 to give a 50 Hz lower cutoff frequency. Calculate the circuit minimum overall voltage gain. Assume that $r_s = 350\ \Omega$.

12-6 DC FEEDBACK PAIR

DC Feedback Pair with Two Amplification Stages

The two-stage amplifier circuit shown in Fig. 12-25a is known as a *dc feedback pair*. The base of transistor Q_2 is directly connected to the collector of Q_1, and the base of Q_1 is biased from the emitter of Q_2 via resistor R_2. Comparing this circuit to the capacitor-coupled circuit of Fig. 12-15 and the direct-coupled circuit of Fig. 12-20, it is seen that a further saving in components has been effected: R_1, R_2, R_4, and C_2 in Fig. 12-20 are now replaced with the single resistor R_2 in Fig. 12-25.

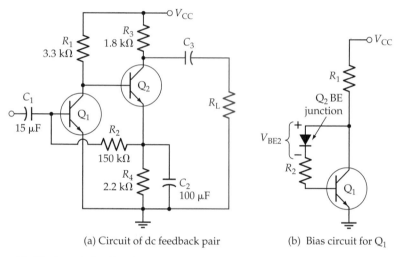

(a) Circuit of dc feedback pair   (b) Bias circuit for Q_1

Figure 12-25   In a dc feedback pair, the base of Q_2 is directly connected to the collector of Q_1, and the base of Q_1 is biased via R_2 from the emitter of Q_2. This arrangement uses fewer components than a capacitor-coupled two-stage circuit.

To understand the bias arrangement for the dc feedback pair, note that there is only a *pn* junction voltage drop (V_{BE2}) between the Q_1 collector and the Q_2 emitter. This is illustrated in Fig. 12-25b, where it is seen that the first stage of the circuit is essentially a collector-to-base bias circuit. Stage 2 bias arrangement is similar to voltage-divider bias, with resistor R_4 stabilizing the transistor emitter current. The base voltage for Q_2 is derived from the collector of Q_1; consequently, the bias stability of the second stage is no better than that of the first stage.

Because collector-to-base bias is not as stable as voltage-divider bias, the bias conditions in this circuit are not as stable as in the circuit of Fig. 12-20. However, the stability is adequate for most purposes, and the component savings can be a major advantage.

The dc feedback pair derives its name from the direct coupling between stages and from the feedback from the collector of Q_1 to its base. As explained for collector-to-base bias in Section 5-3, a change in the level of Q_1 collector voltage (V_{C1}) produces a change in base current (I_{B1}). The I_{B1} change in turn alters I_{C1}, and the voltage drop $I_{C1}R_1$ tends to push V_{C1} back toward its original level.

As well as stabilizing the circuit dc conditions, the feedback can produce ac degeneration (see Section 6-1). Capacitor C_2 shorts the ac feedback to ground to eliminate ac degeneration from stage 1. As the emitter bypass capacitor, C_2 also eliminates ac degeneration from stage 2.

The design approach to the dc feedback pair is similar to those already discussed. The output stage is designed exactly like the output stage for the direct-coupled circuit in Fig. 12-20. The input stage is designed as a collector-to-base bias circuit to have a collector voltage (V_{C1}) equal to the desired level of the stage 2 base voltage (V_{B2}).

The input impedance of the circuit is

$$Z_i = R_2 \| h_{ie} \tag{12-19}$$

So (from Eq. 12-4) the impedance of C_1 at f_1 is

$$X_{C1} = \frac{(R_2 \| h_{ie}) + r_s}{10}$$

Since C_2 is by far the largest capacitor in the circuit, it determines the circuit low 3 dB frequency:

From Eq. 12-3, $\qquad\qquad X_{C2} = \dfrac{h_{ie}}{1 + h_{fe}}$ at f_1

C_3 is once again calculated from Eq. 12-5:

$$X_{C3} = \frac{Z_o + R_L}{10}$$

DC Feedback Pair with Emitter Follower Output

Figure 12-26 shows a dc feedback pair that has an emitter follower output stage. In this case, a bypass capacitor cannot be used at the emitter of Q_2; otherwise the output would be ac-shorted to ground. So now another method of removing the ac feedback must be employed. In Fig. 12-26, the base bias resistor for Q_1 is split into two resistors and the junction of these two is ac-grounded via capacitor C_2.

The design of this circuit differs from other designs only in the calculation of the capacitor values. In this case, X_{C2} should be very much smaller than R_3 at f_1. This is because the ac output voltage at the emitter of Q_2 is divided across R_3 and X_{C2} before being fed back to the Q_1 base. Capacitor values normally depend on the resistance in series with them, and in Fig. 12-26 R_3 is in series

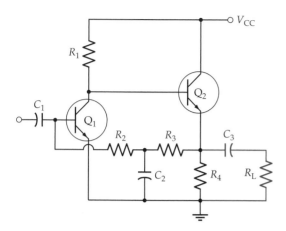

Figure 12-26 DC feedback pair with an emitter follower output stage. In this case, two series-connected feedback resistors (R_2 and R_3) are used, and the junction of the two is grounded via C_2 to eliminate ac degeneration.

with C_2 for ac feedback. Because this capacitor is in the feedback path from output to input, it is calculated at f_1 from the equation

$$X_{C2} = \frac{R_3}{100} \qquad (12\text{-}20)$$

The reasons for using a factor of 100, instead of the usual factor of 10, will be understood when negative feedback is studied (see Chapter 13).

Normally, the largest capacitor in any circuit is employed to determine the circuit low 3 dB frequency (f_1). Where two capacitors are likely to be equally large, they are both used to set f_1. The largest capacitor is always the one in series with the lowest resistance. In the circuit in Fig. 12-26, the external load resistance is likely to be the smallest resistance in series with a capacitor. Consequently, C_3 is calculated from

$$X_{C3} = (Z_o + R_L) \text{ at } f_1 \qquad (12\text{-}21)$$

Example 12-14

Design a dc feedback pair with an emitter follower output, as shown in Fig. 12-26. The circuit is to use 2N3904 transistors with a 12 V supply. The load resistance is 50 Ω, the lower cutoff frequency is 150 Hz, and the output voltage is to be ±50 mV. Assume that $r_s \ll Z_{in}$.

Solution

$$i_p = \frac{v_p}{R_L} = \frac{50 \text{ mV}}{50 \ \Omega}$$

$$= 1 \text{ mA}$$

$$I_{E2} > i_p, \text{ select} \quad I_{E2} = 2 \text{ mA}$$

Select $\qquad\qquad\qquad V_{E2} = 5 \text{ V}$

$$R_4 = \frac{V_{E2}}{I_{E2}} = \frac{5\ V}{2\ mA}$$

$$= 2.5\ k\Omega\ \text{(use 2.2 k}\Omega\ \text{standard value)}$$

$$I_{E2} = \frac{5\ V}{2.2\ k\Omega} = 2.27\ mA$$

$I_{C1} \gg I_{B2}$, select $I_{C1} = 1\ mA$

$$V_{R1} = V_{CC} - (V_{BE2} + V_{E2}) = 12\ V - (0.7\ V + 5\ V)$$

$$= 6.3\ V$$

$$R_1 = \frac{V_{R1}}{I_{C1}} = \frac{6.3\ V}{1\ mA}$$

$$= 6.3\ k\Omega\ \text{(use 5.6 k}\Omega\ \text{and recalculate } I_{C1})$$

$$I_{C1} = \frac{6.3\ V}{5.6\ k\Omega} = 1.13\ mA$$

From data sheet A-5 in Appendix A for the 2N3904: $h_{FE(min)} = 70$, $h_{fe} = 100$

Therefore $$I_{B1} = \frac{I_{C1}}{h_{FE}} = \frac{1.13\ mA}{70}$$

$$= 16.1\ \mu A$$

$$R_2 + R_3 = \frac{V_{E2} - V_{BE}}{I_{B1}} = \frac{5\ V - 0.7\ V}{16.1\ \mu A}$$

$$= 267\ k\Omega$$

Select R_2 equal to 120 kΩ and R_3 equal to 150 kΩ (see Fig. 12-27)

$$h_{ie1} = h_{fe} \times r'_{e1} = 100 \times \frac{26\ mV}{1.13\ mA}$$

$$= 2.3\ k\Omega$$

$$h_{ie2} = h_{fe} \times r'_{e2} = 100 \times \frac{26\ mV}{2.27\ mA}$$

$$= 1.14\ k\Omega$$

Eq. 12-19: $$Z_{i1} = R_2 \| h_{ie1} = 120\ k\Omega \| 2.3\ k\Omega$$

$$= 2.26\ k\Omega$$

$$X_{C1} = \frac{Z_i + r_s}{10} = \frac{2.26\ k\Omega}{10}$$

$$= 226\ \Omega$$

$$C_1 = \frac{1}{2\pi f_1 X_{C1}} = \frac{1}{2\pi \times 150\ Hz}$$

$$= 4.7\ \mu F\ \text{(standard value)}$$

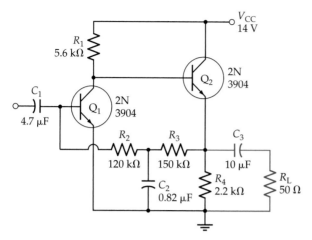

Figure 12-27   Two-stage dc feedback pair designed in Example 12-14.

Eq. 12-20: $X_{C2} = \dfrac{R_3}{100} = \dfrac{150 \text{ k}\Omega}{100}$

$= 1.5 \text{ k}\Omega$

$C_2 = \dfrac{1}{2\pi f_1 X_{C2}} = \dfrac{1}{2\pi \times 150 \text{ Hz} \times 1.5 \text{ k}\Omega}$

$= 0.71 \text{ }\mu\text{F (use 0.82 }\mu\text{F standard value)}$

$Z_o = \left[\dfrac{h_{ie2} + R_1}{h_{fe}}\right] \| R_4 = \left[\dfrac{1.14 \text{ k}\Omega + 5.6 \text{ k}\Omega}{100}\right] \| 2.2 \text{ k}\Omega$

$= 65 \text{ }\Omega$

Eq. 12-21: $X_{C3} = R_L + Z_o = 50 \text{ }\Omega + 65 \text{ }\Omega$

$= 115 \text{ }\Omega$

$C_3 = \dfrac{1}{2\pi f_1 X_{C3}} = \dfrac{1}{2\pi \times 150 \text{ Hz} \times 115 \text{ }\Omega}$

$= 9.2 \text{ }\mu\text{F (use 10 }\mu\text{F standard value)}$

Practice Problems

12-6.1 A dc feedback pair as in Fig. 12-25 is to use $V_{CC} = 15$ V, $R_L = 47$ kΩ, and 2N3903 transistors. Determine suitable resistor values.

12-6.2 The circuit in Problem 12-6.1 is to have $f_1 = 75$ Hz. Determine suitable capacitor values, and calculate the circuit minimum overall voltage gain. Assume that $r_s = 350$ Ω.

12-7 BIFET CIRCUITS

BJT-FET Considerations

Two-stage BJT circuits usually have relatively low input impedances. To increase Z_i, a field effect transistor may be used as the first stage. Circuits that are composed of BJTs and FETs are termed BIFET circuits. Since a FET circuit normally has much lower voltage gain than a BJT circuit, a BIFET circuit can be expected to have lower overall voltage gain than a two-stage BJT circuit. Two-stage FET circuits can be designed and constructed. However, the voltage gain of a two-stage FET circuit is substantially lower than that of a BIFET circuit, and there are no advantages to offset the lower voltage gain.

Capacitor-Coupled BIFET Amplifier

A two-stage, capacitor-coupled BIFET amplifier circuit is shown in Fig. 12-28. The input stage provides a very high input resistance and a voltage gain usually ranging from 5 to 15. Stage 2 has the typical bipolar voltage gain of 200 to 500, depending upon transistor parameters and resistor values. Each stage is designed independently, as explained in Sections 12-1 and 12-2. The two stages of the circuit shown in Fig. 12-28 are those designed in Examples 12-2 to 12-5. In this case, C_3 in Fig. 12-14 is omitted, and C_3 in Fig. 12-28 is the interstage coupling capacitor.

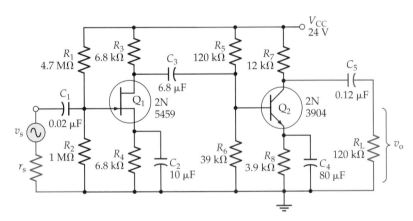

Figure 12-28 Capacitor-coupled two-stage BIFET amplifier. The FET input stage provides a high input impedance, and the BJT produces substantial voltage gain.

Direct-Coupled BIFET Amplifier

In the direct-coupled BIFET circuit in Fig. 12-29a, transistor Q_1 is an *n*-channel JFET and Q_2 is a *pnp* bipolar transistor. An *npn* BJT could be used in the circuit, as illustrated in Fig. 12-29b; however, this arrangement uses a large voltage drop across Q_2 emitter resistor (R_6), leaving a smaller voltage drop across the collector resistor (R_5). Q_2 can also be driven into saturation if the drain current

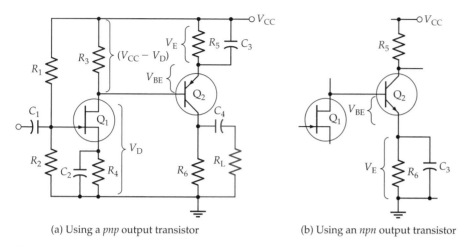

(a) Using a *pnp* output transistor   (b) Using an *npn* output transistor

Figure 12-29 Direct-coupled two-stage BIFET amplifier. When Q_1 is an *n*-channel FET, it is preferable that Q_2 be a *pnp* BJT.

of Q_1 is lower than the design level. For the circuit in Fig. 12-29a, V_{CE2} is increased when I_{D1} is less than the $I_{D(max)}$ design level.

Design Calculations

Example 12-15

Determine suitable resistors for the BIFET amplifier in Fig. 12-29a. The circuit is to use a 2N3906 BJT and a 2N5486 FET. The supply voltage is 20 V, the external load resistance is 80 kΩ, the circuit input impedance is to be approximately 500 kΩ, and the low 3 dB frequency is to be 150 Hz.

Solution

The 2N5486 has:

$$V_{GS(off)(max)} = -6 \text{ V}, I_{DSS(max)} = 20 \text{ mA, and}$$

$$Y_{fs} = 4000 \text{ μS (min)}, 8000 \text{ μS (max)}$$

Select $I_{D(max)} \approx 2 \text{ mA}$

Eq. 10-8: $V_{GS} = V_{GS(off)(max)}[1 - \sqrt{(I_D/I_{DSS})}]$

$$= -6 \text{ V}[1 - \sqrt{(2 \text{ mA}/20 \text{ mA})}]$$

$$\approx -4.1 \text{ V}$$

Eq. 12-8: $V_{DS(min)} = V_{GS(off)(max)} + 1 \text{ V} - V_{GS} = 6 \text{ V} + 1 \text{ V} - 4.1 \text{ V}$

$$= 2.9 \text{ V}$$

$$V_{R3} + V_{R4} = V_{CC} - V_{DS} = 20 \text{ V} - 2.9 \text{ V}$$

$$= 17.1 \text{ V}$$

$$V_{R3} = V_{R4} = \frac{17.1 \text{ V}}{2}$$

$$\approx 8.5 \text{ V}$$

$$R_3 = R_4 = \frac{V_{R4}}{I_D} = \frac{8.5 \text{ V}}{2 \text{ mA}}$$

$$= 4.25 \text{ k}\Omega \text{ (use 3.9 k}\Omega \text{ standard value,}$$
$$\text{and recalculate } V_{R3} \text{ and } V_{R4})$$

$$V_{R3} = V_{R4} = I_D \times R_4 = 2 \text{ mA} \times 3.9 \text{ k}\Omega$$

$$= 7.8 \text{ V}$$

$$V_{R2} = V_G = V_{R4} - V_{GS} = 7.8 \text{ V} - 4.1 \text{ V}$$

$$= 3.7 \text{ V}$$

$$V_{R1} = V_{CC} - V_{R3} = 20 \text{ V} - 3.7 \text{ V}$$

$$= 16.3 \text{ V}$$

For $Z_i \approx 500 \text{ k}\Omega$, select $R_2 = 560 \text{ k}\Omega$

Eq. 12-10: $$R_1 = \frac{V_{R1} \times R_2}{V_{R2}} = \frac{16 \text{ V} \times 560 \text{ k}\Omega}{3.7 \text{ V}}$$

$$\approx 2.5 \text{ M}\Omega \text{ (use 2.7 M}\Omega \text{ standard value)}$$

Select $$R_6 \approx \frac{R_L}{10} = \frac{80 \text{ k}\Omega}{10}$$

$$= 8 \text{ k}\Omega \text{ (use 8.2 k}\Omega \text{ standard value)}$$

$$V_{R5} = V_{R3} - V_{BE2} = 7.8 \text{ V} - 0.7 \text{ V}$$

$$= 7.1 \text{ V}$$

$$V_{R6} = V_{CC} - V_{R5} - V_{CE(min)} = 20 \text{ V} - 7.1 \text{ V} - 3 \text{ V}$$

$$= 9.9 \text{ V}$$

$$I_{C2} = \frac{V_{R6}}{R_6} = \frac{9.9 \text{ V}}{8.2 \text{ k}\Omega}$$

$$= 1.2 \text{ mA}$$

$$R_5 = \frac{V_{R5}}{I_{C2}} = \frac{7.1 \text{ V}}{1.2 \text{ mA}}$$

$$= 5.9 \text{ k}\Omega \text{ (use 6.8 k}\Omega \text{ standard value)}$$

Example 12-16

Determine suitable capacitor values for the BIFET direct-coupled amplifier in Example 12-15.

Solution

$$Z_i = R_1 \| R_2 = 2.7 \text{ M}\Omega \| 560 \text{ k}\Omega$$

$$= 464 \text{ k}\Omega$$

Eq. 12-4:
$$X_{C1} = \frac{Z_i}{10} = \frac{464 \text{ k}\Omega}{10}$$

$$= 46.4 \text{ k}\Omega$$

$$C_1 = \frac{1}{2\pi f_1 X_{C1}} = \frac{1}{2\pi \times 150 \text{ Hz} \times 46.4 \text{ k}\Omega}$$

$$= 0.02 \text{ } \mu\text{F (standard value)}$$

From Eqs. 12-12 and 12-13,

$$X_{C2} = \frac{0.65}{Y_{\text{fs (max)}}} = \frac{0.65}{8000 \text{ } \mu\text{S}}$$

$$= 81 \text{ } \Omega$$

$$C_2 = \frac{1}{2\pi f_1 X_{C2}} = \frac{1}{2\pi \times 150 \text{ Hz} \times 81 \text{ } \Omega}$$

$$= 11 \text{ } \mu\text{F (use 15 } \mu\text{F standard value)}$$

$$r'_{e2} = \frac{26 \text{ mV}}{I_E} = \frac{26 \text{ mV}}{1.2 \text{ mA}}$$

$$\approx 22 \text{ } \Omega$$

$$X_{C3} = 0.65 \text{ } r'_{e2} \text{ at } f_1$$

$$= 0.65 \times 22 \text{ } \Omega$$

$$= 14 \text{ } \Omega$$

$$C_3 = \frac{1}{2\pi f_1 X_{C3}} = \frac{1}{2\pi \times 150 \text{ Hz} \times 14 \Omega}$$

$$= 76 \text{ } \mu\text{F (use 80 } \mu\text{F standard value)}$$

From Eq. 12-5,
$$X_{C4} = \frac{R_6 + R_L}{10} = \frac{8.2 \text{ k}\Omega + 80 \text{ k}\Omega}{10}$$

$$= 8.8 \text{ k}\Omega$$

$$C_4 = \frac{1}{2\pi f_1 X_{C4}} = \frac{1}{2\pi \times 150 \text{ Hz} \times 8.8 \text{ k}\Omega}$$

$$= 0.12 \text{ } \mu\text{F (standard value)}$$

Example 12-17

Analyze the amplifier circuit in Fig. 12-30 to determine its minimum overall voltage gain.

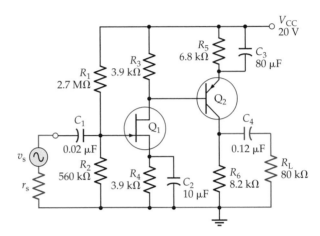

Figure 12-30 BIFET amplifier designed in Examples 12-15 and 12-16.

Solution

For the 2N3906,

$$Z_{i2} = h_{fe} \times r'_{e2} = 100 \times 22 \ \Omega$$

$$= 2.2 \ \text{k}\Omega$$

Eq. 11-7:

$$A_{v1} = -Y_{fs} \left(R_3 \| Z_{i2}\right) = -4000 \ \mu\text{S} \times (3.9 \ \text{k}\Omega \| 2.2 \ \text{k}\Omega)$$

$$= -5.6$$

From Eq. 12-17,

$$A_{v2} = \frac{-h_{fe2}(R_6 \| R_L)}{h_{ie2}} = \frac{-100 \times (8.2 \ \text{k}\Omega \| 80 \ \text{k}\Omega)}{2.2 \ \text{k}\Omega}$$

$$= -338$$

From Eq. 12-18,

$$A_v = A_{v1} \times A_{v2} = (-5.6) \times (-338)$$

$$= 1893$$

Practice Problems

12-7.1 A BIFET circuit, as in Fig. 12-28, is to use the circuits designed in Examples 12-2 to 12-5. Make the necessary design modifications to give a 100 Hz lower cutoff frequency.

12-7.2 Calculate the minimum voltage gain for the circuit in Problem 12-7.1.

12-8 DIFFERENTIAL AMPLIFIER

Differential Amplifier Circuit

The *differential amplifier* is widely applied in integrated circuitry, because it has both good bias stability and good voltage gain without the use of large bypass capacitors. Its use in integrated circuits is studied further in Chapter 14. Differential amplifiers can also be constructed as discrete component circuits.

Figure 12-31a shows that a basic differential amplifier circuit consists of two voltage-divider bias circuits with a single emitter resistor. The circuit is also known as an *emitter-coupled amplifier,* because the transistors are coupled at the emitter terminals. If transistors Q_1 and Q_2 are assumed to be identical in all respects, and if $V_{B1} = V_{B2}$, then the emitter currents are equal and the total emitter current is

$$I_E = I_{E1} + I_{E2} \qquad\qquad (12\text{-}22)$$

Also

$$V_E = V_B - V_{BE}$$

Therefore,

$$I_E = \frac{V_B - V_{BE}}{R_E}$$

and

$$I_{E1} = I_{E2} = \frac{I_E}{2}$$

Like the emitter current in a single-transistor voltage-divider bias circuit, I_E in the differential amplifier remains virtually constant regardless of the transistor h_{FE} value. As a result, I_{E1}, I_{E2}, I_{C1}, and I_{C2} all remain substantially constant, and the constant collector current levels keep V_{C1} and V_{C2} stable. So, the

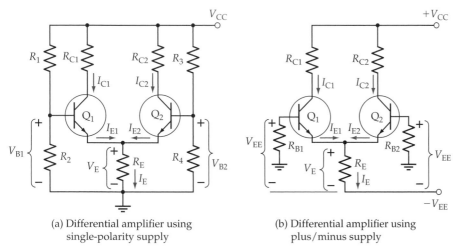

(a) Differential amplifier using single-polarity supply

(b) Differential amplifier using plus/minus supply

Figure 12-31 A differential amplifier has two emitter-coupled transistors with a single emitter resistor. The emitter current tends to split evenly between Q_1 and Q_2.

differential amplifier has the same excellent bias stability as a single-transistor voltage-divider bias circuit.

The circuit of a differential amplifier using a plus/minus supply is shown in Fig. 12-31b. In this case, the voltage across the emitter resistor is $(V_{EE} - V_{BE})$ as illustrated,

and

$$I_{E1} = I_{E2} = \frac{V_{EE} - V_{BE}}{2R_E}$$ (12-23)

The base resistors (R_{B1} and R_{B2}) are included to bias the transistor bases to ground while offering an acceptable input resistance to a signal applied to one of the bases. (See the emitter current bias circuit discussion in Section 5-8.) The transistor emitter currents (I_{E1} and I_{E2}) are exactly equal only if the devices are perfectly matched. To allow for some differences in transistor parameters, a small-value potentiometer (R_{EE}) is sometimes included between the emitters (see Fig. 12-32). Adjustment of R_{EE} increases the resistance in series with the emitter of one transistor and reduces the emitter resistance for the other transistor. This reduces the I_E for one transistor and increases it for the other, while the total emitter current remains constant.

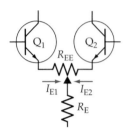

Figure 12-32 A potentiometer connected between the emitters of Q_1 and Q_2 allows I_{E1} and I_{E2} to be equalized.

AC Operation

Consider what happens when the ac input voltage (v_i) at the base of Q_1 is positive-going, as illustrated in Fig. 12-33. Q_1 emitter current (I_{E1}) increases. Also, I_{E2} decreases because the total emitter current ($I_{E1} + I_{E2}$) remains constant. This means that I_{C1} increases and I_{C2} decreases, and consequently, v_{c1} falls and v_{c2} rises, as shown. So, the ac output voltage at Q_1 collector is in antiphase to v_i at Q_1 base, and the output at Q_2 collector is in phase with v_i.

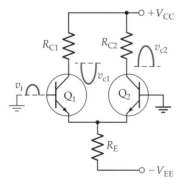

Figure 12-33 A positive-going ac input applied to the base of Q_1 produces a negative-going output at Q_{1C} and a positive-going output at Q_{2C}.

Voltage Gain

The voltage gain of a single-stage amplifier with an unbypassed emitter resistor and no external load is given by Eq. 6-21:

$$A_v = \frac{-h_{fe}R_C}{h_{ie} + (1 + h_{fe})R_E}$$

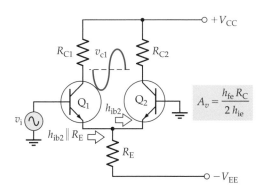

Figure 12-34 The voltage gain of a differential amplifier is half the gain of a common-emitter circuit using similar components. Note the absence of an emitter bypass capacitor.

Referring to Fig. 12-34, it is seen that the resistance *looking into* the emitter of Q_2 is h_{ib}; so $h_{ib} \| R_E$ behaves like an unbypassed resistor in series with the emitter of Q_1. If we ignore R_E because it is very much larger than h_{ib}, the voltage gain from the base of Q_1 to its collector is

$$A_v = \frac{-h_{fe}R_C}{h_{ie} + (1 + h_{fe})h_{ib}}$$

This reduces to

$$A_v = \frac{-h_{fe}R_C}{2\,h_{ie}} \tag{12-24}$$

Equation 12-24 gives the voltage gain from one input terminal to one output of a differential amplifier. It is seen to be half the voltage gain of a similar single-transistor CE amplifier with R_E bypassed; but note that *the differential amplifier requires no bypass capacitor.* This is an important advantage, because bypass capacitors are usually large and expensive.

Another way to contemplate the operation of the differential amplifier circuit is to think of the input voltage being equally divided between Q_1 base-emitter and Q_2 base-emitter. This is illustrated in Fig. 12-35, where it is seen that (for a positive-going input) $v_i/2$ is applied *positive on the base* of Q_1,

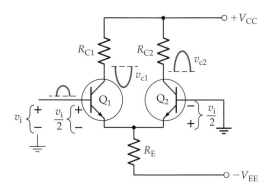

Figure 12-35 A signal at the base of Q_1 is divided equally between the BE junctions of Q_1 and Q_2. Half of v_i appears at Q_1 base-emitter circuit, and the other half is applied to Q_2 base-emitter as a common-base circuit.

while the other half of v_i appears *positive on the emitter* of Q_2. Thus, for v_i at Q_{1B}, transistor Q_1 behaves as a common-emitter circuit, and because Q_2 receives the input at its emitter, Q_2 behaves as a common-base circuit. Consequently,

$$v_{c1} = \frac{v_i}{2} \times \frac{-h_{fe}R_C}{h_{ie}}$$

and

$$v_{c2} = \frac{v_i}{2} \times \frac{h_{fe}R_C}{h_{ie}}$$

Note the minus sign in the v_{c1} equation.

Input and Output Impedances

From Eq. 6-20, the input impedance at the base of a CE circuit with an unbypassed emitter resistor is

$$Z_b = h_{ie} + (1 + h_{fe})R_E$$

Referring to Fig. 12-34, we see that the differential amplifier has $h_{ib}\|R_E$ as an unbypassed resistor in series with the emitter of Q_1. If R_E is ignored (because $R_E \gg h_{ib}$), the input resistance at Q_{1B} is

$$Z_b = h_{ie} + (1 + h_{fe})h_{ib}$$

This reduces to

$$Z_b = 2\,h_{ie} \qquad\qquad (12\text{-}25)$$

or $\qquad\qquad Z_b = 2\,(Z_b$ for a single-stage CE circuit)

Note that there are usually bias resistors in parallel with Z_b, so that (from Equation 6-12) the circuit input impedance is

$$Z_i = R_B\|Z_b$$

As in the case of CE and CB circuits, the output impedance at the transistor collector terminals is given by Equation 6-14:

$$Z_o = R_C\|(1/h_{oe})$$

DC Amplification

When one transistor base is grounded in a differential amplifier and an input is applied to the other one, as already discussed, v_i is amplified to produce the outputs at the collector terminals. In this case v_i is the voltage *difference* between the two base terminals. Figure 12-36 shows a differential amplifier with dc input voltages V_{i1} and V_{i2} applied to the transistor bases. If the voltage gain from the base to the collector is A_v, the dc voltage changes at the collectors are

$$\Delta V_{o1} = A_v(V_{i2} - V_{i1})$$

and $\qquad\qquad \Delta V_{o2} = A_v(V_{i1} - V_{i2})$

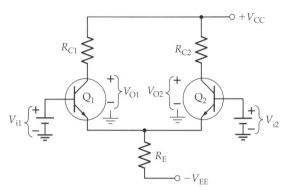

Figure 12-36 When dc voltages are applied to the two inputs of a differential amplifier, the outputs at the transistor collectors are proportional to the differences between the input voltages.

It is seen that the differential amplifier can be employed as a *direct-coupled amplifier* or *dc amplifier*. The term *difference amplifier* is also used for this circuit.

Design Calculations

Design procedures for a differential amplifier are similar to those for voltage-divider bias circuits. Because there is no bypass capacitor in a differential amplifier, one of the coupling capacitors determines the circuit lower cutoff frequency (f_1). The capacitor with the smallest resistance in series with it is normally the largest capacitor, and in the case of a differential amplifier this is usually the input coupling capacitor. So, the input coupling capacitor determines the circuit lower cutoff frequency.

Consider the capacitor-coupled differential amplifier in Fig. 12-37. The circuit uses a plus/minus supply and a single collector resistor (R_C). No output is taken from Q_1 collector, so there is no need for a collector resistor. R_C is selected in the usual way for a small-signal amplifier: $R_C \ll R_L$. The collector-emitter voltage should be a minimum of 3 V, as always. Then, I_C is calculated from R_C and the selected voltage drop across R_C.

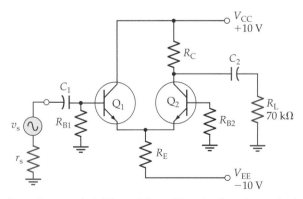

Figure 12-37   Capacitor-coupled differential amplifier circuits using a plus/minus supply.

The total emitter current is determined as

$$I_E \approx I_{C1} + I_{C2} \approx 2I_C$$

and

$$R_E = \frac{V_{EE} - V_{BE}}{I_E} \qquad (12\text{-}26)$$

The base bias resistors are determined from Eq. 5-17:

$$R_B = \frac{V_{BE}}{10\,I_{B(max)}}$$

Since capacitor C_1 sets the lower cutoff frequency,

$$X_{C1} = Z_i \text{ at } f_1 \qquad (12\text{-}27)$$

and C_2 is determined from Eq. 12-5:

$$X_{C2} = R_L/10 \text{ at } f_1$$

Example 12-18

Determine suitable resistor values for the differential amplifier circuit in Fig. 12-37. The transistor parameters are $h_{FE} = 60$, $h_{fe} = 60$, and $h_{ie} = 1.4\ \text{k}\Omega$.

Solution

$$R_{C2} < R_L$$

Select

$$R_{C2} \approx \frac{R_L}{10} \approx \frac{70\ \text{k}\Omega}{10}$$

$$= 7\ \text{k}\Omega \text{ (use } 6.8\ \text{k}\Omega \text{ standard value)}$$

Select

$$V_{CE(min)} = 3\ \text{V}$$

$$V_{RC2} = V_{CC} + V_{BE} - V_{CE(min)} = 10\ \text{V} + 0.7\ \text{V} - 3\ \text{V}$$

$$= 7.7\ \text{V}$$

$$I_C = \frac{V_{RC2}}{R_C} = \frac{7.7\ \text{V}}{6.8\ \text{k}\Omega}$$

$$= 1.13\ \text{mA}$$

Eq. 12-26:

$$R_E = \frac{V_{EE} - V_{BE}}{I_E} \approx \frac{10\ \text{V} - 0.7\ \text{V}}{2 \times 1.13\ \text{mA}}$$

$$= 4.1\ \text{k}\Omega \text{ (use } 4.7\ \text{k}\Omega \text{ standard value)}$$

From Eq. 5-17, $R_B = \dfrac{V_{BE}}{10\, I_{B(max)}} = \dfrac{V_{BE}}{10\,(I_C/h_{FE})} = \dfrac{0.7\text{ V}}{10 \times (1.3\text{ mA}/60)}$

$$= 3.7\text{ k}\Omega\ (\text{use }3.9\text{ k}\Omega\text{ standard value})$$

$$R_{B1} = R_{B2} = 3.9\text{ k}\Omega$$

Example 12-19

Determine suitable capacitor values for the amplifier in Ex. 12-18 to give $f_1 = 60$ Hz. Calculate the circuit voltage gain. Assume that $r_s \ll Z_i$.

Solution

See Fig. 12-38.

$$r'_e = \frac{26\text{ mV}}{I_E} \approx \frac{26\text{ mV}}{1.13\text{ mA}}$$

$$= 20\ \Omega$$

$$h_{ie} = h_{fe}r'_e = 60 \times 20\ \Omega$$

$$= 1.2\text{ k}\Omega$$

Eq. 12-25: $Z_b = 2h_{ie} = 2 \times 1.2\text{ k}\Omega$

$$= 2.4\text{ k}\Omega$$

$$Z_i = R_B \| Z_b = 3.9\text{ k}\Omega \| 2.4\text{ k}\Omega$$

$$= 1.5\text{ k}\Omega$$

From Eq. 12-27, $C_1 = \dfrac{1}{2\pi f_1 Z_i} = \dfrac{1}{2\pi \times 60\text{ Hz} \times 1.5\text{ k}\Omega}$

$$= 1.77\ \mu\text{F}\ (\text{use }1.8\ \mu\text{F standard value})$$

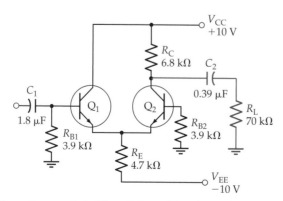

Figure 12-38 Capacitor-coupled differential amplifier circuit designed in Examples 12-18 and 12-19.

From Eq. 12-5, $C_2 = \dfrac{1}{2\pi f_1(R_L/10)} = \dfrac{1}{2\pi \times 60 \text{ Hz} \times (70 \text{ k}\Omega/10)}$

$$= 0.38 \ \mu\text{F (use 0.39} \ \mu\text{F standard value)}$$

From Eq. 12-24, $A_v = \dfrac{h_{fe}(R_{C2}\|R_L)}{2 \ h_{ie}} = \dfrac{60 \times (6.8 \text{ k}\Omega\|70 \text{ k}\Omega)}{2 \times 1.2 \text{ k}\Omega}$

$$= 155$$

Practice Problems

12-8.1 The differential amplifier in Fig. 12-31a is to use a 2N3904 transistor and a 20 V supply and is to have a 100 kΩ load capacitor-coupled to Q_2 collector. Determine suitable resistor values.

12-8.2 Determine suitable coupling capacitor values for the circuit in Problem 12-9.1 to give a 75 Hz lower cutoff frequency. Calculate the overall voltage gain. Assume that $r_s = 200 \ \Omega$.

12-9 SMALL-SIGNAL HIGH-FREQUENCY AMPLIFIERS

Common-Base Amplifier

A practical CB amplifier (using a plus/minus supply) is shown in Fig. 12-39. (Note the polarity of the input coupling capacitor (C_1).) The CB circuit is analyzed in Section 6-7, where it is found to have a very low input impedance, as well as a voltage gain and output impedance similar to a CE circuit. Because the CB circuit has no phase inversion between the input and output terminals, there is no Miller-effect amplification of the collector-base capacitance (see Sections 8-3 and 8-4). Consequently, a CB circuit is capable of operating at much higher frequencies than a CE circuit.

The procedure for designing a CB circuit is exactly the same as that for a CE circuit (see Section 12-1). Resistor calculation is performed just as if a CE circuit were being designed. The output coupling capacitor (C_2 in Fig. 12-39) is determined in the usual way. Because of the very low input impedance of a CB circuit, the input coupling capacitor (C_1) is the largest capacitor in the circuit. So, C_1 determines the circuit low cutoff frequency, and its capacitance is calculated as for an emitter bypass capacitor in a CE circuit.

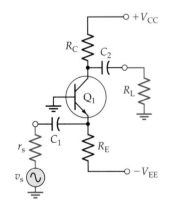

Figure 12-39 A capacitor-coupled CB circuit can be designed in exactly the same way as a CE circuit.

Cascode Amplifier

The very low input impedance of a CB circuit (typically 25 Ω) is a major disadvantage. To increase the circuit input impedance while retaining its high frequency performance, the *cascode* amplifier (Fig. 12-40) uses a CE input stage that drives a CB output stage. Transistor Q_1 and its associated components in Fig. 12-40 operate as a CE input stage, while the circuit of Q_2 functions as a CB output stage.

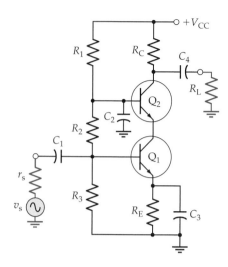

Figure 12-40 A cascode amplifier is like a CB circuit with an additional input transistor (Q_1) that behaves like a CE stage with a gain of one.

The circuit dc voltage and current levels are illustrated in Fig. 12-41. It is seen that I_{E1} is set by V_{B1} and R_E. Collector current I_{C1} approximately equals I_{E1}, and I_{E2} is the same current as I_{C1}. So, I_{C2} approximately equals I_{E1}. The current levels remain constant so long as V_{CE1} and V_{CE2} are large enough for satisfactory transistor operation.

A positive-going ac signal at the base of Q_1 produces an increase in I_{E1}, and this results in an I_{C2} increase. Thus, the voltage drop across R_C is increased, producing a decrease in the output voltage. So there is an ac phase inversion between the input and output.

The input impedance at the emitter of Q_2 constitutes the load for the collector of Q_1. Therefore, from Eq. 6-15, the voltage gain from the circuit input to Q_1 collector is

$$A_{v1} = \frac{-h_{fe} \times (Z_i \text{ to } Q_2)}{h_{ie}} = \frac{-h_{fe} \times [h_{ie}/(1 + h_{fe})]}{h_{ie}}$$

$$\approx -1$$

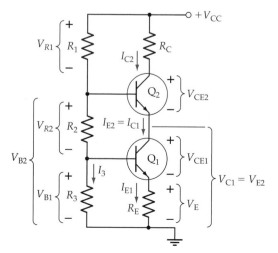

Figure 12-41 DC voltage levels in a cascode circuit.

With an input stage gain of -1, there is no significant Miller effect at the circuit input.

From Eq. 6-32, the voltage gain produced by the Q_2 CB circuit is

$$A_{v2} = \frac{h_{fb} \times (R_C \| R_L)}{h_{ib}}$$

Converting to CE parameters, the overall voltage gain for the cascode amplifier is the same as that for a CE circuit:

$$A_v = \frac{-h_{fe} \times (R_C \| R_L)}{h_{ie}}$$

Design Calculations

The design procedure for a cascode amplifier is easily derived from the CE design process. The voltage across R_E should typically be 5 V, and the V_{CE} for each transistor should be a minimum of 3 V. The emitter voltage for Q_2 is

$$V_{E2} = (V_{E1} + V_{CE1})$$

The transistor base voltages are

$$V_{B1} = (V_{E1} + V_{BE})$$

and
$$V_{B2} = (V_{E2} + V_{BE})$$

When the voltage and current levels have been chosen, the resistor values are easily calculated.

The base bypass capacitor (C_2) in Fig. 12-40 should have an impedance equal to one-tenth of the h_{ie} of Q_2. This ensures that the impedance 'looking into' Q_2 emitter terminal is not significantly affected by R_1 and R_2. Also, no portion of the input signal (at Q_1 base) is developed at Q_2 base.

Example 12-20

Determine suitable resistor values for the cascode amplifier circuit in Figs. 12-40 and 12-41 if $V_{CC} = 20$ V and $R_L = 90$ kΩ. Assume that the transistors have the following parameters: $h_{fe} = 50$, $h_{ie} = 1.2$ kΩ, and $h_{ib} = 24\ \Omega$.

Solution

Select $\qquad R_C \approx \dfrac{R_L}{10} \approx \dfrac{90\text{ k}\Omega}{10}$

$$= 9 \text{ k}\Omega \text{ (use 8.2 k}\Omega \text{ standard value)}$$

Select $\qquad V_{CE1(min)} = V_{CE2(min)} = 3$ V

and $\qquad V_E = 5$ V

$$V_{RC} = V_{CC} - V_{CE1} - V_{CE2} - V_E = 20 \text{ V} - 3 \text{ V} - 3 \text{ V} - 5 \text{ V}$$

$$= 9 \text{ V}$$

$$I_C = \frac{V_{RC}}{R_C} = \frac{9\text{ V}}{8.2\text{ k}\Omega}$$

$$= 1.1\text{ mA}$$

$$R_E \approx \frac{V_E}{I_C} = \frac{5\text{ V}}{1.1\text{ mA}}$$

$$= 4.5\text{ k}\Omega \text{ (use } 4.7\text{ k}\Omega \text{ standard value)}$$

Select $\qquad R_3 = 10\,R_E = 10 \times 4.7\text{ k}\Omega$

$$= 47\text{ k}\Omega$$

$$V_{B1} = V_E + V_{BE} = 5\text{ V} + 0.7\text{ V}$$

$$= 5.7\text{ V}$$

$$I_3 = \frac{V_{B1}}{R_3} = \frac{5.7\text{ V}}{47\text{ k}\Omega}$$

$$= 121\ \mu\text{A}$$

$$V_{B2} = V_E + V_{CE1} + V_{BE2} = 5\text{ V} + 3\text{ V} + 0.7\text{ V}$$

$$= 8.7\text{ V}$$

$$V_{R2} = V_{B2} - V_{B1} = 8.7\text{ V} - 5.7\text{ V}$$

$$= 3\text{ V}$$

$$R_2 = \frac{V_{R2}}{I_3} = \frac{3\text{ V}}{121\ \mu\text{A}}$$

$$= 24.8\text{ k}\Omega \text{ (use } 22\text{ k}\Omega + 2.7\text{ k}\Omega)$$

$$R_1 = \frac{V_{CC} - V_{B2}}{I_3} = \frac{20\text{ V} - 8.7\text{ V}}{121\ \mu\text{A}}$$

$$= 93.4\text{ k}\Omega \text{ (use } 47\text{ k}\Omega + 47\text{ k}\Omega)$$

Example 12-21

Determine suitable capacitor values for the cascode circuit in Example 12-20 to give a 25 Hz low cutoff frequency. Assume that $r_s \ll Z_i$.

Solution

See Fig. 12-42.

$$Z_i = h_{ie1}\|R_3\|R_2 = 1.2\text{ k}\Omega\|47\text{ k}\Omega\|24.7\text{ k}\Omega$$

$$= 1.1\text{ k}\Omega$$

From Eq. 12-4, $\quad C_1 = \dfrac{1}{2\pi f_1(Z_i/10)} = \dfrac{1}{2\pi \times 25\text{ Hz} \times (1.1\text{ k}\Omega/10)}$

$$= 57.9\ \mu\text{F (use } 56\ \mu\text{F standard value)}$$

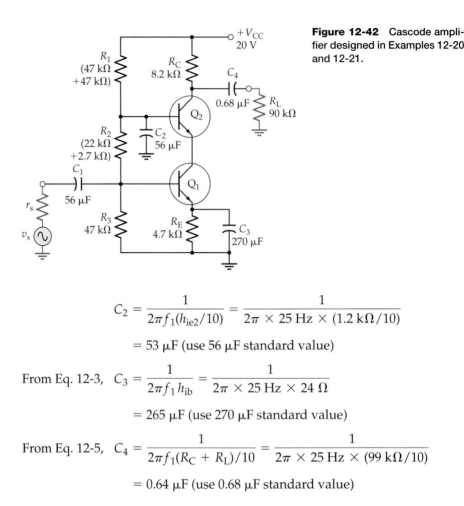

Figure 12-42 Cascode amplifier designed in Examples 12-20 and 12-21.

$$C_2 = \frac{1}{2\pi f_1 (h_{ie2}/10)} = \frac{1}{2\pi \times 25\ \text{Hz} \times (1.2\ \text{k}\Omega/10)}$$

$$= 53\ \mu\text{F (use 56 } \mu\text{F standard value)}$$

From Eq. 12-3, $\quad C_3 = \dfrac{1}{2\pi f_1 h_{ib}} = \dfrac{1}{2\pi \times 25\ \text{Hz} \times 24\ \Omega}$

$$= 265\ \mu\text{F (use 270 } \mu\text{F standard value)}$$

From Eq. 12-5, $\quad C_4 = \dfrac{1}{2\pi f_1 (R_C + R_L)/10} = \dfrac{1}{2\pi \times 25\ \text{Hz} \times (99\ \text{k}\Omega/10)}$

$$= 0.64\ \mu\text{F (use 0.68 } \mu\text{F standard value)}$$

Differential Amplifier as a High-Frequency Amplifier

The differential amplifier discussed in Section 12-8 can have the same high-frequency performance as a CB circuit. The differential amplifier circuit shown in Fig. 12-43 has a voltage gain of zero from Q_1 base to Q_1 collector, because there is no collector resistor. Consequently, there is no Miller effect on the

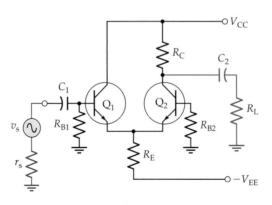

Figure 12-43 A differential amplifier can have a good high-frequency response when the input stage has a voltage gain of 1 because of the absence of a collector resistor.

input capacitance. Also, the amplifier input is applied (via Q_1) to the emitter of Q_2, as in the case of a CB circuit. So there is no Miller effect associated with Q_2. The input impedance (at the base of Q_1) is reasonably high, like a CE circuit and unlike the case of a CB amplifier.

Practice Problems

12-9.1 A cascode amplifier circuit as in Fig. 12-40 is to use 2N3904 BJTs and a 16 V supply. The external load resistor is 33 kΩ. Determine suitable resistor values for the circuit.

12-9.2 Determine suitable capacitor values for the circuit in Problem 12-9.1 to give a 100 Hz lower cutoff frequency. Calculate the circuit voltage gain.

12-10 AMPLIFIER TESTING

Preparation

Transistor amplifiers and other circuits should be tested in a methodical fashion; otherwise the results obtained may be useless. The circuit should first be carefully constructed in breadboard form, with the components placed to resemble the circuit diagram as closely as possible. This makes testing and troubleshooting much easier than when the layout is untidy. Figure 12-44a shows a CE circuit, and Fig. 12-44b illustrates a suitable circuit layout on a plug-in type breadboard. The common errors listed in Section 5-6 should be referred to once again.

Instability

After the circuit has been constructed, the power supply should be set to the appropriate voltage and then connected and switched on. An oscilloscope

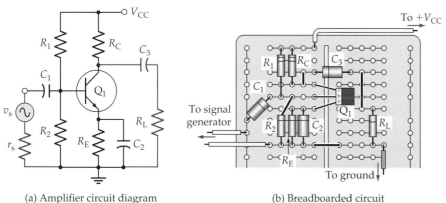

(a) Amplifier circuit diagram   (b) Breadboarded circuit

Figure 12-44   When a circuit is constructed for laboratory testing, the layout of the breadboard should resemble the circuit diagram as closely as possible.

should be connected to monitor the output of the amplifier to check that the circuit is not oscillating. This is especially important for a multi-stage circuit. If the circuit is oscillating, the oscillations must be stopped before proceeding further. DC voltage measurements made on an oscillating circuit have absolutely no value.

Oscillations at the output of an amplifier circuit could be the result of high-frequency noise pickup at the circuit input. If the signal source is not connected, directly grounding the input of a capacitor-coupled circuit should eliminate the problem. When a direct-coupled circuit is under test, the input should be grounded via a 0.1 μF capacitor.

Amplifier instability can be the result of incorrect design or poor circuit layout. It can also be caused by feedback along the conductors from the power supply to the circuit. To stabilize an unstable amplifier, start by connecting a 0.01 μF (or 0.1 μF, if necessary) decoupling capacitor from the positive supply line to ground (Fig. 12-45). Where a plus/minus supply is used, connect capacitors from each supply line to ground. *The capacitors must be installed right on the breadboard.* It is useless to place them at power supply terminals.

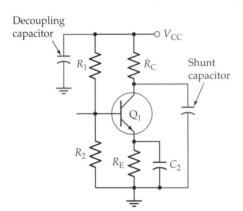

Decoupling capacitor

Shunt capacitor

Figure 12-45 Oscillations may be eliminated in an unstable circuit by using power supply decoupling capacitors and/or by employing small shunt capacitors from the transistor collector to ground.

If the circuit is still unstable, small shunt capacitors should be connected from the transistor collector terminals to ground, or between collector and base (Fig. 12-45). Start with 50 pF capacitors for this purpose, and increase the capacitance in steps to about 300 pF, as necessary. Normally, these capacitors should not be part of the final circuit. They are usually installed temporarily (if they are required) only until the source of the circuit instability is located.

DC Voltage Measurements

Once it is established that the circuit is stable, the next step is to measure the dc voltage levels at all transistor terminals. An electronic voltmeter should be used, and every dc voltage level should be measured with respect to ground. Some electronic voltmeters have their common terminal capacitor-coupled to ground. So, it is best to keep the voltmeter common terminal grounded at all times. If the dc voltages are not satisfactory, they must be corrected before proceeding further. Even when the voltage levels are not exactly as designed, it might be possible to proceed.

The measured dc voltage levels should at least show that every BJT base-emitter junction is forward-biased by the appropriate amount (approximately 0.7 V for silicon transistors and 0.3 V for germanium transistors). The measurements should also show a minimum of 2 V for each collector-emitter voltage drop. BJT and FET bias circuit troubleshooting procedures are discussed in Sections 5-6 and 10-6, respectively.

AC Testing

When satisfactory dc levels are established throughout the circuit, ac tests may proceed. Connect a signal generator to the input, and connect a two-channel oscilloscope to monitor simultaneously the input and output voltage waveforms (Fig. 12-46). Adjust the signal frequency to somewhere around the middle of the amplifier bandwidth, and adjust the signal amplitude to give maximum undistorted output voltage. If the signal generator amplitude cannot be set low enough to give an undistorted output, a resistive voltage divider should be employed to further reduce the signal amplitude (see Fig. 12-47).

There are several reasons why an oscilloscope is preferable to the use of voltmeters for measuring circuit input and output voltages. With the waveforms displayed on the oscilloscope, phase shifts between input and output can be monitored and waveform distortion can be observed. It should also be

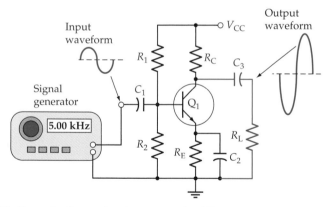

Figure 12-46 For ac testing, a signal generator should be connected at the circuit input and an oscilloscope should be used to monitor the input and output waveforms.

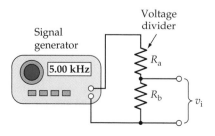

Figure 12-47 A resistive voltage divider might be required to reduce the signal generator amplitude to a low enough level for the circuit input voltage.

noted than many hand-held ac multimeters have a frequency range of only 40 Hz to 500 Hz. Even for high-performance portable instruments, the typical frequency range is 40 Hz to 30 kHz.

The amplifier input and output voltage amplitudes (v_i and v_o) can be measured and the voltage gain can be calculated as $A_v = v_o/v_i$. When a two-stage circuit is being tested, one oscilloscope probe may be moved to monitor the output voltage (v_{o1}) of the first stage. The gain of the first stage is $A_{v1} = v_{o1}/v_i$, and that of the second stage is $A_{v2} = v_o/v_{o1}$.

With the oscilloscope monitoring the input and output waveforms, the frequency response of the circuit may be investigated. The signal voltage level should be maintained constant, and the signal frequency adjusted in convenient steps. At each frequency, v_o should be measured and A_v calculated. The upper and lower cutoff frequencies are found where v_o drops to 0.707 of its normal (mid-frequency) level with v_i still at it normal level.

Input and Output Impedances

The amplifier input and output impedances should normally be measured with the signal frequency at the middle of the bandwidth. The simplest way to measure the input impedance is to connect a resistor (not a decade box) in series with the circuit input terminal and the signal source (see Fig. 12-48a). When the ac output voltage drops to half its normal level, the series-connected resistor equals the input impedance. If the output voltage change is other than half, the input resistance can be calculated with the voltage-divider equation.

The output impedance may be determined by connecting a resistor as a capacitor-coupled load at the circuit output (Fig. 12-48b). The ac output voltage is once again halved when the load resistance equals the circuit output impedance.

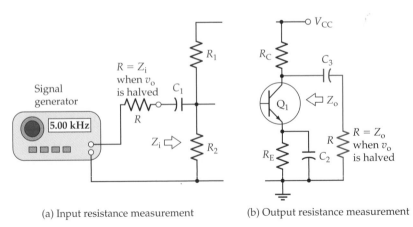

(a) Input resistance measurement   (b) Output resistance measurement

Figure 12-48 Circuit input impedance is measured by connecting a resistor in series with the circuit input. Output resistance is measured by the use of a capacitor-coupled load resistor.

Section 12-10 Review

12-10.1 Prepare a point-form list of the correct test procedure for a two-stage transistor amplifier circuit.

Practice Problem

12-10.1 Refer to the circuit designed in Examples 12-2 and 12-3. Calculate the percentage change in output voltage when a 1 kΩ resistor is connected in series with the signal source and the circuit input. Determine the effect on the output voltage when a 27 kΩ resistor is connected in place of the 120 kΩ load.

Review Questions

Section 12-1

12-1 Discuss the factors involved in the selection of I_C, R_C, and R_E for a single-stage common-emitter BJT amplifier circuit using voltage-divider bias. Explain the process for determining suitable bias resistors.

12-2 Derive the equation for calculating the value of the emitter bypass capacitor for a single-stage common-emitter BJT amplifier using voltage-divider bias.

12-3 Write an equation for calculating the coupling capacitors in a capacitor-coupled common-emitter amplifier. Show that, when this equation is used, the coupling capacitors have very little effect on the performance of the circuit.

12-4 Show how a capacitor may be used to obtain a desired upper cutoff frequency in a transistor common-emitter amplifier. Write the equation for determining the capacitor value.

Section 12-2

12-5 Discuss the factors involved in the choice of I_D, R_D, and R_S for a single-stage common-source FET amplifier circuit using voltage-divider bias. Explain the process for determining suitable bias resistors.

12-6 Derive the equation for calculating the value of the source bypass capacitor for a single-stage common-source amplifier using voltage-divider bias.

Section 12-3

12-7 Sketch the circuit of a two-stage capacitor-coupled common-emitter BJT amplifier. Briefly explain its operation.

12-8 Discuss the lower cutoff frequency for each stage of a two-stage common-emitter amplifier and the lower cutoff frequency for the complete circuit. Write equations for calculating the bypass capacitor values for each stage.

12-9 Write equations for calculating the coupling capacitors in a two-stage capacitor-coupled common-emitter amplifier. Explain.

12-10 Sketch the h-parameter equivalent circuit for a two-stage common-emitter amplifier. Write equations for Z_i, Z_o, A_{v1}, A_{v2}, and A_v.

Section 12-4

12-11 Sketch the circuit of a two-stage direct-coupled common-emitter amplifier using npn BJTs. Discuss the advantages of direct coupling between stages.

12-12 Sketch the h-parameter equivalent circuit for the amplifier in Question 12-11. Write equations for Z_i, Z_o, and A_v for the circuit.

12-13 Sketch a two-stage direct-coupled BJT amplifier using an npn input stage and a pnp output stage. Explain the advantages of using complementary transistors.

Section 12-5

12-14 Sketch the circuit of a two-stage direct-coupled BJT amplifier using a common-emitter input stage and an emitter follower output stage. Discuss the circuit operation.

12-15 Write equations for Z_i, Z_o, and A_v for the circuit in Question 12-14.

Section 12-6

12-16 Sketch the circuit of a dc feedback pair with two stages of amplification. Explain the circuit operation, and discuss its advantages and disadvantages.

12-17 Sketch the circuit of a dc feedback pair with an amplifier input stage and an emitter follower stage. Explain the capacitor bypassing for this circuit.

Section 12-7

12-18 Sketch the circuit of a two-stage capacitor-coupled amplifier using an FET common-source input stage and a BJT common-emitter output stage. Discuss the advantages of this type of BIFET circuit.

12-19 Write equations for Z_i, Z_o, and A_v for the circuit in Question 12-18.

12-20 Sketch the circuit of a two-stage direct-coupled amplifier using an n-channel FET common-source input stage and a pnp BJT common-emitter output stage. Explain the advantages of this arrangement.

Section 12-8

12-21 Draw the circuit diagram for a BJT differential amplifier: (a) using a single-polarity supply, and (b) using a plus/minus supply. Explain the dc and ac operation of each circuit, and discuss the advantages and disadvantages of differential amplifiers.

12-22 Write equations for voltage gain and impedance for a differential amplifier circuit.

Section 12-9

12-23 Sketch the circuit of a single-stage, capacitor-coupled common-base amplifier, and then show the modifications required to convert it to a cascode amplifier. Explain the operation of the cascode amplifier.

12-24 Derive an equation for a cascode amplifier voltage gain. Write equations for input and output impedance.

12-25 Sketch a differential amplifier circuit for high-frequency signal amplification use. Explain its advantages in this application.

Section 12-10

12-26 Briefly explain the measures that should be taken to stop a laboratory-constructed circuit from oscillating.

12-27 List the correct procedure for testing a laboratory-constructed amplifier circuit for voltage gain, frequency response, input impedance, and output impedance.

Problems

Section 12-1

12-1 A single-stage, BJT, common-emitter amplifier with voltage-divider bias is to be designed to use a 2N3903 transistor. A 15 V supply is to be used, and 70 kΩ load resistor is to be capacitor-coupled to the output. Sketch the circuit and determine suitable resistor values.

12-2 Determine suitable capacitor values for the circuit in Problem 12-1 to give a 60 Hz lower cutoff frequency and a 60 kHz upper cutoff frequency. The signal source resistance is 300 Ω.

12-3 Analyze the circuit designed for Problems 12-1 and 12-2 to determine A_v, Z_i, and Z_o.

12-4 A single-stage common-emitter amplifier using a 2N3906 transistor with voltage-divider bias is to be designed to have $V_{CC} = 18$ V and $R_L = 56$ kΩ. Sketch the circuit and determine suitable resistor values.

12-5 The circuit in Problem 12-4 is to have $f_1 = 80$ Hz and $f_2 = 75$ kHz. The signal source resistance is 500 Ω. Determine suitable capacitor values.

12-6 Calculate A_v, Z_i, and Z_o for the circuit designed for Problems 12-4 and 12-5.

12-7 A single-stage common-emitter amplifier with voltage divider bias is to use a 2N3904 transistor with $V_{CC} = 12$ V. The circuit is to be designed so that $Z_o \approx 3.9$ kΩ. Sketch the circuit and determine suitable resistor values.

12-8 The circuit in Problem 12-7 is to have $f_1 = 50$ Hz, $R_L = 47$ kΩ, and a signal source resistance $r_s = 600$ Ω. Determine suitable capacitor values.

12-9 Determine A_v, Z_i, and Z_o for the circuit designed for Problems 12-7 and 12-8.

Section 12-2

12-10 A single-stage common-source amplifier with voltage-divider bias is to use a 2N5457 FET with $V_{DD} = 22$ V and $R_L = 70$ kΩ. Sketch the circuit and determine suitable resistor values.

12-11 Determine suitable capacitor values for the circuit in Problem 12-10 to give a 60 Hz lower cutoff frequency and a 60 kHz upper cutoff frequency. The signal source resistance is 300 Ω.

12-12 Analyze the circuit designed for Problems 12-10 and 12-11 to determine A_v, Z_i, and Z_o.

12-13 In a single-stage common-source amplifier using a 2N5458 FET with voltage-divider bias, $V_{DD} = 25$ V and $R_L = 56$ kΩ. Sketch the circuit and determine suitable resistor values.

12-14 The circuit in Problem 12-13 to have $f_1 = 75$ Hz and $f_2 = 90$ kHz. The signal source resistance is 500 Ω. Determine suitable capacitor values.

12-15 Calculate A_v, Z_i, and Z_o for the circuit designed for Problems 12-13 and 12-14.

12-16 A single-stage common-source amplifier with voltage divider bias is to use a 2N5459 FET with $V_{DD} = 20$ V. The circuit is to be designed so that $Z_i \approx 500$ kΩ and $Z_o \approx 5$ kΩ. Draw the circuit diagram and determine suitable resistor values.

12-17 The circuit in Problem 12-16 is to have $f_1 = 80$ Hz, $R_L = 47$ kΩ, and a signal source resistance $r_s = 600$ Ω. Determine suitable capacitor values.

12-18 Determine A_v, Z_i, and Z_o for the circuit designed for Problems 12-16 and 12-17.

Section 12-3

12-19 A two-stage capacitor-coupled common-emitter amplifier is to be designed to have $Z_o \approx 3.9$ kΩ. The circuit is to use 2N3903 transistors with $V_{CC} = 12$ V. Draw the complete circuit diagram and calculate suitable resistor values.

12-20 The circuit in Problem 12-19 is to have $f_1 = 100$ Hz, $R_L = 47$ kΩ, and $r_S = 600$ Ω. Determine suitable capacitor values.

12-21 Calculate A_v, Z_i, and Z_o for the circuit designed for Problems 12-19 and 12-20.

12-22 A two-stage capacitor-coupled common-emitter amplifier is to be designed to use 2N3906 *pnp* transistors with $V_{CC} = 18$ V and $R_L = 60$ kΩ. Draw the circuit diagram and calculate suitable resistor values.

12-23 The circuit in Problem 12-22 is to have $f_1 = 120$ Hz and $f_2 = 82$ kHz. The signal source resistance is $r_s = 600$ Ω. Determine suitable capacitor values.

12-24 Calculate A_v, Z_i, and Z_o for the circuit designed for Problems 12-22 and 12-23.

Section 12-4

12-25 A two-stage direct-coupled common-emitter amplifier is to be designed to use 2N3903 transistors with $V_{CC} = 16$ V. The capacitor-coupled load is $R_L = 75$ kΩ, and the signal source resistance is $r_s = 200$ Ω. Draw the circuit diagram and calculate suitable resistor values.

12-26 Determine suitable capacitor values for the circuit in Problem 12-25. The lower cutoff frequency is to be 90 Hz.

12-27 Calculate A_v, Z_i, and Z_o for the circuit designed for Problems 12-25 and 12-26.

12-28 A two-stage direct-coupled common-emitter amplifier is to be designed to use a 2N3904 *npn* transistor as stage 1, and a 2N3906 *pnp* device as stage 2. The circuit is to have $V_{CC} = 17$ V and $R_L = 85$ kΩ. Draw the circuit diagram and calculate suitable resistor values.

12-29 The circuit in Problem 12-28 is to have $f_1 = 55$ Hz. Determine suitable capacitor values if $r_s = 600$ Ω.

12-30 Calculate A_v, Z_i, and Z_o for the circuit designed for Problems 12-28 and 12-29.

12-31 A two-stage direct-coupled common-emitter amplifier is to use a 2N3905 *pnp* transistor as stage 1, and a 2N3903 *npn* device as stage 2. The circuit is to have $V_{CC} = 11$ V, $R_L = 75$ kΩ, $f_1 = 80$ Hz, and $r_s = 300$ Ω. Sketch the circuit diagram and determine suitable resistor and capacitor values.

Section 12-5

12-32 A two-stage direct-coupled common-emitter amplifier with an emitter follower output is to be designed to use 2N3904 transistors. The circuit is to have $V_{CC} = 16$ V, $R_L = 65$ Ω, $f_1 = 50$ Hz, and $v_o \approx \pm 75$ mV. Sketch the circuit and determine suitable resistor and capacitor values. Take $r_s = 300$ Ω.

12-33 Analyze the circuit designed for Problem 12-32 to calculate the overall voltage gain.

12-34 Design a two-stage direct-coupled common-emitter amplifier with an emitter follower output. The circuit is to use 2N3906 BJTs with $V_{CC} = 18$ V. The load is $R_L = 85$ Ω, the lower cutoff frequency is $f_1 = 65$ Hz, the output voltage is $v_o \approx \pm 55$ mV, and $r_s = 100$ Ω. Draw the circuit diagram and determine suitable resistor and capacitor values.

12-35 Determine the overall voltage gain for the amplifier circuit in Problem 12-34.

Section 12-6

12-36 A dc feedback pair with two stages of amplification is to be designed to use 2N3904 transistors. The circuit is to have $V_{CC} = 18$ V, $R_L = 50$ kΩ, $f_1 = 75$ Hz, and $r_s = 100$ Ω. Sketch the circuit and determine suitable component values.

12-37 Analyze the circuit designed for Problem 12-36 to calculate the overall voltage gain.

12-38 A dc feedback pair with two stages of amplification is to be designed to use 2N3906 BJTs and to have $V_{CC} = 15$ V, $R_L = 60$ kΩ, $f_1 = 150$ Hz, and $r_s = 220$ Ω. Draw the circuit diagram and determine suitable component values.

12-39 A dc feedback pair with an emitter follower output stage is to use 2N3904 BJTs and to have $V_{CC} = 14$ V, $R_L = 40$ Ω, $r_s = 200$ Ω, $f_1 = 60$ Hz, and $v_o \approx \pm 65$ mV. Determine suitable component values.

12-40 Analyze the circuit designed for Problem 12-39 to calculate A_v, Z_i, and Z_o.

12-41 A dc feedback pair with an emitter follower output stage is to use 2N3906 BJTs. The circuit is to have $V_{CC} = 9$ V, $R_L = 100$ Ω, $f_1 = 75$ Hz, and $v_o \approx \pm 75$ mV. Sketch the circuit and determine suitable resistor and capacitor values.

12-42 Calculate A_v, Z_i, and Z_o for the circuit in Problem 12-41.

Section 12-7

12-43 A two-stage capacitor-coupled BIFET amplifier using a common-source input stage and a common-emitter output stage is to be designed to use a 2N5459 FET and a 2N3904 BJT. The external load resistance is 120 kΩ, the supply is 22 V, and the circuit input impedance is to be 800 kΩ. Draw the circuit diagram and calculate suitable resistor values.

12-44 The circuit in Problem 12-43 is to have $f_1 = 140$ Hz. Assuming that $r_s = 220$ Ω, determine suitable capacitor values.

12-45 Calculate A_v, Z_i, and Z_o for the circuit designed for Problems 12-43 and 12-44.

12-46 A two-stage direct-coupled BIFET amplifier using a common-source input stage with a 2N5457 FET and a common-emitter output stage with a 2N3906 *pnp* BJT is to be designed. The external load resistance is 100 kΩ, the supply is 25 V, and the circuit input impedance is to be approximately 300 kΩ. Draw the circuit diagram and calculate suitable resistor values.

12-47 The circuit in Problem 12-46 is to have $f_1 = 80$ Hz. Determine suitable capacitor values. Assume that $r_s = 800$ Ω.

12-48 Calculate A_v, Z_i, and Z_o for the circuit designed for Problems 12-46 and 12-47.

12-49 Design a two-stage capacitor-coupled BIFET amplifier circuit using a 2N5486 FET with self-bias for Q_1 and a voltage-divider biased 2N3904 BJT for Q_2. The circuit specifications are $V_{CC} = 23$ V, $R_L = 200$ kΩ, and $f_1 = 80$ Hz. For the 2N5486, $Y_{fs} = 6000$ μS typical, 8000 μS maximum, $I_{DSS(max)} = 20$ mA, and $V_{GS(off)(max)} = -6$ V. Assume that $r_s \ll Z_i$.

12-50 Calculate A_v, Z_i, and Z_o for the circuit in Problem 12-49.

12-51 Redesign the circuit in Problem 12-49 to use a 2N3905 *pnp* BJT for the second stage with direct coupling between stages.

Section 12-8

12-52 A differential amplifier is to be designed to use 2N3904 BJTs. The external capacitor-coupled load resistance is 120 kΩ, and the supply is ± 12 V. Draw the circuit diagram and calculate suitable resistor values.

12-53 Determine suitable capacitor values for the circuit in Problem 12-52 so that $f_1 = 100$ Hz. Assume that $r_s = 600$ Ω.

12-54 Determine A_v, Z_i, and Z_o for the circuit designed for Problems 12-52 and 12-53.

12-55 A differential amplifier is to use 2N3903 BJTs with a 15 V supply. The external load resistance is 66 kΩ, the signal source resistance is 600 Ω, and the lower cutoff frequency is to be 75 Hz. Draw the circuit diagram and determine suitable resistor and capacitor values. Assume that $r_s \ll Z_i$.

12-56 Determine A_v, Z_i, and Z_o for the circuit in Problem 12-55.

Section 12-9

12-57 Design a common-base BJT amplifier to use a 2N3904 transistor. The specifications are $V_{CC} = 9$ V, $R_L = 27$ kΩ, and $f_1 = 75$ Hz. Sketch the circuit and determine suitable resistor and capacitor values.

12-58 Calculate A_v, Z_i, and Z_o for the circuit in Problem 12-57.

12-59 Design a cascode amplifier to use 2N3904 transistors. The specifications are $V_{CC} = 19$ V, $R_L = 47$ kΩ, and $f_1 = 85$ Hz. Draw the circuit diagram and calculate suitable resistor and capacitor values.

12-60 Analyze the circuit in Problem 12-59 to determine A_v, Z_i, and Z_o.

Section 12-10

12-61 The output voltage of a transistor amplifier under test falls from ±2 V to ±1.5 V when a 4.7 kΩ resistor is connected in series with the 600 Ω source resistance and the circuit input. Calculate the circuit input resistance.

12-62 Determine the output resistance of a transistor amplifier if v_o changes from ±1 V to ±1.2 V when the capacitor-coupled 60 kΩ load resistance is disconnected.

12-63 The voltage gain of the first stage of a two-stage BJT amplifier drops from 300 to 66 when the second stage is connected. The first stage uses a 6.8 kΩ collector resistor. Calculate the input impedance of the second stage.

Practice Problem Answers

12-1.1	150 Ω, 33 kΩ, 6.8 kΩ, 3.3 kΩ
12-1.2	30 μF, 240 μF, 0.56 μF
12-2.1	3.3 MΩ, 1 MΩ, 8.2 kΩ, 8.2 kΩ
12-2.2	0.068 μF, 30 μF, 0.5 μF
12-3.1	100 kΩ, 56 kΩ, 8.2 kΩ, 5.6 kΩ, 100 kΩ, 56 kΩ, 8.2 kΩ, 5.6 kΩ
12-3.2	10 μF, 150 μF, 15 μF, 150 μF, 0.25 μF
12-3.3	9700

12-4.1   (150 kΩ + 15 kΩ), 56 kΩ, 5.6 kΩ, 5.6 kΩ, 1.8 kΩ, 5.6 kΩ

12-4.2   20 μF, 330 μF, 1000 μF, 1 μF

12-4.3   16 907

12-5.1   47 kΩ (82 kΩ + 1.5 kΩ), 4.7 kΩ, 6.8 kΩ, 2.7 kΩ

12-5.2   10 μF, 200 μF, 39 μF, -132

12-6.1   10 kΩ, 220 kΩ, 4.7 kΩ, 3.3 kΩ

12-6.2   20 μF, 180 μF, 0.68 μF, 7582

12-7.1 $C_2 = 15$ μF, $C_3 = 1.8$ μF, $C_4 = 150$ μF

12-7.2   1481

12-8.1   56 kΩ, 22 kΩ, 10 kΩ, 2.2 kΩ, 10 kΩ, 56 kΩ, 22 kΩ

12-8.2   0.68 μF, 0.2 μF, 206

12-9.1   (39 kΩ + 3.3 kΩ), (15 kΩ + 2.2 kΩ), 33 kΩ, 3.3 kΩ, 3.3 kΩ

12-9.2   10 μF, 10 μF, 100 μF, 0.47 μF, 171

12-10.1   28%, 23.9%

CHAPTER 13
Amplifiers with Negative Feedback

CONTENTS

Objectives

You will be able to:

1 Using diagrams, explain series voltage negative feedback, list its effects on amplifier performance, and write voltage gain and impedance equations.

2 Sketch the following amplifier circuits using series voltage negative feedback, and explain the operation of each:
  • two-stage, BJT capacitor-coupled
  • complementary BJT direct-coupled
  • dc feedback pair
  • two-stage BIFET
  • differential-input

3 Analyze each of the circuit types listed in Objective 2 to determine its performance characteristics.

4 Design each of the circuit types listed in Objective 2 to fulfill a given specification.

5 Explain emitter current feedback, list its effects, and write voltage gain and impedance equations.

6 Design and analyze emitter current feedback circuits.

7 Explain parallel current negative feedback, list its effects, and write current gain and impedance equations.

8 Design and analyze parallel current negative feedback circuits.

9 Explain the effects of negative feedback on circuit bandwidth, distortion, and phase shift, and make related calculations.

INTRODUCTION

Negative feedback is produced by feeding a portion of an amplifier output back to the input, where it behaves as an additional signal. The feedback quantity is applied in opposition to the signal, so that the effective input is reduced. This results in stabilized amplifier gain, extended bandwidth, reduced distortion, and modified input and output impedances. *Series voltage feedback* increases input impedance; *parallel current feedback* reduces input impedance. The most important advantage of negative feedback is that it produces amplifiers with stable, predictable voltage gains.

13-1 SERIES VOLTAGE NEGATIVE FEEDBACK

Negative Feedback Concept

In a negative feedback amplifier, a small portion of the output voltage is fed back to the input. When the feedback voltage is applied in series with the signal voltage, the arrangement is *series voltage feedback*. The instantaneous polarity of the feedback voltage is normally opposite to the signal voltage polarity (they are in *series opposition*). So the feedback voltage is *negative* with respect to the signal voltage; hence the term *negative feedback (NFB)*.

Consider the illustration in Fig. 13-1a. An amplifier with two input terminals and one output is shown (in triangular representation). The amplifier has a voltage gain (A_v), and its output voltage (v_o) is applied to a feedback network that reduces v_o by a factor (B) to produce a feedback voltage (v_f). The feedback network may be as simple as the resistive voltage divider shown in Fig. 13-1b. At the amplifier input, the instantaneous level of v_f is applied negative with respect to v_s, so that the amplifier input terminal voltage is

$$v_i = v_s - v_f$$

Because the amplifier input voltage is lower than the signal voltage, the output voltage is lower than that produced when negative feedback is not used.

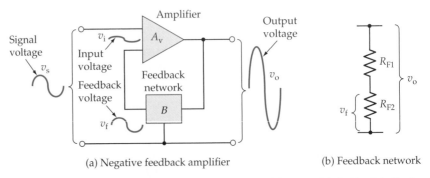

(a) Negative feedback amplifier   (b) Feedback network

Figure 13-1   In a negative feedback amplifier, a portion of the output is fed back to the input. The instantaneous polarity of the feedback voltage (v_f) is negative with respect to the signal voltage (v_s).

This means, of course, that the overall voltage gain (v_o/v_i) is reduced by negative feedback. However, as will be demonstrated, the stability of the voltage gain is greatly improved with negative feedback.

Voltage Gain

Consider the feedback amplifier illustrated in Fig. 13-2. Recall that v_i is amplified to produce v_o, and that v_o is divided by the feedback network to produce v_f. Remember too that v_f is applied (along with v_s) to the amplifier input. It is seen that there is a *closed loop* from the amplifier input to the output, and then back to the input. Because of this closed loop, the overall voltage gain with negative feedback is termed the *closed-loop gain* (A_{CL}).

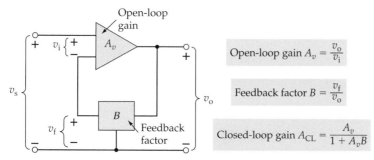

Open-loop gain $A_v = \dfrac{v_o}{v_i}$

Feedback factor $B = \dfrac{v_f}{v_o}$

Closed-loop gain $A_{CL} = \dfrac{A_v}{1 + A_v B}$

Figure 13-2 Open-loop voltage gain, feedback factor, and closed-loop voltage gain for a negative feedback amplifier.

When the feedback network is disconnected, the loop is opened, and the gain without feedback is referred to as the *open-loop gain* (A_v). Alternative symbols sometimes used for the open-loop gain are A_{OL} and $A_{v(OL)}$.

The overall voltage gain (closed-loop gain) of the amplifier in Fig. 13-2 is

$$A_{CL} = \frac{v_o}{v_s}$$

and the feedback factor is

$$B = \frac{v_f}{v_o} \tag{13-1}$$

The input voltage is

$$v_i = v_s - v_f$$

So

$$v_i = v_s - Bv_o$$

and

$$v_o = A_v v_i = A_v(v_s - Bv_o)$$

$$= A_v v_s - A_v B v_o$$

or

$$v_o(1 + A_v B) = A_v v_s$$

giving

$$\frac{v_o}{v_s} = \frac{A_v}{1 + A_v B}$$

Therefore the equation for overall voltage gain with negative feedback is

$$A_{CL} = \frac{A_v}{1 + A_vB} \qquad (13\text{-}2)$$

Analysis of any transistor amplifier not using negative feedback shows that the voltage gain has a wide range, depending on the actual h_{fe} values of the transistors used. Consequently, the open-loop gain of a negative feedback amplifier can vary significantly. However, as demonstrated in Ex. 13-1, the closed loop gain can be a very stable quantity.

Example 13-1

Calculate the closed-loop gain for the negative feedback amplifier shown in Fig. 13-3. Also calculate the closed-loop gain when the open-loop gain is changed by ±50%. Note that $A_v = 100\,000$ and $B = 1/100$.

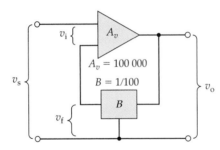

Figure 13-3 Negative feedback amplifier for Ex. 13-1.

Solution

When $A_v = 100\,000$:

Eq. 13-2: $\qquad A_{CL} = \dfrac{A_v}{1 + A_vB} = \dfrac{100\,000}{1 + (100\,000/100)}$

$\qquad\qquad\qquad = 99.9$

When $A_v = 150\,000$:

Eq. 13-2: $\qquad A_{CL} = \dfrac{A_v}{1 + A_vB} = \dfrac{150\,000}{1 + (150\,000/100)}$

$\qquad\qquad\qquad = 99.93$

When $A_v = 50\,000$:

Eq. 13-2: $\qquad A_{CL} = \dfrac{A_v}{1 + A_vB} = \dfrac{50\,000}{1 + (50\,000/100)}$

$\qquad\qquad\qquad = 99.8$

Example 13-1 shows that when the open-loop gain changes by ±50%, the closed-loop gain of the negative feedback amplifier remains stable within ±0.1%. Thus,

series voltage negative feedback stabilizes amplifier voltage gain.

Amplifier gain stabilization is the most important advantage of negative feedback amplifiers. Amplifiers that are required to have stable voltage gain are always designed as negative feedback amplifiers. Examination of Eq. 13-2 reveals that if $A_vB \gg 1$, then

$$A_{\text{CL}} \approx \frac{1}{B} \qquad\qquad (13\text{-}3)$$

From Eq. 13-3, it is seen that to design a negative feedback amplifier with a particular closed-loop gain, it is necessary only to design the feedback network so that $A_{\text{CL}} \approx 1/B$. For example, if $A_{\text{CL}} = 100$, $B \approx 1/100$. It must be remembered that for Eq. 13-3 to be correct, $A_vB \gg 1$, or

$$A_v \gg \frac{1}{B}$$

which means that

$$A_v \gg A_{\text{CL}}$$

Thus, for satisfactory operation of a negative feedback amplifier,

the open-loop voltage gain must be much greater than the required closed-loop gain.

In general, for A_{CL} to be stable within ±1%, A_v should be equal to or greater than 100 A_{CL}.

Input Impedance

The input impedance of a BJT amplifier without negative feedback is usually the impedance 'looking into' the base of a transistor. If no negative feedback is present in the amplifier in Fig. 13-4, the input impedance is given by

$$Z_b = \frac{v_i}{i_i}$$

This gives

$$i_i = \frac{v_i}{Z_b}$$

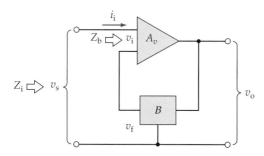

Figure 13-4 The input impedance of an amplifier is increased by a factor of $(1 + A_vB)$ when series voltage negative feedback is employed.

With negative feedback, the input impedance is

$$Z_i = \frac{v_s}{i_i} = \frac{v_s \times Z_b}{v_i} \qquad (1)$$

and

$$v_i = v_s - v_f$$
$$= v_s - Bv_o$$
$$= v_s - A_v Bv_i$$

or

$$v_i(1 + A_v B) = v_s \qquad (2)$$

Substituting for v_s from Eq. 2 into Eq. 1 gives

$$Z_i = (1 + A_v B)Z_b \qquad (13\text{-}4)$$

where Z_i is the input impedance with negative feedback, and Z_b is the input impedance without negative feedback. It is seen that

series voltage feedback increases the input impedance of an amplifier by a factor of $(1 + A_v B)$.

This is the same factor involved in gain reduction.

Input Impedance with Bias Resistors

The bias resistors at the input of a transistor circuit (BJT or FET) are not normally affected by series voltage negative feedback (see Fig. 13-5). This is because the bias resistors are outside the feedback loop. The circuit input impedance in Fig. 13-5 is

$$Z_{in} = Z_i \| R_1 \| R_2 \qquad (13\text{-}5)$$

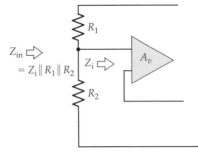

Figure 13-5 The bias resistors at the input of an amplifier are not affected by negative feedback

Example 13-2

The input impedance of the amplifier in Fig. 13-6 (Z_b) is 1 kΩ when negative feedback is not used, and the open-loop voltage gain is 100 000. Calculate the circuit input impedance with negative feedback.

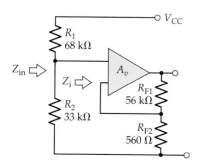

Figure 13-6 Negative feedback amplifier for Example 13-2.

Solution

From Eq. 13-1,

$$B = \frac{v_f}{v_o} = \frac{R_{F2}}{R_{F1} + R_{F2}} = \frac{560\ \Omega}{560\ \Omega + 56\ k\Omega}$$

$$= \frac{1}{101}$$

Eq. 13-4:

$$Z_i = (1 + A_v B)Z_b = \left[1 + \left(\frac{100\ 000}{101}\right)\right] \times 1\ k\Omega$$

$$= 991\ k\Omega$$

Eq. 13-5:

$$Z_{in} = Z_i \| R_1 \| R_2 = 991\ k\Omega \| 68\ k\Omega \| 33\ k\Omega$$

$$= 21.7\ k\Omega$$

Output Impedance

To derive an equation for the output impedance of a negative feedback amplifier, refer to Fig. 13-7. This circuit is similar to that in Fig. 13-1, but it shows the
internal impedance at the amplifier output terminal. The input terminals are
short-circuited, so that v_f is the only voltage at the amplifier input. The output
impedance without feedback is identified as Z_c (the output impedance at a BJT
collector terminal, normally equal to $1/h_{oe}$).

The output voltage (v_o) produces the feedback voltage (v_f), and this generates a voltage $A_v v_f$ in series with Z_c, as illustrated. Assuming an instantaneous
positive polarity for v_o, the output current (i_o) occurs in the direction shown.

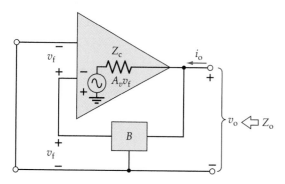

Figure 13-7 The output impedance of an amplifier is reduced by a
factor of $(1 + A_v B)$ by series voltage
negative feedback.

Writing an equation for the voltage drops around the output circuit of the negative feedback amplifier gives

$$v_o = i_o Z_c - A_v v_f$$

or

$$i_o Z_c = v_o + A_v v_f$$

$$= v_o + A_v B v_o$$

$$= v_o(1 + A_v B)$$

The output impedance of the negative feedback amplifier is

$$Z_o = \frac{v_o}{i_o}$$

Therefore

$$Z_o = \frac{Z_c}{1 + A_v B} \qquad (13\text{-}6)$$

In Eq. 13-6, Z_o is the output impedance with negative feedback, and Z_c is the output impedance without negative feedback. Thus,

series voltage feedback reduces the output impedance of an amplifier by a factor of $(1 + A_v B)$.

This is the same factor involved in the reduction of amplifier gain and in the increase of the input impedance.

Output Impedance with a Collector Resistor

The collector resistor in the second stage of a two-stage circuit is not normally affected by series voltage negative feedback (see Fig. 13-8). This is because (as in the case of input bias resistors) the collector resistor is outside the feedback loop.

Recall from Section 6-4 that, for a CE circuit, $Z_c = 1/h_{oe}$ and $Z_o = R_C \| (1/h_{oe})$. Because $1/h_{oe}$ is usually much larger than R_C, Z_o is usually taken as equal to R_C when there is no feedback involved. However, with negative feedback, the collector output impedance is much lower than $1/h_{oe}$.

From Eq. 13-6,

$$Z_o = \frac{1/h_{oe}}{1 + A_v B}$$

So the circuit output impedance should be calculated as

$$Z_{out} = R_C \| Z_o \qquad (13\text{-}7)$$

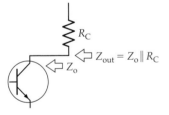

Figure 13-8 The output impedance at the collector of Q_2 is $R_C \| Z_o$; resistor R_C is not affected by negative feedback.

Example 13-3

The amplifier circuit in Fig. 13-9 (reproduced from Fig. 12-21) is designed and analyzed in Examples 12-8 to 12-10. The input impedance Z_b at the base of Q_1 is 1 kΩ, and Q_2 has $1/h_{oe} = 50$ kΩ. Calculate the circuit input and output impedances when negative feedback is used with a feedback factor of 1/100.

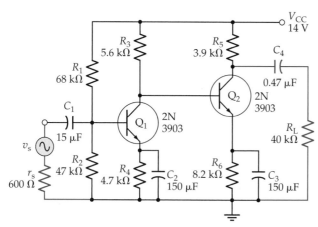

Figure 13-9 Amplifier circuit for Example 13-3.

Solution

Eq. 13-4:
$$Z_i = (1 + A_v B)Z_b = \left[1 + \left(\frac{5562}{100}\right)\right] \times 1 \text{ k}\Omega$$

$$\approx 57 \text{ k}\Omega$$

Eq. 13-5:
$$Z_{in} = Z_i \| R_1 \| R_2 = 57 \text{ k}\Omega \| 68 \text{ k}\Omega \| 47 \text{ k}\Omega$$

$$= 18.7 \text{ k}\Omega$$

Eq. 13-6:
$$Z_o = \frac{1/h_{oe}}{1 + A_v B} = \frac{50 \text{ k}\Omega}{1 + (5562/100)}$$

$$= 883 \text{ }\Omega$$

Eq. 13-7:
$$Z_{out} = R_C \| Z_o = 3.9 \text{ k}\Omega \| 883 \text{ }\Omega$$

$$= 720 \text{ }\Omega$$

Practice Problem

13-1.1 A BJT amplifier with an open-loop gain of 350 000 has the following transistor parameters: $h_{ie} = 1.5$ kΩ, $h_{fe} = 100$, and $1/h_{oe} = 73$ kΩ. The input bias resistors are $R_1 = 82$ kΩ and $R_2 = 39$ kΩ, and the output collector resistor is $R_C = 5.6$ kΩ. Negative feedback is applied with $B = 1/135$. Calculate Z_{in} and Z_{out} without negative feedback and with negative feedback. Determine the closed-loop gain.

13-2 TWO-STAGE CE AMPLIFIER WITH SERIES VOLTAGE NEGATIVE FEEDBACK

Negative Feedback Amplifier Circuit

A two-stage, capacitor-coupled BJT amplifier is shown in Fig. 13-10. This is the same two-stage CE circuit discussed in Section 12-3 with the addition of feedback components R_{F2}, R_{F1}, and C_{F1}. These components constitute the feedback network referred to in Section 13-1. The output voltage is divided across R_{F2} and R_{F1} to produce a feedback voltage in series with the signal at the base of Q_1.

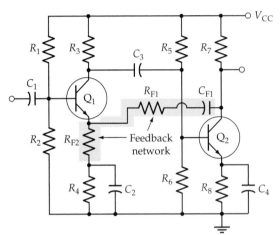

Figure 13-10 Two-stage capacitor-coupled BJT amplifier using series voltage negative feedback. R_{F1}, R_{F2}, and C_{F1} constitute the feedback network.

C_{F1} (in Fig. 13-10) is a dc-blocking capacitor that prevents the dc voltage at Q_2 collector from affecting the Q_1 bias conditions. C_{F1} behaves as an open-circuit to dc and a short-circuit to ac. Consequently, C_{F1} is not included in the mid-frequency ac equivalent circuit of the feedback network in Fig. 13-11. Capacitor C_2 behaves as a short-circuit at middle and higher frequencies, and so C_2 and R_4 (shorted by C_2) are also absent from the mid-frequency ac equivalent circuit.

The signal voltage (v_s) is applied between Q_1 base and ground, as illustrated in Fig. 13-11, and the output voltage (v_o) is developed between Q_2 collector and ground. The feedback voltage (v_f) is developed across resistor R_{F2},

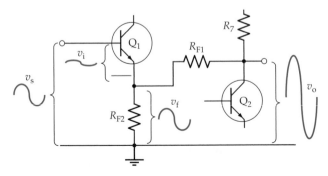

Figure 13-11 Mid-frequency ac equivalent circuit for the feedback network in the circuit of Fig. 13-10.

between the Q_1 emitter terminal and ground. The input voltage (v_i) appears across the Q_1 base-emitter terminals as the difference between v_s and v_f.

From previous studies of a two-stage amplifier circuit (see Section 12-3), it is known that the circuit output voltage is in phase with the signal voltage. When the instantaneous level of v_s is positive-going, v_o is also positive-going, and consequently v_f is in phase-opposition with v_s. When v_s goes up (positively), v_f also goes up, and so the voltage (v_i) at the base-emitter of Q_1 is reduced from v_s to

$$v_i = v_s - v_f$$

Note that the Q_1 bias resistors (R_1 and R_2) in Fig. 13-10 are outside the feedback loop and are unaffected by feedback (see Section 13-1). Since the impedance 'looking into' the transistor base is within the feedback loop, it is altered by feedback. Also recall that the Q_2 collector resistor (R_7) is unaffected by negative feedback and that the impedance 'looking into' the transistor collector terminal is changed by negative feedback.

The feedback factor for the circuit in Figs. 13-10 and 13-11 is given by

$$B = \frac{R_{F2}}{R_{F1} + R_{F2}} \tag{13-8}$$

From Eq. 13-3,

$$A_{CL} \approx \frac{R_{F1} + R_{F2}}{R_{F2}} \tag{13-9}$$

Since usually $R_{F1} \gg R_{F2}$,

$$A_{CL} \approx \frac{R_{F1}}{R_{F2}} \tag{13-10}$$

Example 13-4

Determine the voltage gain, input impedance, and output impedance for the amplifier in Fig. 13-12. Without the feedback network, the circuit has an open-loop voltage gain of 58 000, an input impedance at Q_1 base of $Z_b = 1 \ k\Omega$, and a Q_2 collector resistance of $1/h_{oe} = 85 \ k\Omega$.

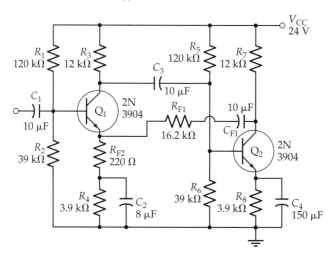

Figure 13-12 Negative feedback amplifier analyzed in Example 13-4.

Solution

Eq. 13-8:
$$B = \frac{R_{F2}}{R_{F1} + R_{F2}} = \frac{220\ \Omega}{16.2\ k\Omega + 220\ \Omega}$$

$$= \frac{1}{74.6}$$

Eq. 13-2:
$$A_{CL} = \frac{A_v}{(1 + A_vB)} = \frac{58\ 000}{1 + (58\ 000/74.6)}$$

$$= 74.5$$

Or, using Eq. 13-9,
$$A_{CL} \approx \frac{R_{F1} + R_{F2}}{R_{F2}} = \frac{16.2\ k\Omega + 220\ \Omega}{220\ \Omega}$$

$$= 74.6$$

Eq. 13-4:
$$Z_i = Z_b(1 + A_vB) = 1\ k\Omega[1 + (58\ 000/74.6)]$$

$$= 778\ k\Omega$$

Eq. 13-5:
$$Z_{in} = Z_i \,\|\, R_1 \,\|\, R_2 = 778\ k\Omega \,\|\, 120\ k\Omega \,\|\, 39\ k\Omega$$

$$= 28.4\ k\Omega$$

Eq. 13-6:
$$Z_o = \frac{1/h_{oe}}{(1 + A_vB)} = \frac{85\ k\Omega}{1 + (58\ 000/74.6)}$$

$$= 109\ \Omega$$

Eq. 13-7:
$$Z_{out} = R_7 \,\|\, Z_o = 12\ k\Omega \,\|\, 109\ \Omega$$

$$= 108\ \Omega$$

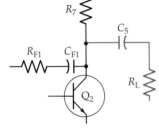

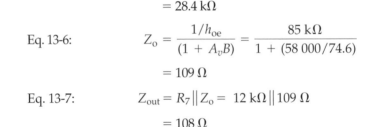

Figure 13-13 The circuit open-loop gain can be reduced by the presence of feedback resistor R_{F1}.

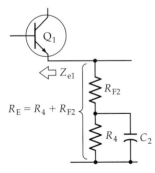

Figure 13-14 Resistor R_{F2} is in parallel with the impedance Z_{e1} 'looking into' the emitter of Q_1. The total (dc) emitter resistance (R_E) is also increased by the presence of R_{F2}.

Negative Feedback Amplifier Design

A negative feedback amplifier, like the one shown in Fig. 13-12, is best designed first as an amplifier without feedback. The feedback component values are then determined, and the circuit is modified as necessary.

Like R_L, R_{F1} is an additional load at the collector of the second stage in Fig. 13-12. As illustrated in Fig. 13-13, the voltage gain of the second stage is proportional to the total load $(R_L \,\|\, R_7 \,\|\, R_{F1})$; consequently, R_{F1} has the effect of reducing the second-stage gain. A large open-loop gain is required for good gain stability, so R_{F1} should be selected to be as large as possible.

Resistor R_{F2} is an unbypassed resistor in series with the emitter of Q_1. As such, it would seem to reduce the first-stage voltage gain, and thus reduce the circuit open-loop gain. However, because v_f is applied to Q_1 emitter, the effective open-loop gain is the voltage gain from the BE terminals of Q_1 to the collector of Q_2. Thus, R_{F2} *does not alter the overall open-loop gain*. (R_{F2} is omitted when calculating the circuit open-loop gain.)

Another consideration for R_{F2} is that it is in parallel with the impedance (Z_{e1}) at the emitter of Q_1 (see Fig. 13-14). Without negative feedback, Z_{e1} would usually equal the transistor r'_e value, which is typically 26 Ω when $I_C = 1$ mA. With negative feedback, Z_{e1} can be shown to range from about 2.5 kΩ to 30 kΩ, depending upon A_v and B. Resistor R_{F2} should be much smaller than Z_{e1}. However, R_{F2} should be as large as possible to give the largest possible value for R_{F1} (from Eq. 13-10, $R_{F1} \approx A_{CL} \times R_{F2}$).

A reasonable starting point for feedback network design is to choose an R_{F2} much lower than $Z_{e1}/10$, in the range of 100 Ω to 470 Ω. R_{F2} could also alter the dc bias conditions for Q_1. Consequently, it might be necessary to reduce R_4 so that R_4 and R_{F2} add up to the original resistance selected for R_4 (or R_E) (see Fig. 13-14).

Figure 13-15 shows a low-frequency ac equivalent circuit for the feedback network. Note that since capacitor C_{F1} is still taken as an ac short circuit, X_{CF1} is assumed to be very much smaller than R_{F1}. The impedance of capacitor C_2 is show, and because X_{C2} is assumed to be very much smaller than resistor R_4, R_4 is omitted.

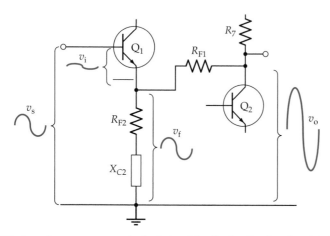

Figure 13-15 Low-frequency ac equivalent circuit for the feedback network in the circuit of Fig. 13-12. Note that X_{C2} is in series with R_{F2}.

From Fig. 13-15, v_o is divided across R_{F1} and $(R_{F2} + jX_{C2})$ to give v_f; therefore the feedback factor is

$$B = \frac{R_{F2} - jX_{C2}}{(R_{F2} - jX_{C2}) + R_{F1}}$$

Assuming that $(R_{F2} - jX_{C2}) \ll R_{F1}$,

$$B \approx \frac{R_{F2} - jX_{C2}}{R_{F1}}$$

From Eq. 13-3:

$$A_{CL} \approx \frac{R_{F1}}{R_{F2} - jX_{C2}}$$

or

$$|A_{CL}| \approx \frac{R_{F1}}{\sqrt{(R_{F2}^2 + X_{C2}^2)}}$$

When $X_{C2} = R_{F2}$,

$$|A_{CL}| \approx \frac{R_{F1}}{\sqrt{2}R_{F2}} \approx 0.707 \times \text{(mid-frequency gain)}$$

To set a desired frequency lower cutoff frequency (f_1),

$$X_{C2} = R_{F2} \text{ at } f_1 \qquad (13\text{-}11)$$

As already mentioned, X_{CF1} must be very much smaller than R_{F1} at the low cutoff frequency for the amplifier. It can be shown that, if $X_{CF1} = R_{F1}$ at f_1, the closed-loop gain would actually increase at frequencies below f_1. C_{F1} is normally calculated from

$$X_{CF1} = \frac{R_{F1}}{100} \text{ at } f_1 \qquad (13\text{-}12)$$

Capacitor C_{F1} can often be eliminated. If R_{F1} is large enough, direct connection of R_{F1} from Q_{2C} to Q_{1E} in Fig. 13-12 will have a negligible effect on the circuit bias conditions. In fact, with a direct-coupled circuit, direct connection of the feedback resistors might enhance the circuit bias stability.

Since the circuit input impedance is increased by negative feedback, the input coupling capacitor C_1 can be recalculated to be smaller than without NFB ($X_{C1} = Z_{in}/10$ at f_1). Recall that the bias resistors are not affected by feedback. Regardless of how high the circuit input impedance might become, C_1 should normally be selected as a minimum of 0.1 µF, to ensure that it is much larger than any stray capacitance at the circuit input.

The emitter bypass capacitor C_4 (in Fig. 13-12) was calculated (in Example 12-6) to have X_{C4} equal to 0.65 r'_e at f_1. So, X_{C4} will reduce the circuit open-loop gain (A_v) at f_1. However, even with this gain reduction, A_v should still be much larger than A_{CL}; so C_4 can normally be left as calculated for an amplifier without feedback.

Example 13-5

Modify the two-stage capacitor-coupled BJT amplifier designed in Example 12-6 to use series voltage feedback, as in Fig. 13-16. The closed-loop gain is to be 75, and the lower cutoff frequency is to be 100 Hz.

Solution

Select $R_{F2} = 220 \ \Omega$ ($R_{F2} < Z_{E1}$)

R_4 becomes $R_4 = \text{(original } R_4) - R_{F2} = 3.9 \text{ k}\Omega - 220 \ \Omega$

                    $= 3.68 \text{ k}\Omega$ (use 3.9 kΩ standard value)

From Eq. 13-9, $R_{F1} = (A_{CL} - 1) R_{F2} = (75 - 1) \times 220 \ \Omega$

                    $= 16.3 \text{ k}\Omega$ (use 15 kΩ + 1.2 kΩ)

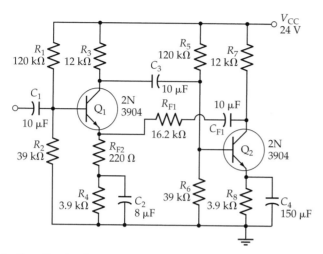

Figure 13-16 Negative feedback amplifier for Example 13-5. Note that this is the circuit analyzed in Example 13-4.

Eq. 13-11:
$$X_{C2} = R_{F2} \text{ at } f_1$$

$$C_2 = \frac{1}{2\pi f_1 R_{F2}}$$

$$= \frac{1}{2\pi \times 100 \text{ Hz} \times 220 \ \Omega}$$

$$= 7.2 \ \mu F \text{ (use 8 } \mu F)$$

Eq. 13-12:
$$X_{CF1} = \frac{R_{F1}}{100} \text{ at } f_1$$

$$C_{F1} = \frac{1}{2\pi f_1 R_{F1}/100}$$

$$= \frac{1}{2\pi \times 100 \text{ Hz} \times 16.2 \text{ k}\Omega/100}$$

$$= 9.8 \ \mu F \text{ (use 10 } \mu F)$$

Practice Problems

13-2.1 Modify the circuit in Fig. 12-18 to use series voltage negative feedback (as in Fig. 13-16) to give $A_{CL} = 100$ and $f_1 = 50$ Hz.

13-2.2 Analyze the negative feedback circuit in Problem 13-2.1 to determine the input and output impedances. The transistor parameters are as follows: $h_{ie} = 2$ kΩ, $h_{oe} = 40$ μS, and $h_{fe} = 100$.

13-3 MORE AMPLIFIERS WITH SERIES VOLTAGE NEGATIVE FEEDBACK

Complementary Direct-Coupled Circuit with Feedback

The two-stage, direct-coupled BJT amplifier in Fig. 13-17 is reproduced from Fig. 12-22 and modified to include feedback components R_{F1} and R_{F2}. The design procedure for the feedback network is exactly as discussed in Section 13-2. Other direct-coupled circuits can be converted into negative feedback amplifiers by following the same procedure. Similarly, BIFET circuits, both direct-coupled and capacitor-coupled, can be designed for overall negative feedback. As always, the best approach is first to design the circuit as a non-feedback amplifier and then to determine the feedback component values.

Note that there is no coupling capacitor for the feedback network in the circuit shown in 13-17. This is because (as discussed in Section 13-2) the omission of the capacitor can have a negligible effect on the circuit dc conditions if feedback resistor R_{F1} is large enough.

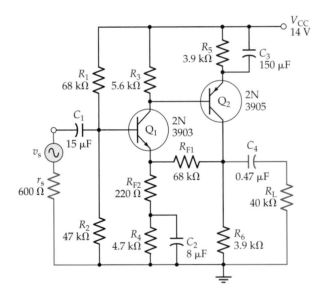

Figure 13-17 Complementary, direct-coupled amplifier using series-voltage negative feedback. The amplifier closed-loop voltage gain is $A_{CL} = (R_{F1} + R_{F2})/R_{F2}$.

Example 13-6

Modify the direct-coupled amplifier in Fig. 12-22 to use series-voltage negative feedback as in Fig 13-17. The amplifier closed-loop voltage gain is to be approximately 300, and the lower cutoff frequency is to be 75 Hz.

Solution

$$R_{F2} < Z_{e1}$$

Select $\qquad R_{F2} = 220 \ \Omega$

R_4 becomes $\qquad R_4 = (\text{original } R_4) - R_{F2} = 4.7 \text{ k}\Omega - 220 \ \Omega$

$$= 4.48 \text{ k}\Omega \text{ (use 4.7 k}\Omega \text{ standard value)}$$

From Eq. 13-9, $R_{F1} = (A_{CL} - 1) R_{F2} = (300 - 1) \times 220\ \Omega$

$$\approx 66\ \text{k}\Omega \text{ (use 68 k}\Omega \text{ standard value)}$$

Eq. 13-11: $X_{C2} = R_{F2}\,\text{at}\,f_1$

$$C_2 = \frac{1}{2\pi f_1 R_{F2}} = \frac{1}{2\pi \times 100\ \text{Hz} \times 220\ \Omega}$$

$$= 7.2\ \mu\text{F (use 8 }\mu\text{F standard value)}$$

Analysis of the circuit in Example 13-6 reveals that the transistor dc collector currents are approximately 1 mA and that the direct current through R_{F1} is about 4.4 μA.

DC Feedback Pair Using Negative Feedback

Figure 13-18a shows a dc feedback pair with ac feedback components C_{F1}, R_{F1}, and R_{F2}. This is the most economical of all two-stage BJT amplifier circuits, because it has the fewest components. It can have just as high a voltage gain (open-loop and closed-loop) as any other two-stage circuit. As a negative feedback amplifier, its input resistance is normally higher than that for a BJT circuit using voltage divider bias (R_2 is usually larger than the parallel resistance of voltage divider resistors). As always, the feedback components are determined after the circuit is designed for the largest possible open-loop voltage gain.

The lower cutoff frequency for the circuit in Fig. 13-18a is unpredictable. The modification shown in Fig. 13-18b gives a definite low 3 dB frequency (when $X_{C4} = R_{F2}$). In this case, emitter resistor R_5 affects the circuit dc bias

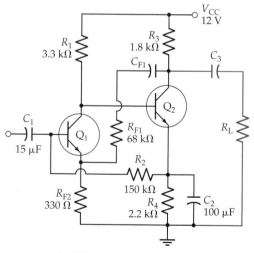

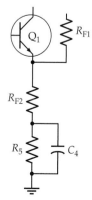

(a) DC feedback pair using series-voltage negative feedback

(b) Circuit modification for predictable low 3 dB frequency

Figure 13-18   DC feedback pair as a negative feedback amplifier. The closed-loop voltage gain is $A_{CL} = (R_{F1} + R_{F2})/R_{F2}$.

conditions; so it must be designed into the circuit from the start. As in other circuits, it might be possible to omit the feedback network coupling capacitor (C_{F1} in Fig. 13-18). The direct current through R_{F1} should be calculated to ensure that it does not significantly affect the bias conditions of the circuit.

BIFET Two-Stage Circuit with Negative Feedback

A two-stage, direct-coupled BIFET circuit with series voltage negative feedback is shown in Fig. 13-19. This is a modified version of the circuit in Fig. 12-30, designed in Examples 12-15 and 12-16.

The open-loop voltage gain of a BIFET circuit is usually 50 to 100 times smaller than that for a two-stage BJT amplifier (see Example 12-17). Therefore, BIFET negative feedback amplifiers must be designed for relatively small closed-loop voltage gains, usually a maximum of around 30. The single major advantage of BIFET circuits, very high input impedance, is largely unchanged by negative feedback. The design procedure for the feedback network in the BIFET circuit is exactly as already discussed.

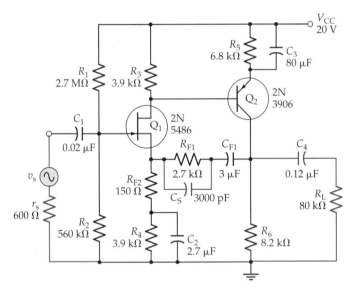

Figure 13-19 BIFET circuit using series voltage negative feedback. The closed-loop voltage gain is $A_{CL} = (R_{F1} + R_{F2})/R_{F2}$.

Setting the Circuit Upper Cutoff Frequency

The upper cutoff frequency of any of the series voltage negative feedback amplifiers already discussed can be set by the simple process of connecting a capacitor (C_S) across feedback resistor R_{F1}, as illustrated in Fig. 13-19. It can be shown that f_2 occurs when $X_{CS} = R_{F1}$, so long as the upper cutoff frequency without C_S is much higher than the desired f_2 value. Therefore, to determine C_S,

$$X_{CS} = R_{F1} \text{ at } f_2 \qquad (13\text{-}13)$$

13-4 DIFFERENTIAL AMPLIFIER WITH NEGATIVE FEEDBACK

Two-Stage Differential Amplifier

The circuit shown in Fig. 13-20 has a differential-amplifier input stage with *npn* BJTs and a direct-coupled *pnp* transistor differential-amplifier second stage. (The differential amplifier is discussed in detail in Section 12-8.) The circuit uses a plus/minus supply voltage, and the base terminals of Q_1 and Q_2 are biased to ground via resistors R_1 and R_6, respectively. Transistors Q_3 and Q_4 have the bases directly connected to the collector terminals of Q_1 and Q_2. So bias voltage for Q_3 and Q_4 bases is provided by the voltage drops across the Q_1 and Q_2 collector resistors (R_2 and R_4).

The amplifier input terminal is the base of Q_1, and the output is taken from Q_3 collector. No collector resistor is provided for Q_4, because no output or feedback is taken from Q_4. Note that the feedback network (R_5 and R_6) is connected from the output at the collector of Q_3 back to the base of Q_2. Because the emitter of Q_2 is directly coupled to Q_1 emitter, applying the feedback voltage (v_f) to Q_2 base is similar to applying it to Q_1 emitter.

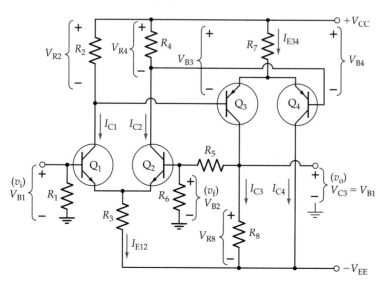

Figure 13-20 Two-stage direct-coupled differential amplifier with series voltage negative feedback.

DC Bias Conditions

The circuit in Fig. 13-20 is designed to have the dc bias voltage at Q_2 base equal to Q_1 base bias voltage, in this case, ground level. This means that the dc voltage at the collector of Q_3 must also equal the Q_1 base voltage, because Q_3 collector is directly connected to Q_2 base via R_5. Consider what would happen if V_{C3} (at the output terminal) is not exactly equal to V_{B1} (at the input).

If V_{C3} goes lower than its normal level:

- V_{B2} is reduced below the level of V_{B1}, and this reduces I_{C2} and increases I_{C1}.
- The increased level of I_{C1} causes an increased voltage drop across R_2, and the reduced level of I_{C2} decreases the voltage drop across R_4. Thus V_{B3} is increased and V_{B4} is decreased.
- This raises the level of I_{C3} and lowers I_{C4}. So, V_{R8} is increased (by the I_{C3} increase) to drive V_{C3} back up to its normal voltage.

Similarly, if V_{C3} somehow drifts to a higher than normal level:

- V_{B2} is raised above the level of V_{B1}, thus increasing I_{C2} and reducing I_{C1}.
- The decreased level of I_{C1} and increased level of I_{C2} produces voltage drops across R_2 and R_4 that reduce V_{B3} and increase V_{B4}.
- This change reduces I_{C3}, thus decreasing V_{R8} to drive V_{C3} back down to its normal voltage.

It is seen that there is *dc negative feedback* that stabilizes the circuit bias conditions.

AC Operation

Consider the circuit waveforms shown in Fig. 13-21. A positive-going input signal (v_i) at the base of Q_1 produces a positive-going output (v_o) at Q_3 collector. As in the case of the dc bias conditions, the instantaneous ac voltage at Q_2 base follows the instantaneous level of the ac signal voltage at Q_1 base.

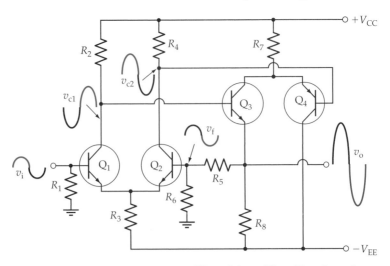

Figure 13-21 AC waveforms in a two-stage differential amplifier with series voltage negative feedback.

With $v_{b2} = v_{b1}$ or $v_i = v_f$,

$$v_f = Bv_o = v_o \times \frac{R_6}{R_5 + R_6}$$

So

$$v_o = v_f \times \frac{R_5 + R_6}{R_6} = v_i \times \frac{R_5 + R_6}{R_6}$$

giving

$$A_{CL} = \frac{R_5 + R_6}{R_6}$$

Resistors R_5 and R_6 in Figs. 13-20 and 13-21 are comparable to R_{F1} and R_{F2}, respectively, in the other negative feedback circuits already discussed. So the closed-loop gain equation is the same equation that applies in the case of all series voltage negative feedback amplifiers. As always, the equation is true only when the open-loop gain is very much larger than the closed-loop gain.

The input and output impedances for a circuit with a differential input stage are calculated exactly as discussed for other negative feedback circuits.

Note that there are no bypass capacitors in the circuit in Figs. 13-20 and 13-21. If the signal and load are capacitor-coupled to the circuit, the coupling capacitors determine the lower cutoff frequency. If the signal and load are direct-coupled, then the circuit is a *dc amplifier*, one that amplifies direct voltage signals.

Circuit Design

The design procedure for the input stage is similar to that discussed in Section 12-8. The second-stage components are determined in essentially the same way as the input stage components, bearing in mind that the second stage is *upside down* compared to the input stage. Feedback network resistor R_6 at the base of Q_2 is selected equal to bias resistor R_1 at Q_1 base. This is to equalize the resistances at the bases of Q_1 and Q_2 and thus equalize any voltage drops due to I_{B1} and I_{B2}. R_5 is calculated from the specified closed loop gain and the R_6 resistance.

Example 13-7

The two-stage differential amplifier in Fig. 13-22 is to use a ±10 V supply and to have a closed-loop gain of 33. Select suitable direct current and voltage levels, and calculate resistor values for the circuit. Assume that the transistors have $h_{FE} = 100$.

Solution

Select

$$I_{C1} = I_{C2} = I_{C3} = I_{C4} \approx 1 \text{ mA}$$

From Eq. 5-17,

$$R_1 \approx \frac{V_{BE}}{10 \, I_{C1}/h_{FE}} \approx \frac{0.7 \text{ V}}{10 \times 1 \text{ mA}/100}$$

$$= 7 \text{ k}\Omega \text{ (use 6.8 k}\Omega \text{ standard value)}$$

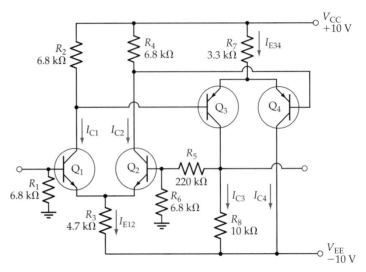

Figure 13-22 Amplifier circuit for Ex. 13-7.

$$R_3 \approx \frac{V_{EE} - V_{BE}}{I_{C1} + I_{C2}} = \frac{10\ V - 0.7\ V}{2\ mA}$$

$$= 4.65\ k\Omega \text{ (use } 4.7\ k\Omega \text{ standard value)}$$

Select $\qquad V_{CE1} \approx 3\ V$

$$V_{R2} = V_{CC} + V_{BE1} - V_{CE1} = 10\ V + 0.7\ V - 3\ V$$

$$= 7.7\ V$$

$$R_2 = R_4 = \frac{V_{R2}}{I_{C1}} = \frac{7.7\ V}{1\ mA}$$

$$= 7.7\ k\Omega \text{ (use } 6.8\ k\Omega \text{ standard value)}$$

$$R_7 = \frac{V_{R2} - V_{BE}}{I_{C3} + I_{C4}} = \frac{7.7\ V - 0.7\ V}{2\ mA}$$

$$= 3.5\ k\Omega \text{ (use } 3.3\ k\Omega \text{ standard value)}$$

$$R_8 = \frac{V_{EE}}{I_{C3}} = \frac{10\ V}{1\ mA}$$

$$= 10\ k\Omega \text{ (standard value)}$$

Select $\qquad R_6 = R_1 = 6.8\ k\Omega$

From Eq. 13-9, $\qquad R_5 = (A_{CL} - 1)\ R_6 = (33 - 1) \times 6.8\ k\Omega$

$$= 217.6\ k\Omega \text{ (use } 220\ k\Omega \text{ standard value)}$$

Example 13-8

Assuming an open-loop gain of 25 000, $h_{ie} = 2\ k\Omega$, $h_{ib} = 25\ \Omega$, and $1/h_{oe} = 100\ k\Omega$ for the circuit in Example 13-7, calculate A_{CL}, Z_{in}, and Z_{out}.

Solution

From Eq. 13-8,
$$B = \frac{R_6}{R_5 + R_6} = \frac{6.8 \text{ k}\Omega}{220 \text{ k}\Omega + 6.8 \text{ k}\Omega}$$

$$= \frac{1}{33.4}$$

Eq. 13-2:
$$A_{\text{CL}} = \frac{A_v}{1 + A_v B} = \frac{25\,000}{1 + (25\,000/33.4)}$$

$$= 33.4$$

Eq. 13-4:
$$Z_i = 2\,h_{\text{ie}}(1 + A_v B) = 2 \text{ k}\Omega[1 + (25\,000/33.4)]$$

$$= 3 \text{ M}\Omega$$

Eq. 13-5:
$$Z_{\text{in}} = Z_i \| R_1 = 3 \text{ M}\Omega \| 6.8 \text{ k}\Omega$$

$$\approx 6.8 \text{ k}\Omega$$

Eq. 13-6:
$$Z_o = \frac{1/h_{\text{oe}}}{1 + A_v B} = \frac{100 \text{ k}\Omega}{1 + (25\,000/33.4)}$$

$$= 133 \,\Omega$$

Eq. 13-7:
$$Z_{\text{out}} = R_8 \| Z_o = 10 \text{ k}\Omega \| 133 \,\Omega$$

$$= 131 \,\Omega$$

Modification for Reduced Z_{out}

Although the application of series voltage negative feedback substantially reduces the output impedance of a circuit, further reduction of Z_{out} is sometimes
required. Figure 13-23 shows the circuit of Fig. 13-22 with the addition of an

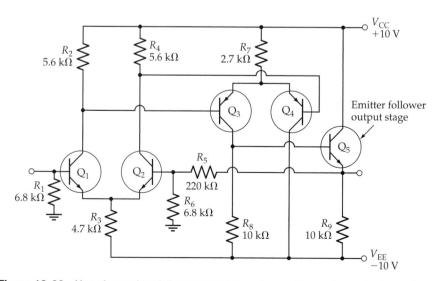

Figure 13-23   Use of an emitter follower output stage to provide very low output impedance.

emitter follower output stage (Q_5). Note that the feedback network is now connected at the emitter follower output terminal.

The output impedance at Q_5 emitter terminal in Fig. 13-23 is found from the derivation of Eq. 6-24 (for a common-collector circuit).

$$Z_e = \frac{h_{ic} + (Z \text{ in series with } Q_5 \text{ base})}{h_{fc}}$$

This gives

$$Z_e = \frac{h_{ic} + R_8}{h_{fc}} \qquad (13\text{-}14)$$

This quantity is modified by negative feedback according to (a rewritten) Eq. 13-6:

$$Z_o = \frac{Z_e}{1 + A_v B}$$

Then, as always, the circuit output impedance is given by Equation 13-7, which must also be rewritten for the circuit in Fig. 13-23:

$$Z_{out} = Z_o \| R_9$$

An emitter follower output stage can be added to any of the amplifiers previously discussed. However, the combination of an emitter follower output stage with a differential amplifier has a particular significance that will be explained in Chapter 14.

Example 13-9

Calculate the output impedance for the circuit modification in Fig. 13-23, assuming that $h_{ic} = h_{ie} = 2 \text{ k}\Omega$, and $h_{fc} = h_{fe} = 100$. Also assume that $A_v = 25\,000$ and $B = 1/33.4$, as for Examples 13-7 and 13-8.

Solution

Eq. 13-14:
$$Z_e = \frac{h_{ic} + R_8}{h_{fc}} = \frac{2 \text{ k}\Omega + 10 \text{ k}\Omega}{100}$$

$$= 120 \ \Omega$$

Eq. 13-6:
$$Z_o = \frac{Z_e}{1 + A_v B} = \frac{120 \ \Omega}{1 + (25\,000/33.4)}$$

$$= 0.16 \ \Omega$$

Eq. 13-7:
$$Z_{out} = R_5 \| Z_o = 10 \text{ k}\Omega \| 0.16 \ \Omega$$

$$= 0.16 \ \Omega$$

Practice Problems

13-4.1 A two-stage differential amplifier with negative feedback, as in Fig. 13-22, is to use 2N3904 and 2N3906 BJTs with a ±15 V supply. The closed-loop gain is to be 50. Determine suitable resistor values.

13-4.2 Calculate the minimum open-loop gain for the circuit in Problem 13-4.1. Determine Z_{in} and Z_{out}.

13-5 EMITTER CURRENT FEEDBACK

Emitter Current Feedback Circuit

Emitter current feedback is produced by connecting an unbypassed resistor (termed a *swamping resistor*) in series with the emitter terminal of a transistor, as shown in Fig. 13-24a. The effects of an unbypassed emitter resistor in a CE circuit were analyzed in Section 6-5, where it was shown that the circuit impedances and voltage gains are as follows:

Eq. 6-20:
$$Z_b = h_{ie} + R_{E1}(1 + h_{fe})$$

Eq. 6-12:
$$Z_{in} = R_1 \| R_2 \| Z_b$$

Eq. 6-14:
$$Z_{out} = (1/h_{oe}) \| R_C \approx R_C$$

Eq. 6-21:
$$A_v = \frac{-h_{fe}(R_C \| R_L)}{h_{ie} + R_{E1}(1 + h_{fe})}$$

Eq. 6-22:
$$A_v \approx \frac{R_C \| R_L}{R_{E1}}$$

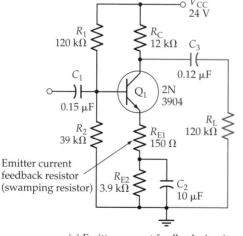

(a) Emitter current feedback circuit

(b) Signal, input, and feedback voltages

Figure 13-24   Single-stage CE amplifier using emitter current feedback. The voltage gain is $A_v \approx (R_C \| R_L)/R_{E1}$.

Equation 6-20 shows that the input impedance at the base of a transistor with an unbypassed emitter resistor is considerably increased above its normal value without feedback (h_{ie}). This raises the circuit input impedance, as defined by Eq. 6-12. Therefore,

emitter current feedback increases the circuit input impedance.

Figure 13-24b shows that the circuit feedback loop is from the transistor base to the emitter, then back to the base again. The collector resistor is outside the feedback loop. The circuit output impedance remains approximately equal to R_C (see Eq. 6-14). So,

emitter current feedback has no effect on the circuit output impedance (see Section 6-5).

The precise voltage gain for the circuit in Fig. 13-24a is given by Equation 6-21, and Equation 6-22 can be used for calculation of the approximate voltage gain. Equation 6-22 is derived from Equation 6-21 by assuming that $h_{ie} \ll R_{E1}(1 + h_{fe})$. If we use $R_{E1} = 100 \ \Omega$ and $h_{fe} = 100$,

$$R_{E1}(1 + h_{fe}) \approx 10 \ \text{k}\Omega$$

This is 10 times the typical value for h_{ie}. So Equation 6-22 gives a reasonably close approximation to the circuit voltage gain.

It is apparent that

emitter current feedback stabilizes single-stage amplifier voltage gain.

Example 13-10

The emitter current feedback circuit in Fig. 13-24 uses a transistor with $h_{fe(min)} = 100$, $h_{fe(max)} = 400$, $h_{ie(min)} = 2 \ \text{k}\Omega$, and $h_{ie(max)} = 5 \ \text{k}\Omega$. Calculate the precise value of the circuit voltage gain at the h_{fe} extremes, and then calculate the approximate voltage gain by using Eq. 6-22.

Solution

Eq. 6-21:
$$A_v = \frac{h_{fe}(R_C \| R_L)}{h_{ie} + R_{E1}(1 + h_{fe})}$$

$$A_{v(max)} = \frac{h_{fe(max)}(R_C \| R_L)}{h_{ie(max)} + R_{E1}(1 + h_{fe(max)})} = \frac{400(12 \ \text{k}\Omega \| 120 \ \text{k}\Omega)}{5 \ \text{k}\Omega + 150 \ \Omega(1 + 400)}$$

$$= 67$$

$$A_{v(min)} = \frac{h_{fe(min)}(R_C \| R_L)}{h_{ie(min)} + R_{E1}(1 + h_{fe(min)})} = \frac{100(12 \ \text{k}\Omega \| 120 \ \text{k}\Omega)}{2 \ \text{k}\Omega + 150 \ \Omega(1 + 100)}$$

$$= 63.6$$

Eq. 6-22:
$$A_v \approx \frac{R_C \| R_L}{R_{E1}} = \frac{12 \ \text{k}\Omega \| 120 \ \text{k}\Omega}{150 \ \Omega}$$

$$= 72.7$$

Circuit Design

As in the case of other negative feedback amplifiers, an emitter-current feed-back circuit should first have its resistor values calculated as for a circuit without feedback. Resistor R_{E1} is then determined from Equation 6-22, and if R_{E1} is large enough to affect the circuit bias conditions, R_{E2} is reduced to its original value minus R_{E1}. The input and output coupling capacitors are calculated in the usual way, but remember that Z_{in} is increased by the presence of R_{E1}. The equation for the emitter bypass capacitor (Eq. 12-3) is rewritten to take R_{E1} into account. Equation 12-3 was derived from the fact that X_{C2} must equal the impedance in series with C_2 at the lower cutoff frequency for the circuit. For the circuit in Fig. 13-24a,

$$X_{C2} = (r'_e + R_{E1}) \text{ at } f_1$$

Usually, $R_{E1} \gg r'_e$, so the capacitor can be calculated from

$$X_{C2} = R_{E1} \text{ at } f_1 \qquad\qquad (13\text{-}15)$$

Example 13-11

Modify the CE amplifier circuit designed in Examples 12-2 and 12-3 to use emitter current feedback to give $A_v = 70$ (see Fig. 13-25). The lower cutoff frequency is 100 Hz, and the signal source resistance is 600 Ω.

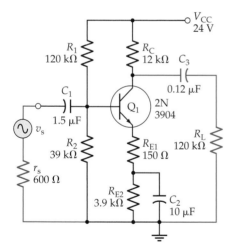

Figure 13-25   Emitter current feedback circuit for Ex. 13-11.

Solution

From Eq. 6-22,

$$R_{E1} \approx \frac{R_C \| R_L}{A_v} = \frac{12 \text{ k}\Omega \| 120 \text{ k}\Omega}{70}$$

$$\approx 156 \ \Omega \text{ (use 150 } \Omega \text{ standard value)}$$

$$R_{E2} = R_E - R_{E1} = 3.9 \text{ k}\Omega - 150 \ \Omega$$

$$= 3.75 \text{ k}\Omega \text{ (use 3.9 k}\Omega)$$

Eq. 6-20: $Z_{b(min)} = h_{ie} + R_{E1}(1 + h_{fe}) = 2\text{ k}\Omega + 150\text{ }\Omega(1 + 100)$

$= 17.2\text{ k}\Omega$

Eq. 6-12: $Z_{in} = R_1 \| R_2 \| Z_b = 120\text{ k}\Omega \| 39\text{ k}\Omega \| 17.2\text{ k}\Omega$

$= 10.8\text{ k}\Omega$

Eq. 12-4: $C_1 = \dfrac{1}{2\pi f_1 (Z_{in} + r_s)/10}$

$= \dfrac{1}{2\pi \times 100\text{ Hz} \times (10.8\text{ k}\Omega + 600\text{ }\Omega)/10}$

$= 1.4\text{ }\mu\text{F}$ (use 1.5 µF standard value)

Eq. 13-15: $X_{C2} = R_{E1}$ at f_1

$C_2 = \dfrac{1}{2\pi f_1 R_{E1}} = \dfrac{1}{2\pi \times 100\text{ Hz} \times 150\text{ }\Omega}$

$= 10.6\text{ }\mu\text{F}$ (use 10 µF standard value)

Two-Stage Circuit with Emitter Current Feedback

All of the two-stage amplifiers discussed in Chapter 12 can be designed to use emitter current feedback for voltage gain stabilization and increased input impedance. The design procedures are essentially those already discussed.

Consider the two-stage, capacitor-coupled circuit in Fig. 13-26. The voltage gain for stage 1 is calculated as follows:

From Eq. 6-22,
$$A_{v1} \approx \frac{R_3 \| Z_{in2}}{R_4} \tag{13-16}$$

and for stage 2,
$$A_{v2} = \frac{R_8 \| R_L}{R_9} \tag{13-17}$$

The overall voltage gain is

$$A_v = A_{v1} \times A_{v2}$$

When designing a two-stage circuit with emitter current feedback, the gain of each stage is determined as

$$A_{v1} = A_{v2} = \sqrt{A_v} \tag{13-18}$$

Although the two stages can be designed for identical gains, $R_3 \| Z_{in2}$ is unlikely to equal $R_8 \| R_L$. So R_4 and R_9 are likely to be unequal.

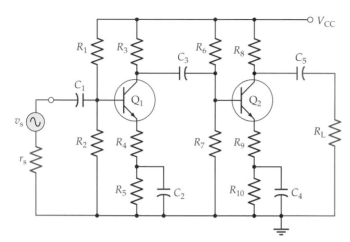

Figure 13-26   Two-stage, capacitor-coupled BJT amplifier using emitter current feedback in each stage.

For the reasons discussed in Section 12-3, the emitter bypass capacitors are determined from

$$X_{C2} = 0.65\ R_4 \text{ at } f_1 \tag{13-19}$$

and

$$X_{C4} = 0.65\ R_9 \text{ at } f_1 \tag{13-20}$$

The circuit input impedance is, once again, given by Eqs. 6-20 and 6-12, and the output impedance is simply the collector resistor in the second stage (see Eq. 6-14).

Example 13-12

The circuit in Fig. 13-25 is to be modified for use as each stage of the circuit in Fig. 13-27. The overall voltage gain is to be 1000, and the lower cutoff frequency is to be 100 Hz. Determine suitable emitter resistor values. Assume transistor parameters of $h_{ie} = 2\ \text{k}\Omega$ and $h_{fe} = 100$.

Solution

Eq. 13-18:

$$A_{v1} = A_{v2} = \sqrt{A_v} = \sqrt{1000}$$

$$= 31.6$$

From Eq. 13-17,

$$R_9 \approx \frac{R_8 \| R_L}{A_{v2}} = \frac{12\ \text{k}\Omega \| 120\ \text{k}\Omega}{31.6}$$

$$\approx 345\ \Omega \text{ (use } 330\ \Omega \text{ standard value)}$$

$$R_{10} = R_E - R_9 = 3.9\ \text{k}\Omega - 330\ \Omega$$

$$= 3.57\ \text{k}\Omega \text{ (use } 3.9\ \text{k}\Omega)$$

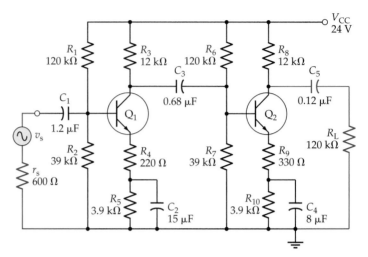

Figure 13-27   Emitter current feedback amplifier circuit for Examples 13-12 and 13-13.

A_{v2} becomes	$A_{v2} \approx \dfrac{R_8 \| R_L}{R_9} = \dfrac{12 \text{ k}\Omega \| 120 \text{ k}\Omega}{330 \ \Omega}$

$$\approx 33$$

Therefore,	$A_{v1} \approx \dfrac{A_v}{A_{v2}} = \dfrac{1000}{33}$

$$= 30.3$$

From Eq. 6-20,	$Z_{b2} = h_{ie} + R_9(1 + h_{fe})$

$$= 2 \text{ k}\Omega + 330 \ \Omega(1 + 100)$$

$$= 35.3 \text{ k}\Omega$$

From Eq. 6-12,	$Z_{in2} = R_6 \| R_7 \| Z_{b2} = 120 \text{ k}\Omega \| 39 \text{ k}\Omega \| 35.3 \text{ k}\Omega$

$$= 16 \text{ k}\Omega$$

From Eq. 13-16,	$R_4 \approx \dfrac{R_3 \| Z_{in2}}{A_{v1}} = \dfrac{12 \text{ k}\Omega \| 16 \text{ k}\Omega}{30.3}$

$$\approx 220 \ \Omega \ (\text{use } 220 \ \Omega \text{ standard value})$$

$$R_5 = R_E - R_4 = 3.9 \text{ k}\Omega - 220 \ \Omega$$

$$= 3.68 \text{ k}\Omega \ (\text{use } 3.9 \text{ k}\Omega)$$

Example 13-13

Calculate suitable capacitor values for the two-stage circuit in Example 13-12.

Solution

Eq. 6-20:	$Z_{b1} = h_{ie} + R_4(1 + h_{fe}) = 2 \text{ k}\Omega + 220 \ \Omega(1 + 100)$

$$= 24.2 \text{ k}\Omega$$

Eq. 6-12: $Z_{in1} = R_1 \| R_2 \| Z_{b1} = 120 \text{ k}\Omega \| 39 \text{ k}\Omega \| 24.2 \text{ k}\Omega$

$$= 13.3 \text{ k}\Omega$$

Eq. 12-4: $C_1 = \dfrac{1}{2\pi f_1 (Z_{in1} + r_s)/10} = \dfrac{1}{2\pi \times 100 \text{ Hz} \times (13.3 \text{ k}\Omega + 600 \ \Omega)/10}$

$$= 1.1 \ \mu\text{F (use 1.2 } \mu\text{F standard value)}$$

Eq. 13-19: $X_{C2} = 0.65 \ R_4$ at f_1

$$C_2 = \dfrac{1}{2\pi \ f_1 \ 0.65 \ R_4} = \dfrac{1}{2\pi \times 100 \text{ Hz} \times 0.65 \times 220 \ \Omega}$$

$$= 11.1 \ \mu\text{F (use 15 } \mu\text{F standard value)}$$

Eq. 12-4: $C_3 = \dfrac{1}{2\pi f_1 (Z_{in2} + R_3)/10} = \dfrac{1}{2\pi \times 100 \text{ Hz} \times (16 \text{ k}\Omega + 12 \text{ k}\Omega)/10}$

$$= 0.56 \ \mu\text{F (use 0.68 } \mu\text{F standard value)}$$

Eq. 13-20: $X_{C4} = 0.65 \ R_9$ at f_1

$$C_4 = \dfrac{1}{2\pi f_1 \ 0.65 \ R_9} = \dfrac{1}{2\pi \times 100 \text{ Hz} \times 0.65 \times 330 \ \Omega}$$

$$= 7.4 \ \mu\text{F (use 8 } \mu\text{F standard value)}$$

Eq. 12-5: $C_5 = \dfrac{1}{2\pi f_1 (R_8 + R_L)/10} = \dfrac{1}{2\pi \times 100 \text{ Hz} \times (12 \text{ k}\Omega + 120 \text{ k}\Omega)/10}$

$$= 0.12 \ \mu\text{F (standard value)}$$

Other two-stage circuits can be designed to use emitter current bias (for example, see Fig. 13-28). However, series voltage feedback produces more

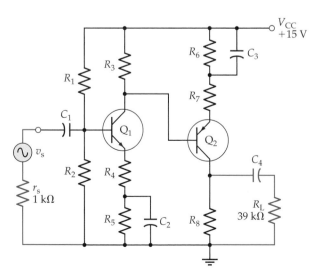

Figure 13-28   Two-stage direct-coupled BJT amplifier circuit using emitter current feedback.

stable (closed-loop) voltage gains than single-stage emitter current feedback. Series voltage feedback also produces reduced output impedance, as well as increased input impedance. Emitter current feedback increases the circuit input impedance, but leaves the output impedance unaffected. Consequently, overall series voltage negative feedback is normally preferable to single-stage emitter current feedback.

Practice Problems

13-5.1 Calculate resistor values for the emitter current feedback circuit in Fig. 13-28 to give $A_v = 600$. The transistor parameters are as follows: $h_{ie} = 1.5 \text{ k}\Omega$ and $h_{fe} = 60$; and the cutoff frequency is to be 80 Hz.

13-5.2 Determine suitable capacitor values for the circuit in Problem 13-5.1.

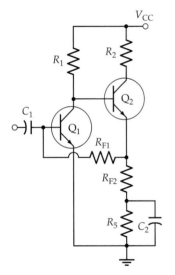

Figure 13-29 DC feedback pair with parallel current negative feedback. The circuit current gain is $A_{i(CL)} \approx (R_{F1} + R_{F2})/R_{F2}$.

13-6 PARALLEL CURRENT NEGATIVE FEEDBACK

Current Feedback Circuit

In series voltage feedback, a portion of the output voltage is fed back in series with the signal source. In *parallel current feedback*, a portion of the output current is fed back in parallel with the signal source. Just as series voltage feedback stabilizes the voltage gain of a circuit, so parallel current feedback stabilizes the current gain.

Consider the dc feedback pair circuit in Fig. 13-29. An unbypassed resistor (R_{F2}) is connected in the emitter circuit of transistor Q_2, and Q_1 bias resistor (R_{F1}) is connected at the emitter terminal of Q_2. With this arrangement, any ac voltage developed across R_{F2} is applied to R_{F1}.

Look at the instantaneous polarities of the input voltage (v_i) and feedback voltage (v_f), as illustrated in the ac equivalent circuit in Fig. 13-30. When v_i

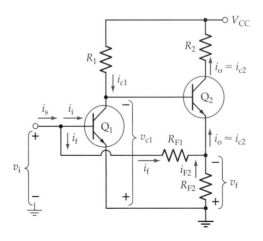

Figure 13-30 AC equivalent for the current feedback circuit Fig. 13-29, showing ac voltages and currents.

is moving in a positive direction, v_{c1} is negative-going. Because the emitter of Q_2 follows the voltage (v_{c1}) at its base, v_f (across R_{F2}) is also negative-going at this time.

Now look at the instantaneous ac current directions indicated in Fig. 13-30. When v_i goes positive, the signal current (i_s) flows into the circuit and an input current (i_i) flows into the base of Q_1, as shown. Because v_f goes negative when v_i is positive-going, a feedback current (i_f) flows in R_{F1} in the direction indicated (from Q_1 base toward Q_2 emitter). Thus, some of the signal current is diverted away from the base of Q_1. This means that the output current (i_{c2}) is less than it would be if there were no current feedback; consequently, the circuit current gain is reduced.

Current Gain

To determine the effect of negative feedback on the circuit current gain, first investigate the current gain without feedback (the open-loop current gain). Suppose that R_{F2} in Fig. 13-30 has a bypass capacitor connected across it so that there is no negative feedback. Assuming that the resistance of R_{F1} is very much larger than the Q_1 input impedance (h_{ie1}), virtually all of i_s enters the base of Q_1 (see Fig. 13-30).

So
$$i_s \approx i_1$$

and
$$i_{c1} \approx h_{fe1}\, i_1$$

Current i_{c1} divides between R_1 and Z_{i2}. Using the current divider equation,

$$i_{b2} = \frac{i_{c1} \times R_1}{R_1 + Z_{i2}}$$

$$i_o = i_{c2} = h_{fe2}\, i_{b2}$$

or
$$i_o = \frac{h_{fe2}\, i_{c1}\, R_1}{R_1 + Z_{i2}} = \frac{h_{fe1}\, h_{fe2}\, i_i\, R_1}{R_1 + Z_{i2}}$$

The open-loop current gain is

$$A_i = \frac{i_o}{i_i}$$

$$A_i = \frac{h_{fe1}\, h_{fe2}\, R_1}{R_1 + Z_{i2}} \tag{13-21}$$

Now derive an equation for the current gain with negative feedback. Referring to Fig. 13-30, it is seen that i_o (in the emitter circuit of Q_2) divides between R_{F1} and R_{F2}, giving

$$i_f = \frac{i_o \times R_{F2}}{R_{F1} + R_{F2}} = i_o B$$

where
$$B = \frac{R_{F2}}{R_{F1} + R_{F2}}$$

Also, $i_s = i_i + i_f = i_i + i_oB$

and $i_o = A_i i_i$

Therefore $i_s = i_i + (A_i i_i B) = i_i(1 + A_i B)$

The closed-loop current gain is

$$A_{i(CL)} = \frac{i_o}{i_s} = \frac{A_i i_i}{i_i(1 + A_iB)}$$

$$A_{i(CL)} = \frac{A_i}{1 + A_iB} \qquad\qquad (13\text{-}22)$$

If $A_iB \gg 1$, then

$$A_{i(CL)} \approx \frac{1}{B} \approx \frac{R_{F1} + R_{F2}}{R_{F2}} \qquad\qquad (13\text{-}23)$$

It is seen that

parallel current negative feedback stabilizes circuit current gain.

It should be noted that, in the current gain equations, the ac output current is taken as the current in the collector circuit of transistor Q_2. How this current divides between a capacitor-coupled external load (R_L) and the Q_2 collector resistor (R_3) depends on the values of R_L and R_3.

Output and Input Impedances

In the circuit shown in Figs. 13-29 and 13-30, the Q_2 collector resistor (R_2) is outside the feedback loop. Therefore, like the case of emitter current feedback, the output impedance of this circuit is largely unaffected by feedback. (It can be shown that the impedance 'looking into' the collector of Q_2 is increased by current feedback.) So the circuit output impedance is

$$Z_o \approx R_2$$

Assuming that $R_{F1} \gg h_{ie1}$, the circuit input impedance without feedback is given by

$$Z_i \approx h_{ie1} = \frac{v_i}{i_s}$$

As already shown, with negative feedback,

$$i_s = i_i + i_f = i_i(1 + A_iB)$$

So, $Z_i = \dfrac{v_i}{i_s} = \dfrac{v_i}{i_i(1 + A_iB)}$

or $Z_i = \dfrac{h_{ie1}}{1 + A_iB} \qquad\qquad (13\text{-}24)$

Therefore,

> *parallel-current negative feedback reduces circuit input impedance by a factor of* $(1 + A_i B)$.

Another way of looking at the circuit input impedance is in terms of the feedback resistor R_{F1} and the voltage gain of the first stage. It can be shown that

$$Z_i = h_{ie1} \| \frac{R_{F1}}{A_{v1}} \qquad (13\text{-}25)$$

Circuit Design

As with other negative feedback circuits, the designing of a parallel current negative feedback amplifier is approached by first ignoring the ac negative feedback components. Resistors R_{F2} and R_5 in Fig. 13-29 are calculated as a single emitter resistor ($R_E = R_{F2} + R_5$) to fulfil the desired dc conditions. Also, R_{F1} is calculated (as for other dc feedback bias circuits) to provide the required base current to Q_1. R_{F2} is then calculated from Eq. 13-23. Usually, the resistance of R_{F1} is very large, and this results in a calculated resistance for R_{F2} larger than ($R_{F2} + R_5$)! In this case, the base resistor for Q_1 should be made up of two resistors, one of which is bypassed: see ($R_2 = R_{F1} + R_7$) in Fig. 13-31. R_{F1} is the unbypassed portion of R_2, and this is calculated in relation to R_{F2}.

Bypass capacitor C_2 (in Fig. 13-31) is calculated to give $X_{C2} = R_{F2}$ at the desired lower cutoff frequency. Because the signal is derived from a current source, the impedance of the input coupling capacitor (C_1) at f_1 should be

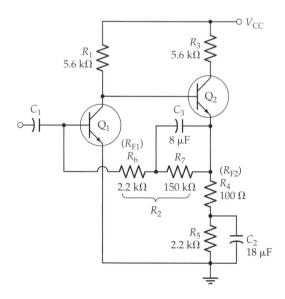

Figure 13-31 Current feedback circuit with part of R_2 bypassed. The current gain is given by $A_{i(CL)} \approx (R_6 + R_4)/R_4$.

much smaller than the (normally high) impedance of the signal source. Capacitor C_3 is determined by making X_{C3} very much smaller than R_{F1} at f_1.

Example 13-14

Analyze the circuit in Fig. 13-31 to determine the current gain and the input impedance. Assume transistor parameters of $h_{fe} = 100$ and $h_{ie} = 2 \text{ k}\Omega$.

Solution

From Eq. 6-20, $Z_{i2} = h_{ie2} + (1 + h_{fe2})R_4 = 2 \text{ k}\Omega + (1 + 100)100 \text{ }\Omega$

$$= 12.1 \text{ k}\Omega$$

Open-loop current gain:

Eq. 13-21: $A_i = \dfrac{h_{fe1}\, h_{fe2}R_1}{R_1 + Z_{i2}} = \dfrac{100 \times 100 \times 5.6 \text{ k}\Omega}{5.6 \text{ k}\Omega + 12.1 \text{ k}\Omega}$

$$= 3164$$

From Eq. 13-23, $B = \dfrac{R_4}{R_6 + R_4} = \dfrac{100 \text{ }\Omega}{2.2 \text{ k}\Omega + 100 \text{ }\Omega}$

$$= \dfrac{1}{23}$$

Closed-loop current gain:

Eq. 13-22: $A_{i(CL)} = \dfrac{A_i}{1 + A_iB} = \dfrac{3164}{1 + (3164/23)}$

$$= 22.8$$

$$A_{i(CL)} \approx (R_6 + R_4)/R_4.$$

Eq. 13-24: $Z_i = \dfrac{h_{ie1}}{1 + A_iB} = \dfrac{2 \text{ k}\Omega}{1 + (3164/23)}$

$$= 14.6 \text{ }\Omega$$

Practice Problems

13-6.1 Modify the dc feedback pair circuit shown in Fig. 12-25a to give $A_{i(CL)} = 55$ and $f_1 = 80$ Hz. Assume that $r_s = 12 \text{ k}\Omega$.

13-6.2 Assuming transistor parameters of $h_{ie} = 1 \text{ k}\Omega$ and $h_{fe} = 50$, calculate Z_i and $A_{i(CL)}$ for the circuit in Problem 13-6.1.

13-7 ADDITIONAL EFFECTS OF NEGATIVE FEEDBACK

Decibels of Feedback

Negative feedback can be measured in decibels. A statement that 40 dB of feedback has been applied to an amplifier means that the amplifier gain has been reduced by 40 dB (that is, by a factor of 100). Thus,

$$A_{CL} = A_v - 40 \text{ dB} = \frac{A_v}{100}$$

Bandwidth

Consider the typical gain-frequency response of an amplifier, as illustrated in Fig. 13-32. Without negative feedback, the amplifier open-loop gain (A_v) falls off to its lower 3 dB frequency ($f_{1(OL)}$), as illustrated. This is usually due to the impedance of bypass capacitors increasing as the frequency decreases. Similarly, the open-loop upper cutoff frequency ($f_{2(OL)}$) is produced by transistor cutoff, by shunting capacitance, or by a combination of both. As discussed in Section 8-2, the circuit open-loop bandwidth is given by

$$BW_{OL} = f_{2(OL)} - f_{1(OL)}$$

Now look at the typical frequency response for the same amplifier when negative feedback is used. The closed-loop gain (A_{CL}) is much smaller than the

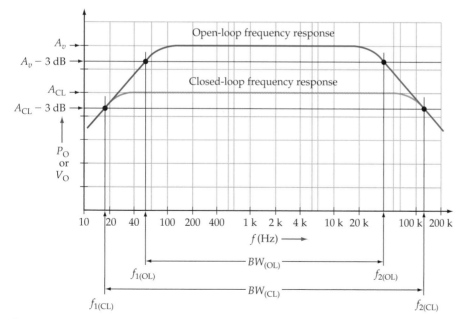

Figure 13-32 Amplifier frequency response with and without negative feedback. Negative feedback extends the amplifier bandwidth.

open-loop gain, and A_{CL} does not begin to fall off (at high or low frequencies) until A_v (open-loop gain) falls substantially. Consequently, $f_{1(CL)}$ is much lower than $f_{1(OL)}$, and $f_{2(CL)}$ is much higher than $f_{2(OL)}$. So the circuit bandwidth with negative feedback (the closed-loop bandwidth) is much greater than the bandwidth without negative feedback.

$$BW_{CL} = f_{2(CL)} - f_{1(CL)}$$

From Eq. 13-2,

$$A_{CL} = \frac{A_v}{1 + A_v B}$$

It can be shown that there is a 90° phase shift associated with the open-loop gain at frequencies below $f_{1(OL)}$ and above $f_{2(OL)}$. Thus, Equation 13-2 must be rewritten as

$$A_{CL} = \frac{-jA_v}{1 - jA_v B}$$

or

$$|A_{CL}| = \frac{A_v}{\sqrt{[1 + (A_v B)^2]}}$$

When $A_v = 1/B$,

$$|A_{CL}| = \frac{1/B}{\sqrt{[1 + 1]}} = \frac{A_{CL}}{\sqrt{2}}$$

$$= A_{CL} - 3 \text{ dB}$$

Thus, for a negative feedback amplifier designed to have the widest possible bandwidth, the cutoff frequencies would occur when the open-loop gain falls to the equivalent of $1/B$. Thus, $f_{2(CL)}$ occurs when

$$A_v = 1/B \approx A_{CL} \tag{13-26}$$

So, for example, the cutoff frequencies for a negative feedback amplifier designed for a closed-loop gain of 100 would occur when the open-loop gain falls to 100. It is seen that

negative feedback increases amplifier bandwidth.

The upper cutoff frequency for an amplifier is usually greater than 20 kHz, and the lower cutoff frequency is around 100 Hz or lower. So $f_2 \gg f_1$, and consequently,

$$BW = f_2 - f_1 \approx f_2$$

This means that the amplifier bandwidth is essentially equal to the upper cutoff frequency.

Now refer to Fig. 13-32 once again. The amplifier gain multiplied by the upper cutoff frequency is a constant quantity. This is known as the *gain-bandwidth product.* Therefore,

$$A_{CL} \times f_{2(CL)} = A_v \times f_{2(OL)}$$

or

$$f_{2(CL)} = \frac{A_v f_{2(OL)}}{A_{CL}} \qquad\qquad (13\text{-}27)$$

Thus the closed-loop upper cutoff frequency for a negative feedback amplifier can be calculated from the open-loop upper cutoff frequency, the open-loop gain, and the closed-loop gain.

Harmonic Distortion

Harmonic distortion occurs when a transistor or other device is driven beyond the linear range of its characteristics. Figure 13-33 shows the I_C/V_{BE} characteristics of a transistor biased to a base-emitter voltage V_B. As illustrated, when the base-emitter voltage changes ($\pm v_{be}$) are very small, the collector current changes by equal positive and negative amounts ($\pm i_c$). When $\pm v_{be}$ is large, the change in $+i_c$ is greater than the change in $-i_c$. This is because of the non-linearity in the I_C/V_{BE} characteristics. The result is *harmonic distortion* (or *non-linear distortion*) in the waveform of i_c and in the amplifier output.

The distorted waveform can be shown to consist of a *fundamental* frequency waveform and a number of smaller amplitude *harmonic* components (see Fig. 13-34). The fundamental waveform is the amplified signal, or amplifier output voltage (v_o), and the harmonics are unwanted voltage components (v_H).

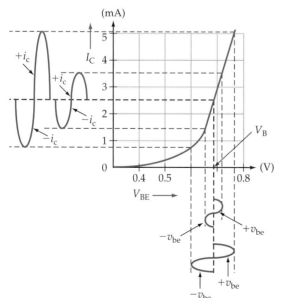

Figure 13-33 Harmonic distortion in the output of an amplifier is caused by non-linearity in transistor characteristics. Negative feedback reduces this type of distortion by a factor of $(1 + A_v B)$.

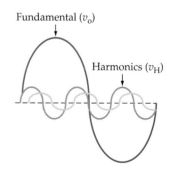

Figure 13-34 Harmonic distortion in an amplifier is reduced by negative feedback.

The *harmonic distortion* is the rms value of v_H expressed as a percentage of the rms value of v_o. The harmonics (generated within the feedback loop) are reduced by negative feedback by a factor of $(1 + A_v B)$. Thus,

$$\% \text{ distortion with } NFB = \frac{\% \text{ distortion without } NFB}{(1 + A_v B)} \tag{13-28}$$

Therefore,

negative feedback reduces harmonic distortion.

Attenuation Distortion

Attenuation distortion occurs when different frequencies are amplified by different amounts. This type of distortion is the result of the amplifier open-loop gain being frequency-dependent (see Fig. 13-35). When negative feedback is used, the closed-loop gain is a constant quantity largely independent of signal frequency within the circuit bandwidth. Like harmonic distortion, attenuation distortion is reduced by a factor of $(1 + A_v B)$ by negative feedback. Equation 13-28 applies. Thus,

negative feedback reduces attenuation distortion.

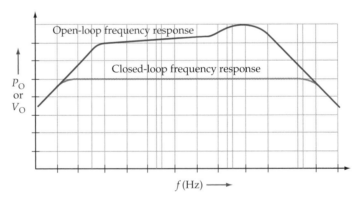

Figure 13-35   Attenuation distortion, resulting from different signal frequencies being amplified by different amounts, is reduced by negative feedback.

Phase Shift

Input signals to an amplifier go through phase shifts from the input to the output. The waveforms in Fig. 13-36 show that, in a two-stage circuit, there is a 180° phase shift from v_s to v_{o1}, and a further 180° shift from v_{o1} to v_{o2}. In some cases, an amplifier can produce different phase shifts for different signal frequencies. For example, higher frequency signals might be shifted by more than 180° at each stage, resulting in an undesirable phase difference (ϕ) between input and output, as illustrated.

Audio (and other) signals at an amplifier input are usually made up of a combination of component waveforms with different amplitudes and different frequencies. Consequently, any variation in amplifier-introduced phase shift will create distortion in the output waveform that is not present in the input.

To investigate the effect of negative feedback on amplifier phase shift, assume that the open-loop gain has a phase shift angle of ϕ, and that the closed-loop phase shift is ϕ_{CL}. Then,

$$\text{open-loop gain} = A_v\underline{/\phi}$$

and

$$A_{CL}\underline{/\phi_{CL}} = \frac{A_v\underline{/\phi}}{1 + A_vB\underline{/\phi}} = \frac{A_vB \sin \phi}{1 + A_vB \cos \phi + jA_vB \sin \phi}$$

giving

$$\phi_{CL} = \phi - \tan^{-1}\frac{A_vB \sin \phi}{1 + A_vB \cos \phi} \qquad \text{(13-29)}$$

Thus,

negative feedback reduces amplifier phase shift

by an angle of

$$\tan^{-1}\frac{A_vB \sin \phi}{1 + A_vB \cos \phi}$$

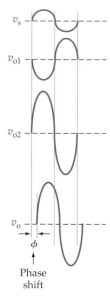

Figure   13-36 Undesirable phase shifts in amplifier ac waveforms are reduced by negative feedback.

Example 13-15

A negative feedback amplifier has an open-loop gain of 60 000 and a closed-loop gain of 300. If the open-loop upper cutoff frequency is 15 kHz, estimate the closed-loop upper cutoff frequency. Also, calculate the total harmonic distortion with feedback if there is 10% harmonic distortion without feedback.

Solution

Eq. 13-27:

$$f_{2(CL)} = \frac{A_v f_{1(OL)}}{A_{CL}} = \frac{60\,000 \times 15 \text{ kHz}}{300}$$

$$= 3 \text{ MHz}$$

Eq. 13-28:

$$\% \text{ distortion with } NFB = \frac{\% \text{ distortion without } NFB}{1 + (A_vB)} = \frac{10\%}{1 + (60\,000/300)}$$

$$= 0.05\%$$

Noise

Circuit *noise* generated within the feedback loop of an amplifier is reduced by a factor of $(1 + A_vB)$ in the same way that unwanted harmonics are reduced. It should be noted that only the noise generated within the feedback loop is reduced by feedback. Noise produced by bias resistors outside the feedback loop will not be affected by feedback.

Negative feedback reduces circuit noise.

Circuit Stability

In Chapter 16, it is explained that, for a circuit to oscillate, the *loop phase shift* must be 360° when the *loop gain* is 1. The *loop phase shift* is the phase shift around the loop from the amplifier input to the output, and back via the feedback network to the input. Similarly, the loop gain is the product of the amplifier open-loop gain and the feedback network attenuation (A_vB).

The loop phase shift of an amplifier can approach 360° as the gain falls off at the high end of the bandwidth. In this case, the amplifier might oscillate, and it is said to be *unstable.* The usual method of combating this kind of instability is to include small shunting capacitors from the transistor collector terminals to ground. These tend to cause the loop gain to fall below 1 before the loop phase shift approaches 360°.

Amplifier instability and compensation techniques are further covered in Chapter 15.

Practice Problems

13-7.1 The amplifier in Example 13-15 has a 15° open-loop phase shift at high signal frequencies. Calculate the phase shift with negative feedback.

13-7.2 A negative feedback amplifier with a closed-loop gain of 250 has a 4 MHz upper cutoff frequency. Calculate the open-loop cutoff frequency if the open-loop gain is 200 000.

Review Questions

Section 13-1

13-1 Draw a sketch to illustrate the principle of series voltage negative feedback, and briefly explain. List the major effects of negative feedback on an amplifier.

13-2 Derive an equation for the voltage gain of an amplifier that uses series voltage negative feedback.

13-3 Derive an equation for the input impedance of an amplifier that uses series voltage negative feedback.

13-4 Derive an equation for the output impedance of an amplifier that uses series voltage negative feedback.

13-5 Discuss the effect of input bias resistors on the input impedance of a negative feedback amplifier, and the effect of a collector resistor at the output of a BJT negative feedback amplifier.

Section 13-2

13-6 Sketch the circuit of a two-stage, capacitor-coupled BJT amplifier that uses series voltage negative feedback. Briefly explain how the feedback operates, and write an equation for the voltage gain in terms of the feedback components.

13-7 Briefly discuss the procedure for determining the negative feedback components for the circuit in Question 13-6.

13-8 For the circuit in Question 13-6, write equations for the input impedance and output impedance in terms of the circuit components and transistor parameters.

Section 13-3

13-9 Sketch the circuit of a direct-coupled, two-stage BJT amplifier that uses series voltage negative feedback. Q_1 should be an *npn* device, and Q_2 should be a *pnp* transistor. Briefly explain the feedback operation, and write the equation for voltage gain in terms of the circuit components.

13-10 For the circuit in Question 13-9, write equations for Z_{in} and Z_{out} in terms of the transistor parameters and the circuit components.

13-11 Briefly discuss the design procedures for the circuit in Question 13-9.

13-12 Sketch the circuit of a dc feedback pair that uses series voltage negative feedback. Explain the operation of the feedback network, and write an equation for the voltage gain in terms of the circuit components.

13-13 For the circuits in Question 13-12, write equations for Z_{in} and Z_{out} in terms of the transistor parameters and the circuit components.

13-14 Briefly discuss the design procedures for the circuit in Question 13-12.

13-15 Sketch the circuit of a direct-coupled, two-stage BIFET amplifier that uses series voltage negative feedback. Explain the operation of the circuit, and write an equation for the voltage gain in terms of the circuit components.

13-16 For the circuits in Question 13-15, write equations for Z_{in} and Z_{out} in terms of the transistor parameters and the circuit components.

13-17 Briefly discuss the design procedures for the circuit in Question 13-15.

Section 13-4

13-18 Sketch the circuit of a two-stage, direct-coupled negative feedback amplifier that uses a two-transistor, emitter-coupled input stage. Explain the operation of the circuit.

13-19 Write equations for $A_{v(CL)}$, Z_{in}, and Z_{out} for the circuit in Question 13-18.

13-20 Briefly discuss the design procedures for the circuit in Question 13-18.

Section 13-5

13-21 Sketch the circuit of a single-stage common emitter amplifier that uses emitter current feedback. Briefly explain the operation of the circuit.

13-22 For the single-stage circuit sketched for Question 13-21, write equations for $A_{v(CL)}$, Z_{in}, and Z_{out}.

13-23 Briefly explain the design procedure for the circuit in Question 13-21.

13-24 Sketch a two-stage, capacitor-coupled BJT amplifier that uses emitter current feedback in each stage. Briefly explain the operation of the circuit.

13-25 Discuss the design procedure for the two-stage circuit in Question 13-24.

13-26 Sketch the circuit of a direct-coupled, two-stage, BJT amplifier that uses emitter current feedback. Q_1 should be an *npn* device, and Q_2 should be a *pnp* transistor.

13-27 Write equations for $A_{v(CL)}$, Z_{in}, and Z_{out} for the circuit in Question 13-26.

Section 13-6

13-28 Sketch a direct-coupled dc feedback pair amplifier that uses parallel current negative feedback. Briefly explain the operation of the circuit, and discuss its performance characteristics.

13-29 Write equations for $A_{i(CL)}$, Z_{in}, and Z_{out} for the circuit in Question 13-28.

13-30 Discuss the design procedure for the two-stage circuit in Question 13-28.

Section 13-7

13-31 Using illustrations, explain the effects of negative feedback on the bandwidth of an amplifier. Discuss the effect of open-loop gain reduction on the closed-loop gain.

13-32 Draw sketches to explain how harmonic distortion occurs in an amplifier. Discuss the effects of negative feedback on harmonic distortion.

13-33 Explain attenuation distortion, and discuss the effects of negative feedback on attenuation distortion.

13-34 Explain how phase shift distortion occurs in an amplifier, and discuss how negative feedback affects phase shift.

13-35 Discuss the effect of negative feedback on circuit noise.

13-36 Briefly discuss amplifier stability with negative feedback.

Problems

Section 13-1

13-1 A two-stage BJT amplifier has an open-loop voltage gain of 224 000 when the transistors have a maximum h_{fe} value, and 14 000 when the transistor h_{fe} values are a minimum. Calculate the closed-loop gain for both cases when negative feedback is used with a feedback factor of 1/125.

13-2 The amplifier in Problem 13-1 has an input impedance of 1.2 kΩ at the base of the input transistor when negative feedback is not used. Voltage divider bias with $R_1 = 150$ kΩ and $R_2 = 56$ kΩ is used at the input transistor. Calculate the circuit input impedance (with negative feedback) at both h_{fe} extremes.

13-3 The output transistor in the amplifier in Problem 13-1 has $1/h_{oe}$ 80 kΩ and a 5.6 kΩ collector resistor. Calculate the circuit output impedance at both h_{fe} extremes.

13-4 The closed-loop gain of a negative feedback amplifier is measured as 99.7 when the feedback factor is $1/100$, and as 297 when a feedback factor of $1/300$ is used. Determine the open-loop gain.

Section 13-2

13-5 The circuit in Fig. 13-16 is to be modified to have an overall voltage gain of 50 and a lower cutoff frequency of 150 Hz. Calculate the new values for R_{F1}, R_{F2}, C_2 , and C_{F1}.

13-6 Analyze the negative feedback amplifier in Problem 13-5 to determine the input and output impedances. Assume that the transistor parameters are $h_{ie} = 800$ Ω, $h_{fe} = 60$, and $1/h_{oe} = 90$ kΩ. The open-loop gain is 97 000.

13-7 A two-stage, capacitor-coupled BJT amplifier as in Fig. 12-15 has the following components: $R_1 = 150$ kΩ, $R_2 = 56$ kΩ, $R_3 = 4.7$ kΩ, $R_4 = 5.6$ kΩ, $R_5 = 150$ kΩ, $R_6 = 56$ kΩ, $R_7 = 4.7$ kΩ, $R_8 = 5.6$ kΩ, $R_L = 47$ kΩ, $C_1 = 10$ μF, $C_2 = 180$ μF, $C_3 = 10$ μF, and $C_4 = 180$ μF. Modify the circuit to use series voltage negative feedback. The voltage gain is to be 180, and the required lower cutoff frequency is 90 Hz.

13-8 Calculate the input and output impedances for the circuit in Problem 13-7. Assume that the transistor parameters are $h_{ie} = 1.2$ kΩ, $h_{fe} = 90$, and $1/h_{oe} = 75$ kΩ.

Section 13-3

13-9 A two-stage, direct-coupled BJT amplifier as in Fig. 12-20 has the following components: $R_1 = 100$ kΩ, $R_2 = 39$ kΩ, $R_3 = 3.9$ kΩ, $R_4 = 3.9$ kΩ, $R_5 = 2.7$ kΩ, $R_6 = 3.9$ kΩ, $R_L = 33$ kΩ, $C_1 = 8$ μF, $C_2 = 150$ μF, $C_3 = 150$ μF. Modify the circuit to use series voltage negative feedback to give a gain of 75 and a lower cutoff frequency of 60 Hz.

13-10 Calculate the input and output impedances for the circuit in Problem 13-9 if the transistor parameters are $h_{ie} = 1$ kΩ, $h_{fe} = 120$, and $1/h_{oe} = 100$ kΩ.

13-11 The two-stage, direct-coupled BJT amplifier designed in Examples 12-8 and 12-9 (and analyzed in Example 12-10) is to be converted into a negative feedback amplifier with a voltage gain of 120 and a lower cutoff frequency of 20 Hz. Make the necessary modifications and determine the new component values.

13-12 A two-stage, direct-coupled amplifier with complementary transistors is to be designed to use negative feedback, as in Fig. 13-17. The overall voltage gain is to be 150, and the lower 3 dB frequency is to be 200 Hz. Calculate suitable values for R_{F1}, R_{F2}, and C_2.

13-13 A dc feedback pair using negative feedback (as in Fig. 13-18) is to have $A_{v(CL)} = 80, f_1 = 120$ Hz, and $f_2 = 50$ kHz. Calculate suitable component values for the feedback network.

13-14 Calculate Z_{in} and Z_{out} for the circuit in Fig. 13-18. Use the component values shown and the transistor parameters $h_{ie} = 1.2$ kΩ, $h_{fe} = 90$, and $1/h_{oe} = 95$ kΩ.

13-15 A two-stage, negative feedback BIFET amplifier, as in Fig. 13-19, is to have $A_{v(CL)} = 19$, and $f_1 = 400$ Hz. Calculate suitable values for R_{F1}, R_{F2}, and C_{F1}.

13-16 Calculate the direct current through the feedback networks in the circuits in Figures 13-17 and 13-19 when the feedback network is directly connected instead of capacitor-coupled.

Section 13-4

13-17 The circuit in Fig. 13-21 is to be designed to have $A_{v(CL)} = 175$. Using $V_{CC} = \pm12.5$ V, select suitable current and voltage levels, and calculate resistor values. Use 2N3904 and 2N3906 transistors.

13-18 Calculate the minimum values of $A_{v(CL)}$, Z_{in}, and Z_{out} for the circuit in Problem 13-17.

13-19 A two-stage differential amplifier circuit (as in Fig. 13-21) is to be designed to amplify a 10 mV input to 750 mV. The amplifier supply is ±9 V, and the transistors to be used have $h_{FE} = h_{fe} = 150$, $h_{ie} = 1.5$ kΩ, and $1/h_{oe} = 150$ kΩ. Determine suitable resistor values.

13-20 Calculate Z_i and Z_o for the circuit in Problem 13-19.

Section 13-5

13-21 Design a single-stage BJT common-emitter amplifier with emitter current feedback. The circuit has the following specifications: $Q_1 = 2N3906$, $V_{CC} = -18$ V, $R_L = 150$ kΩ, $A_v = 45$, and $f_1 = 50$ Hz.

13-22 Analyze the circuit designed for Problem 13-21 to determine A_v, Z_{in}, Z_{out}, and f_1.

13-23 A common-emitter amplifier as in Fig. 12-1 has the following components: $R_1 = 120$ kΩ, $R_2 = 33$ kΩ, $R_C = 6.8$ kΩ, $R_E = 3.3$ kΩ, and $R_L = 68$ kΩ. Modify the circuit to use emitter current feedback. The voltage gain is to be 23, and the required lower cutoff frequency is 60 Hz. The transistor parameters are the following: $h_{ie} = 1.2$ kΩ, $h_{fe} = 90$, $1/h_{oe} = 95$ kΩ and $r'_e = 13\Omega$.

13-24 Calculate A_v, Z_{in}, and Z_{out} for the circuit in Problem 13-23.

13-25 Two stages of the circuit in Problem 13-23 are to be connected in cascade to give an overall voltage gain of approximately 1600 and a lower cutoff frequency of 50 Hz. Determine the necessary modifications.

13-26 An amplifier is to be designed using two stages of the circuit designed for Problem 13-21. The overall voltage gain is to be 900 and the lower cutoff frequency is to be 100 Hz. Determine the necessary modifications.

13-27 Analyze the circuit designed for Problem 13-25 to determine A_v, Z_{in}, and Z_{out}.

13-28 Analyze the circuit designed for Problem 13-26 to determine A_v, Z_{in}, and Z_{out}.

13-29 A dc feedback pair (as in Fig. 12-25a) is to be modified to use emitter current feedback. The circuit specifications are $V_{CC} = 15$ V, $R_L = 80$ kΩ, $A_v = 400$, and $f_1 = 70$ Hz. Design the circuit to use 2N3904 BJTs.

13-30 Analyze the circuit designed for Problem 13-29 to determine A_v, Z_{in}, and Z_{out}.

Section 13-6

13-31 A dc feedback pair is to be designed to use parallel current negative feedback, as in Fig. 13-29. Using $V_{CC} = 22$ V and transistors with $h_{FE} = h_{fe} = 75$ and $h_{ie} = 2.2$ kΩ, design the circuit to have $I_C = 0.9$ mA, $A_i = 22$, and $f_1 = 200$ Hz.

13-32 Calculate A_i and Z_{in} for the circuit in Problem 13-31.

13-33 A dc feedback pair as in Fig. 12-25a has the following components: $R_1 = 12$ kΩ, $R_2 = 330$ kΩ, $R_3 = 4.7$ kΩ, $R_4 = 2.7$ kΩ, and $C_2 = 240$ µF. Determine the necessary modification to give $A_i = 30$ and $f_1 = 75$ Hz.

13-34 Analyze the circuit designed for Problem 13-33 to determine A_i and Z_{in}. Assume that $h_{fe1} = h_{fe2} = 100$ and that $h_{ie1} = h_{ie2} = 1$ kΩ.

Section 13-7

13-35 The amplifier in Example 13-4 has 5% total harmonic distortion in its output waveform when negative feedback is not used. Calculate the new distortion content with negative feedback.

13-36 If the amplifier in Example 13-4 has a 17° open-loop phase shift at high frequencies, determine the phase shift with negative feedback.

13-37 The negative feedback amplifier in Examples 13-7 and 13-8 has a 30 kHz open-loop upper cutoff frequency. Calculate the upper cutoff frequency with negative feedback.

13-38 A negative feedback amplifier with $B = 1/105$ produces a 2 V output when the input is 20 mV. Distortion content in the output is 1%. Find the new signal level to give 2 V output when the negative feedback is disconnected. Also, determine the new distortion content in the output.

Practice Problem Answers

13-1.1 1.42 kΩ, 5.2 kΩ, 26.3 kΩ, 28 Ω, 134.9
13-2.1 100 Ω, 3.9 kΩ, 33 μF, 10 kΩ, 33 μF
13-2.2 27.5 kΩ, 113 Ω
13-3.1 330 Ω, 68 kΩ, 171 μA
13-3.2 150 Ω, 2.7 kΩ, 5.6 μF, 30 μF, 3000 pF
13-4.1 4.7 kΩ, 10 kΩ, 6.8 kΩ, 10 kΩ, (220 kΩ + 10 kΩ), 4.7 kΩ, 4.7 kΩ, 15 kΩ
13-4.2 4.7 kΩ, 34.9 Ω
13-5.1 (68 kΩ + 10 kΩ), 47 kΩ, 4.7 kΩ, 120 Ω, 4.7 kΩ, 2.7 kΩ, 150 Ω, 3.9 kΩ
13-5.2 2.7 μF, 27 μF, 22 μF, 0.47 μF
13-6.1 100 Ω, 5.6 kΩ, 1.8 μF, 20 μF, 39 μF
13-6.2 61 Ω, 878
13-7.1 0.08°
13-7.2 5 kHz

CHAPTER 14
IC Operational Amplifiers and Basic Op-Amp Circuits

CONTENTS

Objectives

You will be able to:

1 Sketch and briefly explain an operational amplifier circuit symbol and the basic input and output stages, and identify all terminals.

2 List and discuss the most important op-amp parameters.

3 Design op-amp bias circuits, and show how to null the output offset.

4 Sketch the following linear op-amp circuits in direct-coupled and capacitor-coupled configuration and explain the operation of each:
   • voltage follower
   • non-inverting amplifier
   • inverting amplifier

5 Analyze and design circuits of the type listed in Objective 4 above.

6 Sketch the following linear op-amp circuits and explain the operation of each:
   • summing amplifier
   • difference amplifier
   • instrumentation amplifier

7 Analyze and design circuits of the type listed in Objective 6 above.

8 Sketch the following nonlinear op-amp circuits and explain the operation of each:
   • zero crossing detector
   • inverting Schmitt trigger circuit
   • non-inverting Schmitt trigger

9 Analyze and design circuits of the type listed in Objective 8 above.

INTRODUCTION

An operational amplifier is a high-gain amplifier circuit with two high-impedance input terminals and one low-impedance output. The inputs are identified as *inverting input* and *non-inverting input*. The basic circuit consists of a differential amplifier input stage and an emitter follower output stage. The voltage gains of integrated circuit (IC) operational amplifiers are extremely high, typically 200 000. Because of this high voltage gain, externally connected resistors must be employed to provide negative feedback. Design of IC op-amp circuits involves determination of suitable values for the external components. The design calculations are usually much simpler than for discrete-component circuits.

14-1 INTEGRATED CIRCUIT OPERATIONAL AMPLIFIERS

Circuit Symbol and Packages

Figure 14-1a shows the triangular circuit symbol for an operational amplifier (op-amp). As illustrated, there are two input terminals, one output terminal, and two supply terminals. The inputs are identified as the *inverting input* ($-$ sign) and the *non-inverting input* ($+$ sign). A positive-going voltage at the inverting input produces a negative-going (inverted) voltage at the output terminal. Conversely, a positive-going voltage at the non-inverting input generates an output which is also positive-going (non-inverted). Plus/minus supplies are normally used with op-amps, and so the supply terminals are identified as $+V_{CC}$ and $-V_{EE}$.

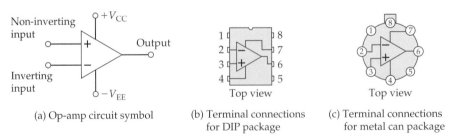

(a) Op-amp circuit symbol   (b) Terminal connections for DIP package   (c) Terminal connections for metal can package

Figure 14-1 Operational amplifier circuit symbol and terminal connections. The $+$ and $-$ signs identify the non-inverting and inverting input terminals respectively.

Two typical op-amp packages are illustrated in Fig. 14-1b and c. For both the dual-in-line plastic (DIP) package and the metal can package, terminals 2 and 3 are the inverting and non-inverting inputs respectively, terminal 6 is the output, and terminals 7 and 4 are the $+$ and $-$ supply terminals.

Basic Internal Circuit

The basic circuit of an IC operational amplifier is essentially as shown in Fig. 13-23 and further illustrated in Fig. 14-2a and b. (The actual op-amp internal

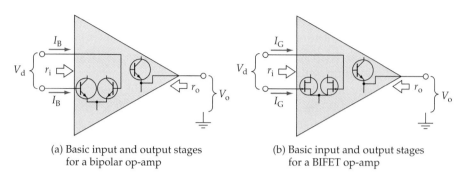

(a) Basic input and output stages
for a bipolar op-amp

(b) Basic input and output stages
for a BIFET op-amp

Figure 14-2   Input and output stages for bipolar and BIFET operational amplifiers. The FET gate current is very much smaller than the BJT base current.

circuitry is much more complex.) The differential-amplifier input stage has two (inverting and non-inverting) high-impedance input terminals, and the emitter follower output stage has a low impedance. An intermediate stage between the input and output stages is largely responsible for the op-amp high voltage gain. Figure 14-2a shows a BJT input stage, while Fig. 14-2b has a FET input stage. Op-amps that exclusively use BJTs are termed bipolar op-amps, and those that use a FET input stage with BJT additional stages are referred to as BIFET op-amps.

Important Parameters

Some of the most important parameters for the 741 operational amplifier (one of the most commonly used) are listed in the data sheet portion shown in Fig. 14-3.

The *large-signal voltage gain* (A_v), also called the *open-loop gain* $(A_{v(OL)})$, for the 741 is listed as as 50 000 minimum and 200 000 typical. This means that a signal applied as a voltage difference between the two input terminals is

Electrical characteristics ($V_S = \pm15$ V, $T_A = 25°C$ unless otherwise specified)

Parameters	Conditions	Min	Typ	Max	Unit
Large signal voltage gain (A_v)	$R_L \geq 2$ kΩ, $V_{out} = \pm10$ V	50 000	200 000		
Input bias current (I_{IB})			80	500	nA
Input offset voltage (V_{IO})	$R_S \leq 10$ kΩ		1.0	5.0	mV
Input offset current (I_{IO})			20	200	nA
Input resistance (r_i)		0.3	2		MΩ
Output resistance (r_o)			75		Ω

Figure 14-3   Portion of data sheet for a 741 op-amp.

amplified by a factor of at least 50 000. The typical amplification is 200 000, and the maximum amplification can be considerably greater. This is similar to the situation with the h_{fe} of a BJT, where the manufacturer specifies minimum, typical, and maximum values.

The *input bias current* (I_{IB}) is the base current for the op-amp input stage transistors. In Fig. 14-3 I_{IB} is listed as 80 nA typical and 500 nA maximum.

Ideally, the input stage transistors should be perfectly matched, so that zero voltage difference between the two input (base) terminals should produce a zero output voltage. In practice, there is always some difference between the base-emitter voltages of the transistors, and this results in an *input offset voltage* (V_{IO}), that might be as large as 5 mV for a 741 op-amp. Similarly, because of mismatch of input transistors, there is also an *input offset current* (I_{IO}). For the 741 I_{IO} is a maximum of 200 nA.

The *input resistance* (r_i) and the *output resistance* (r_o) are the resistances at the op-amp terminals when no feedback is involved. Since virtually all op-amp applications use negative feedback, these (resistance) quantities are modified in practical circuits.

The dc supply voltage for electronic circuits can vary when the supply current changes, and in some cases there is an *ac ripple voltage* superimposed on the dc. The supply voltage changes can be passed to the output of an amplifier. Ideally, the supply voltage variations should have no effect on the circuit output. How good the amplifier is at attenuating supply voltage variations is determined by the *power supply voltage rejection ratio (PSRR)*. This is usually specified in decibel form.

Common mode inputs are voltage changes that are applied simultaneously to both input terminals of an operational amplifier. Common-mode input voltages can be passed to the output. The *common-mode rejection ratio (CMRR)* (expressed in dB) defines how good the amplifier is at attenuating common-mode input voltages.

Practice Problems

14-1.1 Determine the following parameter values from the LM108 op-amp data sheet A-14 in Appendix A: large signal voltage gain, input bias current, input offset voltage, and input resistance.

14-1.2 Repeat Problem 14-1.1 for the LF353 op-amp data sheet A-15 in Appendix A.

14-2 BIASING OPERATIONAL AMPLIFIERS

Biasing Bipolar Op-amps

Like other electronic devices, operational amplifiers must be correctly biased if they are to function properly. As already explained, the inputs of an operational amplifier are the base terminals of the transistors in a differential amplifier.

Base currents must flow into these terminals for the transistors to operate. Consequently, the input terminals must be directly connected to suitable *dc* bias voltage sources.

The most appropriate dc bias voltage for op-amp inputs is approximately halfway between the positive and negative supply voltages. One of the two input terminals is usually connected in some way to the op-amp output to facilitate negative feedback. Where a plus/minus supply is used, the other input might be biased directly to ground via a signal source (see Fig. 14-4). Base current I_{B1} flows into the op-amp non-inverting input via the signal source, while I_{B2} flows from the output into the inverting input, as illustrated.

Figure 14-5 shows a situation in which one input is connected via resistor R_1 to ground, and the other is connected via R_2 to the op-amp output. Once again base current for the input stage transistors flows into both input terminals. Resistors R_1 and R_2 should have equal values, so that voltage drops $I_{B1}R_1$ and $I_{B2}R_2$ are approximately equal. Any difference in these two voltage drops appears as an op-amp dc input voltage, which may be amplified to produce a dc offset at the output.

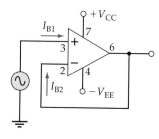

Figure 14-4 Op-amp circuit with one input terminal grounded via a signal generator, and the other input directly connected to the output.

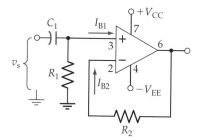

Figure 14-5 Op-amp circuit with one input grounded via resistor R_1, and the other input connected to the output via R_2.

If very small resistance values are selected for R_1 and R_2 in the circuit in Fig. 14-5, then a very small input resistance (R_1) would be offered to a capacitor-coupled signal source. On the other hand, if R_1 and R_2 are very large, the voltage drops $I_{B1}R_1$ and $I_{B2}R_2$ may be ridiculously high. An acceptable maximum voltage drop across these resistors must be very much smaller than the typical V_{BE} level for a forward-biased base-emitter junction. (This is also discussed in Section 5-8.)

Select
$$V_{R1} \approx V_{BE}/10$$

and
$$R_{1(max)} = V_{R1}/I_{B(max)}$$

Therefore,
$$R_{1(max)} = \frac{V_{BE}}{10\,I_{B(max)}} \qquad \text{(14-1)}$$

Figure 14-6 shows a voltage divider (R_1 and R_2) providing an input terminal bias voltage derived from the supply voltages. The voltage-divider current I_2 should be very much larger than the input bias current. This is to ensure that

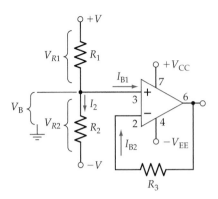

Figure 14-6 Op-amp circuit using a voltage-divider bias circuit.

I_B has a negligible effect upon the bias voltage. Usually, the minimum I_2 is selected as

$$I_{2(min)} = 100\ I_{B(max)} \tag{14-2}$$

Then R_1 and R_2 are simply calculated as V_{R1}/I_2 and V_{R2}/I_2 respectively.

Resistor R_3 in Fig. 14-6 should be approximately equal to $R_1 \| R_2$ to minimize any difference between $I_{B1}(R_1 \| R_2)$ and $I_{B2}R_3$ that could act as a dc input voltage.

Example 14-1

Calculate the maximum resistance for R_1 and R_2 in Fig. 14-7a if a 741 op-amp is used. Determine suitable resistances for R_1, R_2, and R_3 in Fig. 14-7b to give $V_B = 0$ when the supply voltages are ± 15 V.

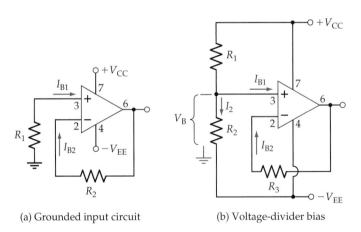

(a) Grounded input circuit   (b) Voltage-divider bias

Figure 14-7   Op-amp circuits for Example 14-1.

Solution

For Fig. 14-7a:

$$V_{BE} \approx 0.7 \text{ V (for op-amps with BJT input stages)}$$

From the 741 data sheet A-13 in Appendix A,

$$I_{B(max)} = 500\ nA$$

Eq. 14-1:
$$R_{1(max)} = \frac{V_{BE}}{10\ I_{B(max)}} = \frac{0.7\ V}{10 \times 500\ nA}$$
$$= 140\ k\Omega\ (use\ 120\ k\Omega\ standard\ value)$$
$$R_2 = R_1 = 120\ k\Omega$$

For Fig. 14-7b:

Eq. 14-2:
$$I_2 = 100\ I_{B(max)} = 100 \times 500\ nA$$
$$= 50\ \mu A$$
$$V_{R1} = V_{R2} = 15\ V$$
$$R_1 = R_2 = \frac{V_{R1}}{I_2} = \frac{15\ V}{50\ \mu A}$$
$$= 300\ k\Omega\ (use\ 270\ k\Omega\ standard\ value)$$
$$R_3 = R_1 \| R_2 = 270\ k\Omega \| 270\ k\Omega$$
$$= 135\ k\Omega\ (use\ 120\ k\Omega\ standard\ value)$$

A single-polarity supply voltage can be used with an operational amplifier. For example, a 741 could use a $+30$ V (as illustrated in Fig. 14-8) instead of a ±15 V supply. Resistors R_1 and R_2 are normally selected to set $V_B = V_{CC}/2$.

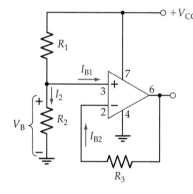

Figure 14-8 Op-amp circuit using a single-polarity supply.

The input offset voltage referred to in Section 14-1 and the resulting output offset can be reduced or eliminated in the circuit shown in Fig. 14-6 by making one of the resistors adjustable. Similarly, if R_3 in Fig. 14-8 is a variable resistor, it could be adjusted to reduce the op-amp output offset voltage. The process is termed *voltage offset nulling*.

Figure 14-9 shows another method of voltage offset nulling with a 741 op-amp. A 10 kΩ potentiometer is connected to the differential amplifier input stage via terminals 1 and 5, as illustrated. With the moving contact connected to $-V_{EE}$, the potentiometer can be adjusted to null the output offset voltage.

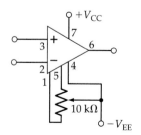

Figure 14-9 Output offset nulling can be accomplished in a 741 op-amp with a 10 kΩ potentiometer connected to terminals 1 and 5.

Biasing BIFET Op-amps

The input bias current for a BIFET op-amp is typically 50 pA, which is very much smaller that that for a bioplar op-amp. So the bias resistor selection method already discussed would produce very high resistor values. This is undesirable because electric charges can accumulate at the FET gates and thus make the bias levels unstable. Also, stray capacitance becomes more effective with high-value bias resistors, possibly resulting in unwanted circuit oscillations. Furthermore, large resistors at an amplifier input can produce unacceptable levels of thermal noise output (see Section 8-6). To combat these effects, *the largest resistor in a BIFET op-amp bias circuit should normally not exceed 1 MΩ*. Because there is virtually no input current with a BIFET op-amp, there is no need for equal resistance values at the inverting and non-inverting input terminals.

Example 14-2

Determine suitable resistor values for the type of circuit shown in Fig. 14-7b when a BIFET op-amp is used. The supply is ±9 V and the output is to be biased to +3 V.

Solution

Select

$$R_2 = 1 \text{ M}\Omega \text{ (standard value)}$$

$$V_B = V_o = +3 \text{ V}$$

$$V_{R2} = V_B - V_{EE} = 3 \text{ V} - (-9 \text{ V})$$
$$= 12 \text{ V}$$

$$V_{R1} = V_{CC} - V_B = 9 \text{ V} - 3 \text{ V}$$
$$= 6 \text{ V}$$

$$I_2 = \frac{V_{R2}}{R_2} = \frac{12 \text{ V}}{1 \text{ M}\Omega}$$
$$= 12 \text{ }\mu\text{A}$$

$$R_1 = \frac{V_{R1}}{I_2} = \frac{6 \text{ V}}{12 \text{ }\mu\text{A}}$$
$$= 500 \text{ k}\Omega \text{ (use 470 k}\Omega + 33 \text{ k}\Omega\text{)}$$

$$R_3 = 0$$

Practice Problems

14-2.1 A 741 op-amp is used with a +24 V supply in the type of circuit shown in Fig. 14-8. Determine suitable bias resistor values.

14-2.2 Repeat Problem 14-2.1 to bias the input to +10 V.

14-3 VOLTAGE FOLLOWER CIRCUITS

Direct-Coupled Voltage Follower

The IC operational amplifier can be employed for an infinite variety of applications. The very simplest application is the direct-coupled *voltage follower* shown in Fig. 14-10a. The output terminal is connected directly to the inverting input terminal, the signal is applied to the non-inverting input, and the load is directly coupled to the output.

Figure 14-10b shows the basic op-amp circuit connected as a voltage follower. With the differential amplifier input stage (see Sections 12-8 and 13-4), V_2 (which equals V_o) must equal V_i. When V_2 is higher or lower than V_i, the voltage difference is amplified to move the output back to equality with the input. Suppose V_o were slightly higher than V_i. The voltage at Q_2 base (terminal 2) would be higher than that at Q_1 base (terminal 3); therefore I_{C2} would be increased above its normal level. This would cause an increase in the voltage drop across R_C, thus reducing V_{B3} and driving V_o back to equality with V_i. Similarly, when V_o is lower than V_i, Q_2 base voltage is lower than that at Q_1 base and I_{C2} is reduced, thus reducing V_{RC}, increasing V_{B3}, and driving V_o back up toward V_i. This is, of course, negative feedback as described in Section 13-4.

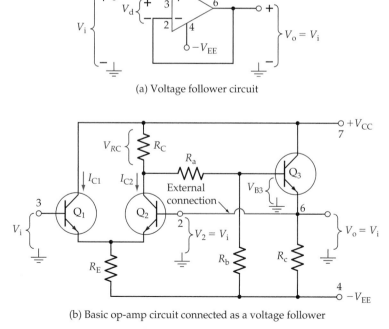

(a) Voltage follower circuit

(b) Basic op-amp circuit connected as a voltage follower

Figure 14-10 In a voltage follower circuit, the output terminal is directly connected to the inverting input terminal. As the input voltage changes, the output follows the input to keep the inverting input terminal voltage equal to the non-inverting terminal voltage.

When the input voltage (at terminal 3) is increased or decreased, the feedback effect causes the output to follow the input almost perfectly. The actual difference between the input and output voltage is easily calculated from the output voltage level and the op-amp open-loop voltage gain.

Like an emitter follower, the voltage follower has a high input impedance, a low output impedance, and a voltage gain of 1. However, the voltage follower performance is superior to that of the emitter follower. As demonstrated in Example 14-3, the typical difference between input and output voltage with a voltage follower is only 5 μV when the input amplitude is 1 V. With an emitter follower, there is a 0.7 V dc voltage difference between input and output. The voltage follower also has a much higher input impedance and a much lower output impedance than the emitter follower. The actual values of Z_{in} and Z_{out} can be calculated from the negative feedback equations derived in Chapter 13.

Example 14-3

Calculate the typical difference between the input and output voltage for a voltage follower using a 741 op-amp with a 1 V input, as in Fig. 14-11. Also, determine typical Z_{in} and Z_{out} values.

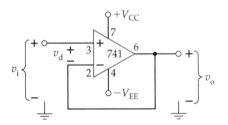

Figure 14-11 Voltage follower circuit for Example 14-3.

Solution

From data sheet A-13 in Appendix A, typical parameters are:

open-loop voltage gain (A_{OL} or A_v) = 200 000, r_i = 2 MΩ, r_o = 75 Ω

$$V_d = V_o - V_i$$

$$V_d = \frac{V_o}{A_v} = \frac{1\ V}{200\ 000}$$

$$= 5\ \mu V$$

From Eq. 13-4, $Z_i = (1 + A_v B)r_i = [1 + (200\ 000 \times 1)] \times 2\ M\Omega$

$$= 400\ 000\ M\Omega$$

From Eq. 13-6, $Z_o = \dfrac{r_o}{1 + A_v B} = \dfrac{75\ \Omega}{1 + (200\ 000 \times 1)}$

$$= 0.375 \times 10^{-3}\ \Omega$$

Capacitor-Coupled Voltage Follower

When a voltage follower is to have capacitor-coupled input and output terminals, the non-inverting input must be grounded via a resistor (R_1 in Fig. 14-12). As explained in Section 14-2, the resistor is required for passing bias current to the non-inverting input terminal. It also offers an input resistance to the signal source, rather than a short-circuit. Because the load is capacitor-coupled, the dc output offset voltage might seem to be unimportant. However, the dc offset voltage at the op-amp output terminal can limit the amplitude of the ac output voltage. So a resistor (R_2) equal to R_1 should be included in series with the inverting input terminal to equalize the $I_B R_B$ voltage drops and thus minimize output offset voltage.

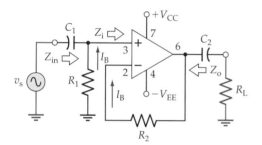

Figure 14-12 A capacitor-coupled voltage follower must have the op-amp non-inverting terminal grounded via a resistor to provide a path for input bias current.

Designing the capacitor-coupled circuit in Fig. 14-12 involves the calculation of R_1, C_1, and C_2. The circuit input impedance is $R_1 \| Z_i$; however, Z_i is always much larger than R_1. So the circuit input impedance is simply taken as

$$Z_{in} = R_1 \tag{14-3}$$

Normally, the load resistance R_L is much smaller R_1, and consequently, the smallest capacitor values are calculated when C_2 is selected to set the circuit lower cutoff frequency, f_1. Therefore,

$$X_{C2} = (Z_o + R_L) \text{ at } f_1 \tag{14-4}$$

Capacitor C_1 should then be calculated from the equation $X_{C1} = (Z_{in} + r_s)/10$ at f_1, so that it has no significant effect on the circuit lower cutoff frequency:

$$X_{C1} = (R_1 + r_s)/10 \text{ at } f_1 \tag{14-5}$$

Coupling capacitor equations are discussed in Section 12-1.

Example 14-4

Design a capacitor-coupled voltage follower using a 741 op-amp. The circuit lower cutoff frequency is to be 70 Hz, the load resistance is 4 kΩ, and the signal source resistance is much smaller than R_1.

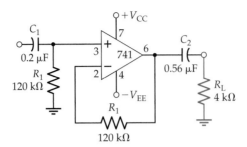

Figure 14-13 Capacitor-coupled voltage follower for Example 14-4.

Solution

See Fig. 14-13. From data sheet A-13 in Appendix A,

$$I_{B(max)} = 500 \text{ nA}$$

Eq. 14-1: $\quad R_{1(max)} = \dfrac{V_{BE}}{10\, I_{B(max)}} = \dfrac{0.7 \text{ V}}{10 \times 500 \text{ nA}}$

$$= 140 \text{ k}\Omega \text{ (use 120 k}\Omega \text{ standard value)}$$

$$R_2 = R_1 = 120 \text{ k}\Omega$$

From Eq. 14-4, $\quad C_2 = \dfrac{1}{2\pi f_1 R_L} = \dfrac{1}{2\pi \times 70 \text{ Hz} \times 4 \text{ k}\Omega}$

$$\approx 0.56 \text{ μF (standard value)}$$

From Eq. 14-5, $\quad C_1 = \dfrac{1}{2\pi f_1 R_1/10} = \dfrac{1}{2\pi \times 70 \text{ Hz} \times 120 \text{ k}\Omega/10}$

$$\approx 0.19 \text{ μF (use 0.2 μF standard value)}$$

Practice Problems

14-3.1 A capacitor-coupled voltage follower with $Z_{in} = 22 \text{ k}\Omega$ and $R_L = 10 \text{ k}\Omega$ is to have $f_1 = 50$ Hz. Determine suitable capacitor values.

14-3.2 Determine coupling capacitor values for the circuit in part (b) of Example 14-1 if $r_s = 600 \ \Omega$, $R_L = 8.2 \text{ k}\Omega$, and $f_1 = 20$ Hz.

14-3.3 Calculate the maximum dc offset voltage that might be produced at the output of the circuit designed in part (a) of Ex. 14-1 when R_2 is replaced with a short-circuit.

14-4 NON-INVERTING AMPLIFIERS

Direct-Coupled Non-Inverting Amplifier

The non-inverting amplifier circuit in Fig. 14-14 behaves similarly to a voltage follower circuit with one major difference. Instead of all of the output voltage being fed directly back to the inverting input terminal (as in a voltage follower),

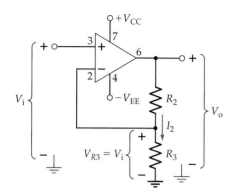

Figure 14-14 Op-amp non-inverting amplifier circuit. The signal is applied to the non-inverting input terminal. The circuit voltage gain is $A_{CL} = (R_2 + R_3)/R_3$.

only a portion of V_o is fed back. The output voltage is divided by resistors R_2 and R_3, and the voltage across R_3 is applied to the inverting input terminal. As in the case of the voltage follower, the output voltage changes as necessary to keep the inverting input terminal voltage equal to that at the non-inverting input. Thus, the voltage V_{R3} always equals V_i, and the output voltage is then determined by the resistances of R_2 and R_3.

Because the signal voltage is applied to the op-amp non-inverting input terminal, the output always has the same polarity as the input. A positive-going input produces a positive-going output, and vice versa. Thus, the input is not inverted (at the output), and the circuit is identified as a *non-inverting amplifier.*

The voltage-divider current (I_2) is always selected to be very much larger than the operational amplifier input bias current, and

$$V_{R3} = V_i = I_2 R_3 \qquad (14\text{-}6)$$

Also, in Fig. 14-14 it is seen that

$$I_2 = \frac{V_o}{R_2 + R_3} \qquad (14\text{-}7)$$

The circuit voltage gain is

$$A_{CL} = \frac{V_o}{V_i} = \frac{I_2(R_2 + R_3)}{I_2 R_3}$$

Therefore,

$$A_{CL} = \frac{R_2 + R_3}{R_3} \qquad (14\text{-}8)$$

Figure 14-15 shows a non-inverting amplifier circuit with a resistor (R_1) connected in series with the non-inverting input terminal. As in the case of other op-amp circuits, this is done to equalize the resistor voltage drops produced by the input bias currents. In this case, R_1 is chosen to be approximately equal to the resistance 'seen looking out of' the inverting input terminal. Thus,

$$R_1 = R_2 \| R_3 \qquad (14\text{-}9)$$

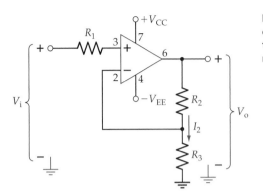

Figure 14-15 Non-inverting amplifier circuit with a resistor R_1 in series with the input. R_1 is selected to be approximately equal to $R_2 \| R_3$.

The input and output impedances for a non-inverting amplifier are easily determined from the negative feedback equations in Chapter 13.

Design of a non-inverting amplifier mostly involves determining suitable voltage-divider resistors (R_2 and R_3). So, as explained in Section 14-2, design begins with the selection of a voltage divider current that is much larger than the op-amp input bias current.

Example 14-5

Design a direct-coupled non-inverting amplifier (as in Fig. 14-15) to use a 741 op-amp. The output voltage is to be 2 V when the input is 50 mV.

Solution

From the 741 data sheet A-13 in Appendix A,

$$I_{B(max)} = 500 \text{ nA}$$

From Eq. 14-2,

$$I_{2(min)} = 100 \, I_{B(max)} = 100 \times 500 \text{ nA}$$
$$= 50 \, \mu A$$

From Eq. 14-6,

$$R_3 = \frac{V_i}{I_2} = \frac{50 \text{ mV}}{50 \, \mu A}$$
$$= 1 \text{ k}\Omega \text{ (standard resistor value)}$$

From Eq. 14-7,

$$R_2 + R_3 = \frac{V_o}{I_2} = \frac{2 \text{ V}}{50 \, \mu A}$$
$$= 40 \text{ k}\Omega$$

$$R_2 = (R_2 + R_3) - R_3 = 40 \text{ k}\Omega - 1 \text{ k}\Omega$$
$$= 39 \text{ k}\Omega \text{ (standard value)}$$

From Eq. 14-9,

$$R_1 = R_2 \| R_3 = 39 \text{ k}\Omega \| 1 \text{ k}\Omega$$
$$\approx 1 \text{ k}\Omega \text{ (use 1 k}\Omega \text{ standard value)}$$

Example 14-6
Calculate typical input and output impedances for the non-inverting amplifier designed in Example 14-5.

Solution

From data sheet A-13 in Appendix A, typical parameters are

$$A_v = 200\ 000,\ r_i = 2\ \text{M}\Omega,\ r_o = 75\ \Omega$$

Also,

$$B = \frac{R_3}{R_2 + R_3} = \frac{1\ \text{k}\Omega}{39\ \text{k}\Omega + 1\ \text{k}\Omega}$$

$$= \frac{1}{40}$$

From Eq. 13-4, $Z_i = (1 + A_vB)\ r_i = \left[1 + \dfrac{200\ 000}{40}\right] \times 2\ \text{M}\Omega$

$$\approx 10\ 000\ \text{M}\Omega$$

From Eq. 13-6, $Z_o = \dfrac{r_o}{1 + A_vB} = \dfrac{75\ \Omega}{1 + (200\ 000/40)}$

$$\approx 0.015\ \Omega$$

Capacitor-Coupled Non-Inverting Amplifier
When a non-inverting amplifier is to have a signal capacitor coupled to its input, the op-amp non-inverting input terminal must be grounded via a resistor to provide a path for the input bias current. This is illustrated in Fig. 14-16, where R_1 allows for the passage of I_{B1}. As in the case of the capacitor-coupled voltage follower, the input resistance is essentially equal to R_1 for the capacitor-coupled non-inverting amplifier.

Resistors R_1, R_2, and R_3 in the capacitor-coupled circuit are determined exactly as for a direct-coupled non-inverting amplifier. The capacitor values are calculated in the same way as for a capacitor-coupled voltage follower.

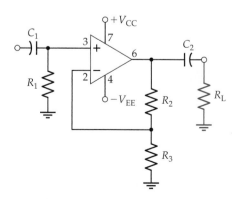

Figure 14-16 A capacitor-coupled non-inverting amplifier circuit must have the non-inverting input terminal grounded via a resistor to provide a path for the input bias current.

Example 14-7

Calculate the voltage gain and lower cutoff frequency for the capacitor-coupled non-inverting amplifier circuit in Fig. 14-17.

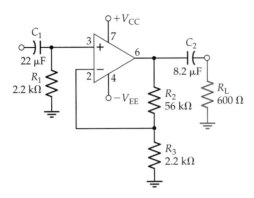

Figure 14-17 Capacitor-coupled non-inverting amplifier circuit for Example 14-7.

Solution

Eq. 14-8:
$$A_{CL} = \frac{R_2 + R_3}{R_3} = \frac{50\text{ k}\Omega + 2.2\text{ k}\Omega}{2.2\text{ k}\Omega}$$

$$= 26.5$$

From Eq. 14-4,
$$f_1 = \frac{1}{2\pi C_2 R_L} = \frac{1}{2\pi \times 8.2\text{ }\mu\text{F} \times 600\text{ }\Omega}$$

$$= 32.3\text{ Hz}$$

Practice Problems

14-4.1 A direct-coupled non-inverting amplifier using a 741 op-amp is to have a 100 mV input voltage amplitude and a voltage gain of 40. Select suitable resistor values.

14-4.2 Design a capacitor-coupled non-inverting amplifier using a 741 op-amp to have a ±5 V output amplitude when the input is ±250 mV. The load resistance is 820 Ω, and the lower cutoff frequency is to be 20 Hz.

14-4.3 Redesign the circuit in Problem 14-4.1 to use a BIFET op-amp.

14-5 INVERTING AMPLIFIERS

Direct-Coupled Inverting Amplifier

The circuit in Fig. 14-18 is termed an *inverting amplifier* because, with V_i applied via R_1 to the inverting input terminal, the output goes negative when the input goes positive, and vice versa. Note that the non-inverting input terminal is grounded via resistor R_3. With the non-inverting terminal grounded, the voltage at the op-amp inverting input terminal remains close to ground. Any

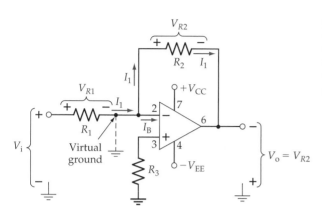

Figure 14-18 Op-amp inverting amplifier. The signal voltage is applied via resistor R_1 to the inverting input terminal. The circuit voltage gain is $A_{CL} = -R_2/R_1$.

increase or decrease in the (very small) difference in voltage between the two input terminals is amplified by the op-amp open-loop gain and fed back via R_2 and R_1 to correct the change. Because the inverting input terminal is not grounded but remains close to ground, the inverting input terminal in this application is termed a *virtual ground* or *virtual earth*.

The circuit input current in Fig. 14-18 can be calculated as

$$I_1 = \frac{V_{R1}}{R_1}$$

The input voltage is applied to one end of R_1, and the other end of R_1 is at ground level. Consequently,

$$V_{R1} = V_i$$

and
$$I_1 = \frac{V_i}{R_1} \tag{14-10}$$

I_1 is always selected to be very much larger than the op-amp input bias current (I_B). So virtually all of I_1 flows through resistor R_2 (see Fig. 14-18), and the voltage drop across R_2 is

$$V_{R2} = I_1 R_2$$

The left side of R_2 is connected to the op-amp inverting input terminal, which, as discussed, is always at ground level. This means that the right side of R_2 (the output terminal) is V_{R2} below ground. So

$$V_o = -I_1 R_2 \tag{14-11}$$

From Eq. 14-10, $V_i = I_1 R_1$

Therefore the circuit voltage gain is

$$A_{CL} = \frac{V_o}{V_i} = \frac{-I_1 R_2}{I_1 R_1}$$

or
$$A_{CL} = \frac{-R_2}{R_1} \tag{14-12}$$

If the input of the inverting amplifier is grounded, the circuit is seen to be exactly the same as a non-inverting amplifier with R_3 as the input resistor and zero input voltage. Thus, negative feedback occurs (as in a non-inverting amplifier) to maintain the op-amp inverting input terminal at the same voltage level as the non-inverting input terminal.

The output impedance of an inverting amplifier is calculated in exactly the same way as for a non-inverting amplifier (using Equation 13-6). As in the case of the non-inverting amplifier, the output impedance is very low for an inverting amplifier.

The input impedance of an inverting amplifier is easily determined by recalling that the right side of R_1 (in Fig. 14-18) is always at ground level and that the signal is applied to the left side of R_1. Therefore,

$$Z_i = R_1 \qquad (14\text{-}13)$$

Design of an inverting amplifier is very simple. Voltage-divider current I_1 is selected to be very much larger than the op-amp input bias current. Resistors R_1 and R_2 are calculated from Equations 14-10 and 14-11, or 14-12, and R_3 is chosen to be approximately equal to $R_1 \| R_2$.

Example 14-8

Design a direct-coupled inverting amplifier, as in Fig. 14-19, to use a 741 op-amp. The input voltage amplitude is 20 mV and the voltage gain is to be 144.

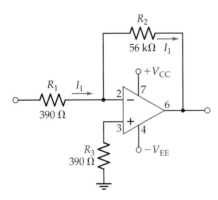

Figure 14-19 Inverting amplifier for Example 14-8.

Solution

For the 741, $\qquad\qquad I_{B(max)} = 500\ \text{nA}$

Select $\qquad\qquad I_{1(min)} = 100\ I_{B(max)} = 100 \times 500\ \text{nA}$
$$= 50\ \mu\text{A}$$

From Eq. 14-10, $\qquad R_1 = \dfrac{V_i}{I_1} = \dfrac{20\ \text{mV}}{50\ \mu\text{A}}$
$$= 400\ \Omega\ (\text{use } 390\ \Omega\ \text{standard value})$$

From Eq. 14-12,
$$R_2 = A_{CL}R_1 = 144 \times 390\ \Omega$$
$$= 56.2\ k\Omega \text{ (use 56 k}\Omega \text{ standard value)}$$

$$R_3 = R_1 \| R_2 = 390\ \Omega \| 56\ k\Omega$$
$$\approx 390\ \Omega \text{ (use 390 }\Omega)$$

Capacitor-Coupled Inverting Amplifier

A capacitor-coupled inverting amplifier circuit is shown in Fig. 14-20. In this case, the bias current to the op-amp inverting input terminal flows (from the output) via resistor R_2, so that the input coupling capacitor does not interrupt the input bias current. A voltage drop $I_B R_2$ is produced at the inverting input by the bias current flow, and this must be equalized by the $I_B R_3$ voltage drop at the non-inverting input. Therefore, for the capacitor-coupled non-inverting circuit,

$$R_3 = R_2 \qquad\qquad (14\text{-}14)$$

Apart from the choice of R_3, the circuit is designed exactly as the direct-coupled circuit, and the capacitor values are calculated in the same way as for a capacitor-coupled non-inverting amplifier.

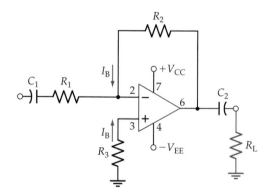

Figure 14-20 Capacitor-coupled inverting amplifier circuit. Note that the flow of bias current is not interrupted by capacitor C_1.

Practice Problems

14-5.1 A direct-coupled inverting amplifier using a 741 op-amp is to have a 300 mV input voltage amplitude and a voltage gain of 15. Calculate suitable resistor values.

14-5.2 The circuit in Problem 14-5.1 is to have a 1 kΩ capacitor-coupled load and a 600 Ω capacitor-coupled signal source. Determine suitable capacitor values for a 20 Hz lower cutoff frequency.

14-5.3 Redesign the circuit in Problem 14-5.1 to use a BIFET op-amp.

14-6 SUMMING AMPLIFIER

The *summing amplifier* circuit in Fig. 14-21 is simply a direct-coupled inverting amplifier with two inputs applied to two resistors (R_1 and R_1). With two input voltages (V_{i1} and V_{i2}) there are two input currents (I_1 and I_2). Also, because the op-amp inverting input terminal behaves as a virtual ground (as for an inverting amplifier), the input currents are

$$I_1 = \frac{V_{i1}}{R_1} \qquad I_2 = \frac{V_{i2}}{R_2}$$

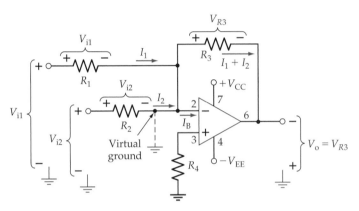

Figure 14-21   An op-amp summing amplifier amplifies the sum of two, or more, input voltages.

All of I_1 and I_2 flows through resistor R_3, giving

$$V_o = -(I_1 + I_2)\, R_3$$

Or

$$V_o = -\left[\frac{V_{i1}}{R_1} + \frac{V_{i2}}{R_2}\right] R_3$$

With $R_1 = R_2$,

$$V_o = \frac{-R_3}{R_1}(V_{i1} + V_{i2}) \tag{14-15}$$

When $R_3 = R_1 = R_2$, the output voltage is the direct (inverted) sum of the two inputs. When R_3 is greater than R_1 and R_2, the output is an amplified version of the sum of the inputs.

A summing amplifier is not limited to two inputs. There can be almost any number of inputs, and the output remains the sum of the inputs. Summing amplifiers are designed in the same way as ordinary inverting amplifiers.

Example 14-9

Design a three-input summing amplifier, as in Fig. 14-22, to use a BIFET op-amp and to have a voltage gain of 3. Calculate the resistor currents and the output voltage when all three inputs are 1 V.

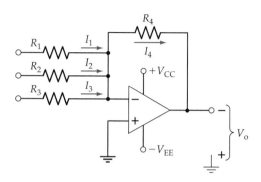

Figure   14-22   Three-input   summing amplifier for Example 14-9.

Solution

Select $R_4 = 1\ \text{M}\Omega$

From Eq. 14-12, $R_1 = R_2 = R_3 = \dfrac{R_4}{A_{CL}} = \dfrac{1\ \text{M}\Omega}{3}$

$$= 333\ \text{k}\Omega\ (\text{use }330\ \text{k}\Omega\ \text{standard value})$$

$$I_1 = I_2 = I_3 = \frac{V_i}{R_1} = \frac{1\ \text{V}}{330\ \text{k}\Omega}$$

$$= 3.03\ \mu\text{A}$$

$$I_4 = I_1 + I_2 + I_3 = 3.03\ \mu\text{A} + 3.03\ \mu\text{A} + 3.03\ \mu\text{A}$$
$$= 9.09\ \mu\text{A}$$

$$V_o = -I_4\,R_4 = -9.09\ \mu\text{A} \times 1\ \text{M}\Omega$$
$$= -9.09\ \text{V}$$

Or, from Eq. 14-15,

$$V_o = \frac{-R_4}{R_1}(V_{i2} + V_{i2} + V_{i3}) = \frac{-1\ \text{M}\Omega}{330\ \text{k}\Omega}(1\ \text{V} + 1\ \text{V} + 1\ \text{V})$$

$$= -9.09\ \text{V}$$

Practice Problems

14-6.1 A two-input summing amplifier is to be designed to produce a 5 V output from two 0.25 V inputs. Calculate suitable resistor values for a circuit using a 741 op-amp.

14-6.2 If the circuit designed in Example 14-9 has 0.75 V, 1.5 V, and 1.25 V inputs, determine V_o, I_1, I_2, I_3, and I_4.

14-7 DIFFERENCE AMPLIFIER

Circuit Operation

A *difference amplifier* amplifies the difference between two inputs. The circuit shown in Fig. 14-23 is a combination of inverting and non-inverting amplifiers. Resistors R_1, R_2, and the op-amp constitute an inverting amplifier for a voltage (V_{i1}) applied to R_1. The same components (R_1, R_2, and the op-amp) also function as a non-inverting amplifier for a voltage (V_{R4}) at the non-inverting input terminal. It is seen that V_{R4} is derived from input voltage V_{i2} by the voltage divider R_3 and R_4. To understand the circuit operation, consider the output produced by each input voltage when the other input is zero:

With $V_{i2} = 0$,
$$V_{o1} = \frac{-R_2}{R_1} \times V_{i1}$$

With $V_{i1} = 0$,
$$V_{o2} = \frac{R_1 + R_2}{R_1} \times V_{R4}$$

and
$$V_{R4} = \frac{R_4}{R_3 + R_4} \times V_{i2}$$

Therefore
$$V_{o2} = \frac{R_1 + R_2}{R_1} \times \frac{R_4}{R_3 + R_4} \times V_{i2}$$

With $R_3 = R_1$ and $R_4 = R_2$,
$$V_{o2} = \frac{R_2}{R_1} \times V_{i2}$$

When both inputs are present,
$$V_o = V_{o2} + V_{o1}$$
$$= \frac{R_2}{R_1} V_{i2} + \frac{-R_2}{R_1} V_{i1}$$

Therefore
$$V_o = \frac{R_2}{R_1} (V_{i2} - V_{i1}) \qquad (14\text{-}16)$$

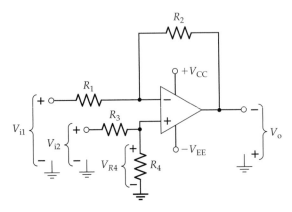

Figure 14-23 An op-amp diference amplifier amplifies the difference between two input voltages.

When $R_2 = R_1$, the output voltage (as calculated by Equation 14-16) is the direct difference between the two inputs. With R_2 greater than R_1, the output becomes an amplifier version of $(V_{i2} - V_{i1})$.

Input Resistances

Consider the input portion of the difference amplifier circuit reproduced in Fig. 14-24. The resistance at input *terminal 1* is the same as the input impedance for an inverting amplifier: $Z_{i1} = R_1$. The input resistance at the op-amp non-inverting input terminal is very high (as in the case of a non-inverting amplifier), and this is in parallel with resistor R_4. So the input impedance at *terminal 2* in Fig. 14-24 is $Z_{i2} = R_3 + R_4$.

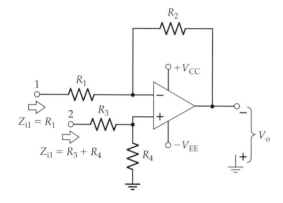

Figure 14-24 Difference amplifier circuit input impedances.

Equation 14-16 was derived by assuming that $R_3 = R_1$ and $R_4 = R_2$. It can be shown that the same result is obtained when the *ratio R_4/R_3 equals R_2/R_1*, so that the actual resistor values do not have to be equal. For equal resistances at the two input terminals, select

$$R_3 + R_4 = R_1 \qquad\qquad (14\text{-}17)$$

Then, calculate the resistances of R_3 and R_4 from

$$\frac{R_4}{R_3} = \frac{R_2}{R_1} \qquad\qquad (14\text{-}18)$$

A simple rule of thumb can be used for determining suitable resistance values for R_3 and R_4 when the two input resistances do not have to be exactly equal. Select $R_4 = R_2/A_{CL}$, which always makes $R_4 = R_1$. Then, calculate R_3 as $R_3 = R_1/A_{CL}$.

Example 14-10

A difference amplifier is to be designed to amplify the difference between two voltages by a factor of 10. The inputs each approximately equal 1 V. Determine suitable resistor values for a circuit using a 741 op-amp (see Fig. 14-25).

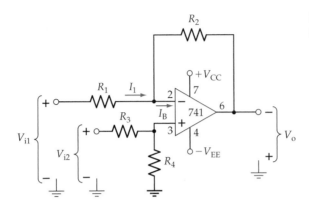

Figure 14-25 Difference amplifier circuit for Example 14-10.

Solution

Select
$$I_1 \approx 100\, I_B = 100 \times 500 \text{ nA}$$
$$= 50\ \mu\text{A}$$

$$R_1 = \frac{V_{i1}}{I_1} = \frac{1\ \text{V}}{50\ \mu\text{A}}$$
$$= 20\ \text{k}\Omega\ (\text{use } 18\ \text{k}\Omega\ \text{standard value})$$

$$R_2 = A_{CL}\,R_1 = 10 \times 18\ \text{k}\Omega$$
$$= 180\ \text{k}\Omega\ (\text{standard values})$$

$$R_4 = R_1 = 18\ \text{k}\Omega$$

$$R_3 = \frac{R_1}{A_{CL}} = \frac{18\ \text{k}\Omega}{10}$$
$$= 1.8\ \text{k}\Omega\ (\text{standard value})$$

Common-Mode Voltages

A common-mode input voltage is a signal voltage (dc or ac) applied to both input terminals at the same time. This is illustrated in Fig. 14-26, where V_{i1}, V_{i2}, and the common-mode voltage (V_n) are all represented as inputs from dc

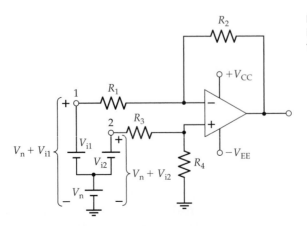

Figure 14-26 Common-mode input voltage applied to a difference amplifier.

sources. As shown, the input voltages at *terminals 1* and *2* are changed from V_{i1} and V_{i2} to $(V_n + V_{i1})$ and $(V_n + V_{i2})$.

Equation 14-16 shows that the output voltage is the amplified difference between the two input voltages. So,

$$V_o = \frac{R_2}{R_1}[(V_{i2} + V_n) - (V_{i1} + V_n)]$$

$$= \frac{R_2}{R_1}(V_{i2} - V_{i1})$$

This shows that the common-mode input is completely cancelled. However, recall that the gain equation depends upon the resistor ratios (R_4/R_3 and R_2/R_1) being equal. If the ratios are not exactly equal, one input will undergo a larger amplification than the other. Moreover, the common-mode voltage at one input terminal will be amplified by a larger amount than that applied to the other input terminal. In this case, common-mode inputs will not be completely cancelled.

Because it is difficult to match resistor ratios perfectly (especially for standard-value components), some common-mode output voltage is almost certain to be produced where a common-mode input exists. Figure 14-27 shows a circuit modification for minimizing common-mode outputs from a difference amplifier. Resistor R_4 is made up of a fixed-value resistor and a small-value adjustable resistor connected in series. This provides adjustment of the ratio R_4/R_3 to match R_2/R_1, so that common-mode outputs can be nulled to zero.

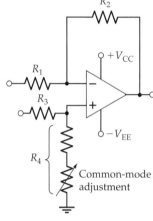

Figure 14-27 Common-mode outputs may be nulled by adjustment of R_4.

Practice Problems

14-7.1 If the circuit designed in Example 14-10 has $V_{i1} = 0.75$ V and $V_{i2} = 1.5$ V, determine V_o, I_{R1}, I_{R2}, I_{R3}, and V_{R4}.

14-7.2 A 500 Ω resistor is connected in series with R_4 in the circuit designed in Example 14-10. Calculate the common-mode gain.

14-7.3 A difference amplifier is to be designed to produce a 3 V output when the inputs are 1.5 V and 2 V. Determine suitable resistor values for a circuit using a BIFET op-amp.

14-8 INSTRUMENTATION AMPLIFIER

Circuit Operation

At first glance, the *instrumentation amplifier* circuit in Fig. 14-28 looks complex, but when considered section by section it is found to be quite simple. First, note that the second stage (consisting of op-amp A_3 and resistors R_4 to R_7) is a difference amplifier that operates exactly as described in Section 14-7. Next, look at A_1 and resistors R_1 and R_2; this is a non-inverting amplifier. Similarly, A_2, combined with resistors R_2 and R_3, constitutes another non-inverting

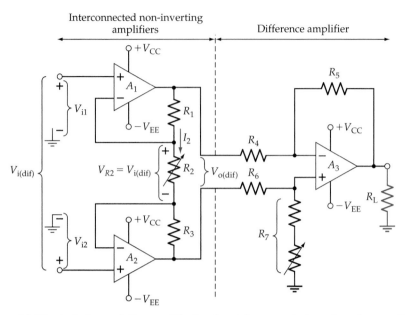

Figure 14-28 An instrumentation amplifier has two interconnected non-inverting amplifiers as the first stage, and a difference amplifier as the second stage.

amplifier. Because the first-stage circuits share a single resistor, their operation is slightly different from the usual non-inverting amplifier operation.

The first stage accepts a *differential input voltage* ($V_{i(dif)}$), and produces a *differential output voltage* ($V_{o(dif)}$). The differential input could be the difference between two grounded inputs (V_{i1} and V_{i2}), as illustrated. But often a differential ungrounded input voltage is derived, for example, from two voltage monitoring electrodes connected to a human body for medical purposes. In this case, there could be a large common-mode input voltage, which can be shown to pass to the output of the first stage without amplification.

The difference amplifier second stage accepts $V_{o(dif)}$ from the first stage as an input and produces an output to a grounded load, as shown on the circuit diagram. As explained in Section 14-7, the difference amplifier tends to reject common-mode voltages, and the circuit can also have an adjustment for reducing common-mode outputs to zero.

The instrumentation amplifier is now seen to be a circuit with two high-impedance input terminals and one low-impedance output. The differential input voltage is amplified and converted to a single-ended output, and common-mode inputs are attenuated.

Voltage Gain

Recall that, with a non-inverting amplifier, the feedback voltage to the op-amp inverting input terminal always equals the input voltage to the non-inverting input terminal. Therefore, the voltage at the $R_1 R_2$ junction equals V_{i1}, and that at the R_2R_3 junction equals V_{i2}. Consequently, the voltage drop across R_2 equals the difference between the two input voltages. That also means that V_{R2} equals the differential input voltage ($V_{i(dif)}$). The current through R_2 can now be calculated as

$$I_2 = \frac{V_{i(dif)}}{R_2}$$

The voltage drop across R_1, R_2, and R_3 is the differential output voltage of the first stage ($V_{o(dif)}$):

$$V_{o(dif)} = I_2 (R_1 + R_2 + R_3)$$

$$= \frac{V_{i(dif)}}{R_2} (R_1 + R_2 + R_3)$$

The (closed-loop) voltage gain of the *differential input-differential output* first stage is

$$A_{CL1} = \frac{V_{o(dif)}}{V_{i(dif)}} = \frac{R_1 + R_2 + R_3}{R_2}$$

Normally, R_1 and R_3 are always equal. So, the first stage gain can be written as

$$A_{CL1} = \frac{2R_1 + R_2}{R_2} \qquad (14\text{-}19)$$

From Equation 14-16, the second-stage gain is

$$A_{CL2} = \frac{R_5}{R_4}$$

The overall voltage gain is

$$A_{CL} = A_{CL1} \times A_{CL2}$$

The second stage is often designed for a gain of one, so that the overall voltage gain can be calculated from Equation 14-19. Note that, as shown in Fig. 14-28, R_2 can be a variable resistor for adjustment of the circuit overall voltage gain.

Example 14-11

Calculate the overall voltage gain for the instrumentation amplifier in Fig. 14-29. Determine the current and voltage levels throughout the circuit when a $+1$ V common-mode input (V_n) is present along with ± 10 mV signals.

Solution

From Eqs. 14-16 and 14-19,

$$A_{CL} = \frac{2R_1 + R_2}{R_2} \times \frac{R_5}{R_4}$$

$$= \frac{(2 \times 33 \text{ k}\Omega) + 300 \text{ }\Omega}{300 \text{ }\Omega} \times \frac{15 \text{ k}\Omega}{15 \text{ k}\Omega}$$

$$= 221$$

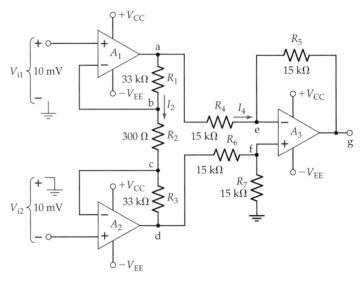

Figure 14-29 Instrumentation amplifier circuit for Example 14-11.

At the junction of R_1 and R_2,

$$V_b = V_{i1} + V_n = 10\text{ mV} + 1\text{ V}$$
$$= +1.01\text{ V}$$

At the junction of R_2 and R_3,

$$V_c = V_{i2} + V_n = -10\text{ mV} + 1\text{ V}$$
$$= +0.99\text{ V}$$

The current through R_2,

$$I_2 = \frac{V_b - V_c}{R_2} = \frac{1.01\text{ V} - 0.99\text{ V}}{300\text{ }\Omega}$$
$$= 66.67\text{ }\mu\text{A}$$

At the output of A_1,

$$V_a = V_b + (I_2 \times R_1) = 1.01\text{ V} + (66.67\text{ }\mu\text{A} \times 33\text{ k}\Omega)$$
$$= +3.21\text{ V}$$

At the output of A_2,

$$V_d = V_c - (I_2 \times R_3) = 0.99\text{ V} - (66.67\text{ }\mu\text{A} \times 33\text{ k}\Omega)$$
$$= -1.21\text{ V}$$

At the junction of R_6 and R_7,

$$V_f = V_d \times \frac{R_7}{R_6 + R_7} = -1.21\text{ V} \times \frac{15\text{ k}\Omega}{15\text{ k}\Omega + 15\text{ k}\Omega}$$
$$= -0.605\text{ V}$$

At the junction of R_4 and R_5,

$$V_e = V_f = -0.605 \text{ V}$$

The current through R_4,

$$I_4 = \frac{V_a - V_e}{R_4} = \frac{3.21 \text{ V} - (-0.605 \text{ V})}{15 \text{ k}\Omega}$$

$$= 254.3 \text{ }\mu\text{A}$$

At the output of A_3,

$$V_g = V_e - (I_4 \times R_5) = -0.605 \text{ V} - (254.3 \text{ }\mu\text{A} \times 15 \text{ k}\Omega)$$

$$= -4.42 \text{ V}$$

Practice Problem

14-8.1 An instrumentation amplifier is to be designed to produce a 4.75 V output when the differential input is 50 mV. Using 741 op-amps, determine suitable resistor values.

14-9 VOLTAGE LEVEL DETECTORS

Op-amps in Switching Applications

Operational amplifiers are often used in circuits in which the output is switched between the positive and negative saturation voltages $+V_{o(sat)}$ and $-V_{o(sat)}$. The actual voltage change that occurs is known as the *output voltage swing*. For many op-amps (such as a 741, for example), the output saturation voltages are typically the supply voltage levels minus 1 V. Thus, as illustrated in Fig. 14-30, the typical output voltage swing is

$$\Delta V \approx (V_{CC} - 1 \text{ V}) - (V_{EE} + 1 \text{ V}) \qquad (14\text{-}20)$$

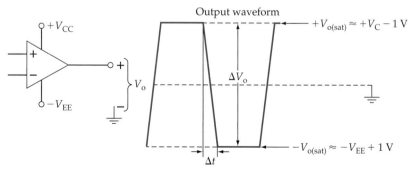

Figure 14-30 An operational amplifier used in a switching application produces an output voltage that switches between positive and negative saturation levels.

For many op-amps the output can be switched from one supply level to the other (there is no 1 V drop), so that

$$\Delta V \approx V_{CC} - V_{EE} \qquad \text{(14-21)}$$

This is referred to as *rail-to-rail* operation.

The switching speed, or rate of change, of the op-amp output voltage is termed the *slew rate (SR)*. Refering to Fig. 14-30,

$$SR = \frac{\Delta V}{\Delta t} \qquad \text{(14-22)}$$

Suppose the output changes from -10 to $+10$ V; the voltage change ΔV is 20 V. If the transition time (or *rise time*, see Section 8-5) is $\Delta t = 1$ μs, the slew rate is 20 V/μs.

Another integrated circuit known as a *voltage comparator* is often used in place of an operational amplifier in switching applications. Voltage comparators are similar to op-amps in that they have two input terminals (inverting and non-inverting) and one output terminal. However, comparators are designed exclusively for switching, and they have slew rates much faster that those available with op-amps.

Zero Crossing Detector

The *zero crossing detector* circuit in Fig. 14-31a is seen to be simply an operational amplifier with the inverting input grounded and the signal applied to the non-inverting input. When the input is above ground level, the output is saturated at its positive maximum, and when the input is below ground, the output is at its negative maximum level. This is illustrated by the input and output waveforms, which show that the output voltage changes from one extreme to the other each time the input voltage crosses zero. The input waveform could have any shape (sinusoidal, pulse, ramp, and so on), but the output will always be a rectangular-type wave.

The actual input voltage that causes the output to switch is not precisely zero, but some very small voltage above or below zero, depending upon the op-amp open-loop gain.

If the op-amp non-inverting input is grounded and the signal is applied to the inverting input in Fig. 14-31a, the output is negative when the input is above ground, and vice versa. Because of the waveform inversion, this circuit is often termed an *inverter.*

The circuit in Fig. 14-31b has the op-amp inverting input biased via a voltage divider (R_1 and R_2). The bias voltage level (V_{R2}) could be positive or negative. In the circuit illustrated, the bias voltage is a positive quantity. The waveforms show that the output voltage changes when the input voltage crosses the bias level. This circuit is appropriately named a *voltage level detector.*

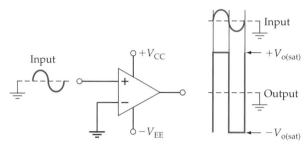

(a) Non-inverting zero crossing detector

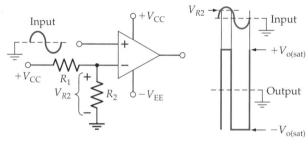

(b) Voltage level detector

Figure 14-31 Operation amplifiers can be connected to function as zero crossing detectors or as voltage level detectors.

Example 14-12

For the zero crossing detector in Fig. 14-32, determine the typical output voltage swing and the typical input voltage level above and below ground at which the output switches. Calculate the rise time of the output voltage.

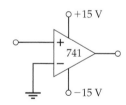

Figure 14-32 Zero crossing detector circuit for Example 14-12.

Solution

Eq. 14-20: $\Delta V \approx (V_{CC} - 1\text{ V}) - (V_{EE} + 1\text{ V}) \approx (15\text{ V} - 1\text{ V}) - (-15\text{ V} + 1\text{ V})$

≈ 28 V

From data sheet A-13 in Appendix A,

$$A_v = 200\,000 \text{ (typical), and SR} = 0.5 \text{ V/}\mu\text{s}$$

$$V_i = \frac{V_o}{A_v} = \frac{\pm 14\text{ V}}{200\,000}$$

$$= \pm 70\ \mu\text{V}$$

From Eq. 14-22, $\Delta t = \dfrac{\Delta V}{\text{SR}} = \dfrac{28\text{ V}}{0.5\text{ V/}\mu\text{s}}$

$$= 56\ \mu\text{s}$$

Practice Problem

14-9.1 A 353 op-amp with a ±18 V supply is used as a zero crossing detector. Determine the typical output voltage swing and the typical input voltage level required to switch the output. Calculate the rise time of the output voltage. Data sheet A-15 in Appendix A shows a partial specification for the 353.

14-10 SCHMITT TRIGGER CIRCUITS

Inverting Schmitt Trigger

A *Schmitt trigger circuit* is a fast-operating voltage level detector. When the input voltage arrives at a level determined by the circuit components, the output voltage switches rapidly between its maximum positive level and its maximum negative level.

An op-amp inverting Schmitt trigger circuit is shown in Fig. 14-33, together with input and output waveforms. At first glance the circuit looks like a non-inverting amplifier. But note that (unlike a non-inverting amplifier) the input voltage (V_i) is applied to the inverting input terminal, and the feedback voltage goes to the non-inverting input. The waveforms show that the output switches rapidly from the positive saturation $(+V_{o(sat)})$ voltage to the negative saturation level $(-V_{o(sat)})$ when the input exceeds a certain positive level: the *upper trigger point (UTP)*. Similarly, the output voltage switches from low to high when the input goes below a negative triggering point: the *lower trigger point (LTP)*.

Note that after V_i has increased to the *UTP* and V_o has switched to $-V_{o(sat)}$, the output remains at $-V_{o(sat)}$ even when V_i falls below the *UTP*. Switch-over from $-V_{o(sat)}$ to $+V_{o(sat)}$ does not occur until $V_i = LTP$. Similarly, after V_i has been reduced to the *LTP* and V_o has switched to $+V_{o(sat)}$, the output remains at $+V_{o(sat)}$ when V_i is increased above the *LTP*. Switch-over from $+V_{o(sat)}$ to $-V_{o(sat)}$ does not occur again until $V_i = UTP$.

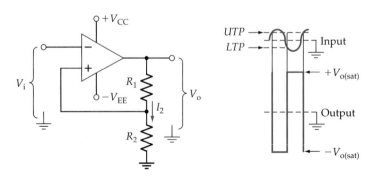

Figure 14-33 A Schmitt trigger circuit is a fast-operating voltage level detector that switches between negative and positive output voltage levels. In an inverting Schmitt trigger circuit, the output goes negative when the input goes positive, and vice versa.

Triggering Points

If the output voltage for the circuit in Fig. 14-33 is high, the voltage at the non-inverting terminal is

$$V_{R2} = \frac{+V_o \times R_2}{R_1 + R_2}$$

If the input voltage (at the inverting input terminal) is below V_{R2} (at the non-inverting input), the output voltage is kept at its high positive level. For the output to switch to its low level, the input voltage must exceed V_{R2} by a very small amount (approximately 70 μV for a 741 op-amp). So the *UTP* essentially equals V_{R2}:

$$UTP = \frac{+V_o \times R_2}{R_1 + R_2} \qquad (14\text{-}23)$$

When the output is negative, the LTP can be calculated as

$$LTP = \frac{-V_o \times R_2}{R_1 + R_2} \qquad (14\text{-}24)$$

Input/Output Characteristic

A graph of output voltage (V_o) versus input voltage (V_i) can be plotted for an inverting Schmitt trigger circuit, as shown in Fig. 14-34. This is the circuit input/output characteristic. To understand the characteristic, consider the voltages at each of the numbered points on the graph:

- At point 1, $V_o = +V_{o(sat)}$ and $V_i < LTP$, thus keeping $V_o = +V_{o(sat)}$.
- From point 1 through points 2 and 3, V_o remains at $+V_{o(sat)}$ as V_i is increased through the *LTP* and through zero, until $V_i = UTP$.
- From point 3 to point 4, V_o switches rapidly from $+V_{o(sat)}$ to $-V_{o(sat)}$ when $V_i = UTP$.

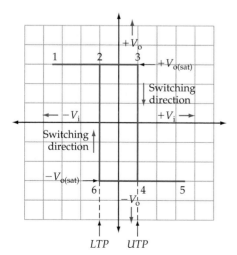

Figure 14-34 Output/input characteristics for an inverting Schmitt trigger circuit.

- From point 4 to point 5, V_o remains at $-V_{o(sat)}$ as V_i is increased above the *UTP*.
- From point 5 through points 4 and 6, V_o remains at $-V_{o(sat)}$ as V_i is reduced through the *UTP* and through zero, until $V_i = LTP$.
- From point 6 to point 2, V_o switches rapidly from $-V_{o(sat)}$ to $+V_{o(sat)}$ when $V_i = LTP$.
- From point 2 to point 1, V_o remains at $+V_{o(sat)}$ as V_i is reduced below the *LTP*.

The difference between the *UTP* and the *LTP* is termed *hysteresis*. Some applications require a small amount of hysteresis, and for other applications a large amount of hysteresis is essential.

Circuit Design

The design procedure for a Schmitt trigger circuit is similar to that for an op-amp amplifier. A voltage-divider current (I_2 in Fig. 14-33) is selected much larger than the op-amp input bias current. The resistor values are then calculated as

$$R_2 = \frac{UTP}{I_2} \qquad (14\text{-}25)$$

and

$$R_1 = \frac{V_{o(sat)} - UTP}{I_2} \qquad (14\text{-}26)$$

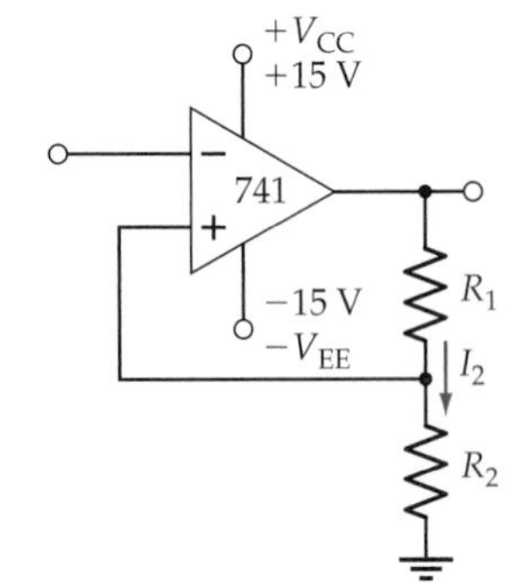

Figure 14-35 Inverting Schmitt trigger circuit for Example 14-13.

Example 14-13

Using a 741 op-amp, calculate resistor values for the Schmitt trigger circuit in Fig. 14-35 to give triggering points of ±5 V.

Solution

Select

$$I_1 \approx 100\, I_B = 100 \times 500 \text{ nA}$$
$$= 50\ \mu\text{A}$$

Eq. 14-25:

$$R_2 = \frac{UTP}{I_2} = \frac{5 \text{ V}}{50\ \mu\text{A}}$$
$$= 100 \text{ k}\Omega \text{ (standard value)}$$

Eq. 14-26:

$$R_1 = \frac{V_{o(sat)} - UTP}{I_2} \approx \frac{(15 \text{ V} - 1 \text{ V}) - 5 \text{ V}}{50\ \mu\text{A}}$$
$$= 180 \text{ k}\Omega \text{ (standard value)}$$

Adjusting the Trigger Points

Many Schmitt trigger circuit applications require *UTP* and *LTP* levels that are not equal in magnitude. This is usually achieved by the use of diodes, as illustrated in Fig. 14-36.

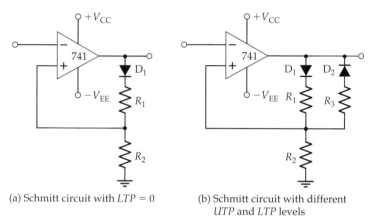

(a) Schmitt circuit with $LTP = 0$

(b) Schmitt circuit with different UTP and LTP levels

Figure 14-36   Diodes can be used with inverting Schmitt trigger circuits to produce different levels of positive and negative triggering voltages.

The circuit shown in Fig. 14-36a simply has a diode (D$_1$) connected in series with resistor R_1. The diode is forward-biased only when the op-amp output is a positive quantity. At this time, the *UTP* is V_{R2}, as before. When V_o is negative, D$_1$ is reverse-biased, making I_2 equal zero. Consequently, there is no voltage drop across R_2, and so the non-inverting terminal is grounded via R_2. This gives a zero level for the *LTP*. Thus, this circuit has a positive *UTP* and a zero-voltage *LTP*.

Figure 14-36b shows a circuit with two different-level trigger points. When V_o is positive, D$_1$ is forward-biased and D$_2$ is reversed and the *UTP* is set by resistors R_1 and R_2. With V_o negative, D$_2$ is forward-biased and D$_1$ is reversed. The *LTP* is now determined by the resistances of R_3 and R_2.

The diode-forward voltage drop (V_F) must be accounted for when calculating the trigger points for both of the circuits in Fig. 14-36. This is done simply by replacing $V_{o(sat)}$ with ($V_{o(sat)} - V_F$) in Equation 14-26. Another important design consideration is that the voltage-divider current (I_2) should normally be a minimum of 100 μA for satisfactory diode operation.

Non-Inverting Schmitt Trigger

A non-inverting Schmitt trigger circuit is shown in Fig. 14-37. This circuit looks like an inverting amplifier, but note that (unlike an inverting amplifier) the inverting input is grounded and the non-inverting input is connected to the junction of R_1 and R_2. The waveforms in Fig. 14-37 show that V_o switches rapidly from $-V_{o(sat)}$ to $+V_{o(sat)}$ when V_i arrives at the *UTP*, and that V_o switches back to $-V_{o(sat)}$ when V_i falls to the *LTP*.

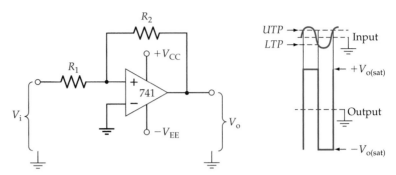

Figure 14-37 Non-inverting Schmitt trigger circuit. The output goes positive when the input arrives at the *UTP*, and negative when $V_i = LTP$.

Consider the situation when $V_i = UTP$ and V_o has just switched to $+V_{o(sat)}$. The voltage at the junction of R_1 and R_2 is pulled up far above the ground-level voltage at the op-amp inverting input terminal. So the positive voltage at the non-inverting input keeps the output at its positive saturation level. To switch the output to $-V_{o(sat)}$, the voltage at the junction of R_1 and R_2 must be pulled down to the (ground-level) voltage at the inverting input terminal. This occurs when V_i is at the *LTP* (see Fig. 14-38a). So switching occurs when one end of R_1

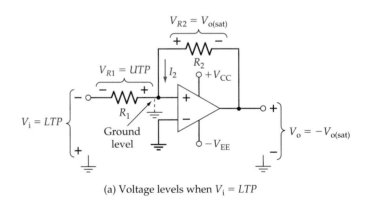

(a) Voltage levels when $V_i = LTP$

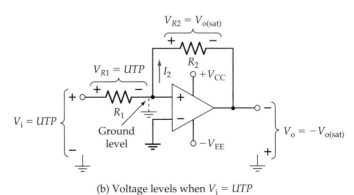

(b) Voltage levels when $V_i = UTP$

Figure 14-38 At the instant of switching in a non-inverting Schmitt trigger circuit, the op-amp non-inverting input terminal voltage must be at ground level.

(the right end) is at ground and the other (left) end is at $V_i = LTP$, that is, when $V_{R1} = LTP$. Similarly, switching in the other direction occurs when $V_{R1} = UTP$, (Fig. 14-38b). So, at the trigger points,

$$V_i = UTP = LTP = I_2 R_1$$

Figure 14-38a and b show that the output voltage is at one of its saturation levels at the instant of triggering. This means that one end of R_2 is at ground (left end) and the other (right) end is at $V_{o(sat)}$. So, $V_{R2} = V_{o(sat)}$ when triggering occurs, giving

$$I_2 = \frac{V_{o(sat)}}{R_2}$$

Also,
$$UTP = I_2 R_1$$

So
$$UTP = V_{o(sat)} \times \frac{R_1}{R_2} \qquad (14\text{-}27)$$

and
$$LTP = -V_{o(sat)} \times \frac{R_1}{R_2} \qquad (14\text{-}28)$$

The design procedure for a non-inverting Schmitt trigger circuit is just as simple as for the inverting circuit. Voltage-divider current I_2 is again selected to be much larger than the op-amp input bias current. Then the resistor values are

$$R_1 = \frac{UTP}{I_2} \qquad (14\text{-}29)$$

and
$$R_2 = \frac{V_{o(sat)}}{I_2} \qquad (14\text{-}30)$$

Non-inverting Schmitt trigger circuits can be designed for different upper and lower trigger point voltages by the use of diodes, as in the case of the inverting circuit. Figure 14-39a and b show two possible circuits. The diode-forward voltage drop (V_F) must be included in the UTP and LTP calculations.

Example 14-14

Determine the upper and lower trigger points for the non-inverting Schmitt trigger circuit shown in Fig. 14-39b. Assume that the op-amp used can be rail-to-rail operated, and that the supply is ±15 V.

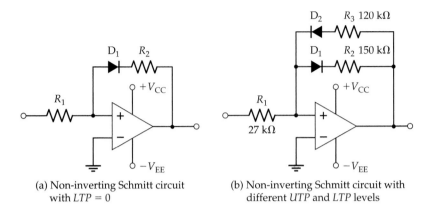

(a) Non-inverting Schmitt circuit   (b) Non-inverting Schmitt circuit with
    with $LTP = 0$ different UTP and LTP levels

Figure 14-39   Diodes can be used to produce different UTP and LTP levels for non-inverting Schmitt trigger circuits.

Solution

With rail-to-rail operation,

$$V_{o(sat)} = \pm V_{CC} = \pm 15 \text{ V}$$

UTP calculation (including V_F):

From Eq. 14-30, $I_2 = \dfrac{V_{o(sat)} - V_F}{R_2} = \dfrac{15 \text{ V} - 0.7 \text{ V}}{150 \text{ k}\Omega}$

$$= 95.3 \text{ }\mu\text{A}$$

From Eq. 14-29, $UTP = I_2 R_1 = 95.3 \text{ }\mu\text{A} \times 27 \text{ k}\Omega$

$$= 2.57 \text{ V}$$

LTP calculation (including V_F):

From Eq. 14-30, $I_3 = \dfrac{V_{o(sat)} - V_F}{R_3} = \dfrac{15 \text{ V} - 0.7 \text{ V}}{120 \text{ k}\Omega}$

$$= 119 \text{ }\mu\text{A}$$

From Eq. 14-29, $LTP = -I_3 R_1 = 119 \text{ }\mu\text{A} \times 27 \text{ k}\Omega$

$$= -3.2 \text{ V}$$

Practice Problems

14-10.1 Using a 741 op-amp with a ±12 V supply, design an inverting Schmitt trigger circuit (as in Fig. 14-36b) to have $UTP = 3$ V and $LTP = -2$ V.

14-10.2 Design a non-inverting Schmitt trigger circuit to have approximately ±1 V trigger points. Use a 353 op-amp with a ±15 V supply.

14-11 TESTING AND TROUBLESHOOTING OP-AMP CIRCUITS

DC Voltages

Section 5-6 lists common errors that apply to all laboratory-constructed circuits, and these should be considered once again when one is preparing to test an op-amp circuit.

Before making dc measurements on a linear op-amp circuit, the input should be grounded, assuming that plus/minus supply voltages are used. Where a single-polarity supply is employed, capacitor-coupled inputs can be grounded, and direct-coupled inputs should be connected to a $+V_{CC}/2$ voltage level. Also, before one proceeds with dc level measurements, an oscilloscope must be connected to check that the circuit is not oscillating. Methods of ensuring circuit stability discussed in Section 12-10 should be considered. Op-amp circuit stability is explored further in Section 15-6.

If the dc output voltage is found to be near $+V_{CC}$ or $-V_{EE}$ (for a linear circuit) the following are possible fault sources:

- The input is not grounded.
- One or more components are incorrectly connected.
- An incorrectly identified component value is used.
- The op-amp inverting and non-inverting input terminals are connected incorrectly.

AC Gain and Frequency Response

To investigate the frequency response of a linear op-amp circuit, the procedure described in the 'AC Testing' part of Section 12-10 is applicable.

When an op-amp circuit appears to function correctly but the voltage gain is much smaller than expected, it is probable that two resistors are interchanged (that is, the components are in the wrong places). Another possibility is that an incorrectly identified resistor is used in the circuit (for example, a 1 kΩ instead of a 100 kΩ). If an op-amp circuit has a higher voltage gain than expected, the most likely cause is incorrectly identified resistor values.

Testing more complex linear circuits, such as the instrumentation amplifier in Section 14-8, might require checking individual parts of the overall circuit. For example, the non-inverting circuit consisting of A_1, R_1, and R_2 in Fig. 14-28 can be tested by grounding the bottom of R_2. An appropriate input voltage can then be applied, and the A_1 output voltage measured. Similarly, the A_2, R_2, R_3 non-inverting amplifier can be tested if the top of R_2 is grounded. If the A_1A_2 input stage is found to function correctly, it can be used to apply appropriate test levels to the difference amplifier output stage. When the input stage is not functioning correctly, it might be necessary to disconnect the input and output stages to test them individually.

Switching Circuits

Op-amp switching circuits are unlikely to self-oscillate, except when the inverting and non-inverting input terminals are incorrectly connected. In this situation, the circuit might behave as an amplifier rather than a switching circuit. When the output of a switching circuit will not move from a near $+V_{CC}$ or $-V_{EE}$ saturated level, the most likely cause is an open-circuited connection. Incorrect triggering levels for a circuit, such as the Schmitt trigger in Fig. 14-33, are most likely due to two resistors that are interchanged or to incorrectly identified resistor values.

Review Questions

Section 14-1

14-1 Sketch the circuit symbol for an operational amplifier, identify all terminals, and briefly explain.

14-2 Sketch the input and output stages of a bipolar operational amplifier, explain the circuit operation, and discuss the input and output impedances.

14-3 Repeat Question 14-2 for a BIFET operational amplifier.

14-4 List the most important parameters for an operational amplifier, and state typical values for each parameter.

Section 14-2

14-5 Draw a complete circuit diagram to show an operational amplifier with a signal source capacitor-coupled to the non-inverting input, and 100% feedback from the output to the inverting input terminal. Explain the requirement for all components.

14-6 Draw a complete circuit diagram for an operational amplifier with a plus/minus supply, its non-inverting input terminal biased to a positive voltage level, and 100% feedback from the output to the inverting input terminal. Explain the requirement for all components, and discuss the component current levels.

14-7 An operational amplifier is to use a single-polarity supply and to have its non-inverting input terminal biased to a positive voltage level; 100% feedback is to be provided from the output to the inverting input terminal. Draw the circuit diagram, and explain the function of each component.

14-8 Explain output voltage offset nulling, and show how it can be performed in a 741 op-amp.

Section 14-3

14-9 Draw a circuit diagram for a direct-coupled voltage follower, identifying all terminals. Sketch a basic op-amp (internal) circuit with the external connection for voltage follower operation. Explain the circuit operation.

14-10 Discuss the performance of a voltage follower, and compare it to an emitter follower.

14-11 Sketch a capacitor-coupled voltage follower circuit, and explain the effects of capacitor-coupling on the circuit input impedance.

Section 14-4

14-12 Draw a circuit diagram for a direct-coupled non-inverting amplifier. Identify all terminals and explain the circuit operation. Derive the closed-loop voltage gain equation, and write equations for the input and output impedances.

14-13 Sketch a basic operational amplifier (internal) circuit with the external connection for non-inverting amplifier operation. Explain how the circuit amplifies a signal.

14-14 Sketch a capacitor-coupled non-inverting amplifier circuit. Explain the circuit operation, and discuss the effects of capacitor-coupling on the circuit input impedance.

Section 14-5

14-15 Sketch a circuit diagram for a direct-coupled inverting amplifier. Identify all terminals and explain the circuit operation. Derive the closed-loop voltage gain equation, and write equations for the input and output impedances.

14-16 Sketch a circuit for a capacitor-coupled inverting amplifier. Explain the amplifier operation, and discuss the effects of capacitor-coupling on the circuit input impedance.

Section 14-6

14-17 Sketch a three-input summing amplifier circuit and explain its operation.

14-18 For a three-input summing amplifier, derive an equation for the output voltage in terms of the input voltage and the circuit components.

Section 14-7

14-19 Sketch the circuit of an op-amp difference amplifier. Explain its operation, and derive an equation for the output voltage in terms of the input and the circuit components.

14-20 Identify the input resistances for the difference amplifier in Question 14-19. Also, discuss the effects of a common-mode input voltage, and show how common-mode voltages can be nulled.

Section 14-8

14-21 Draw the circuit diagram for an instrumentation amplifier. Identify each section of the circuit, and explain its operation.

14-22 Derive an equation for the overall voltage gain of the instrumentation amplifier circuit drawn for Question 14-21. Discuss the effects of common-mode input voltages.

Section 14-9

14-23 Sketch circuit diagrams for an op-amp used as (a) a zero crossing detector and (b) a voltage level detector. Show typical input and output waveforms and briefly explain. Define slew rate, and show its effect on an op-amp output waveform.

Section 14-10

14-24 Draw the circuit diagram of an op-amp inverting Schmitt trigger circuit. Sketch typical input and output waveforms, and explain the circuit operation. Write equations for the circuit upper and lower trigger points.

14-25 Sketch typical output/input characteristics for an inverting Schmitt trigger circuit. Identify the upper and lower trigger points, and explain the characteristics.

14-26 Draw circuit diagrams for inverting Schmitt trigger circuits with different levels of positive and negative trigger points. Explain.

14-27 Draw the circuit diagram of an op-amp non-inverting Schmitt trigger circuit. Sketch typical input and output waveforms, and explain the circuit operation. Write equations for the circuit upper and lower trigger points.

14-28 Draw circuit diagrams for non-inverting Schmitt trigger circuits with different levels of positive and negative trigger points. Explain.

Problems

Section 14-1

14-1 Referring to the 741 operational amplifier specification in data sheet A-13 in Appendix A, determine typical values of the following parameters: input resistance, input bias current, output resistance, output short-circuit current, supply current, and power dissipation.

14-2 Referring to data sheet A-14 in Appendix A, determine typical values of the following parameters for a LM308 operational amplifier: input resistance, input bias current, supply current, and power supply rejection ratio.

14-3 From the LF353A op-amp specification in data sheet A-15 in Appendix A, determine typical values of the following parameters: supply current, common-mode rejection ratio, output voltage swing, and input offset voltage.

Section 14-2

14-4 A 741 op-amp with a ±9 V supply uses voltage-divider bias to provide +3 V at the non-inverting input terminal. There is also 100% feedback from the output to the inverting input. Determine suitable resistor values.

14-5 Determine resistor values for a 741 op-amp with a +18 V supply, +9 V voltage-divider bias at the non-inverting input terminal, and 100% feedback from the output to the inverting input.

14-6 Redesign the circuit in Problem 14-4 to use a LM108 op-amp. Note that, because the LM108 has a very low bias current, it should be treated like a BIFET op-amp.

14-7 Redesign the circuit in Problem 14-5 to use a LF353 op-amp. Then investigate the effect of using a 741 in place of the LF353 without changing the resistor values.

Section 14-3

14-8 An operational amplifier with an open-loop gain of 200 000 is used as a voltage follower. Calculate the precise level of the output voltage if the input is exactly 8 V.

14-9 Calculate the input and output resistance of a voltage follower circuit that uses an op-amp with the following parameters: $A_{v(OL)} = 100\ 000$, $r_i = 1\ M\Omega$, and $r_o = 50\ \Omega$.

14-10 Design a capacitor-coupled voltage follower to use a 741 op-amp. The load resistance is 2 kΩ, the signal source resistance is 600 Ω, and the lower cutoff frequency is to be 100 Hz.

14-11 A capacitor-coupled voltage follower uses an op-amp with a 1 μA input bias current. The load resistance is 1 kΩ, and the signal source resistance is 400 Ω. Select suitable component values for the circuit to give a 120 Hz lower cutoff frequency.

Section 14-4

14-12 Design a direct-coupled non-inverting amplifier to produce a ±5 V output when the input is ±75 mV. Use a 741 op-amp.

14-13 A direct-coupled non-inverting amplifier that uses an op-amp with $I_{B(max)} = 750$ nA is to have a maximum output of ±10 V and a voltage gain of 120. Design the circuit.

14-14 A direct-coupled non-inverting amplifier (as in Fig. 14-15) uses a BIFET op-amp and has the following resistor values: $R_1 = 22$ kΩ, $R_2 = 1$ MΩ, $R_3 = 22$ kΩ. Calculate the circuit voltage gain.

14-15 Redesign the circuit in Problem 14-12 to use an LF353 BIFET op-amp.

14-16 The non-inverting amplifier designed for Problem 14-12 is to be modified to operate as a capacitor-coupled circuit with $f_1 = 80$ Hz, $R_L = 100$ kΩ, and $r_s = 300$ Ω. Calculate suitable capacitor values.

14-17 Calculate the input impedance of the direct-coupled circuit in Problem 14-12 and of the capacitor-coupled circuit in Problem 14-16.

14-18 A capacitor-coupled non-inverting amplifier is to be designed to produce a 7 V output from a 70 mV input. The supply voltage is ±18 V, the lower cutoff frequency is 60 Hz, the load is 82 kΩ, and the signal source resistance is 600 Ω. Design the circuit to have an input impedance of 12 kΩ.

14-19 Design a direct-coupled non-inverting amplifier using a BIFET op-amp with a 100 000 open-loop voltage gain. The closed-loop gain is to be 45, and the maximum output is to be ± 2 V.

14-20 Modify the circuit for Problem 14-19 to make it capacitor-coupled with $R_L = 2$ kΩ and $f_1 = 150$ Hz.

Section 14-5

14-21 A direct-coupled inverting amplifier using a 741 op-amp is to have a 45 mV input signal and a voltage gain of 200. Calculate suitable resistor values and determine the circuit input impedance.

14-22 A direct-coupled inverting amplifier is to have a 1 kΩ input impedance and is to produce a ± 3.3 V output from a ± 100 mV input. Determine suitable resistor values.

14-23 Redesign the circuit in Problem 14-21 to use a BIFET op-amp.

14-24 The inverting amplifier designed for Problem 14-21 is to be modified to operate as a capacitor-coupled circuit with $f_1 = 50$ Hz, $R_L = 150$ kΩ, and $r_s = 300$ Ω. Determine suitable capacitor values.

14-25 A capacitor-coupled inverting amplifier is to have a 680 Ω input impedance and a voltage gain of 49. The load resistance is 12 kΩ, and the lower cutoff frequency is to be 40 Hz. Determine suitable component values.

14-26 The inverting amplifier designed for Problem 14-22 is to function as a capacitor-coupled circuit with $f_1 = 30$ Hz, $R_L = 33$ kΩ, and $r_s = 600$ Ω. Determine suitable capacitor values.

14-27 A capacitor-coupled inverting amplifier is to use a 741 op-amp with a +30 V single-polarity supply, and have $R_L = 5$ kΩ, $A_{CL} = 50$, $V_o = \pm 3$ V, and $f_1 = 75$ Hz. Sketch the circuit and determine suitable component values.

Section 14-6

14-28 Design a two-input summing amplifier for a voltage gain of 10 and input levels of 100 mV. Use an operational amplifier with $I_{B(max)} = 1$ μA and $A_{v(OL)} = 300\ 000$.

14-29 The circuit designed for Problem 14-28 has inputs of 60 mV and 80 mV. Analyze the circuit to determine all voltage and current levels.

14-30 A three-input summing amplifier using a 741 op-amp is to be designed for a voltage gain of 1. The input voltages are approximately 1 V. Calculate suitable component values.

Section 14-7

14-31 The difference amplifier circuit in Fig. 14-27 has the components $R_1 = 15$ kΩ, $R_2 = 33$ kΩ, $R_3 = 6.8$ kΩ, and $R_4 = (12$ kΩ $+ 2.7$ kΩ $+ 500$ Ω variable). The input voltages are V_{i1} (to R_1) = 1.47 V and V_{i2} (to R_3) = 1.23 V. Calculate V_o when R_4 is adjusted to 14.96 kΩ.

14-32 A difference amplifier is to be designed for a voltage gain of 100 and input levels of 100 mV. Using an operational amplifier with $I_{B(max)} = 1 \ \mu A$ and $A_{v(OL)} = 300 \ 000$, determine suitable resistor values.

14-33 If the difference amplifier in Problem 14-31 has a 500 mV common-mode input, calculate the common-mode output voltage: (a) when R_4 is set to its maximum resistance, (b) when R_4 is at its minimum value, and (c) when $R_4 = 14.96 \ k\Omega$.

Section 14-8

14-34 Design an instrumentation amplifier to produce a 3 V output when the inputs are +450 mV and +150 mV. Use 741 op-amps with a ±12 V supply.

14-35 Analyze the circuit designed for Problem 14-34 to determine all current and voltage levels.

Section 14-9

14-36 A zero crossing detector uses an op-amp with $V_{CC} = \pm 9$ V, $SR = 3$ V/μs, and $A_{v(OL)} = 100 \ 000$. Calculate the typical output voltage swing, the output rise time, and the input voltage level required to switch the output.

Section 14-10

14-37 Using a 741 op-amp with $V_{CC} = \pm 18$ V, design an inverting Schmitt trigger circuit to have ±2 V triggering points.

14-38 Plot the output/input characteristics for the circuit in Problem 14-37.

14-39 Design a non-inverting Schmitt trigger circuit to fulfil the specification in Problem 14-37.

14-40 An inverting Schmitt trigger circuit is to be designed to have $UTP = 0$ V and $LTP = -1.5$ V. Sketch the circuit and determine suitable component values if a 741 op-amp is used with a ±15 V supply.

14-41 An non-inverting Schmitt trigger circuit is to be designed to have $UTP = 1$ V and $LTP = -0.5$ V. Sketch the circuit and determine suitable component values if an LF353 op-amp is used with a ±12 V supply.

14-42 An non-inverting Schmitt trigger circuit using a 741 op-amp with a ±10 V supply is to have trigger points adjustable from ±0.5 V to ±0.75 V. Sketch the circuit and determine suitable component values.

Practice Problem Answers

14-1.1   300 000, 0.8 nA, 0.7 mV, 70 MΩ
14-1.2   100 000, 50 pA, 5 mV, $10^{12} \ \Omega$
14-2.1   220 kΩ, 220 kΩ, 100 kΩ
14-2.2   180 kΩ, (220 kΩ + 33 kΩ), 100 kΩ
14-3.1   1.5 μF, 0.33 μF
14-3.2   0.68 μF, 1 μF
14-3.3   60 mV

14-4.1   1.8 kΩ, (68 kΩ + 2.2 kΩ), 1.8 kΩ

14-4.2   4.7 kΩ, (82 kΩ + 6.8 kΩ), 4.7 kΩ, 18 μF, 10 μF

14-4.3   180 kΩ, 1 MΩ, (220 kΩ + 33 kΩ)

14-5.1   5.6 kΩ, (82 kΩ + 2.2 kΩ), 5.6 kΩ

14-5.2   8 μF, 15 μF

14-5.3   (33 kΩ + 33 kΩ), 1 MΩ, 68 kΩ

14-6.1   4.7 kΩ, 4.7 kΩ, 47 kΩ, 2.2 kΩ

14-6.2   −10.6 V, 2.27 μA, 4.55 μA, 3.79 μA, 10.6 μA

14-7.1   +7.5 V, −33.9 μA, −33.9 μA, 75.76 μA, 1.36 V

14-7.2 25×10^{-3}

14-7.3   (100 kΩ + 68 kΩ), 1 MΩ, 27 kΩ, (100 kΩ + 68 kΩ)

14-8.1   47 kΩ, 1 kΩ, 47 kΩ, 47 kΩ, 47 kΩ, 47 kΩ, 47 kΩ

14-9.1   33 V, ±330 μV, 2.5 μS

14-10.1   (68 kΩ + 27 kΩ), 39 kΩ, (150 kΩ + 12 kΩ)

14-10.2   (56 kΩ + 18 kΩ), 1 MΩ

CHAPTER 15
Operational Amplifier Frequency Response and Compensation

CONTENTS

Objectives

You will be able to:

1   Show how feedback can produce instability in op-amp circuits.

2   Sketch and explain typical gain/frequency and phase/frequency response graphs for uncompensated and compensated operational amplifiers.

3   Define loop gain, loop phase shift, and phase margin.

4   Discuss compensation methods for stabilizing op-amp circuits, and calculate component values for compensating circuits.

5   Explain how the bandwidth of an op-amp circuit is affected by closed-loop gain.

6   Define gain-bandwidth product, slew rate, and full-power bandwidth.

7   Determine the bandwidths of various op-amp circuits using frequency response graphs, gain-bandwidth product, and slew rate.

8   Explain how stray and load capacitance can affect op-amp circuit stability, sketch appropriate compensating circuits, and calculate suitable component values.

9   List precautions that should be observed to ensure op-amp circuit stability.

INTRODUCTION

Signals applied to operational amplifiers experience phase shifts as they pass from input to output. These phase shifts are greatest at high frequencies, and at some particular frequency the total loop phase shift (from the inverting input terminal to the output and back to the input via the feedback network) can add up to 360°. When this occurs, the amplifier circuit can go into a state of unwanted oscillation. The conditions that produce oscillation are a loop voltage gain greater than or equal to unity when the loop phase shift approaches 360°. Measures taken to combat instability include the use of capacitors and resistors to reduce the total phase shift. Most operational amplifiers have compensating components included in the circuitry to ensure stability.

15-1 OPERATIONAL AMPLIFIER CIRCUIT STABILITY

Loop Gain and Loop Phase Shift

Consider the inverting amplifier circuit and waveforms in Fig. 15-1a. The signal voltage (v_s) is amplified by a factor R_2/R_1, and phase shifted through $-180°$. The circuit is redrawn in Fig. 15-1b to illustrate the fact that the output voltage (v_o) is divided by the feedback network to produce the feedback voltage (v).

For an ac voltage (v) at the op-amp inverting input terminal (in Fig. 15-1b), the amplified output is $v_o = A_v v$, as shown. The output is divided by the feedback factor $[B = R_1/(R_1 + R_2)]$ and is fed back to the input. An additional $-180°$ of phase shift can occur within the op-amp at high frequencies, and this causes v to be in phase with v_o, as illustrated by the waveforms. Thus, the feedback voltage can be exactly equal to and in phase with the voltage (v) at the inverting input. In this case, the circuit is supplying its own ac input voltage and a state of continuous oscillation exists.

Because of the feedback network, high-frequency oscillations can occur in many operational amplifier circuits, and when this happens the circuit is termed *unstable.* Measures taken to combat circuit instability are referred to as *frequency compensation.*

Two conditions normally have to be fulfilled for a circuit to oscillate: the *loop gain* must be equal to or greater than 1, and the *loop phase shift* must equal 360°. The *loop gain* is the voltage gain around the loop from the inverting input terminal to the amplifier output, and back to the input via the feedback network. The *loop phase shift* is the total phase shift around the loop from the inverting input terminal to the amplifier output, and back to the input via the feedback network.

The gain from the inverting input terminal to the output is the op-amp open-loop gain (A_v). For the feedback network, the gain from the amplifier

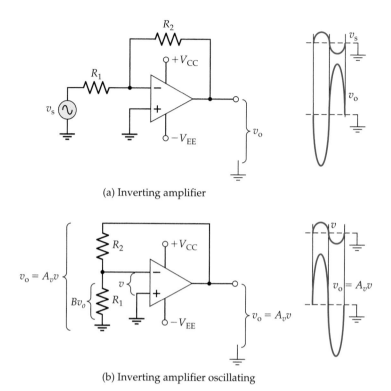

(a) Inverting amplifier

(b) Inverting amplifier oscillating

Figure 15-1 Because an inverting amplifier consists of an operational amplifier and a feedback network, the circuit can supply its own ac input (v derived from v_o), and a state of continuous oscillations can occur.

output back to the input is actually an attenuation. So,

$$\text{loop gain} = (\text{amplifier gain}) \times (\text{feedback network attenuation})$$
$$= A_vB$$

Assuming that the feedback network is purely resistive, it adds nothing to the loop phase shift. The loop phase shift is essentially the amplifier phase shift. The phase shift from the inverting input terminal to the output is normally $-180°$. (The output goes negative when the input goes positive, and vice versa.) At high frequencies there is additional phase shift caused by circuit capacitances, and the total can approach $-360°$. When this occurs, the circuit is virtually certain to oscillate. Most currently available operational amplifiers have internal *compensating components* to prevent oscillations. In some cases, compensating components must be connected externally to stabilize a circuit.

Uncompensated Gain and Phase Response

Straight-line approximations of the gain/frequency and phase/frequency response graphs for a typical operational amplifier *without any compensating components* are shown in Fig. 15-2. Note that the overall open-loop voltage gain

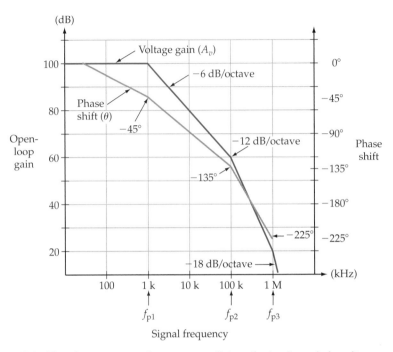

Figure 15-2 The three stages of an op-amp (internal) circuit each has its own gain/ frequency response with a 6 dB/octave fall-off, and its own phase shift/frequency response with a maximum phase shift of 90°. These responses combine to give the overall op-amp A_v/f and θ/f responses.

(A_v) initially falls off (from its 100 dB level) at 6 dB/octave (-20 dB/decade) from f_{p1} (pole frequency #1). From f_{p2}, the rate of decline is 12 dB/octave (-40 dB/ decade), and from f_{p3}, the fall-off rate of A_v is 18 dB/octave (-60 dB/decade). The reason there are three different fall-off rates is that the three internal stages of the op-amp (input, intermediate, and output) have different upper cutoff frequencies.

The phase/frequency response graph in Fig. 15-2 shows that the phase shift (θ) is approximately $-45°$ at f_{p1}, $-135°$ at f_{p2}, and $-225°$ at f_{p3}. This open-loop phase shift is in addition to the $-180°$ phase shift that normally occurs from the op-amp inverting input terminal to the output. Thus, the total loop phase shift (ϕ_L) at f_{p1} is $(-45° -180°) = -225°$; at f_{p2}, $\phi_L = (-135° -180°) = -315°$; and at f_{p3}, $\phi_L = (-225° -180°) = -405°$.

As already pointed out, oscillations occur when the loop gain equals or exceeds 1 and the loop phase shift is 360°. In fact, the phase shift does not have to be exactly 360° for oscillation to occur. A phase shift of 330° at $A_vB = 1$ can make the circuit unstable. To avoid oscillations, the total loop phase shift must not be greater than 315° when $A_vB = 1$. The difference between 360° and the actual loop phase shift at $A_vB = 1$ is referred to as the *phase margin* (ϕ_m). Thus, for circuit stability, the phase margin should be a minimum of

$$\phi_m = 360° - 315° = 45°$$

Compensated Op-amp Gain and Phase Response

The open-loop gain/frequency and phase/frequency responses for two internally compensated operational amplifiers are shown in Figs. 15-3 and 15-4. The 741 frequency response graph in Fig. 15-3 shows that the gain starts at 100 dB and falls by 20 dB/decade over most of its frequency range. The phase shift remains $-90°$ or less for most of the frequency range. The open-loop gain falls off to 1 (0 dB) at a frequency of approximately 800 kHz. The 741 is known as a general purpose operational amplifier for use in relatively low-frequency applications.

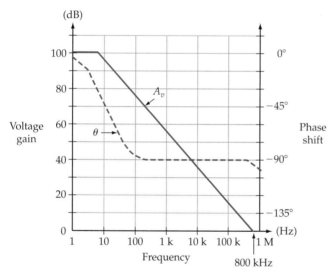

Figure 15-3   Approximate gain/frequency and phase/frequency responses for a 741 op-amp.

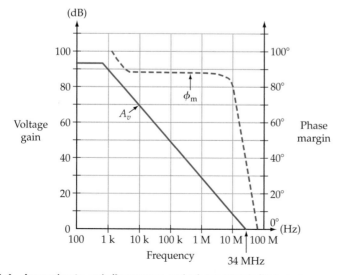

Figure 15-4   Approximate gain/frequency and phase-margin/frequency response for an AD843 op-amp.

The AD843 frequency response in Fig. 15-4 shows an open-loop gain of 90 dB at low frequencies, falling off at 20 dB per decade to $A_v = 1$ at 34 MHz. Instead of the open-loop phase shift, the phase margin is plotted versus frequency. The phase margin is close to 90° over much of the frequency range, starts to become smaller around 3 MHz, and falls to approximately 40° at $f = 34$ MHz.

Amplifier Stability and Gain

From Eq. 13-3, the overall voltage gain of an amplifier with negative feedback is

$$A_{CL} \approx \frac{1}{B}$$

and the loop gain is

$$A_v B \approx \frac{A_v}{A_{CL}}$$

So the loop gain $(A_v B)$ equals 1 when

$$A_{CL} \approx A_v$$

This is one of the conditions required for circuit oscillation. To determine if oscillation will occur in a given circuit, it is necessary first to find the frequency at which $A_{CL} \approx A_v$, and then to determine the op-amp phase margin at that frequency.

Example 15-1

The inverting amplifier in Fig. 15-5 is to be investigated for stability. Determine the frequency at which the loop gain equals 1, and estimate the phase margin if the operational amplifier is (a) one with the gain/frequency characteristics in Fig. 15-2, (b) a 741, and (c) an AD843.

Solution

(a) Refer to the A_v/f and θ/f graphs reproduced in Fig. 15-6 (from Fig. 15-2).

$$A_{CL} = \frac{R_2}{R_1} = \frac{560 \text{ k}\Omega}{1.8 \text{ k}\Omega}$$

$$= 311$$

Or

$$A_{CL} = 20 \log 311$$

$$\approx 50 \text{ dB}$$

Draw a horizontal line on the frequency response graph at $A_v = A_{CL} = 50$ dB (Fig. 15-6). Draw a vertical line where the horizontal line intersects the A_v/f graph. The frequency at this point is identified as f_2.

$$f_2 \approx 150 \text{ kHz (logarithmic scale)}$$

From the θ/f graph, the op-amp phase shift at f_2 is

$$\theta \approx -165°$$

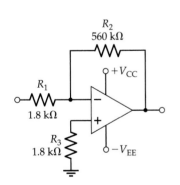

Figure 15-5 Inverting amplifier circuit for Example 15-1.

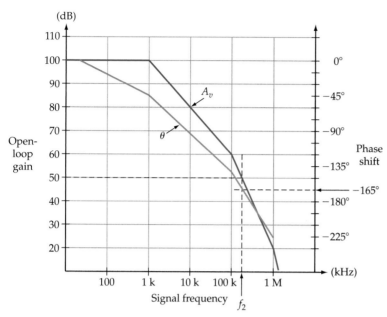

Figure 15-6 A_v/f and θ/f characteristics for the op-amp in Example 15-1a.

The loop phase shift is

$$\phi_L = \theta - 180° = -165° - 180°$$
$$= -345°$$

and

$$\phi_m = 360° - \phi_L = 360° - 345°$$
$$= 15°$$

Because the phase margin is less than 45°, the circuit is likely to be unstable.

(b) For a 741:

A horizontal line at 50 dB on the frequency response in Fig. 15-3 gives

$$f_2 \approx 1.5 \text{ kHz} \quad \text{and} \quad \theta \approx -90°$$
$$\phi_L = -90° - 180°$$
$$= -270°$$

and

$$\phi_m = 360° - 270°$$
$$= 90° \text{ (stable circuit)}$$

(c) For an AD843:

A horizontal line at 50 dB on the frequency response in Fig. 15-4 gives

$$f_2 \approx 90 \text{ kHz}$$

and

$$\phi_m \approx 90° \text{ (stable circuit)}$$

   A circuit with the frequency response in Fig. 15-6 and with $A_{CL} = 50$ dB was shown to be unstable. If the amplifier had $A_{CL} = 70$ dB, reconsideration shows that it is stable. That is, an amplifier with a high closed-loop gain is more likely to be stable than one with the lower gain. Low-gain amplifiers are more difficult to stabilize than high-gain circuits. The voltage follower (with a closed-loop gain of 1) can be one of the most difficult circuits to stabilize.

Some internally compensated op-amps are specified as being stable to closed-loop gains as low as 5. In this case, external compensating components must be used with lower gain circuits.

Practice Problem

15-1.1 Investigate the stability of an inverting amplifier with a closed-loop gain of 60 dB if the operational amplifier is (a) one with the gain/frequency characteristics in Fig. 15-2, (b) a 741, and (c) an AD843.

15-2 FREQUENCY COMPENSATION METHODS

Phase-Lag and Phase-Lead Compensation

Lag compensation and *lead compensation* are two methods often employed to stabilize op-amp circuits. The phase-lag network in Fig. 15-7a introduces additional phase lag at some low frequency where the op-amp phase shift is still so small that additional phase lag has no effect. It can be shown that at frequencies

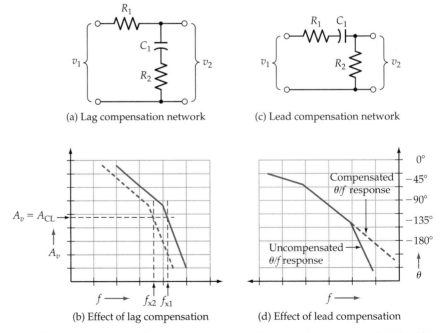

(a) Lag compensation network   (c) Lead compensation network

(b) Effect of lag compensation   (d) Effect of lead compensation

Figure 15-7   A phase-lag network reduces an amplifier open-loop gain, so that the phase shift at $A_v B = 1$ is too small for instability. A phase-lead network cancels phase lag.

where $X_{C1} \gg R_2$, the voltage v_2 lags behind v_1 by as much as 90°. At higher frequencies where $X_{C1} \ll R_2$, no significant phase lag occurs, and the lag network merely introduces some attenuation. The effect of this attenuation is that the A_v/f graph is moved to the left, as illustrated in Fig. 15-7b. Thus, the frequency (f_{x1}) at which $A_vB = 1$ [for a given closed-loop gain (A_{CL})] is moved to a lower frequency (f_{x2}), as shown. Because f_{x2} is less than f_{x1}, the phase shift at f_{x2} is less than that at f_{x1} and the circuit is likely to be stable.

The network in Fig 15-7c introduces a phase lead. In this network, when $X_{C1} \gg R_1$, the voltage v_2 leads v_1. This phase lead cancels some of the unwanted phase lag in the operational amplifier θ/f graph (see Fig. 15-7d), thus rendering the circuit more stable. Phase-lag and phase-lead networks are both used internally to compensate op-amp circuits. Both types of circuit can also be used externally.

Manufacturer's Recommended Compensation

Most currently available operational amplifiers contain internal compensating components and do not require additional external components. Some have internal compensating resistors and need only a capacitor connected externally to complete a compensating network. For those that require compensation, IC manufacturers list recommended component values and connection methods on the op-amp data sheet. An example of this is illustrated in Fig. 15-8 for the LM108.

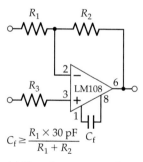

$$C_f \geq \frac{R_1 \times 30\ \text{pF}}{R_1 + R_2}$$

(a) Phase lag compensation

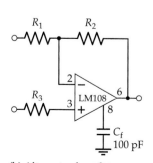

(b) Alternate phase lag compensation

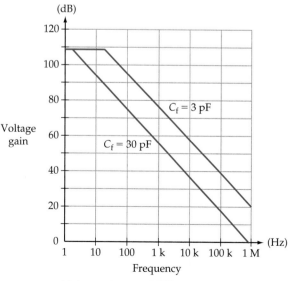

(c) Approximate gain/frequency response

Figure 15-8 Manufacturer's recommended compensation methods and gain/frequency response for the LM108 op-amp. (Reproduced with permission of National Semiconductor Corp.)

When standard value compensating capacitors are selected, the next larger values should be used. This is termed *over-compensation* and it results in better amplifier stability, but it also produces a smaller circuit bandwidth.

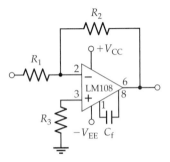

Figure 15-9 Op-amp circuit for Example 15-2.

Example 15-2

The inverting amplifier in Fig. 15-9 is required to amplify a 200 mV input by a factor of 4.5. Determine suitable component values.

Solution

Because the LM108 has a very low input bias current (see data sheet A-14 in Appendix A), it should be treated as a BIFET op-amp.

Select $\qquad\qquad\qquad\qquad R_2 = 1 \text{ M}\Omega$

$$R_1 = \frac{R_2}{A_{CL}} = \frac{1 \text{ M}\Omega}{4.5}$$

$$= 222 \text{ k}\Omega \text{ (use 220 k}\Omega \text{ standard value)}$$

$$R_3 = R_1 \| R_2 = 220 \text{ k}\Omega \| 1 \text{ M}\Omega$$

$$= 180 \text{ k}\Omega \text{ (standard value)}$$

From Fig. 15-8, $\qquad C_f = \dfrac{R_1 \times 30 \text{ pF}}{R_1 + R_2} = \dfrac{220 \text{ k}\Omega \times 30 \text{ pF}}{220 \text{ k}\Omega + 1 \text{ M}\Omega}$

$$= 5.4 \text{ pF (use 10 pF standard value}$$
$$\text{for over-compensation)}$$

Connect C_f between terminals 1 and 8, as shown in Fig. 15-9.

Miller-Effect Compensation

Miller effect (discussed in Section 8-3) involves capacitance between the output and input terminals of an inverting amplifier. Miller-effect compensation of an op-amp circuit is very simple, and it is often the only external method available for stabilizing a circuit where the op-amp is internally compensated. A capacitor (C_f) is connected across the feedback resistor (R_f), as shown in Fig. 15-10a and b. The capacitor value is calculated to have an impedance equal to the feedback resistor value at the desired signal cutoff frequency (f_2):

$$X_{Cf} = R_f \text{ at } f_2 \qquad\qquad (15\text{-}1)$$

This reduces the closed-loop gain by 3 dB at the selected frequency. So long as the op-amp is stable at this frequency, the circuit will not oscillate. The op-amp used should have an upper cutoff frequency much higher than f_2.

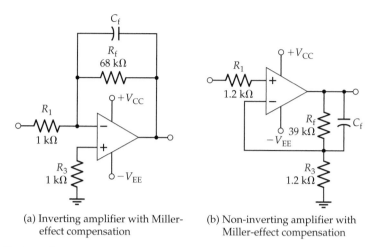

(a) Inverting amplifier with Miller-effect compensation

(b) Non-inverting amplifier with Miller-effect compensation

Figure 15-10   Miller-effect frequency compensation for amplifier circuits.

Example 15-3

Calculate a suitable Miller-effect capacitor to stabilize the circuit in Fig. 15-10a at $f_2 = 35$ kHz.

Solution

From Eq. 15-1, $$C_f = \frac{1}{2\pi f_2 R_f} = \frac{1}{2\pi \times 35 \text{ kHz} \times 68 \text{ k}\Omega}$$

$$\approx 67 \text{ pF (use 68 pF standard value)}$$

Practice Problems

15-2.1 The components of the lag and lead compensation network in Fig. 15-7 are $R_1 = 6.8$ kΩ, $R_2 = 390$ Ω, and $C_1 = 500$ pF. Calculate the approximate phase lag and phase lead at a frequency of 50 kHz.

15-2.2 Calculate a suitable Miller-effect capacitor to stabilize the circuit in Fig. 15-10b at $f_2 = 50$ kHz.

15-3 OP-AMP CIRCUIT BANDWIDTH AND SLEW RATE

Low Cutoff Frequency

Operational amplifiers are direct-coupled internally, and so where they are employed in direct-coupled applications, the circuit lower cutoff frequency (f_1) is zero. In capacitor-coupled circuits, the lower cutoff frequency is determined by the selection of coupling capacitors. The circuit high cutoff frequency (f_2) is, of course, dependent on the frequency response of the operational amplifier.

High Cutoff Frequency

In Section 13-7 it is shown that for a negative feedback amplifier the high cut-off frequency occurs when the amplifier open-loop gain approximately equals the circuit closed-loop gain:

Eq. 13-26: $A_v \approx A_{CL}$

So, the circuit high cutoff frequency (f_2) can be found simply by drawing a horizontal line at $A_v \approx A_{CL}$ on the op-amp open-loop gain/frequency response graph. Because the op-amp low cutoff frequency (f_1) is zero (as explained above), the circuit bandwidth is

$$BW = f_2 - f_1$$
$$= f_2$$

Consequently, the op-amp high cutoff frequency is often referred to as the circuit bandwidth.

The frequency response graphs published on manufacturer's data sheets are typical for each particular type of operational amplifier. Like all typical device characteristics, the precise frequency response differs from one op-amp to another. All frequencies derived from the response graphs should be taken as typical quantities. The process of determining circuit cutoff frequency from the op-amp frequency response graph is demonstrated in Example 15-4.

Example 15-4

Determine the typical upper cutoff frequency for the inverting amplifier in Fig. 15-11 when the compensating capacitor (C_f) value is (a) 30 pF and (b) 3 pF. The A_v/f graph for the LM108 is shown in Fig. 15-12.

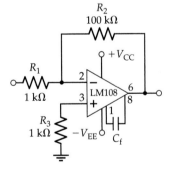

Figure 15-11 Amplifier circuit for Example 15-4.

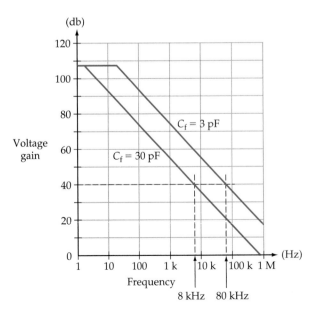

Figure 15-12 Straight-line approximation of LM108 gain/frequency response.

Solution

$$A_{CL} = \frac{R_2}{R_1} = \frac{100\ \text{k}\Omega}{1\ \text{k}\Omega}$$

$$= 100 = 40\ \text{dB}$$

f_2 occurs at $A_v = A_{CL} = 40\ \text{dB}$

(a) For $C_f = 30$ pF:

Draw a horizontal line on the A_v/f graph at $A_v = 40$ dB.

Where the line cuts the A_v/f characteristic for $C_f = 30$ pF, read

$$f_2 \approx 8\ \text{kHz}$$

(b) For $C_f = 3$ pF:

Where the $A_v = 40$ dB line cuts the A_v/f characteristic for $C_f = 3$ pF, read

$$f_2 \approx 80\ \text{kHz}$$

Gain-Bandwidth Product

The *gain-bandwidth product (GBW)*, or *unity-gain bandwidth,* of an operational amplifier is the open-loop gain at a given frequency multiplied by the frequency. Referring to the A_v/f response for the 741 reproduced in Fig. 15-13, it is seen that at $A_{v(a)} = 10^4$, the frequency is $f_{(a)} \approx 80$ Hz. Thus,

$$GBW = A_{v(a)} \times f_{(a)} = 10^4 \times 80\ \text{Hz}$$

$$= 8 \times 10^5$$

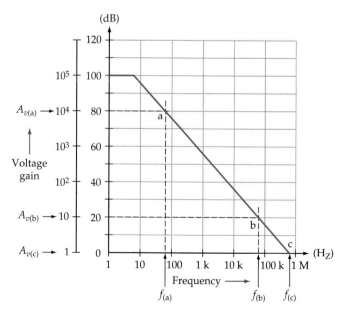

Figure 15-13   The gain-bandwidth product (*GBW*) for an operational amplifier can be used to determine the cutoff frequency for any closed-loop gain.

Similarly, at $A_{v(b)} = 10$, $f_{(b)} \approx 80$ kHz, again giving $GBW = 8 \times 10^5$. Also, at $A_{v(c)} = 1$, $f_{(c)} \approx 800$ kHz, once more giving $GBW = 8 \times 10^5$. This last determination explains the term unity-gain bandwidth, because the GBW is simply equal to the frequency at which $A_v = 1$.

Because the high cutoff frequency for an op-amp circuit occurs when the closed-loop gain equals the open-loop gain, the circuit upper cutoff frequency can be calculated by dividing the gain-bandwidth product by the closed-loop gain:

$$f_2 = \frac{GBW}{A_{CL}} \qquad\qquad (15\text{-}2)$$

It is important to note that Eq. 15-2 applies only to operational amplifiers that have a gain/frequency response that falls off to the unity-gain frequency at 20 dB/decade. Where the A_v/f response falls off at some other rate, Eq. 15-2 cannot be used.

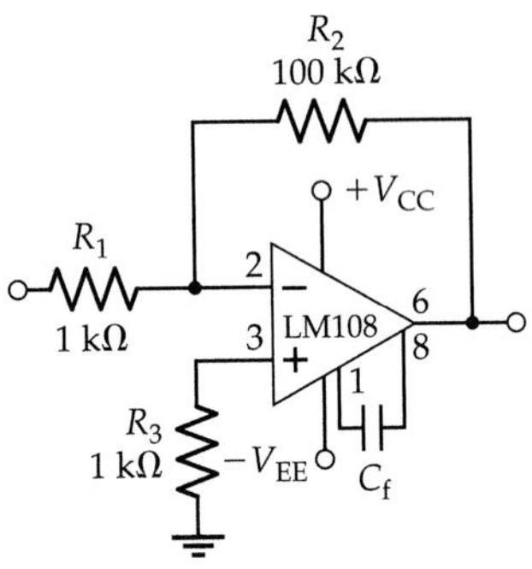

Figure 15-14 Amplifier circuit for Example 15-5.

Example 15-5

Using the gain-bandwidth product, determine the cutoff frequencies for the circuit in Example 15-4 (reproduced in Fig. 15-14) when the compensating capacitor is (a) 30 pF and (b) 3 pF.

Solution

(a) For $C_f = 30$ pF:

Referring to the LM108 A_v/f graph for $C_f = 30$ pF in Fig. 15-12, we see that

$$GBW = f \text{ at } A_v = 1$$
$$\approx 800 \text{ kHz}$$

Eq. 15-2: $f_2 = \dfrac{GBW}{A_{CL}} = \dfrac{800 \text{ kHz}}{100}$

$$= 8 \text{ kHz}$$

(b) For $C_f = 3$ pF:

Referring to the LM108 A_v/f graph for $C_f = 3$ pF in Fig. 15-12, we see that at $A_v = 20$ dB $= 10$, $f \approx 800$ kHz

$$GBW = f \times A_v = 800 \text{ kHz} \times 10$$
$$= 8 \text{ MHz}$$

Eq. 15-2: $f_2 = \dfrac{GBW}{A_{CL}} = \dfrac{8 \text{ MHz}}{100}$

$$= 80 \text{ kHz}$$

Full-Power Bandwidth and Slew Rate

The A_v/f response graphs, upper cutoff frequencies, and *GBW* specified on op-amp data sheets normally refer to the operational amplifier performance as a small-signal amplifier. In this case, the measurements are usually made *for output amplitudes not exceeding 100 mV peak-to-peak.* Where an amplifier circuit has to produce a large output voltage, the op-amp *full-power bandwidth* (f_p) must be used. The AD843 operational amplifier, for example, is specified as having a typical unity gain bandwidth of 34 MHz for an output amplitude of 90 mV peak-to-peak, and a typical full power bandwidth of 3.9 MHz when the output amplitude is 20 V p-to-p.

The op-amp slew rate (SR) (see Section 14-9) can be used to calculate the full-power bandwidth for a given output amplitude. For a sinusoidal voltage waveform, the fastest rate-of-change of voltage occurs at the point where the waveform crosses from its negative half-cycle to its positive half-cycle, and vice versa. This is illustrated in Fig. 15-15a. It can be shown that the voltage rate-of-change at this point is

$$\Delta V/\Delta t = 2\pi f V_p \text{ (volts/second)}$$

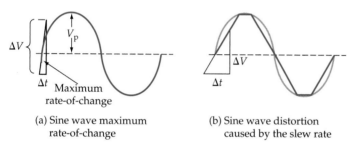

(a) Sine wave maximum
rate-of-change

(b) Sine wave distortion
caused by the slew rate

Figure 15-15 The op-amp slew-rate limits the upper cutoff frequency of an op-amp circuit and the output amplitude at a given frequency.

The maximum rate-of-change of the waveform is limited by the maximum slew rate of the op-amp used. Where the waveform amplitude or frequency is higher than the limits imposed by the slew rate, distortion will occur as illustrated in Fig. 15-15b.

The slew rate can be equated to the sine wave rate-of-change:

$$SR = 2\pi f_p V_p \qquad\qquad (15\text{-}3)$$

where f_p is the slew-rate limited frequency, or full-power bandwidth, and V_p is the peak level of the circuit output voltage. Equation 15-3 can be used to determine the full-power bandwidth of an op-amp circuit for a given output voltage amplitude. Sometimes Equation 15-3 gives an f_p value greater than that determined from the A_v/f graph or the *GBW* product. In these cases, the circuit bandwidth is still dictated by the A_v/f graph or the *GBW* product.

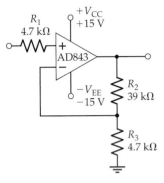

Figure 15-16 Amplifier circuit for Example 15-6.

Example 15-6

(a) Calculate the full-power bandwidth for an AD843 op-amp circuit (Fig. 15-16), given a 1 V peak input and op-amp slew rate of 250 V/μs.

(b) Determine the maximum peak output voltage obtainable from a 741 op-amp circuit with a 100 kHz signal frequency. (SR = 0.5 V/μs for a 741.)

Solution

(a) For the AD843:

$$V_{o(p)} = \frac{R_2 + R_3}{R_3} \times V_{i(p)} = \frac{39 \text{ k}\Omega + 4.7 \text{ k}\Omega}{4.7 \text{ k}\Omega} \times 1 \text{ V}$$

$$= 9.3 \text{ V}$$

From Eq. 15-3, $$f_P = \frac{SR}{2\pi V_P} = \frac{250 \text{ V}/\mu s}{2\pi \times 9.3 \text{ V}}$$

$$\approx 4.2 \text{ MHz}$$

(b) For a 741:

From Eq. 15-3, $$V_P = \frac{SR}{2\pi f_P} = \frac{0.5 \text{ V}/\mu s}{2\pi \times 100 \text{ kHz}}$$

$$= 0.79 \text{ V}$$

Practice Problems

15-3.1 Determine the typical upper cutoff frequency for an inverting amplifier with a closed-loop gain of 15 using a 741 op-amp. The A_v/f graph for the 741 is shown in Fig. 15-13.

15-3.2 Using the gain-bandwidth product, calculate the cutoff frequencies for an inverting amplifier with a closed-loop gain of 30 when the op-amp used is (a) 741 and (b) an AD843.

15-3.3 Calculate the full-power bandwidth for an LF353 op-amp circuit with a 14 V peak-to-peak output voltage.

15-4 STRAY CAPACITANCE EFFECTS

Stray capacitance (C_s) at the input terminals of an operational amplifier effectively introduces an additional phase-lag network in the feedback loop (see Fig. 15-17), thus making the op-amp circuit unstable. Stray capacitance problems can be avoided by good circuit construction techniques that keep the stray to a minimum. The effects of stray capacitance also depend upon the resistor values used in the feedback network. High resistance values make it easier for small stray capacitances to produce phase lag. With low resistances, small stray capacitances normally have little effect on the circuit stability.

Analysis of an RC phase-lag circuit shows that the capacitor voltage lags behind the input voltage by 45° when the capacitor impedance (X_C) equals the

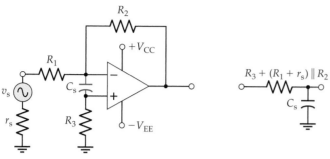

(a) Stray capacitance (C_s) at amplifier input   (b) C_s and its series resistance

Figure 15-17 Stray capacitance can cause instability in an op-amp circuit by introducing additional phase lag in the feedback network.

series resistance (R). Also, when $X_C = 10R$, the phase lag is approximately $10°$; it is this $10°$ of additional phase lag that might make the circuit oscillate if its phase margin is already close to the minimum for stability. If the phase margin is known to be large at the frequency where $A_vB = A_{CL}$ (the frequency at which the circuit is likely to oscillate), the stray capacitance may be unimportant. Where the phase margin is small, for circuit stability the op-amp input stray capacitance should normally be much less than

$$C_s = \frac{1}{2\pi f(10R)} \qquad (15\text{-}4)$$

where R is the equivalent resistance in series with the stray capacitance. In Fig. 15-17, $R = R_3 + (R_1 + r_s)\|R_2$.

From Equation 15-4 it is seen that (as already mentioned) the larger the resistor values, the smaller the stray capacitance that can produce circuit instability. If the signal source is disconnected from the circuit, R becomes equal to $(R_2 + R_3)$, which is much larger than $[R_3 + (r_s + R_1)\|R_2]$. In this situation, an extremely small stray capacitance can make the circuit unstable.

Miller-effect compensation can be used to compensate for stray capacitance at an op-amp input, as shown in Fig. 15-18. To eliminate the phase shift introduced by the stray capacitance, the division of the output voltage produced by C_S and C_2 in series should be equal to the division produced by R_1 and R_2. Therefore,

$$\frac{X_{CS}}{X_{C2}} = \frac{R_1}{R_2}$$

This gives

$$C_2R_2 = C_SR_1 \qquad (15\text{-}5)$$

Note that Equation 15-5 does not allow for r_s or R_3 in Fig. 15-17. Where r_s is not very much smaller than R_1, it must be added to R_1. Also, resistor R_3 could be bypassed with another capacitor to reduce the total series resistance.

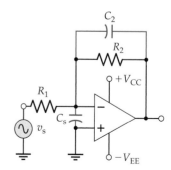

Figure 15-18 Use of Miller-effect compensation for stray capacitance at the input terminals of an op-amp.

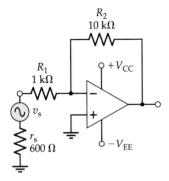

Figure 15-19 Op-amp amplifier circuit for Example 15-7.

Example 15-7

Calculate the op-amp input terminal stray capacitance that might cause instability in the circuit of Fig. 15-19 if the amplifier cutoff frequency is 800 kHz. Determine a suitable Miller-effect compensating capacitor value.

Solution

Stray capacitance:

Eq. 15-4: $C_s = \dfrac{1}{2\pi f \times 10[(r_s + R_1)\,\|\,R_2]}$

$= \dfrac{1}{2\pi \times 800 \text{ kHz} \times 10[(600\ \Omega + 1\text{ k}\Omega)\,\|\,10\text{ k}\Omega]}$

$= 14.4 \text{ pF}$

Compensation:

Eq. 15-5: $C_2 = \dfrac{C_s(r_s + R_1)}{R_2} = \dfrac{14.4 \text{ pF} \times (600\ \Omega + 1\text{ k}\Omega)}{10\text{ k}\Omega}$

$= 2.3 \text{ pF}$

Practice Problems

15-4.1 Determine the op-amp input stray capacitance that might cause instability in an inverting amplifier with $R_1 = 1.8$ kΩ, $R_2 = 560$ kΩ, and $f_2 = 600$ kHz (a) when the signal source is open-circuited, and (b) when $r_s = 600\ \Omega$ and R_1 and R_2 are reduced by a factor of 10.

15-4.2 Determine a suitable Miller-effect compensating capacitor value for the circuit in part (b) of Problem 15-4.1.

15-5 LOAD CAPACITANCE EFFECTS

Capacitance connected at the output of an operational amplifier is termed load capacitance (C_L). Figure 15-20 shows that C_L is in series with the op-amp output resistance (r_o), so that C_L and r_o constitute a phase-lag circuit in the feedback network. As in the case of stray capacitance, another 10° of phase lag introduced by C_L and r_o could cause circuit instability where the phase margin is already small. The equation for calculating the load capacitance that might cause instability is similar to that for stray capacitance:

$$C_L = \dfrac{1}{2\pi\, f(10r_o)} \qquad (15\text{-}6)$$

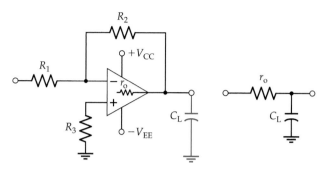

Figure 15-20   Load capacitance at an op-amp output can cause instability by introducing additional phase lag in the feedback network.

In Eq. 15-6, f is the frequency at which $A_v B = A_{CL}$. If r_o is reduced, Eq. 15-6 gives a larger C_L value. Thus, *an op-amp with a low output resistance can tolerate more load capacitance than one with a higher output resistance.*

One method often used to counter instability caused by load capacitance is shown in Fig. 15-21a. A resistor (R_x), usually ranging from 12 Ω to 400 Ω, is connected in series with the load capacitance. The presence of R_x (with R_2

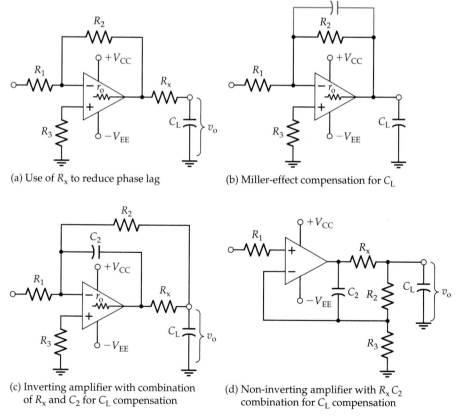

(a) Use of R_x to reduce phase lag

(b) Miller-effect compensation for C_L

(c) Inverting amplifier with combination of R_x and C_2 for C_L compensation

(d) Non-inverting amplifier with $R_x C_2$ combination for C_L compensation

Figure 15-21   Compensation methods for load capacitance.

connected at the op-amp output) can severely reduce the phase lag produced by r_o and C_L. However, R_x also has the undesirable effect of increasing the circuit output impedance to approximately the resistance of R_x.

A Miller-effect capacitor (C_2) connected across feedback resistor R_2 may be used to compensate for the load capacitance (see Fig. 15-21b). In this case, C_2 introduces some phase lead in the feedback network to counter the phase lag. The equation for calculating a suitable capacitance for C_2 is, once again, similar to that for stray capacitance:

$$C_2 R_2 = C_L r_o \qquad (15\text{-}7)$$

A modified form of Miller-effect compensation for load capacitance is shown in Fig. 15-21c. An additional resistor (R_x) is included in series with C_L to reduce the phase lag. But now, R_2 is connected at the junction of R_x and C_L, so that (because of feedback) R_x has no significant effect on the circuit output impedance. Also, C_2 is connected from the op-amp output terminal to the inverting input. With this arrangement, Equation 15-7 is modified to

$$C_2 R_2 = C_L (r_o + R_x) \qquad (15\text{-}8)$$

It should be noted from Equations 15-7 and 15-8 that, as for stray capacitance, smaller resistance values for R_2 give larger, more convenient compensating capacitor values.

Example 15-8

Calculate the load capacitance that might cause instability in the circuit of Fig. 15-22a if the amplifier cutoff frequency is 2 MHz and its output resistance is 25 Ω. Determine a suitable compensating capacitor value for the circuit as modified in Fig. 15-22b with a 0.1 µF load capacitance.

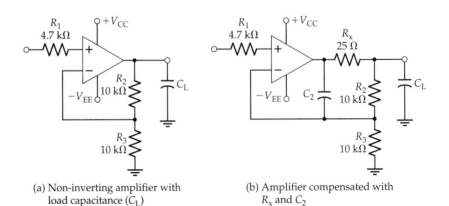

(a) Non-inverting amplifier with load capacitance (C_L)

(b) Amplifier compensated with R_x and C_2

Figure 15-22 Circuits for Example 15-8.

Solution

Load capacitance:

Eq. 15-6: $$C_L = \frac{1}{2\pi f(10\, r_o)} = \frac{1}{2\pi \times 2\,\text{MHz} \times 10 \times 25\,\Omega}$$

$$= 318\,\text{pF}$$

Compensation:

Eq. 15-8: $$C_2 = \frac{C_L(r_o + R_x)}{R_2} = \frac{0.1\,\mu\text{F} \times (25\,\Omega + 25\,\Omega)}{10\,\text{k}\Omega}$$

$$= 500\,\text{pF (standard value)}$$

Practice Problems

15-5.1 Calculate the load capacitance that might cause instability in the circuit in Ex. 15-7 if the op-amp output resistance is 20 Ω. Determine a suitable Miller-effect compensating capacitor value.

15-5.2 The circuit in Problem 15-5.1 is modified as in Fig. 15-21c with $R_x = 5\,R_o$ and $C_L = 0.5\,\mu\text{F}$. Calculate the required C_2 value.

15-6 CIRCUIT STABILITY PRECAUTIONS

Power Supply Decoupling

Feedback along supply lines is another source of op-amp circuit instability. This can be minimized by connecting 0.01 μF high-frequency capacitors from each supply terminal to ground (see Fig. 15-23). The capacitors must be connected as closely as possible to the IC terminals. Sometimes larger-value capacitors are required.

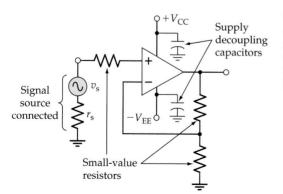

Signal source connected

Small-value resistors

Figure 15-23 For op-amp circuit stability, keep resistor values to a minimum, use the recommended compensating components, bypass the supply terminals to ground, and keep the signal source connected.

Stability Precautions

The following precautions should be observed for circuit stability:

1. Where low-frequency performance is required, use an internally compensated op-amp. Alternatively, use Miller-effect compensation to give the lowest acceptable cutoff frequency.
2. Use small-value resistors in the feedback network if possible, instead of using the largest possible resistor values.
3. With an op-amp that must be compensated, use the methods and components recommended by the IC manufacturer.
4. Keep all component leads as short as possible, and take care with component placement. A resistor connected to an op-amp input terminal should have the resistor body placed close to the input terminal.
5. Use 0.01 μF capacitors (or 0.1 μF capacitors if necessary) to bypass the supply terminals of the op-amp (or groups of op-amps) to ground. Connect these capacitors close to the ICs.
6. Always have a signal source connected to a circuit being tested. Alternatively, ground the circuit input. With an open-circuited input, very small stray capacitances can cause instability.
7. Do not connect oscilloscopes or other instrument at the op-amp input terminals. Instrument input capacitance can cause instability.
8. If a circuit is unstable after all of the above precautions have been observed, reduce the value of all circuit resistors (except compensating resistors). Also, reduce the signal source resistance if possible.

Review Questions

Section 15-1

15-1 Show how feedback in an op-amp inverting amplifier can produce instability. Explain the conditions necessary for oscillations to occur in an op-amp circuit.

15-2 Show how feedback in an op-amp non-inverting amplifier circuit can produce instability.

15-3 Define loop gain, loop phase shift, and phase margin.

15-4 Sketch typical gain/frequency response and phase/frequency response graphs for an uncompensated operational amplifier. Identify the pole frequencies and rates of fall of voltage gain, and show the typical phase shift at each pole frequency.

15-5 Sketch typical gain/frequency response and phase/frequency response graphs for a compensated operational amplifier. Briefly explain.

15-6 Derive an equation for the open-loop gain of an operational amplifier when the loop gain equals 1.

Section 15-2

15-7 Sketch a lag compensation circuit. Explain its operation, and show how it affects the frequency and phase response graphs of an operational amplifier.

15-8 Sketch a lead compensation circuit. Explain its operation, and show how it affects the frequency and phase response graphs of an operational amplifier.

15-9 Show how Miller-effect compensation can be applied to an op-amp circuit. Briefly explain.

Section 15-3

15-10 Define bandwidth, gain-bandwidth product, and full-power bandwidth for an operational amplifier. Explain the circuit conditions that apply in each case.

15-11 Define slew rate, and explain its effect on the output waveform from an operational amplifier.

Section 15-4

15-12 Discuss the effects of stray capacitance at the input terminals of an operational amplifier. Explain the precautions that should be observed to deal with input stray capacitance problems.

15-13 Show how Miller-effect compensation can be used to counter the effects of op-amp input stray capacitance.

Section 15-5

15-14 Discuss the effects of op-amp load capacitance.

15-15 Show how op-amp instability due to load capacitance can be countered by means of an additional resistor, Miller-effect compensation, and a combination of both.

Section 15-6

15-16 List precautions that should be observed for operational amplifier circuit stability. Briefly explain in each case.

Problems

Section 15-1

15-1 Investigate the stability of the circuit in Fig. 15-11 if the LM108 is replaced with an op-amp that has the gain/frequency and phase/frequency responses in Fig. 15-2.

15-2 Investigate the stability of the inverting amplifier circuit in Fig. 15-10a if the op-amp used is (a) a 741, and (b) an AD843. Use the response graphs in Fig. 15-3 and 15-4.

15-3 Investigate the stability of an amplifier circuit with $A_{CL} = 70$ dB if the op-amp has the gain/frequency and phase/frequency responses in Fig. 15-2.

Section 15-2

15-4 The phase-lag network in Fig. 15-7a has $R_1 = 8.2$ kΩ, $R_2 = 470$ Ω, and $C_1 = 3300$ pF. Calculate the approximate phase lag at 7.5 kHz and at 750 kHz.

15-5 The phase-lead network in Fig. 15-7c has $R_1 = 560$ Ω, $R_2 = 27$ kΩ, and $C_1 = 1000$ pF. Calculate the approximate phase lead at 6 kHz and 300 kHz.

15-6 If the circuit in Fig. 15-22a uses an LM108 op-amp, determine suitable compensation capacitor values. Refer to the LM108 information in Fig. 15-8.

15-7 Calculate a suitable Miller-effect compensating capacitor to stabilize the circuit in Fig. 15-19 at an 80 kHz cutoff frequency.

15-8 The circuit in Fig. 15-10b is to be stabilized at $f_2 = 50$ kHz. Determine the value of a suitable Miller-effect compensating capacitor.

Section 15-3

15-9 Determine the bandwidth of the circuit in Fig. 15-10a if the op-amp has the gain/frequency response graph in Fig. 15-2. Determine the bandwidth of the circuit in Fig 15-10b if the op-amp is a 741.

15-10 Find the upper cutoff frequency for the circuit in Fig. 15-19 if the op-amp is (a) a 741, and (b) an AD843. Use the response graphs in Fig. 15-3 and 15-4.

15-11 Using the gain-bandwidth product, determine the upper cutoff frequencies for the circuits in Problem 15-10.

15-12 Use the gain-bandwidth product to determine the upper cutoff frequencies for the circuits in Figs. 15-10a and b if they both have LM108 op-amps with $C_f = 30$ pF.

15-13 The circuits in Examples 14-5 and 14-8 use 741 op-amps. Use the gain-bandwidth product to determine the upper cutoff frequency for each circuit.

15-14 If the circuit in Fig. 15-11 has the LM108 replaced with an LF353, use the gain-bandwidth product to determine the upper cutoff frequency.

15-15 Calculate the full-power bandwidth for an amplifier using a 741 op-amp if the output voltage is to be (a) 5 V peak-to-peak, and (b) 1 V peak-to-peak.

15-16 Recalculate the full-power bandwidth in each case in Problem 15-15 if the 741 is replaced with an LF353.

15-17 Calculate the full-power bandwidth for the circuit in Example 14-5 if the peak output is to be 2 V. Determine the maximum peak output voltage that can be produced by the circuit at the cutoff frequency calculated in Problem 15-13.

15-18 Calculate the slew-rate limited cutoff frequency for the circuit in Example 14-8 if the peak input is 20 mV. Determine the maximum peak output voltage at the circuit cutoff frequency calculated in Problem 15-13.

Section 15-4

15-19 A circuit as in Fig. 15-10a with C_f removed has a cutoff frequency of 600 kHz. Determine the op-amp input stray capacitance that might cause instability (a) when the signal source is open-circuited, and (b) when a 300 Ω signal source is connected.

15-20 Determine the input stray capacitance that might make the circuit in Fig. 15-10b become unstable when a 300 Ω signal source is connected. Assume that the circuit cutoff frequency is 30 kHz and that C_f is removed.

15-21 Calculate the Miller-effect capacitor value required to compensate for 250 pF of input stray capacitance in the circuitry of Problem 15-19b.

15-22 Calculate the Miller-effect capacitor value required to compensate for 90 pF of stray capacitance in the circuit of Problem 15-20.

15-23 An inverting amplifier (as in Fig. 15-19) uses an LF353 op-amp and has $r_s = 600\ \Omega$, $R_1 = 220\ k\Omega$, $R_2 = 2.2\ M\Omega$, $f_2 = 18\ kHz$. Calculate the input stray capacitance that might make the circuit unstable; (a) when the signal source is connected, and (b) when the signal source is open-circuited.

15-24 Repeat Problem 15-23 when R_1 and R_2 are each reduced by a factor of 10.

Section 15-5

15-25 Determine the load capacitance that might cause instability in the circuit in Fig. 15-10a with C_f removed, if the circuit cutoff frequency is 600 kHz and $r_o = 100\ \Omega$.

15-26 An inverting amplifier (as in Fig. 15-19) uses an op-amp with a 300 Ω output resistance, and has $R_1 = 220\ k\Omega$ and $R_2 = 2.2\ M\Omega$. Calculate the load capacitance that might make the circuit unstable.

15-27 Calculate the Miller-effect capacitor value required to compensate for a 0.1 μF load capacitance in the circuit of Fig. 15-10a if $f_2 = 600\ kHz$.

15-28 Calculate the Miller-effect capacitor value required to compensate for the load capacitance in Problem 15-26.

15-29 Determine the load capacitance that might cause instability in the circuit in Fig. 15-22a if the cutoff frequency is 400 kHz, and the op-amp has an r_o of 150 Ω.

15-30 The circuit in Problem 15-29 is rearranged as in Fig. 15-22b with $R_x = 10\ r_o$. Calculate the required C_2 value to compensate for a 5000 pF load capacitance.

Practice Problem Answers

15-1.1   45°, 90°, 90°
15-2.1   −48.5°, 41.5°
15-2.2   82 pF
15-3.1   50 kHz
15-3.2   27 kHz, 1.13 MHz
15-3.3   296 kHz
15-4.1   0.05 pF, 34.5 pF
15-4.2   0.5 pF
15-5.1   995 pF, 2 pF
15-5.2   6000 pF

CHAPTER 16
Signal Generators

CONTENTS

Objectives

You will be able to:

1 Draw the following types of sine wave oscillator circuits and explain the operation of each: phase-shift, Colpitts, Hartley, and Wein bridge.

2 Analyze each of the above oscillator circuits to determine the oscillation frequency.

3 Design each of the above oscillator circuits to produce a specified output frequency.

4 Sketch oscillator amplitude stabilization circuits and explain their operation.

5 Design circuits to limit oscillator outputs to a specified amplitude.

6 Draw square-wave and triangular-wave generator circuits. Sketch the circuit waveforms, and explain the operation of each circuit.

7 Analyze square and triangular waveform generators to determine the amplitude and frequency of the output waveform.

8 Design square and triangular waveform generators to produce a specified output amplitude and frequency.

9 Analyze and design pulse generator circuits using 555 IC timers.

10 Explain piezoelectric crystals and sketch the crystal equivalent circuit and the crystal impedance/frequency graph.

11 Show how crystals may be used for oscillator frequency stabilization, and design crystal-controlled oscillators.

INTRODUCTION

A sinusoidal oscillator usually consists of an amplifier and a phase-shifting network. The amplifier receives the output from the network, amplifies it, phase-shifts it by 180°, and applies it to the network input. The network phase shifts the amplifier output by a further 180° and attenuates it before feeding it back to the amplifier input. When the amplifier gain equals the inverse of the network attenuation, and the amplifier phase shift equals the network phase shift, the circuit is amplifying an input to produce an output which is attenuated to become the input. The circuit is generating its own input signal, and a state of oscillation exists.

Some signal generators produce square or triangular waveforms. These normally use non-linear circuits and resistor-capacitor charging circuits.

16-1 PHASE SHIFT OSCILLATORS

Op-Amp Phase Shift Oscillator

Figure 16-1 shows the circuit of a *phase shift oscillator,* which consists of an inverting amplifier and an *RC* phase-shifting network. The amplifier phase-shifts its input by −180°, and the *RC* phase-lead network phase-shifts the amplifier output by a +180°, giving a total loop phase shift of zero. The attenuated feedback signal (at the amplifier input) is amplified to reproduce the output. In this condition the circuit is generating its own input signal; consequently, it is oscillating. The output and feedback voltage waveforms in Fig. 16-1 illustrate the circuit operation.

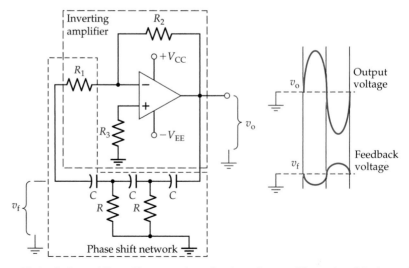

Figure 16-1   A phase shift oscillator consists of an inverting amplifier and an *RC* phase shifting feedback network. The *RC* network attenuates the output and phase-shifts it by 180°. The amplifier amplifies the network output and phase-shifts it through a further 180°.

For a state of oscillation to be sustained in any sinusoidal oscillator circuit, certain conditions, known as the *Barkhausen criteria,* must be fulfilled:

The loop gain must be equal to (or greater than) one.
The loop phase shift must be zero.

The *RC* phase-lead network in Fig. 16-1 consists of three equal-value resistors and three equal-value capacitors. Resistor R_1 functions as the last resistor in the *RC* network and as the amplifier input resistor. A phase-lag network would give a total loop phase shift of $-360°$, and so it would work just as well as the phase-lead network.

The frequency of the oscillator output depends upon the component values in the *RC* network. The circuit can be analyzed to show that the phase shift is $180°$ when

$$X_C = \sqrt{6}\,R$$

This gives an oscillation frequency:

$$f = \frac{1}{2\pi RC\sqrt{6}} \qquad\qquad (16\text{-}1)$$

As well as phase-shifting the amplifier output, the *RC* network attenuates the output. It can be shown that, when the required $180°$ phase shift is produced, the feedback factor (B) is always $1/29$. This means that the amplifier must have a closed-loop voltage gain (A_{CL}) of at least 29 to give a loop gain (BA_{CL}) of one; otherwise the circuit will not oscillate. For example, if the amplifier output voltage is 10 V, the feedback voltage is

$$v_f = Bv_o = 10\text{ V}/29$$

To reproduce the 10 V output, v_f must be amplified by 29:

$$v_o = A_{CL}v_f = 29 \times (10\text{ V}/29)$$
$$= 10\text{ V}$$

If the amplifier voltage gain is much greater than 29, the output waveform will be distorted. When the gain is slightly greater than 29, a reasonably pure sine wave output can be expected. The gain is usually designed to be just over 29 to ensure that the circuit oscillates. The output voltage amplitude normally peaks at $\pm(V_{CC} - 1\text{ V})$ unless a rail-to-rail op-amp is used (see Section 14-9).

Op-Amp Phase Shift Oscillator Design

Design of a phase shift oscillator begins with design of the amplifier to have a closed-loop gain just greater than 29. The resistor values for the *RC* network are then selected to be equal to the amplifier input resistor (R_1), and the capacitor values are calculated from Equation 16-1. In some cases, this procedure might produce capacitor values not much larger than stray capacitance. So, alternatively, the design might start with selection of convenient capacitor values. Equation 16-1 is then used to calculate the resistance of R (and R_1). Finally, R_2 is calculated to give the required amplifier gain.

Example 16-1

Using a 741 op-amp with a ±10 V supply, design the phase shift oscillator in Fig. 16-2 to produce a 1 kHz output frequency.

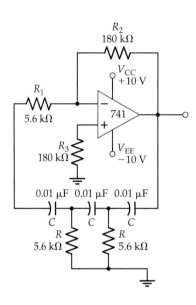

Figure 16-2 Phase shift oscillator circuit for Example 16-1.

Solution

Select

$$I_1 \approx 100 \times I_{B(max)} = 100 \times 500 \text{ nA}$$

$$= 50 \text{ }\mu\text{A}$$

$$v_o \approx \pm(V_{CC} - 1 \text{ V}) \approx \pm(10 \text{ V} - 1 \text{ V})$$

$$\approx \pm 9 \text{ V}$$

$$v_i = \frac{v_o}{A_{CL}} = \frac{\pm 9 \text{ V}}{29}$$

$$= \pm 0.31 \text{ V}$$

$$R_1 = \frac{v_i}{I_1} = \frac{0.31 \text{ V}}{50 \text{ }\mu\text{A}}$$

$$= 6.2 \text{ k}\Omega \text{ (use 5.6 k}\Omega \text{ standard value)}$$

$$R_2 = A_{CL}R_1 = 29 \times 5.6 \text{ k}\Omega$$

$$\approx 162 \text{ k}\Omega \text{ (use 180 k}\Omega \text{ to give } A_{CL} > 29)$$

$$R_3 = R_2 = 180 \text{ k}\Omega \text{ (the dc path through } R_1$$
$$\text{is interrupted by } C)$$

$$R = R_1 = 5.6 \text{ k}\Omega$$

From Eq. 16-1: $C = \dfrac{1}{2\pi R f \sqrt{6}} = \dfrac{1}{2\pi \times 5.6 \ \text{k}\Omega \times 1 \ \text{kHz} \times \sqrt{6}}$

$$\approx 0.01 \ \mu\text{F (standard value)}$$

Although the 741 op-amp used in Ex. 16-1 is likely to be quite suitable for the particular circuit, some care should always be taken when selecting an operational amplifier. It should be recalled (from Chapter 15) that when a large output voltage swing is required, the op-amp full-power bandwidth is involved. This must be considered when selecting an operational amplifier for an oscillator circuit.

BJT Phase Shift Oscillator

A phase shift oscillator using a single BJT amplifier is shown in Fig. 16-3. Once again, the amplifier and phase shift network each produce 180° of phase shift, the BJT amplifies the network output, and the network attenuates the amplifier output.

First thoughts about this circuit (in comparison to the op-amp phase shift oscillator) would suggest that a BJT amplifier with a voltage gain of 29 is required. An attempt to design such a circuit reveals that in many cases the amplifier output is overloaded by the phase shift network, or else the network output is overloaded by the amplifier input. The problem can be solved by including an emitter follower in the circuit. However, the circuit can be made to function satisfactorily without any additional components if the transistor is

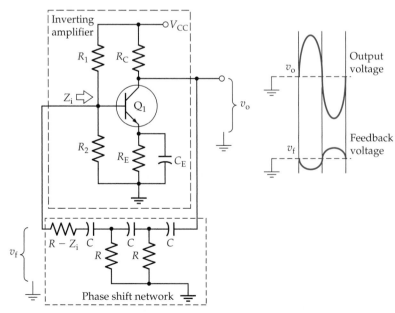

Figure 16-3   Phase shift oscillator using an inverting BJT amplifier and an *RC* feedback network.

treated as a current amplifier, rather than as a voltage amplifier. In this case, circuit analysis gives

$$f = \frac{1}{2\pi RC\sqrt{(6 + 4R_C/R)}} \quad (16\text{-}2)$$

and

$$h_{fe(min)} = 23 + \frac{29R}{R_C} + \frac{4R_C}{R} \quad (16\text{-}3)$$

The circuit oscillates only if the BJT current gain is equal to (or larger than) the minimum value determined from Equation 16-3. When $R = R_C$, a minimum h_{fe} of 56 is required to sustain circuit oscillation. With $R = 10R_C$, $h_{fe(min)}$ must be greater than 300. The output waveform is likely to be distorted if h_{fe} is substantially greater than the calculated $h_{fe(min)}$. Because h_{fe} varies widely from one transistor to another, R_C should be partially adjustable to minimize distortion. Note that in Fig. 16-3, the amplifier input resistance (Z_i) constitutes part of the last resistor in the phase shift network.

BJT Phase Shift Oscillator Design

BJT phase shift oscillator design should be approached by first choosing an R equal to or greater than the estimated amplifier Z_i. Then, an R_C is selected equal to R, C is calculated from Eq. 16-2, and the rest of the component values are determined for the circuit dc conditions. The impedance of C_E should be much lower than $h_{ie}/(1 + h_{fe})$ at the oscillating frequency.

Practice Problems

16-1.1 Using a BIFET op-amp with rail-to-rail operation, design a phase shift oscillator to produce a 6.5 kHz, ±12 V output.

16-1.2 Design a BJT phase shift oscillator, as in Fig. 16-3, to produce a 900 Hz output with a 10 V peak-to-peak amplitude. Assume that the BJT has $h_{fe(min)} \approx 60$ and $h_{ie} \approx 1.5$ kΩ.

16-2 COLPITTS OSCILLATORS

Op-Amp Colpitts Oscillator

The *Colpitts oscillator* circuit shown in Fig. 16-4 is similar to the op-amp phase shift oscillator, except that an *LC* network is used to produce the necessary phase shift in the feedback voltage. In this case, the *LC* network acts as a filter that passes the oscillating frequency and blocks all other frequencies. The filter circuit resonates at the required oscillating frequency. For resonance,

$$X_L = X_{CT}$$

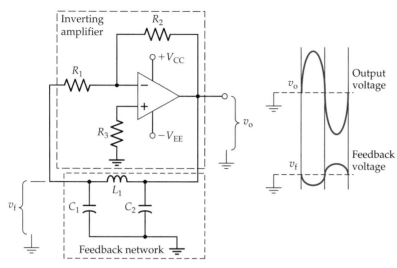

Figure 16-4 A Colpitts oscillator consists of an inverting amplifier and an *LC* phase shifting feedback network.

where X_{CT} is impedance of the total capacitance in parallel with the inductor. This gives the resonance frequency (and oscillating frequency) as

$$f = \frac{1}{2\pi\sqrt{(L_1 C_T)}} \qquad (16\text{-}4)$$

Capacitors C_1 and C_2 are connected in series across L_1; so,

$$C_T = \frac{C_1 C_2}{C_1 + C_2} \qquad (16\text{-}5)$$

Consideration of the *LC* network shows that its attenuation (from the amplifier output to input) is due to the voltage divider effect of L and C_1. This gives

$$B = \frac{X_{C1}}{X_{L1} - X_{C1}}$$

It can be shown that the required 180° phase shift occurs when

$$X_{C2} = X_{L1} - X_{C1}$$

and this gives

$$B = \frac{X_{C1}}{X_{C2}} = \frac{C_2}{C_1}$$

As in the case of all oscillator circuits, the loop gain must be a minimum of one to ensure oscillation. Therefore,

$$A_{CL(min)}B = 1$$

or

$$A_{CL(min)} = \frac{C_1}{C_2} \qquad (16\text{-}6)$$

When deriving the above equations, it was assumed that the inductor coil resistance (R_w) is very much smaller than the inductor impedance, that is, that the coil Q factor ($\omega L/R_w$) is large. This must be taken into consideration when an inductor is selected. It was also assumed that the amplifier input resistance is much greater than the impedance of C_1 at the oscillating frequency. Because of the inductor resistance and the amplifier input resistance, and because of stray capacitance effects when the oscillator operates at a high-frequency, the amplifier voltage gain usually has to be substantially larger than C_1/C_2.

Op-Amp Colpitts Oscillator Design

The designing of a Colpitts oscillator can begin with the choice of the smallest capacitor (C_2) that is much larger than stray capacitance, or with the selection of a convenient value of L. To keep the amplifier input voltage fairly low, the feedback network is often designed to attenuate the output voltage by a factor of 10. This requires that $C_1/C_2 \approx 10$. (It should be recalled that large A_{CL} values require larger op-amp bandwidths.) Also, X_{C2} should be much larger than the amplifier output impedance. Using C_T and the desired oscillating frequency, L can be calculated from Equation 16-4. Amplifier input resistor R_1 must be large enough to avoid overloading the feedback network ($R_1 \gg X_{C1}$). Resistor R_2 is determined from A_{CL} and R_1.

Example 16-2

Design the Colpitts oscillator in Fig. 16-5 to produce a 40 kHz output frequency. Use a 100 mH inductor and an op-amp with a ±10 V supply.

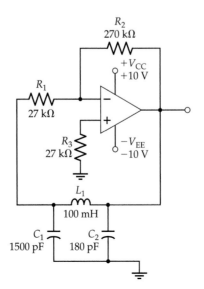

Figure 16-5   Colpitts oscillator circuit for Example 16-2.

Solution

From Eq. 16-4, $C_T = \dfrac{1}{4\pi^2 f^2 L} = \dfrac{1}{4\pi^2 \times (40 \text{ kHz})^2 \times 100 \text{ mH}}$

$$= 153.8 \text{ pF}$$

For $C_1 \approx 10C_2$, $C_1 \approx 10C_T = 10 \times 153.8 \text{ pF}$

$$\approx 1538 \text{ pF (use 1500 pF standard value)}$$

From Eq. 16-5, $C_2 = \dfrac{1}{(1/C_T) - (1/C_1)} = \dfrac{1}{(1/158.3 \text{ pF}) - (1/1500 \text{ pF})}$

$$= 177 \text{ pF (use 180 pF standard value)}$$

$$X_{C2} = \dfrac{1}{2\pi f C_2} = \dfrac{1}{2\pi \times 40 \text{ kHz} \times 180 \text{ pF}}$$

$$= 22 \text{ k}\Omega$$

$$X_{C2} \gg Z_o \text{ of the amplifier}$$

$$X_{C1} = \dfrac{1}{2\pi f C_1} = \dfrac{1}{2\pi \times 40 \text{ kHz} \times 1500 \text{ pF}}$$

$$= 2.65 \text{ k}\Omega$$

$$R_1 \gg X_{C1}$$

Select $R_1 = 10 X_{C1} = 10 \times 2.65 \text{ k}\Omega$

$$= 26.5 \text{ k}\Omega \text{ (use 27 k}\Omega \text{ standard value)}$$

From Eq. 16-6,

$$A_{CL(min)} = \dfrac{C_1}{C_2} = \dfrac{1500 \text{ pF}}{180 \text{ pF}}$$

$$= 8.33$$

$$R_2 = A_{CL} R_1 = 8.33 \times 27 \text{ k}\Omega$$

$$= 225 \text{ k}\Omega \text{ (use 270 k}\Omega \text{ standard value)}$$

$$R_3 = R_1 \| R_2 = 27 \text{ k}\Omega \| 270 \text{ k}\Omega$$

$$= 24.5 \text{ k}\Omega \text{ (use 27 k}\Omega \text{ standard value)}$$

The op-amp full-power bandwidth (f_p) must be a minimum of 40 kHz when $v_o \approx \pm 9$ V and $A_{CL} = 8.33$.

From Eq. 15-2, $f_2 = A_{CL} \times f_p = 8.33 \times 40 \text{ kHz}$

$$= 333 \text{ kHz}$$

From Eq. 15-3, $SR = 2\pi f_p v_p = 2\pi \times 40 \text{ kHz} \times 8 \text{ V}$

$$= 2 \text{ V}/\mu s$$

BJT Colpitts Oscillator

A Colpitts oscillator using a single BJT amplifier is shown in Fig. 16-6a. This is the basic circuit, and its similarity to the op-amp Colpitts oscillator is fairly obvious. A more complex version of the circuit is shown in Fig. 16-6b. Components Q_1, R_1, R_2, R_E, and C_E in (b) are unchanged from (a), but collector resistor R_C is replaced with inductor L_1. A radio frequency choke (RFC) is included in series with V_{CC} and L_1. This allows dc collector current (I_C) to pass but offers a very high impedance at the oscillating frequency, so that the top of L_1 is ac isolated from V_{CC} and ground. The output of the LC network (L_1, C_1, C_2) is coupled via C_C to the amplifier input. The circuit output voltage (v_o) is derived from a secondary winding (L_2) coupled to L_1. As in the case of the BJT phase shift oscillator, the transistor current gain is important. Circuit analysis gives Eq. 16-4 for frequency. For current gain,

$$h_{fe(min)} = \frac{C_1}{C_2} \tag{16-7}$$

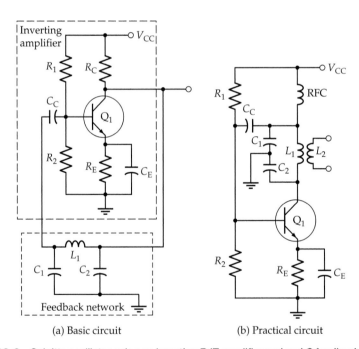

(a) Basic circuit   (b) Practical circuit

Figure 16-6 Colpitts oscillator using an inverting BJT amplifier and an LC feedback network.

Practice Problems

16-2.1 Design a Colpitts oscillator circuit to produce a 12 kHz, ±10 V output. Use a 741 op-amp.

16-2.2 Design the oscillator in Fig. 16-6a to produce a 20 kHz, 6 V p-to-p output. Use a 10 mH inductor and assume that the BJT has $h_{ib} \approx 26 \ \Omega$ and $h_{ie} \approx 1.5 \ k\Omega$.

16-3 HARTLEY OSCILLATORS

Op-Amp Hartley Oscillator

The *Hartley oscillator* circuit is similar to the Colpitts oscillator, except that the feedback network consists of two inductors and a capacitor instead of two capacitors and an inductor. Figure 16-7a shows the Hartley oscillator circuit, and Fig. 16-7b illustrates the fact that L_1 and L_2 may be wound on a single core so that there is mutual inductance (M) between the two windings. In this case, the total inductance is

$$L_T = L_1 + L_2 + 2M \qquad\qquad (16\text{-}8)$$

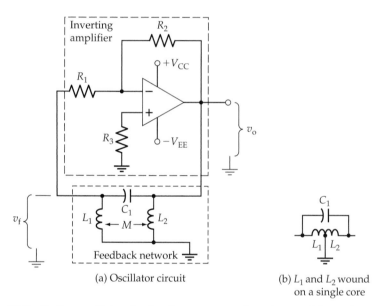

(a) Oscillator circuit

(b) L_1 and L_2 wound on a single core

Figure 16-7   Hartley oscillator circuit using an op-amp inverting amplifier and an *LC* feedback network.

Oscillation occurs at the feedback network resonance frequency:

$$f = \frac{1}{2\pi\sqrt{(C_1 L_T)}} \qquad\qquad (16\text{-}9)$$

The attenuation of the feedback network is

$$B = \frac{X_{L1}}{X_{L1} - X_{C1}}$$

It can be shown that the required 180° phase shift occurs when

$$X_{L2} = X_{L1} - X_{C1}$$

The loop gain must be a minimum of one, giving

$$A_{CL(min)} = \frac{L_2}{L_1} \tag{16-10}$$

Design procedure for a Hartley oscillator circuit is similar to that for a Colpitts oscillator.

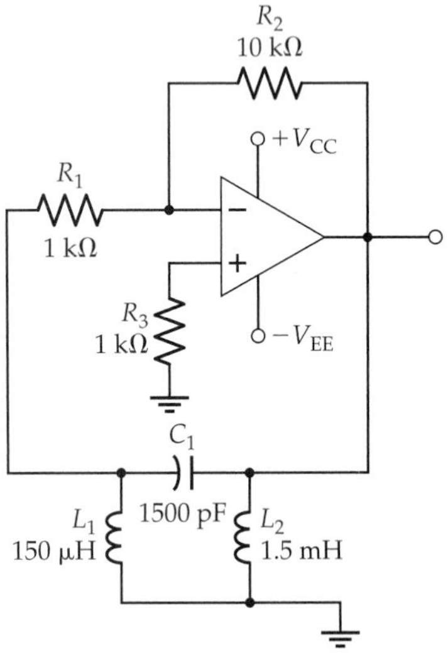

Example 16-3

Design the Hartley oscillator in Fig. 16-8 to produce a 100 kHz output frequency with an amplitude of approximately ±8 V. For simplicity, assume that there is no mutual inductance between L_1 and L_2.

Figure 16-8 Hartley oscillator circuit for Example 16-3.

Solution

$$V_{CC} \approx v_o + 1 \text{ V} = \pm(8 \text{ V} + 1 \text{ V})$$

$$\approx \pm 9 \text{ V}$$

$$X_{L2} \gg Z_o \text{ of the amplifier}$$

Select $\qquad X_{L2} \approx 1 \text{ k}\Omega$

$$L_2 = \frac{X_{L2}}{2\pi f} = \frac{1 \text{ k}\Omega}{2\pi \times 100 \text{ kHz}}$$

$$= 1.59 \text{ mH (use 1.5 mH standard value)}$$

Select $\qquad L_1 = \frac{L_2}{10} = \frac{1.5 \text{ mH}}{10}$

$$= 150 \text{ μH (standard value)}$$

$$L_T = L_1 + L_2 = 1.5 \text{ mH} + 150 \text{ μH (assuming } M = 0)$$

$$= 1.65 \text{ mH}$$

From Eq. 16-9, $\qquad C_1 = \dfrac{1}{4\pi^2 f^2 L_T} = \dfrac{1}{4\pi^2 \times (100 \text{ kHz})^2 \times 1.65 \text{ mH}}$

$$= 1535 \text{ pF (use 1500 pF with additional parallel}$$
$$\text{capacitance, if necessary)}$$

$$C_1 \gg \text{stray capacitance}$$

$$X_{L1} = 2\pi f L_1 = 2\pi \times 100 \text{ kHz} \times 150 \text{ μH}$$

$$= 94.2 \text{ }\Omega$$

$$R_1 \gg X_{L1}$$

Select $\qquad R_1 = 1 \text{ k}\Omega \text{ (standard value)}$

From Eq. 16-10,

$$A_{CL(min)} = \frac{L_2}{L_1} = \frac{1.5\,mH}{150\,\mu H}$$

$$= 10$$

$$R_2 = A_{CL}R_1 = 10 \times 1\,k\Omega$$

$$= 10\,k\Omega \text{ (standard value)}$$

$$R_3 = R_1 \| R_2 = 1\,k\Omega \| 10\,k\Omega$$

$$= 909\,\Omega \text{ (use 1 k}\Omega \text{ standard value)}$$

The op-amp full-power bandwidth (f_p) must be a minimum of 100 kHz when $v_o \approx \pm 8$ V and $A_{CL} = 10$.

From Eq. 15-2, $\quad f_2 = A_{CL} \times f = 10 \times 100\,kHz$

$$= 1\,MHz$$

From Eq. 15-3, $\quad SR = 2\pi f_p v_p = 2\pi \times 100\,kHz \times 8\,V$

$$= 5\,V/\mu s$$

BJT Hartley Oscillator

Figure 16-9 shows the circuit of a Hartley oscillator using a BJT amplifier. The basic circuit in Fig. 16-9a is similar to the op-amp Hartley oscillator, and its operation is explained in the same way as for the op-amp circuit. Note that

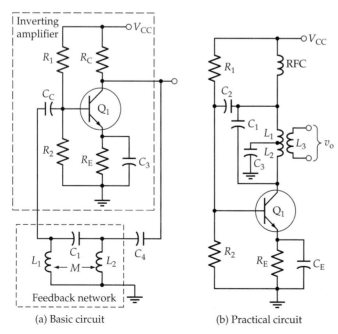

(a) Basic circuit   (b) Practical circuit

Figure 16-9 Hartley oscillator consisting of a BJT inverting amplifier and an *LC* feedback network.

coupling capacitors C_2 and C_4 are required in order to avoid dc-grounding the transistor base and collector terminals through L_1 and L_2.

In the practical BJT Hartley oscillator circuit shown in Fig. 16-9b L_1, L_2, and C_1 constitute the phase shift network. In this case, the inductors are directly connected in place of the transistor collector resistor (R_C). The circuit output is derived from the additional inductor winding (L_3). The radio frequency choke (RFC) passes the direct collector current, but ac isolates the upper terminal of L_1 from the power supply. Capacitor C_2 couples the output of the feedback network back to the amplifier input. Capacitor C_4 at the BJT collector in Fig. 16-9a is not needed in Fig. 16-9b, because L_2 is connected directly to the collector terminal. The junction of L_1 and L_2 must now be capacitor-coupled to ground (via C_3) instead of being direct-coupled.

Practice Problems

16-3.1 A Hartley oscillator circuit using a 741 op-amp is to produce a 7 kHz, ±10 V output. Determine suitable component values.

16-3.2 Analyze the BJT Hartley oscillator in Fig. 16-9b to determine the oscillating frequency. The important component values are: $L_1 = L_2 = 4.7$ mH, $C_1 = 600$ pF, and $C_2 = C_3 = 0.03$ µF. The mutual inductance between L_1 and L_2 is 100 µH.

16-4 WEIN BRIDGE OSCILLATOR

The *Wein bridge* is an ac bridge that balances only at a particular supply frequency. In the *Wein bridge oscillator* (Fig. 16-10), a Wein bridge circuit is used as a feedback network between the amplifier output and input. The bridge is

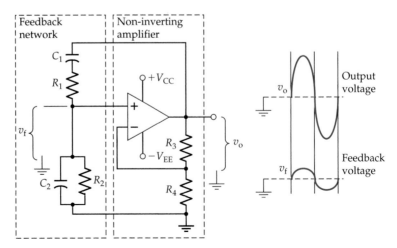

Figure 16-10 The Wein bridge oscillator circuit uses an operational amplifier and a Wein bridge that balances at a particular frequency.

made up of all of the resistors and capacitors. The operational amplifier together with resistors R_3 and R_4 constitute a non-inverting amplifier. The feedback network from the amplifier output to its non-inverting input terminal is made up of components C_1, R_1, C_2, and R_2.

At the balance frequency of the Wein bridge, the feedback voltage is in phase with the amplifier output. This (in-phase) voltage is amplified to reproduce the output. At all other frequencies, the bridge is off balance; that is, the feedback and output voltages do not have the correct phase relationship to sustain oscillations. The Barkhausen requirement for zero loop phase shift is fulfilled in this circuit by the amplifier and feedback network both having zero phase shift at the oscillation frequency.

Analysis of the bridge circuit shows that balance is obtained when two equations are fulfilled:

$$\frac{R_3}{R_4} = \frac{R_1}{R_2} + \frac{C_2}{C_1} \tag{16-11}$$

and
$$2\pi f = \frac{1}{\sqrt{(R_1 C_1 R_2 C_2)}} \tag{16-12}$$

If $R_1 C_1 = R_2 C_2$, Equation 16-12 yields

$$f = \frac{1}{2\pi R_1 C_1} \tag{16-13}$$

For simplicity, the components are often selected as $R_1 = R_2$ and $C_1 = C_2$, causing Equation 16-11 to give

$$R_3 = 2R_4 \tag{16-14}$$

In this case, the amplifier closed-loop gain is $A_{CL} = 3$.

Sometimes it is preferable to have an amplifier voltage gain substantially greater than 3; then the relationship between the component values is determined by Equations 16-11 and 16-12.

Design of a Wein bridge oscillator can be started by selecting a current level for each arm of the bridge. This should be much larger than the op-amp input bias current. Resistors R_3 and R_4 can then be calculated from the estimated output voltage and the closed-loop gain. After that, the other component values can be determined from Equations 16-11 to 16-14.

An alternative design approach is to start by choosing a convenient value for the smallest capacitor in the circuit. The other component values are then calculated from the equations.

Example 16-4

Design the Wein bridge oscillator in Fig. 16-11 to produce a 100 kHz, ±9 V output. Design the amplifier to have a closed-loop gain of 3.

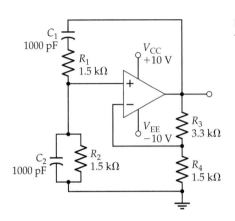

Figure 16-11 Wein bridge oscillator circuit for Example 16-4.

Solution

$$V_{CC} \approx \pm(V_o + 1\ V) = \pm(9\ V + 1\ V)$$

$$= \pm 10\ V$$

For $A_{CL} = 3$, $\quad R_1 = R_2$ and $C_1 = C_2$

Also, $\quad R_3 = 2\ R_4$

Select $\quad C_1 = 1000\ pF$ (standard value)

$$C_2 = C_1 = 1000\ pF$$

From Eq. 16-13, $\quad R_1 = \dfrac{1}{2\pi f C_1} = \dfrac{1}{2\pi \times 100\ kHz \times 1000\ pF}$

$$= 1.59\ k\Omega \text{ (use } 1.5\ k\Omega \text{ standard value)}$$

$$R_2 = R_1 = 1.5\ k\Omega$$

Select $\quad R_4 \approx R_2 = 1.5\ k\Omega$ (standard value)

$$R_3 = 2R_4 = 2 \times 1.5\ k\Omega$$

$$= 3\ k\Omega \text{ (use } 3.3\ k\Omega \text{ standard value)}$$

The op-amp must have a minimum full-power bandwidth (f_p) of 100 kHz when $v_o \approx \pm 9$ V and $A_{CL} = 3$.

From Eq. 15-2, $\quad f_2 = A_{CL} \times f = 3 \times 100\ kHz$

$$= 300\ kHz$$

From Eq. 15-3, $\quad SR = 2\pi f_p v_p = 2\pi \times 100\ kHz \times 9\ V$

$$\approx 5.7\ V/\mu s$$

Practice Problems

16-4.1 Resistors R_1 and R_2 in Fig. 16-11 are switched to (a) 15 kΩ and (b) 5.6 kΩ. Calculate the new oscillating frequency in each case.

16-4.2 A Wein bridge oscillator using an op-amp is to produce a 15 kHz, ±14 V output. Design the circuit with the amplifier having $A_{CL} = 11$.

16-5 OSCILLATOR AMPLITUDE STABILIZATION

Output Amplitude

For all of the oscillator circuits discussed, the output voltage amplitude is determined by the amplifier maximum output swing. The output waveform may also be distorted by the output saturation limitations. To minimize distortion and reduce the output voltage to an acceptable level, *amplitude stabilization* circuitry must be employed. Amplitude stabilization operates by ensuring that oscillation is not sustained if the output exceeds a predetermined level.

Diode Stabilization Circuit for a Phase Shift Oscillator

The phase shift oscillator discussed in Section 16-1 must have a minimum amplifier gain of 29 for the circuit to oscillate. Consider the oscillator circuit in Fig. 16-12a that has part of resistor R_2 bypassed by series-parallel connected diodes. When the output amplitude is low, the diodes do not become forward-biased, and so they have no effect on the circuit. At this time, the amplifier voltage gain is $A_{CL} = R_2 / R_1$. As always for a phase shift oscillator, A_{CL} is designed to exceed the critical value of 29. When the output amplitude becomes large enough to forward-bias either D_1 and D_2, or D_3 and D_4, resistor R_4 is short-circuited and the amplifier gain becomes R_5/R_1. This is designed to be too small to sustain oscillations. So, this circuit cannot oscillate with a high-amplitude output; however it can (and does) oscillate with a low-amplitude output.

In the design of the amplitude stabilization circuit, the inverting amplifier is designed in the usual manner with one important difference. The current (I_1) used in calculating the resistor values must be large enough to forward-bias the diodes into the near-linear region of their characteristics. This usually requires a minimum current around 1 mA. The resistor values are calculated as follows:

$$R_1 = \frac{v_o/29}{I_1} \qquad\qquad (16\text{-}15)$$

$$R_2 = 29R_1 \qquad\qquad (16\text{-}16)$$

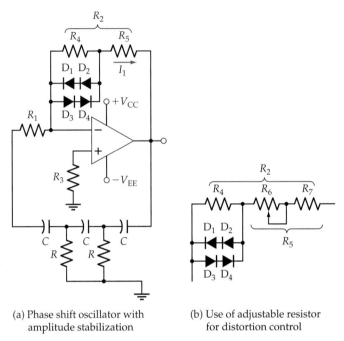

(a) Phase shift oscillator with amplitude stabilization

(b) Use of adjustable resistor for distortion control

Figure 16-12 The output amplitude of a phase shift oscillator can be limited by using diodes to modify the amplifier gain.

The diodes should become forward-biased just when the output voltage is at the desired maximum level. At this time, I_1 produces a voltage drop of $2 V_F$ across R_4.

Therefore

$$R_4 \approx \frac{2V_F}{I_1}$$

(16-17)

and

$$R_5 = R_2 - R_4$$

The resulting component values should give $(R_4 + R_5)/R_1$ slightly greater than 29, and R_5/R_1 less than 29.

Some distortion of the waveform can occur if $(R_4 + R_5)/R_1$ is much larger than 29; however, attempts to make the gain close to 29 can cause the circuit to stop oscillating. Making a portion of R_5 adjustable, as illustrated in Fig. 16-12b, provides for gain adjustment to give the best possible output waveform. Usually R_6 should be approximately 40% of the calculated value of R_5, and R_7 should be 80% of R_5. This gives a ±20% adjustment of R_5.

The diodes should be low-current switching devices. The diode reverse breakdown voltage should exceed the circuit supply voltage, and the maximum reverse recovery time ($t_{rr(max)}$) should be about one-tenth of the time period of the oscillation frequency:

$$t_{rr(max)} = \frac{T}{10}$$

(16-18)

Example 16-5

Design the phase shift oscillator in Fig. 16-13 to produce a 5 kHz, ±5 V output waveform.

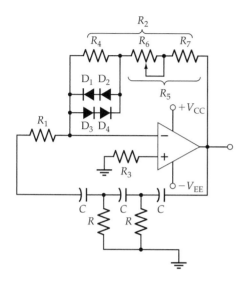

Figure 16-13   Amplitude-controlled phase shift oscillator circuit for Example 16-5.

Solution

Select
$$I_1 \approx 1 \text{ mA when } v_{o(peak)} = 5 \text{ V}$$

Eq. 16-15:
$$R_1 = \frac{v_o/29}{I_1} = \frac{5\text{ V}/29}{1\text{ mA}}$$
$$= 170 \ \Omega \ (\text{use } 150 \ \Omega)$$

Eq. 16-16:
$$R_2 = 29\, R_1 = 29 \times 150 \ \Omega$$
$$\approx 4.4 \text{ k}\Omega$$

Eq. 16-17:
$$R_4 = \frac{2\,V_F}{I_1} = \frac{2 \times 0.7\text{ V}}{1\text{ mA}}$$
$$= 1.4 \text{ k}\Omega \ (\text{use } 1.5 \text{ k}\Omega \text{ standard value})$$
$$R_5 = R_2 - R_4 = 4.4 \text{ k}\Omega - 1.5 \text{ k}\Omega$$
$$= 2.9 \text{ k}\Omega$$
$$R_6 = 0.4\, R_5 = 0.4 \times 2.9 \text{ k}\Omega$$
$$= 1.16 \ \Omega \ (\text{use } 1 \text{ k}\Omega \text{ adjustable})$$
$$R_7 = 0.8\, R_5 = 0.8 \times 2.9 \text{ k}\Omega$$
$$= 2.32 \text{ k}\Omega \ (\text{use } 2.7 \text{ k}\Omega \text{ standard value})$$
$$R_3 \approx R_2 = 4.4 \text{ k}\Omega \ (\text{use } 4.7 \text{ k}\Omega \text{ standard value})$$
$$R = R_1 = 150 \ \Omega$$

From Eq. 16-1, $$C = \frac{1}{2\pi R f \sqrt{6}} = \frac{1}{2\pi \times 150\ \Omega \times 5\ \text{kHz} \times \sqrt{6}}$$

$$= 0.087\ \mu\text{F (use } 0.082\ \mu\text{F standard value)}$$

Diode Stabilization Circuit for a Wein Bridge Oscillator

Figures 16-14 and 16-15 show two output amplitude stabilization methods that can be used with a Wein bridge oscillator. These can also be applied to other oscillator circuits because they all operate by limiting the amplifier voltage gain.

The circuit in Fig. 16-14 uses diodes and operates in the same way as the amplitude control for the phase shift oscillator. Resistor R_6 becomes shorted by

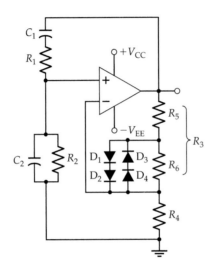

Figure 16-14   Wein bridge oscillator with its output amplitude stabilized by a diode circuit that modifies the amplifier gain.

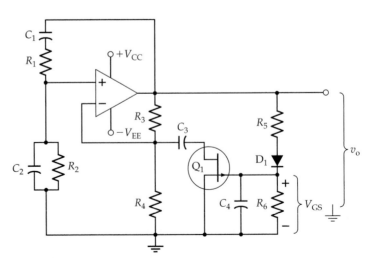

Figure 16-15   Wein bridge oscillator with the output amplitude stabilized by a FET voltage-controlled resistance circuit.

the diodes when the output amplitude exceeds the design level, thus rendering the amplifier gain too low to sustain oscillations.

FET Stabilization Circuit for a Wein Bridge Oscillator

The circuit in Fig. 16-15 is slightly more complex than the diode circuit; however, like other circuits, it stabilizes the oscillator output amplitude by controlling the amplifier gain. The channel resistance (r_{DS}) of the p-channel FET (Q_1) is in parallel with resistor R_4. Capacitor C_3 ensures that Q_1 has no effect on the amplifier dc conditions. The amplifier voltage gain is

$$A_{CL} = \frac{R_3 + R_4 \| r_{DS}}{R_4 \| r_{DS}} \qquad (16\text{-}19)$$

The FET gate-source bias voltage is derived from the amplifier ac output. The output voltage is divided across resistors R_5 and R_6 and rectified by diode D_1. Capacitor C_4 smooths the rectified waveform to give the FET dc bias voltage (V_{GS}). The polarity shown on the circuit diagram reverse-biases the gate-source of the p-channel device. When the output amplitude is low, V_{GS} is low and this keeps the FET drain-source resistance (r_{DS}) low. When the output gets larger, V_{GS} is increased, causing r_{DS} to increase. The increase in r_{DS} reduces A_{CL}, thus preventing the circuit from oscillating with a high output voltage. It is seen that the FET is behaving as a *voltage variable resistance (VVR)*.

Design of a FET Stabilization Circuit

To design the FET amplitude stabilization circuit, knowledge of a possibly suitable FET is required; in particular, the channel resistance at various gate-source voltages must be known.

Suppose the circuit is to oscillate when $r_{DS} = 6\ k\Omega$ at $V_{GS} = 1\ V$, and that the peak output is to be $V_{o(pk)} = 6\ V$. Resistors R_5 and R_6 should be selected to give $V_{GS} = 1\ V$ when $V_{o(pk)} = 6\ V$, allowing for V_F across the diode. Capacitor C_4 smooths the half-wave rectified waveform and discharges via R_6 during the time interval between peaks of the output waveform. The capacitance of C_4 is calculated to allow perhaps a 10% discharge during the time period of the oscillating frequency. The voltage-divider current (I_5) should be a minimum of about 100 μA for satisfactory diode operation.

C_3 is a coupling capacitor; its impedance at the oscillating frequency should be much smaller than the r_{DS} of the FET. Resistors R_3 and R_4 are calculated from Eq. 16-19 to give the required amplifier voltage gain when $r_{DS} = 6\ k\Omega$.

Example 16-6

Design the FET output amplitude stabilization circuit in Fig. 16-16 to limit the output amplitude of the Wein bridge oscillator in Example 16-4 to ±6 V. Assume that the $r_{ds} = 500\ \Omega$ at $V_{GS} = 1\ V$, and 800 Ω at $V_{GS} = 3\ V$.

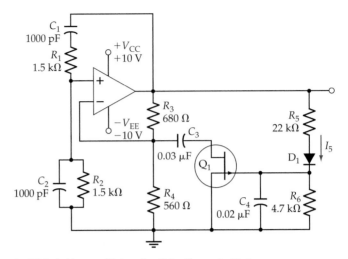

Figure 16-16   Wein bridge oscillator circuit for Example 16-6.

Solution

Select $R_4 \approx r_{DS}$ at $V_{GS} = 1$ V

$$\approx 600 \ \Omega \text{ (use } 560 \ \Omega)$$

$$R_4 \| r_{DS} = 560 \ \Omega \| 600 \ \Omega$$

$$\approx 290 \ \Omega$$

For $A_{CL} = 3$, $R_3 = 2\,(R_4 \| r_{DS}) = 2 \times 290 \ \Omega$

$$= 580 \ \Omega \text{ (use } 680 \ \Omega)$$

Select $I_5 \approx 200 \ \mu A$ when $V_{o(peak)} = 6$ V

$$R_6 = \frac{V_{GS}}{I_5} = \frac{1 \text{ V}}{200 \ \mu A}$$

$$= 5 \ k\Omega \text{ (use } 4.7 \ k\Omega)$$

$$R_5 = \frac{V_{o(peak)} - (V_{GS} + V_{D1})}{I_5} = \frac{6 \text{ V} - (1 \text{ V} + 0.7 \text{ V})}{200 \ \mu A}$$

$$= 21.5 \ k\Omega \text{ (use } 22 \ k\Omega)$$

C_4 discharge voltage:

$$\Delta V_C = 0.1 \ V_{GS} = 0.1 \times 1 \text{ V} = 0.1 \text{ V}$$

C_4 discharge time:

$$T = \frac{1}{f} = \frac{1}{100 \text{ kHz}}$$

$$= 10 \ \mu s$$

$$I_C \approx \frac{V_{R6}}{R_6} \approx I_5 \approx 200 \ \mu A$$

$$C_4 = \frac{I_C T}{\Delta V_C} = \frac{200 \ \mu A \times 10 \ \mu s}{0.1 \ V}$$

$$= 0.02 \ \mu F \ (\text{standard value})$$

$$X_{C3} = \frac{r_{DS}}{10} \ \text{at the oscillating frequency}$$

$$C_3 = \frac{1}{2\pi f r_{DS}/10} = \frac{1}{2\pi \times 100 \ kHz \times 500 \ \Omega/10}$$

$$= 0.032 \ \mu F \ (\text{use } 0.03 \ \mu F)$$

D_1 should be a low-current switching diode with a $t_{rr} \ll T$.

16-6 SQUARE WAVE GENERATOR

A *square wave generator* can be constructed by adding a resistor and capacitor to an inverting Schmitt trigger circuit (see Section 14-10). Figure 16-17 shows the circuit, which is also known as an *astable multivibrator*. The operational amplifier together with resistors R_2 and R_3 constitute the inverting Schmitt trigger circuit. Capacitor C_1 controls the voltage at the Schmitt input, and resistor R_1 charges and discharges C_1 from the Schmitt output.

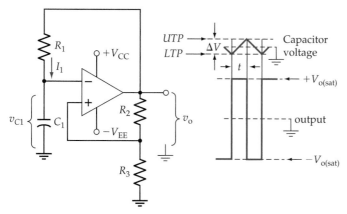

Figure 16-17 Square wave generator consisting of an inverting Schmitt trigger circuit and a series *RC* circuit.

The circuit waveforms in Fig. 16-17 illustrate the square wave generator operation. When the output is *high* (at the op-amp output saturation level), current flows through R_1, charging C_1 positively until v_{C1} equals the Schmitt *UTP*. The Schmitt output then switches to the op-amp negative saturation level. Current now starts to flow out of the capacitor via R_1, causing v_{C1} to decrease until it arrives at the *LTP* of the Schmitt circuit. At this point, the Schmitt output switches to its positive level once again, and the cycle begins again.

Design of this very simple square wave generator involves design of the Schmitt trigger circuit and determination of suitable R_1 and C_1 values for the trigger voltage levels and the required charge and discharge times. Selecting the *UTP* and *LTP* to be very much smaller than the op-amp output levels keeps the voltage drop across R_1 approximately constant. This means that the capacitor charging current is also kept fairly constant, and so the simple constant-current equation can be used for the capacitor:

$$C_1 = \frac{I_1 \times t}{\Delta V} \tag{16-20}$$

In Eq. 16-20, I_1 is the average charging current to the capacitor (through R_1), t is the charging time (see Fig. 16-17), and ΔV is the capacitor voltage change between the *UTP* and *LTP*, as illustrated. Charging current I_1 should be selected to be much larger than the op-amp input bias current; then R_1 and C_1 are calculated. Alternatively, a convenient value of C_1 can be chosen first. The level of I_1 is then determined from Eq. 16-20.

If a Schmitt trigger circuit with a large difference between the *UTP* and *LTP* is used, the capacitor changing equation is

$$e_c = E - (E - E_o)\,\varepsilon^{-t/RC} \tag{16-21}$$

Example 16-7

Design the square wave generator in Fig. 16-18 to produce a 1 kHz square wave with an amplitude of approximately ±14 V. Use a 741 op-amp.

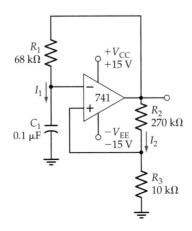

Figure 16-18   Square wave generator circuit for Example 16-7.

Solution

$$V_{CC} \approx \pm(V_o + 1 \text{ V}) \approx \pm(14 \text{ V} + 1 \text{ V})$$

$$\approx \pm 15 \text{ V}$$

$$V_{R3} = UTP = -LTP \ll V_{CC}$$

Select $V_{R3} \approx 0.5 \text{ V}$

Select $I_2 \approx 100 \times I_{B(\max)} = 100 \times 500 \text{ nA}$

$$= 50 \text{ }\mu\text{A}$$

$$R_3 = \frac{V_{R3}}{I_2} = \frac{0.5 \text{ V}}{50 \text{ }\mu\text{A}}$$

$$= 10 \text{ k}\Omega \text{ (standard value)}$$

$$R_2 = \frac{V_o - V_{R3}}{I_2} = \frac{14 \text{ V} - 0.5 \text{ V}}{50 \text{ }\mu\text{A}}$$

$$= 270 \text{ k}\Omega \text{ (standard value)}$$

$$t = \frac{T}{2} = \frac{1}{2f} = \frac{1}{2 \times 1 \text{ kHz}}$$

$$= 0.5 \text{ ms}$$

$$\Delta V = UTP - LTP = 0.5 \text{ V} - (-0.5 \text{ V})$$

$$= 1 \text{ V}$$

Select $C_1 = 0.1 \text{ }\mu\text{F (convenient value)}$

From Eq. 16-20, $I_1 = \dfrac{C_1 \Delta V}{t} = \dfrac{0.1 \text{ }\mu\text{F} \times 1 \text{ V}}{0.5 \text{ ms}}$

$$= 200 \text{ }\mu\text{A}$$

$$V_{R1(ave)} \approx 14 \text{ V}$$

$$R_1 = \frac{V_{R1}}{I_1} = \frac{14 \text{ V}}{200 \text{ }\mu\text{A}}$$

$$= 70 \text{ k}\Omega \text{ (use 68 k}\Omega \text{ standard value)}$$

Practice Problems

16-6.1 The square wave generator designed in Example 16-7 is to be modified to make the output frequency adjustable. Determine the maximum and minimum values for R_1 to produce a frequency range of 500 Hz to 5 kHz.

16-6.2 A Schmitt trigger circuit with ± 0.8 V trigger points and a ± 9 V supply is to be used in a 9 kHz square wave generator (as in Fig. 16-17). Determine suitable R_1 and C_1 values.

16-7 555 PULSE GENERATOR

Timer Block Diagram

The 555 IC timer has many applications, one of which is as a pulse waveform generator. To explain the operation of the 555 timer in any circuit, it is first necessary to understand the block diagram in Fig. 16-19. The component parts are

- voltage divider R_A, R_B, R_C
- voltage comparator 1 (CP_1)
- voltage comparator 2 (CP_2)
- set-reset flip-flop
- low impedance (buffer) output stage
- *npn* transistor Q_1
- *pnp* transistor Q_2

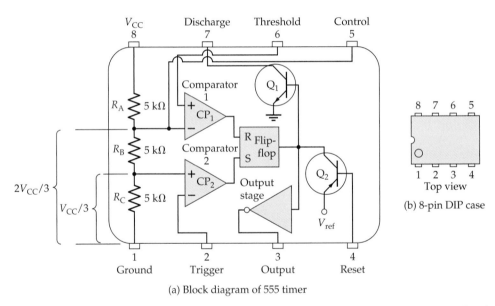

(a) Block diagram of 555 timer

Figure 16-19 Block diagram and DIP package for 555 IC timer. The components consist of a voltage divider (R_A, R_B, R_C), two voltage comparators (CP_1 and CP_2), a set-reset flip-flop, a buffer output stage, and two transistors.

Supply voltage V_{CC} is applied to one end of the voltage divider at *terminal 8*, and the other end is grounded at *terminal 1*. V_{CC} and ground are also connected to the other circuitry, but for simplicity these connections are not shown in the block diagram.

Voltage comparators (discussed in Section 14-9) are voltage level detectors that *compare* the voltages at their inverting and non-inverting inputs. When the non-inverting input voltage is lower than that at the inverting input, the output is *low*. The output rapidly switches from *low* to *high* at the instant the non-inverting input voltage goes above the inverting input level. The inverting input of CP_1 is biased by the voltage divider to a level of $2V_{CC}/3$, and the non-inverting input of CP_2 is biased to $V_{CC}/3$.

The *set-reset flip-flop* is a switching circuit that maintains a constant output voltage (*high* or *low*) depending on the last *triggering* input it received. A positive-going input applied to its R terminal (from CP_1) *resets* its output to *high*. Similarly, a positive-going input applied to its S terminal (from CP_2) *sets* the flip-flop output to *low*. The flip-flop output is applied to the low-impedance *output stage*, which performs like a voltage follower, except that it inverts its input voltage. The small circle at the output of the symbol indicates voltage inversion.

Transistor Q_1 is controlled by the flip-flop output voltage. Because Q_1 is an *npn* device with its emitter terminal grounded, a *high* input at its base drives it into saturation, thus pulling *terminal 7* close to ground level. A *low* base voltage keeps Q_1 in a cutoff state. Transistor Q_2 is a *pnp* device with its emitter biased to an internal reference voltage (V_{ref}). Q_2 is not used in most 555 applications, and it is normally biased into an off state by having its base (*terminal 4*) connected to V_{CC}.

Pulse Generator

Figure 16-20a shows a 555 timer connected to function as a pulse waveform generator, and Fig. 16-20b shows the important circuit waveforms. First note that V_{CC} is applied to *terminal 8* and to *terminal 4*, as discussed, and that *terminal 1* is grounded. C_1 is a small-value capacitor solely for coupling the inverting input of CP_1 to ground via *terminal 5* to minimize the possibility that unwanted noise signals will affect CP_1 when there is no other connection to

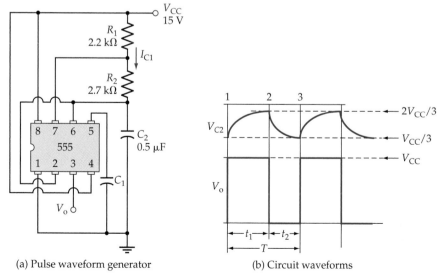

(a) Pulse waveform generator   (b) Circuit waveforms

Figure 16-20   555 pulse wave generator circuit. Capacitor C_2 is charged via resistors R_1 and R_2 to a voltage of $2V_{CC}/3$, which causes (discharge) *terminal 7* to switch to ground level. C_2 then discharges via R_2 until its voltage is $V_{CC}/3$, at which point *terminal 7* becomes an open circuit. Then the cycle begins again.

terminal 5. Capacitor C_2 is charged from V_{CC} via resistors R_1 and R_2. *Terminal 2* and *terminal 6* (CP_2 inverting input and CP_1 non-inverting input respectively) are connected directly to C_2. Finally, note that *terminal 7* is connected to the junction of R_1 and R_2.

Assume that the flip-flop output is low, which means that Q_1 is off and that V_o at *terminal 3* is high. Also, assume that C_2 is charged to $V_{CC}/3$ and is starting to charge toward V_{CC}. These are the conditions at point 1 on the circuit waveforms.

Output voltage V_o remains high (during time t_1) until V_{C2} arrives at $2V_{CC}/3$. At this point V_{C2} (applied to the CP_1 non-inverting input via *terminal 6*) causes CP_1 output to switch to *high*. The *high* output from CP_1 resets the flip-flop output to *high*, and produces a *low* V_o level (via the inverting output stage) at *terminal 3* (point 2 on the waveforms). The *high* level at the flip-flop output drives Q_1 into saturation, thus shorting *terminal 7* and the junction of R_1 and R_2 to ground level (approximately). C_2 now starts to discharge via R_2, and this continues (during time t_2) until V_{C2} falls to $V_{CC}/3$ once again. This voltage level (applied to the CP_2 inverting input via *terminal 2*) caused CP_2 output to switch to *high*, thus setting the flip-flop output to a *low* level once again. The *low* flip-flop output switches Q_1 *off* and produces a *high* V_o level (point 3 on the waveforms).

It is seen that C_2 charges from $V_{CC}/3$ to $2V_{CC}/3$ and that the charging time is t_1. Also, C_2 discharges from $2V_{CC}/3$ to $V_{CC}/3$, and the discharge time is t_2. The timer output voltage (V_o) is *high* during t_1 and *low* for the duration of t_2. Although the same capacitor voltage change occurs during charge and discharge of C_2, the charging resistance ($R_1 + R_2$) is larger than the discharge resistance (R_2). Consequently, charging time t_1 is always longer than discharge time t_2, making V_o a pulse waveform rather than a square wave.

Component Values

Capacitor C_1 is normally selected as 0.01 µF for ac grounding *terminal 5*, as discussed. The total resistance ($R_1 + R_2$) is determined by first choosing a minimum charging current ($I_{C1(min)}$). This should be much larger than the *threshold current* (I_{th}) into *terminal 6* and the *trigger current* (I_{trig}) into *terminal 2*, as specified on the 555 data sheet. The minimum level of I_{C1} occurs when $V_{R1} + V_{R2}$ is a minimum, that is, when C_2 is charged to $2V_{CC}/3$. At this point,

$$V_{R1} + V_{R2} = V_{CC}/3$$

so that
$$R_1 + R_2 = \frac{V_{CC}/3}{I_{C1(min)}} \qquad (16\text{-}22)$$

From the basic equation for a capacitor charged via a resistor from a constant voltage source, the following equations can be derived:

$$t_1 = 0.693\, C_2(R_1 + R_2) \qquad (16\text{-}23)$$

$$t_2 = 0.693\, C_2 R_2 \qquad (16\text{-}24)$$

The above equations can be used to analyze a 555 pulse generator or to determine capacitor and resistor values to suit a given specification.

Example 16-8

Analyze the circuit in Fig. 16-20 to determine t_1, t_2, and the pulse frequency. Calculate the minimum level of capacitor charging current.

Solution

Eq. 16-23:
$$t_1 = 0.693\, C_2\, (R_1 + R_2)$$
$$= 0.693 \times 0.5\ \mu F \times (2.2\ k\Omega + 2.7\ k\Omega)$$
$$= 1.7\ ms$$

Eq. 16-24:
$$t_2 = 0.693\, C_2\, R_2 = 0.693 \times 0.5\ \mu F \times 2.7\ k\Omega$$
$$= 0.94\ ms$$

$$T = t_1 + t_2 = 1.7\ ms + 0.94\ ms$$
$$= 2.64\ ms$$

$$f = \frac{1}{T} = \frac{1}{2.64\ ms}$$
$$= 379\ Hz$$

$$I_{C1(min)} = \frac{V_{CC}/3}{R_1 + R_2} = \frac{15\ V/3}{2.2\ k\Omega + 2.7\ k\Omega}$$
$$= 1.02\ mA$$

Duty Cycle

Consider the V_o waveform in Fig. 16-20b once again. Time t_1 can be referred to as the *pulse width (PW)* and t_2 as the *space width (SW)*. The *duty cycle* is the percentage of time period T occupied by the pulse width. In the waveforms illustrated, PW is greater than SW, and so in this case the duty cycle is greater than 50%. A square wave has a 50% duty cycle, which means that PW and SW are equal.

Square Wave Generator

The 555 pulse generator can be modified to produce a duty cycle of 50% or less by including diodes D_1 and D_2 in the circuit, as illustrated in Fig. 16-21. In this arrangement, D_2 is reverse-biased during the charging time (t_1), and charging current (I_{ch}) to C_2 flows through R_1 and D_1. During the discharge time (t_2), the discharge current (I_{dis}) flows from C_2 via D_2 and R_2 to *terminal 7*. If resistors R_1 and R_2 are equal, t_1 and t_2 are equal and the output is a square wave. The

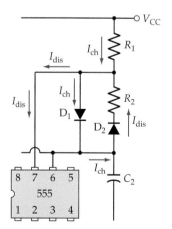

Figure 16-21 Use of diodes to convert a 555 pulse generator into a square wave generator (50% duty cycle). C_2 charging current (I_{ch}) flows through R_1 and D_1, and the discharge current (I_{dis}) flows through R_2 and D_2.

circuit equations must be modified for the different charge and discharge paths and to take account of the diode voltage drops (V_F).

$$R_1 = \frac{(V_{CC}/3) - V_F}{I_{C1(min)}} \qquad (16\text{-}25)$$

Equation 16-23 becomes

$$t_1 = 0.693 C_2 R_1 \qquad (16\text{-}26)$$

while Eq. 16-24 remains

$$t_2 = 0.693 C_2 R_2$$

Example 16-9

A 555 square wave generator operating from a 12 V supply is to have a 1 kHz frequency. Determine suitable component values.

Solution

$$I_{C1(min)} \gg I_{trig} \text{ and } I_{th}$$

From the 555 data sheet,

$$I_{trig} = 0.5 \text{ μA and } I_{th} = 0.25 \text{ μA}$$

Select $\quad I_{C1(min)} = 100 \text{ μA}$

Eq. 16-25: $\quad R_1 = \dfrac{(V_{CC}/3) - V_F}{I_{C1(min)}} = \dfrac{(12 \text{ V}/3) - 0.7 \text{ V}}{100 \text{ μA}}$

$$= 33 \text{ k}\Omega$$

$$R_2 = R_1 = 33 \text{ k}\Omega$$

$$T = \frac{1}{f} = \frac{1}{1 \text{ kHz}}$$

$$= 1 \text{ ms}$$

$$t_1 = t_2 = T/2 = 0.5 \text{ ms}$$

From Eq. 16-26, $\quad C_2 = \dfrac{t_1}{0.693 \, R_1} = \dfrac{0.5 \text{ ms}}{0.693 \times 33 \text{ k}\Omega}$

$$= 0.022 \text{ μF (use 0.02 μF standard value)}$$

Voltage-Controlled Oscillator

The frequency of the output from a 555 square wave generator can be altered by applying a direct voltage (V_i) to *terminal 5* (the control terminal) (see Fig. 16-22). The voltage at the non-inverting input of CP_1 is now V_i, instead of $2V_{CC}/3$, and C_2 has to charge to V_i before discharge can begin. When V_i is higher than $2V_{CC}/3$, capacitor charge and discharge times are increased, and so the output

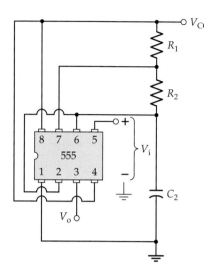

Figure 16-22   An adjustable voltage (V_i) applied to (control) terminal 5 of a 555 pulse generator can increase or decrease the output frequency, thus converting the circuit into a voltage-controlled oscillator (VCO).

frequency is decreased. When V_i is lower than $2V_{CC}/3$, charge and discharge times are decreased and there is an increase in output frequency. The circuit is now a *voltage-controlled oscillator (VCO)*.

Practice Problems

16-7.1 A 555 square wave generator has $V_{CC} = 18$ V, $C_2 = 0.1$ μF, and $R_1 = R_2 = 18$ kΩ. Calculate the maximum and minimum levels of charge and discharge currents, and determine the output frequency.

16-7.2 A 555 pulse wave generator operating from a 15 V supply is to have a 1.5 kHz output frequency with a 60% duty cycle. Determine suitable resistor and capacitor values.

16-8 TRIANGULAR WAVE GENERATOR

Integrator

The circuit in Fig. 16-23 is an *integrator*; its output amplitude can be shown to be directly proportional to the area of the input pulse (amplitude × time). The circuit is similar to an op-amp inverting amplifier, except that capacitor C_1 replaces the resistor usually connected between the output and inverting input terminals. As in the case of the inverting amplifier, the op-amp inverting input terminal remains at ground level (a virtual ground) because the non-inverting input is grounded. The output voltage depends upon the capacitor charge:

$$V_o = V_{C1}$$

If the capacitor charge is $V_{C1} = 0$, then the output is $V_o = 0$. If $V_{C1} = -1$ V (negative on the right, as illustrated), $V_o = -1$ V. If $V_{C1} = +1$ V (positive on the right), $V_o = +1$ V.

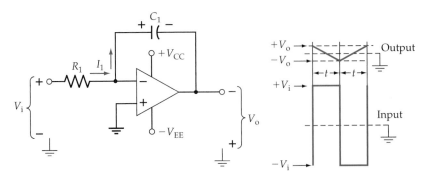

Figure 16-23   An integrator circuit produces a triangular wave output from a square wave input.

The circuit input current is

$$I_1 = \frac{V_i}{R_1}$$

With V_i constant, I_1 is a constant current flowing into C_1. The capacitor constant current charging equation can be used to calculate the capacitor voltage:

$$V_{C1} = \frac{I_1 t}{C_1}$$

or

$$V_o = \frac{I_1 t}{C_1} \tag{16-27}$$

In Eq. 16-27, t is the time duration of I_1, or the input pulse width, as illustrated in Fig. 16-23.

When the input voltage is positive $(+V_i)$, I_1 is a positive quantity flowing through R_1 and into C_1 in the direction shown on the circuit diagram. This causes C_1 to charge, positive on the left, negative on the right. With the input voltage constant, I_1 is a constant current, and V_o increases constantly in a negative direction. When the polarity of the input voltage is reversed (to $-V_i$), the direction of I_1 is reversed, and the charging direction of C_1 is also reversed. Thus, V_o begins to grow in a positive direction. So a square wave input to the integrating circuit produces a triangular wave output.

Integrator Combined with Schmitt Trigger

Figure 16-24 shows an integrator combined with a non-inverting Schmitt trigger circuit (see Section 14-10). As will be explained, this combination constitutes a *triangular waveform generator*. The Schmitt trigger output is applied to the integrator input, and the integrator output functions as the Schmitt circuit input. The waveforms illustrate the operation of the circuit.

During the time from instant t_1 to instant t_2, the Schmitt output is positive (at $+V_{o(sat)}$), and the integrater output is changing at a constant rate in a negative-going direction. The integrator output change is ΔV, from $+V_o$ to $-V_o$. The Schmitt circuit is designed to have upper and lower trigger points (UTP and

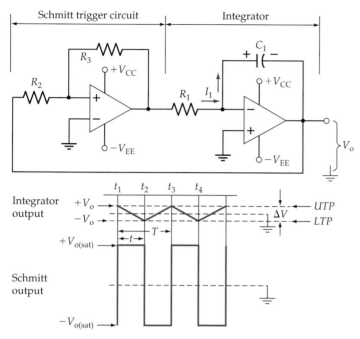

Figure 16-24 Triangular waveform generator consisting of a non-inverting Schmitt trigger circuit and an integrator. The Schmitt output is the integrator input, and the integrator output is applied as the Schmitt input.

LTP) equal to the desired levels of $+V_o$ and $-V_o$. Thus, when the integrator output arrives at the LTP, the Schmitt output (the integrator *input*) switches from $+V_{o(sat)}$ to $-V_{o(sat)}$. The integrator input is now a constant negative voltage, so that the integrator output direction is reversed. From t_2 to t_3, the integrator output increases linearly from $-V_o$ to $+V_o$, that is, from the Schmitt LTP to its UTP. At the UTP the Schmitt output reverses again, causing the integrator output to reverse direction once more. The cycle is repeated again and again, producing a triangular waveform at the integrator output terminal.

The frequency of the triangular output wave can be varied by altering the charging rate of capacitor C_1. This is done by making R_1 adjustable, so that I_1 can be increased or decreased. Reducing the resistance of R_1 increases the level of I_1 and causes C_1 to be charged faster. This reduces the time (t) between $+V_o$ and $-V_o$, and thus increases the output frequency. An increase in the resistance of R_1 reduces I_1, increases t, and results in a lower output frequency. The output amplitude of the triangular wave can be varied by altering the Schmitt UTP and LTP. This can be done by making one of the Schmitt resistors (R_2 or R_3) partially adjustable.

Circuit Design

A waveform generator is normally designed to produce a specified output amplitude and frequency. This means that the Schmitt circuit trigger points must be equal to the required positive and negative output peaks of the triangular wave (see Section 14-10). Also, the integrator has to accept the Schmitt output as its input, and its capacitor charging time should equal half the time period of the specified output frequency. Resistor R_1 might be made partially variable to set the output precisely to the required frequency.

Example 16-10

Design a triangular wave generator to produce a ± 3 V, 500 Hz output. Use 741 op-amps with a ± 9 V supply. (See Fig. 16-25.)

Solution

Integrator design:

$$V_i \approx \pm(V_{CC} - 1\text{ V}) = \pm(9\text{ V} - 1\text{ V})$$

$$= \pm 8\text{ V}$$

$$\Delta V = +V_o - (-V_o) = 3\text{ V} - (-3\text{ V})$$

$$= 6\text{ V}$$

$$I_1 \gg I_{B(max)}\text{ for the op-amp}$$

Select

$$I_1 = 1\text{ mA}$$

$$R_1 = \frac{V_i}{I_1} = \frac{8\text{ V}}{1\text{ mA}}$$

$$= 8\text{ k}\Omega\text{ (use 8.2 k}\Omega)$$

$$t = \frac{1}{2f} = \frac{1}{2 \times 500\text{ Hz}}$$

$$= 1\text{ ms}$$

From Eq. 16-27,

$$C_1 = \frac{I_1 t}{\Delta V} = \frac{1\text{ mA} \times 1\text{ ms}}{6\text{ V}}$$

$$= 0.16\ \mu\text{F (use 0.15 standard value)}$$

Schmitt design:

Select

$$I_2 = 1\text{ mA}$$

$$R_2 = \frac{UTP}{I_2} = \frac{3\text{ V}}{1\text{ mA}}$$

$$= 3\text{ k}\Omega\text{ (use 3.3 k}\Omega)$$

$$R_3 = \frac{V_{CC} - 1\text{ V}}{I_2} = \frac{8\text{ V}}{1\text{ mA}}$$

$$= 8\text{ k}\Omega\text{ (use 8.2 k}\Omega)$$

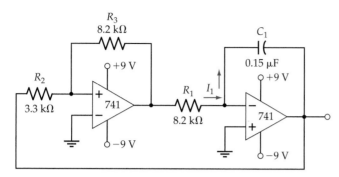

Figure 16-25 Triangular wave generator designed in Example 16-10.

Practice Problems

16-8.1 A triangular waveform generator circuit (as in Fig. 16-25) has the following component values: $R_1 = 4.7\,k\Omega$, $R_2 = 3.9\,k\Omega$, $R_3 = 22\,k\Omega$, and $C_1 = 0.05\,\mu F$. If the supply is $V_{CC} = \pm 12$ V, determine the amplitude and frequency of the output.

16-8.2 Design a triangular wave generator to have $V_o = \pm 2.5$ V, and an output frequency adjustable from 200 Hz to 400 Hz. Use 741 op-amp with $V_{CC} = 15$ V.

16-9 OSCILLATOR FREQUENCY STABILIZATION

Frequency Stability

The output frequency of oscillator circuits is normally not as stable as necessary for a great many applications. The component values all have tolerances, so that the actual oscillating frequency may easily be 10% higher or lower than the desired frequency. But by making a capacitor or resistor partially adjustable, the frequency of most oscillator circuits can be set fairly precisely. However, such adjustments may not be convenient in production circuits.

A further problem is that component values vary with the changes in temperature, and this causes changes in oscillating frequency. The component temperature changes might be due to variations in the ambient temperature or the result of component power dissipation. A well-designed circuit (of any type) normally uses as little power as possible to avoid component heating and to reduce the power supply load. Oscillator frequency stability can be dramatically improved by the use of *piezoelectric crystals.*

Piezoelectric Crystals

If a mechanical stress is applied to a wafer of quartz crystal, a voltage proportional to the pressure appears at the surfaces of the crystal (see Fig. 16-26a). The

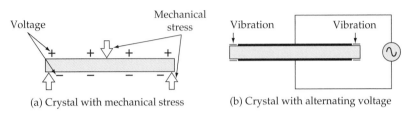

(a) Crystal with mechanical stress　　(b) Crystal with alternating voltage

Figure 16-26　A piezoelectric crystal under stress produces a surface voltage. It also vibrates when an ac voltage is applied to its surfaces.

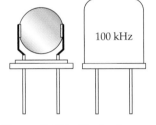

100 kHz

Figure 16-27　Electronic crystal contained in a metal can.

crystal also vibrates, or *resonates,* when an alternating voltage with the natural resonance frequency of the crystal is applied to its surfaces (Fig. 16-26b). All materials with this property are termed *piezoelectric.* Because the crystal resonance frequency is extremely stable, piezoelectric crystals are used to stabilize the frequency of oscillators.

Quartz crystals for electronics applications are cut from the natural material in several different shapes. The crystals are ground to precise dimensions, and silver or gold electrodes are plated on to opposite sides for electrical connections. The crystal is usually mounted inside a vacuum-sealed glass envelope or a hermetically sealed metal can. Figure 16-27 shows a typical crystal contained in a metal can. Crystals are also available in surface-mount and other types of enclosures.

Crystal Equivalent Circuit

The electrical equivalent circuit for a crystal is shown in Fig. 16-28a. The crystal behaves as a series *RLC* circuit (R_s, L_s, C_s) in parallel with the capacitance of the connecting terminals (C_p). The series *RLC* components are referred to as the *motional resistance* (R_s), the *motional inductance* (L_s), and the *motional capacitance* (C_s), because they represent the piezoelectric performance of the crystal.

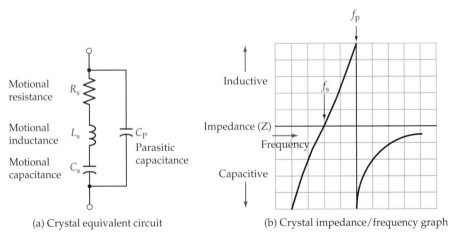

(a) Crystal equivalent circuit　　　　　(b) Crystal impedance/frequency graph

Figure 16-28　The equivalent circuit for a piezoelectric crystal is a series *RLC* circuit with a parallel capacitance. The circuit has two resonance frequencies: (f_s) and (f_p).

C_p is sometimes referred to as a *parasitic capacitance*. Because of the presence of C_p, the crystal has a *parallel resonance frequency* (f_p) when C_p resonates with the series circuit reactance, as well as a *series resonance frequency* (f_s) when L_s and C_s resonate. At series resonance the device impedance is reduced to R_s, and at parallel resonance the impedance is very high.

Figure 16-28b shows that, like all *RLC* circuits, the impedance of the crystal equivalent circuit is capacitive below the series resonance frequency and inductive above f_s. At frequencies greater than f_s, the series *RLC* circuit becomes inductive until it resonates with the parallel capacitance at f_p. The impedance of the complete circuit then becomes capacitive (with increasing frequency) as the reactance of C_p is reduced. The two resonance frequencies (f_s and f_p) are very close together.

A measure of the quality of a resonance circuit is the ratio of reactance to resistance, termed the *Q factor*. Because the resistive component of the crystal equivalent circuit is relatively small, crystals have very large *Q* factors. Crystal *Q* factors range approximately from 2000 to 100 000, compared to a maximum of about 400 for an actual *LC* circuit. Resonance frequencies of available crystals are usually 10 kHz to 200 MHz.

When a series *RLC* circuit is operating at its resonance frequency, the inductive and capacitive reactances cancel each other, and the power supplied is dissipated in the resistance. If the power dissipation increases the temperature of a crystal, the resonance frequency can drift by a small amount. Most crystals maintain their frequency to within a few cycles of the resonance frequency at 25°C. For greater frequency stability, crystals are sometimes enclosed in an insulated, thermostatically controlled *crystal oven*.

Crystals Control of Oscillators

The frequency of an oscillator may be stabilized by using a crystal operating at either its series or parallel resonance frequency. The circuit is then usually referred to as a *crystal oscillator*.

In many circuits the crystal is connected in series with the feedback network. The crystal offers a low impedance at its series resonance and a high impedance at all other frequencies, so that oscillation occurs only at the crystal series resonance frequency. The *Pierce oscillator* in Fig. 16-29 would appear to be such a circuit, but it actually operates as a Colpitts oscillator. Capacitors C_1 and C_2 are present at the input and output of the inverting amplifier (compare to Fig. 16-6a). A Colpitts oscillator also has an inductor connected between the amplifier input and output. In Fig. 16-29 the crystal behaves as an inductance by operating at a frequency slightly above f_s (see Fig. 16-28).

The circuit in Fig. 16-29 is often used as a square wave generator or *clock oscillator* for digital circuit applications. In this situation, the amplifier gain is made as large as possible, so that Q_1 is driven between saturation and cutoff to produce a square wave output.

Figure 16-30 shows one method of modifying a Wein bridge oscillator for frequency stabilization. The crystal is connected in series with C_1 and R_1, and

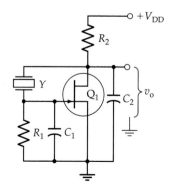

Figure 16-29 Pierce oscillator circuit using a crystal that operates at its series resonance frequency.

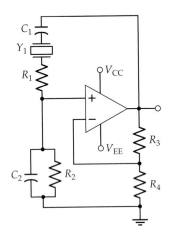

Figure 16-30 Use of a series-connected crystal to stabilize the frequency of a Wein bridge oscillator circuit.

in this situation it offers a low resistance at its series resonance frequency (f_s) and a relatively high impedance at all other frequencies. Feedback voltage from the output via C_1 and R_1 is severely attenuated at all frequencies except f_s, so the circuit can oscillate only at f_s.

A crystal-controlled Wein bridge oscillator circuit is first designed without the crystal. A crystal is selected with f_s equal to the desired frequency. Then the circuit is modified by subtracting the crystal series resonance resistance (R_s) from the calculated value of the series resistor (R_1 in Fig. 16-30). Alternatively, circuit design could start by selection of a crystal. R_1 is then made larger than R_s, and $R_2 = (R_1 + R_s)$. C_1 and C_2 are calculated from f_s and R_2.

The crystal power dissipation must be kept below the specified maximum. Too much power can overheat the crystal and cause drift in the resonance frequency. As with other devices, too much power dissipation can destroy a crystal. Typical maximum crystal drive powers are 1 mW to 10 mW, but manufacturers usually recommend operating at $1/10$ of the specified maximum.

The dc insulation resistance of crystals ranges from 100 MΩ to 500 MΩ. This allows them to be directly connected into a circuit without the need for coupling capacitors. Care must be taken in any crystal oscillator circuit to ensure that the crystal does not interrupt the flow of a necessary direct current.

The typical frequency stability of oscillators that do not use crystals is about 1 in 10^4. This means, for example, that the frequency of a 1 MHz oscillator might be 100 Hz higher or lower than 1 MHz. If a crystal is used, the frequency stability can be improved to better than 1 in 10^6, which gives a ±1 Hz variation in the output of a 1 MHz oscillator.

Example 16-11

Redesign the Wein bridge oscillator in Example 16-4 to use a 100 kHz crystal with $R_s = 1.5$ kΩ.

Solution

Select
$$R_1 = 2R_s = 2 \times 1.5 \text{ k}\Omega$$
$$= 3 \text{ k}\Omega \text{ (use 2.7 k}\Omega \text{ standard value)}$$
$$R_2 = R_1 + R_s = 2.7 \text{ k}\Omega + 1.5 \text{ k}\Omega$$
$$= 4.2 \text{ k}\Omega \text{ (use 3.9 k}\Omega \text{ standard value)}$$

From Eq. 16-13,
$$C_1 = \frac{1}{2\pi f R_2} = \frac{1}{2\pi \times 100 \text{ kHz} \times 3.9 \text{ k}\Omega}$$
$$= 408 \text{ pF (use 390 pF standard value)}$$
$$R_4 \approx R_2 = 3.9 \text{ k}\Omega \text{ (standard value)}$$
$$R_3 = 2R_4 = 2 \times 3.9 \text{ k}\Omega$$
$$= 7.8 \text{ k}\Omega \text{ (use 8.2 k}\Omega \text{ standard value)}$$

Example 16-12

The Pierce oscillator in Fig. 16-31 has a crystal with $f_s = 1$ MHz and $R_s = 700\ \Omega$. Calculate the inductance offered by the crystal at the circuit oscillating frequency. Estimate the peak power dissipated in the crystal.

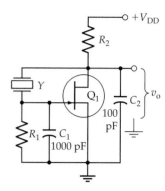

Figure 16-31 Crystal oscillator circuit for Example 16-12.

Solution

Eq. 16-5:
$$C_T = \frac{C_1 \times C_2}{C_1 + C_2} = \frac{1000\ \text{pF} \times 100\ \text{pF}}{1000\ \text{pF} + 100\ \text{pF}}$$

$$= 90.9\ \text{pF}$$

At resonance, $X_L = X_{CT}$

$$2\pi f L = \frac{1}{2\pi f C_T}$$

or
$$L = \frac{1}{(2\pi f)^2\, C_T} = \frac{1}{(2\pi \times 1\ \text{MHz})^2 \times 90.9\ \text{pF}}$$

$$= 279\ \mu\text{H}$$

$$i_p = \frac{V_{DD}}{R_1 + R_2 + R_s} = \frac{5\ \text{V}}{1\ \text{M}\Omega + 10\ \text{k}\Omega + 700\ \Omega}$$

$$\approx 5\ \mu\text{A}$$

$$P_D = (0.707\, i_p)^2\, R_S = (0.707 \times 5\ \mu\text{A})^2 \times 700\ \Omega$$

$$= 9\ \text{nW}$$

Practice Problems

16-9.1 A 40 kHz crystal with $R_s = 3$ kΩ is to be used with the Colpitts oscillator in Fig. 16-5. Determine the necessary circuit modifications, and estimate the peak power dissipation in the crystal.

16-9.2 A Pierce oscillator (as in Fig. 16-31) has a crystal with $L = 500$ μH at $f_s = 1$ MHz. Select suitable capacitor values and calculate the minimum amplifier voltage gain.

16-9.3 Design a 50 kHz, crystal-controlled, Wein bridge oscillator to use a crystal with $R_s = 2$ kΩ. The available supply voltage is $V_{CC} = \pm 12$ V. Specify the op-amp bandwidth.

Review Questions

Section 16-1

16-1 State the Barkhausen criteria for a sinewave oscillator, and explain why they must be fulfilled to sustain oscillations.

16-2 Draw the circuit diagram of an op-amp phase shift oscillator. Sketch the circuit waveforms, and briefly explain the oscillator operation.

16-3 Write the frequency equation for a phase shift oscillator. Discuss the phase shift network attenuation and the amplifier gain requirements.

16-4 Sketch the circuit diagram for a phase shift oscillator that uses a single-stage BJT amplifier. Briefly explain the circuit operation.

Section 16-2

16-5 Draw the circuit diagram of an op-amp Colpitts oscillator. Sketch the oscillator waveforms, and briefly explain the circuit operation.

16-6 Write the frequency equation for a Colpitts oscillator. Discuss the phase shift network attenuation and the amplifier gain requirements.

16-7 Sketch circuit diagrams for a Colpitts oscillator using a single-stage BJT amplifier. Briefly explain the circuit operation.

Section 16-3

16-8 Draw the circuit diagram of an op-amp Hartley oscillator. Sketch the oscillator waveforms, and explain the circuit operation.

16-9 Write the frequency equation for a Hartley oscillator. Discuss the phase shift network attenuation and the amplifier gain requirements.

16-10 Sketch circuit diagrams for a Hartley oscillator using a single-stage BJT amplifier. Briefly explain the circuit operation.

Section 16-4

16-11 Draw the circuit diagram of an op-amp Wein bridge oscillator. Sketch the oscillator waveforms, and explain the circuit operation.

16-12 Write the frequency equation for a Wein bridge oscillator. Discuss the phase shift network attenuation and the amplifier gain requirements.

Section 16-5

16-13 Show how the output amplitude of a phase shift oscillator can be stabilized by means of a diode circuit. Explain the circuit operation.

16-14 Sketch a diode amplitude stabilization circuit for a Wein bridge oscillator, and explain its operation.

16-15 Draw a FET circuit diagram for stabilizing the output amplitude of a Wein bridge oscillator. Explain the circuit operation.

Section 16-6

16-16 Draw the circuit diagram of a square wave generator that uses an inverting Schmitt trigger circuit. Sketch the circuit waveforms, and explain the operation of the circuit.

Section 16-7

16-17 Referring to the 555 timer block diagram in Fig. 16-19a, identify each component and explain its function.

16-18 Draw a circuit diagram and waveforms for a pulse wave generator using a 555 timer. Explain its operation.

16-19 Sketch circuits to show how a 555 pulse generator can be modified for (a) square wave output, (b) operation as a voltage controlled oscillator.

Section 16-8

16-20 Sketch an op-amp integrating circuit together with the circuit wave-forms. Explain the circuit operation.

16-21 Draw the circuit diagram of a triangular wave generator that uses an integrator circuit and a non-inverting Schmitt trigger. Sketch the wave-forms and explain the circuit operation.

16-22 Discuss how the output amplitude and frequency can be made adjustable in a triangular wave generator.

Section 16-9

16-23 Describe a piezoelectric crystal as used with electronic circuits. Sketch the crystal equivalent circuit and impedance/frequency graph. Explain the behaviour of electronic crystals.

16-24 Show how piezoelectric crystals are employed for oscillator stabiliza-tion. Explain.

Problems

Section 16-1

16-1 Design a phase shift oscillator to have a 3 kHz output frequency. Use a 741 op-amp with $V_{CC} = \pm12$ V.

16-2 A phase shift oscillator is to use three 0.05 μF capacitors and an op-amp with $V_{CC} = \pm9$ V. Design the circuit to have $f = 7$ kHz. Select a suitable operation amplifier.

16-3 Redesign the circuit in Problem 16-1 to use a single-stage BJT amplifier. Use a 2N3904 transistor with $V_{CC} = 15$ V.

16-4 The phase shift oscillator circuit in Fig. 16-1 has the following compo-nent values: $R_1 = 3.9$ kΩ, $R_2 = 120$ kΩ, $R_3 = 120$ kΩ, $R = 3.9$ kΩ, and $C = 0.025$ μF. Calculate the circuit oscillating frequency.

16-5 Determine new capacitor values for the phase shift oscillator in Prob-lem 16-1 to switch its frequency to (a) 700 Hz and (b) 5 kHz.

Section 16-2

16-6 Using a 741 op-amp with $V_{CC} = \pm10$ V, design a Colpitts oscillator to have a 30 kHz oscillating frequency.

16-7 Design a Colpitts oscillator to have $f = 55$ kHz, and to use a 20 mH inductor and a 741 op-amp with $V_{CC} = \pm18$ V.

16-8 The Colpitts oscillator circuit in Fig. 16-4 has $R_1 = 18$ kΩ, $R_2 = 180$ kΩ, $R_3 = 18$ kΩ, $L = 75$ mH, $C_1 = 3000$ pF, and $C_2 = 300$ pF. Calculate the circuit oscillating frequency.

16-9 Design a 20 kHz Colpitts oscillator using a 10 mH inductor and a single-stage BJT amplifier. Use a 2N3904 transistor with $V_{CC} = 20$ V.

Section 16-3

16-10 Design a 6 kHz Hartley oscillator circuit to use a 741 op-amp, and an inductor with $L_1 = 10$ mH and $L_2 = 100$ mH.

16-11 Calculate the oscillating frequency for the Hartley oscillator in Fig. 16-7 if the components are $R_1 = 3.3$ kΩ, $R_2 = 56$ kΩ, $R_3 = 3.3$ kΩ, $L_1 = 3$ mH, $L_2 = 50$ mH, and $C_1 = 1500$ pF.

16-12 An op-amp Hartley oscillator has two inductors with $L_1 = 5$ mH, $L_2 = 40$ mH, and a total inductance of $L_T = 50$ mH. Determine the capacitor value to give $f = 2.25$ kHz. Calculate the required amplifier voltage gain.

Section 16-4

16-13 Design a 15 kHz Wein bridge oscillator to use an op-amp with $V_{CC} = \pm14$ V. Specify the operational amplifier.

16-14 A Wein bridge oscillator uses two 5000 pF capacitors and a 741 op-amp with $V_{CC} = \pm12$ V. Complete the circuit design to produce a 9 kHz output frequency.

16-15 The Wein bridge oscillator in Fig. 16-10 has the following components: $R_1 = R_2 = R_4 = 5.6$ kΩ, $R_3 = 12$ kΩ, and $C_1 = C_2 = 2700$ pF. Calculate the oscillating frequency.

16-16 The oscillator in Problem 16-15 has its capacitors changed to 0.05 μF, and resistors R_1 and R_2 are adjustable from 4.7 kΩ to 6.7 kΩ. Calculate the maximum and minimum output frequencies.

Section 16-5

16-17 The phase shift oscillator in Problem 16-1 is to have its output amplitude stabilized to ±7 V. Design a suitable diode amplitude stabilization circuit.

16-18 Design a diode amplitude stabilization circuit to limit the output of the Wein bridge oscillator in Example 16-4 to a maximum of ±4 V.

16-19 Design a 3 kHz Wein bridge oscillator using a diode circuit to stabilize the output to ±10 V. A ±20 V supply is to be used, and a suitable op-amp is to be selected.

16-20 Modify the circuit for Problem 16-19 to use a FET stabilization circuit, as in Fig. 16-15. Use a FET with $r_{DS} = 200$ Ω at $V_{GS} = 3$ V, and $r_{DS} = 600$ Ω at $V_{GS} = 5$ V.

Section 16-6

16-21 The square wave generator circuit in Fig. 16-17 has the following quantities: $R_1 = 33$ kΩ, $R_2 = 56$ kΩ, $R_3 = 2.2$ kΩ, $C_1 = 0.15$ μF, $V_{CC} = \pm12$ V. A rail-to-rail op-amp is used. Determine the output amplitude and frequency.

16-22 Design a square wave generator circuit (as in Fig. 16-17) to produce a 3 kHz, ±9 V output. Select a suitable op-amp.

16-23 A square wave generator circuit is to use a BIFET op-amp with $V_{CC} = \pm18$ V. Design the circuit to produce an output frequency adjustable from 500 Hz to 5 kHz.

16-24 Modify the circuit designed for Problem 16-22 to make the output adjustable from 2.5 kHz to 3.5 kHz.

Section 16-7

16-25 A 555 pulse generator circuit as in Fig. 16-20 has the following components: $R_1 = R_2 = 1.8$ kΩ, and $C_2 = 0.3$ μF. The supply voltage is $V_{CC} = 12$ V. Calculate the output frequency and duty cycle.

16-26 Design a 555 pulse generator circuit (as in Fig. 16-20) to produce an output pulse waveform with PW = 350 μs and SW = 200 μs. The supply voltage is $V_{CC} = 15$ V.

16-27 A 555 square wave generator circuit as in Fig. 16-21a has: $R_1 = R_2 = 2.7$ kΩ, and $C_2 = 0.6$ μF. Calculate the output frequency.

16-28 The square wave generator circuit in Problem 16-27 uses a 12 V supply. Determine the output frequency if a voltage $V_i = 7$ V is applied as illustrated in Fig. 16-22.

Section 16-8

16-29 Design a triangular waveform generator to produce a ±1 V, 1 kHz output. Use 741 op-amps with $V_{CC} = \pm12$ V.

16-30 Modify the circuit designed for Problem 16-29 to make the output adjustable from 500 Hz to 1.5 kHz.

16-31 Determine the amplitude and frequency of the output from a triangular waveform generator (as in Fig. 16-24) that has the following components and supply voltage: $R_1 = 10$ kΩ, $R_2 = 3.9$ kΩ, $R_3 = 22$ kΩ, $C_1 = 0.5$ μF, and $V_{CC} = \pm12$ V.

16-32 Determine the output frequency range from the circuit in Problem 16-31 if R_1 is replaced with a 4.7 kΩ resistor in series with a 10 kΩ potentiometer.

Section 16-9

16-33 The oscillating frequency of the Hartley oscillator circuit in Fig. 16-8 is to be stabilized by means of a 100 kHz crystal with $R_s = 700$ Ω. Determine the necessary modifications, and estimate the peak power dissipation in the crystal.

16-34 Select suitable capacitor values for a Pierce oscillator (as in Fig. 16-31) that uses a 3 MHz crystal. The crystal has $L = 390$ μH at the oscillating frequency.

16-35 A 200 kHz crystal with $R_s = 700$ Ω is used in the Wein bridge oscillator circuit in Fig. 16-30. Design the circuit using $A_{CL} \approx 3$ for the amplifier. Assume that $V_{CC} = \pm15$ V and that the output is limited to ±5 V.

Practice Problem Answers

16-1.1   10 kΩ, 10 kΩ, 330 kΩ, 1000 pF

16-1.2 $V_{CC} = 18$ V, $R_1 = (56$ kΩ $+ 2.2$ kΩ$)$, $R_2 = 27$ kΩ, $R_C = 3.3$ kΩ, $R_E = 2.7$ kΩ, $(R - Z_{in}) = 1.8$ kΩ, $R = 3.3$ kΩ, $C = (0.15$ μF$\|0.02$ μF$)$, $C_E = 75$ μF

16-2.1   12 kΩ, 120 kΩ, 12 kΩ, 0.01 μF, 220 mH, 1000 pF, ±12 V

16-2.2 $V_{CC} = 15$ V, $R_1 = (68$ kΩ $+ 8.2$ kΩ$)$, $R_2 = 47$ kΩ, $R_C = 4.7$ kΩ, $R_E = 4.7$ kΩ, $C_1 = 0.06$ μF, $C_2 = 6000$ pF , $C_C = 0.05$ μF, $C_E = 3$ μF

16-3.1   ±12 V, 1 kΩ, 10 kΩ, 1 kΩ, 2.2 mH, 22 mH, 0.02 μF

16-3.2   66.3 kHz

16-4.1   10.6 kHz, 28.4 kHz

16-4.2   10 kΩ, 2.2 kΩ, 27 kΩ, 2.2 kΩ, 1000 pF, 5000 pF

16-5.1   220 Ω, 5.6 kΩ, 1.5 kΩ, 2 kΩ, 3.9 kΩ, 220 Ω, 0.082 μF

16-5.2   1.5 kΩ, 2.2 kΩ, 1.5 kΩ

16-6.1   12 kΩ, 150 kΩ

16-6.2   0.1 μF, 2.7 kΩ

16-7.1   628 μA, 294 μA, 400 Hz

16-7.2   11 kΩ, 39 kΩ, 0.01 μF

16-8.1   ±1.42 V, 6.13 kHz

16-8.2   (6.8 kΩ + 10 kΩ pot.), 0.5 μF, 2.2 kΩ, 12 kΩ

16-9.1   Crystal in series with $R_1 = 22$ kΩ, 2.8 μW

16-9.2   270 pF, 62 pF, 4.4

16-9.3   3.9 kΩ, 5.6 kΩ, 560 pF, 560 pF, 12 kΩ, 5.6 kΩ, 150 kHz

CHAPTER 17
Active Filters

CONTENTS

Objectives

You will be able to:

1   Sketch the circuits of first-order and second-order low-pass and high-pass active filters. Explain the operation of each circuit.
2   Sketch and explain the typical gain/frequency and phase/frequency response graphs for each of the above circuits. Discuss the various frequency response shapes possible.
3   Design each of the above circuits to meet a given specification.
4   Analyze each of the above circuits to determine its performance.
5   Show how third-order active filters are constructed. Design and analyze third-order filters.

6   Sketch various band-pass and notch filter block diagrams and circuits and their gain/frequency responses. Explain the operation of each circuit.
7   Design and analyze band-pass and notch filters.
8   Sketch a state-variable filter circuit and explain its operation.
9   Using appropriate circuit diagrams, explain how a resistor is simulated by a switched capacitor.
10 Sketch and explain the block diagram of an IC switched-capacitor filter, and determine suitable external resistor values and clock frequency for a given application.

INTRODUCTION

Filters are circuits that pass only a certain range of signal frequencies and that attenuate unwanted frequencies. Filters are usually classified according to the band of frequencies that they pass: *low-pass, high-pass,* and so on. Passive filters employ only passive components such as resistors, capacitors, and inductors. Active filters use amplifiers, together with passive components. Analysis of filter circuits can be extremely complex, but the operation of most commonly used filters can be readily understood, and simple design techniques are available.

17-1 FILTER TYPES AND CHARACTERISTICS

The functions of various types of filters are illustrated by the waveforms and typical gain/frequency response graphs in Fig. 17-1. The gain/frequency response graphs show that filters are classified according to the band of frequencies that they pass. In all cases the signal frequency (f) is plotted to a logarithmic scale, and the circuit gain ($A_v = v_o/v_i$) has a decibel scale, just as in the case of an amplifier frequency response graph (see Section 8-2).

Low-Pass

In Fig. 17-1a the input waveforms consist of a low-frequency signal (v_s) and an unwanted high-frequency noise (v_n). The low-frequency signal is reproduced at the output with very little attenuation, while the high-frequency noise voltage is severely attenuated. The A_v/f response graph shows that the circuit gain remains constant from $f = 0$ up to the *cutoff frequency* (f_c), defined as the frequency at which A_v is down by 3 dB from its normal level in the pass band, as in the case of amplifiers. Ideally, the gain should fall steeply to zero, as shown by the straight-line *ideal response* graph. In practice, of course, an ideal response is impossible to produce. In the actual response graph, A_v begins to fall off slowly as the signal frequency approaches f_c, and then falls off at a constant rate above f_c, as illustrated. Because all signal frequencies from zero to f_c are passed, and those above f_c are attenuated, this is a *low-pass filter* (only low frequencies are passed). The filter pass band ranges from $f = 0$ to $f = f_c$. The circuit bandwith (*BW*) is ($f_c - 0$), which equals f_c. As will be shown, $BW = f_c$ only in the case of a low-pass filter.

High-Pass

Figure 17-1b shows a *high-pass filter* with input waveforms again consisting of a combination of low and high frequencies. In this case, the low-frequency input is severely attenuated and the high-frequency signal is passed to the output. On the gain/frequency response graph, f_c is once again the frequency

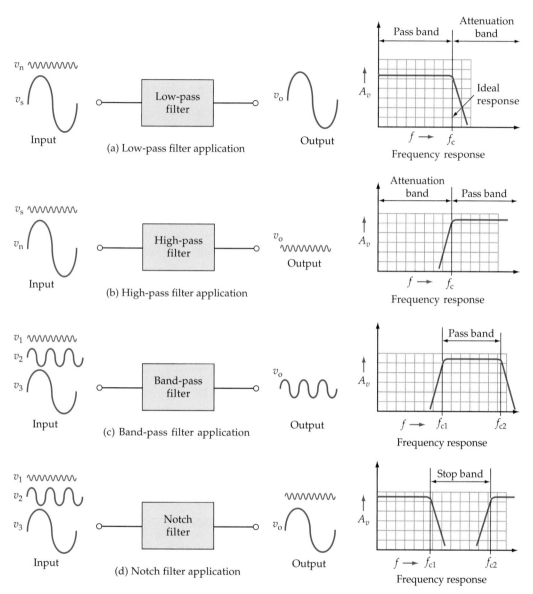

Figure 17-1 Typical input and output waveforms and gain/frequency response graphs for various filters.

at which the circuit gain is 3 dB below the gain in the pass band. The *pass band* extends from f_c to the upper cutoff frequency for the amplifier used (normally an op-amp). The amplifier upper cutoff frequency must be higher than the highest signal frequency to be passed.

Band-Pass

As the name implies, a *band-pass filter* (Fig. 17-1c) passes a band of signal frequencies ranging from a low cutoff frequency (f_{c1}) to a high cutoff frequency (f_{c2}).

The high- and low-frequency input waveforms are rejected while middle-frequency signals are passed to the output. The cutoff frequencies are again defined as those signal frequencies at which the gain falls to 3 dB below the gain in the pass band. As shown on the A_v/f response graph, the *pass band* is the range of frequencies between f_{c1} and f_{c2}, and the bandwith is $BW = (f_{c2} - f_{c1})$.

Notch

A *notch filter* (Fig. 17-1d), also known as a *band-stop filter*, attenuates an unwanted band of signal frequencies and passes all other frequencies above and below the filter cutoff frequencies. The A_v/f graph resembles a combination of the low-pass and high-pass response graphs. Signal frequencies up to f_{c1} are passed, frequencies between f_{c1} and f_{c2} are attenuated, and frequencies above f_{c2} are passed. The bandwidth is classified as the width of the *stop band*; $BW = (f_{c2} - f_{c1})$.

Filter Characteristics

Figure 17-2a shows a basic RC low-pass filter circuit together with a signal source (v_s and r_s) and load resistor (R_L). The circuit is referred to as a *passive filter* because it has no active components (transistors or operational amplifiers). Resistance R_1 and capacitive reactance X_{C1} constitute a voltage divider, as shown in Fig. 17-2b. The capacitive reactance $[1/(2\pi f C_1)]$ is very high at low frequencies; consequently, there is very little attenuation of low-frequency signals. As X_{C1} becomes progressively smaller with increasing signal frequency, input voltages are increasingly attenuated.

Gain/frequency and phase/frequency response graphs for a low-pass filter are shown in Fig. 17-2c. As already discussed, the output voltage (across C_1) remains substantially constant for all frequencies up to the cutoff frequency (f_c) (the frequency at which A_v is down by 3 dB from its normal level in the pass band). For signal frequencies above f_c, A_v falls off at *20 dB/decade*, which is a gain reduction of 20 dB for each frequency increase by a factor of 10 (a *decade*). This can also be stated as *6 dB/octave*: a 6 dB gain reduction for each doubling (an *octave*) of the signal frequency. As shown, the A_v/f response is divided into *pass band* and *attenuation band*, and the attenuation band can be further subdivided into the *transition band* (during which the output voltage is falling) and the *stop band* (where the unwanted signals are sufficiently attenuated). The cutoff frequency for the simple low-pass circuit in Fig. 17-2a depends upon the capacitance and upon the combination of r_s, R_1, and R_L. Source resistance r_s can be neglected if it is very much smaller than R_1, and R_L can also be neglected when it is very much larger than R_1. In these circumstances the filter cutoff frequency occurs when $X_{C1} = R_1$.

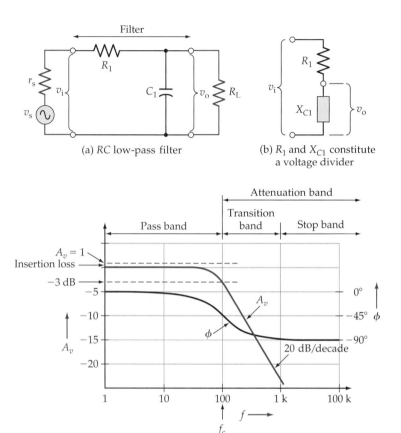

(a) RC low-pass filter

(b) R_1 and X_{C1} constitute a voltage divider

(c) Gain/frequency response and phase/frequency response

Figure 17-2 Resistor and capacitor connected to function as a low-pass filter. The A_v/f graph shows that the filter passes all signal frequencies up to the cutoff frequency f_c, and the ϕ/f graph shows a 45° phase lag at f_c.

Thus,
$$R_1 = \frac{1}{2\pi f_c C_1}$$

which gives
$$f_c = \frac{1}{2\pi R_1 C_1} \tag{17-1}$$

When R_L is not very much larger than R_1, the two resistors appear in parallel across the capacitor and the equation for the cutoff frequency becomes

$$f_c = \frac{1}{2\pi(R_1 \| R_L)C_1} \tag{17-2}$$

For signal frequencies within the filter pass band there is some attenuation of the input, known as *insertion loss* (see Fig. 17-2c). The insertion loss is defined as the loss of signal due to the presence of the filter, and it is the result of R_1 and R_L acting as a voltage divider. The *transfer function* is the ratio of the output voltage to the input. For the simple RC low-pass circuit in Fig. 17-2 with no load resistor connected, the transfer function is easily shown to be

$$\frac{v_o}{v_i} = \frac{-jX_C}{R_1 - jX_C}$$

or

$$\frac{v_o}{v_i} = \frac{X_C}{\sqrt{(R_1^2 + X_C^2)}} \qquad (17\text{-}3)$$

Because the filter is an RC circuit, there is a signal waveform phase shift from input to output, as well as an attenuation. The phase shift is illustrated by the ϕ/f graph in Fig. 17-2c. The output is in phase with the input at low signal frequencies (when $X_{C1} \gg R_1$) and increasingly lags behind the input as the frequency rises. At the cutoff frequency (when $X_{C1} = R_1$), there is a phase lag of 45°, and the lag increases above f_c to a maximum of 90° (when $X_{C1} \ll R_1$).

Example 17-1

A low-pass filter circuit, as in Fig. 17-2a, has $r_s = 60\ \Omega$, $R_1 = 12\ \text{k}\Omega$, $R_L = 100\ \text{k}\Omega$, and $C_1 = 0.013\ \mu\text{F}$. Calculate the cutoff frequency (a) when R_L is not connected, and (b) when R_L is connected. Calculate the attenuation when the signal frequency is $4f_c$.

Solution

(a) When R_L is not connected:

Eq. 17-1: $\qquad f_c = \dfrac{1}{2\pi R_1 C_1} = \dfrac{1}{2 \times \pi \times 12\ \text{k}\Omega \times 0.013\ \mu\text{F}}$

$\qquad\qquad = 1.02\ \text{kHz}$

(b) When R_L is connected:

Eq. 17-2: $\quad f_c = \dfrac{1}{2\pi(R_1 \| R_L)C_1} = \dfrac{1}{2 \times \pi \times (12\ \text{k}\Omega \| 100\ \text{k}\Omega) \times 0.013\ \mu\text{F}}$

$\qquad\qquad = 1.14\ \text{kHz}$

at f_c, $\qquad\qquad\qquad$ attenuation $= 3\ \text{dB}$

$\qquad\qquad\qquad\qquad$ falloff rate $= 6\ \text{dB}$ for each frequency doubling

$\qquad\qquad\qquad$ attenuation at $2f_c = 3\ \text{dB} + 6\ \text{db}$

and $\qquad\qquad\qquad$ attenuation at $4f_c = 3\ \text{dB} + 6\ \text{dB} + 6\ \text{dB}$

$\qquad\qquad\qquad\qquad = 15\ \text{dB}$

Practice Problems

17-1.1 A basic *RC* low-pass filter has $R_1 = 27$ kΩ and $C_1 = 300$ pF. Calculate the frequency at which the output is 3 dB below the normal level in the pass band. Estimate the signal attenuation at a frequency of $10 f_c$.

17-1.2 Calculate the insertion loss in decibels for the filter in Problem 17-1.1 when a 200 kΩ load is connected, and determine the effect of the load on the filter cutoff frequency.

17-2 FIRST-ORDER ACTIVE FILTERS

First-Order Low-Pass Filter

A passive *RC* low-pass filter circuit is shown in Fig. 17-3 with an op-amp voltage follower connected to its output. The voltage follower functions as a *buffer amplifier* to isolate the passive filter circuit from the load. A non-inverting amplifier can be used instead of the voltage follower if some voltage gain is required. The typical input impedance for a voltage follower is extremely high (see Example 14-3), and so the voltage follower eliminates the possibility of load effect on the filter performance. The presence of the op-amp converts the circuit into a *first-order low-pass active filter*. The A_v/f and ϕ/f responses for the circuit in Fig. 17-3 are exactly the same as those of the unloaded passive filter. The circuit is analyzed exactly as for a passive low-pass filter without a load.

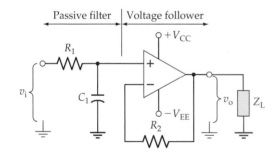

Figure 17-3   A first-order active filter uses an operational amplifier to isolate the RC passive filter from the load.

The design of a basic low-pass active filter circuit is very simple. Resistor R_1 should be selected to be small enough that no significant voltage drop occurs across it because of the op-amp input bias current, as explained in Section 14-2. It is also desirable to keep R_1 as large as possible for maximum filter input impedance and for minimum power dissipation. With the resistance of R_1 determined, capacitor C_1 can be calculated from Equation 17-1. The capacitance of C_1 must be very much larger than stray capacitance, so that the filter cutoff frequency is not affected by stray capacitance. When the calculated capacitance is *not* very much larger than possible stray capacitance, a suitable value of C_1 should be selected first, and the resistance of R_1 should then be calculated from Equation 17-1. Selecting C_1 first is the normal procedure when using a BIFET op-amp or any op-amp with a very low input bias current. The op-amp used in the filter circuit must have a suitable frequency response; its unity gain upper cutoff frequency must be much higher than the highest signal frequency to be passed. Resistor R_2 is made equal to R_1 to equalize the resistances at the op-amp input terminals (to avoid dc offset).

Example 17-2

Using a 741 op-amp, design a first-order active low-pass filter to have a cutoff frequency of 1 kHz. (See Fig. 17-4.)

Solution

Eq. 14-1: $R_1 = \dfrac{70\ \text{mV}}{I_{B(max)}} = \dfrac{70\ \text{mV}}{500\ \text{nA}}$

$= 140\ \text{k}\Omega$ (use 121 kΩ ±1% standard value)

$R_2 \approx R_1 = 121\ \text{k}\Omega$ (use 120 kΩ ±10% standard value)

Eq. 17-1: $C_1 = \dfrac{1}{2\pi R_1 f_c} = \dfrac{1}{2 \times \pi \times 121\ \text{k}\Omega \times 1\ \text{kHz}}$

$= 1315\ \text{pF}$ (use 1300 pF standard value)

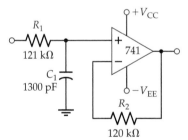

Figure 17-4 First-order low-pass active filter designed in Example 17-2.

First-Order High-Pass Filter

The *first-order active high-pass filter* circuit in Fig. 17-5a is made up of a passive high-pass circuit (C_1 and R_1) and a voltage follower (*buffer amplifier*). The A_v/f and ϕ/f responses shown in Fig. 17-5b are exactly the same as those for an unloaded passive high-pass filter. At high signal frequencies, X_{C1} is very much smaller than R_1, so that there is virtually zero attenuation of the input signal. When the signal frequency is reduced to the cutoff frequency (f_c), X_{C1} equals R_1 and there is a 3 dB drop in signal voltage across C_1. The voltage gain rolls off at 20 dB per decade at frequencies below f_c. The ϕ/f response shows a 90° phase lead at low signal frequencies, decreasing to a maximum of +45° at f_c, then approaching zero in the pass band. The bandwidth for a high-pass active filter is calculated as BW = ($f_2 - f_c$), where f_2 is the amplifier upper cutoff frequency. In the case of a voltage follower, f_2 is the op-amp unity gain frequency (f_u).

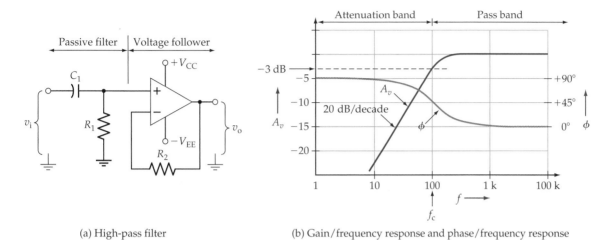

(a) High-pass filter

(b) Gain/frequency response and phase/frequency response

Figure 17-5 First-order active high-pass filter using an operational amplifier as a voltage follower to isolate the RC passive filter from the load. Note that the A_v/f response graph shows that the filter passes all signal frequencies above the cutoff frequency f_c, and the ϕ/f graph shows a 45° phase lead at f_c.

The procedure for designing a first-order active high-pass filter circuit is similar to that for a low-pass filter, except that a low cutoff frequency (f_c) is involved instead of a high cutoff frequency.

Example 17-3

Design a first-order high-pass active filter circuit to have a cutoff frequency of 5 kHz. Use a 108 op-amp, and determine the filter bandwidth.

Solution

Because the 108 has an extremely low input bias current (see data sheet A-14 in Appendix A), it should be treated as a BIFET op-amp. Start by selecting a capacitance value for C_1 very much larger than stray capacitance. (See Fig. 17-6.)

Select $\qquad C_1 = 1000 \text{ pF}$

From Eq. 17-1, $\quad R_1 = \dfrac{1}{2\pi f_c C_1} = \dfrac{1}{2 \times \pi \times 5 \text{ kHz} \times 1000 \text{ pF}}$

$\qquad\qquad = 31.8 \text{ k}\Omega \text{ (use } 31.6 \text{ k}\Omega \pm 1\% \text{ standard value)}$

$\qquad\qquad R_2 \approx R_1 = 31.6 \text{ k}\Omega \text{ (use } 33 \text{ k}\Omega \pm 10\% \text{ standard value)}$

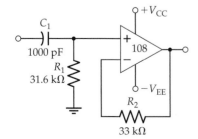

Figure 17-6 First-order high-pass active filter designed in Example 17-3.

The 108 gain/frequency response graph in data sheet A-14 in Appendix A shows that the op-amp unity gain frequency $f_u \approx 1$ MHz when $C_f = 30$ pF. (Maximum frequency compensation must be used for a voltage follower, see Section 15-1.)

$$BW = f_u - f_c = 1 \text{ MHz} - 5 \text{ kHz}$$
$$= 995 \text{ kHz}$$

Recall from Chapter 15 that the above determination of unity gain frequency refers only to op-amps used in small-signal applications. Usually, the peak-to-peak output should be less than 1 V. When larger output voltages are involved, the op-amp full power bandwidth must be used (see Section 15-3).

Practice Problems

17-2.1 Design a first-order active low-pass filter to have a cutoff frequency of 15 kHz. Use a BIFET op-amp.

17-2.2 Calculate the bandwidth for a first-order active high-pass filter that uses a 741 op-amp and has $R_1 = 330 \ \Omega$ and $C_1 = 0.01 \ \mu F$.

17-3 HIGHER-ORDER ACTIVE FILTERS

Falloff Rate

As well as classification according to the pass band, higher-order filters are distinguished by the falloff rate of the gain/frequency response graph. Figure 17-7 shows three different A_v/f response graphs for a low-pass filter, which fall off at different rates above the cutoff frequency. As already discussed, the falloff

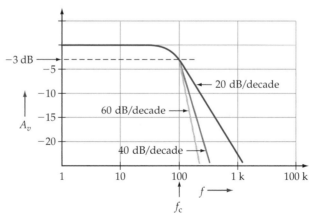

Figure 17-7 A_v/f graphs for first-order, second-order, and third-order active low-pass filters. The gain of a first-order filter falls off (or rolls off) at 20 dB/decade in the transition band. A second-order filter gain falls off at 40 dB/decade. The rate for a third-order filter is 60 dB/decade.

rate for a first-order filter is *20 dB/decade (6 dB/octave)*. Because a first-order filter has only one reactive component (a capacitor or inductor) in its circuit, it is
referred to as a *single-pole filter*. Second-order filters are termed *two-pole filters*
because they have two reactive components, and this gives a falloff rate of
40 dB/decade (12 dB/octave). For a *third-order* filter (*three-pole*), the falloff rate is
60 dB/decade (18 dB/octave). A steeper falloff rate means that the A_v/f response
graph approaches the ideal (vertical straight line) more closely, so that the filter
is better at rejecting unwanted frequencies.

Filter Design Categories

The design and analysis of high-performance filter circuits can be extremely
complex, and it is the subject of many books. There are several major approaches to filter circuit design which give different performance characteristics, each with advantages and disadvantages. These produce filters that are
variously known as *Bessel, Butterworth, Chebyshev,* and *Elliptic* (also termed
Cauer). The amplitude/frequency responses in Fig. 17-8 illustrate the major
differences between the filter types. The Bessel filter has a response that rolls
off relatively slowly from the pass band into the transition band. The Butterworth filter has a flat pass band response and a sharper corner than the Bessel
at the beginning of the transition band. Chebyshev and Elliptic filter responses
move sharply from the pass band into the transition band, and both have
steeper initial falloff rates than the other two. However, as illustrated, the
Chebyshev and Elliptic filter have *ripple*, or uneven amplitude response within
the pass band. The Elliptic also has ripple within the stop-band. Such filters
can be designed for a specified ripple amplitude.

Phase/frequency response can be a major consideration in the selection of a
filter circuit. When input signals containing many frequency components
are involved, it is usually very important that the phase relationships between
the component frequencies remain substantially unchanged by the filter. The
Bessel filter gives the most linear phase response.

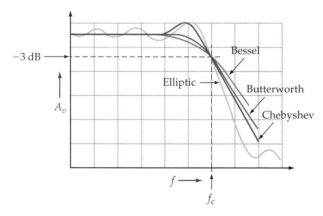

Figure 17-8   Different approaches to the design of active filters yield different shapes of A_v/f
response graph. The Elliptic filter response rolls off most steeply in the transition band but is
not flat in the pass band.

As already mentioned, filter design can be extremely complex. However, designing for a Butterworth response can be amazingly simple for circuits that use a voltage follower. This is demonstrated in following sections.

Second-Order Low-Pass Active Filter

Figure 17-9a shows a *second-order low-pass filter* circuit (*two-pole*) that has an A_v/f response falloff rate of 40 dB per decade above the cutoff frequency (Fig. 17-9b). This is achieved by using the C_1R_2 section, together with feedback from the output via capacitor C_2 to the junction of R_1 and R_2. The circuit is know as a *Sallen-Key* filter and also as a *voltage controlled voltage source (VCVS)* filter.

At low frequencies, X_{C1} and X_{C2} are very much larger than R_1 and R_2; consequently, they have no significant effect on the circuit. Input signals are reproduced at the output with zero attenuation. At high frequencies, when the capacitive impedances are very much smaller than the resistance values, C_1 effectively shorts the voltage follower input to ground, producing ground

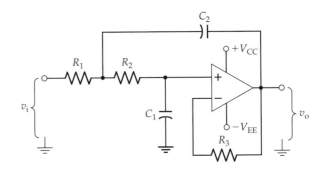

(a) −40 dB/decade low-pass filter

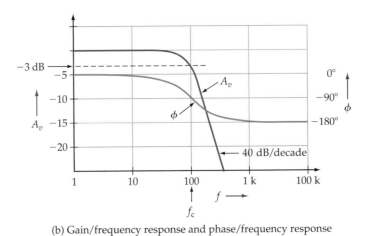

(b) Gain/frequency response and phase/frequency response

Figure 17-9 Second-order low-pass active filter with the A_v/f and ϕ/f response graphs. The gain falls off at 40 dB/decade above the cutoff frequency, and there is a 90° phase lag at f_C.

level at the output, and C_2 shorts the junction of R_1 and R_2 to the (ground-level) output. Consequently, input signals are severely attenuated.

In the transition band of the A_v/f response graph, C_1 and R_2 produce a 20 dB per decade falloff rate as the frequency increases, and feedback via C_2 and $R_1 \| R_2$ produces a further falloff (or *roll-off*) of 20 dB per decade. The two attenuating effects combine to produce a (*two-pole*) falloff rate of 40 dB per decade, as illustrated in Fig. 17-9b. Note that the phase shift from input to output decreases from zero at low frequencies to a 90° lag at f_c, and to a maximum of $-180°$ at frequencies above f_c.

The cutoff frequency for the a second-order low-pass filter can be shown to be

$$f_c = \frac{1}{2\pi \sqrt{(R_1 R_2 C_1 C_2)}} \tag{17-4}$$

A very simple approach can be taken to the design of a -40 dB per decade Butterworth low-pass filter. The resistance of $R_1 + R_2$ is first determined in the usual way (for minimum voltage drop caused by the op-amp input bias current). Alternatively, the capacitance of C_2 can first be selected to be much larger than stray capacitance. Then, either R_1 or C_2 is calculated as follows:

$$X_{C1} = \sqrt{2}\, R_2 \text{ at } f_c \tag{17-5}$$

Next,

$$C_1 = 2C_2 \tag{17-6}$$

and

$$R_1 = R_2 = 0.5(R_1 + R_2) \tag{17-7}$$

which gives

$$R_1 = \sqrt{2}\, X_{C2} \text{ at } f_c \tag{17-8}$$

and

$$R_3 \approx R_1 + R_2 \tag{17-9}$$

The effect of selecting the components in this way ($X_{C1} = \sqrt{2}\,R_2$ and $R_1 = \sqrt{2}\,X_{C2}$) produces a total attenuation of 3 dB at the desired cutoff frequency. It should be noted that *the above design procedure applies only to circuits that use a voltage follower, not to circuits that have the op-amp connected as a noninverting amplifier*. As always, the op-amp selected must have a suitable frequency response, and the capacitance values must be much larger than stray capacitance.

Example 17-4

Design a Butterworth second-order low-pass filter, as in Fig. 17-10, to have a cutoff frequency of 1 kHz. Use the selected components to calculate the actual cutoff frequency for the circuit.

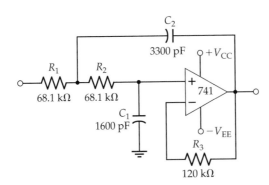

Figure 17-10 Second-order lowpass active filter circuit designed in Example 17-4.

Solution

The frequency response of the 741 op-amp extends to almost 800 kHz for unity gain (see Fig. 15-3). The 741 is suitable.

Eq. 14-1: $R_1 + R_2 = \dfrac{70 \text{ mV}}{I_{B(max)}} = \dfrac{70 \text{ mV}}{500 \text{ nA}}$

$= 140 \text{ k}\Omega$

Eq. 17-7: $R_1 = R_2 = 0.5(R_1 + R_2) = 0.5 \times 140 \text{ k}\Omega$

$= 70 \text{ k}\Omega$ (use 68.1 kΩ ±1% standard value)

Eq. 17-9: $R_3 \approx R_1 + R_2 = 2 \times 68.1 \text{ k}\Omega$

$= 136.2 \text{ k}\Omega$ (use 120 kΩ ±10% standard value)

Eq. 17-5: $X_{C1} = \sqrt{2}\, R_2$ at f_c

or $C_1 = \dfrac{1}{2\pi f_c \sqrt{2}\, R_2} = \dfrac{1}{2 \times \pi \times 1 \text{ kHz} \times \sqrt{2} \times 68.1 \text{ k}\Omega}$

$= 1653 \text{ pF}$ (use 1600 pF standard value)

Eq. 17-6: $C_2 = 2C_1 = 2 \times 1600 \text{ pF}$

$= 3200 \text{ pF}$ (use 3300 pF standard value)

Eq. 17-4: $f_c = \dfrac{1}{2\pi \sqrt{(R_1 R_2 C_1 C_2)}}$

$= \dfrac{1}{2\pi \sqrt{(68.1 \text{ k}\Omega \times 68.1 \text{ k}\Omega \times 1600 \text{ pF} \times 3300 \text{ pF})}}$

$= 1.02 \text{ kHz}$

Because the selected components values in any filter design are unlikely to be exactly as calculated, the circuit cutoff frequency cannot be expected to be exactly as specified. Even if all the calculations produce standard value components, the component tolerances are likely to affect the cutoff frequency. To achieve a desired cutoff frequency more precisely in any filter circuit, low-tolerance components must be used, and some of the components may have to be partially adjustable.

Practice Problems

17-3.1 A second-order active low-pass filter is to have a 1.5 kHz cutoff frequency. Design a Butterworth circuit using a BIFET op-amp, and calculate the actual cutoff frequency after the components are selected.

17-3.2 A filter is required to attenuate all signal frequencies above 12 kHz by a minimum of 15 dB. Design a suitable second-order Butterworth filter, and use the selected components to calculate the actual cutoff frequency.

17-4 SECOND-ORDER HIGH-PASS ACTIVE FILTERS

The second-order (-40 dB per decade) highpass Sallen-Key filter circuit in Fig. 17-11a is similar to the low-pass filter in Fig. 17-9a, except that resistor and capacitor positions are interchanged. At high frequencies, X_{C1} and X_{C2} are so

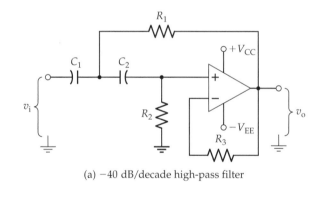

(a) -40 dB/decade high-pass filter

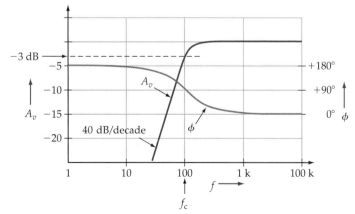

(b) Gain/frequency response and phase/frequency response

Figure 17-11 Second-order high-pass active filter with the A_v/f and ϕ/f response graphs. The gain falls off at 40 dB/decade at frequencies below the cutoff frequency, and there is a 90° phase lead at f_c.

much smaller than R_1 and R_2 that they behave like short-circuits, passing the input signal unattenuated to the voltage follower. At low frequencies, when the capacitive impedances are very much larger than the resistance values, the combined effects of C_1R_1 and C_2R_2 produce severe attenuation of the input. In the transition band, C_2 and R_2 cause a 20 dB per decade falloff rate as the frequency decreases below f_c, and feedback via R_1 and $C_1\|C_2$ produces a further falloff of 20 dB per decade. The two attenuating effects combine to produce a 40 dB per decade (two-pole) falloff rate, as illustrated in Fig. 17-11b. The phase shift from input to output approximates $+180°$ at low frequencies, decreases to a 90° lead at f_c, and approaches zero at frequencies above f_c.

Like the low-pass two-pole circuit, a very simple design procedure can be used for a -40 db per decade Butterworth high-pass filter. Resistor R_2 is first determined from Equation 14-1 or else capacitor C_2 is chosen to be much larger than stray capacitance. Then the other component values are calculated from the following equations:

$$R_2 = \sqrt{2}\, X_{C2} \text{ at } f_c \tag{17-10}$$

$$C_1 = C_2 \tag{17-11}$$

$$R_1 = 0.5R_2 \tag{17-12}$$

and

$$R_3 \approx R_2 \tag{17-13}$$

When the component values are known, Equation 17-4 can be used to calculate the filter cutoff frequency.

As in the case of the first-order high-pass filter, the bandwidth for a second-order high-pass filter is calculated as $BW = f_2 - f_c$. Here f_2 is the amplifier upper cutoff frequency, which for a voltage follower is the op-amp unity gain frequency (f_u).

Example 17-5

Using a BIFET op-amp, design a Butterworth second-order high-pass filter circuit, as in Fig. 17-12, to have a cutoff frequency of 12 kHz. Using the selected component values, calculate the actual cutoff frequency for the circuit.

Solution

Select $C_1 = 1000$ pF

Eq. 17-11: $C_2 = C_1 = 1000$ pF

Eq. 17-10: $R_2 = \dfrac{\sqrt{2}}{2\pi f_c C_1} = \dfrac{\sqrt{2}}{2 \times \pi \times 12 \text{ kHz} \times 1000 \text{ pF}}$

$= 18.75 \text{ k}\Omega \text{ (use 18.7 k}\Omega \pm 1\% \text{ standard value)}$

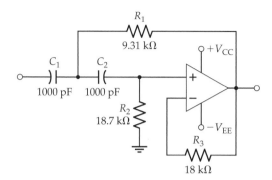

Figure 17-12   Second-order highpass active filter circuit designed in Example 17-5.

Eq. 17-12: $R_1 = 0.5\,R_2 = 0.5 \times 18.7\text{ k}\Omega$

   $= 9.35\text{ k}\Omega$ (use 9.31 kΩ ±1% standard value)

Eq. 17-13: $R_3 \approx R_2 = 18.7\text{ k}\Omega$ (use 18 kΩ ±10% standard value)

Eq. 17-4: $f_c = \dfrac{1}{2\pi \sqrt{(R_1 R_2 C_1 C_2)}}$

   $= \dfrac{1}{2\pi \sqrt{(9.31\text{ k}\Omega \times 18.7\text{ k}\Omega \times 1000\text{ pF} \times 1000\text{ pF})}}$

   $= 12.06\text{ kHz}$

Practice Problems

17-4.1 A second-order active high-pass filter is to have a 600 Hz cutoff frequency. Design a suitable Butterworth circuit using a 741 op-amp, and calculate the actual cutoff frequency from the selected components.

17-4.2 A filter is to attenuate all signal frequencies below 6.25 kHz by a minimum of 27 dB. Design a suitable second-order Butterworth filter.

17-5 THIRD-ORDER ACTIVE FILTERS

Third-Order Low-Pass filter

The low-pass filter circuit in Fig. 17-13 consists of two stages: a −20 dB per decade (first-order) stage and a −40 dB per decade (second-order) stage. The complete circuit is known as a *third-order low-pass filter*. The overall gain of the circuit falls off at a rate of −60 dB per decade at frequencies above the cutoff frequency. Each stage of the circuit is designed independently in the usual way. However, if the gain of *stage 1* is down by 3 dB at f_c, and the gain of *stage 2* is also down by 3 dB at f_c, then the overall circuit gain is 6 dB below its pass band level. This cannot be allowed, of course, because f_c is defined as the frequency at which the circuit gain is 3 dB down from the pass band gain. So, each stage

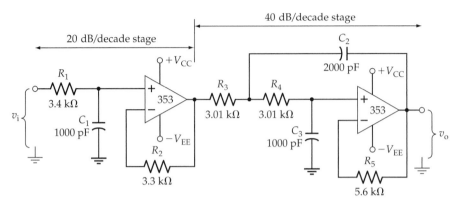

Figure 17-13 A third-order low-pass filter can be constructed by cascading first-order and second-order stages. The gain falloff rate above the cutoff frequency is 60 dB/decade.

must be designed to have an attenuation of 1.5 dB at the specified cutoff frequency. For a low-pass filter, this can be achieved by using the following cutoff frequencies:

For the first-order stage:
$$f_{c1} = \frac{f_c}{0.65}$$
(17-14)

For the second-order stage:
$$f_{c2} = \frac{f_c}{0.8}$$
(17-15)

Example 17-6

Design a third-order low-pass filter, as in Fig. 17-13, to have a cutoff frequency of 30 kHz. Use BIFET op-amps.

Solution

−20 dB per decade stage:

Select $\qquad C_1 = 1000$ pF

Eq. 17-14: $\qquad f_{c1} = \dfrac{f_c}{0.65}$

Eq. 17-1: $\qquad R_1 = \dfrac{1}{2\pi f_{c1} C_1} = \dfrac{0.65}{2 \times \pi \times 30 \text{ kHz} \times 1000 \text{ pF}}$

$\qquad\qquad\qquad = 3.4 \text{ k}\Omega \text{ (use } 3.4 \text{ k}\Omega \pm 1\% \text{ standard value)}$

$\qquad\qquad R_2 \approx R_1 = 3.4 \text{ k}\Omega \text{ (use } 3.3 \text{ k}\Omega \pm 10\% \text{ standard value)}$

−40 dB per decade stage:

Select $\qquad C_3 = 1000$ pF

From Eq. 17-6, $\qquad C_2 = 2C_3 = 2 \times 1000$ pF

$\qquad\qquad\qquad = 2000 \text{ pF (standard value)}$

Eq. 17-15: $f_{c2} = \dfrac{f_c}{0.8}$

From Eq. 17-5, $X_{C3} = \sqrt{2}\, R_4$ at f_{c2}

This gives $R_4 = \dfrac{1}{2\pi f_{c2} C_3 \sqrt{2}} = \dfrac{0.8}{2 \times \pi \times 30 \text{ kHz} \times 1000 \text{ pF} \times \sqrt{2}}$

$= 3 \text{ k}\Omega$ (use 3.01 kΩ ±1% standard value)

$R_3 = R_4 = 3.01 \text{ k}\Omega$

$R_5 \approx R_3 + R_4$

$= 6.02 \text{ k}\Omega$ (use 5.6 kΩ ±10% standard value)

Third-Order High-Pass Filters

Figure 17-14 shows *third-order high-pass filter* (−60 db per decade) consisting of a −20 dB per decade stage and a −40 dB per decade stage. As in the case of the −60 dB per decade low-pass circuit, each stage should be designed to have an attenuation of 1.5 dB at the desired cutoff frequency. For a high-pass filter, this is achieved by designing each stage in the usual way, and using the following stage frequencies:

For the first-order stage, $f_{c1} = 0.65 f_c$ (17-16)

For the second-order stage, $f_{c2} = 0.8 f_c$ (17-17)

Example 17-7

Design a third-order high-pass filter as in Fig. 17-14 to have a 20 kHz low cut-off frequency. Use 741 op-amps.

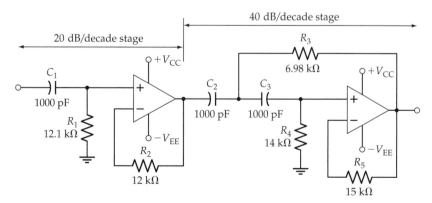

Figure 17-14 Third-order high-pass filter consisting of cascaded first-order and second-order stages. The circuit gain falls off at 60 dB/decade below the cutoff frequency.

Solution

−20 dB per decade stage:

Select $R_1 = 121\ \text{k}\Omega$ (as in Example 17-2)

Eq. 17-16: $f_{c1} = 0.65\,f_c$

Eq. 17-1: $C_1 = \dfrac{1}{2\pi f_{c1}R_1} = \dfrac{1}{2 \times \pi \times 0.65 \times 20\ \text{kHz} \times 120\ \text{k}\Omega}$

$= 102\ \text{pF}$

This is so small that it might be affected by stray capacitance. Redesign, first choosing a suitable capacitance for C_1.

Select $C_1 = 1000\ \text{pF}$

Eq. 17-1: $R_1 = \dfrac{1}{2\pi f_{c1}C_1} = \dfrac{1}{2 \times \pi \times 0.65 \times 20\ \text{kHz} \times 1000\ \text{pF}}$

$= 12.2\ \text{k}\Omega$ (use 12.1 kΩ ±1% standard value)

$R_2 \approx R_1 = 12.2\ \text{k}\Omega$ (use 12 kΩ ±10% kΩ standard value)

−40 dB per decade stage:

Select $C_3 = 1000\ \text{pF}$

Eq. 17-17: $f_{c2} = 0.8\,f_c$

Eq. 17-12: $R_4 = \dfrac{\sqrt{2}}{2\pi\,0.8\,f_c\,C_3} = \dfrac{\sqrt{2}}{2 \times \pi \times 0.8 \times 20\ \text{kHz} \times 1000\ \text{pF}}$

$= 14.06\ \text{k}\Omega$ (use 14 kΩ ±1% standard value)

Eq. 17-13: $C_2 = C_3 = 1000\ \text{pF}$

Eq. 17-10: $R_3 = 0.5\,R_4 = 0.5 \times 14\ \text{k}\Omega$

$= 7\ \text{k}\Omega$ (use 6.98 ±1% standard value)

$R_5 \approx R_4 = 14\ \text{k}\Omega$ (use 15 kΩ ±10% standard value)

Practice Problems

17-5.1 A third-order active low-pass filter is to have a 10 kHz cutoff frequency. Design a suitable Butterworth circuit using 741 op-amps.

17-5.2 A third-order active high-pass filter is to have a 6.6 kHz cutoff frequency. Design a suitable Butterworth circuit using BIFET op-amps.

17-6 BAND-PASS FILTERS

Multi-Stage Band-Pass Filter

A band-pass filter can be constructed simply by connecting low-pass and high-pass filters in cascade, as in Fig. 17-15a. For example, suppose a low-pass filter with f_c = 100 kHz is cascaded with a high-pass circuit which has f_c = 10 kHz, as illustrated by the A_v/f graph in Fig. 17-15b. The low-pass circuit will pass all frequencies up to 100 kHz, while the high-pass circuit will block all frequencies below 10 kHz. Consequently, the combination gives a filter with a pass band from 10 kHz to 100 kHz.

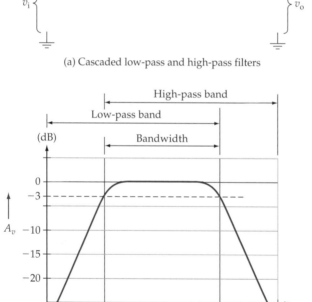

(a) Cascaded low-pass and high-pass filters

(b) Band-pass frequency response

Figure 17-15   A band-pass filter can be constructed by cascading low-pass and high-pass filters.

Single-Stage Band-Pass Filter

A single-stage band-pass filter circuit is shown in Fig. 17-16. If the capacitors were not present, the circuit would look like an inverting amplifier. The capacitors are selected to have X_{C2} large enough to be neglected at low frequencies, and X_{C1} small enough to be neglected at high frequencies.

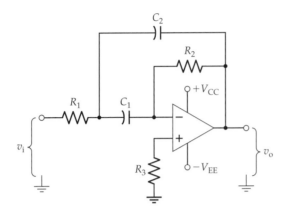

Figure 17-16 Single-stage band-pass filter circuit. Capacitor C_1 and resistor R_1 determine the filter low cutoff frequency (f_1), while R_2 and C_2 set the high cutoff frequency (f_2).

At low frequencies, X_{C2} is so large that it can be eliminated from the low-frequency equivalent circuit. As shown in Fig. 17-17a, the circuit is an inverting amplifier with a voltage gain of

$$A_v = \frac{R_2}{Z_1} = \frac{R_2}{\sqrt{(R_1^2 + X_{C1}^2)}} \tag{17-18}$$

At signal frequencies in the pass band of the circuit, X_{C1} becomes very much smaller than R_1. So the circuit voltage gain becomes

$$A_v \approx \frac{R_2}{R_1}$$

Equation 17-18 shows that for the voltage gain to be down by 3 dB (from the mid-frequency gain),

$$X_{C1} = R_1 \text{ at } f_1 \tag{17-19}$$

where f_1 is the low cutoff frequency.

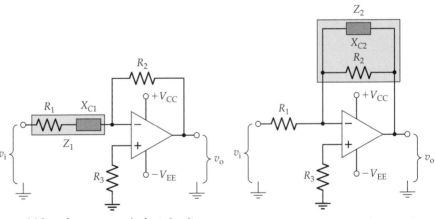

(a) Low-frequency equivalent circuit   (b) High-frequency equivalent circuit

Figure 17-17 Low-frequency and high-frequency equivalent circuits for a single-stage band-pass filter, showing how X_{C1} and X_{C2} affect the circuit gain.

At high frequencies, X_{C1} becomes so small compared to R_1 that C_1 can be eliminated from the high-frequency equivalent circuit. But X_{C2} is no longer large enough to be neglected. So the high frequency equivalent circuit is as shown in Fig. 17-17b. Once again, the circuit functions as an inverting amplifier, and its voltage gain is

$$A_v = \frac{X_{C2} \| R_2}{R_1} = \frac{R_2}{R_1\sqrt{[1 + (R_2/X_{C2})^2]}} \qquad \text{(17-20)}$$

At frequencies in the pass band, X_{C2} is very much larger than R_2, and so the circuit voltage gain is once again

$$A_v \approx \frac{R_2}{R_1}$$

From Equation 17-20, the voltage gain is down by 3 dB from its mid-frequency value when

$$X_{C2} = R_2 \text{ at } f_2 \qquad \text{(17-21)}$$

where f_2 is the high cutoff frequency.

It is seen that the circuit behaves as a high-pass filter for low signal frequencies, as an inverting amplifier for frequencies in the pass band, and as a low-pass filter for high frequencies. The low cutoff frequency is dictated by Equation 17-19, and the high cutoff frequency depends on Equation 17-21. This circuit is essentially a wide-band filter. The high cutoff frequency should typically be 10 or more times the low cutoff frequency.

The normal design approach of determining resistor values first is likely to give unacceptably small capacitance values whether bipolar or BIFET op-amps are involved. So it is usually best to begin by choosing the value of the smallest capacitor (C_2). Then R_2 can be determined from Equation 17-21, and R_1 can be calculated in relation to R_2 for the desired voltage gain. The value of R_1 can be substituted into Equation 17-19 to calculate C_1. Because direct current through R_1 is interrupted by C_1, resistor R_3 should be made equal to R_2 for resistance equality at the op-amp input terminals.

Example 17-8

Design a single-stage band-pass filter, as in Fig. 17-18, to have unity voltage gain and a pass band from 300 Hz to 30 kHz.

Solution

Select $C_2 = 1000 \text{ pF}$

Eq. 17-21: $X_{C2} = R_2 \text{ at } f_2$

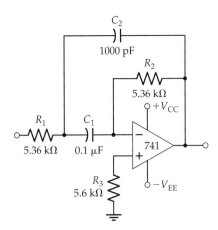

Figure 17-18 Single-stage band-pass filter circuit designed in Example 17-8.

giving
$$R_2 = \frac{1}{2\pi f_2 C_2} = \frac{1}{2 \times \pi \times 30 \text{ kHz} \times 1000 \text{ pF}}$$

$$= 5.3 \text{ k}\Omega \text{ (use 5.36 k}\Omega \pm 1\% \text{ standard value)}$$

For $A_v = 1$, $R_1 = R_2 = 5.36 \text{ k}\Omega$

Eq. 17-19: $X_{C1} = R_1$ at f_1,

giving
$$C_1 = \frac{1}{2\pi f_1 R_1} = \frac{1}{2 \times \pi \times 300 \text{ Hz} \times 5.36 \text{ k}\Omega}$$

$$= 0.1 \ \mu\text{F (standard value)}$$

$$R_3 \approx R_2 = 5.36 \text{ k}\Omega \text{ (use 5.6 k}\Omega \pm 10\% \text{ standard value)}$$

Bandwidth

As already stated, the band-pass filter discussed above is essentially a wide-band circuit. A typical wide-band frequency response is illustrated in Fig. 17-19a. Narrow-band band-pass filters have the kind of frequency response shown in Fig. 17-19b. In all cases the bandwidth is

$$BW = f_2 - f_1 \qquad\qquad (17\text{-}22)$$

The circuit Q *factor* is a figure of merit for a filter circuit. It defines the selectivity of the filter in passing the centre frequency and rejecting other frequencies. The Q factor is the relationship of the centre frequency f_o to the bandwidth:

$$Q = \frac{f_o}{BW} \qquad\qquad (17\text{-}23)$$

Figure 17-19b shows that a filter with a Q of 10 has a much narrower bandwidth than a filter with Q equal to 1. Narrow-band filters are usually classified as those with a Q factor greater than 5, while wide-band circuits have a Q less than 5. The centre frequency of the filter can be determined from

$$f_o = \sqrt{(f_1 f_2)} \qquad\qquad (17\text{-}24)$$

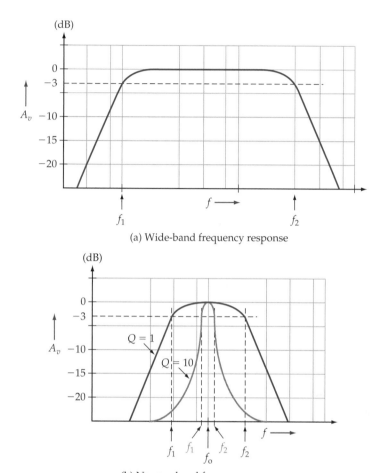

(a) Wide-band frequency response

(b) Narrow-band frequency responses

Figure 17-19   Wide-band and narrow-band frequency responses for band-pass filters. The narrowest bands are achieved by filters with the highest *Q* factor.

Example 17-9

Calculate the Q factor for the wide-band filter designed in Example 17-8.

Solution

Eq. 17-24:
$$f_o = \sqrt{(f_1 f_2)} = \sqrt{(300 \text{ Hz} \times 30 \text{ kHz})}$$
$$= 3 \text{ kHz}$$

Eq. 17-22:
$$BW = f_2 - f_1 = 30 \text{ kHz} - 300 \text{ Hz}$$
$$= 29.7 \text{ kHz}$$

Eq. 17-23:
$$Q = \frac{f_o}{BW} = \frac{3 \text{ kHz}}{29.7 \text{ kHz}}$$
$$\approx 0.1$$

Narrow-Band Single-Stage Band-Pass Filter

If the circuit in Fig. 17-16 is reconsidered, it is seen that if f_1 and f_2 are brought closer together to narrow the bandwidth, C_2 will affect the low cutoff frequency, and C_1 will interfere with the high cutoff frequency. To counter this, another resistor (R_4) is included at the circuit input, as shown in Fig. 17-20. This modification produces larger Q factors and narrower bandwidths. The additional resistor makes the circuit analysis quite complex; however, the following equations can be shown to relate the component values to the circuit performance:

$$C_1 = C_2 = C \tag{17-25}$$

$$R_1 = 0.5\, R_2 \tag{17-26}$$

$$R_2 = 2Q\, X_C \text{ at } f_o \tag{17-27}$$

$$Q^2 = \frac{R_1 + R_4}{2R_4} \tag{17-28}$$

Design of the circuit is approached in the usual way by first selecting a suitable resistance value for R_2 (considering the op-amp input bias current), and then calculating C. As always, if this produces a capacitance value that might be affected by stray capacitance, then a suitable value for C is first selected, and R_2 is determined from C. Other components are then determined from the appropriate equations.

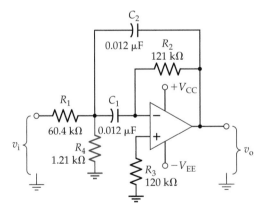

Figure 17-20 Modified single-stage band-pass filter. The addition of resistor R_4 allows the circuit to be designed for a narrow-band A_v/f response.

Example 17-10

Calculate the centre frequency and bandwidth of the band-pass filter in Fig. 17-20.

Solution

Eq. 17-28:

$$Q^2 = \frac{R_1 + R_4}{2R_4} = \frac{60.4\text{ k}\Omega + 1.21\text{ k}\Omega}{2 \times 1.21\text{ k}\Omega}$$

$$= 25.46$$

$$Q = \sqrt{25.46} = 5.05$$

From Eq. 17-27, $f_o = \dfrac{Q}{\pi C R_2} = \dfrac{5.05}{\pi \times 0.012\ \mu F \times 121\ k\Omega}$

$$= 1.1\ kHz$$

Eq. 17-23: $BW = \dfrac{f_o}{Q} = \dfrac{1.1\ kHz}{5.05}$

$$= 218\ Hz$$

Practice Problems

17-6.1 A single-stage band-pass filter, as in Fig. 17-16, has the following component values: $R_1 = R_2 = 7.5\ k\Omega$, $R_3 = 6.8\ k\Omega$, $C_1 = 8200\ pF$, and $C_2 = 750\ pF$. Determine the circuit bandwidth, centre frequency, and Q factor.

17-6.2 The band-pass circuit in Fig. 17-20 is to have a 1 kHz centre frequency and a 200 Hz bandwidth. Using a 741 op-amp, determine the required component values.

17-7 NOTCH FILTERS

A *notch filter* is the inverse of a band-pass filter; its function is to block a band of signal frequencies. Other names are *band-stop* and *band-reject filter*.

Figure 17-21a shows how a notch filter can be constructed by the use of low-pass and high-pass filters. The circuit inputs are connected in parallel, and the outputs are applied to a summing circuit (see Section 14-6). The summing circuit is necessary because the filters would overload each other if the outputs were directly connected in parallel. Suppose that the low-pass circuit has a cutoff frequency of 10 kHz and that the cutoff frequency of the high-pass circuit is 100 kHz. The low-pass circuit will attenuate signal frequencies above 10 kHz, while the high-pass circuit will attenuate frequencies below 100 kHz. The result is a stop band of 10 kHz to 100 kHz, as shown in Fig. 17-21b. Note that the frequency limits of the stop band are those frequencies at which the signal is -3 dB from its normal output level. The low-pass and high-pass filters can be first-order, second-order, or higher, for any desired roll-off rate of the notch frequency response.

Another method of creating a notch filter is to sum the output of a band-pass filter with its own input signal, as shown in Fig. 17-22. In this case, the band-pass filter must have a voltage gain of 1, and it must invert the input signal (or else an inverting amplifier must be included). During the pass band of the band-pass circuit, its output v_{BP} equals $-v_i$, and so the two inputs to the summing circuit cancel each other, giving zero output voltage. Above and below the pass band of the band-pass circuit, v_{BP} is very much smaller than v_i. This causes the summing circuit output to equal v_i, and the combination has a notch frequency response.

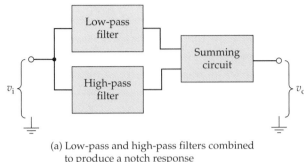

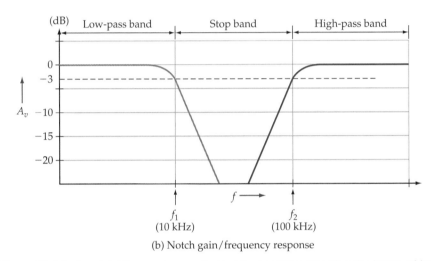

(a) Low-pass and high-pass filters combined
to produce a notch response

(b) Notch gain/frequency response

Figure 17-21 A notch filter can be constructed by parallel-connecting the inputs of low-pass and high-pass filters, and summing the outputs.

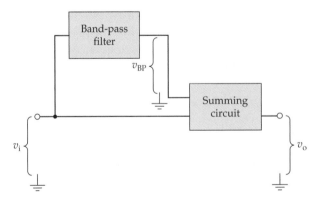

Figure 17-22 A notch filter can be constructed by summing the input and output of a band-pass filter. The notch filter cutoff frequencies are the same as those of the band-pass filter.

The band-pass filter circuit in Fig. 17-18 (designed in Example 17-8) can be converted into a single-stage notch filter by the addition of resistor R_4, as shown in Fig. 17-23. R_4 should be chosen to be equal to R_1 and R_2, and resistor R_3 should also be made precisely equal to R_2 (instead of approximately equal). The circuit now functions as a *difference amplifier* (see Section 14-7) with only

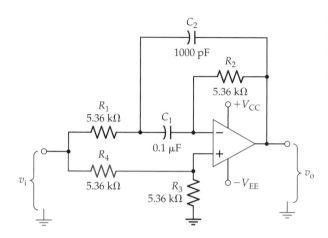

Figure 17-23 A single-stage notch filter can be constructed by the addition of a resistor (R_4) to a single-stage band-pass filter.

one input (v_i). For signal frequencies between f_1 and f_2, the input voltage (applied to both R_1 and R_4) produces positive-going and negative-going equal outputs that cancel each other. For signals above and below the cutoff frequencies, input voltages applied via R_1 are attenuated, and those applied via R_4 are passed to the output. Thus, the circuit functions as a notch filter.

Practice Problem

17-7.1 A single-stage notch filter, as in Fig. 17-23, is to be designed to have a stop band ranging from 200 Hz to 20 kHz. Determine suitable component values for the circuit.

17-8 ALL-PASS FILTERS

Phase-Lag Circuit

An *all-pass filter* is a circuit that will pass all frequencies without attenuation but with phase shifts that vary with the signal frequency. A phase lag can also be termed a phase delay, so an all-pass phase-lag circuit is also known as a *delay filter*. Figure 17-24a shows an all-pass circuit, and Fig. 24b shows its phase/frequency response graph. The circuit is similar to a difference amplifier (see Section 14-7) except that capacitor C_1 replaces a resistor. Also, the two input terminals are connected together so that v_i is applied to both. The procedure used in the analysis of the difference amplifier can be used to investigate the all-pass filter. With the inputs separated, terminal 2 grounded, and v_i applied to terminal 1:

$$v_{o1} = -\frac{R_2}{R_1} \times v_i$$

With $R_1 = R_2$,
$$v_{o1} = -v_i$$

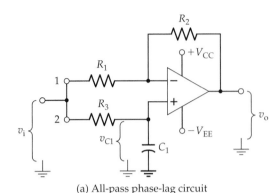

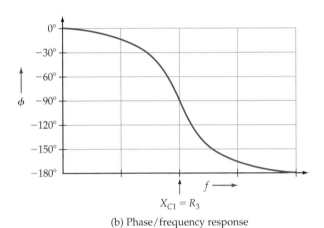

(a) All-pass phase-lag circuit

Figure 17-24 An all-pass phase-lag circuit passes the input signals with no attenuation but with a phase lag dependent on the signal frequency.

(b) Phase/frequency response

With terminal 1 grounded, and v_i applied to terminal 2,

$$v_{o2} = \frac{R_2 + R_1}{R_1} \times v_{C1}$$

With $R_1 = R_2$, $v_{o2} = 2v_{C1}$

$$v_{C1} = \frac{v_i(-jX_{C1})}{R_3 - jX_{C1}} = \frac{v_i}{1 + j(R_3/X_{C1})}$$

and $v_o = v_{o1} - v_{o2}$

which gives $v_o = v_i\underline{/-2\,\tan^{-1}\,(R_3/X_{C1})}$ (17-29)

It is seen that the output voltage amplitude is always equal to the input, regardless of the signal frequency. Also, the output lags behind the input by an angle of

$$\phi = -2\,\tan^{-1}\,(R_3/X_{C1})\qquad(17\text{-}30)$$

When $X_{C1} = R_3$, $\phi = -90°$

At high signal frequencies, when $X_{C1} \ll R_3$,

$$\phi \approx -180°$$

At low signal frequencies, when $X_{C1} \gg R_3$,

$$\phi \approx 0°$$

The above quantities are illustrated by the ϕ/f response graph in Fig. 17-24b.

   An all-pass circuit is designed by first choosing suitable values for resistors R_1 and R_2 as for an inverting amplifier with a gain of 1. This requires $R_1 = R_2$. Resistor R_3 is then selected to be approximately equal to $R_1 \| R_2$, and C_1 is calculated to give $X_{C1} = R_3$ at the desired 90° frequency. If an uncompensated op-amp is used, it should be compensated as an inverting amplifier with a gain of 1.

Example 17-11

Using a 741 op-amp, design an all-pass filter to have a phase lag adjustable from 80° to 100°. The signal amplitude is 1 V, and its frequency is 5 kHz.

Solution

See Fig. 17-25.

$$I_1 \gg I_{B(max)}$$

Select
$$I_1 = 50 \ \mu A$$

$$R_1 = \frac{v_i}{I_1} = \frac{1 \ V}{50 \ \mu A}$$

$$= 20 \ k\Omega \ (\text{use } 20 \ k\Omega \pm 1\% \text{ standard value})$$

$$R_2 = R_1 = 20 \ k\Omega$$

$$R_3 \approx R_1 \| R_2 = 10 \ k\Omega$$

For a 90° phase shift,

$$X_{C1} = R_3$$

or
$$C_1 = \frac{1}{2\pi f R_3} = \frac{1}{2 \times \pi \times 5 \ kHz \times 10 \ k\Omega}$$

$$= 3183 \ pF \ (\text{use } 3300 \ pF \text{ standard value})$$

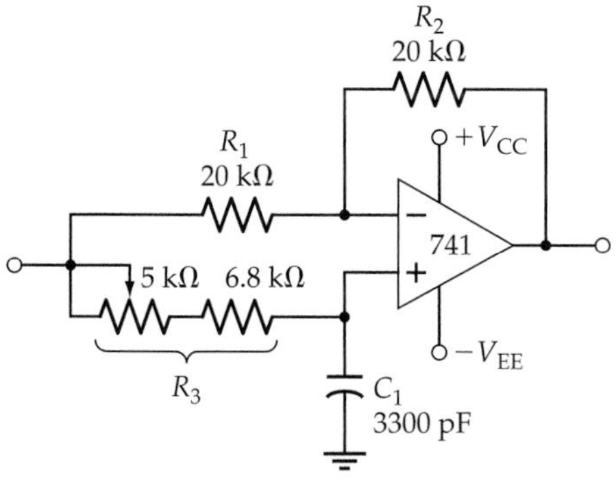

Figure 17-25 All-pass phase-lag circuit designed in Example 17-11.

For an 80° phase shift,

Eq. 17-30: $\phi = -2 \arctan (R_3/X_{C1})$

or $R_3 = \dfrac{\tan(\phi/2)}{2\pi f C_1} = \dfrac{\tan(80°/2)}{2 \times \pi \times 5 \text{ kHz} \times 3300 \text{ pF}}$

$\approx 8.1 \text{ k}\Omega$

For a 100° phase shift,

$R_3 = \dfrac{\tan(\phi/2)}{2\pi f C_1} = \dfrac{\tan(100°/2)}{2 \times \pi \times 5 \text{ kHz} \times 3300 \text{ pF}}$

$\approx 11.5 \text{ k}\Omega$

For R_3, use a 6.8 kΩ fixed value ±10% resistor in series with a 5 kΩ variable resistor to give a total resistance adjustable from 6.8 kΩ to 11.8 kΩ.

Phase-Lead Circuit

The phase-lead circuit in Fig. 17-26 is similar to the phase-lag circuit, except that R_3 and C_1 are interchanged. This has the effect of causing the output to lead the input by a phase angle dependent on the relationship between X_{C1} and R_3. The design procedure for this circuit is similar to that for the phase-lag circuit.

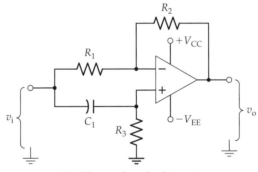

(a) All-pass phase-lead circuit

Figure 17-26 An all-pass phase-lead circuit passes the input signals with no attenuation but with a phase lead dependent on the signal frequency.

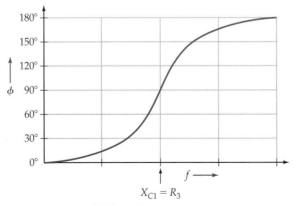

$X_{C1} = R_3$

(b) Phase/frequency response

Practice Problem

17-8.1 An all-pass filter using a BIFET op-amp is to have a phase lead adjustable from 75° to 90°. The input signal has 1 V amplitude and a frequency of 1 kHz. Determine suitable component values for the circuit.

17-9 STATE-VARIABLE FILTERS

The circuit shown in Fig. 17-27, known as a *state-variable filter,* consists of amplifier A_1 and two integrators, A_2 and A_3. Another arrangement of the circuit is named a *biquad filter.* Both are termed *universal filters* because they can function as low-pass, high-pass, band-pass, all-pass, or notch filters. The circuit in Fig. 17-27 is intended to operate as a band-pass filter with the output taken from A_2.

Amplifier A_1, together with resistors R_1, R_2, R_3, and R_4, functions as a difference amplifier. (In a slightly different circuit configuration, A_1 is connected as a summing amplifier.) A_2 and A_3 and their associated components are integrators, which introduce 90° phase shifts. The input to A_1 is applied via R_1 to its non-inverting terminal, but A_1 output is inverted by A_2 and fed back to A_1 via resistor R_2. Consequently, A_1 and A_2 together with R_1 and R_2 constitute an inverting amplifier, which is effective only for signal frequencies close to the filter centre frequency.

At low frequencies, capacitors C_1 and C_2 have high impedances compared to the resistances of R_5 and R_6, and so the stage gains of A_2 and A_3 are very large. In this case, the negative feedback from A_3 to R_4 tends to reduce the A_1 and A_2 outputs to nearly zero. At high-signal frequencies, the impedances of C_1 and C_2 are small compared to R_5 and R_6, so that the A_2 and A_3 gains are very small (an attenuation). In this condition, the output from A_2 is again quite small.

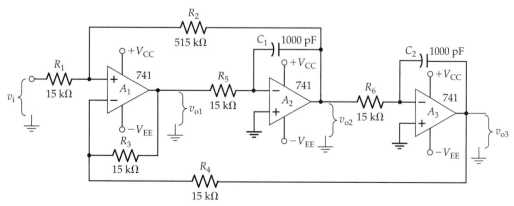

Figure 17-27 State-variable filter circuit consisting of amplifier stage A_1 and integrator stages A_2 and A_3. The circuit functions as a band-pass filter when the output is taken from A_2.

At the centre frequency, or *critical frequency* (f_o), for the circuit, $X_{C1} = R_5$ and $X_{C2} = R_6$. The A_2 and A_3 stages each have a voltage gain of 1 at f_o, as well as a 90° phase shift. So, the voltage fed from A_3 to R_4 is equal to, and in antiphase with, the output of A_1. Thus, A_1 has a very large gain from its non-inverting input terminal to its output, A_2 has a gain of 1, and resistors R_2 and R_1 become effective as a feedback network. The voltage gain from v_i to v_{o2} at the centre of the pass band is then R_2/R_1.

Analysis of the state-variable filter circuit produces a very simple equation for Q:

$$Q = \frac{R_1 + R_2}{2R_1} \tag{17-31}$$

which gives
$$R_2 = R_1(2Q - 1) \tag{17-32}$$

Because the circuit is designed so that at the centre frequency of the bandwidth (f_o), $X_{C1} = X_{C2} = R_5 = R_6$,

$$f_o = \frac{1}{2\pi C_1 R_5} \tag{17-33}$$

Designing this circuit begins with the selection of a convenient capacitance value for C_1 and C_2. Then R_5 and R_6 are calculated from Equation 17-33. Resistors R_3 and R_4 should give an A_1 stage gain of 1. So R_3 equals R_4, and these can also be made equal to R_5 and R_6. Furthermore, R_1 can be made equal to the common value of the other resistors, leaving R_2 to be determined from Equation 17-32. Only R_5 and R_6 need to have precise resistance values in relation to C_1 and C_2 (to set f_o).

The design procedure is seen to consist of only three steps: (1) select a convenient capacitance value for C_1 and C_2, (2) use Equation 17-33 to determine the common value for all resistors except R_2, (3) calculate R_2 from Equation 17-32.

Example 17-12

Design a state-variable bandpass filter to have $f_1 = 10.3$ kHz and $f_2 = 10.9$ kHz.

Solution

Select
$$C_1 = C_2 = 1000 \text{ pF}$$

Eq. 17-24:
$$f_o = \sqrt{(f_1 f_2)} = \sqrt{(10.3 \text{ kHz} \times 10.9 \text{ kHz})}$$
$$= 10.6 \text{ kHz}$$

Eq. 17-33:
$$R_5 = \frac{1}{2\pi f_o C_1} = \frac{1}{2 \times \pi \times 10.6 \text{ kHz} \times 1000 \text{ pF}}$$
$$= 15.01 \text{ k}\Omega \text{ (use 15 k}\Omega \pm 1\% \text{ standard value)}$$

$$R_1 = R_3 = R_4 = R_5 = R_6 = 15 \text{ k}\Omega$$

From Eqs. 17-22 and 17-23,

$$Q = \frac{f_o}{f_2 - f_1} = \frac{10.6 \text{ kHz}}{10.9 \text{ kHz} - 10.3 \text{ kHz}}$$

$$= 17.7$$

Eq. 17-32: $R_2 = R_1(2Q - 1) = 15 \text{ k}\Omega[(2 \times 17.7) - 1]$

$$= 515 \text{ k}\Omega \text{ (use 301 k}\Omega \text{ and 215 k}\Omega \pm 1\% \text{ in series)}$$

Practice Problem

17-9.1 A state-variable band-pass filter (as in Fig. 17-27) has the following components: $R_1 = R_3 = R_4 = R_5 = R_6 = 12$ kΩ, $R_2 = 180$ kΩ, and $C_1 = C_2 = 1500$ pF. Determine the centre frequency and bandwidth for the circuit.

17-10 IC SWITCHED-CAPACITOR FILTERS

Switched-Capacitor Resistor Simulation

A major problem with constructing filters in integrated circuit form is that size restrictions allow only very small capacitance values to be fabricated: a maximum of approximately 100 pF. Small-value capacitances require relatively large-value resistors for the critical frequencies in filters. For example, a critical frequency of 1 kHz with a 100 pF capacitance requires a 1.59 MΩ resistor. Here again there is a problem because such resistance values require more space than is available on an IC chip. The difficulties are overcome by the use of small-value capacitors and MOSFET switches, which are easily fabricated in IC form. The combination of capacitor and switch can be employed to simulate a resistor.

Figure 17-28a shows an integrator stage as used in a state-variable filter circuit. In Fig.17-28b the integrator resistor (R_2) is replaced with a capacitor (C_1)

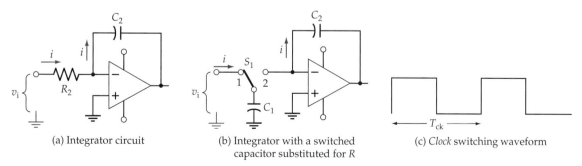

(a) Integrator circuit

(b) Integrator with a switched capacitor substituted for R

(c) *Clock* switching waveform

Figure 17-28   A resistor can be simulated by a switched capacitor, which passes charge at the same rate as the resistor. Switched capacitors are more easily fabricated than resistors in integrated circuits.

and a switch (S_1) that alternately connects C_1 to the input voltage (v_i) at terminal 1 and to the junction of C_2 and the op-amp inverting input at terminal 2. When S_1 is set to terminal 1, capacitor C_1 charges to v_i. When S_1 is switched from terminal 1 to terminal 2, C_1 passes its charge to C_2. Now assume that S_1 is controlled by the *clock* waveform shown in Fig. 17-28c. The charge passed during one cycle of the clock is

$$Q = C_1 v_i$$

or

$$i\, T_{ck} = C_1 v_i$$

which gives

$$i = \frac{C_1 v_i}{T_{ck}}$$

Referring to Fig. 17-28a again, the current passed to C_2 is

$$i = \frac{v_i}{R_2}$$

For equal input current levels via R_2 and via S_1 and C_1,

$$R_2 = \frac{T_{ck}}{C_1}$$

or

$$R_2 = \frac{1}{f_{ck} C_1} \qquad \text{(17-34)}$$

It is seen that resistance R_2 is simulated by C_1 and switch S_1. The resistance value depends upon the capacitance of C_1 and upon the switching (clock) frequency (f_{ck}). Increasing f_{ck} reduces the simulated resistance; reducing f_{ck} increases the resistance. A clock frequency of 50 kHz and a 100 pF capacitance simulate a resistance value of

$$R_2 = \frac{1}{f_{ck}\, C_1} = \frac{1}{50 \text{ kHz} \times 100 \text{ pF}}$$

$$= 200 \text{ k}\Omega$$

Recall from Section 17-9 that the critical frequency of the filter occurs when the integrator resistors equal the capacitive impedance:

$$f_o = \frac{1}{2\pi C_2 R_2}$$

Substituting for R_2,

$$f_o = \frac{f_{ck} C_1}{2\pi C_2} \qquad \text{(17-35)}$$

Equation 17-35 shows that the critical frequency now depends upon the ratio of capacitors C_1 and C_2 as well as on the clock frequency. Although integrated circuit capacitors cannot be fabricated with very accurate absolute values, they can be created with accurate capacitance ratios (usually ±0.1%), and this is good for f_o accuracy. Also, because f_o is directly proportional to f_{ck}, the critical frequency can be altered by variation of f_{ck}.

IC Filter Circuits

Integrated circuit switched-capacitor filters generally use the state-variable (universal) filter circuit configuration (see Section 17-9), because the connections can easily be arranged to give all five filter functions (low-pass, high-pass, band-pass, all-pass, and notch). The internal IC resistive and capacitive component values are not listed on a data sheet. Instead, because the clock frequency determines the (switched-capacitor) resistance values, the filter critical frequency is set by the clock frequency. Two clock connections are usually provided: one for $f_o = f_{ck}/50$, and the other for $f_o = f_{ck}/100$. So, for $f_o = 1$ kHz, for example, a clock of either 50 kHz or 100 kHz might be used. The clock frequency must be much higher than the critical frequency for a given filter design, and it should be at least twice the highest signal frequency. An external clock source is normally required, and these are readily available in IC form.

The MF10 is an integrated circuit containing two switched-capacitor state-variable filter sections which can be operated independently as two second-order filters, or cascaded to make one fourth-order filter. One-half of the MF10 block diagram is shown in Fig. 17-29. The circuit is made up of an amplifier, a

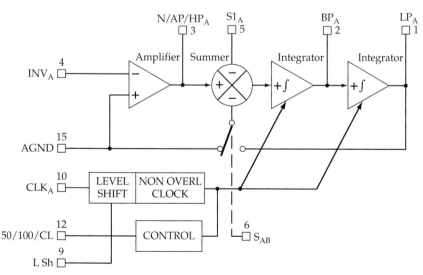

Figure 17-29   Block diagram of one-half of an MF10 switched-capacitor IC filter. (Reprinted with permission of National Semiconductor Corporation. Not authorized for use as critical components in a life support system.)

summing circuit, and two integrators. An external clock is required, and with only two to four external resistors, any filter function can be performed. The connecting pins shown are identified as follows:

L Sh	Shifts the dc level of the clock input
50/100/CL	Sets the ratio of f_{ck}/f_o
CLK$_A$	Clock input
AGND	Analog ground terminal
INV$_A$	Op-amp inverting input terminal
N/AP/HP$_A$	Notch, all-pass, and high-pass output, and summer non-inverting input
S1$_A$	Summer inverting input
BP$_A$	Band-pass output
LP$_A$	Low-pass output
S$_{AB}$	Sets the internal switch to an inverting input of the summer stage

The f_{ck}/f_o ratio is 50/1 when the 50/100/CL pin is connected to the positive supply, and 100/1 when it is connected to ground. When S$_{AB}$ is connected to the positive supply, the switch is set to the right to connect the second integrator output (LP$_A$ output) to the summing stage. With S$_{AB}$ connected to the negative supply, the summing stage inverting input is grounded.

Figure 17-30 shows one-half of an MF10 connected in *Mode 1* to function as a notch, band-pass, or low-pass filter. The design equations are also shown. Six possible modes of operation are listed on the manufacturer's

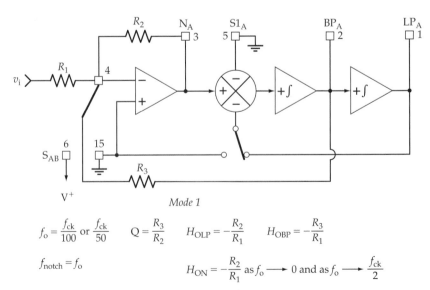

$$f_o = \frac{f_{ck}}{100} \text{ or } \frac{f_{ck}}{50} \qquad Q = \frac{R_3}{R_2} \qquad H_{OLP} = -\frac{R_2}{R_1} \qquad H_{OBP} = -\frac{R_3}{R_1}$$

$$f_{notch} = f_o \qquad\qquad\qquad H_{ON} = -\frac{R_2}{R_1} \text{ as } f_o \longrightarrow 0 \text{ and as } f_o \longrightarrow \frac{f_{ck}}{2}$$

Figure 17-30 One-half of an MF10 connected in *Mode 1* to function as a notch, band-pass, or low-pass filter. (Reprinted with permission of National Semiconductor Corporation. Not authorized for use as critical components in a life support system.)

data sheet, some subdivided into A and B. The H_O equations refer to the voltage gains at the critical frequency (f_o) from the input to each of the outputs. For example, H_{OBP} refers to the gain from v_i to the band-pass output terminal (BP$_A$).

Example 17-13

Determine the required resistance values to operate one-half of an MF10 as a band-pass filter with $f_1 = 10.3$ kHz, $f_2 = 10.9$ kHz, and a voltage gain of 34. (Note that this is the same frequency limit as for Example 17-12.)

Solution

Eq. 17-24:
$$f_o = \sqrt{(f_1 f_2)} = \sqrt{(10.3 \text{ kHz} \times 10.9 \text{ kHz})}$$
$$= 10.6 \text{ kHz}$$

Eqs. 17-22 and 17-23:

$$Q = \frac{f_o}{f_2 - f_1} = \frac{10.6 \text{ kHz}}{10.9 \text{ kHz} - 10.3 \text{ kHz}}$$
$$= 17.7$$

Select $R_3 \approx 120 \text{ k}\Omega$ (use 121 kΩ ±1% standard value)

From the MF10 design equations,

$$R_2 = \frac{R_3}{Q} = \frac{121 \text{ k}\Omega}{17.7}$$
$$= 6.83 \text{ k}\Omega \text{ (use 6.81 k}\Omega \pm 1\% \text{ standard value)}$$

and
$$R_1 = \frac{R_3}{H_{OBP}} = \frac{121 \text{ k}\Omega}{34}$$
$$= 3.6 \text{ k}\Omega \text{ (use 3.57 k}\Omega \pm 1\% \text{ standard value)}$$

$$f_{ck} = 50 \times f_o = 50 \times 10.6 \text{ kHz}$$
$$= 530 \text{ kHz}$$

Practice Problem

17-10.1 A notch filter using one-half of an MF10 is to have $f_1 = 1.7$ kHz, $f_2 = 2.35$ kHz, and a voltage gain of 10. Determine suitable component values and select an appropriate clock frequency.

17-11 FILTER TESTING AND TROUBLESHOOTING

The *common errors* listed in Section 5-6 should be referred to once again. They apply to filter circuits constructed on a laboratory breadboard, just as they do to all circuits. The discussion in Section 14-11 concerning op-amp circuit testing is applicable to active filters, and the op-amp circuit stability treatment in Section 15-6 is very important.

Once it is determined that the circuit is not oscillating and that the dc voltage levels are acceptable, testing for A_v/f and ϕ/f responses can begin. The amplifier *ac testing* procedure in Section 12-10 can be followed. The first signal frequency should be well within the filter-pass band; then fairly large frequency steps can be made while the response is flat. The frequency should be changed in small steps around the cutoff frequencies, to obtain sufficient readings for accurately plotting the graphs. Because of component tolerances, filter cutoff frequencies are unlikely to be precisely as calculated.

If a measured cutoff frequency is far from the expected frequency, the component values should be rechecked, especially the colour-coded resistor values. A common mistake is the selection of a resistor in error by a factor of 10.

When filter falloff rates are less than expected, one possibility is that the feedback portion of the circuit is not functioning correctly. For example, suppose the two-pole filter in Fig. 17-9 behaves like a single-pole filter (20 dB/decade falloff rate instead of 40 db/decade). In this case it is most likely that the feedback capacitor (C_2) is not correctly connected, or perhaps that the capacitance value is much smaller than required.

Higher-order filters consisting of several sections may need to be separated into individual sections for testing if the overall circuit does not perform as expected. The same remarks apply to band-pass and notch filters that contain low-pass and high-pass sections.

Review Questions

Section 17-1

17-1 Sketch ideal and practical gain/frequency response graphs for low-pass, high-pass, band-pass, and notch filters. Explain the function of each type of filter.

17-2 Sketch the circuit of a basic *RC* low-pass filter and explain its operation.

17-3 Sketch typical gain/frequency and phase/frequency response graphs for a low-pass filter. Identify the cutoff frequency, and explain the shape of the graphs.

17-4 Define cutoff frequency, bandwidth, pass band, stop band, attenuation band, transition band, insertion loss, and transfer function.

Section 17-2

17-5 Sketch the circuit of a first-order active low-pass filter, and explain how the circuit operates. Discuss the function of the operational amplifier.

17-6 Sketch the circuit of a first-order active high-pass filter, and explain how the circuit operates.

17-7 Sketch typical gain/frequency and phase/frequency response graphs for a first-order active high-pass filter, and explain the shape of the graphs.

Section 17-3

17-8 Define single-pole, two-pole, and three-pole filters, and describe the performances of each.

17-9 Identify the major types of filter circuit design, and discuss the differences in their gain/frequency and phase/frequency response graphs.

17-10 Sketch the circuit of a second-order low-pass active filter, and explain its operation.

17-11 Sketch typical A_v/f and ϕ/f response graphs for a second-order low-pass active filter. Discuss the shape of the graphs.

Section 17-4

17-12 Sketch the circuit of a second-order high-pass active filter, and explain its operation.

17-13 Sketch typical A_v/f and ϕ/f response graphs for a second-order high-pass active filter. Discuss the shape of the graphs.

Section 17-5

17-14 Sketch a circuit for a third-order low-pass active filter, and explain its operation. Discuss the required cutoff frequency for each stage.

17-15 Sketch a circuit for a third-order high-pass active filter, and explain its operation. Discuss the required cutoff frequency for each stage.

17-16 Sketch typical A_v/f and ϕ/f response graphs for third-order low-pass and high-pass active filters. Discuss the shape of the graphs.

Section 17-6

17-17 Show how a band-pass filter can be constructed by the use of a low-pass and a high-pass filter. Sketch the expected A_v/f response graph and explain the filter operation.

17-18 Sketch the circuit of a single-stage band-pass active filter. Explain the circuit operation.

17-19 Sketch typical A_v/f response graphs for wide-band and narrow-band band-pass filters, and discuss the differences between them. Write equations for bandwidth, Q factor, and centre frequency.

17-20 Show how the circuit of a single-stage wide-band band-pass filter can be modified for narrow-band operation. Briefly explain.

Section 17-7

17-21 Sketch a block diagram to show how a notch filter can be constructed using a low-pass and a high-pass filter. Sketch the expected A_v/f response graph and explain the filter operation.

17-22 Explain by the use of a block diagram how a band-pass filter and a summing circuit can be used as a notch filter. Sketch the expected A_v/f response graph.

17-23 The band-pass circuit in Fig. 17-16 can be modified to convert it to a notch filter. Show the modification and explain how the notch response is obtained.

Section 17-8

17-24 Draw a circuit diagram for an all-pass phase-lag filter. Sketch the typical phase/frequency response and explain the circuit operation.

17-25 Write the input/output equation for an all-pass phase-lag circuit. Explain how the phase lag may be made adjustable.

17-26 Draw a circuit diagram for an all-pass phase-lead filter. Sketch the typical phase/frequency response and explain the circuit operation.

Section 17-9

17-27 Draw the circuit of a state-variable filter. Briefly explain the circuit operation and write equations for Q and f_o.

Section 17-10

17-28 Show how a switched capacitor can be used to simulate a resistor, and discuss the advantages of this process in integrated circuit applications.

17-29 Explain how the critical frequency of an integrator with a switched-capacitor simulated resistor can be controlled by a clock.

17-30 Sketch the basic block diagram for one-half of an MF10 integrated circuit filter. Identify the components and briefly explain the function of each. Show how external resistors may be connected to have the IC function as a band-pass filter.

Section 17-11

17-31 Prepare a point-form list of items to be checked before testing a laboratory-constructed filter circuit.

17-32 Prepare a point-form list of the procedure for testing a filter circuit to determine its A_v/f and ϕ/f responses.

Problems

Section 17-1

17-1 Calculate the cutoff frequency for a basic RC low-pass filter with $R_1 = 18$ kΩ and $C_1 = 1000$ pF. Determine the circuit bandwidth, and estimate the signal attenuation at $f \approx 35.4$ kHz.

17-2 A basic low-pass filter with $C_1 = 0.015$ μF is to attenuate 20 kHz signal frequencies by approximately 23 dB. Determine the required cutoff frequency, and calculate a suitable resistance value for R_1.

17-3 Calculate the new cutoff frequency and insertion loss for the filter in Problem 17-2 when a 60 kΩ load resistance is connected to the output.

17-4 A basic low-pass filter with $C_1 = 0.01$ μF has R_1 consisting of a 3.3 kΩ resistor in series with a 1 kΩ variable resistor. Calculate the maximum and minimum cutoff frequency.

17-5 Calculate the insertion loss for the filter in Problem 17-1 when a 100 kΩ load resistance is connected to the output. Determine the effect of the load on the circuit cutoff frequency.

Section 17-2

17-6 A first-order active low-pass filter is to have a cutoff frequency of 3 kHz. First, design the circuit with a bipolar op-amp, then redesign it with a BIFET op-amp.

17-7 Modify each of the circuits designed for Problem 17-6 to make the cutoff frequency adjustable from 2 kHz to 4 kHz.

17-8 Using a BIFET op-amp, design a first-order active high-pass filter to have a 15 kHz cutoff frequency. Determine the circuit bandwidth if the op-amp has a 1.2 MHz unity gain frequency.

17-9 A first-order active high-pass filter has $R_1 = 56$ kΩ and $C_1 = 600$ pF. If the circuit uses a 741 op-amp, determine the cutoff frequency and the highest signal frequency that can be passed.

Section 17-3

17-10 Using a 741 op-amp, design a second-order active low-pass filter to have a 5 kHz cutoff frequency with a Butterworth response.

17-11 Redesign the filter in Problem 17-10 to use a 108 op-amp.

17-12 A second-order active low-pass Butterworth filter has the following component values: $R_1 = R_2 = 56$ kΩ, $R_3 = 120$ kΩ, $C_1 = 600$ pF, and $C_2 = 1200$ pF. Calculate the circuit cutoff frequency.

17-13 A second-order active low-pass filter is to attenuate 7 kHz signals by 15 dB. Design the circuit to use a BIFET op-amp and to have a Butterworth response.

Section 17-4

17-14 Design a second-order active high-pass Butterworth filter to have a 7 kHz cutoff frequency. Use a 353 op-amp and determine the highest signal frequency that can be passed.

17-15 A second-order active high-pass Butterworth filter has the following component values: $R_1 = 28$ kΩ, $R_2 = 56$ kΩ, and $C_1 = C_2 = 1000$ pF. Calculate the circuit cutoff frequency.

17-16 A second-order active high-pass filter is to attenuate signal frequencies below 1.2 kHz by a minimum of 15 dB. Design the circuit to use a 741 op-amp and to have a Butterworth response. Calculate the circuit bandwidth.

17-17 Redesign the filter in Problem 17-16 to use a BIFET op-amp.

Section 17-5

17-18 A third-order active low-pass filter is to attenuate all signals above 9 kHz by a minimum of 21 dB. Design the circuit to use BIFET op-amps and to have a Butterworth response.

17-19 Using 741 op-amps, design a third-order Butterworth active low-pass filter to have an 18 kHz cutoff frequency.

17-20 A third-order Butterworth active high-pass filter is to pass all signal frequencies above 8 kHz. Design the circuit to use 108 op-amps.

17-21 Determine the bandwidth of the circuit designed for Problem 17-20. Estimate the attenuation of a 2 kHz signal.

Section 17-6

17-22 A bandpass filter using low-pass and high-pass circuits (as in Fig. 17-15) is to have a bandwidth extending from 400 Hz to 30 kHz. Using 741 op-amps, design suitable second-order stages.

17-23 Design a single-stage band-pass filter to have cutoff frequencies of 1 kHz and 50 kHz and a voltage gain of 1.

17-24 A single-stage band-pass filter (as in Fig. 17-16) has the following component values: $R_1 = R_2 = R_3 = 3.9$ kΩ, $C_1 = 0.12$ μF, and $C_2 = 600$ pF. Calculate the circuit cutoff frequencies.

17-25 A single-stage narrow-band band-pass filter (as in Fig. 17-20) is to be designed to use an op-amp with 1 μA input bias current. The centre frequency is to be 3.3 kHz with a bandwidth of approximately 500 kHz. Determine suitable component values.

17-26 Redesign the filter in Problem 17-25 to use a BIFET op-amp.

Section 17-7

17-27 The notch filter in Fig. 17-21 is to have a stop band from 10 kHz to 100 kHz. The signal amplitude is 1 V. Design the complete circuit to use first-order high-pass and low-pass filters.

17-28 The notch filter in Fig. 17-22 is to be used to attenuate an unwanted 120 Hz input. Design the circuit to have a bandwidth off 20 Hz.

17-29 Design a single-stage notch filter (as in Fig. 17-23) to have $f_1 = 1.3$ kHz and $f_2 = 5.4$ kHz.

Section 17-8

17-30 An all-pass filter circuit is to have a phase lag adjustable from 70° to 100°. The 3.3 kHz input voltage has a 3 V amplitude. Design the circuit using a BIFET op-amp.

17-31 An all-pass phase-lead circuit with a 2 V, 1.5 kHz input is to have a phase lead adjustable from 90° to 130°. Design the circuit to use a bipolar op-amp.

Section 17-9

17-32 A state-variable band-pass filter (as in Fig. 17-27) has the following component values: $R_1 = 12\ k\Omega$, $R_2 = 560\ k\Omega$, $R_3 = R_4 = 27\ k\Omega$, $R_5 = R_6 = 33\ k\Omega$, and $C_1 = C_2 = 1200\ pF$. Calculate the width of the pass-band.

17-33 A band-pass filter is to pass 2 kHz to 2.5 kHz signal frequencies. Design a suitable state-variable circuit.

17-34 A state-variable band-pass filter is to have $f_o = 3.3\ kHz$ and BW $\approx 200\ Hz$. Design the circuit using a BIFET op-amp.

Section 17-10

17-35 A band-pass filter with a voltage gain of 20, $f_1 = 3\ kHz$, and $f_2 = 4\ kHz$ is to be constructed using an MF10. Determine suitable resistance values and clock frequency.

17-36 Determine suitable resistance values and clock frequency for an MF10 notch filter which is to have a voltage gain of 15, $f_1 = 2\ kHz$, and $f_2 = 2.9\ kHz$.

Practice Problem Answers

17-1.1 19.65 kHz, 23 dB
17-1.2 1.1 dB, 22.29 kHz
17-2.1 1000 pF, 10.5 kΩ
17-2.2 752 kHz
17-3.1 1000 pF, 2000 pF, 75 kΩ, 75 kΩ, 150 kΩ, 1.5 kHz
17-3.2 1000 pF, 2000 pF, 18.7 kΩ, 18.7 kΩ, 39 kΩ, 6.02 kHz
17-4.1 3000 pF, 3000 pF, 60.4 kΩ, 121 kΩ, 120 kΩ, 621 Hz
17-4.2 1000 pF, 1000 pF, (3.01 kΩ + 1.5 kΩ), 9.09 kΩ, 10 kΩ
17-5.1 1000 pF, 10.2 kΩ, 10 kΩ; 1000 pF, 2000 pF, 9.09 kΩ, 9.09 kΩ, 18 kΩ
17-5.2 1000 pF, 37.4 kΩ, 39 kΩ; 1000 pF, 1000 pF, 21 kΩ, 42.2 kΩ, 39 kΩ
17-6.1 25.7 kHz, 8.56 kHz, 0.33
17-6.2 60.4 kΩ, 121 kΩ, 120 kΩ, 1.21 kΩ, 0.012 μF, 0.012 μF
17-7.1 9100 pF, 1000 pF, (4 $\times$ 26.1 kΩ)
17-8.1 243 kΩ, 243 kΩ, (120 kΩ fixed +50 kΩ variable), 1000 pF
17-9.1 8.84 kHz, 1.1 kHz
17-10.1 3.92 kΩ, 40.2 kΩ, 121 kΩ, 100 kHz

CHAPTER 18
Linear and Switching Voltage Regulators

CONTENTS

Objectives

You will be able to:

1   Sketch transistor series regulator circuits and explain their operation.
2   Show how regulator circuits may be improved by the use of error amplifiers, additional series-pass transistors, preregulation, etc., and how the output voltage may be adjusted.
3   Design transistor series regulator circuits to fulfill a given specification.
4   Analyze transistor series regulators to determine source effect, load effect, line regulation, load regulation, and ripple reduction.
5   Sketch operational amplifier series regulator circuits and explain their operation.
6   Design op-amp series regulator circuits to fulfill a given specification.

7   Sketch and explain the basic circuit of a 723 IC voltage regulator, and design circuits using 723 ICs.
8   Sketch IC regulator block diagrams and determine external component values for various applications.
9   Sketch the block diagram and waveforms for a switching regulator and explain its operation.
10 Sketch circuits and waveforms for step-down, step-up, and inverting converters, and explain their operation.
11 Design LC filter circuits for various switching converters.
12 Show how an IC controller circuit is used with switching converters, and calculate values for the externally connected components.

INTRODUCTION

Almost all electronic circuits require a direct voltage supply. This is usually derived from the standard industrial or domestic ac supply by transformation, rectification, and filtering. The resultant *raw* dc is not stable enough for most purposes, and it usually contains an unacceptably large ac ripple waveform. Voltage regulator circuits are employed to render the voltage more constant and to attenuate the ripple.

Unregulated power supplies and Zener diode regulators are discussed in Chapter 3. Zener diode regulators are normally used only where the load current does not exceed 25 mA. A transistor operating as an emitter follower circuit may be connected to a Zener diode regulator to supply larger load currents. The regulator circuit performance is tremendously improved when an *error amplifier* is included, to detect and amplify the difference between the output and the voltage reference source, and to provide feedback to correct the difference.

A variety of voltage regulator circuits are available in integrated circuit form.

18-1 TRANSISTOR SERIES REGULATOR

Basic Circuit

When a low-power Zener diode is used in the simple regulator circuit described in Section 3-6, the load current is limited by the maximum diode current. A high-power Zener used in such a circuit can supply higher levels of load current, but much power is wasted when the load is light. The emitter follower regulator shown in Fig. 18-1 is an improvement on the simple regulator circuit because it draws a large current from the supply only when required by the load. In Fig. 18-1a, the circuit is drawn in the form of the common collector amplifier (emitter follower) discussed in Section 6-6. In Fig. 18-1b, the circuit is shown in the form usually referred to as a *series regulator*. Transistor Q_1 is termed a *series-pass* transistor.

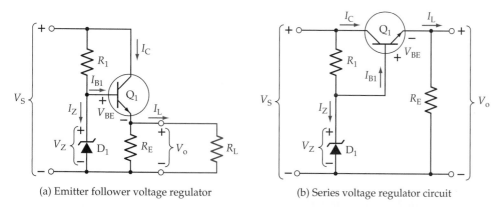

(a) Emitter follower voltage regulator   (b) Series voltage regulator circuit

Figure 18-1   To supply a large output current from a Zener diode voltage regulator, a transistor (Q_1) is connected as an emitter follower. This converts the circuit into a series voltage regulator.

The output voltage V_o from the series regulator in Fig. 18-1 is $V_Z - V_{BE}$, and the maximum load current $I_{L(max)}$ can be the maximum emitter current that Q_1 is capable of passing. For a 2N3055 transistor (specification in data sheet A-8 in Appendix A), I_L could approach 15 A. When I_L is zero, the current drawn from the supply is approximately $I_Z + I_{C(min)}$, where $I_{C(min)}$ is the minimum collector current to keep Q_1 operational. The Zener diode circuit (R_1 and D_1) has to supply only the base current of the transistor. The series voltage regulator is, therefore, much more efficient than a simple Zener diode regulator.

Regulator with Error Amplifier

A series regulator using an additional transistor as an *error amplifier* is shown in Fig. 18-2. The error amplifier improves the line and load regulation of the circuit (see Section 3-5). The amplifier also makes it possible to have an output voltage greater than the Zener diode voltage. Resistor R_2 and diode D_1 are the Zener diode reference source. Transistor Q_2 and its associated components constitute the error amplifier, which controls the series-pass transistor Q_1. The output voltage is divided by resistors R_3 and R_4 and is compared to the Zener voltage level (V_Z). C_1 is a large-value capacitor, usually 50 μF to 100 μF, which is connected at the output to suppress any tendency of the regulator to oscillate.

When the circuit output voltage changes, the change is amplified by transistor Q_2 and fed back to the base of Q_1 to correct the output voltage level. Suppose that the circuit is designed for a V_o of 12 V and that the supply voltage is $V_S = 18$ V. In this case a suitable Zener diode voltage might be $V_Z = 6$ V. For this V_Z level, the base voltage of Q_2 must be $V_{B2} = V_Z + V_{BE2} = 6.7$ V. So resistors R_3 and R_4 are selected to give $V_{B2} = 6.7$ V and $V_o = 12$ V. The voltage at the base of Q_1 is $V_{B1} = V_o + V_{BE1} = 12.7$ V. Also, $V_{R1} = V_S - V_{B1} = 5.3$ V. The current through R_1 is largely the collector current of Q_2.

Now suppose the output voltage drops slightly for some reason. When V_o decreases, V_{B2} decreases. Because the emitter voltage of Q_2 is held at V_Z, any decrease in V_{B2} appears across the base-emitter of Q_2. A reduction in V_{BE2}

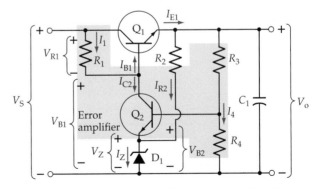

Figure 18-2   Series voltage regulator circuit with an error amplifier. The error amplifier improves the regulator line and load regulation and gives an output voltage greater than the Zener diode voltage.

causes I_{C2} to be reduced. When I_{C2} falls, V_{R1} is reduced, and the voltage at the base of Q_1 rises ($V_{B1} = V_S - V_{R1}$) causing the output voltage to increase. Thus, a decrease in V_o produces a feedback effect which causes V_o to increase back toward its normal level. Similarly, a rise in V_o above its normal level produces a feedback effect that pushes V_o down again toward its normal level.

When the input voltage changes, the voltage across resistor R_1 changes in order to keep the output constant. This change in V_{R1} is produced by a change in I_{C2}, which itself is produced by a small change in V_o. Therefore, a supply voltage change (ΔV_S) produces a small output voltage change (ΔV_o). The relationship between ΔV_S and ΔV_o depends upon the amplification of the error amplifier. Similarly, when the load current (I_L) changes, I_{B1} alters as necessary to increase or decrease I_{E1}. The I_{B1} variation is produced by a change in I_{C2}, which, once again, is the result of an output voltage variation ΔV_o.

Series Regulator Performance

The performance of a series regulator without an error amplifier (Fig. 18-1) is similar to that of a Zener diode regulator (see Section 3-6), except in the case of the load effect. The series-pass transistor tends to improve the regulator load effect by a factor equal to the transistor h_{FE}.

The error amplifier in the regulator in Fig 18-3 (reproduced from Fig. 18-2) improves all aspects of the circuit performance by an amount directly related to the amplifier voltage gain (A_v). When V_S changes by ΔV_S, the output change is

$$\Delta V_o = \frac{\Delta V_S}{A_v} \qquad (18\text{-}1)$$

If ΔV_S is produced by a variation in the ac supply voltage, the power supply source effect is reduced by a factor of A_v. ΔV_S might also be the result of an increase or decrease in load current that causes a change in the average level of the dc supply voltage. Thus, the load effect of the power supply is reduced by a factor of A_v.

Now consider the effect of supply voltage ripple on the circuit in Fig. 18-3. The ripple waveform appears at the collector of transistor Q_1. If there were no

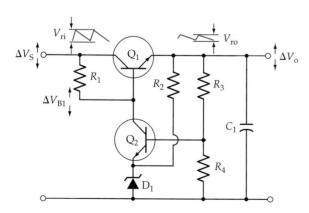

Figure 18-3 Voltage variations at the input of a regulator with an error amplifier are reduced by a factor equal to the amplifier voltage gain.

negative feedback, it would also be present at Q_1 base and at the regulator output. However, like supply voltage changes, the input ripple is reduced by a factor of A_v when it appears at the output. The *ripple rejection ratio* is calculated as the decibel ratio of the input and output ripple voltages.

Example 18-1

The supply voltage for the regulator in Fig. 18-3 has $V_S = 21$ V on no load, and $V_S = 20$ V when $I_{L(max)} = 40$ mA. The output voltage is $V_o = 12$ V, and the regulator has an error amplifier with a gain of 100. Calculate the source effect, load effect, line regulation, and load regulation for the complete power supply. Also determine the ripple rejection ratio in decibels.

Solution

Eq. 3-25: Source effect $= \Delta V_o$ (for $\Delta V_S = 10\%$)

Eq. 18-1: $$\Delta V_o = \frac{\Delta V_S}{A_v} = \frac{10\% \text{ of } 21 \text{ V}}{100}$$

$$= 21 \text{ mV}$$

Eq. 3-27: Load effect $= \Delta V_o$ for $\Delta I_{L(max)}$

Eq. 18-1: $$\Delta V_o = \frac{\Delta V_S}{A_v} = \frac{21 \text{ V} - 20 \text{ V}}{100}$$

$$= 10 \text{ mV}$$

Eq. 3-26: Line regulation $= \dfrac{(\text{Source effect}) \times 100\%}{E_o}$

$$= \frac{21 \text{ mV} \times 100\%}{12 \text{ V}}$$

$$= 0.175\%$$

Eq. 3-28: Load regulation $= \dfrac{(\text{Load effect}) \times 100\%}{E_o}$

$$= \frac{10 \text{ mV} \times 100\%}{12 \text{ V}}$$

$$= 0.08\%$$

Ripple rejection $= 20 \log (1/A_v) = 20 \log (1/100)$

$$= -40 \text{ dB}$$

Regulator Design

To design a series regulator circuit (as in Fig. 18-3), the Zener diode is selected to have V_Z less than the output voltage. A Zener diode voltage approximately equal to $0.75 \, V_o$ is usually suitable. Appropriate current levels are chosen for each resistor, and the resistor values are calculated using Ohm's law. Transistor Q_1 is selected to pass the required load current and to survive the necessary power dissipation. A heat sink (see Section 8-8) is normally required for the series-pass transistor in a regulator that supplies large load currents. As has been explained, a large capacitor is usually connected across the output to ensure amplifier ac stability (C_1 in Fig. 18-3).

The difference between the regulator input and output voltages is the collector-emitter voltage of the series-pass transistor (Q_1), and this voltage must be large enough to keep the transistor operational. The minimum level of V_{CE1} (known as the *dropout voltage*) occurs at the lowest point in the ripple waveform of *raw* dc input (rectified and filtered). If V_{CE1} is too small for correct operation at this point, a large-amplitude ripple waveform appears at the regulator output.

Example 18-2

Design the voltage regulator circuit in Fig. 18-4 to produce $V_o = 12$ V and $I_{L(max)} = 40$ mA. The supply voltage is $V_S = 20$ V.

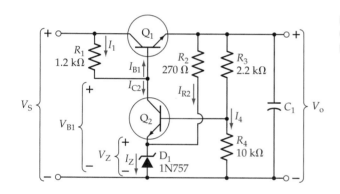

Figure 18-4 Voltage regulator circuit for Example 18-2.

Solution

Select
$$V_Z \approx 0.75\, V_o = 0.75 \times 12 \text{ V}$$
$$= 9 \text{ V}$$

For D_1, use a 1N757 Zener diode with $V_Z = 9.1$ V (see data sheet A-4 in Appendix A)

For minimum D_1 current, select
$$I_{R2} = 10 \text{ mA}$$
$$R_2 = \frac{V_o - V_Z}{I_{R2}} = \frac{12 \text{ V} - 9.1 \text{ V}}{10 \text{ mA}}$$
$$= 290 \text{ } \Omega \text{ (use 270 } \Omega \text{ standard value)}$$
$$I_{E1(max)} \approx I_{L(max)} + I_{R2} = 40 \text{ mA} + 10 \text{ mA}$$
$$= 50 \text{ mA}$$

Specification for Q_1:

$$V_{CE1(max)} \approx V_S = 20 \text{ V}$$

$$I_{C1(max)} \approx I_{E(max)} = 50 \text{ mA}$$

$$P_{D(max)} = (V_S - V_o) \times I_{E1(max)} = (20 \text{ V} - 12 \text{ V}) \times 50 \text{ mA}$$

$$= 400 \text{ mW}$$

Assuming $h_{FE1(min)} = 50$

$$I_{B1(max)} = \frac{I_{E1(max)}}{h_{FE1(min)}} = \frac{50 \text{ mA}}{50}$$

$$= 1 \text{ mA}$$

$$I_{C2} > I_{B1(max)}$$

Select $I_{C2} = 5 \text{ mA}$

$$R_1 = \frac{V_S - V_{B1}}{I_{C2} + I_{B1}} = \frac{20 \text{ V} - (12 \text{ V} + 0.7 \text{ V})}{5 \text{ mA} + 1 \text{ mA}}$$

$$= 1.21 \text{ k}\Omega \text{ (use 1.2 k}\Omega \text{ standard value)}$$

$$I_Z = I_{E2} + I_{R2} = 5 \text{ mA} + 10 \text{ mA}$$

$$= 15 \text{ mA}$$

$$I_4 \gg I_{B1(max)}$$

Select $I_4 = 1 \text{ mA}$

$$R_4 = \frac{V_Z + V_{BE2}}{I_4} = \frac{9.1 \text{ V} + 0.7 \text{ V}}{1 \text{ mA}}$$

$$= 9.8 \text{ k}\Omega \text{ (use 10 k}\Omega \text{ standard value)}$$

$$R_3 = \frac{V_o - V_{R4}}{I_4} = \frac{12 \text{ V} - 9.8 \text{ V}}{1 \text{ mA}}$$

$$= 2.2 \text{ k}\Omega \text{ (standard value)}$$

Practice Problems

18-1.1 A 12 V dc power supply has an 18 V input (V_S) from a rectifier and filter circuit. There is a 1 V drop in V_S from no load to full load. If the regulator has an error amplifier with $A_v = 70$, calculate the source effect, load effect, line regulation, and load regulation.

18-1.2 A voltage regulator circuit as in Fig. 18-4 has $V_S = 25$ V, and is to produce $V_o = 15$ V with $I_{L(max)} = 60$ mA. Design the circuit and specify the series transistor. Assume $h_{FE1(min)} = 100$.

18-2 IMPROVING REGULATOR PERFORMANCE

Error Amplifier Gain

The performance of a regulator is dependent on the voltage gain of the error amplifier (see Eq. 18-1). A higher-gain amplifier gives better line and load regulation. So anything that improves the amplifier voltage gain will improve the regulator performance. Two possibilities for increasing A_v are: use a transistor with a high h_{FE} value for Q_2, and use the highest possible resistance value for R_1.

Output Voltage Adjustment

Regulator circuits such as the one designed in Example 18-2 are unlikely to have V_o exactly as specified. This is because of component tolerances, as well as, perhaps, not finding standard values close to the calculated values. Hence, some form of adjustment is required to enable the output voltage to be set to the desired level. In Fig. 18-5 the potentiometer (R_5) connected between resistors R_3 and R_4 provides output adjustment. The maximum output voltage level is produced when the potentiometer moving contact is at the bottom of R_5:

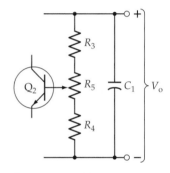

Figure 18-5 Use of a potentiometer to provide regulator output voltage adjustment.

$$V_{o(max)} = \frac{R_3 + R_4 + R_5}{R_4} \times V_{B2} \qquad (18\text{-}2)$$

Minimum V_o occurs when the moving contact is at the top of R_5:

$$V_{o(min)} = \frac{R_3 + R_4 + R_5}{R_4 + R_5} \times V_{B2} \qquad (18\text{-}3)$$

Example 18-3

Modify the voltage regulator circuit in Example 18-2 (as in Fig. 18-6) to make V_o adjustable from 11 V to 13 V.

Figure 18-6 Adjustable output regulator circuit for Example 18-2.

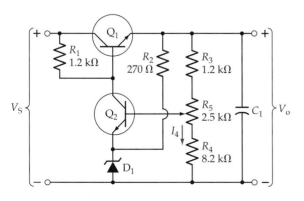

Solution

Select $\quad I_{4(min)} = 1 \text{ mA}$

For $V_o = 11$ V (moving contact at top of R_5),

$$R_3 = \frac{V_o - V_{B2}}{I_{4(min)}} = \frac{11 \text{ V} - 9.8 \text{ V}}{1 \text{ mA}}$$

$$= 1.2 \text{ k}\Omega \text{ (standard value)}$$

$$R_4 + R_5 = \frac{V_{B2}}{I_{4(min)}} = \frac{9.8 \text{ V}}{1 \text{ mA}}$$

$$= 9.8 \text{ k}\Omega$$

For $V_o = 13$ V (moving contact at bottom of R_5),

I_4 becomes $\quad I_4 = \dfrac{V_o}{R_3 + R_4 + R_5} = \dfrac{13 \text{ V}}{1.2 \text{ k}\Omega + 9.8 \text{ k}\Omega}$

$$= 1.18 \text{ mA}$$

$$R_4 = \frac{V_{B2}}{I_4} = \frac{9.8 \text{ V}}{1.18 \text{ mA}}$$

$$= 8.3 \text{ k}\Omega \text{ (use 8.2 k}\Omega \text{ standard value)}$$

$$R_5 = (R_4 + R_5) - R_4 = 9.8 \text{ k}\Omega - 8.2 \text{ k}\Omega$$

$$= 2.6 \text{ k}\Omega \text{ (use 2.5 k}\Omega \text{ standard value potentiometer)}$$

High-Output-Current Circuit

For the circuit in Fig. 18-6 to function correctly, the collector current of Q_2 must be larger than the maximum base current flowing into Q_1. In regulators that supply a large output current, I_{B1} can be too large for I_{C2} to control. In this case, an additional transistor (Q_3) should be connected at the base of Q_1 to form a *Darlington Circuit* (or *Darlington Pair*), as in Fig. 18-7. Transistor Q_1 is usually a high-power BJT requiring a heat sink (see Section 8-8), and Q_3 is a low-power device. The Q_3 base current is

$$I_{B3} = \frac{I_L}{h_{FE1} \times h_{FE3}} \tag{18-4}$$

The current gain for the Darlington is $h_{FE1} \times h_{FE3}$, and I_{B3} is low enough that I_{C2} may easily be made much greater than I_{B3}. In Fig. 18-7 note resistor R_6, which is included to provide a suitable minimum Q_3 operating current when I_L is very low.

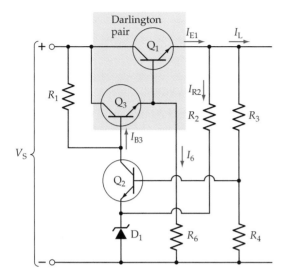

Figure 18-7 An additional transistor (Q_3) connected at the base of Q_1 to constitute a Darlington circuit allows a voltage regulator to supply a higher output current.

Darlington transistors consisting of a pair of BJTs (low-power and high-power) fabricated together and packaged as a single device are available. These are usually referred to as *power Darlingtons*. The 2N6039 is an *npn* power Darlington with an h_{FE} specified as 750 minimum, 18 000 maximum. Resistor R_6 in Fig. 18-7 is not required when a power Darlington is used.

Example 18-4

Modify the voltage regulator circuit in Example 18-2 to change the load current to 200 mA. Use a Darlington circuit (as in Fig. 18-7), and assume that $h_{FE1} = 20$ and that $h_{FE3} = 50$. Specify transistor Q_1.

Solution

$$I_{E1(max)} \approx I_{L(max)} + I_{R2} = 200 \text{ mA} + 10 \text{ mA}$$

$$= 210 \text{ mA}$$

$$I_{B1(max)} = \frac{I_{E1(max)}}{h_{FE1}} = \frac{210 \text{ mA}}{20}$$

$$= 10.5 \text{ mA}$$

$$I_{B3(max)} = \frac{I_{B1(max)}}{h_{FE3}} = \frac{10.5 \text{ mA}}{50}$$

$$= 0.21 \text{ mA}$$

$$I_{C2} > I_{B3(max)}$$

Select $\quad I_{C2} = 1 \text{ mA}$

$$R_1 = \frac{V_S - V_{B3}}{I_{C2} + I_{B3}} = \frac{20 \text{ V} - (12 \text{ V} + 0.7 \text{ V} + 0.7 \text{ V})}{1 \text{ mA} + 0.21 \text{ mA}}$$

$$= 5.45 \text{ k}\Omega \text{ (use 5.6 k}\Omega \text{ standard value)}$$

Select $\qquad I_6 = 0.5$ mA

$$R_6 = \frac{V_o + V_{BE1}}{I_6} = \frac{12\ V + 0.7\ V}{0.5\ mA}$$

$$= 25.4\ k\Omega\ \text{(use 22 k}\Omega\text{ standard value)}$$

Specification for Q_1:

$$V_{CE1(max)} \approx V_S = 20\ V$$

$$I_{C1(max)} \approx I_{E1(max)} = 210\ mA$$

$$P_{D(max)} = (V_S - V_o) \times I_{E1(max)} = (20\ V - 12\ V) \times 210\ mA$$

$$= 1.68\ W$$

Preregulation

Refer to Fig. 18-8, which is a partial reproduction of Fig. 18-2. Note that resistor R_1 is supplied from the regulator input, and consider what happens when the supply voltage drops by 1 V. If I_{C2} does not change, the voltage across R_1 remains constant, and the 1 V drop also occurs at the base of Q_1 and at the output of the regulator. This does not happen, of course. Instead, I_{C2} changes to reduce the voltage across R_1 and thus keep the output voltage close to its normal level. The change in I_{C2} is produced by a small change in the output voltage. If R_1 is connected to a constant-voltage source instead of the input, the change in I_{C2} would not be required when V_S changes, and consequently the output voltage change would not occur.

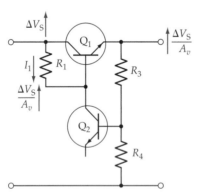

Figure 18-8 A change in a regulator supply voltage (ΔV_S) produces an output change ($\Delta V_o = \Delta V_S/A_V$).

Now look at Fig. 18-9a where R_1 is shown connected to another Zener diode voltage source (R_7 and D_2). This arrangement is called a *pre-regulator*. When V_S changes, the change in V_{Z2} is negligible compared to ΔV_S. Thus, virtually no change is required in I_{C2} and V_o. A preregulator substantially improves the line and the load regulation of a regulator circuit.

The minimum voltage drop across R_1 should typically be 3 V. (Small values of R_1 give low amplifier voltage gain.) A minimum of perhaps 6 V is also required across R_7 to keep a reasonably constant current level through D_2.

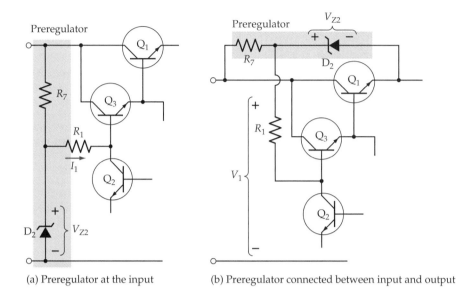

(a) Preregulator at the input   (b) Preregulator connected between input and output

Figure 18-9   A preregulator circuit improves the performance of a voltage regulator.

Figure 18-9b shows another preregulator circuit that has R_7 and D_2 connected across transistor Q_1. This gives an R_1 supply voltage of $V_1 = (V_o + V_{Z2})$.

Example 18-5

Determine suitable component values for a preregulator circuit as in Fig. 18-9a for the voltage regulator modified in Example 18-4.

Solution

Select $V_{R1(min)} = 3$ V

$$R_1 = \frac{V_{R1}}{I_{C2} + I_{B3}} = \frac{3 \text{ V}}{1 \text{ mA} + 0.21 \text{ mA}}$$

$$= 2.5 \text{ k}\Omega \text{ (use 2.7 k}\Omega \text{ standard value)}$$

$$V_{Z2} = V_o + V_{BE1} + V_{BE3} + V_{R1}$$

$$= 12 \text{ V} + 0.7 \text{ V} + 0.7 \text{ V} + 3 \text{ V}$$

$$= 16.4 \text{ V (Use a IN966A with } V_Z = 16 \text{ V. Alternatively,}$$
$$\text{use a 1N753 and a 1N758 connected in series.}$$
$$\text{See data sheet A-4 in Appendix A).}$$

$$I_{R7} \gg I_1 \text{ and } I_{Z2} > (I_{ZK} \text{ for Zener diode } D_2)$$

Select $I_{R7} = 5$ mA

$$R_2 = \frac{V_S - V_{Z2}}{I_{R7}} = \frac{20 \text{ V} - 16 \text{ V}}{5 \text{ mA}}$$

$$= 800 \ \Omega \text{ (use 820 } \Omega \text{ standard value)}$$

Constant-Current Source

The constant-current source shown in Fig. 18-10 may be used in place of resistor R_1 as an alternative to a preregulator circuit. This arrangement passes all the required current to Q_3 base and Q_2 collector, but it behaves as a very high resistance $(1/h_{oe4})$ at the collector of transistor Q_2. Consequently, it increases the error amplifier voltage gain and results in a substantial improvement in the regulator performance. To design the constant-current source, V_{CE4} should typically be a minimum of 3 V. The various voltage drops and current levels are then easily determined for component selection.

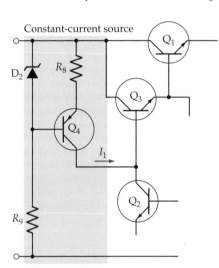

Figure 18-10   A constant-current source can be used as an alternative to a preregulator circuit to improve regulator performance.

Differential Amplifier

Figure 18-11 shows a regulator that uses a differential amplifier or *difference amplifier* (see Section 12-8). In this circuit, Zener diode D_2 is the voltage reference source, and D_1 is used only to provide an appropriate voltage level at the

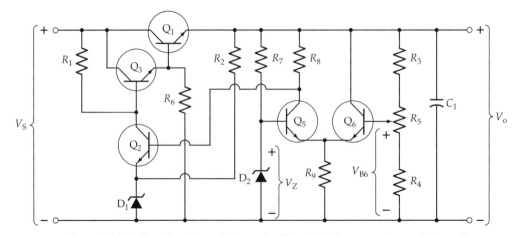

Figure 18-11   Use of a differential amplifier (Q_5 and Q_6) improves the regulator performance by improving the gain of the error amplifier and by using D_2 as a stable voltage reference source.

emitter of transistor Q_2. V_o is divided to provide V_{B6}, which is compared to V_{Z2}. Any difference in these two voltages is amplified by Q_5, Q_6, Q_2, and the associated components, and is then applied to the base of Q_3 to change the output in the required direction.

The performance of the regulator is improved by the increased voltage gain of the error amplifier. However, the performance is also improved in another way. In the regulator circuit in Figs. 18-2 and 18-7, when I_{C1} changes, the current through D_1 is also changed ($I_{Z1} = I_{R2} + I_{E2}$). This causes the Zener voltage to change by a small amount, and this change in V_{Z1} produces a change in output voltage. In the circuit in Fig. 18-11, the current through the reference diode (D_2) remains substantially constant because it is supplied from the regulator output. Therefore, there is no change in output voltage due to a change in the reference voltage.

Example 18-6

Design the differential amplifier stage for the regulator in Fig. 18-11. Use the Q_1, Q_2, Q_3 stage already designed in Examples 18-2 and 18-4. The output voltage is to be adjustable from 10 V to 12 V. (See Fig. 18-12.)

Solution

$$V_{C5} = V_{B2} = 9.8 \text{ V}$$

Select

$$V_{CE5} = 3 \text{ V}$$

$$V_{R9} = V_{C5} - V_{CE5} = 9.8 \text{ V} - 3 \text{ V}$$

$$= 6.8 \text{ V}$$

$$V_{Z2} = V_{R9} + V_{BE5} = 6.8 \text{ V} + 0.7 \text{ V}$$

$$= 7.5 \text{ V (use a 1N755 Zener diode)}$$

$$I_{C5} \gg I_{B2}$$

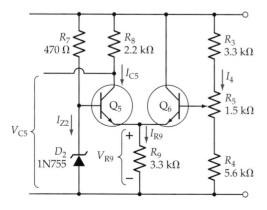

Figure 18-12 Differential amplifier circuit for Example 18-6.

Select
$$I_{C5} = 1 \text{ mA}$$

$$R_8 = \frac{V_o - V_{C2}}{I_{C5}} = \frac{12 \text{ V} - 9.8 \text{ V}}{1 \text{ mA}}$$

$$= 2.2 \text{ k}\Omega \text{ (standard value)}$$

$$I_{R9} \approx 2 \times I_{C5} = 2 \text{ mA}$$

$$R_9 = \frac{V_{R9}}{I_{R9}} = \frac{6.8 \text{ V}}{2 \text{ mA}}$$

$$= 3.4 \text{ k}\Omega \text{ (use 3.3 k}\Omega \text{ standard value)}$$

$$I_{Z2} \gg I_{B5} \text{ and } I_{Z2} > (I_{ZK} \text{ for the Zener diode})$$

Select
$$I_{Z2} = 10 \text{ mA}$$

$$R_7 = \frac{V_o - V_{Z2}}{I_{Z2}} = \frac{12 \text{ V} - 7.5 \text{ V}}{10 \text{ mA}}$$

$$= 450 \ \Omega \text{ (use 470 }\Omega \text{ standard value)}$$

$$I_4 \gg I_{B6}$$

Select
$$I_4 = 1 \text{ mA}$$

$$V_{B6} = V_{Z2} = 7.5 \text{ V}$$

When $V_o = 11$ V (moving contact at the top of R$_5$),

$$R_3 = \frac{V_o - V_{B6}}{I_4} = \frac{11 \text{ V} - 7.5 \text{ V}}{1 \text{ mA}}$$

$$= 3.5 \text{ k}\Omega \text{ (use 3.3 k}\Omega \text{ standard value)}$$

I_4 becomes
$$I_4 = \frac{V_o - V_{B6}}{R_3} = \frac{11 \text{ V} - 7.5 \text{ V}}{3.3 \text{ mA}}$$

$$= 1.06 \text{ mA}$$

$$R_4 + R_5 = \frac{V_{B6}}{I_4} = \frac{7.5 \text{ V}}{1.06 \text{ mA}}$$

$$= 7.07 \text{ k}\Omega$$

When $V_o = 13$ V (moving contact at the bottom of R$_5$),

I_4 becomes
$$I_4 = \frac{V_o}{R_3 + R_4 + R_5} = \frac{13 \text{ V}}{3.3 \text{ k}\Omega + 7.07 \text{ k}\Omega}$$

$$= 1.25 \text{ mA}$$

$$R_4 = \frac{V_{B6}}{I_4} = \frac{7.5 \text{ V}}{1.25 \text{ mA}}$$

$$= 6.25 \text{ k}\Omega \text{ (use 5.6 k}\Omega \text{ standard value)}$$

$$R_5 = (R_4 + R_5) - R_4 = 7.07 \text{ k}\Omega - 5.6 \text{ k}\Omega$$

$$= 1.47 \text{ k}\Omega \text{ (use 1.5 k}\Omega \text{ standard value potentiometer)}$$

Practice Problems

18-2.1 A voltage regulator circuit as in Fig. 18-6 uses a 6.2 V Zener diode. Determine suitable resistor values for R_3, R_4, and R_5 to produce an output adjustable from 9 V to 12 V.

18-2.2 A voltage regulator has an 18 V supply, a 10 V output, and a 150 mA load current. The circuit uses Darlington connected transistors, as in Fig. 18-7. Calculate suitable resistor values, and specify transistor Q_1. Assume that $h_{FE1} = 20$ and $h_{FE3} = 50$.

18-2.3 Determine suitable components for the constant-current circuit in Fig. 18-10. Assume that the supply voltage is 20 V, the output is 12 V, and that I_{B3} is 100 μA.

18-3 CURRENT LIMITING

Short-Circuit Protection

Power supplies used in laboratories are subject to overloads and short circuits. Short-circuit protection by means of current-limiting circuits is necessary in such equipment to prevent the destruction of components when an overload occurs. Transistor Q_7 and resistor R_{10} in Fig. 18-13a constitute a *current-limiting circuit*. When the load current (I_L) flowing through resistor R_{10} is below the maximum design level, the voltage drop V_{R10} is not large enough to forward-bias the base-emitter junction of Q_7. In this case, Q_7 has no effect on the regulator performance.

When the load current reaches the selected maximum ($I_{L(max)}$), V_{R10} biases Q_7 on. Current I_{C7} then produces a voltage drop across resistor R_1 that drives the output voltage down to nearly zero.

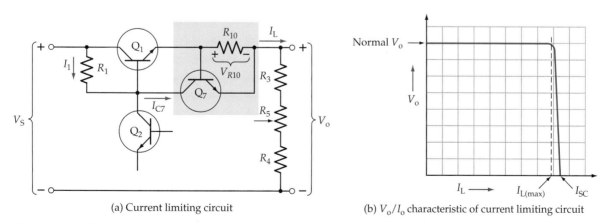

(a) Current limiting circuit   (b) V_o/I_o characteristic of current limiting circuit

Figure 18-13 Regulator overload protection circuit. When I_L increases to a maximum design level, V_{R10} causes Q_7 to conduct, and this pulls Q_1 base down, causing V_o to be reduced to almost zero.

The voltage/current characteristic of the regulator is shown in Fig. 18-13b. It is seen that output voltage remains constant as the load current increases up to $I_{L(max)}$. Beyond $I_{L(max)}$, V_o drops to zero, and a short-circuit current (I_{SC}) slightly greater than $I_{L(max)}$ flows at the output. Under these circumstances, series-pass transistor Q_1 is carrying all of the short-circuit current and has virtually all of the supply voltage developed across its terminals. So the power dissipation in Q_1 is

$$P_1 = V_S \times I_{SC} \qquad \text{(18-5)}$$

Obviously, a Q_1 must be selected to survive this power dissipation.

Design of the current-limiting circuit in Fig. 18-13 is very simple. Assuming that Q_7 is a silicon transistor, it should begin to conduct when $V_{R10} \approx 0.5$ V. Therefore, R_{10} is calculated as

$$R_{10} \approx \frac{0.5\ \text{V}}{I_{L(max)}} \qquad \text{(18-6)}$$

Fold-Back Current Limiting

A problem with the simple short-circuit protection method just discussed is that there is a large amount of power dissipation in the series-pass transistor while the regulator remains short-circuited. The fold-back current-limiting circuit in Fig. 18-14a minimizes this transistor power dissipation. The graph of V_o/I_L in Fig. 18-14b shows that the regulator output voltage remains constant until $I_{L(max)}$ is approached. Then the current reduces (or folds back) to a lower short-circuit current level (I_{SC}). The lower level of I_{SC} produces a lower power dissipation in Q_1.

To understand how the circuit in Fig. 18-14 operates, first note that when the output is shorted, V_o equals zero. Consequently, the voltage drop across

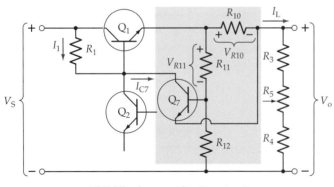

(a) Foldback current limiting circuit

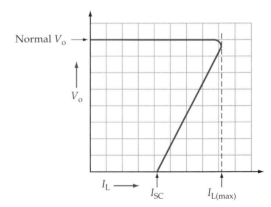

(b) Characteristic of fold-back current-limiting circuit

Figure 18-14   Fold-back current limiting. The short-circuit current (I_{SC}) is less than $I_{L(max)}$, and thus the power dissipation in Q_1 is minimized.

resistor R_{11} is almost zero. The voltage across R_{10} is $I_{SC} \times R_{10}$, and (as in the simple short-circuit protection circuit) this is designed to just keep transistor Q_7 biased on. When the regulator is operating normally with I_L less than $I_{L(max)}$, the voltage drop across R_{11} is

$$V_{R11} \approx \frac{V_o \times R_{11}}{R_{11} + R_{12}}$$

To turn Q_7 on, the voltage drop across R_{10} must become larger than V_{R11} by enough to forward-bias the base-emitter junction of Q_7. Using $V_{BE7} \approx 0.5$ V and making $V_{R11} \approx 0.5$ V gives

$$I_{L(max)} R_{10} = 0.5 \text{ V} + 0.5 \text{ V} = 1 \text{ V}$$

With $I_{SC}R_{10} = 0.5$ V,

$$I_{L(max)} \approx 2 I_{SC} \text{ (see Fig. 18-14 b)}$$

If V_{R11} is selected as 1 V,

$$I_{L(max)} R_{10} = 0.5 \text{ V} + 1 \text{ V} = 1.5 \text{ V}$$

and
$$I_{L(max)} \approx 3 I_{sc}$$

The fold-back current-limiting circuit is designed by first calculating R_{10} to give the desired level of I_{SC}. Then, the voltage-divider resistors (R_{11} and R_{12}) are calculated to provide the necessary voltage drop across R_{11} for the required relationship between I_{SC} and $I_{L(max)}$.

Example 18-7

Design a fold-back current-limiting circuit for a voltage regulator with a 12 V output. The maximum output current is to be 200 mA, and the short-circuit current is to be 100 mA. (See Fig. 18-15.)

Solution

$$I_{SC} = 100 \text{ mA}$$

Select $\qquad V_{R10} \approx 0.5$ V at short circuit

$$R_{10} = \frac{V_{R10}}{I_{SC}} = \frac{0.5 \text{ V}}{100 \text{ mA}}$$

$$= 5 \text{ }\Omega \text{ (use 4.7 }\Omega \text{ standard value)}$$

At $I_{L(max)}$, $\qquad V_{R10} = I_{L(max)} \times R_{10} = 200 \text{ mA} \times 4.7 \text{ }\Omega$

$$= 0.94 \text{ V}$$

$$V_{R11} = (I_{L(max)}R_{10}) - 0.5 \text{ V}$$

$$= 0.44 \text{ V}$$

$$I_{11} \gg I_{B7}$$

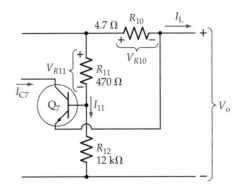

Figure 18-15 Fold-back current-limiting circuit for Example 18-7.

Select $\qquad I_{11} = 1 \text{ mA}$

$$R_{11} = \frac{V_{R11}}{I_{11}} = \frac{0.44 \text{ V}}{1 \text{ mA}}$$

$$= 440 \ \Omega \ (\text{use } 470 \ \Omega \text{ standard value})$$

$$R_{12} = \frac{V_o + V_{R10} - V_{R11}}{I_{11}} = \frac{12 \text{ V} + 0.94 \text{ V} - 0.44 \text{ V}}{1 \text{ mA}}$$

$$= 12.5 \ \text{k}\Omega \ (\text{use } 12 \ \text{k}\Omega \text{ standard value})$$

Practice Problems

18-3.1 A short-circuit protection circuit, as in Fig. 18-13, is to be designed to limit the output of a 15 V regulator to 400 mA. Select a suitable value for R_{10} and specify Q_1. Assume that $V_S = 25$ V.

18-3.2 Modify the circuit designed for Problem 18-3.1 to convert it to a fold-back current-limiting circuit with a 150 mA short-circuit current.

18-4 OP-AMP VOLTAGE REGULATORS

Voltage-Follower Regulator

Refer once again to the voltage regulator circuit in Fig. 18-11. The complete error amplifier has two input terminals at the bases of Q_5 and Q_6 and one output at the collector of Q_2. Transistor Q_6 base is an inverting input, and Q_5 base is a non-inverting input. The error amplifier circuit is essentially an operational amplifier. Thus, IC operational amplifiers with their extremely high open-loop voltage gain are ideal for use as error amplifiers in dc voltage regulator circuits. Normally, an internally compensated op-amp (such as the 741) is quite suitable for most voltage regulator applications.

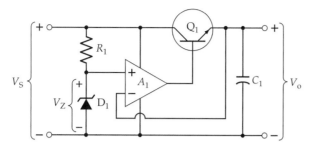

Figure 18-16 Operational amplifier connected as a voltage-follower regulator. The output voltage is held constant at V_z.

A simple voltage-follower regulator circuit is illustrated in Fig. 18-16. In this circuit, the op-amp output voltage always follows the voltage at the non-inverting terminal; consequently, V_o remains constant at V_Z. The only design calculations are those required for the design of the Zener diode voltage reference circuit (R_1 and D_1) and for the specification of Q_1.

Adjustable Output Regulator

The circuit in Fig. 18-17 is that of a variable-output, highly stable dc voltage regulator. As in the transistor circuit in Fig. 18-11, the reference diode in Fig. 18-17 is connected at the amplifier non-inverting input, and the output voltage is divided and applied to the inverting input. The operational amplifier positive supply terminal has to be connected to the regulator supply voltage. If it were connected to the regulator output, the op-amp output voltage (at Q_2 base) would have to be approximately 0.7 V higher than its positive supply terminal, and this is impossible.

Design of the regulator circuit in Fig. 18-17 involves selecting R_1 and D_1, design of the voltage-divider network (R_3, R_4, and R_5), and specification of transistors Q_1 and Q_2. Clearly, an op-amp voltage regulator is more easily designed than a purely transistor regulator circuit.

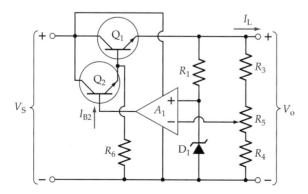

Figure 18-17 Highly stable, adjustable-output voltage regulator using an operational amplifier.

Example 18-8

Design the voltage regulator circuit in Fig. 18-18 to give an output voltage adjustable from 12 V to 15 V. The maximum output current is to be 250 mA, and the supply voltage is 20 V. Assuming that $h_{FE1} = 20$ and $h_{FE2} = 50$, estimate the op-amp maximum output current.

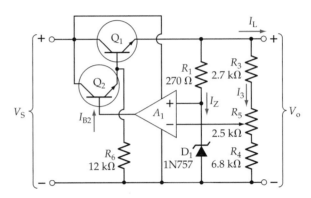

Figure 18-18   Op-amp voltage regulator for Example 18-8.

Solution

$$V_Z \approx 0.75\, V_{o(min)} = 0.75 \times 12\ \text{V}$$

$$= 9\ \text{V}$$

Use a 1N757 diode with $V_Z = 9.1$ V

$$I_Z \gg (\text{op-amp } I_{B(max)}), \text{ and } I_Z > (I_{ZK} \text{ for the diode})$$

Select
$$I_Z = 10\ \text{mA}$$

$$R_1 = \frac{V_{o(min)} - V_Z}{I_{Z1}} = \frac{12\ \text{V} - 9.1\ \text{V}}{10\ \text{mA}}$$

$$= 290\ \Omega \text{ (use 270 } \Omega \text{ standard value)}$$

$$I_{3(min)} \gg (\text{op-amp } I_{B(max)})$$

Select
$$I_{3(min)} = 1\ \text{mA}$$

When $V_o = 12$ V (moving contact at top of R_5)

$$R_3 = \frac{V_o - V_Z}{I_{3(min)}} = \frac{12\ \text{V} - 9.1\ \text{V}}{1\ \text{mA}}$$

$$= 2.9\ \text{k}\Omega \text{ (use 2.7 k}\Omega \text{ standard value)}$$

$$R_4 + R_5 = \frac{V_Z}{I_{3(min)}} = \frac{9.1\ \text{V}}{1\ \text{mA}}$$

$$= 9.1\ \text{k}\Omega$$

When $V_o = 15$ V (moving contact at bottom of R_5),

I_3 becomes
$$I_3 = \frac{V_o}{R_3 + R_4 + R_5} = \frac{15 \text{ V}}{2.7 \text{ k}\Omega + 9.1 \text{ k}\Omega}$$

$$= 1.27 \text{ mA}$$

$$R_4 = \frac{V_Z}{I_3} = \frac{9.1 \text{ V}}{1.27 \text{ mA}}$$

$$= 7.16 \text{ k}\Omega \text{ (use 6.8 k}\Omega \text{ standard value)}$$

$$R_5 = (R_4 + R_5) - R_4 = 9.1 \text{ k}\Omega - 6.8 \text{ k}\Omega$$

$$= 2.3 \text{ k}\Omega \text{ (use 2.5 k}\Omega \text{ potentiometer)}$$

Select
$$I_{R6} = 0.5 \text{ mA}$$

$$R_6 \approx \frac{V_o}{I_{R6}} = \frac{12 \text{ V}}{0.5 \text{ mA}}$$

$$= 24 \text{ k}\Omega \text{ (use 22 k}\Omega \text{ standard value)}$$

Op-amp output current:

$$I_{B2} = \frac{I_{L(max)}}{h_{FE1} \times h_{FE2}} = \frac{250 \text{ mA}}{20 \times 50}$$

$$= 0.25 \text{ mA}$$

Current Limiting with an Op-amp Regulator

When a large output current is to be supplied by an operational amplifier voltage regulator, one of the current-limiting circuits described in Section 18-3 may be used with one important modification. Figure 18-19 shows the modification. A resistor (R_{13}) must be connected between the op-amp output terminal and the junction of Q_{2B} and Q_{7C}. When an overload causes the regulator output voltage to go to zero, the op-amp output goes high (close to V_S) as it

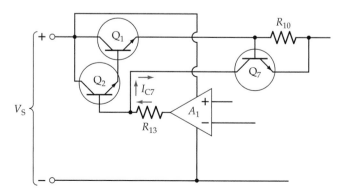

Figure 18-19   When current limiting is used with an op-amp regulator, a resistor (R_{13}) must be included in series with the op-amp output.

attempts to return V_o to its normal level. Consequently, because the op-amp normally has a very low output resistance, R_{13} is necessary to allow I_{C7} to drop the voltage at Q_2 base to near ground level. The additional resistor at the op-amp output is calculated as $R_{13} \approx V_S/I_{C7}$. Resistor R_{13} must not be so large that there is an excessive voltage drop across it when the regulator is supposed to be operating normally.

Practice Problems

18-4.1 Design an op-amp voltage-regulator circuit as in Fig. 18-17 to produce an output adjustable from 15 V to 18 V, with a 300 mA maximum load current. The supply voltage is 25 V.

18-4.2 Design a fold-back current-limiting circuit for the regulator in Problem 18-4.1, to give $I_{L(max)} = 300$ mA and $I_{SC} = 200$ mA when $V_o = 15$ V.

18-5 IC LINEAR VOLTAGE REGULATORS

723 IC Regulator

The basic circuit of a 723 IC voltage regulator in a dual-in-line package is shown in Fig. 18-20. This IC has a voltage reference source (D_1), an error amplifier (A_1), a series pass transistor (Q_1), and a current-limiting transistor (Q_2), all contained in one small package. An additional Zener diode (D_2) is included for voltage dropping in some applications. The IC can be connected to function as a positive or negative voltage regulator with an output voltage ranging from 2 V to 37 V and output current levels up to 150 mA. The maximum supply voltage is 40 V, and the line and load regulations are each specified as 0.01%. A partial specification for the 723 regulator is given in data sheet A-16 in Appendix A.

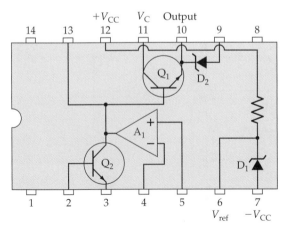

Figure 18-20 The 723 IC voltage regulator contains a reference diode (D_1), an error amplifier (A_1), a series-pass transistor (Q_1), a current limiting transistor (Q_2), and a voltage-dropping diode (D_2).

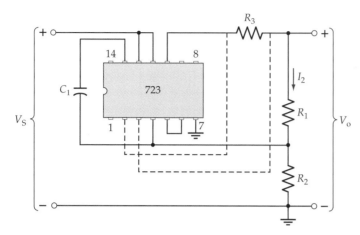

Figure 18-21   Voltage regulator circuit using a 723 IC regulator. The output voltage range is 7 V to 37 V.

Figure 18-21 shows a 723 connected to function as a positive voltage regulator. The complete arrangement (including the internal circuitry shown in Fig. 18-20) is similar to the op-amp regulator circuit in Fig. 18-17. One difference between the two circuits is the 100 pF capacitor (C_1) connected to the error amplifier output and its inverting input terminal. This capacitor is used instead of a large capacitor at the output terminals to prevent the regulator from oscillating. (Sometimes both capacitors are required.) By appropriate selection of resistors R_1 and R_2 in Fig. 18-21, the regulator output can be set to any level between 7.15 V (the reference voltage) and 37 V. A potentiometer can be included between R_1 and R_2 to make the output voltage adjustable.

The dashed lines (in Fig. 18-21) show connections for simple (non-fold-back) current limiting. Fold-back current limiting can also be used with the 723.

It is important to note that, as for all linear regulator circuits, the supply voltage at the lowest point on the ripple waveform should be at least 3 V greater than the regulator output; otherwise a high-amplitude output ripple might occur. The total power dissipation in the regulator should be calculated to ensure that it does not exceed the specified maximum. The specification lists 1.25 W as the maximum power dissipation at a free air temperature of 25°C for a DIP package. This must be derated at 10 mW/°C for higher temperatures. For a metal can package, $P_{D(max)} = 1$ W at 25°C free air temperature, and the derating factor is 6.6 mW/°C. An external series-pass transistor may be Darlington-connected to (internal transistor) Q_1 to enable a 723 regulator to handle a larger load current. This is illustrated in Fig. 18-22.

A regulator output voltage less than the 7.15 V reference level can be obtained by using a voltage divider across the reference source (terminals 6 and 7 in Fig. 18-20). Terminal 5 is connected to the reduced reference voltage, instead of to terminal 6.

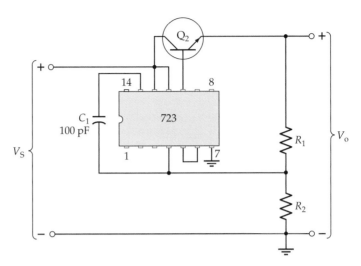

Figure 18-22 An external series-pass transistor may be added to a 723 IC voltage regulator to supply higher load currents than the 723 can normally handle.

Example 18-9

The regulator circuit in Fig. 18-23 is to have an output of 10 V. Calculate resistor values for R_1 and R_2, select a suitable input voltage, and determine the maximum load current that may be supplied if $P_{D(max)} = 1000$ mW.

Solution

$$I_2 \gg \text{(error amplifier input bias current)}$$

Select $\quad I_2 = 1$ mA

$$V_{R2} = V_{ref} = 7.15 \text{ V}$$

$$R_2 = \frac{V_{ref}}{I_2} = \frac{7.15 \text{ V}}{1 \text{ mA}}$$

$$= 7.15 \text{ k}\Omega \text{ (use 6.8 k}\Omega \text{ standard value, and recalculate } I_2)$$

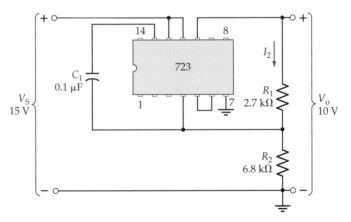

Figure 18-23 Regulator circuit for Example 18-9.

I_2 becomes $I_2 = \dfrac{V_{\text{ref}}}{R_2} = \dfrac{7.15 \text{ V}}{6.8 \text{ k}\Omega}$

$= 1.05 \text{ mA}$

$R_1 = \dfrac{V_{\text{o}} - V_{\text{ref}}}{I_1} = \dfrac{10 \text{ V} - 7.15 \text{ V}}{1.05 \text{ mA}}$

$\approx 2.85 \text{ k}\Omega$ (use 2.7 kΩ standard value)

For satisfactory operation of the series pass transistor,

Select $V_S - V_{\text{o}} = 5 \text{ V}$

$V_S = V_{\text{o}} + 5 \text{ V} = 10 \text{ V} + 5 \text{ V}$

$= 15 \text{ V}$

The internal circuit current is

$$I_{\text{(standby)}} + I_{\text{ref}} \approx 25 \text{ mA}$$

The 723 internal power dissipation on no-load is

$P_{\text{i}} = V_S \times (I_{\text{(standby)}} + I_{\text{ref}}) = 15 \text{ V} \times 25 \text{ mA}$

$= 375 \text{ mW}$

Maximum power dissipated in the (internal) series-pass transistor:

$P_D = (\text{specified } P_{D(\text{max})}) - P_{\text{i}} = 1000 \text{ mW} - 375 \text{ mW}$

$= 625 \text{ mW}$

Maximum load current:

$I_{L(\text{max})} = \dfrac{P_D}{V_S - V_{\text{o}}} = \dfrac{625 \text{ mW}}{5 \text{ V}}$

$= 125 \text{ mA}$

LM317 and LM337 IC Regulators

The LM317 and LM337 IC regulators are three-terminal devices that are extremely easy to use. The 317 is a positive voltage regulator (Fig. 18-24a), and the 337 is a negative voltage regulator (Fig. 18-24b). In each case, input and output terminals are provided for supply and regulated output voltage, and an adjustment terminal (ADJ) is included for output voltage selection. The output voltage range is 1.2 V to 37 V, and the maximum load current ranges from 300 mA to 2 A, depending on the type of device package. Typical line and load regulations are specified as 0.01%/(volt of V_{o}), and 0.3%/(volt of V_{o}), respectively.

The internal reference voltage for the 317 and 337 regulators is typically 1.25 V, and V_{ref} appears across the ADJ and *output* terminals. Consequently, the regulator output voltage is

$$V_{\text{o}} = \dfrac{R_1 + R_2}{R_1} \times V_{\text{REF}} \qquad (18\text{-}7)$$

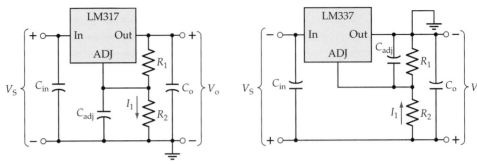

(a) LM317 positive voltage regulator   (b) LM337 negative voltage regulator

Figure 18-24   Application of the LM317 and LM337 integrated circuit voltage regulators. The internal reference voltage appears across resistor R_1.

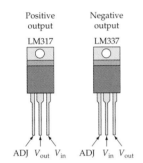

Figure 18-25 Terminal connections for LM317 and LM337 IC regulators contained in 221A packages.

To determine suitable values for R_1 and R_2 for a desired output voltage, first choose a voltage-divider current (I_1) that is much larger than the current that flows in the ADJ terminal of the device. This is specified as 100 μA maximum on the device data sheet. The resistors are calculated from the relationship in Equation 18-7.

Note the capacitors included in the regulator circuits. Capacitor C_{in} is necessary only when the regulator is not located close to the power supply filter circuit. C_{in} eliminates the oscillatory tendencies that can occur with long connecting leads between the filter and regulator. Capacitor C_o improves the transient response of the regulator and ensures ac stability, and C_{adj} improves the ripple rejection ratio.

Figure 18-25 shows the terminal connections for LM317 and LM337 regulators in 221A-type packages.

Example 18-10

An LM317 regulator is to provide a 6 V output from a 15 V supply. The load current is 200 mA. Determine suitable resistance values for R_1 and R_2 (in Fig. 18-24), and calculate the regulator power dissipation.

Solution

$$I_1 \gg I_{ADJ}$$

Select $I_1 = 1 \text{ mA}$

$$R_1 = \frac{V_{ref}}{I_1} = \frac{1.25 \text{ V}}{1 \text{ mA}}$$

$$= 1.25 \text{ k}\Omega \text{ (use 1.2 k}\Omega \text{ standard value)}$$

$$R_2 = \frac{V_o - V_{ref}}{I_1} = \frac{6 \text{ V} - 1.25 \text{ V}}{1 \text{ mA}}$$

$$= 4.75 \text{ k}\Omega \text{ (use 4.7 k}\Omega \text{ standard value)}$$

$$P_D = (V_S - V_o) \times I_{L(max)} = (15 \text{ V} - 6 \text{ V}) \times 200 \text{ mA}$$

$$= 1.8 \text{ W}$$

LM340 Regulators

LM340 devices are three-terminal positive voltage regulators with fixed output voltages ranging from a low of 5 V to a high of 23 V. The regulator is selected for the desired output voltage and then simply provided with a voltage (V_S) from a power supply filter circuit, as illustrated in Fig. 18-26. Here again, capacitor C_1 is required only when the regulator is not close to the filter.

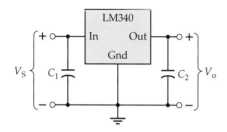

Figure 18-26   The LM340 is a fixed output IC voltage regulator with an output selectable from 5 V to 23 V.

The LM340 data sheet specifies the regulator performance for an output current of 1 A. The tolerance on the output voltage is ±2%, the line regulation is *0.01% per output volt,* and the load regulation is *0.3% per amp of load current.* The IC includes current limiting and a thermal shutdown circuit that protects against excessive internal power dissipation. As with all series regulators, a heat sink must be used when high power dissipation is involved.

Practice Problems

18-5.1   Design the regulator circuit in Fig. 18-22 to have an 18 V, 200 mA output. Include short-circuit protection, and specify Q_2.

18-5.2   Using an LM317, design a 9 V, 150 mA voltage regulator. Select a supply voltage, and calculate the regulator power dissipation.

18-6 SWITCHING REGULATOR BASICS

Switching Regulator Operation

A switching regulator can be thought of as similar to a linear regulator but with the series-pass transistor operating as a switch that is either off or switched on (in a saturated state). The output voltage from the switch is a pulse waveform which is smoothed into a dc voltage by the action of an *LC* filter.

Switching regulators can be classified as

- *step-down converter* (output voltage lower than input);
- *step-up converter* (output higher than input); or
- *inverting converter* (output polarity opposite to input).

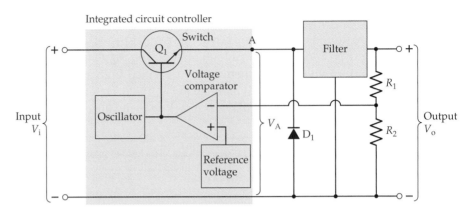

Figure 18-27 A switching regulator has a switch (contained in an IC controller), a filter, and a diode. The switch converts the dc input into a pulse waveform, and the filter smooths the pulse wave into a direct voltage with a ripple waveform.

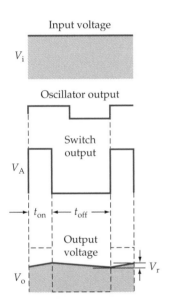

Figure 18-28 Switching regulator waveforms. The switch output is a pulse waveform, and the filter output is a dc voltage with a ripple.

The basic block diagram of a step-down switching regulator in Fig. 18-27 consists of a BJT *switch* (Q_1) (also termed a *power switch*), an *oscillator*, a *voltage comparator*, a *voltage reference source*, a *diode* (D_1), and a *filter*. The switch, oscillator, comparator, and reference source are all usually contained within an integrated circuit *controller*, as illustrated. The filter usually consists of an inductor and capacitor. The operation of the regulator is as follows:

- Refer to the waveforms in Fig. 18-28. The dc input voltage (V_i) is converted into a pulse waveform (V_A) by the action of the switch (Q_1) turning on and off.
- The oscillator switches Q_1 *on*, causing current to flow to the filter and the output voltage to rise.
- The voltage comparator compares the output voltage (divided by R_1 and R_2) to the reference voltage, and it holds Q_1 *on* until V_o equals V_{ref}. Then, Q_1 is turned *off* again.
- A pulse waveform (V_A) is produced at the input of the filter by Q_1 turning *on* and *off*.
- The filter smooths the pulse waveform to produce a dc output voltage (V_o) with a *ripple* waveform (V_r).
- The ripple waveform is the result of the filter capacitor charging via the inductor during t_{on}, and then discharging to the load during t_{off} via D_1. (The is explained further in Section 18-7.)

In the operation described above, the controller can be thought of as a *pulse width modulator*; the *on* time of Q_1 (*pulse width* of its output) is increased or decreased as necessary to supply the output current required. Other systems involve control of the switch *off* time.

Comparison of Linear and Switching Regulators

The power dissipated in the series-pass transistor in a linear regulator is wasted power. This is not very important when the load current is lower than 500 mA. With high current levels, the regulator efficiency becomes important, and there can also be serious heat dissipation problems.

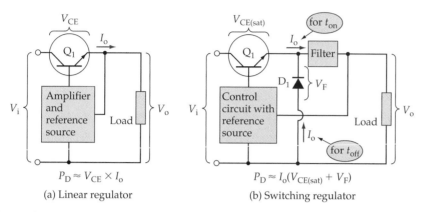

$$P_D \approx V_{CE} \times I_o$$

(a) Linear regulator

$$P_D \approx I_o(V_{CE(sat)} + V_F)$$

(b) Switching regulator

Figure 18-29   Comparison of power dissipation in linear and switching regulators.

In a switching regulator, the power dissipation in the switching transistor (whether it is on or off) is very much smaller than in the series-pass transistor of a linear regulator with a similar output voltage and load current. Thus, a switching regulator is more efficient than a linear regulator.

The approximate efficiency of a linear regulator can be calculated by assuming that the only wasted power (P_D) is that dissipated in the series-pass transistor (see Fig. 18-29a).

$$P_i \approx P_o + (V_{CE} \times I_o) \qquad \text{(18-8)}$$

To estimate the efficiency of a switching regulator, it should be noted that D_1 (in Fig. 18-27) is biased *off* when Q_1 is *on*, and that D_1 is *on* when Q_1 is *off*. Thus, the power dissipated in the switching transistor and diode can be taken as the total wasted power (Fig. 18-29b). The Q_1 and D_1 power dissipations can be calculated in terms of the actual current in each device and the on and off times. This analysis shows that P_D for each device is simply [(average output current) × (device voltage drop)]. Therefore,

$$P_i \approx P_o + I_o(V_{CE(sat)} + V_F) \qquad \text{(18-9)}$$

Since there is much less power dissipation in the transistor in a switching regulator than in the series-pass transistor of a linear regulator with the same output conditions, a lower power rated transistor can be used in the switching regulator. Also, with a switching regulator there is usually no need for the heat sink normally required with a series regulator. Example 18-11 compares switching and linear regulators with similar load requirements.

Example 18-11

Calculate the efficiencies of a linear regulator and a switching regulator that each supply 10 V, 1 A loads. The series-pass transistor of the linear regulator has $V_{CE} = 7$ V. The switching regulator has $V_{CE(sat)} = V_F = 1$ V.

Solution

$$P_o = V_o \times I_o = 10\text{ V} \times 1\text{ A}$$

$$= 10\text{ W}$$

Linear Regulator:

Eq. 18-8: $\qquad P_i \approx P_o + (V_{CE} \times I_o) = 10\text{ W} + (7\text{ V} \times 1\text{ A})$

$$= 17\text{ W}$$

$$\text{Efficiency} = \frac{P_o \times 100\%}{P_i} = \frac{10\text{ W} \times 100\%}{17\text{ W}}$$

$$= 59\%$$

Switching Regulator:

Eq. 18-9: $\qquad P_i \approx P_o + I_o(V_{CE(sat)} + V_F) = 10\text{ W} + 1\text{ A }(1\text{ V} + 1\text{ V})$

$$= 12\text{ W}$$

$$\text{Efficiency} = \frac{P_o \times 100\%}{P_i} = \frac{10\text{ W} \times 100\%}{12\text{ W}}$$

$$= 83\%$$

The efficiency calculations in the above example are approximate, because power dissipations in other parts of the circuits are neglected.

As well as efficiency, there are other considerations in the choice between linear and switching regulators. The output ripple voltage is substantially larger with a switching regulator than with a linear regulator. The source effect can be similar with both types of regulator, but the load effect is usually largest with a switching regulator. For low levels of output power, a switching regulator is usually more expensive than a linear regulator. A linear regulator is usually the best choice for output power levels up to 10 W. A switching regulator should be considered when the output is above 10 W. Table 18-1 compares the two types of voltage regulator.

TABLE 18-1　Comparison of linear and switching regulator performance

	Linear Regulator	Switching Regulator
Typical Efficiency	30% to 70%	85%
Source Effect	<10 mV	<10 mV
Load Effect	<10 mV	>50 mV
Ripple Voltage	<10 mV (120 Hz)	100 mV (10 kHz to 200 kHz)

Practice Problem

18-6.1 Calculate the approximate efficiencies of linear and switching voltage regulators that each supply a 15 V, 750 mA load. The input voltage to both regulators is 21 V. The switching regulator uses a FET with a drain-source resistance of $r_{DS(on)} = 0.6\ \Omega$ as the power switch, and a diode with $V_F = 0.7$ V.

18-7 STEP-DOWN, STEP-UP, AND INVERTING CONVERTERS

Step-Down Converter

A *step-down* switching regulator, or *step-down converter* (also termed a *buck* converter), produces a dc output voltage lower than its input voltage. The basic circuit arrangement for a step-down converter is shown in Fig. 18-30. Note the presence of the *catch diode* (D_1). This is normally reverse-biased when Q_1 is *on*, but becomes forward-biased when Q_1 switches *off*.

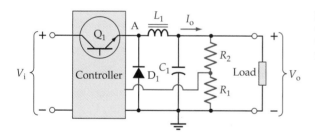

Figure 18-30 Step-down converter circuit using a controller, catch diode (D_1), inductor (L_1), and capacitor (C_1).

Figure 18-31a illustrates the situation when Q_1 is on (in saturation). Inductor current (I_ℓ) flows from V_i, producing an inductor voltage drop (V_ℓ) that is positive on the left and negative on the right. At this time D_1 is reverse-biased. When Q_1 switches off (Fig. 18-31b), the inductor has stored energy, and it opposes any change in current level. In order to maintain inductor current flow (with Q_1 off), the inductor voltage reverses, becoming negative on the left and positive on the right, as illustrated. The reversed polarity of V_ℓ forward-biases D_1 to provide a path for I_ℓ. If D_1 were not in the circuit, the inductor voltage would become large enough to break down the junctions of Q_1. Instead, D_1 *catches* V_ℓ and stops it from increasing; hence the name *catch diode*.

Step-Down Converter Equations

Figure 18-32 shows the circuit waveforms during the Q_1 on and off times (t_{on} and t_{off}). These are switch voltage (V_A) at point A (Q_1 emitter), inductor voltage (V_ℓ), inductor current (I_ℓ), and output voltage (V_o). Equations for calculating component values can be derived by considering the circuit conditions and the waveforms.

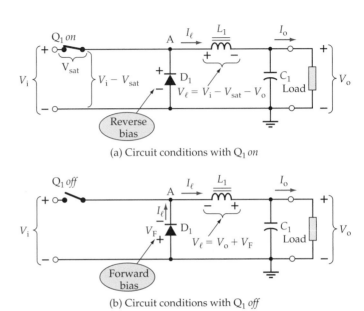

(a) Circuit conditions with Q_1 *on*

(b) Circuit conditions with Q_1 *off*

Figure 18-31   In a step-down converter, inductor current I_ℓ flows while Q_1 is on, producing a voltage drop (V_ℓ). When Q_1 switches off, V_ℓ reverses, and D_1 becomes forward-biased to keep I_ℓ flowing.

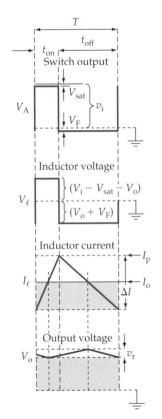

Figure 18-32   Step-down converter waveforms.

The input voltage (V_i) appears at the collector of transistor Q_1. When Q_1 is *on*, the voltage across the inductor and capacitor (in series) is $V_i - V_{sat}$, and the inductor voltage is $V_i - V_{sat} - V_o$, (see Fig. 18-31a). When Q_1 switches *off*, the inductor voltage becomes $V_o + V_F$ (Fig. 18-31b). Typically, $V_F = 0.7$ V and V_{sat} ranges from 0.2 V to 1.5 V, depending on the transistor and the load current. The energy supplied *to the inductor* during t_{on} is proportional to $[t_{on}(V_i - V_{sat} - V_o)]$, and the energy supplied *by the inductor* to the output during t_{off} is proportional to $[t_{off}(V_o + V_F)]$. Because the inductor input energy must equal its output energy,

$$t_{on}(V_i - V_{sat} - V_o) = t_{off}(V_o + V_F)$$

giving
$$\frac{t_{on}}{t_{off}} = \frac{V_o + V_F}{V_i - V_{sat} - V_o} \qquad (18\text{-}10)$$

From Fig. 18-32,

$$t_{on} + t_{off} = T = 1/f \qquad (18\text{-}11)$$

The inductor is charged from the supply via Q_1 during t_{on}, and it discharges to the load and the output capacitor via D_1 during t_{off}. When the output current is at a maximum ($I_{o(max)}$), the inductor current (I_l) can be allowed to change from zero to a peak level.

$$I_p = 2I_{o(max)} \qquad (18\text{-}12)$$

This is the absolute maximum inductor current change ($\Delta I = I_p$) that can occur when the circuit is operating correctly.

An equation for the inductance may be derived from a knowledge of the inductor voltage and the inductor current change during t_{on}.

$$L = \frac{e_L \Delta t}{\Delta I}$$

giving

$$L_{1(min)} = \frac{(V_i - V_{sat} - V_o)t_{on}}{I_p} \tag{18-13}$$

The output voltage waveform (Fig. 18-32) is also the capacitor voltage waveform. As illustrated, the capacitor voltage increases while I_ℓ is greater than I_o, and decreases during the time that I_ℓ is less than I_o. While I_ℓ is less than I_o, the average inductor current is $\Delta I/4$ below I_o, and the capacitor must supply this current to the load. The time involved is $T/2$, and the capacitor voltage change is the peak-to-peak output ripple voltage (V_r). Thus, an equation for the minimum capacitance of C_1 can be derived as follows:

$$C_1 = \frac{It}{\Delta V} = \frac{(\Delta I/4) \times T/2}{V_r}$$

or

$$C_{1(min)} = \frac{I_p}{8fV_r} \tag{18-14}$$

It should be noted that Equations 18-13 and 18-14 give minimum inductor and capacitor values. The use of component values larger than the minimum will result in a lower ripple voltage at the regulator output.

When components are selected, the capacitor *equivalent series resistance* (ESR) and the inductor *winding resistance* (R_W) must be taken into account. ESR can affect the output ripple voltage. The maximum ESR should usually be limited to $(0.1V_r/\Delta I)$. If R_W is too large, it will produce an excessive voltage drop across the inductor. The maximum winding resistive voltage drop should not usually exceed 0.2 V. Also, the peak level of the inductor current must be passed without saturating the core.

Resistance values for R_1 and R_2 (in Fig. 18-30) depend on the output and references voltage. This is treated in Section 18-8.

Example 18-12

A step-down switching regulator with a 30 V input is to have a 12 V, 0.5 A output with a 100 mV maximum ripple. The switching frequency is to be 50 kHz. Specify suitable filter components. Assume that $V_{sat} = 1$ V.

Solution

$$T = \frac{1}{f} = \frac{1}{50 \text{ kHz}}$$

$$= 20 \text{ μs (See Fig. 18-33.)}$$

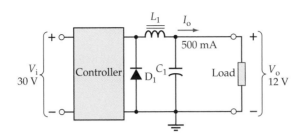

Figure 18-33 Step-down converter circuit for Example 18-12.

Eq. 18-10:
$$\frac{t_{on}}{t_{off}} = \frac{V_o + V_F}{V_i - V_{sat} - V_o} = \frac{12\text{ V} + 0.7\text{ V}}{30\text{ V} - 1\text{ V} - 12\text{ V}}$$

$$= 0.75$$

or
$$t_{on} = 0.75\, t_{off}$$

Eq. 18-11:
$$T = t_{on} + t_{off} = 0.75\, t_{off} + t_{off}$$

or
$$t_{off} = \frac{T}{1.75} = \frac{20\ \mu s}{1.75}$$

$$\approx 11.4\ \mu s$$

$$t_{on} = T - t_{off} = 20\ \mu s - 11.4\ \mu s$$

$$= 8.6\ \mu s$$

Eq. 18-12:
$$I_p = 2\, I_{o(max)} = 2 \times 500\text{ mA}$$

$$= 1\text{ A}$$

Eq. 18-13:
$$L_{1(min)} = \frac{(V_i - V_{sat} - V_o)t_{on}}{I_p} = \frac{(30\text{ V} - 1\text{ V} - 12\text{ V}) \times 8.6\ \mu s}{1\text{ A}}$$

$$= 146\ \mu H$$

Eq. 18-14:
$$C_{1(min)} = \frac{I_p}{8fV_r} = \frac{1\text{ A}}{8 \times 50\text{ kHz} \times 100\text{ mV}}$$

$$= 25\ \mu F$$

Step-Up Converter

A *step-up converter*, or *boost converter*, produces a dc output voltage higher than its supply voltage. In the circuit shown in Fig. 18-34a, L_1 is directly connected to the supply, and D_1 is in series with L_1 and C_1. The collector of Q_1 is connected to the junction of L_1 and D_1 (point A), and its emitter is grounded.

When Q_1 is on (Figure 18-34b), D_1 is reverse-biased, and

$$V_\ell = V_i - V_{sat}$$

When Q_1 is off (Figure 18-34c), V_ℓ reverses polarity to keep I_ℓ flowing. The output voltage is now

$$V_o = V_i + V_\ell - V_F$$

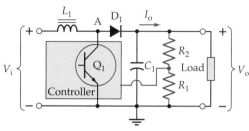

(a) Circuit of step-up converter

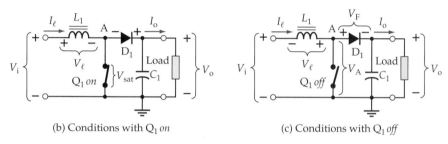

(b) Conditions with Q₁ on   (c) Conditions with Q₁ off

Figure 18-34 In a step-up converter circuit, the inductor current I_ℓ flows while Q_1 is on, producing V_ℓ. When Q_1 switches off, V_ℓ reverses and D_1 becomes forward-biased. The output voltage is $(V_i + V_\ell - V_F)$.

So, the output voltage is larger than the input, and its actual level can be set by selection of R_1 and R_2 (in Fig. 18-34a).

The voltage at point A has the pulse waveform illustrated in Fig. 18-35. During t_{on} (Q_1 on time), $V_A = V_{sat}$; and during t_{off}, $V_A = (V_i + V_\ell)$. The inductor current increases to I_P during t_{on} and decreases during t_{off} as it discharges to C_1 and the load. The output voltage (V_o across C_1) decreases during t_{on} because D_1 is reverse-biased and C_1 is supplying all the load current. During t_{off}, D_1 is forward-biased and C_1 is recharged from L_1. V_o increases while I_ℓ is greater than I_o, and decreases when I_ℓ is less than I_o.

From the circuit conditions and the waveforms, the following equations can be derived:

$$\frac{t_{on}}{t_{off}} = \frac{V_o + V_F - V_i}{V_i - V_{sat}} \tag{18-15}$$

$$I_P = \frac{2\,T\,I_{o(max)}}{t_{off}} \tag{18-16}$$

$$L_{1(min)} = \frac{(V_i - V_{sat})\,t_{on}}{I_P} \tag{18-17}$$

$$C_{1(min)} = \frac{t_{on}\,I_{o(max)}}{V_r} \tag{18-18}$$

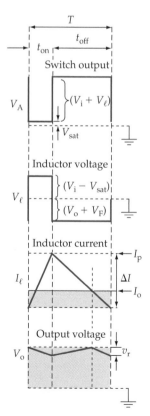

Figure 18-35 Step-up converter circuit waveforms.

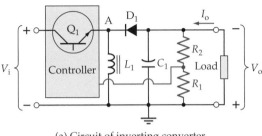

(a) Circuit of inverting converter

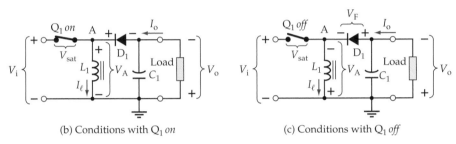

(b) Conditions with Q_1 *on*   (c) Conditions with Q_1 *off*

Figure 18-36 In an inverting converter circuit, the inductor current I_ℓ flows while Q_1 is on, producing V_ℓ. When Q_1 switches off, V_ℓ reverses and D_1 becomes forward-biased. The output voltage is $-(V_\ell - V_F)$.

Inverting Converter

Figure 18-36a shows the circuit of an *inverting converter* (also termed a *flyback converter*). This circuit produces a negative dc output voltage from a positive supply. In this case, L_1 is connected between ground and the emitter terminal of Q_1 (point A). Diode D_1 is in series with Q_1 emitter and C_1. Note the polarity of D_1 and output capacitor C_1.

Figure 18-36b shows that when Q_1 is *on*, D_1 is reverse-biased, and

$$V_A = V_\ell = V_i - V_{sat}.$$

From Fig. 18-36c, when Q_1 is *off*, V_ℓ reverses to keep I_ℓ flowing. Diode D_1 is now forward-biased, giving

$$V_o = -(V_\ell - V_F)$$

The output voltage is negative, and once again the output voltage level is set by the choice of R_1 and R_2.

The voltage at point A in Fig. 18-36 is also the inductor voltage, and it has the pulse waveform shown in Fig. 18-37. During t_{on}, $V_A = (V_i - V_{sat})$; and during t_{off}, $V_A = -(V_o + V_F)$. As in other switching converters, the inductor current increases to I_p during t_{on} and decreases during t_{off}, discharging to C_1 and supplying the load. Also, V_o decreases during t_{on} (D_1 is reverse-biased) as C_1 supplies all of I_o. C_1 is recharged from L_1 during t_{off} (D_1 is forward-biased). So (during t_{off}), V_o increases while I_ℓ exceeds I_o, and decreases while I_ℓ is less than I_o.

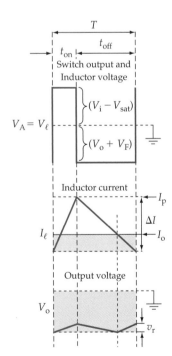

Figure 18-37 Inverting converter circuit waveforms.

For the inverting converter,

$$\frac{t_{on}}{t_{off}} = \frac{V_o + V_F}{V_i - V_{sat}} \qquad \text{(18-19)}$$

Equations 18-16, 18-17, and 18-18 apply for the inverting converter, as well as for the step-up converter.

Practice Problems

18-7.1   A switching regulator with $V_i = 20$ V and $f = 30$ kHz is to produce a 5 V, 1 A output with a 200 mV ripple. Assuming that $V_{sat} = 0.5$ V and $V_F = 0.7$ V, determine the minimum filter component values.

18-7.2   A switching regulator is to have $V_i = 5$ V, $V_o = 9$ V, $I_{o(max)} = 600$ mA, $f = 30$ kHz, and $V_r = 200$ mV. Determine the minimum filter component values. Assume that $V_{sat} = 0.5$ V and that $V_F = 0.7$ V.

18-8 IC CONTROLLER FOR SWITCHING REGULATORS

Function Block Diagram

The functional block diagram of a MC34063 integrated circuit controller is shown in Fig. 18-38. This IC is designed to be used as a variable-off-time switching regulator. The components of the diagram as follows:

- *Switching transistor* (Q_1) controlled by Darlington-connected transistor Q_2.
- *Set-reset flip-flop* with input terminals S and R, and output Q.
- *AND gate* that controls Q_1 and Q_2 by controlling the flip-flop.
- *Oscillator* that generates a square wave output at the desired switching frequency. The oscillator frequency is set by an externally connected timing capacitor (C_T) at terminal 3.
- *Current limiter* terminal (I_{pk} sense). This input to the oscillator permits Q_1 to be turned off via the reset terminal of the flip-flop.
- *Comparator* for comparing the regulator output voltage to the reference voltage.
- *Reference voltage source* (1.25 V).

Step-Down Converter Using an MC35063

Figure 18-39 shows the MC34063 connected to function as a step-down converter. The positive terminal of the supply is connected to terminal 6, and the negative (grounded) terminal is connected to terminal 4. Note that a 100 μF capacitor (C_2) is connected from terminal 6 to ground (right at the IC terminals), to ensure that there is no problem with voltage drops along the supply line when the load current changes.

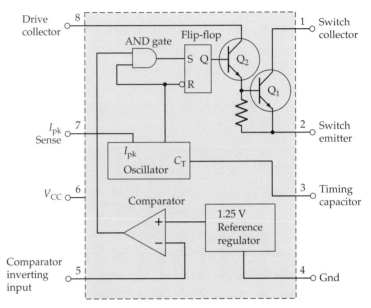

Figure 18-38 Functional block diagram of Motorola MC34063 switching regular controller. (Courtesy of Motorola Inc.)

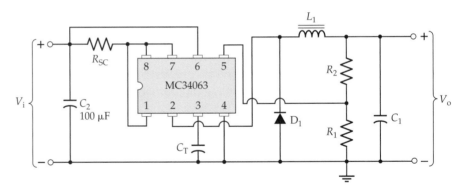

Figure 18-39 Step-down switching converter circuit using an MC34063 controller.

The supply voltage V_i is connected (via R_{SC}) to the collectors of Q_1 and Q_2 at terminals 8 and 1 (see Fig. 18-38). Resistor R_{SC} (connected to terminal 7) senses the peak supply current I_S and provides a voltage drop to turn Q_1 off when the current exceeds the maximum design level. The equation for the current-limiting resistor is given on the IC data sheet as

$$R_{SC} = \frac{0.33\text{ V}}{I_P} \tag{18-20}$$

The output voltage V_o is divided across resistors R_1 and R_2 and applied to the inverting input of the voltage comparator to compare it to the reference voltage (V_{ref}). The voltage across R_1 must be equal to the V_{ref} when the circuit is operating correctly, and the voltage across ($R_1 + R_2$) equals V_o. As in all voltage-divider designs, the resistor current should be much larger than the device or circuit input bias current—in this case, the comparator.

The timing capacitor (C_T), connected from terminal 3 to ground, is selected for the desired on-time of the switching transistor. An equation for the capacitance of C_T is given on the IC data sheet:

$$C_T = 4.8 \times 10^{-5} \times t_{on} \qquad \text{(18-21)}$$

Variable Off-Time Modulator

Switching converter circuits are designed to supply a particular maximum load current. For a converter circuit using an MC35063 controller, the typical circuit waveforms when supplying full load current are as shown in Fig. 18-40a. Transistor Q_1 is switched on at the start of each cycle of the oscillator square wave output. Times t_{on} and t_{off} add up to the time period (T) of the oscillator, as illustrated. When the load current is low, capacitor C_1 discharges more slowly than when supplying full load current. Thus, V_{R1} (in Fig. 18-39) has not dropped below V_{ref} by the beginning of next cycle of the oscillator square wave. Consequently, Q_1 is not switched on again at this point, and so the pulse waveform frequency at Q_1 output is now lower than the oscillator frequency (see Fig. 18-40b). A switching converter controller designed to operate in this way is known as a *variable off-time modulator*.

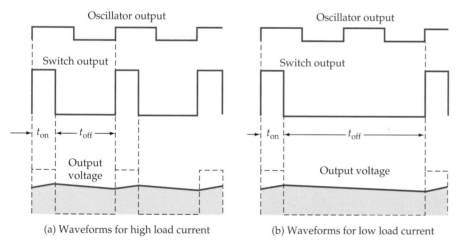

(a) Waveforms for high load current   (b) Waveforms for low load current

Figure 18-40   Waveforms for high-current and low-current conditions in a variable off-time controller for a switching regulator.

Example 18-13

An MC34063 controller is to be used with the step-down switching regulator designed in Example 18-12, as in Fig. 18-39. Determine suitable values for R_1, R_2, R_{SC}, and C_T.

Solution

$$I_1 \gg I_B \text{ (comparator input bias current)}$$

$$I_B = -400 \text{ nA (from the IC data sheet)}$$

Select $I_1 = 1 \text{ mA}$

$$R_1 = \frac{V_{ref}}{I_1} = \frac{1.25 \text{ V}}{1 \text{ mA}}$$

$$= 1.25 \text{ k}\Omega \text{ (use 1.2 k}\Omega \text{ standard value)}$$

I_1 becomes $I_1 = \dfrac{V_{ref}}{R_1} = \dfrac{1.25 \text{ V}}{1.2 \text{ k}\Omega}$

$$= 1.04 \text{ mA}$$

$$R_2 = \frac{V_o - V_{ref}}{I_1} = \frac{12 - 1.25 \text{ V}}{1.04 \text{ mA}}$$

$$= 10.34 \text{ k}\Omega \text{ (use (10 k}\Omega + 330 \text{ }\Omega\text{) standard values)}$$

Eq. 18-20: $R_{SC} = \dfrac{0.33 \text{ V}}{I_P} = \dfrac{0.33 \text{ V}}{1 \text{ A}}$

$$= 0.33 \text{ }\Omega \text{ (special low-value resistor)}$$

Eq. 18-21: $C_T = 4.8 \times 10^{-5} \times t_{on} = 4.8 \times 10^{-5} \times 8.6 \text{ }\mu s$

$$= 413 \text{ pF (use 430 pF standard value)}$$

Catch Diode Selection

The diode power dissipation can be calculated approximately from the output current and the diode voltage drop:

$$P_{D1} \approx I_o \times V_F \tag{18-22}$$

The reverse recovery time (t_{rr}) of the diode should be less than one-tenth of the minimum diode on- or off-time, whichever is smaller:

$$t_{rr} < 0.1 \, (t_{on(min)} \text{ or } t_{off(min)}) \tag{18-23}$$

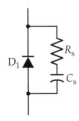

Figure 18-41 A series *RC* circuit connected in parallel with a diode (a diode snubber circuit) is often used to suppress high-frequency ringing.

Diode Snubber

A *diode snubber* is a series *RC* circuit connected across the catch diode to suppress high frequency *ringing* that can result from parasitic inductance and capacitance. This is illustrated in Fig. 18-41. The capacitance of C_s is usually selected in the range of 4 to 10 times the diode junction capacitance, and then the resistance of R_s is calculated from C_s and the ringing frequency (f_r):

$$C_s = (4 \text{ to } 10) \times C_D \tag{18-24}$$

$$R_s C_s = \frac{1}{f_r} \tag{18-25}$$

High-Power Converters

When the load current of a switching regulator is too high for the controller internal transistor, an external device can be employed. Figure 18-42a illustrates how an external BJT should be connected for use with a step-down converter, and Fig. 18-42b shows how an external FET should be connected. Resistor R_B at the base of Q_3 in Fig. 18-42a ensures that Q_3 is biased *off* when Q_1 switches *off*. For the *p*-channel MOSFET (Q_3) in Fig. 18-42b, resistors R_C and R_E are selected to give a gate-channel voltage (V_{RC}) that will switch Q_3 *on* when Q_1 is *on*.

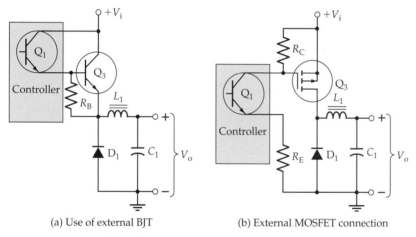

(a) Use of external BJT   (b) External MOSFET connection

Figure 18-42   When a peak current is too high for an IC switching converter controller, an external power device can be connected to pass higher current levels.

Practice Problems

18-8.1 The step-down switching regulator designed for Problem 18-7.1 is to use an MC34063 controller. Determine suitable values for the additional components.

18-8.2 The step-up switching regulator designed for Problem 18-7.2 is to use an MC34063 controller. Determine suitable values for the additional components.

Review Questions

Section 18-1

18-1 Sketch the circuit of an emitter follower voltage regulator. Explain the operation of the circuit, and discuss the effect of the transistor on the performance of the regulator.

18-2 Sketch the circuit of a series regulator with a one-transistor error amplifier. Explain the operation of the circuit, and discuss the effect of the error amplifier on the performance of the regulator.

Section 18-2

18-3 Show how the regulator circuit in Question 18-2 should be modified to produce an adjustable output voltage. Explain.

18-3 Show how the regulator in Question 18-2 should be modified to handle a large load current. Explain.

18-4 Show how voltage regulator performance may be improved by the use of a preregulator. Explain how the circuit performance is improved.

18-5 Show how a constant-current source may be used instead of a preregulator to improve the performance of a voltage regulator. Explain the operation of the circuit, and discuss how the performance of the regulator is improved.

18-6 Sketch the complete circuit of a transistor voltage regulator that uses a differential amplifier. Explain the operation of the circuit and discuss its performance.

Section 18-3

18-7 Sketch a simple current-limiting circuit for short-circuit protection on a voltage regulator. Explain the operation of the circuit, and sketch the V_o/I_L characteristic for the regulator. Discuss the effects of current limiting on the series-pass transistor.

18-8 Sketch a fold-back current-limiting circuit for a voltage regulator. Sketch the V_o/I_L characteristic, and explain the operation of the circuit.

18-9 Write equations for power dissipation in the series-pass transistor in voltage regulators using (a) simple current limiting, and (b) fold-back current limiting. Compare the two methods.

Section 18-4

18-10 Sketch the circuit of a voltage follower regulator using an IC operational amplifier. Explain the circuit operation.

18-11 Compare the performance of an IC op-amp voltage regulator to the performance of an emitter follower voltage regulator.

18-12 Sketch the circuit of a series regulator that uses an IC operational amplifier as an error amplifier. Explain the circuit operation.

18-13 For an IC op-amp series regulator, write equations for V_o in terms of V_Z. Briefly discuss the supply voltage requirements.

Section 18-5

18-14 Sketch the basic circuit of a 723 IC voltage regulator. Briefly explain.

18-15 Sketch a 723 IC voltage regulator connected to function as a positive voltage regulator with V_o greater than V_{ref}. Write an equation for V_o in terms of V_{ref}. Briefly discuss the required supply voltage.

18-16 Sketch a regulator circuit that uses an LM317 IC positive voltage regulator. Briefly explain.

18-17 Sketch a regulator circuit that uses an LM337 IC negative voltage regulator. Briefly explain.

18-18 Sketch a regulator circuit that uses an LM340 IC fixed voltage regulator. Briefly explain.

Section 18-6

18-19 Draw a basic block diagram and waveforms for a switching regulator. Explain its operation.

18-20 Compare switching regulators and linear regulators.

Section 18-7

18-21 Sketch the basic circuit for a step-down switching regulator. Draw current and voltage waveforms, and explain the circuit operation.

18-22 Repeat Question 18-21 for a step-up switching regulator.

18-23 Repeat Question 18-21 for an inverting switching regulator.

Section 18-8

18-24 Sketch the functional block diagram for an MC34063 controller for a switching regulator. Briefly discuss each item in the block diagram.

18-25 Sketch the circuit of a step-up switching regulator using an MC34063 in a similar form to Fig. 18-39.

18-26 Sketch the circuit of an inverting switching regulator using an MC34063 in a similar form to Fig. 18-39.

18-27 Sketch a diode snubber circuit, and briefly explain.

18-28 Draw circuits to show how an external BJT and an external FET should be connected for use with step-down converters. Briefly explain.

Problems

Section 18-1

18-1 Using a 15 V supply and a transistor with $h_{FE(min)} = 50$, design a voltage-follower regulator circuit (as in Fig. 18-1) to give a 9 V output with a 100 mA maximum load current.

18-2 Calculate the line regulation, load regulation, and ripple rejection ratio for the circuit designed for Problem 18-1.

18-3 A regulator circuit as in Fig. 18-1 has $V_Z = 7$ V, $R_1 = 560\ \Omega$, $R_E = 8.2\ k\Omega$, and $R_L = 180\ \Omega$. $V_S = 19$ V when $I_L = 0$, and $V_S = 18$ V when $I_L = 35$ mA. The transistor has $h_{FE} = 60$. Calculate the Q_1 emitter current and the

Zener diode current when the load is connected and when the load is disconnected.

18-4 Determine the line and load regulations for the regulator in Problem 18-3. Assume that $Z_Z = 5\ \Omega$.

18-5 Measurements on a voltage regulator with $V_S = 25$ V and $V_o = 18$ V give the following results: (when I_L changes from zero to 200 mA, $\Delta V_S = 2$ V, and $\Delta V_o = 10$ mV); (when $\Delta V_S = \pm 2.5$ V, $\Delta V_o = \pm 12.5$ mV). Calculate the source effect, load effect, line regulation, and load regulation.

18-6 Determine the voltage gain of the error amplifier in Problem 18-5, and estimate the output ripple voltage amplitude if there is a 2 V peak-to-peak input ripple.

18-7 A series regulator circuit as in Fig. 18-2 is to have a 15 V output and a 50 mA maximum load current. Select a suitable minimum supply voltage and design the circuit.

18-8 Design a voltage regulator circuit as in Fig. 18-2 to produce a 9 V output with a 30 mA maximum load current.

18-9 For the regulator designed for Problem 18-7, calculate the approximate line regulation, load regulation, and ripple reduction, if V_S drops by 1 V from zero to full load current.

18-10 Determine the approximate line regulation, load regulation, and ripple reduction for the regulator designed for Problem 18-8. Assume that $\Delta V_S = 1$ V when $\Delta I_L =$ full load current.

Section 18-2

18-11 Modify the circuit designed for Problem 18-7 to make the output adjustable from 12 V to 15 V.

18-12 Modify the regulator designed for Problem 18-8 to make the output adjustable from 8 V to 10 V.

18-13 Modify the regulator designed for Problem 18-7 to supply a maximum load current of 210 mA. Assume that all transistors have $h_{FE} = 30$.

18-14 Modify the regulator designed for Problem 18-8 to supply a maximum load current of 300 mA. Assume that $h_{FE1} = 20$ and $h_{FE3} = 100$.

18-15 Design a series voltage regulator with a one-transistor error amplifier to provide an output adjustable from 15 V to 18 V. The load current is to be 250 mA. Assume that all transistors have $h_{FE} = 50$.

18-16 Calculate the line and load regulations for the modified regulator circuit in Problem 18-14. Assume that $\Delta V_S = 1$ V when $\Delta I_L =$ full load current.

18-17 Design a preregulator for the circuit designed for Problem 18-7 and modified for Problem 18-13.

18-18 Determine the approximate line regulation, load regulation, and ripple reduction for the modified regulator in Problem 18-17. Assume that $\Delta V_S = 1$ V when $\Delta I_L =$ full load current.

18-19 The preregulator in the circuit referred to in Problem 18-17 is to be replaced with a constant-current source. Design the constant-current source.

18-20 Design a differential amplifier as in Fig. 18-12 to use with the regulator designed for Problem 18-15.

Section 18-3

18-21 Design a current-limiting circuit as in Fig. 18-13a to limit the maximum load current to approximately 220 mA for the regulator in Problem 18-13. Calculate the power dissipation in the series-pass transistor at I_{SC}.

18-22 Design a fold-back current-limiting circuit, as in Fig. 18-14a, to set $I_{L(max)}$ to approximately 220 mA and I_{SC} to approximately 150 mA for the regulator in Problem 18-13. Calculate the power dissipation in the series-pass transistor at $I_{L(max)}$ and at I_{SC}.

18-23 The current-limiting circuit in Fig. 18-14a has $R_{10} = 1.6\ \Omega$, $R_{11} = 2.7\ k\Omega$, and $R_{12} = 47\ k\Omega$. If the normal output level is 20 V, calculate $I_{L(max)}$ and I_{SC}.

Section 18-4

18-24 Design an op-amp voltage regulator to have $V_o = 15$ V and $I_{L(max)} = 120$ mA. Use a 741 IC operational amplifier.

18-25 Design a voltage regulator using a 741 IC operational amplifier to have V_o adjustable from 9 V to 12 V and to deliver a maximum load current of 60 mA.

18-26 Design an op-amp series voltage regulator to provide an output voltage adjustable from 15 V to 18 V. The load current is to be 300 mA. Use a 741 op-amp, and assume that the transistors have $h_{FE} = 60$.

18-27 For the regulator designed for Problem 18-25, calculate the approximate line regulation, load regulation, and ripple reduction if V_S drops by 1 V from zero to full load current.

Section 18-5

18-28 Calculate R_1, R_2, and R_3 for the 723 IC positive voltage regulator circuit in Fig. 18-21. V_o is to be 25 V, and $I_{L(max)}$ is to be approximately 55 mA. Determine a suitable supply voltage.

18-29 A regulator circuit using a 723 IC is to be designed to provide a V_o adjustable from 15 to 20 V. Design the circuit, determine a suitable input voltage, and calculate the maximum load current that can be supplied.

18-30 The LM317 positive voltage regulator in Fig. 18-24a is to produce an 8 V output of with $I_{L(max)} = 100$ mA. Calculate suitable resistances for R_1 and R_2, select an appropriate supply voltage, and determine the device power dissipation.

18-31 The LM317 negative voltage regulator in Fig. 18-24b is to produce a 12 V output with $I_{L(max)} = 80$ mA. Calculate suitable resistances for R_1 and R_2, select an appropriate supply voltage, and determine the IC power dissipation.

Section 18-6

18-32 Calculate the approximate efficiency of the linear regulator in Problem 18-24. Also calculate the approximate efficiency of a switching regulator with similar supply and load conditions. Assume that the switching transistor has $V_{(sat)} = 0.5$ V, and that the switching diode has $V_F = 0.7$ V.

18-33 Calculate the approximate efficiency of the IC series regulator in Problem 18-30. A switching regulator with similar supply and load conditions uses a FET with $r_{DS(on)} = 10\ \Omega$, and a switching diode with $V_F = 0.7\ V$. Calculate the approximate efficiency of the switching regulator.

Section 18-7

18-34 Determine minimum filter component values for a step-down switching regulator to have $f = 28$ kHz, $V_i = 30$ V, $V_o = 15$ V, $I_o = 300$ mA, and $V_r = 250$ mV. Assume that $V_{(sat)} = V_F = 1$ V.

18-35 A step-up switching regulator is to have $f = 28$ kHz, $V_i = 12$ V, $V_o = 30$ V, $I_o = 150$ mA, and $V_r = 250$ mV. Determine suitable minimum filter component values. Assume that $V_{(sat)} = V_F = 1$ V.

18-36 Determine minimum filter component values for an inverting switching regulator to have $f = 28$ kHz, $V_i = 30$ V, $V_o = -12$ V, $I_o = 300$ mA, and $V_r = 250$ mV. Assume that $V_{(sat)} = V_F = 1$ V.

Section 18-8

18-37 The step-down switching regulator in Problem 18-34 is to use an MC34063 IC controller. Determine suitable values for the additional components.

18-38 Calculate values for the additional components for the step-up switching regulator in Problem 18-35 when an MC34063 IC controller is used.

18-39 Calculate the additional component values for the inverting switching regulator in Problem 18-36 to use an MC34063 IC controller.

Practice Problem Answers

18-1.1 25.7 mV, 14.3 mV, 0.21%, 0.12%
18-1.2 1.5 kΩ, 470 Ω, (3.9 kΩ + 390 Ω), 10 kΩ, 1N758, (25 V, 70 mA, 700 mW)
18-2.1 2.2 kΩ, 5.6 kΩ, 2 kΩ
18-2.2 5.6 kΩ, 270 Ω, 1.8 kΩ, 8.2 kΩ, 22 kΩ, (18 V, 160 mA, 1.28 W)
18-2.3 1N749, 3.3 kΩ, 1.5 kΩ
18-3.1 1.25 Ω, (25 V, 400 mA, 10 W)
18-3.2 3.3 Ω, 820 Ω, 15 kΩ
18-4.1 470 Ω, 15 kΩ, 4.7 kΩ, 8.2 kΩ, 2 kΩ, 1N758
18-4.2 2.5 Ω, 220 Ω, (12 kΩ + 1.8 kΩ), 2.2 kΩ
18-5.1 10 kΩ, 6.8 kΩ, 2.5 Ω, 23 V, (200 mA, 23 V, 4.6 W)
18-5.2 270 Ω, 1.5 kΩ, 12 V, 0.45 W
18-6.1 71%, 93%
18-7.1 60 μH, 47 μF
18-7.2 30 μH, 60 μF
18-8.1 1.2 kΩ, (3.3 kΩ + 470 Ω), 0.165 Ω, 360 pF
18-8.2 1.2 kΩ, (6.8 kΩ + 1 kΩ), 0.18 Ω, 560 pF

CHAPTER 19
Audio Power Amplifiers

CONTENTS

Objectives

You will be able to:

1 Sketch and explain Class A, Class B, and Class AB transformer-coupled power amplifier circuits.

2 Design and analyze transformer-coupled power amplifiers and draw the circuit dc and ac load lines.

3 Sketch and explain the basic circuits of Class AB capacitor-coupled and direct-coupled power amplifiers.

4 Show how capacitor-coupled and direct-coupled power amplifiers should be modified for high load currents, output current limiting, and the use of overall negative feedback.

5 Explain complementary and quasi-complementary emitter followers, V_{BE} multiplier, and supply decoupling.

6 Draw and explain complete circuits for direct- and capacitor-coupled power amplifiers using BJT driver stages, op-amp drivers, BJT output stages, and MOSFET output stages.

7 Design and analyze the types of circuits listed in Objective 6 above.

8 Draw and explain complementary common-source power amplifiers, and design and analyze such circuits.

9 Explain the applications of various integrated-circuit power amplifiers.

INTRODUCTION

A *power amplifier*, or *large-signal amplifier*, develops relatively large output voltages across low impedance loads. Audio amplifiers are large-signal amplifiers that supply ac output power to speakers. Power amplifiers may be categorized as *class A* circuits, in which the output transistor is biased to the centre of its load line, or as *Class B* or *Class AB* amplifiers, in which two output transistors are biased at or close to cutoff. The amplifier load may be transformer-coupled, capacitor-coupled, or direct-coupled. Direct coupling usually gives the best performance, but plus-and-minus supply voltages are required. The output stage of the amplifier may use power BJTs or power MOSFETs. IC operational amplifiers may also be used in power amplifiers, and complete power amplifiers are available as integrated circuits.

19-1 TRANSFORMER-COUPLED CLASS A AMPLIFIER

Class A Circuit

Instead of capacitor coupling, a transformer may be used to ac-couple amplifier stages while providing dc isolation between stages. The resistance of the transformer windings is normally very small, so that there is no effect on the transistor bias conditions.

Figure 19-1 shows a load resistance (R_L) transformer-coupled to a transistor collector. The low resistance of the transformer primary winding allows any desired level of (dc) collector current to flow, while the transformer core couples all variations in I_C to R_L via the secondary winding. This circuit is a

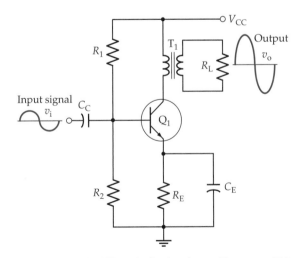

Figure 19-1 Transformer-coupled Class A circuit using emitter current bias. Q_1 is biased to produce maximum equal positive and negative V_{CE} changes. The transistor ac load depends on R_L and the square of the transformer turns ratio.

voltage-divider bias circuit in which resistors R_1 and R_2 determine the transistor base voltage (V_B), and resistor R_E sets the emitter current level.

The circuit in Fig. 19-1 is referred to as a *class A amplifier*, which is defined as one that has the Q-point (bias point) approximately at the centre of the ac load line. This enables the circuit to produce maximum equal positive and negative changes in V_{CE}.

DC and AC Loads

The total dc load for transistor Q_1 in the circuit in Fig. 19-1 is the sum of the emitter resistor (R_E) and the transformer primary winding resistance (R_{PY}):

$$R_{L(dc)} = R_E + R_{PY} \qquad (19\text{-}1)$$

Consider the transformer illustrated in Fig. 19-2. N_1 is the number of turns on the primary winding, and N_2 is the number of secondary turns. The primary ac voltage and current are v_1 and i_1, and the secondary quantities are v_2 and i_2. The (secondary) load resistance can be calculated as

$$R_L = \frac{v_2}{i_2}$$

The ac load resistance measured at the transformer primary terminals (r_L) is calculated as

$$r_L = \frac{v_1}{i_1}$$

From basic transformer theory,

$$\frac{v_1}{v_2} = \frac{N_1}{N_2} \quad \text{and} \quad \frac{i_1}{i_2} = \frac{N_2}{N_1}$$

Those equations give

$$v_1 = \frac{N_1}{N_2} v_2 \quad \text{and} \quad i_1 = \frac{N_2}{N_1} i_2$$

Substituting for v_1 and i_1 in the equation for r_L,

$$r_L = \left[\frac{N_1}{N_2}\right]^2 R_L \qquad (19\text{-}2)$$

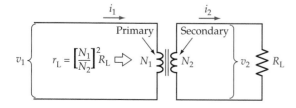

Figure 19-2 The load (R_L) at the secondary terminals of a transformer is reflected into the primary as $R_L(N_1/N_2)^2$.

The load resistance calculated in this way is termed the *reflected load* or the *referred load,* meaning that R_L is reflected or referred from the transformer secondary to the primary as r_L. The total ac load at the transistor collector is the sum of the referred load and the transformer primary winding resistance:

$$r_{L(ac)} = r_L + R_{PY} \qquad (19\text{-}3)$$

Example 19-1

Draw the dc and ac load lines for the circuit in Fig. 19-3 on the transistor common-emitter characteristics in Fig. 19-4. The transformer has $R_{PY} = 40\ \Omega$, $N_1 = 74$, and $N_2 = 14$.

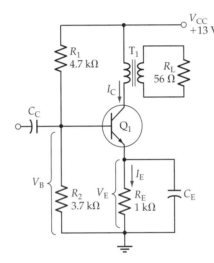

Figure 19-3 Transformer-coupled class A amplifier for Example 19-1.

Solution

Q-point:

$$V_B = V_{CC} \times \frac{R_2}{R_1 + R_2} = 13\text{ V} \times \frac{3.7\text{ k}\Omega}{4.7\text{ k}\Omega + 3.7\text{ k}\Omega}$$

$$\approx 5.7\text{ V}$$

$$I_C \approx I_E = \frac{V_B - V_{BE}}{R_E} = \frac{5.7\text{ V} - 0.7\text{ V}}{1\text{ k}\Omega}$$

$$\approx 5\text{ mA}$$

$$V_{CE} = V_{CC} - I_C(R_{PY} + R_E) = 13\text{ V} - 5\text{ mA}\,(40\ \Omega + 1\text{ k}\Omega)$$

$$\approx 8\text{ V}$$

Plot the *Q*-point on the characteristics at $I_C = 5$ mA and $V_{CE} = 8$ V.

dc load line:

$$V_{CE} = V_{CC} - I_C(R_{PY} + R_E)$$

When $I_C = 0$, $V_{CE} = V_{CC} = 13$ V

Plot point A on the characteristics at $I_C = 0$ and $V_{CE} = 13$ V.

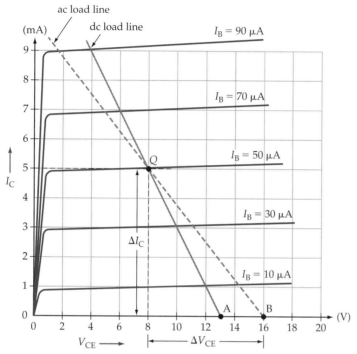

Figure 19-4   The dc load line for a transformer-coupled amplifier is drawn through the Q-point at a slope of $1/R_{L(dc)}$. The ac load line is drawn through the Q-point at a slope of $1/r_{L(ac)}$.

Draw the dc load through point A and point Q.
ac load line:

Eq. 19-2:
$$r_L = \left[\frac{N_1}{N_2}\right]^2 R_L = \left[\frac{74}{14}\right]^2 \times 56 \ \Omega$$
$$= 1565 \ \Omega$$

Eq. 19-3:
$$r_{L(ac)} = r_L + R_{PY} = 1565 \ \Omega + 40 \ \Omega$$
$$\approx 1.6 \ \text{k}\Omega$$

When $\Delta I_C = 5$ mA,

$$\Delta V_{CE} = \Delta I_C \times r_{L(ac)} = 5 \ \text{mA} \times 1.6 \ \text{k}\Omega$$
$$= 8 \ \text{V}$$

Measure ΔI_C and ΔV_{CE} from the Q-point on the characteristics to give point B at $V_{CE} = 16$ V and $I_C = 0$. Draw the ac load line through points Q and B.

Collector Voltage Swing

The ac load line drawn in Example 19-1 is reproduced in Fig. 19-5 to show the effect of an input signal. When the input causes I_B to increase from 50 μA (at I_{BQ}) to 90 μA, the current and voltage become $I_C \approx 9$ mA and $V_{CE} \approx 1.6$ V respectively (point C on the ac load line). The changes are $\Delta I_C = +4$ mA and

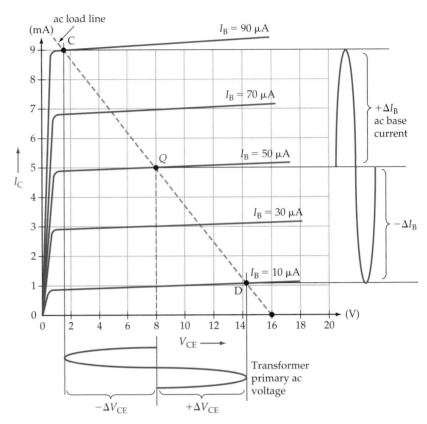

Figure 19-5 A transformer-coupled class A amplifier has its Q-point at the centre of the ac load line. The ac voltage applied to the transformer primary is $\pm\Delta V_{CE}$, which is produced by $\pm\Delta I_{B}$.

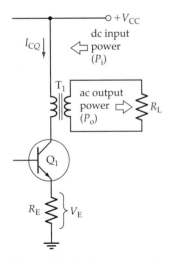

Figure 19-6 A power amplifier converts dc input (supply) power into ac output power.

$\Delta V_{CE} = -6.4$ V. When the input causes I_B to decrease (from I_{BQ}) to 10 μA, I_C changes from 5 mA to 1 mA and V_{CE} changes from 8 V to 14.4 V (point D). The current and voltages changes are now $\Delta I_C = -4$ mA and $\Delta V_{CE} = +6.4$ V.

It is seen that an I_B change of ±40 μA produces a ±4 mA I_C change and a ±6.4 V change in V_{CE}. The V_{CE} variation appears at the primary winding of transformer T_1 (Fig. 19-3), and the I_C variation flows in the primary winding.

Note that, although $V_{CC} = 13$ V, the transistor V_{CE} can actually go to 16 V. This is because of the inductive effect of the transformer primary winding. The transistor used in this type of circuit should have a minimum breakdown voltage approximately equal to $2V_{CC}$.

Efficiency of a Class A Amplifier

Power is delivered to an amplifier from the dc power supply. The amplifier converts the dc power into ac power supplied to the load (see Fig. 19-6). Some of the input power is dissipated in the transistor or in other components. This is wasted power. The efficiency (η) of a power amplifier is a measure of how good the amplifier is at converting the dc input (supply) power (P_i) into ac output power (P_o) dissipated in the load.

$$\eta = \frac{P_o}{P_i} \times 100\% \qquad (19\text{-}4)$$

The dc supply power is

$$P_i = V_{CC} \times I_{ave}$$

In the case of a class A amplifier, $I_{ave} = I_{CQ}$

So
$$P_i = V_{CC} \times I_{CQ}$$

Refer again to the class A circuit in Fig. 19-6, and assume that $V_E \ll V_{CC}$. In this case, V_{CEQ} is approximately equal to V_{CC}, and the peak voltage developed across the transformer primary approaches $\pm V_{CC}$ if the transistor is driven to cutoff and saturation. Also, the peak current developed in the transformer windings approaches $\pm I_{CQ}$. Thus, the maximum ac power delivered to the transformer primary can be calculated as follows:

$$P'_o = V_{rms} \times I_{rms} = (V_p/\sqrt{2}) \times (I_p/\sqrt{2})$$

giving
$$P'_o = 0.5 \, V_p I_p \qquad (19\text{-}5)$$

Using the highest possible current and voltage, and assuming that the transformer is 100% efficient,

$$P_o = 0.5 \, V_{CC} I_{CQ}$$

The maximum theoretical efficiency for a class A transformer-coupled power amplifier can now be determined as

Eq. 19-4:
$$\eta = \frac{P_o}{P_i} \times 100\% = \frac{0.5 \, V_{CC} I_{CQ}}{V_{CC} I_{CQ}} \times 100\%$$

$$= 50\%$$

In a practical class A transformer-coupled power amplifier circuits, 50% efficiency is never approached. Any practical calculation of power amplifier efficiency must take the output transformer efficiency (η_t) into account.

$$\eta_t = \frac{P_o}{P'_o} \times 100\% \qquad (19\text{-}6)$$

A typical transformer efficiency might be 80%. There is also power dissipation in the transistor emitter resistor and in the bias circuit. The practical maximum efficiency for a class A power amplifier is usually about 25%. This means, for example, that 4 W of dc supply power must be provided to deliver 1 W of ac output power to the load.

Example 19-2

Calculate the maximum efficiency of the class A amplifier circuit in Example 19-1 (reproduced in Fig. 19-7). Assume that the transformer has an 80% efficiency.

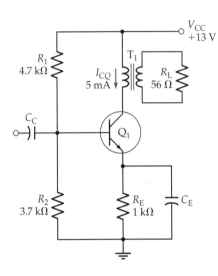

Figure 19-7 Class A power amplifier for Example 19-2.

Solution

$$P_i = V_{CC} \times I_{CQ} = 13 \text{ V} \times 5 \text{ mA}$$

$$= 65 \text{ mW}$$

$$V_p \approx V_{CEQ} = 8 \text{ V}$$

$$I_p \approx I_{CQ} = 5 \text{ mA}$$

Eq. 19-5:
$$P'_o = 0.5 \, V_p I_p = 0.5 \times 8 \text{ V} \times 5 \text{ mA}$$

$$= 20 \text{ mW}$$

Eq. 19-6:
$$P_o = \eta_t \, P'_o = 0.8 \times 20 \text{ mW}$$

$$= 16 \text{ mW}$$

Eq. 19-4:
$$\eta = \frac{P_o}{P_i} \times 100\% = \frac{16 \text{ mW}}{65 \text{ mW}} \times 100\%$$

$$= 24.6\%$$

Practice Problems

19-1.1 The circuit in Ex. 19-1 has the quantities changed as follows: $V_{CC} = 15$ V, $R_1 = 3.9$ kΩ, $R_2 = 2.2$ kΩ, $R_E = 1.5$ kΩ, $R_{PY} = 33$ Ω, $R_L = 100$ Ω, $N_1 = 118$, and $N_2 = 20$. Draw the new dc and ac load lines on Fig. 19-4.

19-1.2 Determine the maximum efficiency for the amplifier in Problem 19-1.1. Assume that the transformer efficiency is 75%.

19-2 TRANSFORMER-COUPLED CLASS B AND CLASS AB AMPLIFIERS

Class B Amplifiers

The inefficiency of class A amplifiers is largely due to the transistor bias conditions. In a class B amplifier, the transistors are biased to cutoff, so that there is no transistor power dissipation when there is no input signal. This gives the class B amplifier a much greater efficiency than the class A circuit.

The output stage of a class B transformer-coupled amplifier is shown in Fig. 19-8. Transformer T_2 couples load resistor R_L to the collector circuits of transistors Q_2 and Q_3. Note that the supply is connected to the centre-tap of the transformer primary, and that Q_2 and Q_3 have grounded emitters. The transistor bases are grounded via resistors R_1 and R_2, so that both are biased *off*.

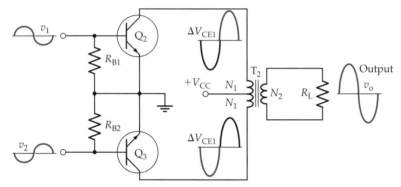

Figure 19-8 In a class B power amplifier, two transistors operate in push-pull. When the input signal turns Q_2 *on*, Q_3 is turned *off*, and vice versa.

The input signals applied to the transistor bases consist of two separate sine waves, which are identical except that they are in antiphase. When v_1 is going positive, v_2 is going negative, so that Q_3 is being biased further *off* as Q_2 is being biased *on*. As the collector current in Q_2 increases from zero, it produces a half sine wave of voltage across the upper half of the transformer primary, as illustrated. When the positive half-cycle of input signal to Q_2 base begins to go negative, the signal at Q_3 base is starting to go positive. Thus, as Q_2 becomes biased *off* again, Q_3 is biased *on*, and a half-cycle of voltage waveform is generated across the lower half of the transformer primary.

The two half-cycles in the separate sections of the transformer primary produce a magnetic flux in the transformer core that flows first in one direction and then in the opposite direction. This flux links with the secondary winding and generates a complete sine wave output which is passed to the load.

In the class B circuit, the two output transistors are said to be operating in *push-pull*. The push-pull action is best illustrated by drawing the ac load line on the *composite characteristics* for Q_2 and Q_3. The composite characteristics are created by drawing the Q_2 characteristics in the normal way and presenting the Q_3 characteristics upside down. This is illustrated in Fig. 19-9.

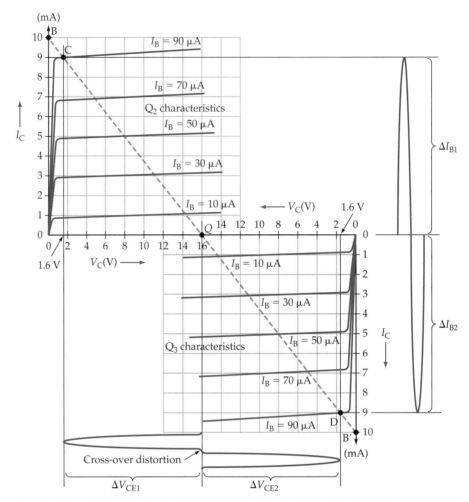

Figure 19-9 AC load line for a class B amplifier drawn on the BJT composite characteristics. Also shown is the effect of I_B changes on V_{CE}.

Example 19-3

Draw the ac load lines for the circuit in Fig. 19-8 on the transistor composite characteristics in Fig. 19-9. The transformer has $R_{PY} = 40\ \Omega$, $N_1 = 74$, and $N_2 = 14$. The supply voltage is $V_{CC} = 16$ V, and the load resistance is $R_L = 56\ \Omega$.

Solution

Q-point:

$$V_B = 0\,\text{V} \quad \text{and} \quad I_C = 0\,\text{mA}$$

$$V_{CE} = V_{CC} - I_C(R_{PY} + R_E) = 16\,\text{V} - 0$$

$$= 16\,\text{V}$$

Plot the Q-point on the characteristics at $I_C = 0$ mA and $V_{CE} = 16$ V.

ac load line:

Eq. 19-2:
$$r_L = \left[\frac{N_1}{N_2}\right]^2 R_L = \left[\frac{74}{14}\right]^2 \times 56 \; \Omega$$

$$= 1565 \; \Omega$$

Eq. 19-3:
$$r_{L(ac)} = r_L + R_{PY} = 1565 \; \Omega + 40 \; \Omega$$

$$\approx 1.6 \; k\Omega$$

When $\Delta I_C = 10$ mA,

$$V_{CE} = \Delta I_C \times r_{L(ac)} = 10 \text{ mA} \times 1.6 \text{ k}\Omega$$

$$= 16 \text{ V}$$

Measure ΔI_C and V_{CE} from the Q-point to give points B and B' at $V_{CE} = 16$ V and $I_C = 0$. Draw the ac load line through points Q, B, and B'.

Now consider the effect of a signal applied to the bases of Q_2 and Q_3 in the circuit in Fig. 19-8. When I_{B1} is increased from zero to 90 μA, Q_3 remains *off* and V_{CE1} falls to 1.6 V (point C on the composite characteristics in Fig. 19-9). At this point the voltage across the upper half of the transformer primary is

$$V_{py} = V_{CC} - V_{CE} = 16 \text{ V} - 1.6 \text{ V}$$

$$= 14.4 \text{ V}$$

When I_{B2} is increased from 0 to 90 μA, Q_2 is *off*, and the Q_3 current and voltage conditions move to point D on the ac load line. This produces 14.4 V across the lower half of the transformer primary. Thus, a full sine wave is developed at the transformer output. When no signal is present, both transistors remain *off* and dissipate zero power. Power is dissipated only while each device is conducting. The wasted power is considerably less with the class B amplifier than with a class A circuit.

Cross-Over Distortion

The waveform delivered to the transformer primary and the resultant output are not perfectly sinusoidal in the class B circuit. *Cross-over distortion* is produced in the output waveform (see Figs. 19-9 and 19-10) because the transistors do not begin to turn on until the input base-emitter voltage is about 0.5 V for a silicon device or 0.15 V for a germanium transistor. To eliminate this effect, the transistors are partially biased *on* instead of being biased *off*. With this modification, the class B amplifier becomes a class AB amplifier.

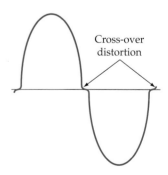

Figure 19-10 Cross-over distortion occurs in class B amplifiers because the transistors are biased off.

Class AB Amplifier

Figure 19-11 shows a class AB transformer-coupled output stage with a class A transformer-coupled *driver stage*. The output transformer (T_2) has a centre-tapped primary winding, with each half of the winding constituting a load for one of the output transistors (Q_2 and Q_3). Resistor R_4 and diode D_1 bias Q_2 and Q_3 partially *on,* and resistors R_5 and R_6 limit the emitter (and collector) currents to the desired bias levels. Transformer T_1, together with transistor Q_1 and the associated components, constitute a class A stage. The secondary of T_1 is centre-tapped to provide the necessary antiphase signals to Q_2 and Q_3.

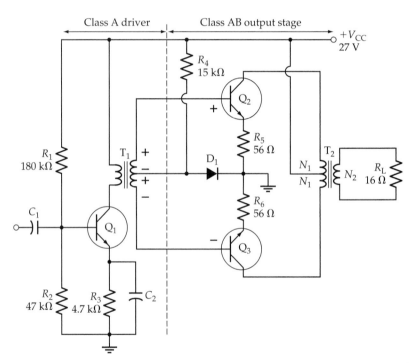

Figure 19-11 Class AB transformer-coupled output stage with a class A driver stage. Transistors Q_2 and Q_3 are biased to a low I_C level.

When the instantaneous polarity of T_1 output is positive at the top, Q_2 base voltage is positive and Q_3 base voltage is negative, as illustrated. At this time Q_2 is *on* and Q_3 is *off.* When the polarity reverses at T_1 output, the base of Q_3 becomes positive and that of Q_2 becomes negative. The output stage functions exactly as for a class B circuit, except that each device commences to conduct just before the signal to its base becomes positive. This eliminates the transistor turn-on delay that creates crossover distortion in a class B amplifier.

The class A portion of the circuit in Fig. 19-11 is referred to as a *driver stage,* simply because it provides the input signals to drive the class AB output stage. Since the input power handled by the driver stage is very much smaller than the circuit output power, the inefficiency of the class A stage is unimportant.

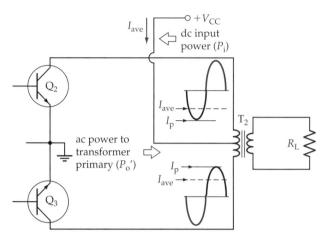

Figure 19-12   The dc input power to a class B amplifier is $P_i = (V_{CC} \times I_{ave})$, and the ac power to the transformer primary is $P'_o = (0.5 \, V_{CC} I_p)$.

Efficiency of Class B and Class AB Amplifiers

For a class B amplifier, the dc supply power (see Fig. 19-12) is calculated as

$$P_{i(dc)} = V_{CC} \times I_{ave}$$

or
$$P_{i(dc)} = V_{CC} \times 0.636 I_p \qquad (19\text{-}7)$$

The ac power input to the transformer primary is given by Eq. 19-5,

$$P'_o = 0.5 V_p I_p \approx 0.5 V_{CC} I_p$$

Assuming a 100% efficiency for the output transformer,

$$P_o = P'_o \approx 0.5 V_{CC} I_p$$

So, the maximum theoretical efficiency for a class B transformer-coupled power amplifier is

Eq. 19-4:
$$\eta = \frac{P_o}{P_i} \times 100\% = \frac{0.5 V_{CC} I_p}{0.636 V_{CC} I_p} \times 100\%$$

$$= 78.6\%$$

Once again, the efficiency of a practical amplifier is lower than the theoretical efficiency. Some power is wasted in the transistors and emitter resistors, and the transformer is never 100% efficient.

The efficiency of a class AB power amplifier is usually a little less than that of a class B circuit because of the additional small amount of power wasted in keeping the output transistors biased in a low-current *on* state. Class B and class AB power amplifiers are employed more often than class A circuits, because of their greater efficiency.

Example 19-4

Calculate the power delivered to the load in the class AB amplifier in Fig. 19-13 (reproduced from Fig. 19-11). Assume that T_2 has a 79% efficiency and that there is a 0.5 V drop across Q_2 and Q_3 at peak output voltage. Assume also that $N_1 = 60$, $N_2 = 10$, and that the primary winding resistance is small enough to neglect.

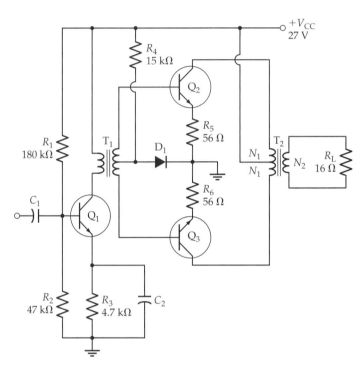

Figure 19-13 Class AB transformer-coupled amplifier for Example 19-4.

Solution

Referred load:

Eq. 19-2:
$$r_L = \left[\frac{N_1}{N_2}\right]^2 \times R_L = \left[\frac{60}{10}\right]^2 \times 16\ \Omega$$
$$= 576\ \Omega$$

Total ac load in series with each of Q_2 and Q_3:
$$R'_L = r_L + R_6 + R_{PY} = 576\ \Omega + 56\ \Omega + 0$$
$$= 632\ \Omega$$

Peak primary current:
$$I_P = \frac{V_{CC} - V_{CE}}{R'_L} = \frac{27\ V - 0.5\ V}{632\ \Omega}$$
$$= 41.9\ mA$$

Peak primary voltage:

$$V_{\mathrm{p}} = V_{\mathrm{CC}} - V_{\mathrm{CE}} - (I_{\mathrm{p}}\,R_6)$$
$$= 27\ \mathrm{V} - 0.5\ \mathrm{V} - (41.9\ \mathrm{mA} \times 56\ \Omega)$$
$$= 24.15\ \mathrm{V}$$

Power delivered to primary:

Eq. 19-5:
$$P'_{\mathrm{o}} = 0.5\ V_{\mathrm{p}}\,I_{\mathrm{p}}$$
$$= 0.5 \times 24.15\ \mathrm{V} \times 41.9\ \mathrm{mA}$$
$$= 506\ \mathrm{mW}$$

Power delivered to the load:

$$P_{\mathrm{o}} = P'_{\mathrm{o}} \times (\text{transformer efficiency})$$
$$= 506\ \mathrm{mW} \times 0.79$$
$$= 400\ \mathrm{mW}$$

Practice Problems

19-2.1 Calculate the bias currents in the circuit in Fig. 19-13, and using the results from Example 19-4, estimate the overall efficiency of the amplifier.

19-2.2 Draw the dc and ac load lines for the output stage in the circuit in Fig. 19-13 on blank composite characteristics.

19-3 TRANSFORMER-COUPLED AMPLIFIER DESIGN

Design of a transformer-coupled amplifier begins with specification of the load resistance and output power. A signal voltage amplitude may also be stated, as well as the upper and lower cutoff frequencies for the amplifier. If a supply voltage is given, then the design must determine a specification for the transformer. Where an available transformer is to be employed, the supply voltage is calculated to suit the transformer. The maximum levels of V_{CE}, I_{C}, and P_{D} must be calculated for each transistor.

The power delivered to the transformer primary is determined from Equation 19-6:

$$P'_{\mathrm{o}} = \frac{P_{\mathrm{o}}}{\eta_t}$$

The power delivered to the primary can also be expressed as

$$P'_{\mathrm{o}} = \frac{(V_{\mathrm{rms}})^2}{r_{\mathrm{L}}}$$

where V_{rms} is the rms primary voltage and r_L is the ac resistance offered by the transformer primary (the referred resistance). Using peak voltages, the equation becomes

$$P'_o = \frac{(V_p/\sqrt{2})^2}{r_L} = \frac{V_p^2}{2r_L}$$

This gives the equation for r_L:

$$r_L = \frac{V_p^2}{2P'_o} \tag{19-8}$$

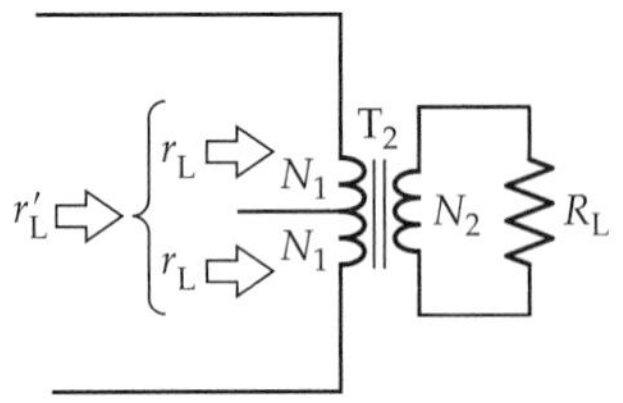

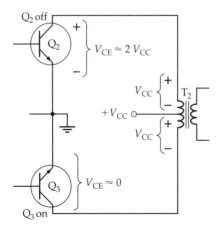

Figure 19-14 The input resistance at each half of the centre-tapped primary winding of an output transformer is r_L, and the input resistance to the whole primary is $r'_L = 4r_L$.

From Eq. 19-2, r_L can be calculated in terms of the load resistance and the transformer turn ratio:

$$r_L = \left[\frac{N_1}{N_2}\right]^2 R_L$$

The resistance seen when 'looking into' the whole winding of a transformer with a centre-tapped primary (see Fig. 19-14) is

$$r'_L = \left[\frac{2N_1}{N_2}\right]^2 R_L = 4\left[\frac{N_1}{N_2}\right]^2 R_L$$

or

$$r'_L = 4r_L \tag{19-9}$$

The transformer can now be specified in terms of the output power (P_o), the load resistance (R_L), and the referred resistance (r'_L).

Referring to Fig. 19-15, note that when Q_2 is *off* and Q_3 is *on*, a voltage with a peak value of approximately $+V_{CC}$ is induced in the half of the transformer primary connected to Q_2. The induced voltage occurs because the other half of the primary has $+V_{CC}$ applied to it via Q_3, and because both windings are on

Figure 19-15 With Q_2 *off* and Q_3 *on*, the voltage across Q_2 collector-emitter is approximately $2V_{CC}$.

the same magnetic core. The induced voltage is superimposed upon the supply, so that the voltage that appears at the collector of Q_2 is

$$V_{CE(max)} = 2V_{CC} \qquad (19\text{-}10)$$

To determine the peak transistor current, the equation for power delivered to the transformer primary is rewritten. From Eq. 19-5,

$$I_p = \frac{2P'_o}{V_p} \qquad (19\text{-}11)$$

The power dissipated in the two output transistors is the difference between the dc supply power to the amplifier and the ac power delivered to the transformer primary:

$$2P_T = P_{i(dc)} - P'_o$$

Each transistor is *on* for half of each cycle of the input signal, so the power dissipated in each transistor is half of $2P_T$.

$$P_T = 0.5(P_{i(dc)} - P'_o) \qquad (19\text{-}12)$$

The transistors are specified in terms of the device power dissipation (P_T), the peak current (I_p), and the maximum collector-emitter voltage ($V_{CE(max)}$). The transistors must also be operated below the maximum power dissipation curve (see Section 8-7).

Example 19-5

The class B circuit in Fig. 19-16 is to dissipate 4 W in the 16 Ω load. Specify the output transformer and transistors. Assume an 80% transformer efficiency.

Solution

Eq. 19-6:
$$P'_o = \frac{P_o}{\eta_t} = \frac{4\,\text{W}}{0.8}$$
$$= 5\,\text{W}$$
$$V_p \approx V_{CC} = 30\,\text{V}$$

Eq. 19-8:
$$r_L = \frac{V_p^2}{2P'_o} = \frac{(30\,\text{V})^2}{2 \times 5\,\text{W}}$$
$$= 90\,\Omega$$
$$r'_L = 4r_L = 4 \times 90\,\Omega$$
$$= 360\,\Omega$$

Transformer specification:

$$P_o = 4\,\text{W},\, R_L = 16\,\Omega,\, r'_L = 360\,\Omega \text{ centre-tapped}$$

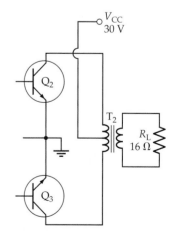

Figure 19-16 Class B power amplifier output stage for Example 19-5.

Eq. 19-10: $\qquad V_{CE(max)} = 2V_{CC} = 2 \times 30 \text{ V}$

$$= 60 \text{ V}$$

Eq. 19-11: $\qquad I_P = \dfrac{2P'_o}{V_P} = \dfrac{2 \times 5 \text{ W}}{30 \text{ V}}$

$$= 333 \text{ mA}$$

Eq. 19-7: $\qquad P_{i(dc)} = V_{CC} \times 0.636 \, I_P = 30 \text{ V} \times 0.636 \times 333 \text{ mA}$

$$= 6.35 \text{ W}$$

Eq. 19-12: $\qquad P_T = 0.5 \, (P_{i(dc)} - P'_o) = 0.5 \, (6.35 \text{ W} - 5 \text{ W})$

$$= 0.68 \text{ W}$$

Transistor specification:

$$P_T = 0.68 \text{ W}, \; V_{CE(max)} = 60 \text{ V}, \; I_P = 333 \text{ mA}$$

Poor frequency response is one disadvantage of transformer-coupled amplifiers, both at the low and high ends of the audio frequency range. This can be improved by the use of overall negative feedback. However, substantial improvement in frequency response can be achieved by eliminating transformers from the circuit. The alternatives are capacitor coupling and direct coupling of the amplifier load.

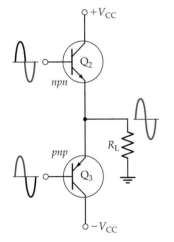

Figure 19-17 A complementary emitter follower uses an *npn* transistor and a *pnp* transistor that have similar characteristics.

Practice Problems

19-3.1 A class A transformer-coupled amplifier is to produce a peak current of 15 mA in a 200 Ω load. The available transformer has $R_L = 200 \; \Omega$, $r_L = 320 \; \Omega$, and efficiency = 90%. Determine a suitable supply voltage and specify the transistor.

19-3.2 A class B transformer-coupled amplifier with $V_{CC} = 25$ V is to supply 6 W to a 12 Ω load. Assuming a 75% transformer efficiency, specify the transistors and transformer.

19-4 CAPACITOR-COUPLED AND DIRECT-COUPLED OUTPUT STAGES

Complementary Emitter Follower

Two BJTs connected to function as emitter followers are shown in Fig. 19-17. Although one is *npn* and the other is *pnp*, the devices are selected to have similar parameters, so they are complementary transistors. The circuit is termed a *complementary emitter follower* or *push-pull emitter follower*.

A single-transistor emitter follower is essentially a small-signal circuit, be-cause large signals can reverse-bias the transistor base-emitter junction when the input polarity is opposite to the transistor V_{BE} polarity. An *npn* emitter follower might not correctly reproduce the negative-going portion of a large signal, while a *pnp* emitter follower might not reproduce the positive-going portion. Complementary emitter followers have similar signals applied simul-taneously to both device bases, as illustrated. Transistor Q_2 conducts during the positive half-cycle of the signal, and it *pulls* the output voltage up to follow the input. During this time, Q_3 base-emitter junction is reverse-biased. For the duration of the negative half-cycle of the input, the Q_2 base-emitter junction is reversed and Q_3 conducts, pulling the output down to follow the input. Thus the complementary emitter follower is a large-signal circuit with the low out-put impedance typical of emitter followers.

Capacitor-Coupled Class AB Output Stage

The basic circuit of a class AB amplifier using a complementary emitter fol-lower output stage and a capacitor-coupled load is shown in Fig. 19-18. The circuit is termed a *complementary symmetry amplifier*. Transistor Q_1 and resistors R_1, R_2, R_C, and R_{E1} constitute a common-emitter amplifier stage that produces all of the circuit voltage gain. The output of Q_1 is developed across R_C and applied to the bases of Q_2 and Q_3. Capacitor C_o ac-couples R_L and dc-isolates R_L to keep it from affecting the circuit bias conditions.

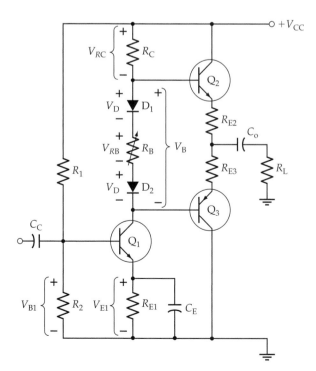

Figure 19-18 A class AB capacitor-coupled power am-plifier that uses a comple-mentary emitter follower as the output stage is known as a complementary symmetry amplifier.

The total voltage drop (V_B) across diodes D_1 and D_2 and resistor R_B forward-biases the base-emitter junctions of Q_2 and Q_3 to prevent cross-over distortion. Emitter resistors R_{E2} and R_{E3} help limit the quiescent current through Q_2 and Q_3. Adjustment of the bias voltage (V_B) is provided by variable resistor R_B. The diodes have voltage drops (V_D) that approximately match the output transistor V_{BE} levels. And since V_D does not change significantly when the diode current changes, the diodes behave like bypassed resistors. The diodes and output transistors can be *thermally coupled* by mounting D_1 and Q_2 on a single heat sink, and D_2 and Q_3 on a single heat sink. In this case, V_D follows the V_{BE} level changes with temperature, thus stabilizing the transistor bias conditions over a wide temperature range. The junction of R_{E2} and R_{E3} must be biased to $V_{CC}/2$ so that the output voltage to C_o can swing by equal amounts in positive- and negative-going directions.

Class AB Capacitor-Coupled Amplifier Design

Designing the type of circuit shown in Fig. 19-18 is largely a matter of selecting appropriate resistor voltage drops and current levels, and then applying Ohm's law to calculate the resistor values. The peak output voltage (V_p) and peak output current (I_p) can be determined from Equations 19-8 and 19-11 respectively. Those equations were developed for the power delivered to a transformer primary, but they apply equally to power delivered to any load resistor.

The voltage drops across R_{E2} and R_{E3} when the peak output current is flowing are typically selected as 5% to 10% of the peak output voltage. This is illustrated in Fig. 19-19a and b where the output capacitor is represented as an ac short-circuit. So,

$$R_{E2} = R_{E3} \approx 0.05 \text{ to } 0.1\ R_L \tag{19-13}$$

It should be remembered that R_{E2} and R_{E3} are included to help stabilize the transistor quiescent currents at a level that eliminates cross-over distortion in the output waveform. For the type of amplifier circuit in Fig. 19-18, without overall negative feedback, it is best if the emitter resistors are as large as possible. Smaller emitter resistors can be used in circuits with dc and ac negative feedback (see Sections 19-5 and 19-6).

When the output is at its negative-going peak, V_{CE1} should be 1 V minimum, to ensure that Q_1 does not go into saturation. V_{E1} should typically be 3 V (see Section 5-7). So the minimum level of V_{C1} is typically 4 V (Fig. 19-19b). Similarly, when the output is at its positive-going peak, there must be an appropriate minimum voltage drop across resistor R_C (Fig. 19-19a). It is not acceptable to set a 1 V minimum for V_{RC}, because the current through R_C would be too small for the required peak base current to Q_2. Therefore it is best to select

$$V_{RC(min)} = V_{C1(min)} = 4 \text{ V} \tag{19-14}$$

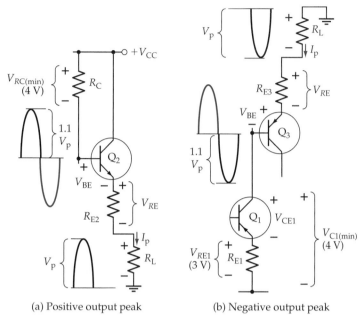

(a) Positive output peak   (b) Negative output peak

Figure 19-19   Supply voltage determination for a class AB capacitor-coupled output stage.

The minimum current through R_C ($I_{RC(min)}$) should typically be selected to be 1 mA larger than the peak base current for the output transistors. R_C is calculated from $V_{RC(min)}$ and $I_{RC(min)}$.

In Fig. 19-19a, it can be seen that the supply required to produce the positive output peak voltage is

$$V_+ = V_p + V_{RE2} + V_{BE2} + V_{RC(min)}$$

Also, from Fig. 19-19b, the negative output peak requires

$$V_- = V_p + V_{RE3} + V_{BE3} + V_{C1(min)}$$

and $$V_+ = V_-$$

So the total supply voltage is

$$V_{CC} = 2(V_p + V_{RE2} + V_{BE2} + V_{RC(min)}) \qquad \textbf{(19-15)}$$

The voltage drop across the diodes and R_B should just bias Q_2 and Q_3 *on* for class AB operation (see Fig. 19-20). The current through R_B is the Q_1 quiescent current (I_{CQ1}), and this is calculated from R_C and the dc voltage drop across R_C. $V_{RC(dc)}$ equals $V_{C1(dc)}$, and the sum of them equals ($V_{CC} - V_B$) (see Fig. 19-20). So,

$$V_{RC(dc)} = V_{C1(dc)} = 0.5(V_{CC} - V_B) \qquad \textbf{(19-16)}$$

The resistance of R_B is now calculated from V_{RB} and I_{CQ1}. Resistors R_1, R_2, and R_{E1} (Fig. 19-18) are determined in the usual manner for a voltage-divider bias circuit.

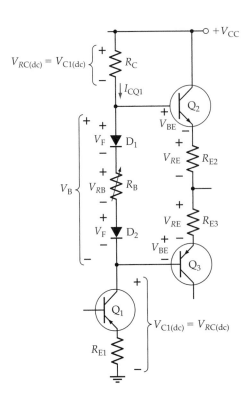

Figure 19-20 The voltage drops across D_1, D_2, and R_B bias Q_2 and Q_3 *on* for class AB operation.

The output transistors should be specified in terms of their maximum voltage, current, and power dissipation. The maximum V_{CE} for Q_2 and Q_3 (in Fig. 19-18) is the total supply voltage (V_{CC}). Maximum current for Q_2 and Q_3 is the peak load current plus the selected quiescent current (I_{Q23}). The maximum is normally 1.1 I_p. Transistor power dissipation is calculated by determining the dc power delivered to the output stage from the power supply and then subtracting the ac load power. The remainder is halved to find the power dissipated in each transistor. Equation 19-12 applies: $P_T = 0.5\,(P_i - P_o)$. Recall that the transistors must be operated within the safe operating area of the characteristics (see Section 8-7).

With a capacitor-coupled load, current is drawn from the power supply during the positive half-cycle of the output, but not during the negative half-cycle. The capacitor acts as an energy reservoir to supply load current when the output is negative-going. Consequently, the supply current has a half-wave rectified waveform (see Fig. 19-21), and so

$$I_{ave} = 0.5 \times 0.636\ 1.1 I_p$$

$$= 0.35 I_p$$

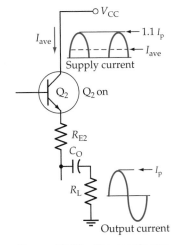

Figure 19-21 The supply current for the output stage of a capacitor-coupled class AB power amplifier has a half-wave rectified waveform.

The dc supply power to the output stage is

$$P_i = V_{CC} \times I_{ave}$$

or

$$P_i = 0.35\ V_{CC} I_p \qquad (19\text{-}17)$$

As always, with the exception of the capacitor chosen to set the circuit low 3 dB frequency (f_1), each capacitor impedance is selected to be one-tenth of the impedance in series with the capacitor. When there is no overall negative feedback, the capacitor with the lowest-value series-connected impedance is normally selected to set f_1. For the circuit shown in Fig. 19-18, the impedance when 'looking into' the emitter of Q_1 is h_{ib1}, and this is in series with C_E. If h_{ib1} is smaller than R_L, then at f_1,

$$X_{CE} \approx h_{ib1}, \quad X_{CC} \approx 0.1 Z_i, \quad X_{CO} \approx 0.1 R_L$$

Most power amplifiers have $R_L = 8\ \Omega$ or $16\ \Omega$. So, the load-coupling capacitor normally sets the circuit low 3 dB frequency.

$$X_{Co} = R_L \text{ at } f_1 \qquad (19\text{-}18)$$

Example 19-6

The class AB amplifier in Fig. 19-22 (reproduced from Fig. 19-18) is to deliver 1 W to a 50 Ω load. Determine the required supply voltage, and calculate resistor values for R_C, R_B, R_{E2}, and R_{E3}. Assume that $h_{FE(min)} = 50$ for Q_2 and Q_3.

Solution

From Eq. 19-8, $\qquad V_p = \sqrt{(2R_L P_o)} = \sqrt{(2 \times 50\ \Omega \times 1\ W)}$

$$= 10\ V$$

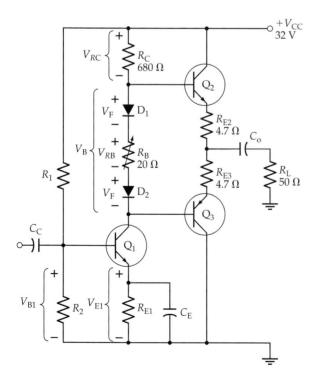

Figure 19-22 Capacitor-coupled class AB power amplifier for Example 19-6.

$$I_{\text{p}} = \frac{V_{\text{P}}}{R_{\text{L}}} = \frac{10 \text{ V}}{50 \ \Omega}$$

$$= 200 \text{ mA}$$

Eq. 19-13:

$$R_{\text{E2}} = R_{\text{E3}} = 0.1 \, R_{\text{L}} = 0.1 \times 50 \ \Omega$$

$$= 5 \ \Omega \text{ (use 4.7 } \Omega \text{ standard value)}$$

Select

$$I_{\text{CQ2}} = 0.1 \, I_{\text{p}} = 0.1 \times 200 \text{ mA}$$

$$= 20 \text{ mA}$$

$$V_{\text{B}} = V_{\text{BE2}} + I_{\text{CQ2}} \, (R_{\text{E2}} + R_{\text{E3}}) + V_{\text{BE3}}$$

$$= 0.7 \text{ V} + 20 \text{ mA } (4.7 \ \Omega + 4.7 \ \Omega) + 0.7 \text{ V}$$

$$= 1.6 \text{ V}$$

Eq. 19-14:

$$V_{\text{C1(min)}} = V_{RC(\text{min})} = 4 \text{ V}$$

$$I_{\text{B2(max)}} = \frac{I_{\text{p}}}{h_{\text{FE2(min)}}} = \frac{200 \text{ mA}}{50}$$

$$= 4 \text{ mA}$$

Select

$$I_{RC(\text{min})} = I_{\text{B2(max)}} + 1 \text{ mA} = 4 \text{ mA} + 1 \text{ mA}$$

$$= 5 \text{ mA}$$

$$R_{\text{C}} = \frac{V_{RC(\text{min})}}{I_{RC(\text{min})}} = \frac{4 \text{ V}}{5 \text{ mA}}$$

$$= 800 \ \Omega \text{ (use 680 } \Omega \text{ standard value)}$$

Eq. 19-15:

$$V_{\text{CC}} = 2(V_{\text{p}} + V_{RE2} + V_{\text{BE2}} + V_{RC(\text{min})})$$

$$= 2[10 \text{ V} + 1 \text{ V} + 0.7 \text{ V} + 4 \text{ V}]$$

$$= 31.4 \text{ V} \text{ (use 32 V)}$$

Eq. 19-16:

$$V_{RC(\text{dc})} = 0.5(V_{\text{CC}} - V_{\text{B}}) = 0.5(32 \text{ V} - 1.6 \text{ V})$$

$$= 15.2 \text{ V}$$

$$I_{\text{C1(dc)}} = \frac{V_{RC(\text{dc})}}{R_{\text{C1}}} = \frac{15.2 \text{ V}}{680 \ \Omega}$$

$$= 22.4 \text{ mA}$$

$$R_{\text{B}} = \frac{V_{\text{B}} - V_{\text{D1}} - V_{\text{D2}}}{I_{\text{C1(dc)}}} = \frac{1.6 \text{ V} - 0.7 \text{ V} - 0.7 \text{ V}}{22.4 \text{ mA}}$$

$$= 8.9 \ \Omega \text{ (use 20 } \Omega \text{ standard value variable}$$
$$\text{resistor to allow for} \pm \text{adjustment)}$$

Example 19-7

Specify the output transistors for the circuit in Example 19-6.

Solution

$$V_{CE(max)} = V_{CC} = 32 \text{ V}$$

$$I_{C(max)} = 1.1\,I_p = 1.1 \times 200 \text{ mA}$$

$$= 220 \text{ mA}$$

Eq. 19-17: $$P_{i(dc)} = 0.35\,V_{CC}\,I_p = 0.35 \times 32 \text{ V} \times 200 \text{ mA}$$

$$= 2.24 \text{ W}$$

From Eq. 19-12, $$P_T = 0.5\,(P_{i(dc)} - P_o) = 0.5\,(2.24 \text{ W} - 1 \text{ W})$$

$$= 0.62 \text{ W}$$

Example 19-8

Calculate capacitor values for C_E and C_o for the circuit in Example 19-6 if the lower cutoff frequency is to be 50 Hz. Assume that $h_{ib1} = 2\ \Omega$.

Solution

Because $h_{ib1} < R_L$, C_E sets f_1:

$$C_E \approx \frac{1}{2\pi f_1 h_{ib1}} = \frac{1}{2 \times \pi \times 50 \text{ Hz} \times 2\ \Omega}$$

$$= 1590\ \mu\text{F (use } 1800\ \mu\text{F standard value)}$$

$$C_o \approx \frac{1}{2\pi f_1 0.1 R_L} = \frac{1}{2 \times \pi \times 50 \text{ Hz} \times 0.1 \times 50\ \Omega}$$

$$= 637\ \mu\text{F (use } 680\ \mu\text{F standard value)}$$

Direct-Coupled Class AB Output Stage

The output capacitor in a capacitor-coupled power amplifier is a large expensive component that should be eliminated if possible. Figure 19-23 shows a class AB amplifier circuit with a direct-coupled load. In this case, the supply voltages must be positive and negative quantities, $+V_{CC}$ and $-V_{EE}$ as shown, so that the dc voltage at the output is zero. This is necessary to avoid a power-wasting direct current through the load. Apart from the positive/negative supply, the direct-coupled circuit operates in the same way as the capacitor-coupled amplifier.

With a few exceptions, the design procedure for a direct-coupled circuit is essentially the same as for a capacitor-coupled circuit. Equation 19-15 gives the total supply voltage, which must be halved to give $+V_{CC}$ and $-V_{EE}$ for the direct-coupled amplifier. The total supply voltage must be used to calculate P_i in Equation 19-17. Also, the transistor maximum V_{CE} is the total supply voltage.

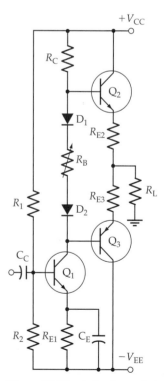

Figure 19-23 A direct-coupled class AB power amplifier must use plus/minus supply voltages.

Practice Problems

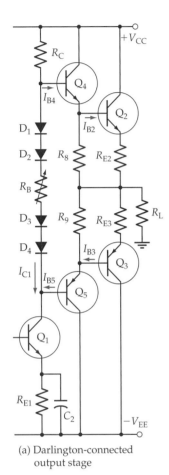

Practice Problems

19-4.1 A class AB power amplifier, as in Fig. 19-22, is to dissipate 200 mW in a 60 Ω load. Calculate V_{CC}, and specify the output transistors.

19-4.2 Determine R_C and R_B for the amplifier in Problem 19-4.1.

19-5 MODIFICATIONS TO IMPROVE POWER AMPLIFIER PERFORMANCE

Darlington-Connected Output Transistors

Because high-power transistors usually have low current gains, relatively large base currents must flow into Q_2 and Q_3 in the circuits in Figs. 19-22 and 19-23 to supply a high-load current. This means that the quiescent current through Q_1 must be large and consequently resistor R_C is small. The small value of R_C keeps the amplifier voltage gain low. To improve on this situation, *Darlington-connected* output transistors may be used, as illustrated in Fig. 19-24a. Transistors Q_4 and Q_5 in Fig. 19-24a are low-power devices that supply base current to output transistors Q_2 and Q_3 respectively. Note the four biasing diodes in Fig. 19-24a for biasing the four transistor base-emitter junctions.

When peak load current (I_p) flows, the peak base current to Q_4 and Q_5 is

$$I_B = \frac{I_p}{h_{fe2} \times h_{fe4}}$$

This reduced base current allows $I_{RC(min)}$ to be smaller, giving a larger resistance for R_C and resulting in a larger voltage gain.

Resistors R_8 and R_9 in Fig. 19-24a are included to bias Q_2 and Q_3 *off* when Q_4 and Q_5 are in cutoff. The largest possible resistance values should normally be selected for R_8 and R_9. When Q_2 and Q_3 are *off*, the collector-base leakage current I_{CBO} flows in R_8 and R_9 (see Fig. 19-24b). The voltage drop across the resistors ($I_{CBO} R_8$) should be much smaller than the normal transistor base-emitter voltage. Selecting $I_{CBO} R_8$ equal to 0.01 V normally gives satisfactory resistor values. R_8 and R_9 are not required when power Darlingtons are used, as illustrated in Fig. 19-25.

(a) Darlington-connected output stage

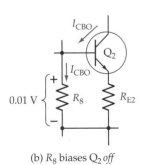

(b) R_8 biases Q_2 *off*

Figure 19-24 Darlington-connected output transistors reduce the current to be supplied by R_C and Q_1, and allow R_C to be increased for a larger voltage gain.

Example 19-9

The class AB amplifier in Example 19-6 is to be redesigned to use power Darlingtons, as in Fig. 19-25. Assuming that the power Darlingtons have $h_{FE(min)} = 2000$, and $V_{BE} = 1.4$ V, determine new values for V_{CC}, R_C, and R_B.

Solution

From Example 19-6: $V_p = 10$ V, $I_p = 200$ mA, $I_{CQ2} = 20$ mA, $R_{E1} = R_{E2} = 4.7\ \Omega$,

$$V_{E1} = 3\ \text{V}, V_{C1(dc)} = V_{RC(dc)} = 15.2\ \text{V}, V_{RC(min)} = 4\ \text{V}$$

$$V_B = V_{BE2} + I_{CQ2}(R_{E2} + R_{E3}) + V_{BE3}$$

$$= 1.4\ \text{V} + 20\ \text{mA}\ (4.7\ \Omega + 4.7\ \Omega) + 1.4\ \text{V}$$

$$= 3\ \text{V}$$

$$V_{CC} = V_{RC(dc)} + V_{C1(dc)} + V_B = 15.2\ \text{V} + 15.2\ \text{V} + 3\ \text{V}$$

$$= 33.4\ \text{V}$$

Use $\qquad V_{CC}/V_{EE} = \pm17\ \text{V}$

$$I_{B2(max)} = \frac{I_P}{h_{FE2(min)}} = \frac{200\ \text{mA}}{2000}$$

$$= 100\ \mu\text{A}$$

Select $\qquad I_{RC(min)} = 1\ \text{mA}$

$$R_C = \frac{V_{RC(min)}}{I_{RC(min)}} = \frac{4\ \text{V}}{1\ \text{mA}}$$

$$= 4\ \text{k}\Omega\ (\text{use } 3.9\ \text{k}\Omega\ \text{standard value})$$

$$I_{C1(dc)} = \frac{V_{RC(dc)}}{R_{C1}} = \frac{15.2\ \text{V}}{3.9\ \text{k}\Omega}$$

$$= 3.9\ \text{mA}$$

$$R_B = \frac{V_B - (4 \times V_D)}{I_{C1(dc)}} = \frac{3\ \text{V} - (4 \times 0.7\ \text{V})}{3.9\ \text{mA}}$$

$$\approx 51\ \Omega\ (\text{use a } 100\ \Omega\ \text{variable resistor for adjustment})$$

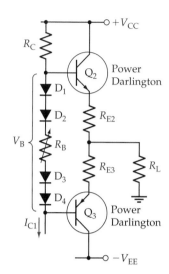

Figure 19-25 Class AB output stage using power Darlingtons.

Quasi-Complementary Output Stage

The *quasi-complementary* circuit was originally developed because complementary high-power transistors were not readily available. Even though such transistors are now available, the quasi-complementary circuit is still widely used.

Consider the arrangement in Fig. 19-26. Q_3 is a high-power *npn* transistor, and Q_5 is a low-power *pnp* device. When Q_5 base current (I_{B5}) flows, the collector current of Q_5 behaves (largely) as base current (I_{B3}) for transistor Q_3. This produces a Q_3 collector current flow (I_{C3}), which combines with I_{E5} to constitute a current flow in the load. Because $I_{C3} \gg I_{E5}$, the output current can be taken to be approximately I_{C3}:

$$I_{C3} = h_{FE3} \times I_{B3}$$

or $\qquad\qquad I_{C3} \approx h_{FE3} \times h_{FE5} \times I_{B5}$

This is the same as the current gain with Darlington-connected transistors.

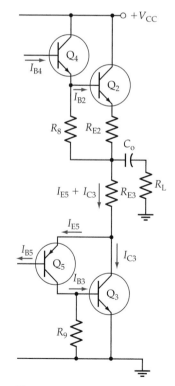

Figure 19-26 A quasi-complementary output stage allows output transistors to be the same type: Q_2 and Q_3 are both *npn* transistors.

Now note that when a negative-going input voltage is applied to the base of Q_5, a negative-going output occurs, because the emitter of Q_5 (and the collector of Q_3) follow the input voltage. Thus, the combination of transistors Q_3 and Q_5 in Fig. 19-26 behaves as a high-power *pnp* emitter follower, just like Darlington-connected transistors Q_3 and Q_5 in Fig. 19-24a. Because transistors Q_2 and Q_3 in Fig. 19-26 are both *npn* devices, they can be the same type of transistor. This eliminates any problem with finding suitable complementary high-power transistors. Resistor R_9 in Fig. 19-26 ensures that Q_3 is biased *off* when Q_5 goes into cutoff.

Output Current Limiting

Because the output transistors can be destroyed by excessive current flow, output current-limiting circuits are often included in a power amplifier. Figure 19-27 shows the typical arrangement for a current-limiting circuit. Emitter resistors R_{E2} and R_{E3} are each made up of two components (R_A and R_B). The current-limiting transistors (Q_4 and Q_5) are connected as shown so that the voltage drop across R_{B2} and R_{B3} (produced by I_L) can turn Q_4 and Q_5 *on*. This occurs only when I_L is at the selected $I_{L(max)}$ level. When Q_4 turns *on*, it pulls the base of Q_2 *down*, so that Q_2 cannot supply current in excess of $I_{L(max)}$.

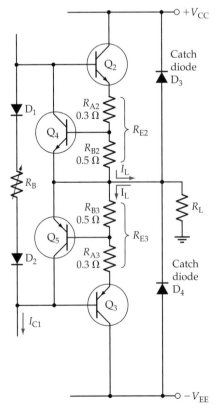

Figure 19-27 Current-limiting transistors (Q_4 and Q_5) turn on when the load current is at the selected maximum. This stops further current increase in Q_2 and Q_3.

Similarly, turning Q_5 *on* causes the base of Q_3 to be pulled *up* toward its emitter, thus again limiting the output current to $I_{L(max)}$.

As already discussed for voltage regulator current-limiting circuits, Q_4 and Q_5 usually turn *on* when $V_{BE} \approx 0.5$ V (see Section 18-3). The component values shown in Fig. 19-27 limit the peak output current to approximately 1 A. *Catch diodes* D_3 and D_4 in Fig. 19-27 are usually included with current limiting to protect the output transistors from the excessive voltage levels generated in an inductive load when the current growth is limited. The diodes prevent the load voltage from exceeding the supply voltage level.

V_{BE} Multiplier

Figure 19-28 shows an alternative to diode biasing for the output stage transistors in a class AB amplifier. This circuit is known as a V_{BE} *multiplier* because it produces a bias voltage ($V_B = V_{CE6}$) that is a multiple of the V_{BE} of transistor Q_6. Referring to the circuit, note that I_{10} is the current through the the voltage divider (R_{10}, R_{11}, and R_{12}) that biases Q_6. Because I_{10} is much smaller than I_{C6}, the Q_1 collector current (I_{C1}) approximately equals I_{C6}.

$$I_{10} = \frac{V_{BE6}}{R_{11} + R_{12}}$$

and $\qquad\qquad V_B = I_{10}(R_{10} + R_{11} + R_{12})$

or $\qquad\qquad V_B = \frac{V_{BE6}(R_{10} + R_{11} + R_{12})}{R_{11} + R_{12}}$ $\qquad$ **(19-19)**

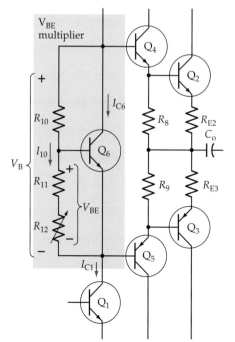

Figure 19-28   A V_{BE} multiplier is an alternative to diode-biasing for the output transistors.

It is seen that the base bias voltage for the output stage transistors can be set by suitable selection of the V_{BE} multiplier resistors. The bias voltage remains constant when I_{C1} changes. It can also be shown that the changes produced in V_B by temperature variations closely match the total V_{BE} temperature changes in the four output stage transistors.

Example 19-10

Design the V_{BE} multiplier in Fig. 19-28 to have $V_B = 3.2$ V adjustable by ± 0.5 V when $I_{C1} = 5$ mA.

Solution

$$V_{B(min)} = 3.2 \text{ V} - 0.5 \text{ V} = 2.7 \text{ V}$$

$$V_{B(max)} = 3.2 \text{ V} + 0.5 \text{ V} = 3.7 \text{ V}$$

Select
$$I_{10} \approx 0.1 \, I_{C1} = 0.1 \times 5 \text{ mA}$$

$$\approx 500 \text{ } \mu\text{A}$$

For $V_{CE} = 3.2$ V,

$$R_{10} = \frac{V_{CE} - V_{BE}}{I_{10}} = \frac{3.2 \text{ V} - 0.7 \text{ V}}{500 \text{ } \mu\text{A}}$$

$$= 5 \text{ k}\Omega \text{ (use 4.7 k}\Omega \text{ standard value)}$$

For $V_{CE} = 3.7$ V,

$$I_{10(max)} = \frac{V_{CE(max)} - V_{BE}}{R_{10}} = \frac{3.7 \text{ V} - 0.7 \text{ V}}{4.7 \text{ k}\Omega}$$

$$= 638 \text{ } \mu\text{A}$$

For $V_{CE} = 2.7$ V,

$$I_{10(min)} = \frac{V_{CE(min)} - V_{BE}}{R_{10}} = \frac{2.7 \text{ V} - 0.7 \text{ V}}{4.7 \text{ k}\Omega}$$

$$= 426 \text{ } \mu\text{A}$$

$$R_{11} + R_{12} = \frac{V_{BE}}{I_{10(min)}} = \frac{0.7 \text{ V}}{426 \text{ } \mu\text{A}}$$

$$= 1.6 \text{ k}\Omega$$

$$R_{11} = \frac{V_{BE}}{I_{10(max)}} = \frac{0.7 \text{ V}}{638 \text{ } \mu\text{A}}$$

$$= 1.1 \text{ k}\Omega \text{ (use 1 k}\Omega \text{ standard value)}$$

$$R_{12} = (R_{11} + R_{12}) - R_{11} = 1.6 \text{ k}\Omega - 1 \text{ k}\Omega$$

$$= 600 \text{ }\Omega \text{ (use a 750 }\Omega \text{ standard}$$
$$\text{value potentiometer)}$$

Power Supply Decoupling

High-power amplifiers require high supply current levels, so unregulated power supplies are often used to save the power wasted in a series regulator. The high-ripple voltage that occurs with unregulated supplies can be amplified to appear at speaker outputs as very unpleasant *power supply hum*. Supply *decoupling* components C_D and R_{15} are employed as shown in Fig. 19-29a to combat hum. Capacitive impedance X_{CD} forms an ac voltage divider with resistor R_{15}, so that the ripple amplitude is attenuated, as illustrated. The resistance of R_{15} is usually selected approximately equal to emitter resistor R_{E1}, and X_{CD} is made very much smaller than R_{15} at the ripple frequency (f_r). If $X_{CD} = R_{15}/100$ at f_r, the ripple voltage will be attenuated by approximately a factor of 100. The dc voltage drop across R_{15} must be taken into account when calculating V_{CC}.

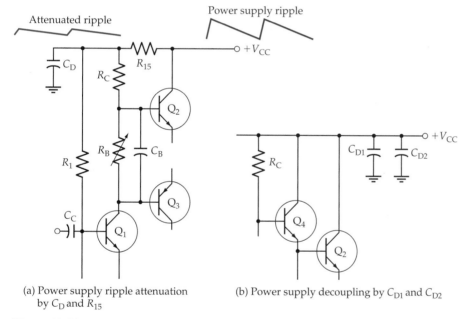

(a) Power supply ripple attenuation by C_D and R_{15}

(b) Power supply decoupling by C_{D1} and C_{D2}

Figure 19-29   Decoupling capacitors are used to minimize power supply ripple voltage and supply line transients.

Figure 19-29b shows power supply decoupling capacitors (C_{D1} and C_{D2}) without any series connected resistors. These are often used even when the supply voltages are regulated and hum is not a problem. When a sudden high output current is switched *on* or *off*, the current change can produce short-lived spike-type voltage drops (*transients*) on the supply lines. These transients may be amplified to produce output distortion if they are allowed to appear at the supply lines for the first or second stages of a circuit. Capacitor

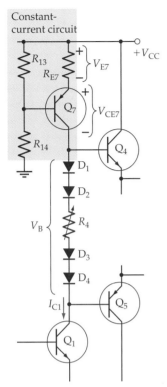

Figure 19-30 A constant-current circuit used as a load for Q_1 will increase the overall amplifier overall gain.

C_{D1} is a relatively high-capacitance component that might normally be expected to perform the necessary decoupling. However, such capacitors usually offer a relatively high impedance to high-frequency variations or fast transients. Consequently, the high-frequency low-capacitance component C_{D2} is necessary to ensure satisfactory decoupling. It is very important that decoupling capacitors be located *right on circuit boards, as close as possible to the terminals of the circuit to be decoupled.*

Increased Voltage Gain and Negative Feedback

The Q_1 collector resistor (R_C) in the circuits discussed so far has its resistance limited by the need to supply base current to the *npn* output transistor. The resistance of R_C also dictates the Q_1 collector current level. This is shown by the $I_{C1(dc)}$ calculation in Example 19-6. A larger resistance for R_C would give a greater voltage gain, which is desirable for negative feedback. In Section 13-7 it is shown that negative feedback reduces distortion and increases circuit bandwidth; so its use is necessary in all power amplifiers.

In Fig. 19-30 R_C is replaced by a constant-current source (see Section 18-2) constituted by transistor Q_7 together with emitter resistor R_{E7}, and base bias resistors R_{13} and R_{14}. The minimum V_{CE7} and the R_{E7} voltage drop are selected to equal the Q_1 levels. This allows the voltage to the output stage to swing positively by the same amount as it can go negatively. The constant-current circuit offers a high ac load resistance ($1/h_{oe7}$) for the Q_1 stage to give the highest possible voltage gain.

Load Compensation

All design and analysis calculations for power amplifiers assume a resistive load with a given resistance value (R_L). For audio amplifiers, the load is usually the coil of a speaker that, as illustrated in Fig. 19-31, combines coil inductance L_c and winding resistance R_c. The load impedance is $Z_L = R_c + j(2\pi f L_c)$, and clearly Z_L increases (from a low of R_c at dc) as the signal frequency increases. An 8 Ω speaker might offer an impedance of 8 Ω only at a frequency around 400 Hz. The fact that the load is inductive means that the load current lags the load voltage, and typically the phase angle could be as high as 60°. Similarly, when capacitive loads are involved, the load current can lead the load voltage. Because the phase difference between load current and voltage can put stress on the output transistors, output compensating components are often included to minimize the phase difference.

Figure 19-32 shows the typical arrangement of the compensating components. Inductor L_x and its parallel-connected resistor R_x are usually recommended by device and IC manufacturers for isolating a capacitive load. Capacitor C_o and series-connected resistor R_o help to correct the lagging phase angle of an inductive load.

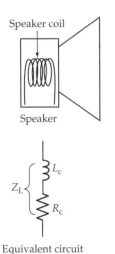

Figure 19-31 A speaker coil has resistance and inductance, so its impedance varies with signal frequency.

Practice Problems

19-5.1 A direct-coupled class AB power amplifier output stage, as in Fig. 19-24a, is to deliver 5 W to an 8 Ω load. The output transistors are power Darlingtons with $h_{FE} = 3000$ and $V_{BE} = 1.5$ V. Calculate the required V_{CC} and the resistance values for R_C and R_B.

19-5.2 Design a V_{BE} multiplier to provide a (4 ± 0.7) V bias for a class AB output stage. The bias current (I_{C1}) is 2 mA.

19-5.3 The constant-current circuit in Fig. 19-30 has $V_{CC} = 25$ V and $I_{C1} = 3$ mA. Determine suitable resistance values for R_{13}, R_{14}, and R_{E7}. Also, calculate the resistance for R_4 to give $V_B = 3.5$ V ± 0.4 V.

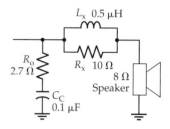

Figure 19-32 Typical arrangement of components to compensate for inductive and capacitive loads.

19-6 BJT POWER AMPLIFIER WITH DIFFERENTIAL INPUT STAGE

Amplifier Circuit

The direct-coupled amplifier in Fig. 19-33 has a differential-amplifier input stage constituted by transistors Q_1 and Q_2 (see Section 12-8). It also has an intermediate stage (Q_3) with a constant-current load (Q_4). Both pairs of output

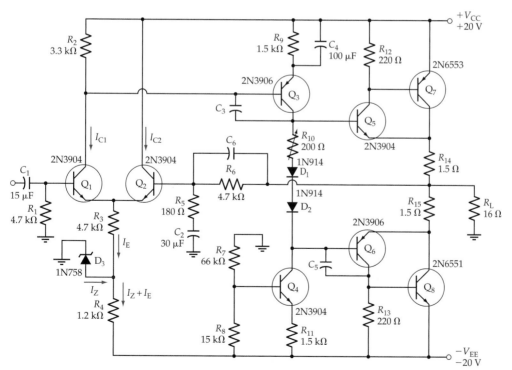

Figure 19-33 Complementary symmetry class AB power amplifier with a quasi-complementary output stage, a differential input stage, and overall negative feedback. The closed-loop gain is $A_{CL} = (R_5 + R_6)/R_5$.

stage transistors (Q_5 and Q_7) and (Q_6 and Q_8) are in quasi-complementary configuration instead of the usual complementary form. This arrangement helps to minimize the required supply voltage by removing the V_{BE} of the power transistors from the V_{CC} equation:

$$V_{CC} = V_p + V_{R14} + V_{BE5} + V_{CE3(min)} + V_{R9} \qquad (19\text{-}20)$$

$$V_{EE} = -(V_p + V_{R15} + V_{BE6} + V_{CE4(min)} + V_{R11})$$
$$= -V_{CC}$$

The differential input stage facilitates negative feedback (NFB), and the whole circuit functions like an operation amplifier. Q_1 base is the non-inverting input, Q_2 base is the inverting input, and the junction of R_{14} and R_{15} is the output terminal. There is 100% dc NFB provided from the output via R_6 to Q_2 base. This keeps the output dc voltage at the same level as Q_1 base (at ground). With C_2 behaving as an ac short-circuit, the ac NFB is divided by R_5 and R_6 to give a closed loop gain: $A_{CL} = (R_5 + R_6)/R_5$.

Zener diode D_3 and resistor R_4 decouple the power supply ripple on the negative supply line. The dynamic impedance of D_3 combined with R_4 functions as an ac voltage divider to attenuate the ripple at the emitters of Q_1 and Q_2. Ripple at this point is amplified just like an input signal. Capacitors C_3 and C_5 are frequency-compensating components (see Section 15-2).

Amplifier Design

The design procedure for the output and intermediate stages of the circuit in Fig. 19-33 is similar to procedures already discussed. Design of the differential stages is very simple. The dc collector currents for Q_1 and Q_2 should be larger than the peak base current for Q_3, and the voltage drop across R_2 is $V_{E3} + V_{BE3}$. Zener voltage V_{Z3} is any convenient level, usually around 0.5 V_{EE}. Resistor R_3 is calculated to pass $I_E \approx (I_{C1} + I_{C2})$, and R_4 must pass $I_Z + I_E$.

Q_1 bias resistor R_1 is determined from Equation 5-17, R_6 is selected equal to R_1, and R_5 is calculated in terms of R_6 to give the desired closed-loop gain. The impedance of C_2 is made equal to R_5 at the desired lower cutoff frequency (f_1), so that C_2 sets f_1. Capacitors C_1 and C_4 are determined in the usual way for capacitors that are not to affect f_1. An additional capacitor (C_6) might be included to set the upper cutoff frequency (f_2): $X_{C6} = R_6$ at f_2.

Example 19-11

The amplifier circuit in Fig. 19-33 is to deliver 6 W to a 16 Ω load. Determine the required supply voltage and specify the output transistors.

Solution

From Eq. 19-8,　　$V_p = \sqrt{(2R_L P_o)} = \sqrt{(2 \times 16\ \Omega \times 6\ W)}$

$$= 13.9\ V$$

$$V_{R14} = V_{R15} = 0.1\ V_{\text{p}}$$

$$\approx 1.4\ \text{V}$$

$$R_{14} = R_{15} = 0.1\ R_{\text{L}}$$

$$= 1.6\ \Omega\ (\text{use } 1.5\ \Omega\ \text{standard value})$$

Select
$$V_{\text{CE3(min)}} = V_{\text{CE4(min)}} = 1\ \text{V}$$

and
$$V_{R9} = V_{R11} = 3\ \text{V}$$

Refer to Fig. 19-34:

$$V_{\text{CC}} = \pm(V_{\text{p}} + V_{R14} + V_{\text{BE5}} + V_{\text{CE3(min)}} + V_{R9})$$

$$= \pm(13.9 + 1.4\ \text{V} + 0.7\ \text{V} + 1\ \text{V} + 3\ \text{V})$$

$$= \pm 20\ \text{V}$$

$$I_{\text{p}} = \frac{V_{\text{p}}}{R_{\text{L}}} = \frac{13.9\ \text{V}}{16\ \Omega}$$

$$= 869\ \text{mA}$$

Dc power input from each supply line:

Eq. 19-17:
$$P_{\text{i(dc)}} = [V_{\text{CC}} - (V_{\text{EE}})] \times 0.35\ I_{\text{p}}$$

$$= [20\ \text{V} + 20\ \text{V}] \times 0.35 \times 869\ \text{mA}$$

$$\approx 12\ \text{W}$$

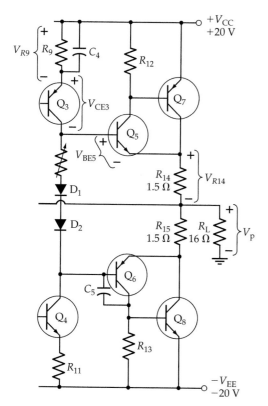

Figure 19-34 Output stage voltage drops for the circuit in Fig. 19-33.

786 Fundamentals of Electronic Devices and Circuits

Output transistor specification:

From Eq. 19-12, $P_T = 0.5(P_{i(dc)} - P_o) = 0.5\,(12\text{ W} - 6\text{ W})$

$$= 3\text{ W}$$

$$V_{CE(max)} = 2V_{CC} = 2 \times 20\text{ V}$$

$$= 40\text{ V}$$

$$I_{C(max)} \approx 1.1 I_p = 1.1 \times 869\text{ mA}$$

$$\approx 956\text{ mA}$$

Example 19-12

Determine suitable resistor values for the output and intermediate stages of the circuit in Example 19-11. Assume that Q_7 and Q_8 have $h_{FE} = 20$ and $I_{CBO(max)} = 50\ \mu A$. Also, assume that Q_5 and Q_6 have $h_{FE} = 70$.

Solution

Refer to Fig. 19-35:

$$R_{12} = R_{13} = \frac{0.01\text{ V}}{I_{CBO}} = \frac{0.01\text{ V}}{50\ \mu A}$$

$$= 200\ \Omega\ \text{(use 220}\ \Omega\ \text{standard value)}$$

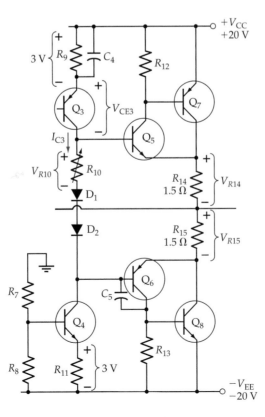

Figure 19-35 Intermediate and output stage dc voltage drops for the circuit in Fig. 19-33.

$$I_{B5(peak)} = \frac{I_P}{h_{FE7} \times h_{FE5}} = \frac{869 \text{ mA}}{20 \times 70}$$

$$\approx 0.62 \text{ mA}$$

$$I_{C3} > I_{B5(peak)}$$

Select
$$I_{C3} = 2 \text{ mA}$$

$$R_9 = R_{11} = \frac{V_{R9}}{I_{C3}} = \frac{3 \text{ V}}{2 \text{ mA}}$$

$$= 1.5 \text{ k}\Omega \text{ (standard value)}$$

$$I_{Q78} \approx Q_7, Q_8 \text{ quiescent current}$$

Select
$$I_{Q78} \approx 0.1 \, I_P = 0.1 \times 869 \text{ mA}$$

$$\approx 86.9 \text{ mA}$$

$$V_{R14(dc)} = V_{R15(dc)} = I_{Q78} \times R_{15}$$

$$= 86.9 \text{ mA} \times 1.5 \ \Omega$$

$$\approx 0.13 \text{ V}$$

$$V_{R10(max)} = (V_{R14(dc)} + V_{R15(dc)}) + 50\%$$

$$= (0.13 \text{ V} + 0.13 \text{ V}) + 50\%$$

$$= 0.39 \text{ V}$$

$$R_{10} = \frac{V_{R10}}{I_{C3}} = \frac{0.39 \text{ V}}{2 \text{ mA}}$$

$$= 195 \ \Omega \text{ (use 200 } \Omega \text{ variable resistor)}$$

Select
$$R_8 = 10 R_{11} = 15 \text{ k}\Omega \text{ (standard value)}$$

$$I_{R8} = \frac{V_{R11} + V_{BE}}{R_8} = \frac{3 \text{ V} + 0.7 \text{ V}}{15 \text{ k}\Omega}$$

$$= 247 \ \mu\text{A}$$

$$R_7 = \frac{V_{EE} - (V_{R11} + V_{BE})}{I_{R8}} = \frac{20 \text{ V} - (3 \text{ V} + 0.7 \text{ V})}{247 \ \mu\text{A}}$$

$$= 66 \text{ k}\Omega \text{ (use 56 k}\Omega + 10 \text{ k}\Omega)$$

Practice Problem

19-6.1 Design the input stage and feedback network for the circuit in Examples 19-11 and 19-12. The signal amplitude is ±0.5 V, and the lower cutoff frequency is to be 30 Hz. Assume that transistors Q_1 to Q_4 have $h_{FE(min)} = 60$.

19-7 COMPLEMENTARY MOSFET COMMON-SOURCE POWER AMPLIFIER

Advantages of MOSFETs

Power MOSFETs (described in Section 9-5) have several advantages over power BJTs for large-signal amplifier applications. One of the most important differences is that MOSFET transfer characteristics (I_D/V_{GS}) are more linear than I_C/V_{BE} characteristics for BJTs. This helps to minimize distortion in the output waveform. Since thermal runaway does not occur with power MOSFETs, the emitter resistors in the BJT output stage (R_{14} and R_{15} in Fig. 19-33) are not needed in a MOSFET amplifier. Thus, the wasted power dissipation in the emitter resistors is eliminated.

Power MOSFETs can be operated in parallel to reduce the total channel resistance and increase the output current. Unlike BJTs operated in parallel, there is no need for resistors to equalize current distribution between parallel-connected MOSFETs. For class AB operation, the MOSFET gate source should be biased to the *threshold voltage* (V_{TH}) for the device, to ensure that it is conducting at a low level when no signal is present.

Power Amplifier with MOSFET Output Stage

The four output transistors in the amplifier circuit in Fig. 19-33 could be replaced with two power MOSFETs operating as source followers. However, the FET gate-source voltage must be included when the supply voltage is calculated, so that a larger supply voltage would be required than with a BJT amplifier. In the power amplifier in Fig. 19-36 the complementary MOSFET output devices operate as common-source amplifiers. As will be seen, this permits the peak output voltage to approach the supply voltage level.

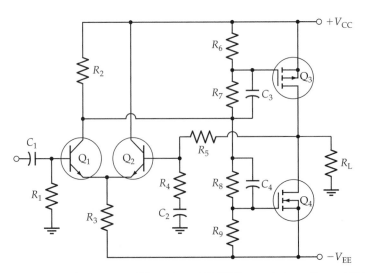

Figure 19-36   Class AB power amplifier using a MOSFET output stage. Power MOSFETs Q_3 and Q_4 operate as common-source amplifiers.

As in the case of the BJT circuit in Fig. 19-33, the differential amplifier input stage in Fig. 19-36 allows the use of dc negative feedback (via R_5) to stabilize the dc output voltage level, and ac negative feedback (via R_4 and R_5) to set the closed-loop voltage gain. Output transistors Q_3 and Q_4 are complementary MOSFETs, and both are operated as common-source amplifiers. Resistors R_6 to R_9 provide bias voltage to set the gate-source voltage of Q_3 and Q_4 to the threshold level (V_{TH}) for the MOSFETs, (see Fig. 19-37). Resistor R_{10} can be included, as shown, to facilitate bias voltage adjustment. Capacitors C_3 and C_4 short-circuit R_7 and R_8 at signal frequencies, so that all the ac voltage from the first stage is applied to the MOSFET gate terminals.

A positive-going voltage at Q_1 collector increases V_{GS4} and decreases V_{GS3}. Thus, Q_4 drain current is increased and Q_3 is turned *off* (see Fig. 19-38). I_{D4} flows through R_L, producing a negative-going load voltage. When the voltage at Q_{1C} is negative-going, V_{GS4} is decreased and V_{GS3} is increased. This causes I_{D3} to increase and Q_4 to be turned *off*. Load current now flows via Q_3 to produce a positive-going load voltage.

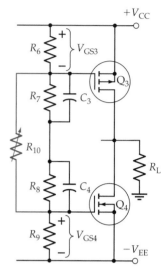

Figure 19-37   MOSFET gate-source bias voltage is provided by resistors R_6 to R_9.

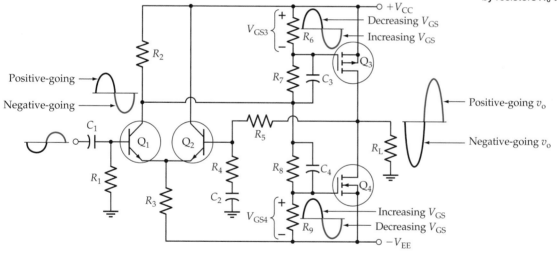

Figure 19-38   A positive-going voltage at Q_1 collector is applied to the gate-source terminals of Q_4 to produce a negative-going load voltage.

MOSFET Power Amplifier Design

In a Class AB MOSFET power amplifier, the FET gate-source voltages should be biased to the minimum specified threshold voltage for the devices. The peak output voltage and current are calculated in the usual way, and the minimum supply voltage is

$$V_{CC} = \pm[V_p + I_{DS} R_{D(on)}] \qquad (19\text{-}21)$$

where $R_{D(on)}$ is the FET channel resistance (see Section 9-3).

The required gate-source voltage swing (ΔV_{GS}) is determined from I_p/g_{FS}. The input stage must provide for $\pm \Delta V_{GS}$ at Q_1 collector (Fig. 19-38). Power dissipation in Q_3 and Q_4 is determined in the same way as for a BJT stage. The selected MOSFETs must survive the total supply voltage and pass a drain current approximately equal to 1.1 I_p. Capacitor values are determined in the usual way.

Example 19-13

The amplifier in Fig. 19-36 is to deliver 2.5 W to a 20 Ω load. The output MOSFETs have $g_{FS} = 250\,\text{mA/V}$, $V_{TH} = 1\,\text{V}$, and $R_{D(on)} = 4\,\Omega$. Calculate the required supply voltage, and determine suitable dc voltage drops across R_2, R_3, and R_6 to R_9.

Solution

From Eq. 19-8, $$V_p = \sqrt{(2R_L P_o)} = \sqrt{(2 \times 20\ \Omega \times 2.5\ \text{W})}$$

$$= 10\ \text{V}$$

$$I_p = \frac{V_p}{R_L} = \frac{10\ \text{V}}{20\ \Omega}$$

$$= 500\ \text{mA}$$

Eq. 19-21: $$V_{CC} = \pm(V_p + I_p R_{D(on)}) = \pm[10\ \text{V} + (500\ \text{mA} \times 4\ \Omega)]$$

$$= \pm 12\ \text{V}$$

$$V_{R6} = V_{R9} = V_{TH} = 1\ \text{V}$$

$$\Delta V_{R6} = \Delta V_{R9} = \frac{I_p}{g_{FS}} = \frac{500\ \text{mA}}{250\ \text{mA/V}}$$

$$= 2\ \text{V}$$

$$V_{R2(min)} = V_{CE1(min)} = \Delta V_{R6} + 1\ \text{V} = 2\ \text{V} + 1\ \text{V}$$

$$= 3\ \text{V}$$

Select $$V_{CE1(dc)} = 3\ \text{V}$$

Then, $$V_{R2(dc)} \approx V_{CC} - V_{CE1(dc)} = 12\ \text{V} - 3\ \text{V}$$

$$= 9\ \text{V}$$

$$V_{R3} = V_{EE} - V_{BE} = 12\ \text{V} - 0.7\ \text{V}$$

$$= 11.3\ \text{V}$$

$$V_{R7} = V_{R2} - V_{R6} = 9\ \text{V} - 1\ \text{V}$$

$$= 8\ \text{V}$$

$$V_{R8} = [V_{CC} - (-V_{EE})] - V_{R6} - V_{R7} - V_{R9}$$

$$= 12\ \text{V} + 12\ \text{V} - 1\ \text{V} - 8\ \text{V} - 1\ \text{V}$$

$$= 14\ \text{V}$$

Example 19-14

Determine resistor values for the MOSFET amplifier circuit in Example 19-13. The input signal is to be ±800 mV.

Solution

$$R_6 = R_9 < 1 \text{ M}\Omega$$

Select

$$R_6 = R_9 = 100 \text{ k}\Omega$$

$$I_6 = \frac{V_{\text{TH}}}{R_6} = \frac{1 \text{ V}}{100 \text{ k}\Omega}$$

$$= 10 \text{ μA}$$

$$R_7 = \frac{V_{R7}}{I_6} = \frac{8 \text{ V}}{10 \text{ μA}}$$

$$= 800 \text{ k}\Omega \text{ (use 820 k}\Omega \text{ standard value)}$$

$$R_8 = \frac{V_{R8}}{I_6} = \frac{14 \text{ V}}{10 \text{ μA}}$$

$$= 1.4 \text{ M}\Omega \text{ (use 1.5 M}\Omega \text{ standard value)}$$

Select

$$I_{C1} = I_{C2} = 1 \text{ mA}$$

$$R_2 = \frac{V_{R2}}{I_{C1}} = \frac{9 \text{ V}}{1 \text{ mA}}$$

$$= 9 \text{ k}\Omega \text{ (use 8.2 k}\Omega \text{ standard value)}$$

$$R_3 = \frac{V_{R3}}{I_{C1} + I_{C2}} = \frac{11.3 \text{ V}}{1 \text{ mA} + 1 \text{ mA}}$$

$$= 5.65 \text{ k}\Omega \text{ (use 5.6 k}\Omega \text{ standard value)}$$

Select

$$R_1 = 4.7 \text{ k}\Omega \text{ (from Eq. 5-17)}$$

$$R_5 = R_1 = 4.7 \text{ k}\Omega$$

$$A_{\text{CL}} = \frac{V_{\text{p(out)}}}{V_{\text{p(in)}}} = \frac{10 \text{ V}}{800 \text{ mV}}$$

$$= 12.5$$

$$R_4 = \frac{R_5}{A_{\text{CL}} - 1} = \frac{4.7 \text{ k}\Omega}{12.5 - 1}$$

$$= 408 \text{ }\Omega \text{ (use 390 }\Omega \text{ standard value)}$$

Practice Problems

19-7.1 Calculate suitable capacitor values for the circuit in Examples 19-13 and 19-14 if the lower cutoff frequency is to be 20 Hz.

19-7.2 A MOSFET class AB amplifier as in Fig. 19-36 is to deliver 1 W to a 100 Ω load. The output devices have g_{FS} = 100 mS, V_{TH} = 1.3 V, and $R_{D(on)}$ = 6 Ω. Determine the supply voltage and the dc voltage drops for the resistors.

19-7.3 Determine resistor and capacitor values for the circuit in Problem 19-7.2. The signal amplitude is ±0.4 V and the lower cutoff frequency is to be 40 Hz.

19-8 BJT POWER AMPLIFIER WITH OP-AMP DRIVER

Circuit Operation

The class AB power amplifier shown in Fig. 19-39 uses an operational amplifier (A_1) for the input stage. Resistors R_4 and R_5 together with the two diodes provide bias for the complementary emitter-follower BJT output stage. There is 100% dc negative feedback via R_3 to keep the dc output at the same level as the op-amp non-inverting input, which is grounded via R_1. Overall ac negative feedback via R_2 and R_3 controls the amplifier ac voltage gain.

No amplification is produced by the intermediate (output-biasing) stage. Instead, resistors R_4 and R_5 provide *active pull-up* for transistors Q_1 and Q_2. When the op-amp output at the junction of D_1 and D_2 is increased in a positive direction, A_1 supplies current through D_2 and R_5. So, the voltage drop across

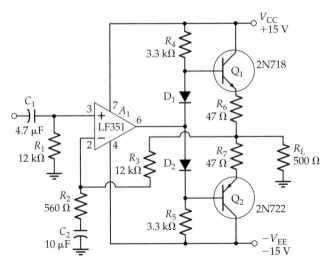

Figure 19-39 Complementary symmetry class AB power amplifier using an operational amplifier and overall negative feedback.

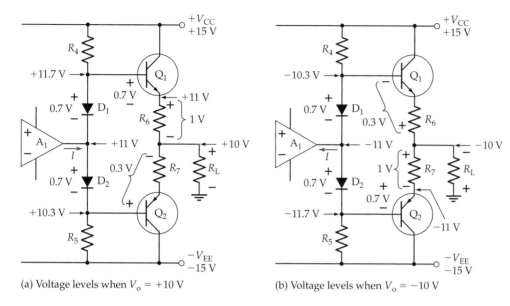

(a) Voltage levels when $V_o = +10$ V

(b) Voltage levels when $V_o = -10$ V

Figure 19-40   Output stage voltage levels for peak output voltages of ±10 V.

R_4 is reduced, allowing it to pull the Q_1 base up to the required level while supplying increased base current to Q_1. This is illustrated by the example of voltage levels shown in Fig. 19-40a. Note that Q_2 is biased *off* when the output voltage is at its positive peak.

Figure 19-40b illustrates the situation when the op-amp output moves in a negative direction. A_1 pulls current through R_4 and D_1, leaving R_5 to pull the base of Q_2 down to the required voltage level while supplying the increased base current. Transistor Q_1 is biased *off* at this time, as indicated by the examples of voltage levels.

The circuit in Fig. 19-39 has no provision for adjusting the bias current in the output transistors. However, the diode voltage drops do bias Q_1 and Q_2, at least into a low-current *on* state. Although this might not seem enough to completely eliminate cross-over distortion, it should be recalled (from Equation 13-28) that overall negative feedback (NFB) reduces distortion by a factor of $(1 + A_v B)$, where A_v is the circuit open-loop gain and B is the feedback factor. Thus, the high open-loop gain of the op-amp severely attenuates the cross-over distortion that would be present without NFB.

Use of Bootstrapping Capacitors

The resistance of (equal resistors) R_4 and R_5 (in Fig. 19-39) is limited by the need to supply base current to the output transistors. The calculation of R_C in Example 19-9 shows the process for determining R_4 and R_5. There is also a need for minimum voltage drop across R_4 and R_5 to produce the base current. Here again, this is shown in Example 19-9, where $V_{RC(min)}$ is the minimum voltage drop across R_C. This minimum resistor voltage requirement keeps the

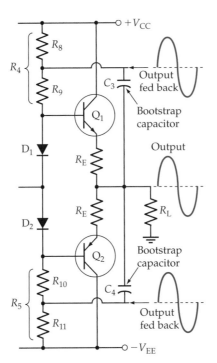

Figure 19-41 Bootstrapping capacitors used with a complementary emitter follower output stage can drive the output transistors close to saturation.

amplifier maximum peak output voltage well below the supply voltage level, and thus limits the amplifier efficiency.

The situation can be substantially improved by the use of the bootstrapping capacitors (C_3 and C_4) shown in Fig. 19-41. Resistors R_4 and R_5 are divided into two equal-value resistors (R_8R_9 and $R_{10}R_{11}$), and the capacitors couple the output voltage back to the junctions of these components.

Consider the example supply voltage and dc bias levels shown in Fig. 19-42a, where the Q_1 emitter resistor is left out for simplicity. The output of A_1 is at 0 V, Q_1 base is at +0.7 V, and the load voltage is 0 V. The supply voltage is +15 V, the voltage at the junction of R_8 and R_9 is +7.5 V, and the voltage across C_3 is 7.5 V. Note that the voltage drop across R_9 is 6.8 V.

The new voltage levels that occur when the op-amp output increases by 3 V are shown in Fig. 19-42b. Q_1 base is at +3.7 V, and the load voltage (V_o) is +3 V. Because C_3 is a large capacitor, its terminal voltage remains substantially constant at 7.5 V, and so the junction of R_8 and R_9 is pushed up to

$$V_o + V_{C3} = 3\,\text{V} + 7.5\,\text{V} = 10.5\,\text{V}$$

With 10.5 V at one end of R_9 and 3.7 V at the other end, the voltage across R_9 is 6.8 V. This is the same level of V_{R9} that occurs when the op-amp output is zero. Thus, C_3 keeps V_{R9} constant. Recall that, without the bootstrapping capacitor, V_{R9} decreases when the op-amp output rises.

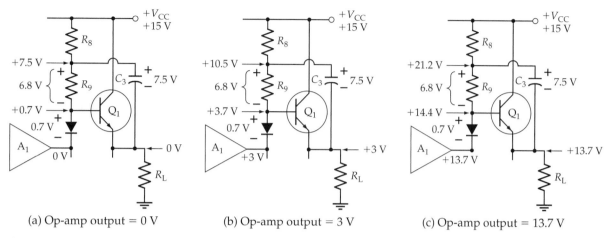

(a) Op-amp output = 0 V   (b) Op-amp output = 3 V   (c) Op-amp output = 13.7 V

Figure 19-42   Output-stage voltage levels for various levels of op-amp output voltage. The voltage on the bootstrapping capacitor remains constant.

Now consider the voltages shown in Fig. 19-42c, where the op-amp output is +13.7 V. Q_1 base voltage is 14.4 V, $V_o = +13.7$, and the voltage at the $R_8 R_9$ junction is

$$V_o + V_{C3} = 13.7 \text{ V} + 7.5 \text{ V} = 21.2 \text{ V}$$

Once again, V_{R9} remains constant; however, note that the bootstrapping capacitor has actually driven the $R_8 R_9$ junction to a level higher than the supply voltage. This allows the output transistors to be driven into saturation, and the voltage drop across $(R_8 + R_9)$ is no longer involved in the supply voltage calculation. In this case, the required supply voltage is

$$V_{CC} = \pm[V_p + (I_p R_E) + V_{CE3(sat)}] \qquad (19\text{-}22)$$

The peak output voltage can also be limited by the output voltage range of the op-amp. For most op-amps the output voltage range is 1 V to 1.5 V less than the positive and negative supply levels. However, *rail-to-rail* op-amps are available with an output that ranges from $+V_{CC}$ to $-V_{EE}$.

Example 19-15
Analyze the amplifier circuit in Fig. 19-43 to determine the bootstrap capacitor terminal voltage (V_{C3}), peak output voltage (V_p), and the peak output power (P_o). The transistors have $V_{CE(sat)} = 1.5$ V.

Solution

When $V_o = 0$ V:

$$I_4 = \frac{V_{CC} - V_{D1}}{R_8 + R_9} = \frac{17 \text{ V} - 0.7 \text{ V}}{1.5 \text{ k}\Omega + 1.5 \text{ k}\Omega}$$

$$= 5.4 \text{ mA}$$

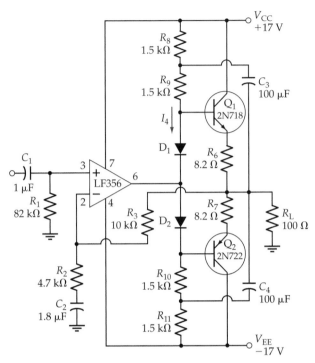

Figure 19-43   Power amplifier circuit for Example 19-15.

$$V_{C3} = V_{CC} - (I_4 R_8)$$

$$= 17\text{ V} - (5.4\text{ mA} \times 1.5\text{ k}\Omega)$$

$$= 8.9\text{ V}$$

$$V_p + V_{R6} = V_{CC} - V_{CE1(sat)} = 17\text{ V} - 1.5\text{ V}$$

$$= 15.5\text{ V}$$

$$I_p = \frac{V_p + V_{R6}}{R_L + R_6} = \frac{15.5\text{ V}}{100\ \Omega + 8.2\ \Omega}$$

$$= 143\text{ mA}$$

$$V_p = I_p R_L = 143\text{ mA} \times 100\ \Omega$$

$$= 14.3\text{ V}$$

$$P_o = \frac{V_p^2}{2\,R_L} = \frac{(14.3\text{ V})^2}{2 \times 100\ \Omega}$$

$$= 1\text{ W}$$

Design Procedure

The amplifier in Fig. 19-44 uses four diodes to forward-bias the base-emitter junctions of the Darlington output transistors. Otherwise, the circuit is exactly the same as in Fig. 19-43.

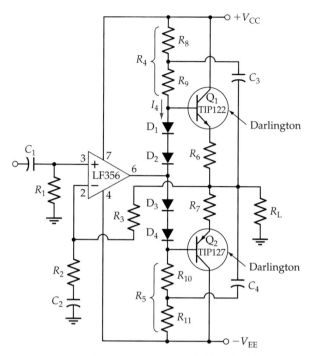

Figure 19-44   Power amplifier circuit for Examples 19-16 and 19-17.

As always, the peak output voltage and current are calculated from the specified output power and load resistance. The supply voltage is determined using Equation 19-22, and the emitter resistors for the output stage are typically selected as $0.1\ R_L$. The bias network current (I_4) should be larger than the peak base current for Q_1 and Q_2. The resistance of R_4 (which equals $R_8 + R_9$) is calculated from I_4 and the circuit dc voltage drops. R_8 should typically be selected as $0.5\ R_4$, and then R_9, R_{10}, and R_{11} are all equal to R_8.

R_1 and R_3 are equal-value resistors that bias the op-amp input terminals. R_2 is calculated from R_3 to give the required voltage gain. C_2 is selected to have its impedance equal to R_2 at the desired lower cutoff frequency (f_1). The bootstrapping capacitors are calculated in terms of the resistance in series with them: $X_{C3} = X_{C4} = 0.1(R_8 \| R_9)$ at f_1. The op-amp must have a suitable full power bandwidth (see Section 15-3) to produce the peak output voltage at the desired upper cutoff frequency for the amplifier.

Example 19-16

The circuit in Fig. 19-44 uses a BIFET op-amp. $R_L = 8\ \Omega$, P_o is to be 6 W, and $v_s = \pm0.1$ V. Q_1 and Q_2 are Darlingtons with $h_{FE(min)} = 1000$ and $V_{CE(sat)} = 2$ V. Determine a suitable supply voltage and resistor values. Also, calculate the minimum op-amp slew rate to give $f_2 = 50$ kHz.

Solution

From Eq. 19-8, $V_p = \sqrt{(2\,R_L\,P_o)} = \sqrt{(2 \times 8\,\Omega \times 6\,\text{W})}$

$$= 9.8\,\text{V}$$

$$I_p = \frac{V_p}{R_L} = \frac{9.8\,\text{V}}{8\,\Omega}$$

$$= 1.2\,\text{A}$$

Select $R_6 = R_7 \approx 0.1 R_L = 0.1 \times 8\,\Omega$

$$\approx 0.8\,\Omega$$

Eq. 19-22: $V_{CC} = \pm[V_p + I_p\,R_6 + V_{CE3(sat)}]$

$$= \pm[9.8\,\text{V} + (1.2\,\text{A} \times 0.8\,\Omega) + 2\,\text{V}]$$

$$\approx \pm 12.8\,\text{V (use }\pm 13\,\text{V)}$$

$$I_{B1(peak)} = \frac{I_p}{h_{FE1}} = \frac{1.2\,\text{A}}{1000}$$

$$= 1.2\,\text{mA}$$

$$I_4 > I_{B3(peak)}$$

Select $I_4 = 2\,\text{mA}$

$$R_4 = \frac{V_{CC} - V_{D1} - V_{D2}}{I_4} = \frac{13\,\text{V} - 0.7\,\text{V} - 0.7\,\text{V}}{2\,\text{mA}}$$

$$= 5.8\,\text{k}\Omega$$

$$R_8 = R_9 = R_{10} = R_{11} = 0.5\,R_4$$

$$= 0.5 \times 5.8\,\text{k}\Omega$$

$$= 2.9\,\text{k}\Omega \text{ (use 2.7 k}\Omega \text{ standard value)}$$

$$A_{CL} = \frac{v_o}{v_s} = \frac{9.8\,\text{V (peak)}}{0.1\,\text{V (peak)}}$$

$$= 98$$

Select $R_1 = R_3 = 100\,\text{k}\Omega$ (see BIFET in Section 14-2)

$$R_2 = \frac{R_3}{A_{CL} - 1} = \frac{100\,\text{k}\Omega}{98 - 1}$$

$$= 1.03\,\text{k}\Omega \text{ (use 1 k}\Omega\text{)}$$

Eq. 15-3: $SR = 2\pi f_2 V_p = 2\pi \times 50\,\text{kHz} \times 9\,\text{V}$

$$= 2.8\,\text{V}/\mu\text{s}$$

Example 19-17

Calculate capacitor values for the circuit in Example 19-16. The lower cutoff frequency is to be 50 Hz.

Solution

$$X_{C1} = 0.1\,R_1 \text{ at } f_1$$

$$C_1 = \frac{1}{2\pi f_1 \times 0.1\,R_1} = \frac{1}{2\pi \times 50\text{ Hz} \times 0.1 \times 100\text{ k}\Omega}$$

$$= 0.318\ \mu\text{F} \ \ (\text{use } 0.33\ \mu\text{F})$$

$$X_{C2} = R_2 \text{ at } f_1$$

$$C_2 = \frac{1}{2\pi f_1 R_2} = \frac{1}{2\pi \times 50\text{ Hz} \times 1\text{ k}\Omega}$$

$$= 3.18\ \mu\text{F} \ \ (\text{use } 3.3\ \mu\text{F})$$

$$X_{C3} = X_{C4} = 0.1(R_8\|R_9) \text{ at } f_1 = 0.1\,(2.7\text{ k}\Omega\|2.7\text{ k}\Omega)$$

$$= 135\ \Omega$$

$$C_3 = C_4 = \frac{1}{2\pi f_1\,X_{CS}} = \frac{1}{2\pi \times 50\text{ Hz} \times 135\ \Omega}$$

$$= 23.6\ \mu\text{F (use } 25\ \mu\text{F})$$

Practice Problems

19-8.1 Calculate $V_{p(out)}$ and P_o for the circuit in Fig. 19-39 when $V_{in} = \pm0.54$ V. Also, determine f_2 when a 741 op-amp is used.

19-8.2 Calculate the supply voltage for an amplifier as in Fig. 19-44 to deliver 3 W to a 12 Ω load. Assume that $h_{FE(min)} = 1500$ and $V_{CE(sat)} = 2$ V for Q_1 and Q_2. Determine the op-amp minimum slew rate to give $f_2 = 65$ kHz.

19-9 MOSFET POWER AMPLIFIER WITH OP-AMP DRIVER STAGE

Basic Circuit Operation

The class AB power amplifier circuit in Fig. 19-45a consists of an operational amplifier (A_1), two MOSFETs (Q_3 and Q_4), and several resistors. The op-amp, together with resistors R_4, R_5, and R_6 and capacitor C_2, constitutes a non-inverting amplifier. The two MOSFETs are a complementary common-source output stage, like the output stage for the circuit discussed in Section 19-7.

The gate-source bias voltages for Q_3 and Q_4 are provided by the voltage drops across resistors R_7 and R_8, which are in series with the op-amp supply

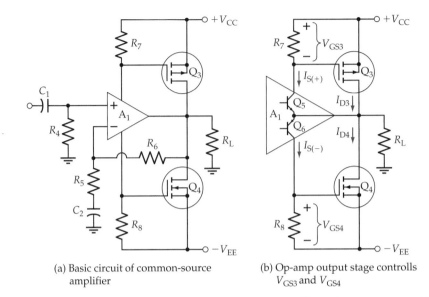

(a) Basic circuit of common-source amplifier

(b) Op-amp output stage controlls V_{GS3} and V_{GS4}

Figure 19-45 Complementary common-source power amplifier using an op-amp driver stage and a MOSFET output stage.

terminals. So the op-amp supply currents ($I_{S(+)}$ and $I_{S(-)}$) determine the levels of V_{GS3} and V_{GS4}:

$$V_{GS3} = I_{S(+)} \times R_7 \tag{19-23}$$

and

$$V_{GS4} = I_{S(-)} \times R_8 \tag{19-24}$$

Suitable gate-source *threshold voltages* ($V_{GS(th)}$) to bias the output transistors for class AB operation, and typical op-amp supply currents (I_S) can be determined from the device data sheets.

Figure 19-45b shows that the op-amp supply currents are largely the collector currents in the op-amp output stage BJTs (Q_5 and Q_6). Thus, R_7 and R_8 are collector resistors for Q_5 and Q_6. When the base voltage for Q_5 and Q_6 is increased in a positive direction, Q_5 collector current ($I_{S(+)}$) increases, causing V_{GS3} to increase and thus increasing the MOSFET drain current I_{D3}. At the same time, the Q_6 collector current ($I_{S(-)}$) decreases, reducing V_{GS4} and drain current I_{D4}. The resulting I_{D3} flows through R_L, producing a positive output voltage swing ($+V_o$). Note that the op-amp output voltage is also positive at this time.

When the base voltage at Q_5 and Q_6 is negative-going, $I_{S(-)}$ increases, causing an increase in V_{GS4} and in I_{D4}. During this time, $I_{S(+)}$ decreases, reducing V_{GS3} and I_{D3}. Thus, I_{D4} flows through R_L, producing a negative output voltage swing ($-V_o$). The op-amp output voltage is also going negative while Q_4 is creating the negative output voltage across R_L.

It is seen that BJTs Q_5 and Q_6 operate as common-emitter amplifier stages and that MOSFETs Q_3 and Q_4 function as common-source circuits. Both stages produce voltage gain, which should be multiplied with the op-amp open-loop gain to determine the total open-loop gain for the circuit. If typical quantities are used, the overall open-loop gain can be shown to be around 4×10^6.

Returning to Fig. 19-45a, the complete (basic) circuit operates as a non-inverting amplifier with a closed-loop voltage gain:

$$A_{\text{CL}} = \frac{R_5 + R_6}{R_5} \qquad\qquad (19\text{-}25)$$

Example 19-18

Determine the MOSFET gate-source bias voltages for the complementary common-source power amplifier in Fig. 19-46. Calculate the peak output voltage, peak output current, and output power if the ac input is ± 100 mV.

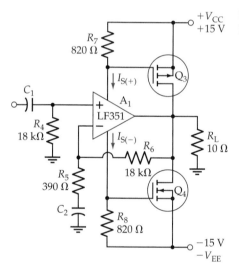

Figure 19-46 Power amplifier circuit for Example 19-18.

Solution

From the LF351 op-amp data sheet:

Supply current: $\qquad\qquad I_S = 1.8 \text{ mA to } 3.4 \text{ mA}$

$$V_{\text{GS3}} = V_{\text{GS4}} = I_S \times R_7$$

$$V_{\text{GS(min)}} = I_{\text{S(min)}} \times R_7 = 1.8 \text{ mA} \times 820 \ \Omega$$

$$\approx 1.5 \text{ V}$$

$$V_{\text{GS(max)}} = I_{\text{S(max)}} \times R_7 = 3.4 \text{ mA} \times 820 \ \Omega$$

$$\approx 2.9 \text{ V}$$

Eq. 19-25: $\qquad\qquad A_{\text{CL}} = \dfrac{R_5 + R_6}{R_5} = \dfrac{390 \ \Omega + 18 \text{ k}\Omega}{390 \ \Omega}$

$$\approx 47.2$$

802   Fundamentals of Electronic Devices and Circuits

$$V_p = A_{CL} \times V_i = 47.2 \times 100 \text{ mV}$$

$$= 4.72 \text{ V}$$

$$I_p = \frac{V_p}{R_L} = \frac{4.72 \text{ V}}{10 \ \Omega}$$

$$= 472 \text{ mA}$$

$$P_o = \frac{V_p I_p}{2} = \frac{4.72 \text{ V} \times 472 \text{ mA}}{2}$$

$$= 1.11 \text{ W}$$

Bias Control

As previously explained, the bias current in the output transistors of a power amplifier should be adjustable. Control over the bias current flowing in transistor Q_3 in Fig. 19-47a can be effected by using a variable resistor for R_7, as illustrated. This allows I_{D3} to be adjusted by varying V_{GS3}. When I_{D3} is increased and the additional drain current flows through R_L, the dc feedback (via R_6 in Fig. 19-46) keeps the V_o equal to zero. The feedback produces the necessary change in V_{GS4} to make I_{D4} closely follow I_{D3}. Thus, adjustment of resistor R_7 controls the level of I_{D4} as well as I_{D3}.

A disadvantage of using R_7 to adjust the bias current is that the voltage gains produced by Q_5 and Q_6 become unequal when R_7 and R_8 have different resistances. This can be overcome by the negative feedback; however, the bias resistors can be kept equal by using the variable current source shown in Fig. 19-47b.

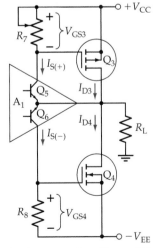

(a) R_7 adjustment alters V_{GS3} and V_{GS4}

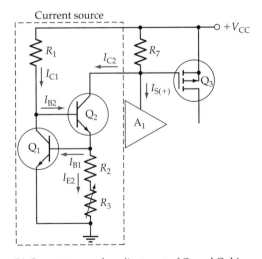

(b) Current source for adjustment of Q_3 and Q_4 bias

Figure 19-47   The gate-source bias voltage for the two output MOSFETs can be adjusted by making R_7 adjustable or by the use of a current source.

In the variable current source, transistor Q_1 is biased from the emitter of Q_2, and the Q_2 base is connected to the collector of Q_1. The collector-emitter voltage of Q_1 is

$$V_{CE1} = V_{BE1} + V_{BE2} = 2V_{BE}$$

Assuming that $I_{B2} \ll I_{C1}$, the Q_1 collector current is

$$I_{C1} = \frac{V_{CC} - 2\,V_{BE}}{R_1} \qquad (19\text{-}26)$$

With $I_{B1} \ll I_{E2}$, the Q_2 collector current is approximately

$$I_{C2} = \frac{V_{BE}}{R_2 + R_3} \qquad (19\text{-}27)$$

Variable resistor R_3 controls the level of I_{C2}.

Also,
$$V_{GS3} = (I_{S(+)} + I_{C2})\,R_7 \qquad (19\text{-}28)$$

Since the dc negative feedback causes V_{GS4} to be always equal to V_{GS3}, R_3 controls the MOSFET gate-source voltages and thus controls the quiescent drain current. The ac load offered to the collector of Q_5 (see Fig. 19-45b) is R_7, regardless of the Q_2 current. Consequently, the Q_5 and Q_6 stage gains are equal.

Example 19-19

Calculate the $V_{GS3(max)}$ and $V_{GS3(min)}$ obtainable by adjusting R_3 for the current source circuit in Fig. 19-48.

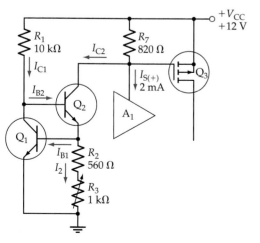

$+V_{CC}$
$+12$ V

Figure 19-48 Current source circuit for Example 19-19.

Solution

Eq. 19-27:
$$I_{C2} = \frac{V_{BE}}{R_2 + R_3}$$

$$I_{C2(max)} = \frac{V_{BE}}{R_2} = \frac{0.7\ V}{560\ \Omega}$$

$$= 1.25\ mA$$

$$I_{C2(min)} = \frac{V_{BE}}{R_2 + R_3} = \frac{0.7\ V}{560\ \Omega + 1\ k\Omega}$$

$$= 449\ \mu A$$

Eq. 19-28:
$$V_{GS3} = (I_{S(+)} + I_{C2})\ R_7$$

$$V_{GS3(min)} = (2\ mA + 449\ \mu A) \times 820\ \Omega$$

$$= 2\ V$$

$$V_{GS3(max)} = (2\ mA + 1.25\ mA) \times 820\ \Omega$$

$$\approx 2.7\ V$$

Output Voltage Swing

One problem with the basic circuit in Fig. 19-45 is that the output voltage swing is limited by the voltage levels at the op-amp supply terminals. As illustrated in Fig. 19-49a, the positive supply voltage to the op-amp is $V_{CC} - V_{R7}$, and the negative supply voltage is $-(V_{EE} - V_{R8})$. Recall too that V_{R7} must be increased by ΔV_{GS3} in order to drive the output in a positive direction, and V_{R8} must be increased by ΔV_{GS4} to drive V_o negative.

$$\Delta V_{GS} = \frac{I_p}{g_{fs}} \tag{19-29}$$

In Equation 19-29, I_p is the peak output current delivered to R_L and g_{fs} is the MOSFET forward transconductance. Thus, the minimum supply voltages at the op-amp terminals are

$$V_s = \pm(V_{CC} - V_{R7} - \Delta V_{GS}) \tag{19-30}$$

The op-amp peak output voltage is normally limited to approximately 1 V below the voltages at the supply terminals (although, as previously noted, rail-to-rail op-amps are available). Consequently, the output voltage swing from the circuit is likely to be less than $\pm(V_{CC} - 4\ V)$ (see Fig. 19-49a). For greatest efficiency, it is necessary to drive the output as closely as possible to $\pm V_{CC}$. This can be done with the circuit modification shown in Fig. 19-49b.

Resistors R_9 and R_{10} in Fig. 19-49b divide V_o, so that the op-amp output can be substantially lower than the peak output voltage developed across R_L. This means that the voltages at the op-amp supply terminals no longer limit the

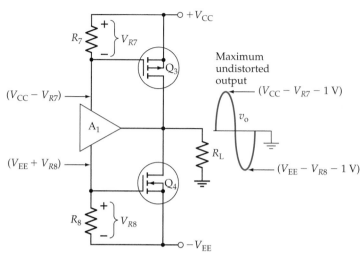

(a) The output voltage swing is limited to approximately $\pm(V_{CC} - V_{R7} - 1\text{ V})$

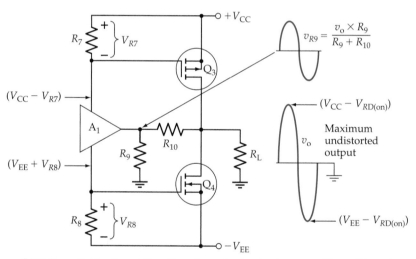

(b) Voltage divider R_9 and R_{10} allows the output swing to go to $\pm(V_{CC} - V_{RD(on)})$

Figure 19-49 The common-source amplifier circuit can be modified to produce an output swing larger than the op-amp maximum output.

amplifier output voltage swing. Now, the largest peak output voltage that can be achieved is limited only by the supply voltage and $R_{D(on)}$.

$$V_P = \frac{V_{CC} \times R_L}{R_{D(on)} + R_L} \qquad (19\text{-}31)$$

R_9 and R_{10} also provide negative feedback that controls the gain of the stage made up of the op-amp output BJTs and the common-source MOSFETs. This is illustrated in Fig. 19-50. A further function of R_9 and R_{10} is that they can be selected to limit the op-amp output current in the event that R_L is short-circuited.

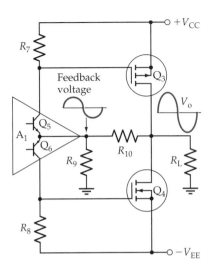

Figure 19-50 Negative feedback to the emitters of Q_5 and Q_6 controls the gain of the $(Q_5 - Q_6) -$ $(Q_3 - Q_4)$ stage.

Example 19-20

The circuit in Fig. 19-51 has MOSFETs with $g_{fs} = 2.5$ S and $R_{D(on)} = 0.5$ Ω. Determine the maximum peak output voltage, the minimum supply voltage at op-amp terminals, and the op-amp peak output voltage when the circuit is producing maximum output power.

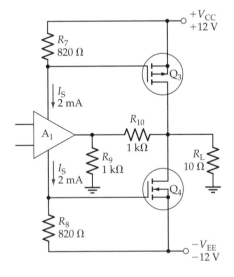

Figure 19-51 Circuit for Example 19-20.

Solution

Eq. 19-31:
$$V_P = \frac{V_{CC} \times R_L}{R_{D(on)} + R_L} = \frac{12 \text{ V} \times 10 \text{ Ω}}{0.5 \text{ Ω} + 10 \text{ Ω}}$$

$$= 11.43 \text{ V}$$

$$I_P = \frac{V_P}{R_L} = \frac{11.43 \text{ V}}{10 \text{ Ω}}$$

$$= 1.14 \text{ A}$$

Eq. 19-29: $\Delta V_{GS} = \dfrac{I_p}{g_{fs}} = \dfrac{1.14\ A}{2.5\ S}$

$= 0.46\ V$

$V_{R7(dc)} = I_S \times R_7 = 2\ mA \times 820\ \Omega$

$= 1.64\ V$

Eq. 19-30: $V_{S(min)} = \pm(V_{CC} - V_{R7(dc)} - \Delta V_{GS}) = \pm(12\ V - 1.64\ V - 0.46\ V)$

$= \pm 9.9\ V$

The op-amp peak output voltage is

$$V_{R9} = \dfrac{V_p \times R_9}{R_9 + R_{10}} = \dfrac{11.43\ V \times 1\ k\Omega}{1\ k\Omega + 1\ k\Omega}$$

$= 5.72\ V$

Complete Amplifier Circuit

The complete circuit of the common-source power amplifier is shown in Fig. 19-52. Note the inclusion of resistors R_{11} and R_{12} and capacitors C_3 and C_4. Resistors R_{11} and R_{12} are typically 100 Ω. They have no effect on the circuit dc conditions, but they help to reduce the possibility of oscillations in the output stage. The additional stage of voltage gain constituted by the MOSFETs and the op-amp (common-emitter) output transistors increases the possibility of circuit instability. Capacitor C_3 helps to ensure frequency stability by acting

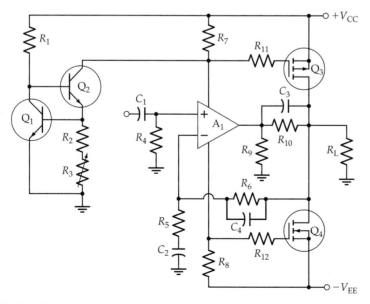

Figure 19-52 Complete circuit of class AB common-source power amplifier.

with resistor R_9 to introduce a phase lead in the output stage feedback loop (see Section 15-2). The phase lead cancels some of the phase lag in the overall circuit.

The additional stage of amplification extends the high cutoff frequency of the amplifier above the cutoff frequency of the op-amp operating alone. If the op-amp (full-power) upper cutoff frequency for an overall voltage gain of 20 (or 26 dB) is 200 kHz, and the additional stage has a gain of 2 (or 6 dB), the circuit cutoff frequency is 400 kHz. For audio applications, it is normal to include capacitor C_4 (see Fig. 19-52), which is usually selected to set the amplifier upper cutoff frequency around 50 kHz or lower.

Example 19-21

Analyse the circuit in Fig. 19-53 to determine the op-amp minimum supply voltage ($V_{S(dc)(min)}$) and the MOSFET maximum gate-source voltage ($V_{GS(max)}$). The op-amp supply current is 0.5 mA.

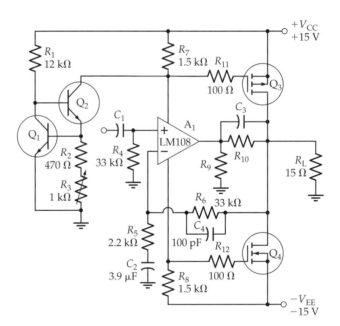

Figure 19-53   Common-source power amplifier circuit for Examples 19-21 and 19-22.

Solution

From Eq. 19-27, $I_{C2(max)} = \dfrac{V_{BE}}{R_2} = \dfrac{0.7 \text{ V}}{470 \text{ }\Omega}$

$$\approx 1.5 \text{ mA}$$

$$I_{C2(min)} = \frac{V_{BE}}{R_2 + R_3} = \frac{0.7 \text{ V}}{470 \text{ }\Omega + 1 \text{ k}\Omega}$$

$$= 476 \text{ }\mu\text{A}$$

Eq. 19-28: $V_{GS(max)} = [I_{S(max)} + I_{C2(max)}] R_7 = (0.5 \text{ mA} + 1.5 \text{ mA}) \times 1.5 \text{ k}\Omega$

$= 3 \text{ V}$

$V_{S(dc)(min)} = \pm(V_{CC} - V_{R7}) = \pm(15 \text{ V} - 3 \text{ V})$

$= \pm12 \text{ V}$

Example 19-22

Analyse the circuit in Fig. 19-53 to determine $P_{o(max)}$, A_{CL}, f_1, and f_2. The op-amp supply current is 0.5 mA, and the MOSFETs have $R_{D(on)} = 0.3 \text{ }\Omega$.

Solution

Power output:

Eq. 19-31: $V_P = \dfrac{V_{CC} \times R_L}{R_{D(on)} + R_L}$

$= \dfrac{15 \text{ V} \times 15 \text{ }\Omega}{0.3 \text{ }\Omega + 15 \text{ }\Omega}$

$= 14.7 \text{ V}$

$I_P = \dfrac{V_P}{R_L} = \dfrac{14.7 \text{ V}}{15 \text{ }\Omega}$

$= 980 \text{ mA}$

$P_{o(max)} = \dfrac{V_P I_P}{2} = \dfrac{14.7 \text{ V} \times 980 \text{ mA}}{2}$

$= 7.2 \text{ W}$

Voltage gain:

$A_v = \dfrac{R_5 + R_6}{R_5} = \dfrac{2.2 \text{ k}\Omega + 33 \text{ k}\Omega}{2.2 \text{ k}\Omega}$

$= 16$

Cutoff frequencies:

$f_1 = \dfrac{1}{2\pi C_2 R_5} = \dfrac{1}{2\pi \times 3.9 \text{ }\mu\text{F} \times 2.2 \text{ k}\Omega}$

$= 18.5 \text{ Hz}$

$f_2 = \dfrac{1}{2\pi C_4 R_6} = \dfrac{1}{2\pi \times 100 \text{ pF} \times 33 \text{ k}\Omega}$

$= 48.2 \text{ kHz}$

Practice Problems

19-9.1 A complementary common-source power amplifier is to deliver 5 W to a 12 Ω load. The available MOSFETs have $R_{D(on)} = 0.6\ \Omega$, $V_{TH} = 1.2$ V, and $g_{fs} = 3$ S. The op-amp to be used has 1 mA supply currents and a maximum output of 20 mA. Design the output stage of the circuit as shown in Fig. 19-51.

19-9.2 Design a BJT current source bias control circuit for the amplifier in Problem 19-9.1 to adjust the V_{GS} of Q_3 and Q_4 by ±20%.

19-9.3 The amplifier in Problem 19-9.1 is to have $f_1 = 20$ Hz and $f_2 = 40$ kHz. If the ac input is ±600 mV, determine suitable values for R_4, R_5, R_6, C_1, C_2, and C_4 (see Fig. 19-52). Use a BIFET op-amp.

19-10 INTEGRATED CIRCUIT POWER AMPLIFIERS

IC Power Amplifier Driver

The LM391 integrated circuit audio power driver contains amplification and driver stages for controlling an externally connected class AB output stage delivering 10 W to 100 W. The voltage gain and bandwidth are set by additional components. Internal circuitry is included for overload and thermal protection and for protection of the (externally connected) amplifier output transistors. The circuit is designed for very low distortion, so that it can be used for high-fidelity amplifiers. Figure 19-54 illustrates the use of the device as an audio amplifier.

The output stage in Fig. 19-54 is seen to be a complementary emitter follower with the low-power and high-power transistor pairs connected in quasi-complementary form. The IC output at terminal 9 is connected to the amplifier output, and the output stage transistors are controlled from *current source* terminal 8 and *current sink* terminal 5. The circuit uses a plus-minus supply, and the non-inverting input terminal of the IC is biased to ground. The inverting input terminal receives feedback from the output, so that the complete circuit operates as a non-inverting amplifier.

Resistors R_A and R_B are connected to an internal transistor (via terminals 5, 6, and 7) to constitute a V_{BE} multiplier for controlling the bias voltage to the amplifier output stage (see Fig. 19-55a). Capacitor C_{AB} by-passes the V_{BE} multiplier circuit to improve the amplifier high-frequency response. Capacitor C_R helps to reject power supply ripple, and C_C is a compensation capacitor for frequency stability. Components R_o, C_o, L_x and R_x are included for load compensation (see Section 19-5).

The LM391 has an internal transistor that can shut the circuit down when turned on by a thermal switch (Fig. 19-55b). This allows the device to be protected from the overheating that might occur with an excessive load current

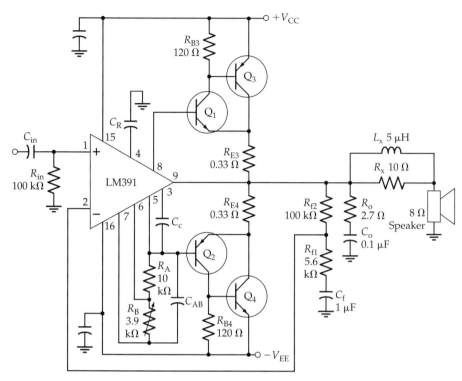

Figure 19-54   Audio power amplifier using an LM391 IC amplifier driver.

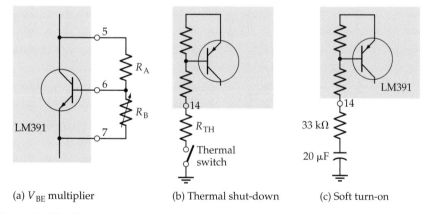

(a) V_{BE} multiplier   (b) Thermal shut-down   (c) Soft turn-on

Figure 19-55   Bias control and shut-down circuits for the LM391.

demand. This same transistor can be employed for *soft turn-on* of the circuit (Fig. 19-55c). If the amplifier supply voltage is switched on at the instant that a peak input signal is applied, a high-level output is passed to the speaker, causing a sharp unpleasant noise. Soft turn-on causes the output to increase slowly, thus eliminating the speaker noise. The circuit in Fig. 19-55c holds the amplifier in a shut-down condition until the capacitor charges.

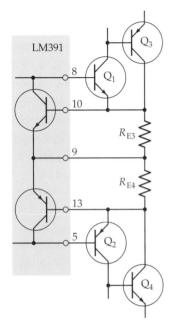

Figure 19-56 Overload protection for the LM391.

Overload protection transistors are included in the LM391, as shown in Fig. 19-56. These transistors turn on when excessive voltage drops occur across the emitter resistors (R_{E3} and R_{E4}) in the output stage (see Section 19-5).

The output stage components in Fig. 19-54 are selected in the same way as for other direct-coupled class AB amplifiers. The minimum supply voltages are calculated by adding 5 V to the peak output voltage.

$$V_{CC} = \pm(V_P + 5\text{ V}) \qquad (19\text{-}32)$$

The input resistance at terminal 1 of the LM391 is extremely high and so the circuit input resistance is set by resistor R_{in}, which is typically selected as 100 kΩ. Feedback resistors R_{f1} and R_{f2} set the amplifier closed-loop voltage gain. The feedback network components are determined in exactly the same way as for other feedback amplifiers. R_{f2} is made equal to R_{in} to minimize output offset, and R_{f1} is calculated from R_{f2} to give the desired voltage gain. Capacitor C_f is determined in terms of R_{f1} to set the low cutoff frequency.

Example 19-23

Determine the maximum output power, the voltage gain, and the low cutoff frequency for the circuit shown in Fig. 19-54, if the supply is ±23 V.

Solution

From Eq. 19-32, $V_P = V_{CC} - 5\text{ V} = 23\text{ V} - 5\text{ V}$

$$= 18\text{ V}$$

$$P_o = \frac{V_P^2}{2R_L} = \frac{(18\text{ V})^2}{2 \times 8\ \Omega}$$

$$\approx 20\text{ W}$$

$$A_{CL} = \frac{R_{f1} + R_{f2}}{R_{f1}} = \frac{100\text{ k}\Omega + 5.6\text{ k}\Omega}{5.6\text{ k}\Omega}$$

$$= 18.9$$

$$f_1 = \frac{1}{2\pi C_f R_{f1}} = \frac{1}{2 \times \pi \times 1\ \mu\text{F} \times 5.6\text{ k}\Omega}$$

$$= 28\text{ Hz}$$

250 mW IC Power Amplifier

The LM386 is a complete power amplifier circuit capable of delivering 250 mW to an 8 Ω load without any additional components. The supply voltage range is 5 V to 18 V, and the (inverting and non-inverting) input terminals are biased to ground or to a negative supply via internal 50 kΩ resistors. The output is automatically centred at half the supply voltage. Feedback resistors are also provided internally to set the voltage gain to 20.

The pin connections for the LM386 are shown in Fig. 19-57a, and the circuit connections for functioning as an amplifier with a gain of 20 are illustrated in

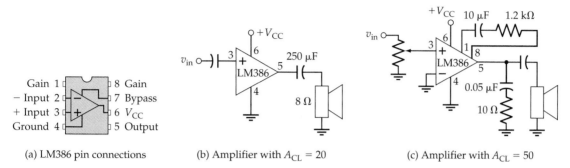

(a) LM386 pin connections   (b) Amplifier with $A_{CL} = 20$ (c) Amplifier with $A_{CL} = 50$

Figure 19-57   The LM386 IC power amplifier can be connected to have a closed-loop gain from 20 to 200.

part b. Figure 19-57c shows how a capacitor and resistor can be connected at pins 1 and 8 to achieve a larger voltage gain. With the 10 μF capacitor alone, a maximum gain of 200 is obtained. The resistor in series with the capacitor allows the voltage gain to be set anywhere between 20 and 200.

Bridge-Tied Load Amplifier

All the power amplifiers already discussed have been *single-ended (SE)*, meaning that they provide power to a load that has one terminal grounded and the other terminal connected to the amplifier output. These amplifiers either use a plus-minus supply with directly-coupled loads or have a capacitor-coupled load and a single-polarity supply. A *bridge-tied load (BTL)* amplifier uses a single-polarity supply and a direct-coupled load.

Figure 19-58a shows the basic circuit of a BTL amplifier. The two op-amps are connected to function as inverting amplifiers, but note from the resistor values that A_1 has a voltage gain of 10 and that A_2 has a gain of 1. Each amplifier has a single-polarity supply (V_{CC}), and a voltage divider (R_5 and R_6) provides a bias voltage of $0.5\,V_{CC}$ to the op-amp non-inverting input terminals.

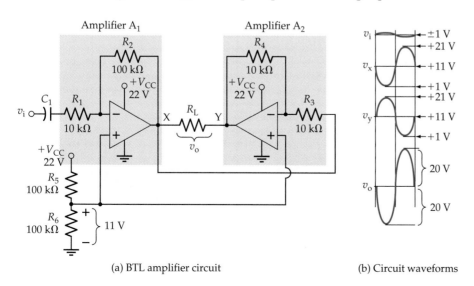

(a) BTL amplifier circuit   (b) Circuit waveforms

Figure 19-58 A bridge-tied load (BTL) amplifier uses two op-amp circuits with their outputs connected to opposite ends of the load.

The load resistor (R_L) is connected from the output of A_1 to the output of A_2. This is the *bridge-tied load* configuration.

When no ac signal is applied, the dc voltage at the load terminals (X and Y) is 0.5 V_{CC}, in this case, 11 V for a 22 V supply. As illustrated by the waveforms in Fig. 19-58(b), a +1 V ac input to A_1 produces a −10 V change at load terminal X. This (−10 V) is also applied to the input of A_2, resulting in a +10 V change at load terminal Y. Thus, a peak of 20 V, negative at X and positive at Y, is developed across the load. When the ac input goes to −1 V, a 20 V peak load voltage is again produced but with the load polarity reversed. So, although a single-polarity +22 V supply is used, the output is 40 V peak-to-peak, and no load-coupling capacitor is required. A (similar performance) single-ended amplifier producing a 40 V peak-to-peak output would require either a ±22 V supply for a direct-coupled load or a +44 V supply for a capacitor-coupled load.

Figure 19-59 shows the pin connections and typical application of a TPA4861, 1 W integrated circuit BTL audio power amplifier. The load is connected across the two output terminals (5 and 8), external resistors R_1 and R_F set the voltage gain, and the signal is coupled to R_1 via C_1. The supply voltage (V_{DD}) is internally divided to bias the op-amp non-inverting terminals to 0.5 V_{DD}. The bias point is externally accessible so that it can be bypassed to ground (via C_B) for soft start-up and to minimize noise.

A TPA4861 using a +5 V supply can dissipate 1 W in an 8 Ω load. The overall voltage gain for the (BTL) amplifier is twice the gain of the inverting amplifier stage:

$$A_{CL} = \frac{2\,R_F}{R_1} \qquad (19\text{-}33)$$

Because the signal is applied to an inverting amplifier, the input resistance is set by resistor R_1. The IC manufacturer recommends that R_1 should be

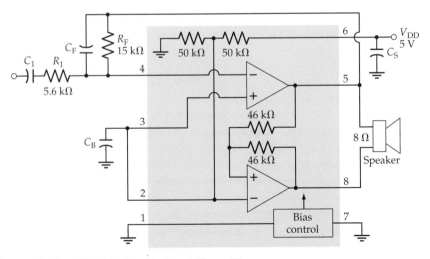

Figure 19-59   TPA4861 bridge-tied load IC amplifier.

selected in the range of 5 kΩ to 20 kΩ. Also, if R_F exceeds 50 kΩ, a small capacitor (C_F = 5 pF) should be connected in parallel with it for ac stability.

The input capacitor (C_1) sets the circuit low cutoff frequency. So,

$$X_{C1} = R_1 \text{ at } f_1 \qquad \text{(19-34)}$$

The internal voltage-divider resistance (50 kΩ‖50 kΩ) is connected in series with the bypass capacitor (C_B). The impedance of C_B should usually be one-tenth of the series resistance:

$$X_{CB} = 2.5 \text{ kΩ at } f_1 \qquad \text{(19-35)}$$

Example 19-24

Analyse the circuit in Fig. 19-59 to determine the load power dissipation when a ±0.5 V signal is applied at the input.

Solution

Eq. 19-33:
$$A_{CL} = \frac{2R_F}{R_1} = \frac{2 \times 15 \text{ kΩ}}{5.6 \text{ kΩ}}$$
$$\approx 5.4$$

$$V_o = A_{CL} \times v_s = \pm 5.4 \times 0.5 \text{ V}$$
$$= \pm 2.7 \text{ V}$$

$$P_o = \frac{V_P^2}{2R_L} = \frac{(2.7 \text{ V})^2}{2 \times 8 \text{ Ω}}$$
$$\approx 0.46 \text{ W}$$

7 W IC Power Amplifier

The LM383 can deliver 7 W to a 4 Ω load. No additional output transistors are required because the amplifier can produce a 3.5 A peak output current. Overload protection circuitry is included, and internal bias is provided for the input terminals. The single-polarity supply voltage ranges from 5 V to 20 V. Amplifier voltage gain can be programmed by means of external components. The circuit bandwidth is 30 kHz at a gain of 40 dB.

Figure 19-60 shows an LM383 (in a 5-pin TO-220 package) connected to function as a non-inverting audio amplifier. Capacitors C_1 and C_4 couple the signal and load. Resistors R_1 and R_2 are feedback components that set the circuit voltage gain, and capacitor C_2 couples the feedback voltage to the inverting input terminal. Other components provide circuit stability. Note that the resistance of R_2 is 2.2 Ω. This is because the inverting input terminal is connected (internally) to a transistor emitter terminal that has an input resistance of about 20 Ω. Capacitor C_2 must be very large to couple the feedback voltage to the low-resistance inverting input. The circuit functions as a non-inverting amplifier.

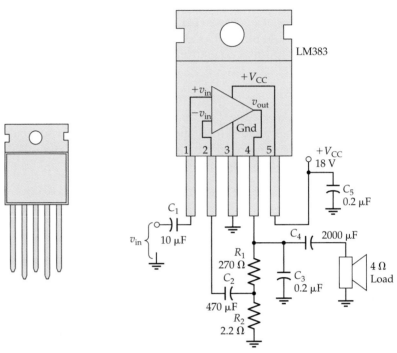

(a) LM383 five-lead TO-220 package   (b) Connection for amplifier with $P_O = 7$ W

Figure 19-60   LM383 IC power amplifier connected to dissipate 7 W in a 4 Ω load.

68 W IC Power Amplifier

Figure 19-61a shows a power amplifier circuit using an LM3886 IC audio amplifier. The LM3886 can deliver 68 W to a 4 Ω load using a ± 28 V supply. Alternatively, it can be used to dissipate 38 W in an 8 Ω load, again using a ± 28 V

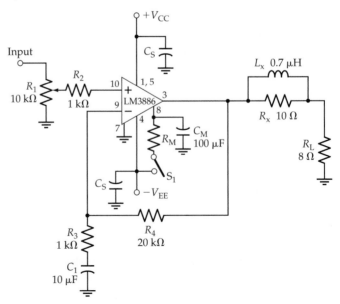

Figure 19-61   LM3886, 68 W audio power amplifier.

supply. The circuit operates as a non-inverting amplifier with the closed-loop gain set by resistors R_3 and R_4 and the low cutoff frequency set by capacitor C_1. Potentiometer R_1 allows the signal amplitude to be adjusted. Switch S_1 is a *mute* control.

Practice Problems

19-10.1 A power amplifier using an LM391 driver (as in Fig. 19-54) has a ±20 V supply, a 50 Ω load, and power Darlington output BJTs with $h_{fe} = 600$. Calculate the maximum output power and the peak output current from the integrated circuit.

19-10.2 Calculate the efficiency of the circuit in Fig. 19-59 when delivering 1 W to the speaker. The quiescent current for the TPA4861 is specified as 2.5 mA.

19-10.3 Determine the signal amplitude for the circuit in Fig. 19-60 to dissipate 7 W in the load. Calculate the low cutoff frequency.

Review Questions

Section 19-1

19-1 Sketch the circuit of a transformer-coupled class A amplifier. Briefly explain the operation of the circuit.

19-2 Derive the equation for the ac load reflected into the primary of the transformer in a class A amplifier. Write the equations for dc and ac load resistances.

19-3 Sketch approximate I_C/V_{CE} transistor characteristics and an ac load line for a class A amplifier. Draw waveforms to show how the transistor output voltage changes with changes in base current. Briefly explain.

19-4 Write equations for a class A transformer-coupled amplifier for dc supply power, ac power to the transformer primary, and circuit efficiency. Show that the maximum theoretical efficiency of a class A amplifier is 50%.

Section 19-2

19-5 Sketch the circuit of a transformer-coupled class B power amplifier output stage. Explain the circuit operation.

19-6 Sketch approximate composite characteristics for a class B output stage and the ac load line. Draw waveforms to show the change in transistor output voltage with change in base current. Briefly explain.

19-7 Sketch the circuit of a class AB transformer-coupled output stage, and explain the difference between class B and class AB amplifiers. Explain the difference in the performance of the two circuits.

19-8 Draw the complete circuit of a class AB transformer-coupled amplifier with a class A driver stage. Explain the operation of the circuit.

19-9 Sketch approximate composite characteristics for a class AB amplifier and the ac load line. Briefly explain.

19-10 Write equations for a class B transformer-coupled amplifier for dc input power to the output stage, ac power delivered to the transformer primary, and circuit efficiency. Show that the maximum theoretical efficiency of a class B amplifier is 78.6%.

Section 19-3

19-11 For a transformer-coupled class B power amplifier, write equations for the resistance seen when 'looking into' one-half of the centre-tapped transformer primary, and the resistance seen when 'looking into' the whole winding of the centre-tapped transformer primary.

19-12 For a transformer-coupled class B power amplifier, write an equation for the power delivered to the transformer primary in terms of peak primary voltage and reflected load. Write an equation for the peak current in terms of primary power and voltage.

19-13 Write equations for maximum voltage, current, and power dissipation for the output transistors in a class B power amplifier.

Section 19-4

19-14 Sketch the circuit of a complementary emitter follower, and explain its operation.

19-15 Sketch the basic circuit of a capacitor-coupled class AB complementary symmetry amplifier. Explain the dc biasing and ac operation of the circuit.

19-16 For the amplifier in Question 19-15, write equations for the average supply current to the output stage, the dc supply power, and the transistor power dissipation.

19-17 Sketch a direct-coupled class AB complementary symmetry amplifier. Explain the circuit operation.

Section 19-5

19-18 Sketch the circuit of a Darlington-connected complementary emitter follower output stage for a class AB power amplifier. Explain the operation of the circuit, and discuss its advantages.

19-19 Sketch the circuit of a quasi-complementary emitter follower output stage. Explain the circuit operation, and discuss its advantages.

19-20 Show how the maximum current can be limited in the output transistors of direct-coupled and capacitor-coupled class AB power amplifiers. Explain.

19-21 Show how a power amplifier can be protected from the effects of power supply ripple and transients. Explain.

Section 19-6

19-22 Sketch the circuit of a power amplifier with a differential BJT input stage, an output driver stage, a complementary symmetry Darlington output with a capacitor-coupled load, and overall negative feedback.

19-23 Explain the biasing arrangement for the two amplification stages of the circuit in Question 19-22, and write an equation for the ac voltage gain.

19-24 Sketch the circuit of a direct-coupled power amplifier that uses an *npn* transistor differential input stage, a class A intermediate stage with a constant-current load, and a quasi-complementary emitter follower output stage.

19-25 Explain the dc and ac operation of the circuit in Question 19-24, and show how a Zener diode may be used to minimize the effect of ripple voltage on the negative supply line.

Section 19-7

19-26 Compare power MOSFETs to power BJTs.

19-27 Draw the circuit of a direct-coupled power amplifier with a differential amplifier BJT input stage, two MOSFETs in the output stage, and overall negative feedback.

19-28 Explain the dc and ac operation of the circuit in Question 19-27.

19-29 Write the supply voltage equation for the circuit in Question 19-27.

Section 19-8

19-30 Draw the circuit of a power amplifier with a direct-coupled complementary symmetry BJT output stage, an op-amp driver stage, and overall negative feedback.

19-31 Explain the dc and ac operation of the circuit in Question 19-30.

19-32 Show how the circuit in Question 19-30 can be modified to use bootstrapping capacitors. Explain the function and advantage of the bootstrap capacitors.

Section 19-9

19-33 Draw the basic circuit of a complementary MOSFET common-source power amplifier that uses an op-amp driver with the op-amp supply currents controlling the MOSFETs.

19-34 Explain the dc and ac operation of the circuit in Question 19-33.

19-35 For the circuit in Question 19-33, show how the MOSFET bias currents can be controlled by a single variable resistor. Explain.

19-36 Show how a current source can be used to control the MOSFET bias currents in the circuit in Question 19-33. Draw the current source circuit and explain its operation.

19-37 Discuss the output voltage swing that can be achieved with the circuit in Question 19-33. Show how the output voltage swing can be made greater than the supply voltage levels at the op-amp terminals.

Section 19-10

19-38 Refer to the IC power amplifier circuit in Fig. 19-54. Explain the function of every component in the circuit.

19-39 Draw a circuit diagram to show how the current in the output transistors of a power amplifier can be limited to a desired maximum level. Explain the circuit operation.

19-40 Sketch the circuit of a bridge-tied load amplifier. Explain the circuit operation and discuss its advantages.

19-41 Explain the function of every component in the IC power amplifier circuit in Fig. 19-60.

Problems

Section 19-1

19-1 A class A transformer-coupled amplifier, as in Fig. 19-1, has $V_{CC} = 20$ V, $R_1 = 3.9$ kΩ, $R_2 = 1$ kΩ, $R_E = 68$ Ω, and $R_L = 23$ Ω. The transformer has $R_{PY} = 32$ Ω, $N_1 = 80$, and $N_2 = 20$. Plot the dc load line and ac load line for this circuit on blank characteristics with vertical axis $I_C = (0$ to 100 mA$)$ and horizontal axis $V_{CE} = (0$ to 40 V$)$.

19-2 Assuming an 85% transformer efficiency, calculate the maximum efficiency for the circuit in Problem 19-1.

19-3 A class A amplifier (as in Fig. 19-1) has the following components: $R_1 = 68$ kΩ, $R_2 = 22$ kΩ, $R_E = 2.2$ kΩ. The supply is $V_{CC} = 25$ V, and the transformer has $R_L = 5$ kΩ, $r'_L = 8$ kΩ, and $R_{PY} = 33$ Ω. Plot the dc load line and ac load line for this circuit on blank characteristics with vertical axis $I_C = (0$ to 5 mA$)$ and horizontal axis $V_{CE} = (0$ to 40 V$)$.

19-4 Calculate the maximum peak load voltage for the circuit in Problem 19-3. Assume a 100% transformer efficiency.

Section 19-2

19-5 A class B transformer-coupled output stage, as in Fig. 19-8, has a load resistance $R_L = 23$ Ω and a supply voltage $V_{CC} = 40$ V. The transformer has $N_1 = 80$, $N_2 = 20$, and a total primary winding resistance $R_{PY} = 64$ Ω. Using blank characteristics with $I_C = (0$ to 100 mA$)$ and $V_{CE} = (0$ to 40 V$)$, plot the complete ac load line.

19-6 Determine the maximum output voltage and power for the circuit in Problem 19-5 if the transistors have $V_{CE(sat)} \approx 0$ V.

19-7 A class B amplifier uses a transformer with 80% efficiency and with $N_p/N_s = 5$, where N_p is the total number of primary turns on the centre-tapped primary. The supply voltage is 45 V, and the load resistance is 8 Ω. Determine the maximum output voltage and power. Assume the transistors have $V_{CE(sat)} \approx 0$ V.

19-8 Using blank composite characteristics with $I_C = $ (0 to 1 A) and $V_{CE} = $ (0 to 45 V), plot the complete ac load line for the circuit in Problem 19-7. Assume $R_{py} \ll r_L$.

19-9 A class AB output stage (as in Fig. 19-11) has $V_{CC} = 30$ V, $R_4 = 6.8$ kΩ, and $R_5 = R_6 = 22$ Ω. The output transformer has $R_L = 24$ Ω and $r'_L = 800$ Ω. Assuming a 75% transformer efficiency, calculate the power delivered to the load. Assume Q_2 and Q_3 have $V_{CE(sat)} \approx 0.5$ V.

19-10 Prepare suitable blank composite characteristics and draw the ac load line for the circuit in Problem 19-9.

Section 19-3

19-11 Specify the maximum transistor voltage, current, and power dissipation for the circuit described in Problem 19-7.

19-12 A class A transformer-coupled amplifier with a 24 V supply is to deliver 1.25 W to a 50 Ω load. Assuming a transformer efficiency of 80%, specify the transformer and transistor.

19-13 Plot dc and ac load lines (on blank characteristics) for the circuit in Problem 19-12.

19-14 A class B amplifier is to supply 8 W to a 12 Ω load. The supply is $V_{CC} = 25$ V. Specify the output transformer and transistors. Assume a transformer efficiency of 75%.

19-15 A class AB transformer-coupled power amplifier (as in Fig. 19-13) is to deliver 0.5 W to a 4 Ω load. The output transformer has $r'_L = 312$ Ω when $R_L = 4$ Ω, and has an efficiency of 75%. Calculate a suitable supply voltage and specify the output transistors.

Section 19-4

19-16 A capacitor-coupled power amplifier as in Fig. 19-18 is to deliver 0.6 W to a 250 Ω load. Specify the supply voltage and the output transistors.

19-17 Determine suitable resistor values for the circuit in Problem 19-16. Assume that the output transistors have $h_{FE(min)} = 40$, $h_{ie} = 1$ kΩ, and $h_{ib} = 1.8$ Ω.

19-18 Calculate suitable capacitance values for the circuit in Problems 19-16 and 19-17 for a 50 Hz lower cutoff frequency.

19-19 A direct-coupled amplifier as in Fig. 19-23 is to deliver 2 W to a 20 Ω load. Specify the supply voltage and the output transistors. Assume that the output transistors have $h_{FE(min)} = 200$.

19-20 Determine resistor values for the circuit in Problem 19-19.

Section 19-5

19-21 A power amplifier is required to deliver 5 W to a 20 Ω load. The output stage is to use Darlington-connected BJTs and is to be direct-coupled, as in Fig. 19-24. Determine a suitable supply voltage, and specify all the output stage transistors. Assume that $h_{FE2} = h_{FE3} = 20$ and that $h_{FE4} = h_{FE5} = 100$.

19-22 Determine suitable resistor values for the amplifier output stage in Problem 19-21.

19-23 A direct-coupled power amplifier using power Darlington BJTs with $h_{fe} = 2000$ and $V_{BE} = 1.5$ V is to dissipate 2 W in a 16 Ω load. Determine the required supply voltage and resistor values for the output stage (see Fig. 19-25).

19-24 Modify the circuit in Problem 19-23 to include current limiting as in Fig. 19-27. The maximum current is to be limited to 20% above the calculated peak level.

19-25 Design a V_{BE} multiplier to replace the diode biasing stage for the output transistors in the circuit for Problem 19-23. Make V_B adjustable by $\pm 20\%$.

19-26 The driver stage collector resistor (R_C) in the circuit designed for Problem 19-23 is to be replaced by a constant-current circuit, as in Fig. 19-30. Design the constant-current circuit.

19-27 The amplifier in Problems 19-16 and 19-17 is to be modified to use power supply decoupling as in Fig. 19-29a. The ripple frequency is 120 Hz. Determine suitable values for R_{15} and C_D, and the new supply voltage level.

Section 19-6

19-28 A direct-coupled amplifier circuit as in Fig. 19-33 is to deliver 7 W to a 22 Ω load. Determine the supply voltage and specify transistors Q_5 to Q_8. Assume that $h_{FE5} = h_{FE6} = 90$ and that $h_{FE7} = h_{FE8} = 15$.

19-29 Determine suitable values for resistors R_9 to R_{15} for the circuit in Problem 19-28. Specify the diodes and transistors Q_3 and Q_4. Assume that $I_{CBO} = 10$ μA for Q_7 and Q_8.

19-30 The input voltage to the circuit in Problems 19-28 and 19-29 is ± 0.5 V. Determine suitable values for resistors R_1 to R_8.

19-31 The circuit in Problems 19-29 and 19-30 is to have a frequency range from 20 Hz to 50 kHz. Determine suitable capacitances for C_1, C_2, and C_6.

19-32 A direct-coupled amplifier as in Fig. 19-33 has $V_{CC} = \pm 20$ V, $V_{R9} = V_{R11} = 3$ V, and $R_L = 16$ Ω. Calculate the maximum power delivered to the load.

19-33 Specify Q_7 and Q_8 for the circuit in Problem 19-32.

19-34 Calculate the approximate efficiency for the circuit in Problem 19-32.

19-35 A direct-coupled amplifier, as in Fig. 19-33, uses 2N3904 and 2N3906 BJTs for Q_5 and Q_6, Q_8 is a 2N3055 and Q_7 is complementary to Q_8. The supply voltage is ± 25 V, and the load resistance is 20 Ω. Determine the maximum output power, and calculate the maximum collector current for Q_7 and Q_8 and the maximum base current for Q_5 and Q_6.

Section 19-7

19-36 A MOSFET power amplifier circuit as in Fig. 19-36 has $V_{CC} = \pm 30$ V and $R_L = 50$ Ω. Transistors Q_3 and Q_4 have $R_{D(on)} = 4$ Ω. Determine the maximum power delivered to the load and the power dissipated in each output transistor.

19-37 A MOSFET amplifier circuit as in Fig. 19-36 has a 25 Ω load resistance and uses output transistors with $R_{D(on)} = 3\ \Omega$. Calculate the required supply voltage to dissipate 8 W in the load. Calculate the power dissipation in the output transistors.

19-38 The MOSFETs in Problem 19-37 have a threshold voltage of $V_{TH} = 1.5$ V and a transconductance of $g_{fs} = 300$ mA/V. Determine suitable dc voltage drops across resistors R_2 and R_3, and R_6 to R_9.

19-39 Calculate resistor and capacitor values for the circuit in Problems 19-37 and 19-38 if $A_{CL} = 15$ and $f_1 = 80$ Hz.

19-40 Calculate the approximate efficiency for the circuit in Problems 19-37 and 19-39.

19-41 Calculate the supply voltage for an amplifier circuit as in Fig. 19-36 to deliver 10 W to a 16 Ω load. The output MOSFETs have $R_{D(on)} = 1\ \Omega$, $V_{TH} = 2$ V, and $g_m = 2$ S.

19-42 Determine suitable dc voltage levels for the circuit in Problem 19-41.

19-43 The amplifier in Problems 19-41 and 19-42 has $v_{in} = \pm0.7$ V, and its low cutoff frequency is to be 40 Hz. Calculate all resistor and capacitor values.

Section 19-8

19-44 A direct-coupled class AB power amplifier using a complementary emitter follower output stage and a operational amplifier driver (as in Fig. 19-39) is to deliver 2.4 W to a 30 Ω load. Calculate the required supply voltage and specify the output transistors in terms of $V_{CE(max)}$, $I_{C(max)}$, and power dissipation.

19-45 Determine suitable dc voltage and current levels for the circuit in Problem 19-44, and calculate all resistor values if $v_{in} = \pm0.5$ V.

19-46 The circuit in Problems 19-44 and 19-45 is to have a frequency range from 30 Hz to 30 kHz. Calculate the capacitor values, and determine the minimum slew rate for the op-amp.

19-47 Modify the circuit in Problems 19-44 to 19-46 to use bootstrapping capacitors, as in Fig. 19-41. Determine suitable values for the bootstrapping capacitors, and calculate the new maximum peak output voltage that can be produced by the modified circuit.

Section 19-9

19-48 The common-source power amplifier circuit in Fig. 19-52 has the following components: $R_7 = R_8 = 680\ \Omega$, $R_9 = R_{10} = 1.2$ kΩ, $R_4 = R_6 = 18$ kΩ, $R_5 = 820\ \Omega$, and $R_L = 32\ \Omega$. The supply voltage is ±30 V, the op-amp supply current is 1 mA, the Q_2 collector current is 1 mA, and the MOSFETs have $R_{D(on)} = 1.5\ \Omega$, and $g_m = 1.2$ S. Determine the gate-source bias voltage for Q_3 and Q_4 and the maximum output power.

19-49 Calculate the op-amp supply terminal voltages and the op-amp peak output voltage for the circuit in Problem 19-48. Determine the required signal voltage to give maximum output.

19-50 Design the bias control circuit for the circuit in Problem 19-48 to give $I_{C2} = 1$ mA $\pm 50\%$.

19-51 A direct-coupled class AB common-source power amplifier, as in Fig. 19-52 without the current source, is to deliver 8 W to a 12 Ω load. Calculate the required supply voltage if the output transistors have $R_{D(on)} = 0.95$ Ω, $V_{th} = 2$ V, and $g_m = 0.9$ S.

19-52 The circuit in Problem 19-51 has $v_{in} = \pm 0.9$ V and an op-amp with $I_s = 1.3$ mA and $I_{o(max)} = 15$ mA. Determine suitable resistor values.

19-53 The circuit in Problems 19-51 and 19-52 is to have a frequency range from 40 Hz to 45 kHz. Calculate the capacitor values, and determine the minimum slew rate for the op-amp.

Section 19-10

19-54 A direct-coupled class AB audio power amplifier is to be designed to dissipate 5 W in a 16 Ω load. The circuit is to use quasi-complementary connected output transistors and a LM391 driver, as in Fig. 19-54. Calculate the required supply voltage, and specify the BJTs. Assume that $h_{FE} = 20$ for Q_3 and Q_4.

19-55 Select suitable resistances for R_{E3} and R_{E4} in the circuit for Problem 19-54 to limit the output current to 20% above the required peak level, as illustrated in Fig. 19-56.

19-56 A TPA4861 BTL amplifier is to be used to dissipate 1 W in a 15 Ω load. The input voltage is ± 0.7 V, and the low cutoff frequency is to be 25 Hz. Calculate the required supply voltage and suitable resistances for R_1 and R_F (in Fig. 19-59). Determine suitable capacitor values.

Practice Problem Answers

19-1.1 [Q-point: $I_C = 3.1$ mA, $V_{CE} = 10.2$ V], [point A: $I_C = 0$, $V_{CE} = 15$ V], [point B: $I_C = 0$, $V_{CE} = 21.1$ V]

19-1.2 25.6%

19-2.1 37.7%

19-2.2 [Q-point: $I_C = 4.8$ mA, $V_{CE} = 26.7$ V], [points A and A': $I_C = 24.8$ mA, $V_{CE} = 25.6$ V], [points B and B': $I_C = 46.7$ mA, $V_{CE} = 2.55$ V]

19-3.1 10 V, (20 V, 26 mA, 67.5 mW)

19-3.2 1.09 W, 50 V, 640 mA

19-4.1 21 V, (21 V, 90 mA, 200 mW)

19-4.2 1.8 kΩ, 50 Ω variable

19-5.1 ± 16 V, 2.7 kΩ, 100 Ω

19-5.2 15 kΩ, 2.2 kΩ, 2 kΩ

19-5.3 10 kΩ, (56 kΩ + 1.5 kΩ), 1 kΩ, 500 Ω variable

19-6.1   4.7 kΩ, 3.3 kΩ, 4.7 kΩ, 1.2 kΩ, 180 Ω, 4.7 kΩ, 15 µF, 30 µF

19-7.1   18 µF, 22 µF, 0.82 µF

19-7.2   ±15 V, 12 V, 14.3 V, 1.3 V, 10.7 V, 17.7 V, 1.3 V

19-7.3   4.7 kΩ, 12 kΩ, 6.8 kΩ, 120 Ω, 4.7 kΩ, 100 kΩ, 820 kΩ, (1.2 MΩ + 150 kΩ), 100 kΩ, 10 µF, 33 µF, 0.47 µF, 0.47 µF

19-8.1   12.1 V, 146 mW, 6.6 kHz

19-8.2   ±12 V, 3.5 V/µs

19-9.1   1.2 kΩ, 1.2 kΩ, 1.2 kΩ, 1.2 kΩ

19-9.2   5.6 kΩ, 390 Ω, 1 kΩ variable, 560 Ω, 560 Ω

19-9.3   100 kΩ, 5.6 kΩ, 100 kΩ, 0.82 µF, 1.5 µF, 39 pF

19-10.1   2.25 W, 0.5 mA

19-10.2   62.5%

19-10.3   ±60 mV, 19.9 Hz

CHAPTER 20
Thyristors

CONTENTS

Objectives

You will be able to:

1 Sketch the basic construction of an SCR and explain its operation. Draw typical SCR characteristics and define the device parameters.

2 Sketch and explain the following SCR circuits: 90° phase control, 180° phase control, zero-point triggering, crowbar, and heater-control.

3 Design and analyze the above types of SCR circuits.

4 Sketch the basic construction of a TRIAC and explain its operation. Draw typical TRIAC characteristics, and define the device parameters.

5 Discuss TRIAC Quadrant I to IV triggering.

6 Sketch and explain TRIAC phase control and zero-point triggering circuits.

7 Design and analyze the above types of TRIAC circuits.

8 Sketch characteristics and graphic symbols for the following devices: DIAC, SUS, SBS, GTO, and SIDAC. Explain the operation and applications of each device.

9 Sketch the basic construction of a UJT and explain its operation. Draw typical UJT characteristics and define the device parameters.

10 Sketch the basic construction of a PUT and explain its operation. Draw typical PUT characteristics and define the device parameters.

11 Sketch and explain relaxation oscillators and thyristor control circuits using UJTs and PUTs.

12 Design and analyze the above types of UJT and PUT circuits.

INTRODUCTION

The *silicon-controlled rectifier (SCR)* can be thought of as an ordinary rectifier with a control element. The current flowing into the control element, which is termed the *gate*, determines the anode-to-cathode voltage at which the device begins to conduct. The SCR is widely applied as an ac power-control device. The gate bias may keep the device off, or it may permit conduction to start at any desired point in the forward half-cycle of a sinusoidal input. Many other devices, such as the DIAC and the TRIAC, are based on the SCR principle. Collectively, SCR-type devices are known as thyristors. This term is derived from *thyratron* and transistor, the thyratron being a gas-filled electron tube that behaves like an SCR.

The *unijunction transistor (UJT)* is a three-terminal device quite different from bipolar and field effect transistors. The device input, called the emitter, has a resistance that rapidly decreases when the input voltage reaches a certain level. This effect is termed a *negative resistance* and it makes the UJT useful in timing and oscillator circuits. The *programmable unijunction transistor (PUT)* is an SCR-type device that behaves like a UJT.

20-1 SILICON-CONTROLLED RECTIFIER (SCR)

SCR Operation

The silicon-controlled rectifier (SCR) consists of four layers of semiconductor material, alternately *p*-type and *n*-type as illustrated in Fig. 20-1a. Because of its construction, the SCR is sometimes referred to as a *four-layer* diode, or a *pnpn device*. The layers are designated p_1, n_1, p_2, and n_2, as shown. There are

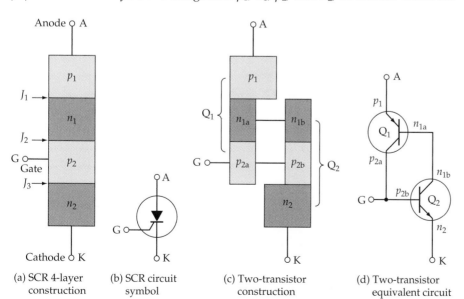

(a) SCR 4-layer construction

(b) SCR circuit symbol

(c) Two-transistor construction

(d) Two-transistor equivalent circuit

Figure 20-1 The silicon-controlled rectifier (SCR) is a four-layer device that can be explained in terms of a two-transistor equivalent circuit.

three junctions: J_1, J_2, and J_3, and three terminals: *anode* (A), *cathode* (K), and *gate* (G). Figure 20-1b shows the SCR circuit symbol.

To understand SCR operation, it is necessary to imagine layers n_1 and p_2 split into n_{1a}, n_{1b}, p_{2a} and p_{2b} as shown in Fig. 20-1c. Since n_{1a} is connected to n_{1b}, and p_{2a} is connected to p_{2b}, nothing is really changed. However, it is now possible to think of p_1, n_{1a}, p_{2a} as a *pnp* transistor, and n_{1b}, p_{2b}, n_2 as an *npn* transistor. Replacing the transistor block representations in Fig. 20-1c with the *pnp* and *npn* BJT circuit symbols gives the *two-transistor equivalent circuit* in Fig. 20-1d. It is seen that the Q_1 collector is connected to the Q_2 base, and the Q_2 collector is commoned with the Q_1 base. The Q_1 emitter is the SCR anode terminal, the Q_2 emitter is the cathode, and the junction of the Q_1 collector and the Q_2 base is the SCR gate terminal.

To forward-bias an SCR, a voltage (V_{AK}) is applied positive on the anode (A) and negative on the cathode (K), as shown in Fig. 20-2a. If the gate (G) is left unconnected, only small leakage currents (I_{CO}) flow, and both transistors remain *off*. Figure 20-1a shows that the leakage currents are the result of junction J_2 being reverse-biased when A is positive and K is negative.

When a negative gate-cathode voltage ($-V_G$) is applied, the Q_2 base-emitter junction is reverse-biased, and only small leakage currents continue to flow, so both Q_1 and Q_2 remain *off*. A positive gate-cathode voltage forward-biases the Q_2 base-emitter junction, causing a gate current ($I_G = I_{B2}$) to flow and producing a Q_2 collector current (I_{C2}). (See Fig. 20-2b.) Because I_{C2} is the same as I_{B1}, Q_1 also switches *on* and I_{C1} starts to flow, providing base current I_{B2}. Each collector current provides much more base current than needed by the transistors, and even when I_G is switched *off*, the transistors remain *on*, conducting heavily with only a small anode-to-cathode voltage drop. The ability of the SCR to remain on when the triggering current is removed is referred to as *latching*.

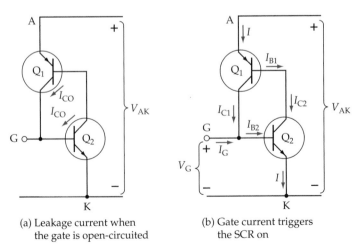

(a) Leakage current when
the gate is open-circuited

(b) Gate current triggers
the SCR on

Figure 20-2   When the SCR gate current is zero, the device normally remains off. The flow of gate current (I_G) triggers the SCR on.

To switch the SCR *on*, only a brief pulse of gate current is required. Once switched on, the gate has no further control and the device remains on until V_{AK} is reduced to near zero.

Consider Fig. 20-1a again. With a forward (anode-to-cathode) bias, junctions J_1 and J_3 are forward-biased, while J_2 is reverse-biased. When V_{AK} is made large enough, J_2 will break down and the resulting current flow across the junction constitutes collector current in each transistor. Each collector current flows into the base of the other transistor, causing both transistors to switch on. Thus, the SCR can be triggered on with the gate open-circuited.

SCR Characteristics and Parameters

Figure 20-3a shows an SCR with a reverse-bias anode-to-cathode voltage $(-V_{AK})$ (negative on A, positive on K). Note that the gate terminal is open-circuited. Figure 20-3b shows that the reverse-bias voltage causes junction J_2 to be forward-biased and J_1 and J_3 to be reverse-biased. When $-V_{AK}$ is small, a *reverse leakage current* (I_{RX}) flows. This is plotted as the reverse characteristic ($-V_{AK}$ versus I_R) on Fig. 20-3c. I_{RX} is typically around 100 μA, and is sometimes referred to as the *reverse blocking current*.

When $-V_{AK}$ is increased, I_{RX} remains approximately constant until the reverse breakdown voltage is reached. At this point the reverse-biased junctions (J_1 and J_3) break down and the reverse current (I_R) increases very rapidly. If I_R is not limited (by additional circuit components), the device will be destroyed by excessive current flow. The region of the reverse characteristics before breakdown is termed the *reverse blocking region*.

An SCR with a forward-bias anode-to-cathode voltage (positive on A, negative on K) is shown in Fig. 20-4a. Here again, the gate terminal is open-circuited. As illustrated in Fig. 20-4b, $+V_{AK}$ forward-biases J_1 and J_3 and

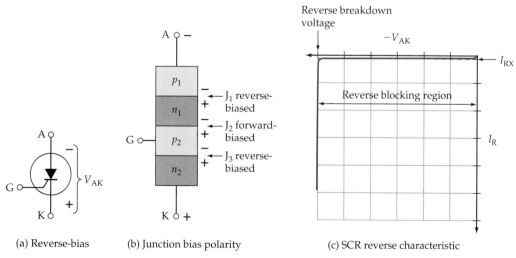

(a) Reverse-bias   (b) Junction bias polarity   (c) SCR reverse characteristic

Figure 20-3   When V_{AK} is negative, J_2 is forward-biased, and J_1 and J_3 are reverse-biased. A small reverse leakage current flows while $-V_{AK}$ is less than the breakdown voltage.

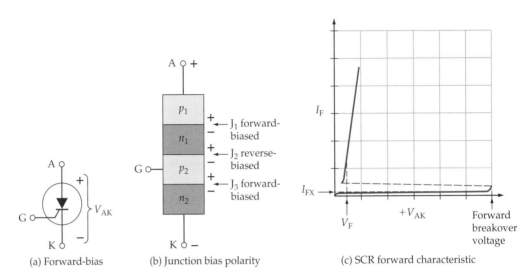

(a) Forward-bias   (b) Junction bias polarity   (c) SCR forward characteristic

Figure 20-4 When V_{AK} is positive, J_2 is reverse-biased and J_1 and J_3 are forward-biased. A small forward leakage current flows while $+V_{AK}$ is less than the forward breakover voltage.

reverse-biases J_2. With low levels of $+V_{AK}$, a small *forward leakage current* (I_{FX}) flows. This is actually the reverse leakage current at junction J_2, and so (like I_{RX}), it is typically around 100 μA. Also like I_{RX}, I_{FX} remains substantially constant until $+V_{AK}$ is made large enough to cause (reverse-biased) J_2 to break down. The applied voltage at this point is termed the forward breakover voltage ($V_{F(BO)}$). This is illustrated by the *forward characteristics* (I_F versus $+V_{AK}$) in Fig. 20-4c. When $V_{F(BO)}$ is reached, the component transistors (Q_1 and Q_2) are immediately switched on into saturation as already explained, and the anode-to-cathode voltage falls rapidly to the *forward conduction voltage* V_F. The device is now into the *forward conduction region*, and I_F must be limited to protect the SCR from excessive current levels.

So far, the SCR forward characteristics have been discussed only for the case of $I_G = 0$. Now consider the effect of I_G levels greater than zero (Fig. 20-5a). As already shown, when $+V_{AK}$ is less than $V_{F(BO)}$ and I_G is zero, a small leakage current flows. This current is too small to have any effect on the level of $+V_{AK}$ that causes SCR switch *on*. When I_G is made just slightly larger than the junction leakage currents, it still has a negligible effect on the level of $+V_{AK}$ for switch-on. Now consider the opposite extreme. When I_G is made larger than the minimum base current required to switch Q_2 on, the SCR switches on when $+V_{AK}$ forward-biases the base-emitter junctions of Q_1 and Q_2 (Fig. 20-5b and Fig. 20-6).

The complete forward characteristics for an SCR are shown in Fig. 20-6. Note that when $I_G = I_{G4}$, switch-*on* occurs with $+V_{AK}$ at a relatively low level (V_4). Gate currents between I_{G0} and I_{G4} permit device switch-on at voltages greater than V_4 and less than $V_{F(BO)}$. The region of the forward characteristics before switch-on is known as the *forward blocking region,* and the region after switch-on is termed the *forward conduction region*, as illustrated. In the forward

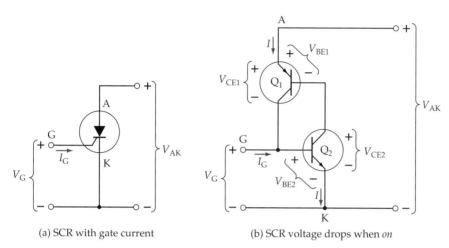

(a) SCR with gate current   (b) SCR voltage drops when *on*

Figure 20-5   A gate current (I_G) can cause the SCR to switch on at a low V_{AK} level.

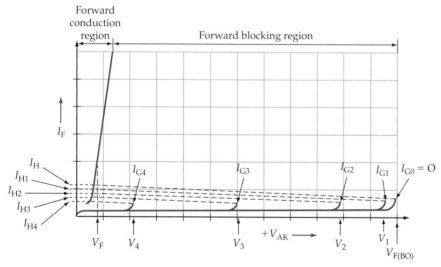

Figure 20-6   Forward characteristics for an SCR. Higher levels of gate current (I_G) cause the SCR to conduct at lower anode-to-cathode voltages ($+V_{AK}$).

conduction region, the SCR behaves as a forward-biased rectifier. The forward (anode-to-cathode) voltage (V_F) when the device is on is typically 1.7 V.

To switch an SCR *off*, the forward current (I_F) must be reduced below the *holding current* (I_H) (see Fig. 20-6). The holding current is the minimum level of I_F that maintains SCR conduction. If a gate current greater than zero is maintained while the SCR is *on*, lower levels of holding current (I_{H1}, I_{H2}, etc.) are possible.

SCR Specification

As in the case of most electronic devices, the SCR maximum voltage and current are important for any given application. The forward breakover voltage and reverse breakdown voltage have already been discussed. The maximum forward voltage that may be applied without causing the SCR to conduct is termed the *forward blocking voltage* (V_{DRM}). Similarly, the maximum reverse voltage that may be applied is the *reverse blocking voltage* (V_{RRM}).

The maximum SCR current is variously specified as the *average current* ($I_{T(AV)}$), the *rms current* ($I_{T(RMS)}$), and the *peak non-repetitive surge current* (I_{TSM}). The first two of these need no explanation. The third is a relatively large current that can normally be permitted to flow for a maximum of a half-cycle of a 60 Hz sine wave. The circuit fusing rating (I^2t) is another parameter that defines the maximum non-repetitive forward current. This can be used to calculate the maximum duration of a given forward current surge. In many circuit applications the SCR current is limited by a series-connected load, so there is usually no need to consider surge-current levels, except in the case of capacitive loads.

Some of the range of available SCRs is illustrated by the partial specifications and packages shown in Fig. 20-7. With 800 mA rms current and 30 V forward and reverse blocking voltage, the 2N5060 is a relatively low-current, low-voltage device. This is packaged in the typical plastic TO-92 transistor-type enclosure. Note that the *peak reverse gate voltage* (V_{GRM}) is 5 V. The 2N6396

	2N5060	2N6396	C35N
Peak forward and reverse voltage (V_{DRM} and V_{RRM})	30 V	200 V	960 V
Maximum rms current ($I_{T(RMS)}$)	0.8 A	12 A	35 A
Forward *on* voltage (V_{TM})	1.7 V	1.7 V	2 V
Holding current (I_H)	5 mA	6 mA	100 mA
Gate trigger current (I_{GT})	200 μA	12 mA	6 mA
Gate trigger voltage (V_{GT})	0.8 V	0.9 V	3 V
Gate reverse voltage (V_{GRM})	5 V	5 V	5 V

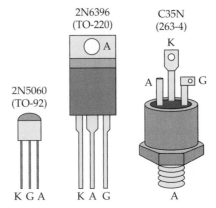

Figure 20-7 Partial specifications and packages for three SCRs for different voltage and current levels.

SCR is capable of handling a maximum rms current of 12 A, and has forward and reverse blocking voltage of 200 V. The package is a TO-220 plastic enclosure with a metal tab for mounting on a heat sink. For the C35N, the peak forward and reverse voltage is 960 V, and maximum rms current is 35 A. The device package is designed for bolt-mounting to a heat sink.

Section 20-1 Review

20-1.1 Sketch the four-layer construction of an SCR and the two-transistor equivalent circuit. Explain the device operation.

20-1.2 Sketch SCR forward and reverse characteristics. Briefly explain.

20-2 SCR CONTROL CIRCUITS

Pulse Control

The simplest of SCR control circuits is shown in Fig. 20-8a. If SCR_1 were an ordinary rectifier, the ac supply voltage would be half-wave-rectified and only the positive half-cycles would appear across the load (R_L). The same would be true if the SCR gate had a continuous bias voltage to keep it *on* when the anode-cathode voltage goes positive. A trigger pulse applied to the gate can switch the device *on* at any time during the positive half-cycle of the supply voltage. The SCR continues to conduct during the rest of the positive half-cycle, and then it switches *off* when the instantaneous level of the supply approaches zero. The resulting load waveform is a portion of the positive half-cycle starting at the instant that the SCR is triggered (Fig. 20-8b). Resistor R_G holds the gate-cathode voltage at zero when no trigger input is present.

Load waveforms that result from the SCR being switched *on* at different points in the positive half-cycle of the supply voltage are shown in Fig. 20-9.

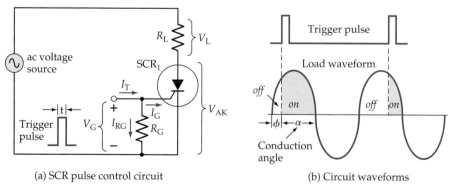

(a) SCR pulse control circuit

(b) Circuit waveforms

Figure 20-8 An SCR can be triggered on by a pulse applied to the gate. Once triggered, the device remains on until the load current falls below the holding current.

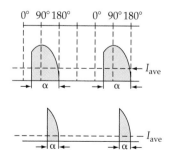

Figure 20-9 The average load current can be varied by controlling the SCR conduction angle.

It is seen that the average load current is determined by the SCR conduction angle. Thus, the load power dissipation can be varied by adjusting the SCR switch-on point. It should be noted that the SCR cannot be triggered precisely at the 0° point in the waveform because the anode-to-cathode voltage must be at least equal to the forward *on* voltage (V_{TM}) for the device. Also, the SCR will switch *off* before the 180° point when the load current falls below the holding current.

The instantaneous level of the load voltage is the instantaneous supply voltage (e_s) minus the SCR forward voltage (V_{TM}):

$$V_L = e_s - V_{TM} \qquad (20\text{-}1)$$

The load current can be calculated from V_L and R_L, and the instantaneous supply voltage ($e_{s(o)}$) that causes the SCR to switch *off* can be determined from V_{TM}, R_L, and the holding current:

$$e_{s(o)} = V_{TM} + (I_H R_L) \qquad (20\text{-}2)$$

For any given application, the SCR selected must have forward and reverse blocking voltages greater than the peak supply voltage. Its specified maximum rms current must also be greater than the rms load current. When designing the circuit, the gate current used should be at least three times the I_G specified for the device. Note that the required triggering current (I_T) for the circuit in Fig. 20-8 is the sum of I_G and the resistor current I_{RG}, as illustrated.

Example 20-1

Select a suitable SCR for the circuit in Fig. 20-8a if the rms supply voltage is 24 V and the load resistance is 25 Ω. Calculate the instantaneous supply voltage that causes the SCR to switch *off*.

Solution

Peak supply voltage:

$$V_{s(pk)} = 1.414 \times V_s = 1.414 \times 25 \text{ V}$$
$$= 33.9 \text{ V}$$

SCR forward and reverse blocking voltage:

$$V_{DRM} \text{ \& } V_{RRM} > 33.9 \text{ V}$$

Reference to the partial specification for the 2N5060 to 2N5064 range of SCRs in Fig. 20-10 shows that the 2N5060 has $V_{DRM} = 30$ V and the 2N5061 has

2N5060 to 2N5064		
Peak forward and reverse voltage (V_{DRM} and V_{RRM})	2N5060	30 V
	2N5061	60 V
	2N5062	100 V
	2N5063	150 V
	2N5064	200 V
Maximum rms current ($I_{T(RMS)}$)		0.8 A
Forward-on voltage (V_{TM})		1.7 V
Holding current (I_H)		5 mA
Gate trigger current (I_G)		200 μA

Figure 20-10   Partial specification for 2N5060 to 2N5064 SCRs.

$V_{DRM} = 60$ V. So the 2N5060 would *not* be suitable, whereas the 2N5061 would seem to be a suitable device.

$$I_{L(pk)} = \frac{V_{s(pk)} - V_{TM}}{R_L} = \frac{33.9\text{ V} - 1.7\text{ V}}{25\ \Omega}$$
$$= 1.29\text{ A}$$

For a half-wave rectified sinusoidal waveform,

$$I_{L(rms)} = 0.5\, I_{L(pk)} = 0.5 \times 1.29\text{ A}$$
$$= 0.64\text{ A}$$

The 2N5061 has $I_{T(rms)} = 0.8$ A.

So, the 2N5061 is a suitable SCR.

Switch-*off* voltage:

From Eq. 20-2,

$$e_{s(o)} = V_{TM} + (I_H \times R_L)$$
$$= 1.7\text{ V} + (5\text{ mA} \times 25\ \Omega)$$
$$\approx 1.8\text{ V}$$

90° Phase Control

In the 90° *phase-control circuit* shown in Fig. 20-11, the gate triggering voltage is derived from the ac supply via resistors R_1, R_2, and R_3. When the moving contact is set to the top of R_2, the SCR can be triggered *on* almost immediately at the beginning of the positive half-cycle of the input. When the moving contact is set to the bottom of R_2, the SCR might not switch *on* until the peak of the positive half-cycle. Between these two extremes, the device can be switched *on* somewhere between the zero level and the peak of the positive half-cycle (between 0° and 90°). If the triggering voltage (V_T) is not large enough to trigger the SCR at 90°, then the device will not trigger *on* at all, because V_T is greatest at the supply voltage peak and falls *off* past the peak.

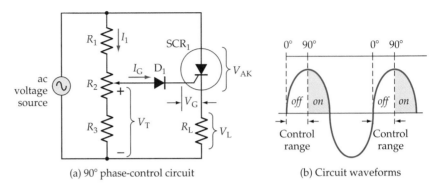

(a) 90° phase-control circuit   (b) Circuit waveforms

Figure 20-11 SCR 90° phase control circuit. The SCR can be triggered on anywhere between 0° and 90°.

Diode D_1 in Fig. 20-11 is included in the circuit to protect the SCR gate from the negative voltage that would otherwise be applied to it during the negative half-cycle of the ac supply.

The load for an SCR phase control circuit could be a permanent magnet motor, so that the circuit controls the motor speed. Or the load may be a heater or a light, and in this case the circuit controls the heater temperature or the light intensity.

The voltage divider ($R_1 R_2 R_3$) in Fig. 20-11 is designed in the usual way for the required range of adjustment of V_T. The voltage-divider current (I_1) is selected much larger than the SCR gate current. The instantaneous triggering voltage at switch-*on* is

$$V_T = V_{D1} + V_G \qquad (20\text{-}3)$$

Figure 20-12 shows a 90° phase-control circuit with its ac voltage source full-wave-rectified. This gives a larger maximum power dissipation in the load than a non-rectified source. Diode D_1 in Fig. 20-11 is not required in Fig. 20-12 because the SCR gate does not become reverse-biased.

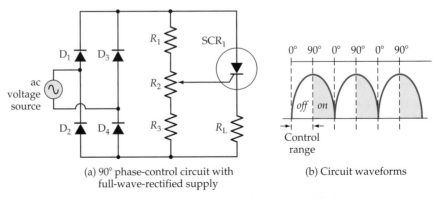

(a) 90° phase-control circuit with full-wave-rectified supply   (b) Circuit waveforms

Figure 20-12 SCR 90° phase-control circuit with a full-wave-rectified supply.

In the circuit in Fig. 20-13a, the two SCRs are connected in inverse parallel and they operate independently as 90° phase-control circuits. SCR₁ controls the load current during the positive half-cycle of the supply voltage, and SCR₂ controls the current during the negative half-cycle. The triggering voltage for each SCR is set by the voltage divider network R_1 to R_4 and adjusted by variable resistor R_3. Diodes D_1 and D_2 protect the gate terminals of each SCR from excessive reverse voltage.

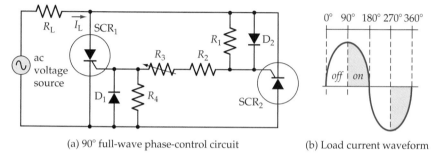

(a) 90° full-wave phase-control circuit   (b) Load current waveform

Figure 20-13   90° full-wave phase control circuit using two inverse-parallel connected SCRs.

During the supply voltage positive half-cycle, D_2 is forward-biased and current flows through R_2, R_3, and R_4. The voltage drop across R_4 triggers SCR₁ at the desired point in the positive half-cycle. When triggered, the SCR forward voltage switches to a low level and remains there until the instantaneous supply voltage approaches zero. During the supply negative half-cycle, D_1 is forward-biased to produce current flow through R_1, R_2, and R_3. With R_1 equal to R_4, the voltage drop across R_1 triggers SCR₂ at the same point in the negative half-cycle as SCR₁ in the positive half-cycle. The resulting 90° full-wave phase-controlled load waveform is shown in Fig. 20-13b.

Example 20-2

The SCR in Fig. 20-14 is to be triggered *on* between 5° and 90° during the positive half-cycle of the 30 V supply. The gate triggering current and voltage are 200 μA and 0.8 V. Determine suitable resistance values for R_1, R_2, and R_3.

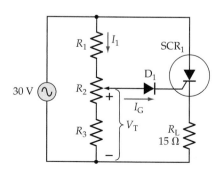

Figure 20-14   SCR 90° phase-control circuit for Example 20-2.

Solution

Peak supply voltage:

$$V_{s(pk)} = 1.414 \times V_s = 1.414 \times 30 \text{ V}$$

$$= 42.4 \text{ V}$$

At 5°,
$$e_s = V_{s(pk)} \sin 5° = 42.4 \text{ V} \sin 5°$$

$$\approx 3.7 \text{ V}$$

At 90°,
$$e_s = V_{s(pk)} = 42.4 \text{ V}$$

Eq. 20-3:
$$V_T = V_{D1} + V_G = 0.7 \text{ V} + 0.8 \text{ V}$$

$$= 1.5 \text{ V}$$

To trigger at $e_s = 3.7$ V, the R_2 moving contact is at the top.

So
$$V_{R2} + V_{R3} = V_T = 1.5 \text{ V}$$

and
$$V_{R1} = e_s - V_T = 3.7 \text{ V} - 1.5 \text{ V}$$

$$= 2.2 \text{ V}$$

$$I_{1(min)} \gg (I_G = 200 \text{ μA})$$

Select
$$I_{1(min)} = 1 \text{ mA}$$

$$R_1 = \frac{V_{R1}}{I_1} = \frac{2.2 \text{ V}}{1 \text{ mA}}$$

$$= 2.2 \text{ kΩ (standard value)}$$

$$R_2 + R_3 = \frac{V_T}{I_1} = \frac{1.5 \text{ V}}{1 \text{ mA}}$$

$$= 1.5 \text{ kΩ}$$

To trigger at $e_s = 42.4$ V, the R_2 moving contact is at the bottom.

So
$$V_{R3} = V_T = 1.5 \text{ V}$$

and
$$I_1 = \frac{e_s}{R_1 + R_2 + R_3} = \frac{42.4 \text{ V}}{2.2 \text{ kΩ} + 1.5 \text{ kΩ}}$$

$$\approx 11.5 \text{ mA}$$

$$R_3 = \frac{V_T}{I_1} = \frac{1.5 \text{ V}}{11.5 \text{ mA}}$$

$$= 130 \text{ Ω (use 120 Ω standard value)}$$

$$R_2 = (R_2 + R_3) - R_3 = 1.5 \text{ kΩ} - 120 \text{ Ω}$$

$$= 1.38 \text{ kΩ (use 1.5 kΩ standard value potentiometer)}$$

180° Phase Control

In the circuit shown in Fig. 20-15, resistor R_1 and capacitor C_1 determine the point in the supply voltage cycle where the SCR switches *on*. During the negative half-cycle of the supply, C_1 is charged via diode D_1 to the negative peak of the supply voltage. When the negative peak is passed, D_1 is reverse-biased because its anode (connected to C_1) is more negative than its cathode. With D_1 reversed, C_1 begins to discharge via R_1. While C_1 voltage remains negative, D_2 is reverse-biased and the gate voltage cannot go positive to trigger the SCR *on*. Depending on the values of C_1 and R_1, the capacitor might be completely discharged at the beginning of the positive half-cycle of the supply, allowing SCR_1 to switch *on*. Alternatively, C_1 might retain some negative charge past the end of positive half-cycle, keeping SCR_1 *off*. Resistor R_2 is included in the circuit to restrict the level of the gate current.

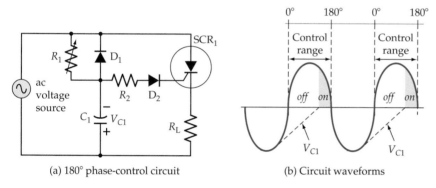

(a) 180° phase-control circuit

(b) Circuit waveforms

Figure 20-15   SCR 180° phase-control circuit. R_1 adjustment allows the SCR triggering point to be set anywhere between 0° and 180° in the positive half-cycle of the ac supply voltage.

Designing the 180° phase-control circuit can begin with selection of a capacitor much larger than stray capacitance. A maximum resistance for R_1 should then be calculated to discharge the capacitor voltage to zero during the time from the negative peak of the supply voltage to the 180° point in the positive half-cycle. The capacitor voltage does not decrease linearly as Fig. 20-16 implies. However, the maximum resistance for R_1 can be most easily calculated by assuming a linear discharge. The average value of the discharging voltage (E) is first determined. Figure 20-16 shows that E is $-0.636\ V_{s(pk)}$ for 0.25 T, and $+0.636\ V_{s(pk)}$ for 0.5 T, which averages out to approximately 0.2 $V_{s(pk)}$ for the total discharge time of 0.75 T. Now the equation for discharge of a capacitor to zero volts via a resistor may be applied:

$$t = RC \ln[(E - E_o)/E]$$

Substituting the appropriate quantities into the equation gives

$$R_1 \approx \frac{0.75\ T}{C_1 \ln 6} \tag{20-4}$$

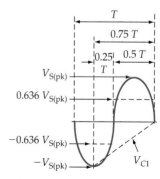

Figure 20-16   Discharge times and voltages for C_1 in the circuit in Fig. 20-15.

20-3 MORE SCR APPLICATIONS

SCR Circuit Stability

An SCR circuit is stable when it operates correctly, switching on and off only at the desired instants. Unwanted triggering (also called *false triggering*) can be produced by noise voltages at the gate, transient voltages at the anode terminal, or very fast voltage changes at the anode (termed *dv/dt triggering*).

Obviously, gate noise voltages may be large enough to forward-bias the gate-cathode junction and cause false triggering. Anode voltage transients (produced by other devices connected to the same ac supply) can exceed the SCR breakover voltage and thus trigger it into conduction. The *dv/dt* effect occurs when the anode voltage changes instantaneously, such as when the supply is switched *on* at its peak voltage level. The SCR capacitance is charged very quickly, and the charging current is sufficient to trigger the device.

Gate noise problems can be minimized by keeping the gate connecting leads short and by the use of a gate bias resistor (R_G in Fig. 20-17a). This should be connected as closely as possible to the SCR gate-cathode terminals, because conductors connected between R_G and the device could pick up noise that

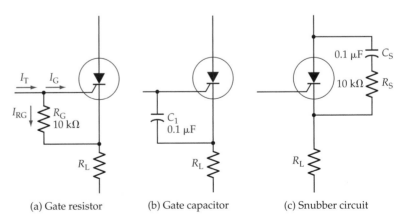

(a) Gate resistor   (b) Gate capacitor   (c) Snubber circuit

Figure 20-17   Unwanted gate noise triggering can be prevented by R_G or C_1 at the gate-cathode terminals. The use of a snubber circuit prevents triggering by transients at the anode terminal.

might cause triggering. Biasing the gate negative with respect to the cathode can also be effective in combating noise. Capacitor C_1 in Fig. 20-17b can be used to short-circuit gate noise voltages. C_1 also operates in conjunction with the anode-gate capacitance as a voltage divider that reduces the possibility of dv/dt triggering. C_1 is usually in the 0.01 μF to 0.1 μF range, and like R_G, it should be connected close to the SCR terminals.

An *RC snubber circuit* can be used to prevent triggering by anode terminal transients (Fig. 20-17c). A snubber is usually necessary for inductive loads and may also be required for resistive loads. With an ac supply, there is a phase difference between an inductive load current and the supply voltage, and this can cause loss of SCR control. Also, the current through an inductor with a dc supply will not go to zero immediately when the SCR switches *off*. A snubber circuit is necessary in both cases.

Zero-Point Triggering

When an SCR is switched on while the instantaneous level of the supply voltage is greater than zero, surge currents occur that generate *electromagnetic interference (EMI)*. The EMI can interfere with other nearby circuits and equipment, and the switching transients can affect control of the SCR. Circuits can be designed to trigger an SCR *on* at the instant the ac supply is crossing the zero voltage point from the negative half-cycle to the positive half-cycle. This is called *zero-point triggering,* and it effectively eliminates the EMI and the switching transients.

The zero-point triggering circuit in Fig. 20-18a shows two inverse-parallel connected SCRs that each have *RC* triggering circuits: C_1 and R_1 for SCR_1, and C_2 and R_2 for SCR_2. SCR_1 is held *off* while switch S_1 is closed, and because capacitor C_2 is uncharged, SCR_2 remains *off*. With S_1 open, positive triggering current (I_{G1}) begins to flow when the supply voltage begins to go positive. As illustrated, I_{G1} flows via C_1 and R_1 to the gate of SCR_1, triggering it into conduction at the zero-crossing point. SCR_1 provides a path for (*positive*) load current (i_{L+}).

With SCR_1 *on*, capacitor C_2 charges (with the polarity shown) almost to the peak of the supply voltage. When the supply voltage crosses zero from the positive half-cycle to the negative half-cycle, SCR_1 switches *off*. D_1 becomes reverse-biased, and the charge on C_1 provides triggering current (I_{G2}) to SCR_2. Thus, SCR_2 is switched on at the start of the supply negative half-cycle, providing a path for (*negative*) load current (i_{L-}).

Both SCRs continue to switch on and off at the zero-crossing points while S_1 remains open, and both stay *off* when S_1 is closed. SCR_2 cannot switch *on* unless SCR_1 has first been *on*, and because of this the arrangement is sometimes termed a *master-slave* circuit, SCR_1 being the *master* and SCR_2 the *slave*. The waveforms in Fig. 20-18b show that power is supplied to the load for several cycles of the supply while S_1 is open, and no load power dissipation occurs for several cycles while S_1 remains closed. The switch might be controlled by a temperature sensor or other device.

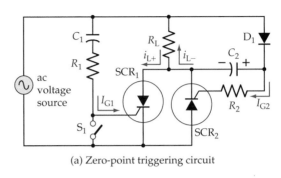

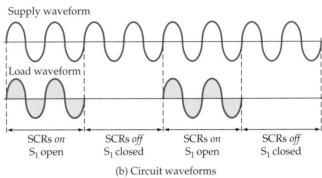

(a) Zero-point triggering circuit

Supply waveform

Load waveform

| SCRs *on* | SCRs *off* | SCRs *on* | SCRs *off* |
| S_1 open | S_1 closed | S_1 open | S_1 closed |

(b) Circuit waveforms

Figure 20-18   In an SCR zero-point triggering circuit the devices are switched on only when the supply waveform crosses the zero-voltage point.

Crowbar Circuit

A *crowbar circuit* (also known as an *overvoltage protection circuit*) is illustrated in Fig. 20-19. This circuit protects a sensitive load against an excessive dc supply voltage. When the supply (V_S) is at its normal voltage level, it is too low to cause the Zener diode (D_1) to conduct. Consequently, there is no current through the gate bias resistor (R_1) and no voltage drop across R_1. The gate voltage (V_G) remains equal to zero, and the SCR remains *off*. When the supply voltage exceeds V_Z, D_1 conducts, and the resultant voltage drop across R_1 triggers the SCR into conduction. The voltage across the load is now reduced to the SCR forward voltage drop. The voltage across V_Z and R_1 is also reduced to the SCR forward voltage, and the dc voltage source is short-circuited by the SCR.

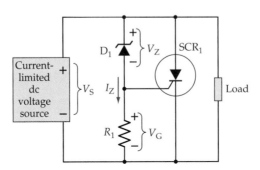

Figure 20-19   An SCR crowbar circuit (or overvoltage protection circuit) short-circuits the load when the supply voltage exceeds a pre-determined level.

The voltage source must have a current-limiting circuit to protect the source and to minimize SCR power dissipation. The supply must be switched *off* for the SCR to cease conducting.

Example 20-3

The dc voltage source in the SCR crowbar circuit in Fig. 20-19 has $V_S = 5$ V and $I_{L(max)} = 300$ mA. The load voltage is not to exceed 7 V. Select suitable components for D_1 and R_1, and specify the SCR. Assume that $V_G = 0.8$ V.

Solution

$$V_Z = V_{L(max)} - V_G = 7\,\text{V} - 0.8\,\text{V}$$
$$= 6.2\,\text{V}$$

For D_1, select a 1N753 with $V_Z = 6.2$ V

Select
$$I_{Z(min)} = 1\,\text{mA}$$

$$R_1 = \frac{V_G}{I_Z} = \frac{0.8\,\text{V}}{1\,\text{mA}}$$
$$= 800\,\Omega\ (\text{use }820\,\Omega\text{ standard value})$$

SCR specification:

$$V_{DRM} > 7\,\text{V},\ I_{T(AV)} > 300\,\text{mA}$$

Heater Control Circuit

The circuit in Fig. 20-20 uses a temperature-sensitive control element (R_2). The resistance of R_2 decreases when the temperature increases, and increases when the temperature falls. Diode D_1 keeps capacitor C_1 charged to the supply voltage peak, and C_1 together with resistor R_1 behaves as a constant-current source for R_2. When R_2 is raised to the desired temperature, V_G drops to a level that keeps the SCR from triggering. When the temperature drops, the resistance of R_2 increases, causing V_G to increase to the SCR triggering level. The result is that the load power is turned *off* when the desired temperature is reached, and turned *on* again when the temperature falls to a predetermined level. Rectifier D_2 might be included, as illustrated, to pass the negative half-cycle of the supply waveform to the load.

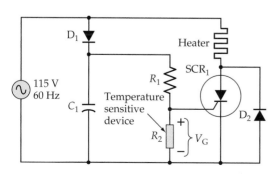

Figure 20-20 SCR heater control circuit. The SCR is triggered on when the temperature is below a specified level and held off when the temperature is satisfactory.

Fundamentals of Electronic Devices and Circuits

20-3.1 The SCR in the circuit in Fig. 20-20 triggers when V_G is 0.8 V or higher, but will not trigger when V_G is 0.6 V. The temperature-sensing element (R_2) has a resistance of 400 Ω at 95°C and 300 Ω at 100°C. Determine suitable values for R_1 and C_1 that will trigger the SCR at 95°C and leave it untriggered at 100°C.

20-4 TRIAC AND DIAC

TRIAC Operation and Characteristics

The basic construction, equivalent circuit, and graphic symbol for a TRIAC are shown in Fig. 20-21. The TRIAC behaves as two inverse-parallel connected SCRs with a single gate terminal. Sections n_1, p_2, n_3, and p_3 in Fig. 20-21a form one SCR that can be represented by transistors Q_1 and Q_2 in Fig. 20-21b. Similarly, p_1, n_2, p_2, and n_4 form another SCR with the transistor equivalent circuit Q_3 and Q_4. Layer p_2, common to the two SCRs, functions as a gate for both sections of the device. The two outer terminals cannot be identified as anode and cathode; instead they are designated *main terminal 1 (MT1)* and *main terminal 2 (MT2)*, as illustrated. The TRIAC circuit symbol is composed of two inverse-parallel connected SCR symbols (Fig. 20-21c).

When MT2 is positive with respect to MT1, transistors Q_3 and Q_4 can be triggered *on* (Fig. 20-21b). In this case the current flow is from MT2 to MT1. When MT1 is positive with respect to MT2, Q_1 and Q_2 can be switched *on*. Now current flow is from MT1 to MT2. It is seen that the TRIAC can be made to conduct in either direction. Regardless of the MT2/MT1 voltage polarity,

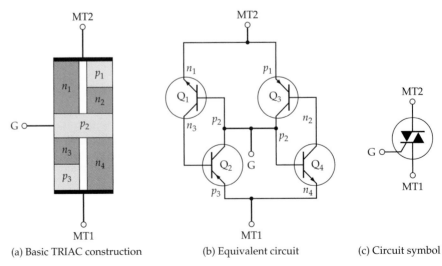

(a) Basic TRIAC construction   (b) Equivalent circuit   (c) Circuit symbol

Figure 20-21 Basic construction, equivalent circuit, and graphic symbol for a TRIAC.

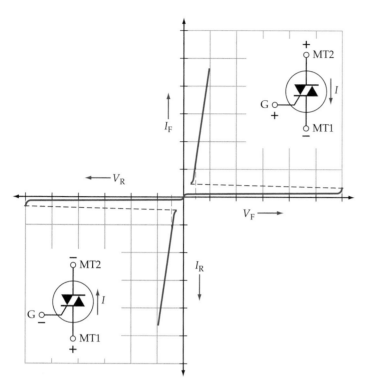

Figure 20-22   TRIAC characteristics. These are similar to the characteristics of two inverse-parallel connected SCRs.

the characteristics for the TRIAC are those of a forward-biased SCR. This is illustrated by the typical TRIAC characteristics shown in Fig. 20-22.

TRIAC Triggering

The characteristics and circuit symbol in Fig. 20-22 show that when MT2 is positive with respect to MT1, the TRIAC can be triggered *on* by application of a positive gate voltage. Similarly, when MT2 is negative with respect to MT1, a negative gate voltage triggers the device into conduction. However, a negative gate voltage can also trigger the TRIAC when MT2 is positive, and a positive gate voltage can trigger the device when MT2 is negative.

Figure 20-23 shows the triggering conditions for a 2N6346, 8 A, 200 V TRIAC. The voltage polarity for MT2 is identified as MT2(+) or MT2(−), and

Figure 20-23   Partial specification showing the triggering conditions for a 2N6346 TRIAC.

2N6346 TRIAC		
	V_{GT}	
	Min	Max
MT2(+), G(+)	0.9 V	2 V
MT2(+), G(−)	0.9 V	2.5 V
MT2(−), G(−)	1.1 V	2 V
MT2(−), G(+)	1.4 V	2.5 V

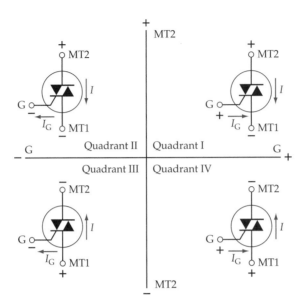

the gate polarity is listed as G(+) or G(−). From the first line of the specifications, it is seen that with MT2 positive, the device gate triggering voltage is +0.9 V minimum and +2 V maximum. From the second line, still with MT2 positive, triggering can be produced by a negative gate voltage: −0.9 V to −2.5 V. The third line shows MT2 negative and the gate trigger voltage as −1.1 V to −2 V. Also, with MT2 negative (fourth line), triggering can be effected by a positive gate voltage: +1.4 V to +2.5 V.

The TRIAC triggering conditions are further illustrated by the diagram in Fig. 20-24. The vertical line identifies MT2 as positive or negative, and the horizontal line shows the gate voltage as positive or negative. The TRIAC is defined as operating in one of the four quadrants: I, II, III, or IV. In quadrant I, MT2 is positive, the gate voltage is positive, and current flow is from MT2 to MT1, as shown. When MT2 is positive and the device is triggered by a negative gate voltage, the TRIAC is operating in quadrant II. In this case, current flow is still from MT2 to MT1. Quadrant III operation occurs when MT2 is negative and the gate voltage is negative. Current flow is now from MT1 to MT2. In quadrant IV, MT2 is again negative, the gate voltage is positive, and current flow is from MT1 to MT2.

Normally, a TRIAC is operated in either quadrant I or quadrant III. When this is the desired condition, it may be necessary to design the circuit to avoid quadrant II or quadrant IV triggering.

The DIAC

A DIAC is simply a low-current TRIAC without a gate terminal. Switch-on is effected by raising the applied voltage to the breakover voltage. Two different DIAC symbols in general use are shown in Fig. 20-25a, and typical DIAC

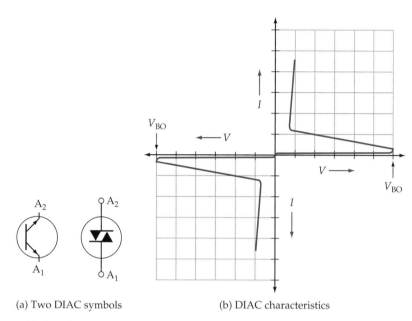

(a) Two DIAC symbols   (b) DIAC characteristics

Figure 20-25 The DIAC is essentially a low-current TRIAC without a gate terminal.

DIACs				
	V_S		$I_{S(max)}$	P_D
	Min	Max		
HS-10	8 V	12 V	400 μA	250 mW
HS-60	56 V	70 V	50 μA	250 mW

Figure 20-26 Partial specifications for two DIACs.

characteristics are illustrated in Fig. 20-25b. Note that the terminals are labelled *anode 2* (A_2) and *anode 1* (A_1). Figure 20-26 shows partial specifications for two DIACs. The HS-10 has a switching voltage that ranges from a minimum of 8 V to a maximum of 12 V. Switching current is a maximum of 400 μA. The HS-60 switching voltage is 56 V to 70 V, and maximum switching current is 50 μA. Both devices have a 250 mW power dissipation, and each is contained in a cylindrical low-current diode-type package. DIACs are most often applied in triggering circuits for SCRs and TRIACs.

Section 20-4 Review

20-4.1 Sketch the construction and transistor equivalent circuit of a TRIAC. Explain the device operation.

20-4.2 Sketch TRIAC characteristics. Briefly explain.

20-5 TRIAC CONTROL CIRCUITS

TRIAC Phase Control Circuit

A TRIAC circuit that allows approximately 180° of phase control is shown in Fig. 20-27a. The waveforms in Fig. 20-27b illustrate the circuit operation. With the TRIAC (Q_1) *off* at the beginning of the supply voltage positive half-cycle, capacitor C_1 is charged positively via resistors R_1 and R_2, as shown. When V_{C1} reaches the DIAC switching voltage plus the Q_1 gate triggering voltage, D_1 conducts passing gate current to trigger Q_1 *on*. C_1 discharges until the discharge current falls below the D_1 holding current level. The TRIAC switches *off* at the end of the supply positive half-cycle, and then the process is repeated during the supply negative half-cycle. The rate of charge of C_1 is set by variable resistor R_1, so that the Q_1 conduction angle is controlled by adjustment of R_1.

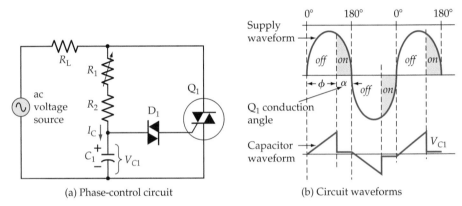

(a) Phase-control circuit (b) Circuit waveforms

Figure 20-27   TRIAC phase-control circuit and circuit waveforms. Q_1 is switched on when the capacitor charges to D_1 breakover voltage.

Example 20-4

Estimate the smallest conduction angle for Q_1 for the circuit in Fig. 20-27a. The supply is 115 V, 60 Hz, and the components are $R_1 = 25$ kΩ, $R_2 = 2.7$ kΩ, and $C_1 = 3$ µF. The D_1 breakover voltage is 8 V, and $V_G = 0.8$ V for Q_1.

Solution

At Q_1 switch-*on*,
$$V_{C1} = V_{D1} + V_G = 8\text{ V} + 0.8\text{ V}$$
$$= 8.8\text{ V}$$

Assume the average charging voltage is
$$E = 0.636 \times V_{ac(pk)} = 0.636 \times 1.414 \times 115\text{ V}$$
$$\approx 103\text{ V}$$

Average charging current: $I_C \approx \dfrac{E}{R_1 + R_2} = \dfrac{103\text{ V}}{25\text{ k}\Omega + 2.7\text{ k}\Omega}$
$$\approx 3.7\text{ mA}$$

Charging time:
$$t \approx \frac{C_1 V_{C1}}{I_C} = \frac{3 \; \mu\text{F} \times 8.8 \; \text{V}}{3.7 \; \text{mA}}$$

$$\approx 7.1 \; \text{ms}$$

$$T = \frac{1}{f} = \frac{1}{60 \; \text{Hz}} = 16.7 \; \text{ms}$$

Q_1 switch-on point:
$$\phi \approx \frac{t \times 360°}{T} = \frac{7.1 \; \text{ms} \times 360°}{16.7 \; \text{ms}}$$

$$= 153°$$

Conduction angle:
$$\alpha = 180° - \phi = 180° - 153°$$

$$= 27°$$

TRIAC Zero-Point Switching Circuit

The TRIAC *zero-point switching circuit* in Fig. 20-28a produces a load waveform similar to that for the SCR zero-point circuit in Fig. 20-18. The load power dissipation is controlled by switching the TRIAC *on* for several cycles of the supply voltage and *off* for several cycles, with switch-*on* occurring only at the

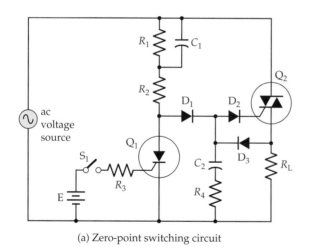

(a) Zero-point switching circuit

Figure 20-28 Zero-point switching circuit for a TRIAC. While Q_1 is on, Q_2 cannot switch on. With Q_1 off, Q_2 starts to conduct at the beginning of the supply positive half-cycle.

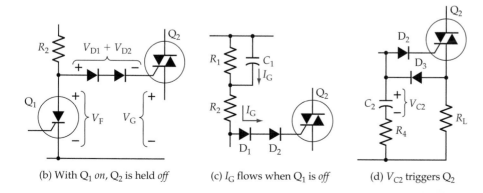

(b) With Q_1 *on*, Q_2 is held *off*   (c) I_G flows when Q_1 is *off*   (d) V_{C2} triggers Q_2

negative-to-positive zero crossing point of the supply waveform, and switch-*off* taking place at the positive-to-negative zero point. Q_1 is a low-current SCR that controls the switching point of Q_2.

With switch S_1 closed, Q_1 is on, and the Q_1 forward voltage drop is below the level required for triggering Q_2 ($V_{G2} + V_{D1} + V_{D2}$) (see Fig. 20-28b). Thus, no gate current flows to Q_2, and no conduction occurs. Q_1 switches *off* when S_1 is opened, so that I_G flows to Q_2 gate via C_1, R_2, D_1, and D_2 to trigger Q_2 into conduction (Fig. 20-28c). With Q_2 conducting, capacitor C_2 is charged via D_3 almost to the positive peak of the load voltage (Fig. 20-28d). The TRIAC switches *off* at the end of the positive half-cycle. Then, the charge on C_2 (applied to the gate via D_2) triggers Q_2 *on* again just after the zero-crossing point into the negative half-cycle. (It should be noted this is quadrant IV triggering.)

The initial Q_2 switch-on occurs only at the beginning of the positive half-cycle of the supply voltage. If S_1 is opened during the supply positive half-cycle, Q_1 continues to conduct until the end of the half-cycle, thus keeping Q_2 *off*. With Q_2 off, C_2 remains uncharged, and so it cannot trigger Q_2 *on* during the supply negative half-cycle. Q_2 triggering now occurs at the beginning of the next positive cycle.

If S_1 is opened during the supply negative half-cycle, Q_2 cannot be triggered into conduction, again because of the lack of charge on C_2. It is seen that Q_2 conduction can start only at the beginning of the positive half-cycle of the supply voltage. Once triggered, Q_2 conduction continues until the end of the cycle.

To design the circuit in Fig. 20-28, the TRIAC is first selected to pass the required load current and to survive the peak supply voltage. Resistor R_2 is a low-resistance component calculated to limit the peak surge current to the Q_2 gate in the event that the peak supply voltage is applied to the circuit without Q_1 being *on*. Capacitor C_1 has to supply triggering current (I_G) to Q_2 at the zero crossing point of the supply waveform when Q_1 is *off*. Usually I_{G2} is chosen to be about three times the specified $I_{G(max)}$ for Q_2, and C_1 is then calculated from the simple equation for capacitor charge: $C = (I \times t)/\Delta V$. In this case $\Delta V/t$ can be replaced by the rate of change of the supply voltage at the zero crossing point, which is $2\pi f V_p$. So the C_1 equation is

$$C_1 = \frac{I_{G2}}{2\pi f V_p} \tag{20-5}$$

Resistor R_1 can now be determined by using the selected gate current for Q_2 (I_{G2}) as the peak anode current for Q_1: $R_1 = V_p/I_{G2}$. The Q_1 gate resistor (R_3) is calculated from the Q_1 triggering current and the voltage of the dc source: $R_1 = (E - V_{G1})/I_{G1}$.

The Q_2 gate current is again used in the calculation of R_4 and C_2. To trigger Q_2 at the start of the supply negative half-cycle, I_{G2} must flow from C_2 into the Q_2 gate; so $R_4 \approx V_p/I_{G2}$. A suitable capacitance for C_2 is now calculated by again using the simple capacitance equation $C_2 = (I_{G2} \times t)/\Delta V$. In this case, time t is selected much larger than the Q_2 turn-on time, and ΔV is approximately $0.1 V_p$.

SCR Q_1 must pass the selected anode current (I_{G2}) and survive the peak supply voltage. The diodes must each survive the peak supply voltage and pass the Q_2 triggering current.

The IC Zero Voltage Switch

The functional block diagram for a typical integrated circuit TRIAC driver, known as a *zero voltage switch*, is shown in Fig. 20-29. The device contains a *voltage limiter* and a *dc power supply*, so that it operates directly from the ac supply to the load to be controlled. There is also a *zero crossing detector* (see Section 14-9) that provides an output pulse each time the supply waveform crosses the zero level. The zero crossing detector output is fed to an AND gate (see Section 3-11), and the AND gate output goes to the TRIAC drive stage that produces the current pulse to the TRIAC gate. An *on-off sensing amplifier* is used to sense the voltage level from an externally connected transducer; for example, a temperature-sensing transducer would be used if the load is a heater. When the temperature drops to a predetermined level, the on-off sensing amplifier provides an input to the AND gate. The gate triggering pulse from the TRIAC drive stage occurs at the supply zero-crossing points only when the temperature is below the desired level. The circuit load waveforms are similar to those shown in Fig. 20-18.

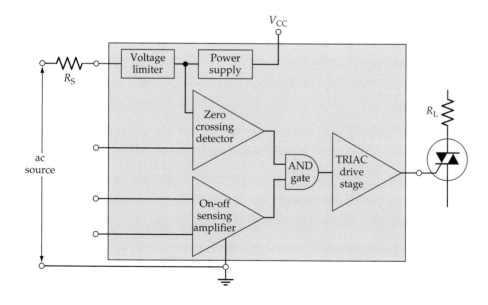

Figure 20-29 Functional block diagram for an integrated circuit zero voltage switch, or TRIAC driver.

Practice Problem

20-5.1 Determine suitable components for the TRIAC zero-point switching circuit in Fig. 20-28, given the following parameters: for Q_1, $I_G = 200\ \mu A$, $V_F = 2$ V; for Q_2, $I_G = 30$ mA, $I_{GM} = 1$ A, $t_{on} = 100\ \mu s$; $E = 6$ V; and ac source = 115 V, 60 Hz.

20-6 SUS, SBS, GTO, AND SIDAC

The SUS

The *silicon unilateral switch (SUS)*, also known as a *four-layer diode* and as a *Schokley diode*, can be treated as a low-current SCR without a gate terminal. Figure 20-30 shows the SUS circuit symbol and typical forward characteristics. The device triggers into conduction when a *forward switching voltage (V_S)* is applied. At this point a minimum *switching current (I_S)* must flow. The voltage falls to a *forward conduction voltage (V_F)* at switch-*on*, and conduction continues until the current level falls below the *holding current (I_H)*. The 2N4988 SUS has V_S ranging from 7.5 V to 9 V, $I_S = 150$ μA, and $I_H = 0.5$ mA. SUS reverse characteristics are similar to SCR reverse characteristics: a very small reverse current flows until the reverse breakdown voltage is reached.

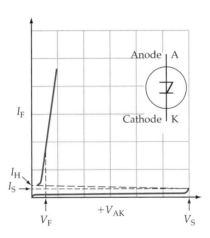

Figure 20-30 SUS circuit symbol and forward characteristics.

The SBS

It is convenient to think of a *silicon bilateral switch (SBS)* as two SUS with a gate terminal, or as a low-current TRIAC. However, the SBS is not simply another four-layer device. Silicon bilateral switches are actually integrated circuits constructed of matched transistors, diodes, and resistors. This produces better parameter stability than is possible with four-layer devices. The SBS equivalent circuit in Fig. 20-31a is similar to the TRIAC equivalent circuit with the addition of resistors R_1 and R_2 and Zener diodes D_1 and D_2. The device circuit symbol in Fig. 20-31b is seen to be composed of inverse-parallel connected SUS symbols with a gate terminal added. Note that the terminals are referred to as *anode 1 (A_1), anode 2 (A_2)*, and *gate (G)*. The typical SBS characteristics shown in Fig. 20-31c are essentially the same shape as TRIAC characteristics.

Returning to the equivalent circuit, the SBS switches *on* when a positive A_1A_2 voltage is large enough to cause D_2 to break down. This produces base current in Q_1, resulting in a Q_1 collector current that switches Q_2 *on*. Similarly, a negative A_1A_2 voltage causes D_1 to break down, producing base current in Q_4 that turns Q_4 and Q_3 *on*. The switching voltage is the sum of the Zener

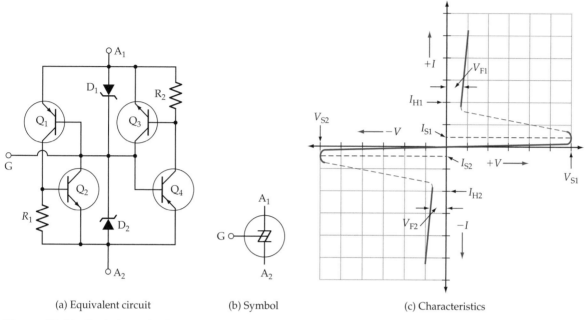

(a) Equivalent circuit   (b) Symbol   (c) Characteristics

Figure 20-31 Equivalent circuit, symbol, and characteristics for the silicon bilateral switch (SBS).

diode voltage and the transistor base-emitter voltage ($V_S = V_Z + V_{BE}$). The Zener diode has a positive *temperature coefficient* (TC) and the transistor base-emitter voltage has a negative TC. This results in a very small TC for the SBS switching voltage.

The partial specification for a MBS4991 SBS in Fig. 20-32 shows a switching voltage that ranges from 6 V to 10 V. Note also that the *switching voltage differential* (the difference between the switching voltages in opposite directions) ($V_{S1} - V_{S2}$), is 0.5 V maximum. The maximum switching current is a 500 μA, and the *switching current differential* ($I_{S1} - I_{S2}$) is 100 μA.

SBS devices are frequently used with the gate open-circuited, so that they simply break down to the forward voltage drop when the applied voltage

MBS4991 SBS			
	Min	Typ	Max
Switching voltage (V_S)	6 V	8 V	10 V
Switching current (I_S)		175 μA	500 μA
Switching voltage differential		0.3 V	0.5 V
Switching current differential			100 μA
Gate trigger current (I_{GF})			100 μA
Forward-on voltage (V_F)		1.4 V	1.7 V

Figure 20-32 Partial specification for the MBS4991 silicon bilateral switch.

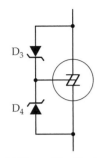

(a) V_S modification by external Zeners

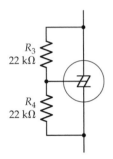

(b) V_S modification by external resistors

Figure 20-33 The switching voltage for an SBS can be modified by externally connected Zener diodes or resistors.

increases to the switching voltage level. The switching voltage can be reduced by connecting Zener diodes with V_Z lower than 6.8 V between the gate and the anodes, as shown in Fig. 20-33a. The new switching voltage is approximately $(V_Z + 0.7 \text{ V})$. The switching voltage can also be modified by the use of external resistors (Fig. 20-33b). Taking the gate current into account, it can be shown that the two 22 kΩ resistors reduce V_S to approximately 3.6 V.

The use of an SBS in a TRIAC phase-control circuit is illustrated in Fig. 20-34. This is essentially the same as the circuit using a DIAC in Fig. 20-27. The SBS turns on and triggers the TRIAC when the capacitor voltage equals the SBS switching voltage plus the TRIAC gate triggering voltage.

For an SBS to switch on, the total resistance in series with it must have a maximum value that allows the switching current to flow. If the resistance is so large that it restricts the current to a level below the SBS switching current, the device will not switch on. The series resistance must not be so small that it allows the holding current to flow when the SBS is supposed to switch off. These restrictions also apply to SCRs, TRIACs, DIACs, and other similar switching devices. Switch-off is usually no problem in thyristor circuits with ac supplies, because the devices normally switch off when the instantaneous supply voltage reduces to zero. With dc supplies, more care must be taken with resistor sizes.

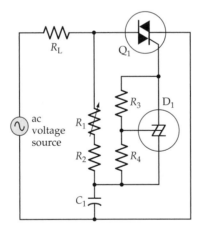

Figure 20-34 Use of an SBS in a TRIAC phase-control circuit. The TRIAC is triggered by the current surge when the SBS switches on.

Figure 20-35 shows a simple circuit that requires careful design to ensure that the SBS switches on and off as required. The circuit is a *relaxation oscillator* that produces an exponential output waveform, as illustrated. Capacitor C_1 is charged via resistor R_1 from the dc supply voltage (E). When the capacitor voltage (V_C) reaches the SBS switching voltage (V_S), D_1 switches *on* and rapidly discharges the capacitor to the D_1 forward voltage (V_F). Then D_1 switches *off*, and the capacitor begins to charge again. The SBS will not switch *off* if the D_1 holding current (I_H) continues to flow through R_1 when V_C equals V_F. SBS switch-*on* will normally occur when V_C equals V_S regardless of the

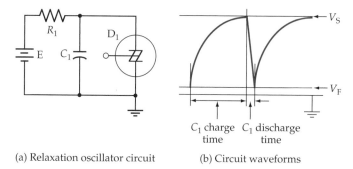

(a) Relaxation oscillator circuit   (b) Circuit waveforms

Figure 20-35   SBS relaxation oscillator. C_1 charges via R_1 to the SBS switching voltage. D_1 switches on at that point and rapidly discharges C_1.

R_1 resistance, because the capacitor discharge should provide the switching current (I_S). However, it is best to select R_1 small enough to allow I_S to flow at D_1 switch-*on*.

The approximate oscillation frequency can be determined from the capacitor charging time (t), and the equation for t is derived from the RC charging equation.

$$t = CR \ln\left[\frac{E - V_F}{E - V_S}\right] \tag{20-6}$$

Example 20-5

The SBS in the circuit in Fig. 20-35 has the following parameters: $V_S = 10$ V, $V_F = 1.7$ V, $I_S = 500$ μA, and $I_H = 1.5$ mA. Calculate the maximum and minimum resistances for R_1 for correct circuit operation when $E = 30$ V. Determine the capacitor charging time when $R_1 = 27$ kΩ and $C_1 = 0.5$ μF.

Solution

$$R_{1(max)} = \frac{E - V_S}{I_S} = \frac{30\ V - 10\ V}{500\ \mu A}$$

$$= 40\ k\Omega$$

$$R_{1(min)} = \frac{E - V_F}{I_H} = \frac{30\ V - 1.7\ V}{1.5\ mA}$$

$$= 18.9\ k\Omega$$

Eq. 20-6: $t = CR \ln\left[\dfrac{E - V_F}{E - V_S}\right] = 0.5\ \mu F \times 27\ k\Omega \ln\left[\dfrac{30\ V - 1.7\ V}{30\ V - 10\ V}\right]$

$$= 4.7\ ms$$

The GTO

When an SCR is triggered into conduction by application of a gate current, the gate loses control and the device continues to conduct until the forward current falls below the holding current. A *gate turn-off (GTO)* device is essentially an SCR designed to be switched on and off by an applied gate signal. The circuit symbol for a GTO is shown in Fig. 20-36a, and the two-transistor equivalent circuit for the device is illustrated in Fig. 20-36b and c. Note that at switch-*on*, the gate current has got to be just large enough to supply base current to transistor Q_2. However, at switch-*off*, the Q_1 collector current has to be diverted through the gate terminal in order to turn Q_2 off. Consequently, for device turn-*off*, relative large levels of gate current are involved, approaching half the GTO forward current.

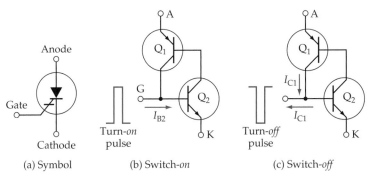

Figure 20-36   The gate turn-off device GTO is effectively an SCR that can be switched off by a voltage applied to the gate.

The SIDAC

The SIDAC is a two-terminal thyristor designed mainly for use in over-voltage protection situations. As a bilateral device with no gate terminal, it simply breaks down to its forward voltage drop when the applied terminal voltage (of either polarity) rises to the breakover voltage level. Like other thyristors, there is a minimum current that must flow to latch the SIDAC into an *on* state. When it is switched *on*, conduction continues until the current falls below a holding current level.

The circuit symbol and typical characteristics for a SIDAC are shown in Fig. 20-37. Available devices have breakover voltages ranging from 110 V to 280 V. Usually, *on* state voltage is 1.1 V, rms current is 1 A, and holding current is 100 mA.

Figure 20-38 shows a SIDAC used to protect a dc power supply from ac line transients. Normally, the SIDAC will behave as an open-circuit. A voltage transient on the ac line will cause it to break down to its forward voltage level, so that it essentially short-circuits the transformer output. This will cause a fuse to blow or a circuit breaker to trip, thus interrupting the ac supply.

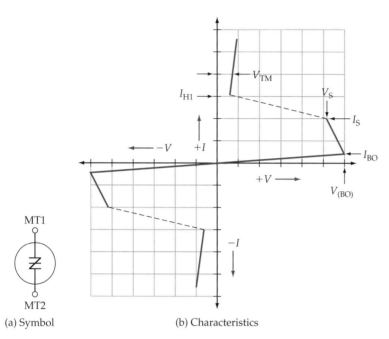

(a) Symbol   (b) Characteristics

Figure 20-37   Circuit symbol and characteristics for a SIDAC.

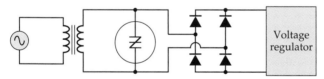

Figure 20-38   SIDAC used for protecting a dc power supply against transients on the supply line.

Practice Problem

20-6.1 The SBS in the circuit in Fig. 20-34 has $V_S \approx 4$ V, $I_S = 500$ μA, and $I_{F(max)} = 200$ mA. The ac supply is 115 V, and $R_3 = R_4 = 22$ kΩ. Determine suitable resistances for R_1 and R_2.

20-7 UNIJUNCTION TRANSISTOR (UJT)

UJT Operation

The *unijunction transistor (UJT)* consists of a bar of lightly doped *n*-type silicon with a block of *p*-type material on one side (see Fig. 20-39a). The end terminals of the bar are identified as *Base 1* (B₁) and *Base 2* (B₂), and the *p*-type block is named the *emitter* (E).

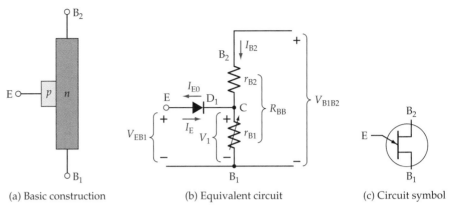

(a) Basic construction (b) Equivalent circuit (c) Circuit symbol

Figure 20-39 A unijunction transistor (UJT) is made up of a p-type emitter joined to a bar of n-type semiconductor.

Figure 20-39b shows the UJT equivalent circuit. The resistance of the n-type silicon bar is represented as two resistors, r_{B1} from B_1 to point C, and r_{B2} from B_2 to C, as illustrated. The sum of r_{B1} and and r_{B2} is called R_{BB}. The p-type emitter forms a pn-junction with the n-type silicon bar, and this junction is shown as a diode (D_1) in the equivalent circuit.

With a voltage V_{B1B2} applied as illustrated, the voltage at the junction point C is

$$V_1 = V_{B1B2} \times \frac{r_{B1}}{R_{BB}}$$

Note that V_1 is also the voltage at the cathode of the diode in the equivalent circuit.

With the emitter terminal open-circuited, the resistor current is

$$I_{B2} = \frac{V_{B1B2}}{R_{BB}} \qquad (20\text{-}7)$$

If the emitter terminal is grounded, the pn-junction is *reverse*-biased and there is a small *emitter reverse current* (I_{E0}).

Now consider what happens when the emitter voltage (V_{EB1}) is slowly increased from zero. When V_{EB1} equals V_1, the emitter current is zero. (With equal voltage levels on each side of the diode, neither reverse nor forward current flows.) A further increase in V_{EB1} forward-biases the pn-junction and causes a forward current (I_E) to flow from the p-type emitter into the n-type silicon bar. When this occurs, charge carriers are injected into the r_{B1} region. Since the resistance of the semiconductor material is dependent on doping, the additional charge carriers cause the resistance of the r_{B1} region to decrease rapidly. The decrease in resistance reduces the voltage drop across r_{B1}, and so the pn-junction is more heavily forward-biased. This in turn results in a greater emitter current and more charge carriers that further reduce the resistance of the r_{B1} region. (The process is termed *regenerative*.) The input voltage is *pulled down*, and the emitter current (I_E) is increased to a limit determined by the V_{EB1} source resistance. The device remains in this *on* condition until the emitter input is open-circuited, or until I_E is reduced to a very low level.

The circuit symbol for a UJT is shown in Fig. 20-39c. As always, the arrowhead points in the conventional current direction for a forward-biased junction. In this case it points from the *p*-type emitter to the *n*-type bar.

UJT Characteristics

A plot of emitter voltage V_{EB1} versus emitter current I_E gives the UJT emitter characteristics. Refer to the UJT terminal voltages and currents shown in Fig. 20-40a and to the equivalent circuit in Fig. 20-39b. Note that when $V_{B1B2} = 0$,

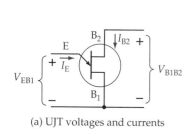

(a) UJT voltages and currents

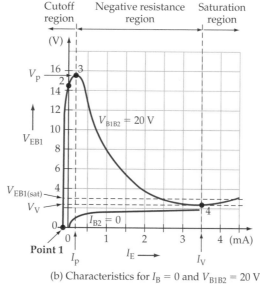

(b) Characteristics for $I_B = 0$ and $V_{B1B2} = 20$ V

Figure 20-40 The UJT characteristics show that the device triggers on at various levels of emitter voltage V_{EB1}, depending upon the level of supply voltage V_{B1B2}.

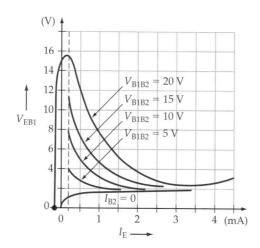

(c) Family of UJT characteristics

$I_{B2} = 0$ and $V_1 = 0$. If V_{EB1} is now increased from zero, the resultant plot of V_{EB1} and I_E is simply the characteristic of a forward-biased diode with some series resistance. This is the characteristic for $I_{B2} = 0$ in Fig. 20-40b.

When V_{B1B2} is 20 V, the level of V_1 (Fig. 20-39b) might be around 15 V, depending on the resistances of r_{B1} and r_{B2}. With $V_{B1B2} = 20$ V and $V_{EB1} = 0$, the emitter junction is reverse-biased and the emitter reverse current I_{E0} flows, as shown at point 1 on the $V_{B1B2} = 20$ V characteristic in Fig. 20-40b. Increasing V_{EB1} until it equals V_1 gives $I_E = 0$; point 2 on the characteristic. A further increase in V_{EB1} forward-biases the emitter junction, and this gives the *peak point* on the characteristic (point 3). At the peak point, V_{EB1} is identified as the *peak voltage* (V_P) and I_E is termed the *peak current* (I_P).

Up until the peak point, the UJT is said to be operating in the *cutoff region* of its characteristics. When V_{EB1} arrives at the peak voltage, charge carriers are injected from the emitter to decrease the resistance of r_{B1}, as already explained. The device enters the *negative resistance region*, r_{B1} falls rapidly to a *saturation resistance* (r_S), and V_{EB1} falls to the *valley voltage* (V_V), (point 4 on the characteristic in Fig. 20-40b). I_E also increases to the *valley current* (I_V) at this time. A further increase in I_E causes the device to enter the *saturation region*, where V_{EB1} equals the sum of V_D and $I_E \times r_S$.

Starting with V_{B1B2} lower than 20 V gives a lower peak point voltage and a different characteristic. Thus, by using various levels of V_{B1B2}, a family of V_{EB1}/I_E characteristics can be plotted for a given UJT, as shown in Fig. 20-40c.

UJT Packages

Two typical UJT packages with the terminal identified are shown in Fig. 20-41. These are similar to low-power BJT packages.

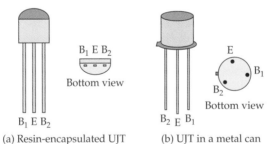

Figure 20-41 UJT packages and terminals.

(a) Resin-encapsulated UJT

(b) UJT in a metal can

UJT Parameters

Interbase Resistance (R_BB): This is the sum of r_{B1} and r_{B2} when I_E is zero. Consider Fig. 20-42, which shows a portion of the manufacturer's data sheet for 2N4949 UJT. R_{BB} is specified as 7 kΩ typical, 4 kΩ minimum, and 12 kΩ maximum. The value of R_{BB}, together with the maximum power dissipation P_D, determines the maximum voltage V_{B1B2} that may be used. When $I_E = 0$,

$$V_{B1B2(\text{max})} = \sqrt{(R_{BB} P_D)} \qquad (20\text{-}8)$$

2N4949 UJT			
	Min	Typ	Max
Interbase resistance (R_{BB})	4 kΩ	7 kΩ	12 kΩ
Intrinsic standoff ratio (η)	0.74		0.86
Emitter saturation voltage ($V_{EB1(sat)}$)		2.5 V	3 V
Peak point current (I_P)		0.6 μA	1 μA
Valley point current (I_V)	2 mA	4 mA	

Figure 20-42 Partial specification for a 2N4949 UJT.

Like all other devices, the P_D of the UJT must be derated for higher temperatures.

Example 20-6
A UJT has $R_{BB(min)} = 4$ kΩ, $P_D = 360$ mW at 25 °C, and a power derating factor $D = 2.4$ mW/°C. Calculate the maximum V_{B1B2} that should be used at a temperature of 100 °C.

Solution

Eq. 8-26:
$$P_{D(100°)} = P_{D(25°)} - [D(T_2 - 25°)]$$
$$= 360 \text{ mW} - [2.4 \text{ mW/°C}(100° - 25°)]$$
$$= 180 \text{ mW}$$

Eq. 20-8:
$$V_{B1B2(max)} = \sqrt{(R_{BB}P_D)} = \sqrt{(4 \text{ kΩ} \times 180 \text{ mW})}$$
$$= 26.8 \text{ V}$$

Intrinsic Standoff Ratio (η): The *intrinsic standoff ratio* is simply the ratio of r_{B1} to R_{BB}. The peak point voltage is determined from η, the supply voltage, and the diode voltage drop:

$$V_P = V_D + \eta V_{B1B2} \tag{20-9}$$

Emitter Saturation Voltage ($V_{EB1(sat)}$): This is the emitter voltage when the UJT is operating in the saturation region of its characteristics; the minimum V_{EB1} level. Because it is affected by the emitter current and the supply voltage, $V_{EB1(sat)}$ is specified for given I_E and V_{B1B2} levels.

Example 20-7
Determine the maximum and minimum triggering voltages for a 2N4949 UJT with $V_{B1B2} = 25$ V.

Solution

From Fig. 20-42, $\quad \eta = 0.74$ minimum, 0.86 maximum

Eq. 20-9: $\quad V_{p(max)} = V_D + (\eta_{max}\ V_{B1B2}) = 0.7\ V + (0.86 \times 25\ V)$

$$= 22.2\ V$$

$$V_{p(min)} = V_D + (\eta_{min}\ V_{B1B2}) = 0.7\ V + (0.74 \times 25\ V)$$

$$= 19.2\ V$$

Peak Point Emitter Current (I_P): I_P is important as a lower limit to the emitter current. If the emitter voltage source resistance is so high that I_E is not greater than I_P, the UJT will simply not trigger into the *on* state. The maximum emitter voltage source resistance is

$$R_{E(max)} = \frac{V_{BB} - V_P}{I_P} \tag{20-10}$$

where V_{BB} is the circuit supply voltage.

Valley Point Current (I_V): I_V is important in some circuits as an upper limit to the emitter current. If the emitter voltage source resistance is so low that I_E is equal to or greater than I_V, the UJT will remain *on* once it is triggered; it will not switch *off*. So the minimum emitter voltage source resistance is

$$R_{E(min)} = \frac{V_{BB} - V_{EB1(sat)}}{I_V} \tag{20-11}$$

UJT Relaxation Oscillator

The relaxation oscillator circuit in Fig. 20-43a consists of a UJT and a capacitor (C_1) charged via resistance R_E. When the capacitor voltage (V_C) reaches V_P, the UJT fires and rapidly discharges C_1 to $V_{EB1(sat)}$. The device then cuts off and the

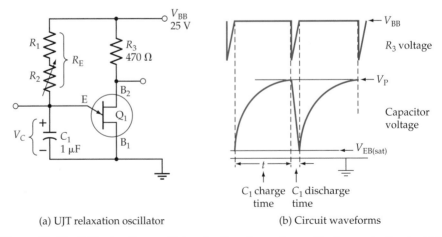

(a) UJT relaxation oscillator　　　　　　(b) Circuit waveforms

Figure 20-43 UJT relaxation oscillator circuit and waveforms. C_1 charges to V_P, at which point the UJT fires, then C_1 is discharged to $V_{BE(sat)}$.

capacitor starts charging again. The cycle is repeated continually, generating a sawtooth waveform across C_1, as illustrated in Fig. 20-43b. The time (t) for the capacitor to charge from $V_{EB1(sat)}$ to V_P can be calculated, and the frequency of the sawtooth determined approximately as $1/t$. The discharge time (t_D) is difficult to calculate because the UJT is in its negative resistance region and its resistance is changing. However, t_D is much less than t, and so it can normally be ignored. Rewriting Equation 20-6 gives

$$t = CR \ln\left[\frac{V_{BB} - V_{EB1}}{V_{BB} - V_P}\right] \qquad (20\text{-}12)$$

Resistor R_3 in the circuit in Fig. 20-43 is included to produce a spike waveform output, as illustrated. When the UJT fires, the current surge through terminal B_2 produces the negative-going voltage spike across R_3. A resistor could also be included in series with terminal B_1 to produce positive-going spikes. Both resistor values should be much lower than the R_{BB} for the UJT.

Example 20-8

Calculate the $R_{E(max)}$ and $R_{E(min)}$ for the relaxation oscillator circuit in Fig. 20-43. The UJT is a 2N4949 with $I_p = 0.6\ \mu A$, $I_V = 2\ mA$, and $V_{EB1(sat)} = 2.5\ V$. Determine the oscillating frequency for the circuit when $R_E = 18\ k\Omega$, $C_1 = 1\ \mu F$, and $V_P = 20\ V$.

Solution

From Example 20-7,

$$V_{P(min)} = 19.2\ V, \text{ and } V_{P(max)} = 22.2\ V$$

Eq. 20-10: $$R_{E(max)} = \frac{V_{BB} - V_{P(max)}}{I_P} = \frac{25\ V - 22.2\ V}{0.6\ \mu A}$$

$$= 4.7\ M\Omega$$

Eq. 20-11: $$R_{E(min)} = \frac{V_{BB} - V_{EB1(sat)}}{I_v} = \frac{25\ V - 2.5\ V}{2\ mA}$$

$$= 11.25\ k\Omega$$

Eq. 20-12: $$t = CR \ln\left[\frac{V_{BB} - V_{EB1(sat)}}{V_{B1B2} - V_P}\right]$$

$$= 1\ \mu F \times 18\ k\Omega \ln\left[\frac{25\ V - 2.5\ V}{25\ V - 20\ V}\right]$$

$$= 27\ ms$$

Eq. 20-12: $$f = \frac{1}{t} = \frac{1}{27\ ms}$$

$$= 37\ Hz$$

UJT Control of an SCR

Unijunction transistors are often used in SCR and TRIAC control circuits. In the typical circuit shown in Fig. 20-44a, diode D_1, resistor R_1, and Zener diode D_2 provide a low-voltage dc supply to the UJT circuit derived from the positive half-cycle of the ac supply voltage. D_1 also isolates the UJT circuit during the supply negative half-cycle. Capacitor C_1 is charged via resistor R_2 to the UJT firing voltage, and the SCR is triggered by the voltage drop across R_3. By adjusting R_2, the charging rate of C_1 and the UJT firing time can be selected. The waveforms in Fig. 20-44b show that 180° of SCR phase control is possible.

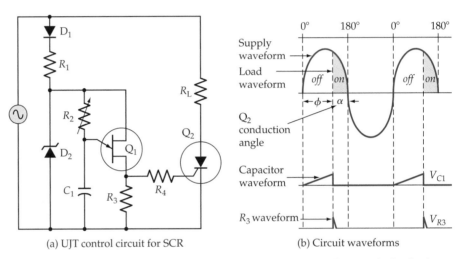

(a) UJT control circuit for SCR   (b) Circuit waveforms

Figure 20-44   UJT 180° phase-control circuit for an SCR. Resistor R_2 controls the C_1 charging rate and the SCR firing point.

Practice Problems

20-7.1 A 2N4870 UJT has the following parameters: P_D = 300 mW at 25°C, D = 3 mW/°C, η = 0.56 to 0.75, R_{BB} = 4 kΩ to 9.1 kΩ, $V_{EB1(sat)}$ = 2.5 V, I_P = 1 μA to 5 μA, and I_V = 2 mA to 5 mA. Determine the maximum V_{B1B2} that may be used at 75°C.

20-7.2 Calculate the $V_{P(max)}$ and $V_{P(min)}$ for a 2N4870 when V_{BB} = 30 V.

20-7.3 A 2N4870 is used in the circuit in Fig. 20-44. If D_2 has V_Z = 30 V, determine the maximum and minimum resistance values for R_2.

20-8 PROGRAMMABLE UNIJUNCTION TRANSISTOR (PUT)

PUT Operation

The *programmable unijunction transistor (PUT)* is actually an SCR-type device used to simulate a UJT. The interbase resistance (R_{BB}) and the intrinsic standoff ratio (η) can be programmed to any desired values by selecting two resistors. This means that the device firing voltage (the peak voltage V_P) can also be programmed.

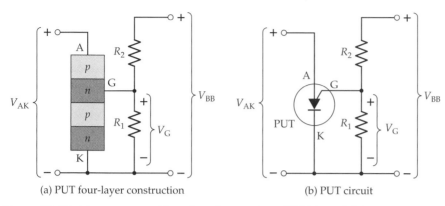

(a) PUT four-layer construction   (b) PUT circuit

Figure 20-45 The programmable unijunction transistor (PUT) is an SCR-type device that can be connected to function like a UJT.

Consider Fig. 20-45a, which shows a four-layer device with its gate connected to the junction of resistors R_1 and R_2. Note that the gate terminal is close to the anode of the device, instead of the cathode as for an SCR. The anode-gate junction becomes forward-biased when the anode is positive with respect to the gate. When this occurs, the device is triggered *on*. The anode-to-cathode voltage then drops to a low level, and the PUT conducts heavily until the current becomes too low to sustain conduction. To simulate the UJT performance, the anode of the device acts as the UJT emitter, and R_1 and R_2 operate as r_{B1} and r_{B2} respectively. Parameters R_{BB}, η, and V_P are programmed by the selection of R_1 and R_2. The four-layer block diagram is replaced with the PUT graphic symbol in Fig. 20-45b. Note that this is the same as the SCR symbol except that the gate terminal is at the anode.

PUT Characteristics

The typical PUT characteristic (V_{AK} plotted versus I_A) shown in Fig. 20-46 is seen to be very similar to UJT characteristics. A small gate reverse current

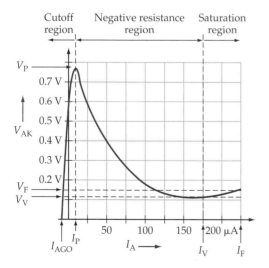

Figure 20-46 The typical V_{AK}/I_A characteristics for a PUT are similar to UJT characteristics.

(I_{AGO}) flows while the anode-gate junction is reverse-biased. At this point the PUT is in the cutoff region of the characteristics. When the anode voltage is raised sufficiently above the gate voltage (V_G in Fig. 20-45), the PUT is triggered into the negative resistance region of its characteristics and the anode-cathode voltage falls rapidly to the valley voltage (V_V). A further increase in I_A causes the device to operate in its saturation region.

PUT Parameters

The intrinsic standoff ratio for the PUT is

$$\eta = \frac{R_1}{R_1 + R_2} \tag{20-13}$$

The gate voltage is simply

$$V_G = \eta V_{BB} \tag{20-14}$$

and the peak voltage is

$$V_P = V_D + \eta V_{BB} \tag{20-15}$$

where V_D is the anode-gate junction voltage, typically 0.7 V.

The gate source resistance (R_G) is an important quantity because it affects the peak current and valley current for the PUT. R_G is the resistance at the junction of voltage divider R_1 and R_2 (Fig. 20-45):

$$R_G = R_1 \| R_2 \tag{20-16}$$

Refer to the partial specification for a 2N6027 PUT in Fig. 20-47, and note the typical quantities. With $R_G = 1\ M\Omega$, $I_P = 1.25\ \mu A$ and $I_V = 18\ \mu A$; with $R_G = 10\ k\Omega$, $I_P = 4\ \mu A$ and $I_V = 150\ \mu A$.

2N6027 PUT			
		Typ	Max
Peak current (I_P)	($R_G = 1\ M\Omega$)	1.25 μA	2 μA
	($R_G = 10\ k\Omega$)	4 μA	5 μA
Valley current (I_V)	($R_G = 1\ M\Omega$)	18 μA	50 μA
	($R_G = 10\ k\Omega$)	150 μA	—
Forward voltage (V_F)	($I_F = 50\ mA$)	0.8 V	1.5 V

Figure 20-47 Partial specification for a programmable unijunction transistor.

PUT Applications

A PUT can be applied in any circuit where a UJT might be used. Figure 20-48 shows a PUT relaxation oscillator used to control an SCR. This circuit operates in essentially the same way as the UJT circuit in Fig. 20-44. It should be noted that there are upper and lower limits to the resistance that can be connected in series with the PUT anode for correct operation of the device. This is similar to the $R_{E(min)}$ and $R_{E(max)}$ requirement for the UJT.

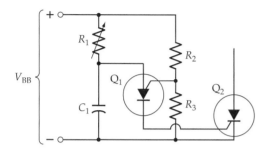

Figure 20-48 SCR phase-control circuit using a PUT relaxation oscillator.

A battery charger circuit using a PUT (Q_1) and an SCR (Q_2) is shown in Fig. 20-49. The ac supply voltage is full-wave-rectified and applied via current-limiting resistor R_5 to the anode of the SCR. The SCR is triggered into conduction by the PUT output coupled via transformer T_1. The PUT gate voltage (V_G) is set by the voltage divider (R_3, R_4, and R_5). While V_G is lower than the Zener diode voltage (V_Z), capacitor C_1 is charged via R_1 to the PUT peak voltage. At this point the PUT fires and triggers the SCR *on*.

As the battery charges, its voltage (E_B) increases, and thus V_G also increases. The increased V_G raises the V_p of the PUT and causes C_1 to take a longer time to charge. Consequently, the SCR is held *off* for a longer portion of the ac supply half-cycle. This means that the average charging current is gradually reduced as the battery approaches full charge. When E_B is fully charged, V_G is raised to

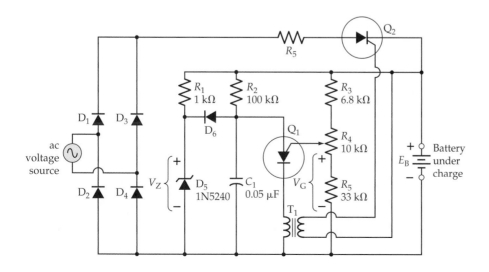

Figure 20-49 SCR battery charger using a PUT control circuit.

the V_Z level, so that D_6 conducts and stops the C_1 voltage from firing the PUT. Thus, the SCR remains *off* and battery charging stops. The circuit will not operate if the battery is connected with the wrong polarity.

Example 20-9

Calculate $V_{P(max)}$ and $V_{P(min)}$ for the PUT in the circuit in Fig. 20-49 when $E_B = 12$ V. Determine the gate bias resistance R_G, and calculate the maximum and minimum resistances for R_2 if the PUT is a 2N6027.

Solution

Eq. 20-13:
$$\eta_{(max)} = \frac{R_4 + R_5}{R_3 + R_4 + R_5} = \frac{10\ k\Omega + 33\ k\Omega}{6.8\ k\Omega + 10\ k\Omega + 33\ k\Omega}$$
$$= 0.86$$

$$\eta_{(min)} = \frac{R_5}{R_3 + R_4 + R_5} = \frac{33\ k\Omega}{6.8\ k\Omega + 10\ k\Omega + 33\ k\Omega}$$
$$= 0.66$$

Eq. 20-15:
$$V_{P(max)} = V_D + (\eta_{(max)}\ V_{BB}) = 0.7\ V + (0.86 \times 12\ V)$$
$$= 11\ V$$

$$V_{P(min)} = V_D + (\eta_{(max)}\ V_{BB}) = 0.7\ V + (0.66 \times 12\ V)$$
$$= 8.6\ V$$

Eq. 20-16:
$$R_G = (R_3 + 0.5\ R_4) \| (R_5 + 0.5\ R_4)$$
$$= (6.8\ k\Omega + 5\ \Omega) \| (33\ k\Omega + 5\ k\Omega)$$
$$= 9\ k\Omega$$

$$R_{2(max)} = \frac{E - V_{P(max)}}{I_P} = \frac{12\ V - 11\ V}{4\ \mu A}$$
$$= 250\ k\Omega$$

$$R_{2(min)} = \frac{E - V_F}{I_V} = \frac{12\ V - 0.8\ V}{150\ \mu A}$$
$$= 74\ k\Omega$$

Practice Problems

20-8.1 The circuit in Fig. 20-48 has $V_{BB} = 15$ V, $R_2 = 12$ kΩ, and $R_3 = 18$ kΩ. The PUT has $I_P = 10\ \mu A$, $I_V = 100\ \mu A$, and $V_F = 1$ V. Calculate R_G, η, V_P, and the maximum and minimum resistances for R_1.

20-8.2 Determine the voltage that the battery will be charged to in Fig. 20-49 when the moving contact is at the middle point on R_4. Assume that the PUT will stop firing when V_{AG} is reduced to 0.5 V.

Review Questions

Section 20-1

20-1 Sketch the construction of a silicon-controlled rectifier. Sketch the two-transistor equivalent circuit and show how it is derived from the SCR construction. Label all terminals and explain how the device operates.

20-2 Sketch typical SCR forward and reverse characteristics. Identify all regions of the characteristics and all important current and voltage levels. Explain the shape of the characteristics in terms of the SCR two-transistor equivalent circuit.

20-3 List the most important SCR parameters and state typical quantities for low-, medium-, and high-current devices.

Section 20-2

20-4 Draw the circuit diagram to show how an SCR can be triggered by the application of a pulse to the gate terminal. Sketch the circuit waveforms and explain its operation.

20-5 Sketch a 90° phase-control circuit for an SCR. Draw the load waveform and explain the operation of the circuit. Show the circuit and load waveforms when the ac supply is full-wave-rectified.

20-6 Draw the diagram for a 90° phase-control circuit using two SCRs for full-wave phase control. Draw the load waveforms and briefly explain the circuit operation.

20-7 Sketch a 180° phase-control for an SCR. Draw the load waveform and explain the circuit operation.

Section 20-3

20-8 Briefly discuss SCR circuit stability and draw diagrams to show methods that can be used to improve stability.

20-9 Draw the diagram for an SCR zero-point triggering circuit. Explain the circuit operation and advantages and draw the load waveform.

20-10 Sketch a circuit that uses an SCR to protect a load from excessive dc supply voltage. Briefly explain.

20-11 Draw a diagram for an SCR heater control circuit using a temperature-sensitive device. Explain the circuit operation.

Section 20-4

20-12 Draw sketches to show the construction, equivalent circuit, and characteristics of a TRIAC. Identify all important voltage and current levels on the characteristics and explain the operation of the device.

20-13 Using diagrams, explain the four quadrant operating conditions for a TRIAC.

20-14 Draw the typical characteristics for a DIAC. Explain the DIAC operation, and sketch the two circuit symbols used for the device.

Section 20-5

20-15 Draw the diagram for a TRIAC 180° phase-control circuit. Draw all waveforms and explain the circuit operation.

20-16 Draw the diagram for an TRIAC zero-point triggering circuit and carefully explain its operation.

20-17 Sketch the functional block diagram for an IC zero voltage switch for TRIAC control. Discuss the components of the block diagram.

Section 20-6

20-18 Using diagrams, briefly explain a silicon unilateral switch (SUS). Draw the device circuit symbol.

20-19 Sketch the equivalent circuit and characteristics for a silicon bilateral switch (SBS). Explain how the device construction differs from that of other thyristors. Discuss the device operation, state typical parameters, and show how the switching voltage can be modified.

20-20 Sketch a relaxation oscillator circuit using an SBS. Draw the output waveform, and explain the circuit operation.

20-21 Draw the circuit symbol and equivalent circuit for a gate turnoff device (GTO) and discuss its operation.

20-22 Draw the circuit symbol and typical characteristics for a SIDAC. Discuss its operation and applications.

Section 20-7

20-23 Draw sketches to show the basic construction and equivalent circuit of a unijunction transistor (UJT). Briefly explain the device operation.

20-24 Sketch typical UJT V_{EB1}/I_E characteristics for $I_{B2} = 0$, $V_{B1B2} = 20$ V, and $V_{B1B2} = 10$ V. Identify each region and all important points on the characteristics, and explain the shape of the characteristics.

20-25 Define the following UJT parameters: intrinsic standoff ratio, interbase resistance, emitter saturation voltage, peak point current, and valley point current.

20-26 Draw the circuit of a UJT relaxation oscillator with provision for frequency adjustment and spike waveform. Show all waveforms, and explain the circuit operation.

20-27 Sketch a UJT circuit for controlling an SCR. Draw all waveforms, and briefly explain how the circuit operates.

Section 20-8

20-28 Draw the basic block diagram and basic circuit for a programmable unijunction transistor (PUT); explain the device operation.

20-29 Sketch typical PUT characteristics, explain how the intrinsic stand-off ratio may be programmed, and identify the most important PUT parameters.

20-30 Draw a basic PUT circuit for controlling an SCR; explain its operation.

20-31 Draw the circuit diagram of a battery charger using a PUT and an SCR. Explain the circuit operation.

Problems

Section 20-1

20-1 Consult 2N6167 and 2N1595 SCR specifications to determine the typical values for V_{DRM}, I_T, V_{TM}, I_H, I_{GT}, and V_{GT}.

Section 20-2

20-2 Choose a suitable SCR from Fig. 20-10 for a circuit with a 115 V ac supply. Calculate the minimum load resistance that can be supplied, and determine the instantaneous voltage level when the SCR switches off.

20-3 A 33 Ω resistor is supplied from an ac source with a 60 V peak level. Current to the load is to be switched on and off by an SCR. Select a suitable device from the specifications in data sheet A-17 in Appendix A, and determine the instantaneous supply voltage at which the SCR switches off.

20-4 An SCR with a 115 V ac supply controls the current through a 150 Ω load resistor. A 90° phase-control circuit (as in Fig. 20-11) is employed to trigger the SCR between 12° and 90°. The gate trigger current is 50 μA, and the trigger voltage is 0.5 V. Calculate suitable resistor values.

20-5 The circuit in Fig. 20-12 has a 50 V ac supply and $R_L = 20\ \Omega$. Determine suitable resistance values for R_1, R_2, and R_3 such that the SCR will be triggered anywhere between 7.5° and 90°. The gate trigger current and voltage are 500 μA and 0.6 V.

20-6 The circuit in Problem 20-5 uses a 2N5170 SCR (see data sheet A-17 in Appendix A). Calculate the instantaneous supply voltage level when the SCR switches *off*.

20-7 Design the circuit in Fig. 20-13 for 10° to 90° phase control; specify the SCRs. The ac supply is 115 V, $R_L = 12\ \Omega$, and the SCRs have $V_G = 0.6$ V and $I_G = 100\ \mu$A.

20-8 A 180° phase-control circuit as in Fig. 20-15 has a 40 V, 400 Hz ac supply and $R_L = 22\ \Omega$. The SCR has $V_G = 0.5$ V, $I_G = 60\ \mu$A, and $I_{GM} = 20$ mA. Determine suitable resistor and capacitor values, and specify the SCRs and the diodes.

Section 20-3

20-9 An SCR crowbar circuit (as in Fig. 20-19) is connected to a 12 V dc supply with a 200 mA current limiter. Design the crowbar circuit to protect the load from voltage levels greater than 13.5 V. Assume that $V_G = 0.7$ V for the SCR.

20-10 A zero-point triggering circuit (as in Fig. 20-18) is to control the power dissipation in a 12 Ω load resistor with a 115 V, 60 Hz ac supply. Assuming that the SCRs have $V_G = 0.5$ V and $I_{G(min)} = 10$ mA, determine suitable capacitor and resistor values.

20-11 An SCR heater control circuit (as in Fig. 20-20) is to switch *on* at 68°C and *off* at 71°C. The circuit is to operate from a 50 V, 60 Hz supply, and the available temperature-sensitive device has a resistance of 500 Ω at 68°C and 350 Ω at 71°C. The load resistance is $R_L = 2.5\ \Omega$ and the SCR has $V_G = 0.6$ V. Determine suitable component values, and calculate the SCR gate voltage at 71°C.

20-12 Specify the SCRs required for the circuits in Problems 20-9, 10, and 11 in terms of maximum anode-cathode voltage and maximum anode current.

Section 20-4

20-13 Consult specifications for 2N6071 and 2N6343 TRIACs to determine the maximum supply voltage, maximum rms current, and the typical quadrant *I* gate triggering voltage.

Section 20-5

20-14 A light dimmer uses a TRIAC 180° phase-control circuit, as in Fig. 20-27. The ac supply is 220 V, 60 Hz, and the total load is 750 W. The TRIAC has triggering current $I_G = 200\ \mu A$, maximum gate current $I_{GM} = 50$ mA, and $V_G = 0.7$ V. The DIAC has $V_S = 9.2$ V and $I_S = 400\ \mu A$. Determine suitable component values.

20-15 The TRIAC control circuit in Fig. 20-27 has the following components: $R_L = 100\ \Omega$, $R_1 = 10$ kΩ, $R_2 = 500\ \Omega$, $C_1 = 3.9\ \mu F$. The DIAC has $V_S = 7$ V, and the TRIAC has $V_G = 1$ V. The ac supply is 60 V, 60 Hz. Determine the minimum conduction angle for the TRIAC.

20-16 Specify the TRIACs required for the circuits in Problems 20-14 and 20-15.

20-17 The TRIAC zero-point switching circuit in Fig. 20-28 uses an SCR with $I_G = 100\ \mu A$ and $V_F = 1.5$ V. The TRIAC has $I_G = 5$ mA, $I_{GM} = 500$ mA, and $t_{on} = 50\ \mu s$. The control voltage is E = 5 V, the ac source is 60 V, 60 Hz, and the load is $R_L = 15\ \Omega$. Determine suitable component values.

Section 20-6

20-18 A relaxation oscillator (as in Fig. 20-35) uses an SBS with $V_S = 8$ V, $V_F = 1$ V, $I_S = 300\ \mu A$, and $I_H = 1$ mA. The dc supply is E = 40 V. Calculate maximum and minimum R_1 values for correct operation of the circuit.

20-19 A relaxation oscillator (as in Fig. 20-35) has a 25 V supply, a 1 μF capacitor, and a 12 kΩ series resistor. The capacitor is to charge up to 15 V and then discharge to approximately 1 V. Specify the required SBS in terms of forward conduction voltage, switching voltage, switching current, and holding current.

20-20 The phase-control circuit in Fig. 20-34 has the following components: $R_L = 18\ \Omega$, $R_1 = 12$ kΩ, $R_2 = 470\ \Omega$, and $C_1 = 10\ \mu F$. Resistors R_3 and R_4 are replaced with 3.3 V Zener diodes, and the TRIAC triggering voltage is $V_G = 1$ V. The ac supply is 115 V, 60 Hz. Determine the TRIAC minimum conduction angle.

Section 20-7

20-21 Determine the maximum power dissipation for a 2N2647 UJT at an ambient temperature of 70°C. Calculate the maximum V_{B1B2} that may be used at 70°C. Data sheet A-18 in Appendix A gives partial specification for the 2N2647.

20-22 Calculate the minimum and maximum V_{EB1} triggering levels for a 2N2647 UJT when $V_{B1B2} = 20$ V.

20-23 A relaxation oscillator (as in Fig. 20-43) uses a 2N2647 UJT with $V_{BB} = 25$ V. Calculate the typical oscillation frequency if $C_1 = 0.5$ µF and $R_E = 3.3$ kΩ.

20-24 Calculate the maximum and minimum charging resistance values that can be used in the circuit of Problem 20-23.

20-25 The UJT phase-control circuit in Fig. 20-44 has a 115 V, 60 Hz ac supply, and an SCR with $V_G = 1$ V and $I_{GM} = 25$ mA. Design the circuit to use a 2N2647 UJT and a Zener diode with $V_Z \approx 15$ V.

Section 20-8

20-26 A PUT operating from a 25 V supply has $V_F = 1.5$ V and $I_G = 50$ µA. Determine values for R_1 and R_2 program η to 0.75. Calculate V_P, V_V, R_{BB}, and R_G.

20-27 A PUT relaxation oscillator (as in Fig. 20-48) has a 20 V supply and a 0.68 µF capacitor. The PUT has $V_G = 1$ V and $I_G = 100$ µA. The peak capacitor voltage is to be 5 V and the oscillating frequency is to be 300 Hz. Determine suitable resistor values.

20-28 The UJT in Problem 20-23 is to be replaced with a PUT. Determine suitable resistance values for the gate bias voltage divider.

20-29 A PUT has a forward voltage of $V_F = 0.9$ V and $I_G = 200$ µA. The device is to be programmed to switch on at $V_G = 15$ V when operating from a 24 V supply. Determine values for R_1 and R_2, and calculate V_P, V_V, R_{BB}, and R_G.

Practice Problem Answers

20-2.1	163 V, 1.6 A, 18 kΩ, 1.5 kΩ, 180 Ω
20-2.2	10 kΩ, 0.82 µF, 1.5 kΩ
20-3.1	(68 kΩ + 12 kΩ), 4 µF
20-5.1	1 kΩ, 180 Ω, 5.6 kΩ, 1 kΩ, 2.5 µF, 1 µF
20-6.1	500 kΩ, 820 Ω
20-7.1	24.5 V
20-7.2	23.2 V, 17.5 V
20-7.3	1.36 MΩ, 6.25 kΩ
20-8.1	7.2 kΩ, 0.6, 9.7 V, 530 kΩ, 140 kΩ
20-8.2	13.4 V

CHAPTER 21
Optoelectronic Devices

CONTENTS

Objectives

You will be able to:

1   Define important illumination units and calculate illumination intensity at a given distance from a source.

2   Explain the fabrication and operation of a light-emitting diode (LED), discuss its parameters, and design LED circuits.

3   Explain the operation of liquid crystal displays (LCDs), show how LCDs and LEDs are used in seven-segment numerical displays, and calculate power dissipations in each type of display.

4   Explain the construction and operation of a photoconductive cell, sketch the device characteristics, and discuss its parameters.

5   Design photoconductive cell circuits to bias BJTs on or off, energize relays, trigger Schmitt circuits, and so on.

6   Explain the construction and operation of a photodiode, sketch photodiode characteristics, and discuss its parameters.

7   Design photodiodes into circuits where they operate as a photoconductive devices and as photovoltaic devices.

8   Explain solar cells and design solar cell battery charger circuits.

9   Explain the operation of a phototransistor, sketch phototransistor characteristics, and discuss its parameters. Explain photodarlingtons and photo-FETs.

10 Design phototransistor circuits for energizing relays, triggering SCRs, and so on.

11 Explain the construction and operation of optocouplers and discuss optocoupler parameters.

12 Design optocoupler circuits for coupling pulse-type and linear signals between systems with different supply voltages.

INTRODUCTION

Optoelectronic devices emit light, modify light, have their resistance affected by light, or produce currents and voltages proportional to light intensity.

Light-emitting diodes (LEDs) produce light and are generally used as indicating lamps and in numerical displays. Liquid crystal displays (LCDs), which modify light, are also used as numerical displays. Photoconductive cells have a resistance that depends upon illumination intensity. They are used in circuits designed to produce an output change when the light level changes. The current and voltage levels in photodiodes and phototransistors are affected by illumination. These devices are also used in circuits that have their conditions altered by changes in light levels. Illumination is converted into electrical energy by means of solar cells, and this energy is often used to charge storage batteries. Optocouplers combine LEDs and phototransistors to provide a means of coupling between circuits that have different supply voltages while maintaining a high level of electrical insolation.

21-1 LIGHT UNITS

The total light energy output, or *luminous flux* (ϕ_s), from a source can be measured in *milliwatts* (mW) or in *lumens* (lm), where 1 lm = 1.496 mW. The *luminous intensity* (E_s) (also termed *illuminance*) of a light source is defined as the luminous flux density per unit solid angle (or cone) emitting from the source (see Fig. 21-1a). This is measured in *candelas* (cd), where one candela is equal to one lumen per unit solid angle (assuming a point source that emits light evenly in all directions).

$$E_s = \frac{\phi_s}{4\pi} \qquad \text{(21-1)}$$

The light intensity (E_A) on an area at a given distance from the source is determined from the surface area of a sphere surrounding the source (Fig. 21-1b).

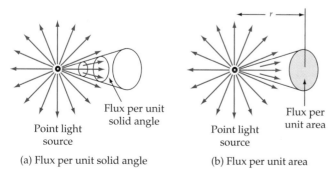

(a) Flux per unit solid angle   (b) Flux per unit area

Figure 21-1   Light intensity can be expressed in flux per unit solid angle or in flux per unit area.

At a distance of r metres, the luminous flux is spread over a spherical area of $4\pi r^2$ square metres. Therefore,

$$E_A = \frac{\phi_s}{4\pi r^2}$$

(21-2)

When the total flux is expressed in lumens, Equation 21-2 gives the luminous intensity in *lumens per square meter* (lm/m^2), also termed *lux* (lx). Comparing Equation 21-2 to Equation 21-1, shows that the luminous intensity per unit area at any distance r from a point source is determined by dividing the source intensity by r^2.

Luminous intensity can also be measured in *milliwatts per square centimetre* (mW/cm^2), or *lumens per square foot* (lm/ft^2), also known as a *foot candle* (fc), where 1 fc = 10.764 lx.

The light intensity of sunlight on the earth at noon on a clear day is approximately 107,640 lx, or 161 W/m^2. The light intensity from a 100 W lamp is approximately 4.8×10^3 cd (allowing for a 90% lamp efficiency). At a distance of 2 m, this is 1.2×10^3 lx. An indicating lamp with a 3 mcd output can be clearly seen at a distance of several metres in normal room lighting conditions.

Example 21-1

Calculate the light intensity 3 m from a lamp that emits 25 W of light energy. Determine the total luminous flux striking an area of 0.25 m^2 at 3 m from the lamp.

Solution

Eq. 21-2: $\qquad E_A = \dfrac{\phi_s}{4\pi r^2} = \dfrac{25 \text{ W}}{4\pi \times (3 \text{ m})^2}$

$$= 0.221 \text{ W/m}^2 = 221 \text{ mW/m}^2$$

Total flux $= E_A \times$ area $= 221 \text{ mW/m}^2 \times 0.25 \text{ m}^2$

$$\approx 55 \text{ mW}$$

Example 21-2

Determine the light intensity at a distance of 2 m from a 10 mcd source.

Solution

$$E_A = \frac{E_s \text{ (cd)}}{r^2} = \frac{10 \text{ mcd}}{2^2}$$

$$= 2.5 \text{ mlx}$$

Light energy is electromagnetic radiation; that is, it is in the form of electromagnetic waves. Thus it can be defined in terms of *frequency* or *wavelength*, as well as intensity. Wavelength, frequency, and velocity are related by the equation

$$c = f\lambda \qquad \text{(21-3)}$$

where c = velocity = 3×10^8 m/s for electromagnetic waves
   f = frequency in Hz
   λ = wavelength in metres

The wavelength of visible light ranges from violet at approximately 380 nm (*nanometres*) to red at 720 nm. From Eq. 21-3 , the frequency extremes are

$$f_{\text{violet}} = \frac{c}{\lambda} = \frac{3 \times 10^8 \text{ m/s}}{380 \text{ nm}}$$

$$\approx 8 \times 10^{14} \text{ Hz}$$

$$f_{\text{red}} = \frac{c}{\lambda} = \frac{3 \times 10^8 \text{ m/s}}{720 \text{ nm}}$$

$$\approx 4 \times 10^{14} \text{ Hz}$$

Practice Problems

21-1.1 A lamp is required to produce a light intensity of 213 lx at a distance of 5 m. Calculate the total light energy output of the lamp in watts.

21-1.2 Calculate the frequency of yellow light with a 585 nm wavelength.

21-1.3 Determine the light intensity 3.3 m from an 8 mcd lamp, and the total luminous flux striking a 4 cm^2 area at that location.

21-2 LIGHT-EMITTING DIODES (LED)

LED Operation and Construction

Charge carrier recombination occurs at a forward-biased *pn*-junction as electrons cross from the *n*-side and recombine with holes on the *p*-side. Free electrons have a higher energy level than holes, and some of this energy is dissipated in the form of heat and light when recombination takes place. If the semiconductor material is translucent, the light is emitted and the junction becomes a light source, that is, a *light-emitting diode (LED)*.

A cross-sectional view of an LED junction is shown in Fig. 21-2a. The semiconductor material is gallium arsenide (GaAs), gallium arsenide phosphide (GaAsP), or gallium phosphide (GaP). An *n*-type epitaxial layer is grown upon a substrate, and the *p*-region is created by diffusion. Since charge carrier recombinations occur in the *p*-region, the *p*-region is kept uppermost to allow

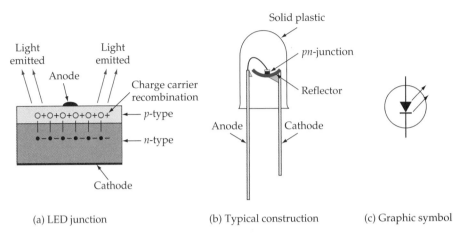

(a) LED junction   (b) Typical construction   (c) Graphic symbol

Figure 21-2   A light-emitting diode (LED) produces energy in the form of light when charge carrier recombination occurs at the *pn*-junction.

the light to escape. The metal film anode connection is patterned to allow most of the light to be emitted. A gold film is applied to the bottom of the substrate to reflect as much light as possible toward the surface of the device and to provide a cathode connection. LEDs made from GaAs emit infrared (invisible) radiation. GaAsP material produces either red light or yellow light, while red or green light can be created by using GaP. Through the use of various materials, LEDs can be manufactured to produce light of virtually any colour.

Figure 21-2b shows the typical construction of a LED. The *pn*-junction is mounted on a cup-shaped reflector, wires are provided for anode and cathode connection, and the device is encapsulated in an epoxy lens. The lens can be colourless or coloured, and (when not energized), the lens colour identifies the LED light colour. The colour of the light emitted by the energized LED is determined solely by the *pn*-junction material. Some LEDs have glass particles embedded in the epoxy lens to diffuse the emitted light and increase the viewing angle of the device. The LED circuit symbol is shown in Fig. 21-2c. Note that the arrow directions indicate emitted light.

Characteristics and Parameters

LED characteristics are similar to those of other semiconductor diodes, except that (as shown in the partial specification in Fig. 21-3) the typical forward voltage drop is 1.6 V. Note also that the reverse breakdown voltage can be as low as 3 V. In some circuits it is necessary to include a diode with a high reverse breakdown voltage in series with a LED. The forward current used with a LED is usually in the 10 mA to 20 mA range, but (depending on the particular device) the peak current can be as high as 90 mA. LED luminous intensity depends on the forward current level; it is usually specified at 20 mA. The peak wavelength of the light output is also normally listed on the specification.

Typical LED Specification			
	Min	Typ	Max
Luminous intensity (I_V at 20 mA)	4 mcd	8 mcd	
Forward voltage (V_F)	1.4 V	1.6 V	2.0 V
Reverse breakdown voltage (V_{FBR})	3 V	10 V	
Peak forward current ($I_{F(max)}$)		90 mA	
Average forward current ($I_{F(av)}$)		20 mA	
Power dissipation (P_D)		100 mW	
Response speed (τ_S)		90 ns	
Peak wavelength (λ_p)		660 nm	

Figure 21-3 Partial specification for a typical light-emitting diode.

LED Circuits

As explained, an LED is a semiconductor diode that emits light when a for-ward current is passed through the device. A single LED might be employed simply as a supply voltage on/off indicator, as illustrated in Fig. 21-4a. A series-connected resistor (R_1) must be included to limit the current to the desired level. Figure 21-4b shows an LED connected at the output of a com-parator to indicate a *high* output voltage. As well as the current-limiting resis-tor (R_1), an ordinary semiconductor diode (D_1) is connected in series with the LED to protect it from an excessive reverse voltage when the comparator output is negative.

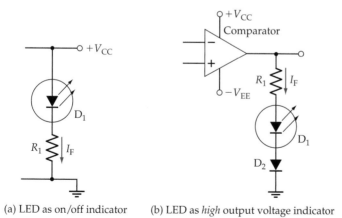

(a) LED as on/off indicator   (b) LED as *high* output voltage indicator

Figure 21-4 Light-emitting diode indicator circuits.

LEDs are often controlled by a BJT, as illustrated in Fig. 21-5a and b. Tran-sistor Q_1 in Fig. 21-5a is switched into saturation by the input voltage (V_B). Resistor R_1 limits the transistor base current, and R_2 limits the LED current. In Fig. 21-5b, the emitter resistor (R_1) limits the LED current to $(V_B - V_{BE})/R_1$.

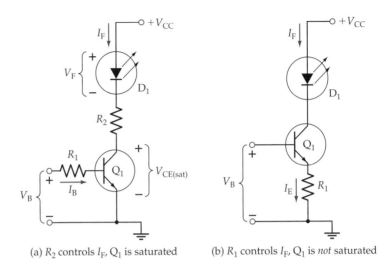

(a) R_2 controls I_F, Q_1 is saturated   (b) R_1 controls I_F, Q_1 is *not* saturated

Figure 21-5   BJT control circuits for light-emitting diodes.

Example 21-3

The LED in Fig. 21-5a is to have a forward current of approximately 10 mA. The circuit voltages are $V_{CC} = 9$ V, $V_F = 1.6$ V, and $V_B = 7$ V; and Q_1 has $h_{FE(min)} = 100$. Calculate suitable resistance values for R_1 and R_2.

Solution

$$R_2 = \frac{V_{CC} - V_F - V_{CE(sat)}}{I_C} = \frac{9\text{ V} - 1.6\text{ V} - 0.2\text{ V}}{10\text{ mA}}$$

$$= 720\ \Omega \text{ (use 680 } \Omega \text{ standard value)}$$

I_C becomes

$$I_C = \frac{V_{CC} - V_F - V_{CE(sat)}}{R_2} = \frac{9\text{ V} - 1.6\text{ V} - 0.2\text{ V}}{680\ \Omega}$$

$$= 10.6\text{ mA}$$

$$I_B = \frac{I_C}{h_{FE(min)}} = \frac{10.6\text{ mA}}{100}$$

$$= 106\ \mu\text{A}$$

$$R_B = \frac{V_B - V_{BE}}{I_B} = \frac{7\text{ V} - 0.7\text{ V}}{106\ \mu\text{A}}$$

$$= 59\text{ k}\Omega \text{ (use 56 k}\Omega \text{ standard value)}$$

Practice Problems

21-2.1 The LED in the circuit in Fig. 21-5b is to pass a 20 mA current. The circuit voltages are $V_{CC} = 15$ V, $V_F = 1.9$ V, and $V_B = 5$ V. Determine a suitable resistance for R_1, and calculate V_{CE} for Q_1.

21-2.2 Determine suitable resistances for the circuits in Fig. 21-4 to give $I_F = 15$ mA. The LEDs have $V_F = 1.8$ V. Also, $V_{CC} = 6$ V in Fig 21-4a, and in Fig. 21-4b the op-amp output is $V_o = \pm 9$ V.

21-3 SEVEN-SEGMENT DISPLAYS

LED Seven-Segment Display

The arrangement of a *seven-segment LED* numerical display is shown in Fig. 21-6a. Since the actual LED devices are very small, solid plastic *light pipes* are often employed to enlarge the lighted surface, as shown. Any numeral from 0 to 9 can be indicated by passing current through the appropriate segments (Fig 21-6b). Part (c) in Fig. 21-6 shows three seven-segment displays together with a two-segment display referred to as a *half-digit*. The whole display, termed a *three-and-a-half digit display*, can be used to indicate numerical values up to a maximum of 1999.

The LEDs in a seven-segment display may be connected in *common-anode* or in *common-cathode* configuration (Fig. 21-6d). When an LED seven-segment display is selected, it is important to determine which of the two connecting arrangements is required.

The relatively large amounts of current consumed by LED seven-segment displays are their major disadvantage. Apart from this, LEDs have the advantage of long life and ruggedness.

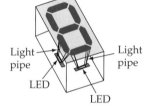

Light pipe — LED — Light pipe

LED

(a) Seven-segment LED display

(b) Seven-segment numeral displays

(c) Three-and-a-half digit display

Example 21-4

Calculate the total power supplied to a $3\frac{1}{2}$ digit LED display when it indicates 1888. A 5 V supply is used, and each LED has a 10 mA current.

Solution

Total segments: $N = [3 \times (\text{segments for } 8)] + [1 \times (\text{segments for } 1)]$

$= (3 \times 7) + (1 \times 2) = 23$

Total current: $I_T = N \times (\text{current per segment}) = 23 \times 10$ mA

$= 230$ mA

Power: $P = I_T \times V_{CC} = 230 \text{ mA} \times 5 \text{ V}$

$= 1.15$ W

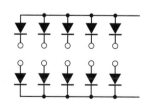

(d) Common-anode and common-cathode connections

Figure   21-6 Light-emitting diodes can be arranged in seven-segment format for numerical displays.

Liquid Crystal Cells

Liquid crystal material is a liquid that exhibits some of the properties of a solid. The molecules in ordinary liquids normally have random orientations. In liquid crystals, however, the molecules are oriented in a definite crystal pattern.

A liquid crystal cell consists of a very thin layer of liquid crystal material sandwiched between glass sheets, as illustrated in Fig. 21-7a. The glass sheets have transparent metal film electrodes deposited on the inside surfaces. In the commonly used *twisted nematic* cell, two thin, polarizing optical filters are placed at the surface of each glass sheet. The liquid crystal material employed twists the light passing through when the cell is not energized. This twisting allows the light to pass through the polarizing filters, so that the cell is semi-transparent. When energized, the liquid molecules are reoriented so that no twisting occurs, and no light can pass through. Thus, the energized cell can appear dark against a bright background (see Fig. 21-7b and c). The cells can also be manufactured to appear bright against a dark background.

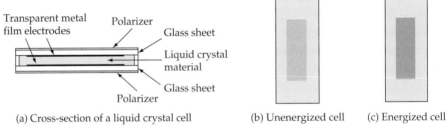

(a) Cross-section of a liquid crystal cell   (b) Unenergized cell   (c) Energized cell

Figure 21-7 A liquid crystal cell consists of a layer of liquid crystal material sandwiched between glass sheets with transparent metal film electrodes and polarizers.

Seven-segment numerical (and other type) displays made from liquid crystal cells are referred to as *liquid crystal displays (LCDs)*. Two types of LCDs are illustrated in Fig. 21-8. The *reflective*-type shown in Fig. 21-8a relies on reflected light. The cell is placed on a reflective surface, so that when not energized it is just as reflective as the surrounding material; consequently, it disappears. When the cell is energized, no light is reflected from it, and it appears dark against the bright background. The *transmittive* cell in Fig. 21-8b allows light to pass through from the back of the cell when it is not energized. When the cell is energized, the light is blocked, and here again the cell appears dark against a bright background. The *trans-reflective* cell is a combination of transmittive and reflective types.

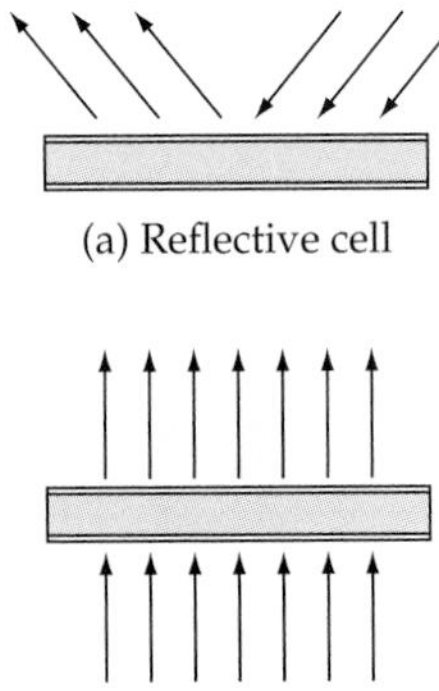

(a) Reflective cell

(b) Transmittive cell

Figure 21-8 Reflective-type liquid crystal cells reflect incident light. Transmittive-type cells pass light from behind.

LCD Seven-Segment Displays

Because liquid crystal cells are light reflectors or transmitters rather than light generators, they consume very small quantities of energy. The only energy required by the cell is that needed to activate the liquid crystal. The total current flow through four small seven-segment LCDs is usually about 20 μA. However, LCDs require an ac voltage supply, in the form of either a sine wave or a square wave. This is because a continuous direct current produces a plating of the cell electrodes that could damage the device. Repeated reversal of the current prevents this problem.

A typical LCD supply is a 3 V to 8 V peak-to-peak square wave with a frequency of 60 Hz. Figure 21-9 illustrates the square-wave drive method. The *back plane*, which is common to all of the cells, is supplied with a square wave (with peak voltage V_p). Similar square waves applied to each of the other terminals are either in phase or in antiphase with the back plane square wave. Those cells with waveforms in phase with the back plane waveform (cells e and f in Figure 21-9) have no voltage developed across them. Since both terminals of the segment are at the same potential, they are not energized. The cells with square waves in antiphase with the back plane input have a square wave with peak voltage $2V_p$ developed across them, and consequently, they are energized.

Unlike LED displays, which are usually quite small, LCDs can be fabricated in almost any convenient size. The major advantage of LCDs is their low power consumption. Perhaps the major disadvantage of the LCD is its decay time of 150 ms (or more). This is very slow compared to the rise and fall times of LEDs. In fact, the human eye can sometimes observe the fading out of LCD segments switching off.

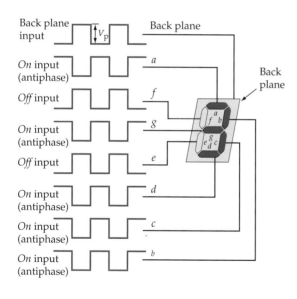

Figure 21-9 Liquid crystal display using a square wave supply. A segment is energized when its input is in antiphase with the back plane input.

21-4 PHOTOCONDUCTIVE CELLS

Cell Construction

Light striking the surface of a material can provide sufficient energy to cause electrons within the material to break away from their atoms. Thus, free electrons and holes (*charge carriers*) are created within the material, and consequently its resistance is reduced. This is known as the *photoconductive effect*.

The construction of a typical photoconductive cell is illustrated in Fig. 21-10a, and the graphic symbol is shown in Fig. 21-10b. Light-sensitive material is arranged in the form of a long strip zigzagged across a disc-shaped base. The connecting terminals are fitted to the conducting material on each side of the strip; they are *not* at the ends of the strip. Thus, the light-sensitive material is actually a short, wide strip between the two conductors. For added protection, a transparent plastic cover is usually included.

Cadmium sulfide (CdS) and cadmium selenide (CdSe) are the two materials normally used in photoconductive cell manufacture. Both respond rather slowly to changes in light intensity. For cadmium selenide, the response time (t_{res}) is about 10 ms, while for cadmium sulfide it may be as long as 100 ms. Temperature sensitivity is another important difference between the two materials. There is a large change in the resistance of a cadmium selenide cell with changes in ambient temperature, but the resistance of cadmium sulfide remains relatively stable. As with all other devices, care must be taken to ensure that the power dissipation is not excessive. The *spectral response* of a cadmium sulfide cell is similar to that of the human eye: it responds to visible light. For a cadmium selenide cell, the spectral response is at the longer wavelength end of the visible spectrum and extends into the infrared region.

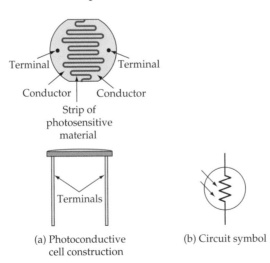

Figure 21-10 A photoconductive cell consists of a strip of light-sensitive material situated between two conductors.

Terminal   Terminal

Conductor   Conductor

Strip of photosensitive material

Terminals

(a) Photoconductive cell construction

(b) Circuit symbol

Characteristics and Parameters

Typical illumination characteristics for a photoconductive cell are shown in Fig. 21-11. It is seen that, when the cell is not illuminated, its resistance can be greater than 100 kΩ. This is known as the *dark resistance* of the cell. When the

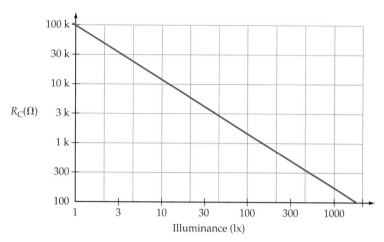

Figure 21-11 Typical photoconductive cell illumination characteristics. The resistance is usually very high when the cell is dark, and relatively low when the cell is illuminated.

cell is illuminated, its resistance may fall to a few hundred ohms. Note that the scales on the illumination characteristic are logarithmic.

A typical photoconductive cell specification is shown in Fig. 21-12. As well as maximum voltage and power dissipation, the cell dark resistance and the resistance at a 10 lx illumination are listed. Note the wide range of cell resistance at 10 lx. The light wavelength that gives peak response (λP) is also given on the specification. Cell *sensitivity* is sometimes used; this is simply the cell current for a given voltage and given level of illumination.

Typical Photoconductive Cell Specification				
PD (mW)	Max ac (V)	Dark resistance	λP (nm)	Resistance at 10 lx
200	180	100 kΩ	550	6 kΩ (min) 18 kΩ (max)

Figure 21-12 Partial specification for a typical photoconductive cell.

Applications
Figure 21-13 shows a photoconductive cell used for relay control. When the cell is illuminated, its resistance is low and the relay current is at its maximum. Thus the relay is energized. When the cell is dark, its high resistance keeps the current too low to energize the relay. Resistance R_1 is included to limit the relay current to the desired level when the cell resistance is low.

Example 21-5
A relay is to be controlled by a photoconductive cell as in Fig. 21-13. The cell has the characteristics shown in Fig. 21-11. The relay is to be supplied with 10 mA from a 30 V supply when the cell is illuminated with about 200 lm/m². Calculate the required series resistance and the level of the dark current. Assume that the coil resistance is much smaller than R_1 and R_C.

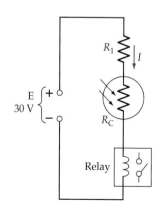

Figure 21-13 Circuit of a relay controlled by a photoconductive cell. The relay is energized when the cell is illuminated.

Solution

From the characteristics, at $200 \ \text{lm/m}^2$: $R_C \approx 1 \ \text{k}\Omega$

When the cell is illuminated,

$$I = \frac{E}{R_1 + R_C}$$

or

$$R_1 = \frac{E}{I} - R_C = \frac{30 \ \text{V}}{10 \ \text{mA}} - 1 \ \text{k}\Omega$$

$$= 2 \ \text{k}\Omega \ \text{(use } 1.8 \ \text{k}\Omega \ \text{standard value)}$$

From the characteristics:

When dark, $R_C \approx 100 \ \text{k}\Omega$

Dark current: $I = \dfrac{E}{R_1 + R_C} = \dfrac{30 \ \text{V}}{1.8 \ \text{k}\Omega + 100 \ \text{k}\Omega}$

$$\approx 0.3 \ \text{mA}$$

Photoconductive cells employed to switch transistors on and off are shown in Fig. 21-14. When the cell in Fig. 21-14a is dark, the cell resistance (R_C) is high. Consequently, the transistor base is biased above its emitter voltage level, and Q_1 is turned *on*. When the cell is illuminated, its resistance is reduced, and the lower cell resistance in series with R_1 biases the transistor base below its emitter voltage level. Thus Q_1 is turned *off* when the cell is illuminated.

In Fig. 21-14b, Q_1 is biased *off* when the cell is dark, because R_C is high. When the cell is illuminated, its reduced resistance causes Q_1 to be biased *on*.

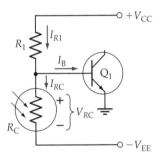

(a) Circuit to switch a BJT *off* when a cell is illuminated

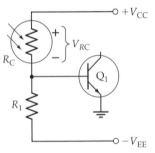

(b) Circuit to switch a BJT *on* when a cell is illuminated

Figure 21-14   BJTs controlled by photoconductive cells. The position of the cell in the circuit determines whether the transistor is switched on or off by an increase in light level.

Example 21-6

The transistor in Fig. 21-14a is to be biased *on* when the photoconductive cell is dark, and *off* when it is illuminated. The supply is ±6 V, and the transistor base current is to be 200 μA when *on*. Design the circuit using the photoconductive cell characteristics in Fig. 21-11. Determine the minimum light level when the transistor is *off*.

Solution

When cell is dark, $R_C \approx 100 \ \text{k}\Omega$

When Q_1 is on, $V_{RC} = V_{EE} + V_{BE} = 6 \ \text{V} + 0.7 \ \text{V}$

$$= 6.7 \ \text{V}$$

$$I_{RC} = \frac{V_{RC}}{R_C} = \frac{6.7 \ \text{V}}{100 \ \text{k}\Omega}$$

$$= 67 \ \mu\text{A}$$

$$I_{R1} = I_{RC} + I_B = 67 \ \mu A + 200 \ \mu A$$

$$= 267 \ \mu A$$

$$V_{R1} = V_{CC} - V_B = 6 \ V - 0.7 \ V$$

$$= 5.3 \ V$$

$$R_1 = \frac{V_{R1}}{I_{R1}} = \frac{5.3 \ V}{267 \ \mu A}$$

$$\approx 20 \ k\Omega \text{ (use 18 k}\Omega \text{ standard value)}$$

When Q_1 is off, $V_{R1} > 6 \ V$

and

$$I_{R1} = \frac{V_{R1}}{R_1} = \frac{6 \ V}{18 \ k\Omega}$$

$$= 333 \ \mu A$$

$$R_C = \frac{V_{RC}}{I_{R1}} = \frac{6 \ V}{333 \ \mu A}$$

$$= 18 \ k\Omega$$

From the characteristics:

$$\text{when } R_C \approx 18 \ k\Omega, \text{ illumination} \approx 7 \ lm/m^2$$

Figure 21-15 shows a photoconductive cell used with an op-amp Schmitt trigger circuit (see Section 14-10). When the cell resistance is low (cell illuminated), the voltage across R_1 is higher than the upper trigger point (UTP) for the Schmitt. Consequently, the op-amp output is *low* (negative). The output switches to a *high* (positive) level when V_{R1} falls to the Schmitt circuit lower trigger point (LTP). This occurs when the cell illumination level falls, causing R_C to rise.

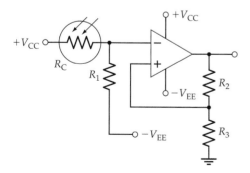

Figure 21-15 Op-amp Schmitt trigger circuit with a photoconductive cell for detecting light level.

The circuit in Fig. 21-16 uses a photoconductive cell to control the current level in an LED. The LED current is low when the ambient light level is low, because the cell resistance is high. The LED current is increased as a result of the decreased resistance of R_C when the ambient light level is high. This increased current gives greater LED brightness so that it can be easily seen.

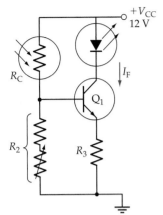

Figure 21-16 Photoconductive cell circuit for controlling LED current level.

Example 21-7

Design the circuit in Fig. 21-16 for a LED current of $I_F \approx 10$ mA at $R_C = 5$ kΩ, and for $I_F \approx 20$ mA at $R_C = 1$ kΩ.

Solution

At $I_F = 10$ mA $\quad V_{B1} = \dfrac{V_{CC} \times R_2}{R_{C1} + R_2} = \dfrac{V_{CC} \times R_2}{5 \text{ k}\Omega + R_2}$

At $I_F = 20$ mA $\quad V_{B2} = \dfrac{V_{CC} \times R_2}{R_{C2} + R_2} = \dfrac{V_{CC} \times R_2}{1 \text{ k}\Omega + R_2} = 2V_{B1}$

To double I_F, $\quad V_{B2} = 2V_{B1}$

Therefore $\quad \dfrac{V_{CC} \times R_2}{1 \text{ k}\Omega + R_2} = \dfrac{2V_{CC} \times R_2}{5 \text{ k}\Omega + R_2}$

giving $\quad R_2 = 3$ kΩ (use 2.2 kΩ + 2 kΩ variable)

$$V_{B1} = \dfrac{V_{CC} \times R_2}{R_{C1} + R_2} = \dfrac{12 \text{ V} \times 3 \text{ k}\Omega}{5 \text{ k}\Omega + 3 \text{ k}\Omega}$$

$$= 4.5 \text{ V}$$

$$V_{B2} = 2V_{B1} = 9 \text{ V}$$

For $I_F = 10$ mA, $\quad R_3 \approx \dfrac{V_{B1} - V_{BE}}{I_C} = \dfrac{4.5 \text{ V} - 0.7 \text{ V}}{10 \text{ mA}}$

$$= 380 \ \Omega \text{ (use 390 } \Omega \text{ standard value)}$$

At V_{B2} $\quad I_F \approx \dfrac{V_{B2} - V_{BE}}{R_3} = \dfrac{9 \text{ V} - 0.7 \text{ V}}{390 \ \Omega}$

$$= 21.3 \text{ mA}$$

Practice Problems

21-4.1 A relay control circuit such as in Fig. 21-13 has $R_1 = 3.3$ kΩ, $E = 9$ V, and a photoconductive cell with the specification in Fig. 21-12. Calculate the maximum and minimum circuit current at a luminous intensity of 10 lx. Determine the current at 30 lx if the cell has the characteristics in Fig. 21-11.

21-4.2 The circuit in Fig. 21-14b has $V_{CC} = +12$ V and $V_{EE} = -9$ V. The photoconductive cell has the characteristics in Fig. 21-11. Calculate the R_1 resistance to have Q_1 *on* at a 3 lx illuminance if $I_{B1} \approx 100$ µA.

21-4.3 The op-amp circuit in Fig. 21-15 has $V_{CC} = \pm 15$ V, $R_1 = 5.6$ kΩ, $R_2 = 18$ kΩ, and $R_3 = 120$ kΩ. The cell characteristics are those in Fig. 21-11. Estimate the light levels that cause the output to switch.

21-5 PHOTODIODES AND SOLAR CELLS

Photodiode Operation

When a *pn*-junction is reverse-biased, there is a small reverse saturation current because thermally generated holes and electrons are being swept across the junction as minority charge carriers (see Section 1-6). Increasing the junction temperature generates more hole-electron pairs, and so the minority carrier (reverse) current is increased. The same effect occurs if the junction is illuminated (see Fig. 21-17). Hole-electron pairs are generated by the incident light energy, and minority charge carriers are swept across the junction to produce a reverse current flow. Increasing the junction illumination increases the number of charge carriers generated and, thus, increases the reverse current level. Diodes designed to be sensitive to illumination are known as *photodiodes*.

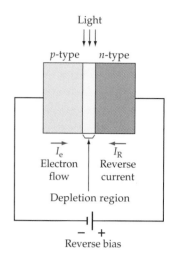

Figure 21-17 A photodiode has a reverse-biased *pn*-junction designed to be light-sensitive. There is a minority charge carrier current when the junction is illuminated.

Characteristics

Consider the typical photodiode illumination characteristics in Fig. 21-18. When the junction is dark, the *dark current* (I_D) would seem to be zero. Typically, I_D is around 2 nA. A 20 mW/cm^2 illumination level produces a reverse current of approximately 60 μA. Increasing the reverse voltage does not increase I_R significantly. So each characteristic is approximately a horizontal line.

Figure 21-19 shows a simple photodiode circuit using a 2 V reverse bias. (Note the device circuit symbol.) Assuming that D_1 has the characteristics in Fig. 21-18, the current at a 5 mW/cm^2 illumination level is approximately 13 μA. At 20 mW/cm^2 the diode current is around 60 μA. The device resistance at each illumination level is readily calculated: (at 5 mW/cm^2, $R = 2\,V/13\,μA = 154\,kΩ$), (at 20 mW/cm^2, $R = 2\,V/60\,μA = 33\,kΩ$). The resistance changes by

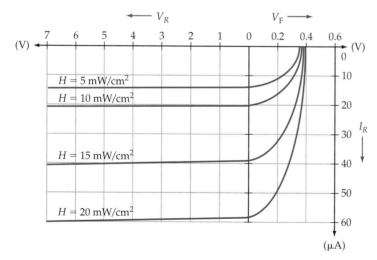

Figure 21-18 Typical photodiode characteristics. The reverse current remains substantially constant for each level of illumination.

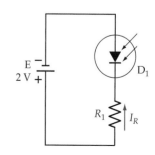

Figure 21-19 Photodiode circuit with a reverse-bias voltage.

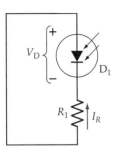

Figure 21-20 An illuminated photodiode without an external bias operates as a photovoltaic device.

a factor of approximately 5 from the low to the high illumination level, showing that a photodiode can be used as a photoconductive device.

When the reverse-bias voltage across a photodiode is removed, minority charge carriers continue to be swept across the junction while the diode is illuminated. With an external circuit connected across the diode terminals, the minority carriers flow back to their original sides. The electrons that crossed the junction from *p* to *n* will now flow out through the *n*-terminal and into the *p*-terminal. This means that the device is behaving as a voltage cell, with the *n*-side as the negative terminal and the *p*-side the positive terminal, as illustrated in Fig. 21-20. In fact, a voltage can be measured at the photodiode terminal, positive on the *p*-side and negative on the *n*-side. So the photodiode is a photovoltaic device as well as a photoconductive device. The characteristics in Figure 21-18 show that, when illuminated, the photodiode actually has to be forward-biased to reduce the reverse current to zero.

It should be noted that V_R and V_F have different scales on the photodiode characteristics shown in Fig. 21-18. A dc load line that crosses between the forward- and reverse-biased regions cannot be drawn on these characteristics. Equal scales must be used for each part of the characteristics to draw such a load line.

Specification

A partial specification for a typical photodiode is shown in Fig. 21-21. The *light current* (I_L) is listed as 10 µA at an illumination level of 5 mW/cm² when the reverse bias is 2 V. This is sometimes defined as a *short-circuit current* (I_{SC}). The *dark current* (I_D) is specified as 2 nA maximum when the reverse voltage is 20 V, and the *open-circuit terminal voltage* (V_{OC}) is given as 350 mV. Note that the typical response time (t_{res}) of 2 ns for a photodiode is very much superior to that for a photoconductive cell. The diode *sensitivity* (S) is the change in diode current produced by a given change in light intensity. The power dissipation, reverse breakdown voltage, and peak output wavelength are also listed.

				Typical Photodiode Specification				
P_D	V_{OC}	$BV_{R(max)}$	$I_{D(max)}$ (dark)	I_L [V_R = 2 V, H = 5 mW/cm²]	t_{res}		S	λP
100 mW	350 mV	100 V	2 nA	10 µA	2 ns		7 µA/mW/cm²	900 nm

Figure 21-21 Partial specification for a low-current photodiode.

Construction

Figure 21-22a shows the cross-section of a diffused photo diode. It is seen that a thin, heavily-doped *p*-type layer is situated at the top, where it is exposed to incident light. The junction depletion region penetrates deeply into the lightly-doped *n*-type layer. This is in contact with a lower, heavily-doped *n*-type layer,

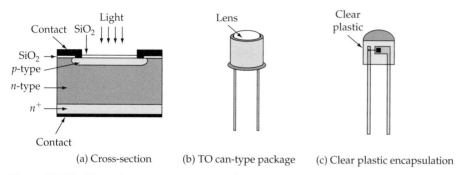

(a) Cross-section (b) TO can-type package (c) Clear plastic encapsulation

Figure 21-22 Photodiode cross-section and typical packages.

which connects to a metal film contact. A ring-shaped contact is provided at the top of the *p*-type layer. Low-current photodiodes (also called *signal photodiodes*) are usually contained in a TO-type can with a lens at the top (see Fig. 21-22b). Transparent plastic encapsulation is also used (Fig. 21-22c).

Photodiode Applications

Photodiodes can be used as photoconductive devices in the type of circuits discussed in Section 21-4. They can also be used in circuits, where they function as photovoltaic devices. Figure 21-23 shows typical photodiode characteristics plotted in the first and second quadrants for convenience. When the device is operated with a reverse voltage, it functions as a photoconductive device. When operating without the reverse voltage, it operates as a photovoltaic device. In some circuits the photodiode can change between the photoconductive mode and the photovoltaic mode.

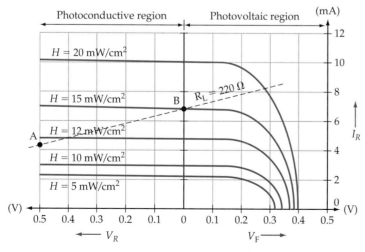

Figure 21-23 Photodiode characteristics have a photoconductive region and a photovoltaic region.

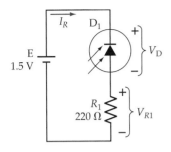

Figure 21-24 Photodiode with a reverse bias and a load resistor.

Example 21-8

The circuit in Fig. 21-24 uses a photodiode with the illumination characteristics in Fig. 21-23. Draw the dc load line, and determine the diode currents and voltages at light levels of 12, 15 and 20 mW/cm^2.

Solution

When $V_D = 0.5$ V, $V_{R1} = E - V_D = 1.5$ V $- 0.5$ V

$$= 1 \text{ V}$$

$$I_D = \frac{V_{R1}}{R_1} = \frac{1 \text{ V}}{220 \text{ }\Omega}$$

$$\approx 4.5 \text{ mA}$$

Plot point A on the characteristics at $I_D = 4.5$ mA and $V_D = 0.5$ V

When $V_D = 0$, $V_{R1} = E = 1.5$ V

Therefore, $I_D = \dfrac{E}{R_1} = \dfrac{1.5 \text{ V}}{220 \text{ }\Omega}$

$$= 6.8 \text{ mA}$$

Plot point B on the characteristics at $V_D = 0$ and $I_D = 6.8$ mA
Draw the dc load line through points A and B.

From the load line:

$$\text{at } 12 \text{ mW/cm}^2, \quad I_D \approx -5 \text{ mA}, \quad V_D \approx -0.4 \text{ V}$$
$$\text{at } 15 \text{ mW/cm}^2, \quad I_D \approx -6.8 \text{ mA}, \quad V_D \approx 0 \text{ V}$$
$$\text{at } 20 \text{ mW/cm}^2, \quad I_D \approx -8.2 \text{ mA}, \quad V_D \approx +0.28 \text{ V}$$

Note the V_D polarity change at the highest illumination level.

Solar Cells

The *solar cell*, or *solar energy converter*, is essentially a large photodiode designed to operate solely as a photovoltaic device and to give as much output power as possible. To provide maximum output current, the surface areas of solar cells are much larger than those of signal photodiodes. Typical solar cell output characteristics are illustrated in Fig. 21-25. Consider the characteristic for a 100 mW/cm^2 illumination level. If the cell is short-circuited, the output current (I_o) is 50 mA. Because the cell voltage (V_o) is zero at this point, the output power (P_o) is zero. Open-circuiting the cell gives $V_o \approx 0.55$ V, but $I_o = 0$. So P_o is again zero. At the *knee* of the characteristic, $V_o \approx 0.44$ V and $I_o \approx 45$ mA, giving $P_o \approx 20$ mW. Therefore, for maximum output power, the device must be operated on the knee of the characteristic. As in the case of all other devices, the power must be derated at high temperatures.

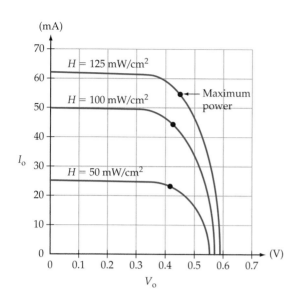

Figure 21-26 shows a group of series-parallel connected solar cells operating as a battery charger. Several cells must be series-connected to produce the required output voltage, and several of these series-connected groups must be connected in parallel to provide the necessary output current.

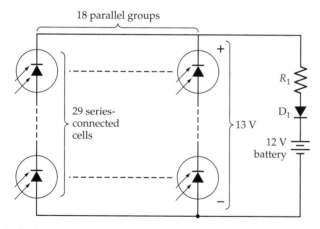

Figure 21-26 Series-parallel arrangement of solar cells connected to function as a solar battery charger.

Example 21-9

An earth satellite has 12 V batteries that supply a continuous current of 0.5 A. Solar cells with the characteristics shown in Fig. 21-25 are employed to keep the batteries charged. If the illumination from the sun for 12 hours in every 24 is 125 mW/cm^2, determine approximately the total number of cells required.

Solution

From Fig. 21-25, maximum output power at 125 mW/cm^2 is achieved when each cell is operated at approximately $V_o = 0.45$ V and $I_o = 57$ mA.

Allowing for the voltage drop across the rectifier, a maximum charging voltage of approximately $V_{CH} = 13$ V is required.

Number of series-connected cells:

$$N_S = \frac{V_{CH}}{V_o} = \frac{13 \text{ V}}{0.45 \text{ V}}$$

$$\approx 29$$

The charge taken from the batteries over a 24-hour period is

$$Q = I_L \times t = 0.5 \text{ A} \times 24 \text{ hours}$$

$$= 12 \text{ Ah}$$

Therefore, the charge delivered by the solar cells must be 12 Ah.

The solar cells deliver current only while they are illuminated, that is, for 12 hours in every 24. Therefore, the charging current from the solar cells is

$$I_{CH} = \frac{Q}{t} = \frac{12 \text{ Ah}}{12 \text{ h}}$$

$$= 1 \text{ A}$$

Number of parallel-connected groups of cells:

$$N_P = \frac{I_{CH}}{I_o/\text{cell}} = \frac{1 \text{ A}}{57 \text{ mA}}$$

$$\approx 18$$

The total number of cells required is

$$N_T = N_P \times N_S = 18 \times 29$$

$$= 522$$

Practice Problems

21-5.1 A photodiode with the characteristics in Fig. 21-23 is connected in series with a 330 Ω resistor and a (reverse-biasing) 3 V battery. Draw the dc load line for the circuit, and determine the diode voltage at an illumination level of 20 mW/cm^2.

21-5.2 A solar voltage source is to be designed to produce a 3 V, 20 mA output from a 15 mW/cm^2 illumination level. Calculate the required number of cells if the available devices have the characteristics shown in Fig. 21-23.

21-6 PHOTOTRANSISTORS

Phototransistor (BJT)

A *phototransistor* is similar to an ordinary BJT, except that its collector-base junction is constructed like a photodiode. Instead of a base current, the input to the transistor is in the form of illumination at the junction. Consider an ordinary BJT with its base terminal open-circuited (Fig. 21-27a). The collector-base leakage current (I_{CBO}) acts as a base current, giving a collector current: $I_C = (h_{FE} + 1) I_{CBO}$. In the case of the photodiode, it was shown that the reverse saturation current is increased by the light energy at the junction. Similarly, in the phototransistor, I_{CBO} is proportional to the collector-base illumination (Fig. 21-27b). This results in I_C also being proportional to the illumination.

For a given amount of illumination on a very small area, the phototransistor provides a much larger output current than that available from a photodiode. Thus the phototransistor is the most sensitive of the two devices. The phototransistor circuit symbol shows a base terminal, and this is often left unconnected, but is sometimes used to provide stable bias conditions.

The cross-section in Fig. 21-28 illustrates the construction of a phototransistor. The emitter area is seen to be quite small, to allow incident illumination to pass to the collector-base junction. Phototransistor packages are similar to the photodiode packages in Fig. 21-22, except that three terminals are provided.

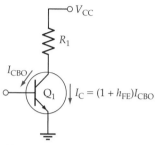

(a) BJT currents with open-circuited base

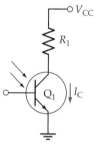

(b) Phototransistor circuit

Figure 21-27   In a phototransistor, the collector current depends upon the illumination level at the CB junction.

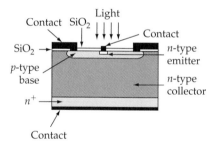

Figure 21-28   Phototransistor cross-section.

Characteristics and Specification

Typical phototransistor output characteristics are shown in Fig. 21-29. These are seen to be similar to BJT characteristics except that the base current levels are replaced with illumination levels. A dc load line can be drawn on the characteristics in the usual way.

The partial specification for a phototransistor in Fig. 21-30 shows a 3 mA minimum current at 5 mW/cm^2, and a sensitivity of 500 μA/mW/cm^2. Comparing this to the photodiode specification (Fig. 21-21) shows (as stated above) that a phototransistor is very much more sensitive than a photodiode. However, the rise and fall times (usually 5 μs and 8 μs) for a phototransistor are very much slower than the 2 ns response time for a photodiode.

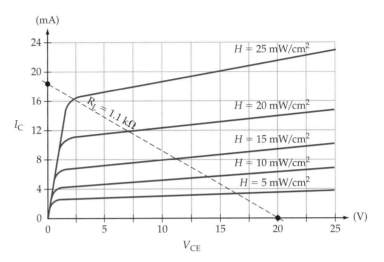

Figure 21-29 The I_C/V_{CE} characteristics for a phototransistor are similar to those of an ordinary BJT except that base current level is replaced by illumination level.

			Typical Phototransistor Specification					
P_D	$V_{CE(max)}$	$I_{CEO(max)}$ (dark) [$V_{CE} = 10$ V]	$I_{C(min)}$ [$V_{CE} = 5$ V] [$H = 5$ mW/cm^2]	t_r [$I_{CE} = 1$ mA]	t_f	S [$H = 5$ mW/cm^2]	λP (nm)	
200 mW	40 V	100 nA	3 mA	5 μs	8 μs	500 μA/mW/cm^2	900 nm	

Figure 21-30 Partial specification for a phototransistor.

Applications

Two phototransistor applications are shown in Fig. 21-31 and 21-32. The relay in Fig. 21-31 is energized when the incident light on the phototransistor is raised to a particular level. This occurs when the Q_1 emitter current produces

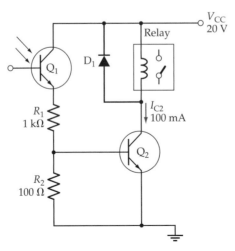

Figure 21-31 Use of a phototransistor to energize a relay when the illumination increases to a predetermined level.

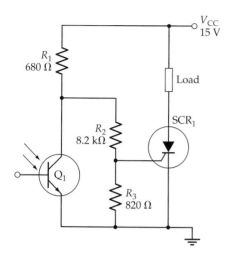

Figure 21-32 Phototransistor circuit for triggering an SCR at low light levels.

sufficient voltage drop across R_2 to forward-bias the BE junction of Q_2. The relay current falls again when Q_2 turns *off* as the light level decreases. Note the presence of the catch diode (D_1) for short-circuiting the relay coil when Q_2 switches *off* (see Section 18-7).

In Fig. 21-32, SCR_1 remains untriggered while the illumination keeps Q_1 in saturation. If the light fails, Q_1 turns *off* and V_{R3} triggers the SCR *on*. This kind of circuit can be used to switch on an emergency lighting system when the normal lighting fails.

Example 21-10

Transistor Q_1 in the circuit in Fig. 21-31 has the characteristics in Fig. 21-29, and Q_2 has $h_{FE} = 80$. Determine the light level required to energize the relay.

Solution

When Q_2 is on,
$$V_{R2} = V_{BE2} = 0.7 \text{ V}$$

$$I_{R2} = \frac{V_{R2}}{R_2} = \frac{0.7 \text{ V}}{100 \text{ }\Omega}$$

$$= 7 \text{ mA}$$

$$I_{B2} = \frac{I_{C2}}{h_{FE}} = \frac{100 \text{ mA}}{80}$$

$$= 1.25 \text{ mA}$$

$$I_{E1} = I_{B2} + I_{R2} = 1.25 \text{ mA} + 7 \text{ mA}$$

$$= 8.25 \text{ mA}$$

The dc load resistance for Q_1 is
$$R_L = R_1 + R_2 = 1 \text{ k}\Omega + 100 \text{ }\Omega$$

$$= 1.1 \text{ k}\Omega$$

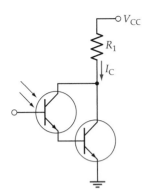

Figure 21-33 A photo-darlington is made up of a phototransistor connected in Darlington with another BJT.

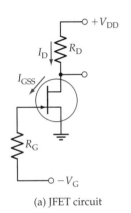

(a) JFET circuit

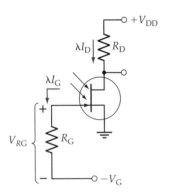

(b) *Photo*-FET circuit

Figure 21-34 In a photo-FET, the gate-channel leakage current depends upon the illumination level. This current produces a voltage drop across the bias resistor to control the gate-source voltage.

Draw the dc load line on the characteristics for $R_L = 1.1 \text{ k}\Omega$. At $I_C = 8.25 \text{ mA}$ on the load line,

$$H \approx 15 \text{ mW/cm}^2$$

Photo-Darlingtons

The *photo-darlington* (Fig. 21-33) consists of a phototransistor connected in Darlington arrangement with another transistor. This device is capable of producing much higher output currents than a phototransistor, and so it has a greater sensitivity to illumination levels than either a phototransistor or a photodiode. With the additional transistor involved, the photo-darlington has a considerably longer switching time than a phototransistor.

Photo-FET

A *photo-FET* is a JFET designed to have its gate-channel junction illuminated. The illumination controls the level of the device drain current. Consider the n-channel JFET and the photo-FET in Fig. 21-34. The gate-source leakage current (I_{GSS}) is the reverse saturation current at a *pn*-junction. The voltage drop across R_G produced by I_{GSS} is normally too small to affect the JFET circuit. In the photo-FET, the junction reverse current (λI_G) is susceptible to light. Illumination on the junction generates additional minority charge carriers, thus increasing λI_G. This current flows through the bias resistance (R_G) and produces a voltage drop (V_{RG}), as illustrated. If the gate bias voltage ($-V_G$) is just sufficient to bias the device *off* when the junction is dark, then when the junction is illuminated, V_{RG} can raise the level of the gate voltage to bias the photo-FET *on*.

The external bias voltage ($-V_G$) may be selected at a level that biases the device on, so that light level variations cause I_D to increase and decrease. In a photo-FET, λI_G is termed the *gate current*, and the normal I_{GSS} at the junction is the *dark gate-leakage current*. The light-controlled drain current is designated λI_D.

Practice Problems

21-6.1 A phototransistor with the characteristics in Fig. 21-29 is connected in series with a resistor $R_1 = 890 \ \Omega$ and $V_{CC} = 18$ V. Determine V_{R1} at illumination levels of 5 mW/cm² and 25 mW/cm².

21-6.2 The SCR in Fig. 21-32 triggers at $V_G = 0.8$ V and $I_G = 100 \ \mu$A. Determine the approximate light level that will trigger the SCR *on* if Q_1 has the characteristics in Fig. 21-29.

21-7 OPTOCOUPLERS

Operation and Construction

An *optocoupler* (*optoelectronic coupler*) is essentially a phototransistor and an LED combined in one package. Figure 21-35 shows the typical circuit and terminal arrangement for one such device contained in a DIP package. When current flows in the LED, the emitted light is directed to the phototransistor, producing current flow in the transistor. The coupler may be operated as a switch, in which case both the LED and the phototransistor are normally *off*. A pulse of current through the LED causes the transistor to be switched *on* for the duration of the pulse. Linear signal coupling is also possible. Because the coupling is optical, there is a high degree of electrical isolation between the input and output terminals, and so the term *optoisolator* is sometimes used. The output (detector) stage has no effect on the input, and the electrical isolation allows a low-voltage dc source to control high-voltage circuits.

The cross-section diagram in Fig. 21-35c illustrates the construction of an optocoupler. The emitter and detector are contained in a transparent insulating material that allows the passage of illumination while maintaining electrical isolation.

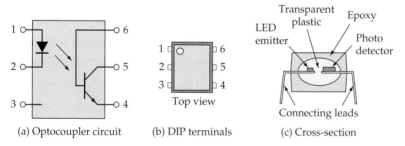

(a) Optocoupler circuit   (b) DIP terminals   (c) Cross-section

Figure 21-35   An octocoupler is composed of an LED and a phototransistor. The input and output are electrically isolated.

Specification

The partial specification for an optoelectronic coupler in Fig. 21-36 has three parts. The first part specifies the current and voltage conditions for the input (LED) stage. The second deals with the output (phototransistor) stage. The third part defines the coupling parameters. The transistor collector current is listed as 5 mA (typical) when its $V_{CE} = 10$ V and the LED has $I_F = 10$ mA. In this particular case, the ratio of output current to input current is 50%. This is known as the *current transfer ratio (CTR)*, and for an optoelectronic coupler with a transistor output it can range from 10% to 150%.

Typical Optocoupler Specification

Input Stage

$I_{F(max)}$	$V_{F(max)}$ [$I_F = 20$ mA]	V_R
60 mA	1.5 V	3 V

Output Stage

$V_{CE(max)}$	$I_{C(max)}$	P_D	$V_{CE(sat)}$	I_{CEO} (dark)
30 V	150 mA	150 mW	0.2 V	50 nA

Coupled

$I_{C(out)}$ [$I_F = 10$ mA]	$t_{on(max)}$	$t_{off(max)}$	Isolation voltage
5 mA	2.5 µs	4 µs	7500 V

Figure 21-36   The partial specification for an octocoupler is made up of three parts: input, output, and coupling.

Applications

The circuit of an optocoupler in a dc or pulse-type coupling application is shown in Fig. 21-37. The diode current is switched on and off by the action of transistor Q_1 operating from a 24 V supply. Transistor Q_2 is turned on into saturation when D_1 is energized. The collector current of Q_2 provides the load (*sinking*) current and the current through resistor R_2. *Pull-up* resistor R_2 is necessary to ensure that the load terminal is held at the 5 V supply level when Q_2 is *off*.

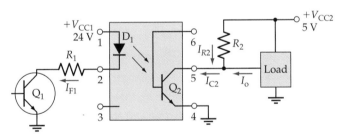

Figure 21-37   Octocoupler used for coupling a signal from a 24 V system to a 5 V system.

Example 21-11

The optocoupler in Fig. 21-37 is required to sink a 2 mA load current when Q_2 is in saturation and $I_{F1} = 10$ mA. The optocoupler has the specification in Fig. 21-36. Determine suitable resistor values.

Solution

From the specification: $I_{C2} = 5$ mA when $I_{FD1} = 10$ mA

$$I_{R2} = I_{C2} - I_o = 5 \text{ mA} - 2 \text{ mA}$$

$$= 3 \text{ mA}$$

$$R_2 = \frac{V_{CC2} - V_{CE(sat)}}{I_{R2}} = \frac{5 \text{ V} - 0.2 \text{ V}}{3 \text{ mA}}$$

$$= 1.6 \text{ k}\Omega \text{ (use 1.8 k}\Omega \text{ to ensure } Q_2 \text{ saturation)}$$

$$R_1 = \frac{V_{CC1} - V_{FD1} - V_{CE(sat)}}{I_{FD1}} = \frac{24 \text{ V} - 1.5 \text{ V} - 0.2 \text{ V}}{10 \text{ mA}}$$

$$= 2.23 \text{ k}\Omega \text{ (use 2.2 k}\Omega \text{)}$$

A linear application of an optocoupler is shown in Fig. 21-38. The 5 V supply provides a dc bias current to D_1 via R_2, and the ac signal coupled via C_1 and R_1 increases and decreases the diode current. Transistor Q_1 is biased into an *on* state by the direct current through D_1, and its emitter current is increased and decreased by the variation in light level produced by the alternating current in D_1. An output voltage is developed across R_3.

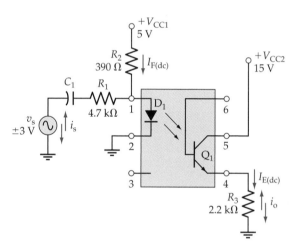

Figure 21-38 Linear signal coupling by means of an octocoupler.

Other Optocouplers

Other types of optocouplers involve different types of output stage. The three types illustrated in Fig. 21-39 are (a) Darlington-output type, (b) SCR-output, and (c) TRIAC-output. In (a), the photo-darlington output stage provides much higher CTR than a BJT phototransistor output stage (typically 500%),

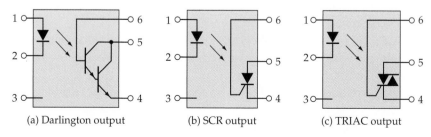

(a) Darlington output   (b) SCR output   (c) TRIAC output

Figure 21-39 Octocouplers are available with various types of output stage.

but it also has a slower response time. The output stages in (b) and (c) are a *light-activated SCR* and a *light-activated TRIAC* respectively. They are used with the kind of control circuits discussed in Chapter 20, where high electrical isolation between the triggering circuit and the control device is an additional requirement. CTR does not apply to SCR and TRIAC output stages; instead, the LED current needed to trigger the thyristor is of interest.

Optocoupler output stages are *not* designed for high load currents. Maximum current levels for Darlington outputs are about 150 mA, and 300 mA is typical for SCR and TRIAC outputs. When high-load currents are to be switched, the optocoupler output stage is used as a trigger circuit for a high-power device.

Practice Problems

21-7.1 An octocoupler with the specification in Fig. 21-36 is to control a 10 mA relay with a 30 V supply. The input stage is connected via a resistor (R_1) to a 5 V supply. Calculate a suitable resistance for R_1.

21-7.2 Analyze the circuit in Fig. 21-38 to determine the dc bias current through D_1 and the ac signal current peaks. Calculate the maximum and minimum dc and ac output voltages. The octocoupler used has a diode with $V_F = 1.5$ V, and a CTR ranging from 20% to 70%.

Review Questions

Section 21-1

21-1 State measurement units for luminous flux and luminous intensity. Using diagrams, explain flux per unit solid angle.

21-2 Define candela, lumen, and foot candle.

Section 21-2

21-3 Sketch diagrams to show the operation and construction of a LED. Briefly explain.

21-4 For a LED, state typical values of forward current, forward voltage, and reverse breakdown voltage.

21-5 Draw circuit diagrams showing LEDs used to indicate (a) a dc supply voltage switched on and (b) a high output level from an op-amp. Explain each circuit.

21-6 The current level in a LED is to be controlled by the use of a BJT. Sketch two possible circuits, and explain the operation of each.

21-7 An op-amp is to be used to control the current level in a LED. Draw a suitable circuit diagram and explain its operation.

Section 21-3

21-8   Sketch a seven-segment LED display. Explain common-anode and common-cathode connections. Discuss total current requirements for a LED four-numeral, seven-segment display.

21-9   Using illustrations, explain the operation of liquid crystal cells. Discuss the difference between reflective-type and transmittive-type cells.

21-10   Sketch a seven-segment LCD and show the waveforms involved in controlling the cells. Explain.

Section 21-4

21-11   Sketch the typical construction and illumination characteristics for a photoconductive cell. Explain its operation.

21-12   Draw circuit diagrams to show how a photoconductive cell can be used for (a) biasing a *pnp* transistor *off* when the cell is illuminated, and (b) biasing an *npn* transistor *on* when the cell is illuminated. Explain how each circuit operates.

21-13   Draw circuit diagrams to show a photoconductive cell used for (a) triggering an op-amp Schmitt trigger circuit, and (b) energizing a relay when the cell is illuminated. Explain the operation of each circuit.

Section 21-5

21-14   Sketch the cross-section of a typical photodiode and explain its operation. Sketch typical photodiode characteristics and discuss their shape.

21-15   For a photodiode, define dark current, light current, and sensitivity. State typical values for each quantity.

21-16   Explain how a solar cell differs from a photodiode. Sketch typical solar cell characteristics, and discuss the best operating point on the characteristics.

21-17   Sketch the circuit diagram for an array of solar cells employed as a battery charger. Briefly explain.

Section 21-6

21-18   Sketch characteristics for a phototransistor, and explain how the device operates.

21-19   Draw a circuit diagram to show how a phototransistor can be used to energize a relay when the incident illumination is increased to a given level. Explain the circuit operation.

21-20   Modify the circuit drawn for Question 21-19 to have the relay energized until the illumination is increased to a given level. Explain.

21-21   Draw a circuit diagram for phototransistor control of an SCR that triggers *on* when the incident illumination falls to a low level. Explain how the circuit operates.

21-22   Sketch a circuit diagram for a photo-darlington. Compare the performance of photo-darlingtons to phototransistors.

21-23   Sketch a circuit diagram to show the operation of a photo-FET circuit. Briefly explain the principle of the device.

Section 21-7

21-24 Draw the circuit diagram of an optocoupler with a BJT output stage. Sketch a cross-section to show the construction of an optocoupler. Explain the device operation.

21-25 Discuss the most important parameters of optocouplers.

21-26 Draw a circuit diagram to show how an optocoupler can use a pulse signal from a low-voltage source to control a circuit with a high-voltage supply, or vice versa. Explain how the circuit operates.

21-27 Draw a circuit diagram to show how an optocoupler can be used to pass a linear signal between two circuits with different supply voltages. Explain how the circuit operates.

21-28 Sketch circuit diagrams for optocouplers with Darlington, SCR, and TRIAC outputs. Briefly discuss each optocoupler.

Problems

Section 21-1

21-1 The total luminous flux striking a 4 cm^2 photocell at 7 m from a lamp is to be 80 mlm. Determine the required energy output from the lamp in watts.

21-2. Calculate the total luminous flux striking the surface of a solar cell located 4.5 m from a lamp with a 509 W output. The surface area of the solar cell is 5 cm^2.

21-3 Calculate the frequency of the light output from red, yellow, and green LEDs with the following peak wavelengths: 635 nm, 583 nm, 565 nm.

Section 21-2

21-4 An LED with $I_F = 20$ mA current and $V_F = 1.4$ V is to indicate when a 25 V supply is switched on. Sketch a suitable circuit and make all necessary calculations.

21-5 Two series-connected LEDs are to be controlled by a 2N3903 transistor with a 12 V supply and $V_B = 5$ V. The diode current is to be approximately 15 mA. Design a suitable circuit.

21-6 An op-amp Schmitt trigger circuit with $V_{CC} = \pm 15$ V is to have the state of its output indicated by LEDs. A green LED is to indicate high, and a red LED is to indicate low. Design the circuit for 10 mA diode currents. Include reverse-voltage protection diodes in series with each LED.

21-7 The BJT-LED circuit in Fig. 21-5a has $V_B = 5$ V, $V_{CC} = 20$ V, and $h_{FE(min)} = 40$ for Q_1. Design the circuit to give a 20 mA LED current with $V_F = 2$ V.

Section 21-3

21-8 Calculate the maximum power used by a three-and-a-half digit seven-segment LED display with a 5 V supply and 10 mA LED currents. Determine the power dissipated in each LED series resistor if the LEDs have $V_F = 1.4$ V.

21-9 Determine the maximum power consumed by a three-and-a-half digit seven-segment LCD display with a 15 V peak square-wave supply and 1 μA LCD segment currents.

Section 21-4
21-10 A *pnp* BJT is to be biased on when the level of illumination on a photo-conductive cell is greater than 100 lx, and off when the cell is dark. A ±5 V supply is to be used, and the BJT collector current is to be 10 mA when on. Design a suitable circuit to use a BJT with $h_{FE} = 50$ and a photo-conductive cell with the characteristics in Fig. 21-11.

21-11 An inverting Schmitt trigger circuit has $V_{CC} = ±12$ V and $UTP/LTP = ±5$ V. The Schmitt output is to switch positively when the illumination level exceeds 30 lx on a photo-conductive cell with the characteristics in Fig. 21-11. Design the circuit, and estimate the light level that causes the output to switch negatively.

21-12 A photoconductive cell with the characteristics in Fig. 21-11 is connected in series with an 820 Ω resistor and a 12 V supply. Determine the illumination level when the circuit current is approximately 6.5 mA, and when it is 1.1 mA.

21-13 A photoconductive cell circuit for controlling the current in a LED (as in Fig. 21-16) has $V_{CC} = 9$ V, $R_2 = 3.3$ kΩ, $R_3 = 270$ Ω. The photoconductive cell has a dark resistance of 100 kΩ, and $R_C = 3$ kΩ at 10 lx. Determine the LED current at light levels of 3 lx and 30 lx.

21-14 The circuit in Fig. 21-14a has $V_{CC} = ±5$ V, $R_1 = 12$ kΩ, and a photoconductive cell with the specification in Fig. 21-12. Calculate the transistor maximum and minimum base voltage at 10 lx.

Section 21-5
21-15 A photodiode with the illumination characteristics in Fig. 21-23 is connected in series with a resistance and a 1 V reverse-bias supply. The diode is to produce a +0.2 V output when illuminated with 20 mW/cm^2. Calculate the required series resistance value, and determine the device voltage and current at a 15 mW/cm^2 illumination level.

21-16 A photodiode with the characteristics in Fig. 21-23 is connected in series with a 1.2 V reverse-bias supply and a 100 Ω resistance. Determine the resistance offered by the photodiode at illumination levels of 15 mW/cm^2 and 20 mW/cm^2.

21-17 Two photodiodes that each have a 100 Ω series resistor are connected to a 0.5 V reverse-bias supply. A voltmeter is connected to measure the voltage difference between the diode cathodes. Assuming that each photodiode has the characteristics illustrated in Fig. 21-23, determine the voltmeter reading when the illumination level is 10 mW/cm^2 on one diode and 15 mW/cm^2 on the other.

21-18 Six photodiodes with the characteristics in Fig. 21-23 are connected in series. Determine the maximum output current and voltage at illumination levels of 15 mW/cm^2 and 12 mW/cm^2.

21-19 A highway sign uses 6 V rechargeable batteries that supply an average current of 50 mA. The batteries are recharged from an array of solar cells, each with the characteristics in Fig. 21-25. The average level of sunshine is 50 mW/cm^2 for 10 hours of each 24-hour period. Calculate the number of solar cells required.

21-20 The roof of a house has an area of 200 m^2 and is covered with solar cells that are each 2 cm $\times$ 2 cm. If the cells have the output characteristics shown in Fig. 21-25, determine how they should be connected to provide an output voltage of approximately 120 V. Take the average daytime level of illumination as 100 mW/cm^2. If the sun shines for an average of 12 hours in every 24 hours, calculate the energy in kilowatt-hours generated by the solar cells each day.

Section 21-6

21-21 The phototransistor circuit in Fig. 21-27b has a 25 V supply, and the device has the output characteristics in Fig. 21-29. Determine the collector resistance required to give $V_{CE} = 10$ V when the illumination level is 20 mW/cm^2.

21-22 Estimate V_{CE} for the circuit in Problem 21-21 at a 5 mW/cm^2 illumination level. If the phototransistor has the specification in Fig. 21-30, calculate the V_{CE} variation produced by a ±0.5 mW/cm^2 illumination change.

21-23 A phototransistor with the characteristics in Fig. 21-29 is connected in series with a 600 Ω relay coil. The coil current is to be 8 mA when the illumination level is 15 mW/cm^2. Determine the required supply voltage. Estimate the coil current at 10 mW/cm^2.

21-24 A phototransistor circuit for controlling an SCR (as in Fig. 21-32) is to be designed. The SCR has triggering conditions of $V_G = 0.7$ V and $I_G = 50$ μA, and the phototransistor has the specification in Fig. 21-30. Calculate suitable resistor values if the SCR is to switch on when the light level drops to 5 mW/cm^2. The supply voltage is $V_{CC} = 12$ V.

Section 21-7

21-25 An optocoupler with the specification in Fig. 21-36 is to control a 12 mA load that has a 6 V supply. The input is a 10 V square wave connected via a resistor (R_1). Determine a suitable resistance for R_1.

21-26 A 25 V, 0.5 W lamp is to be switched on and off by an BJT circuit with $V_{CC} = 9$ V and $I_C = 6$ mA. Design a suitable optocoupler circuit and estimate the required CTR.

21-27 An optocoupler circuit has its input connected via a 820 Ω resistor (R_1) to a pulse source. Its output transistor has a 5 V collector supply and a 470 Ω emitter resistor (R_2). The gate-cathode terminals of an SCR are connected across R_2. The SCR requires $V_G = 1.1$ V and $I_G = 500$ μA for triggering. If the optocoupler has CTR = 40%, calculate the required amplitude of the pulse input to trigger the SCR.

21-28 A optocoupler linear circuit, such as in Fig. 21-38, has $V_{CC1} = 15$ V, $V_{CC2} = 25$ V, $v_s = \pm 0.1$ V, $R_1 = 100\ \Omega$, $R_2 = 1.2\ k\Omega$, $R_3 = 1.5\ k\Omega$. Calculate the dc and ac output voltages and the overall voltage gain. The optocoupler has CTR = 30%.

21-29 An optocoupler switching circuit, as in Fig. 21-37, has $V_{CC1} = 18$ V, $V_{CC2} = 3$ V, $R_1 = 1.8\ k\Omega$, $R_2 = 820\ \Omega$, and $I_o = 1$ mA. Analyze the circuit to determine I_{F1}, I_{C2}, and CTR.

Practice Problem Answers

21-1.1	100 W
21-1.2	5.13×10^{14} Hz
21-1.3	0.73 mlx, 0.29 μlm
21-2.1	220 Ω, 8.4 V
21-2.2	270 Ω, 390 Ω
21-3.1	1.84 mW
21-4.1	422 μA, 968 μA, 1.2 mA
21-4.2	33 kΩ + 2.2 kΩ
21-4.3	30 lx, 15 lx
21-5.1	+0.19 V
21-5.2	48
21-6.1	2.5 V, 13.9 V
21-6.2	13 mW/cm^2
21-7.1	150 Ω
21-7.2	9 mA, ±638mA, (3.96 V to 13.9 V), (±0.27 V to ±0.98 V)

CHAPTER 22
Miscellaneous Devices

CONTENTS

Objectives

You will be able to:

1   Explain the construction and operation of voltage-variable capacitor diodes (VVCs). Sketch typical VVC voltage/capacitance characteristics, draw the equivalent circuits, and discuss typical VVC parameters.

2   Design and analyze resonance circuits using VVCs for frequency tuning.

3   Discuss the construction and operation of thermistors, sketch typical thermistor resistance/temperature characteristics, and discuss typical thermistor parameters.

4   Calculate thermistor resistance at various temperatures from the data sheet information.

Design and analyze circuits using thermistors for temperature level detection.

5   Explain the construction and operation of tunnel diodes. Sketch typical forward and reverse characteristics for a tunnel diode, explain their shape, and identify the important points and regions of the characteristics.

6   Draw tunnel diode piecewise linear characteristics and analyze tunnel diode parallel amplifier circuits.

7   Explain the operation, characteristics, parameters, and applications of Schottky, PIN, and current-limiting diodes. Design and analyze circuits involving these devices.

INTRODUCTION

VVCs are *pn*-junction devices designed to produce a substantial change in junction capacitance when the reverse-bias voltage is adjusted. They can be applied to tune resonant circuits over a range of frequencies. The resistance of a thermistor changes significantly with change in temperature, so its major application is control of circuits that must respond to temperature change. The tunnel diode is a two-terminal negative-resistance device that can function as an oscillator, an amplifier, or a switch. Low forward voltage and fast turn-off time are features of the Schottky diode that make it suitable for preventing saturation in switching BJTs. The PIN diode behaves as a variable resistance at high frequencies, and the current-limiting diode is used in constant-current applications.

22-1 VOLTAGE-VARIABLE CAPACITOR DIODES

VVC Operation

Voltage-variable capacitor diodes (VVCs) are also known as *varicaps*, *varactors*, and *tuning diodes*. Basically, a VVC is a reverse-biased diode, and its capacitance is the junction capacitance. Recall that the width of the depletion region at a *pn*-junction depends upon the reverse-bias voltage (Fig. 22-1). A large reverse-bias produces a wide depletion region, and a small reverse-bias gives a narrow depletion region. The depletion region acts as a dielectric between two conducting plates; thus the junction behaves as a capacitor. The depletion layer capacitance (C_{pn}) is proportional to the junction area and inversely proportional to the width of the depletion region. Because the width of the depletion region is proportional to the reverse-bias voltage, C_{pn} is inversely proportional to the reverse-bias voltage. This is not a direct proportionality; instead C_{pn} is proportional to $1/V^n$, where V is the reverse-bias voltage and n depends upon doping density.

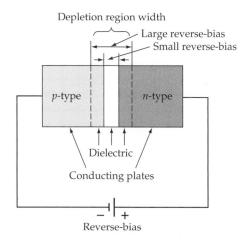

Figure 22-1   A voltage variable capacitor diode (VVC) is essentially a reverse-biased *pn*-junction. Increasing the reverse voltage widens the depletion region and reduces the capacitance.

Figure 22-2 shows the doping profiles for two types of VVC classified as *abrupt junction* and *hyperabrupt junction* devices. In the abrupt junction VVC, the semiconductor material is uniformly doped, and it changes abruptly from *p*-type to *n*-type at the junction. The hyperabrupt junction device has the doping density increased close to the junction. This increasing density produces a narrower depletion region, and so it results in a larger junction capacitance. It also causes the width of the depletion region to be more sensitive to bias voltage variations; thus it produces the largest capacitance change for a given voltage variation. VVCs are packaged just like ordinary low-current diodes.

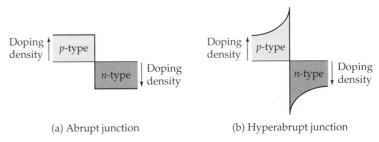

(a) Abrupt junction   (b) Hyperabrupt junction

Figure 22-2 Doping profiles for abrupt junction and hyperabrupt junction VVCs.

Equivalent Circuit

The complete equivalent circuit for a VVC is shown in Fig. 22-3a, and a simplified version is given in Fig. 22-3b. In the complete circuit, the junction capacitance (C_J) is shunted by the junction reverse leakage resistance (R_J). The resistance of the semiconductor material is represented by R_S, the terminal inductance is L_S, and the capacitance of the terminals (or the device package) is C_C. Because L_S is normally very small and R_J is very large, the equivalent circuit can be simplified (Fig. 22-3b) to R_S in series with C_T, where C_T is the sum of the junction and terminal capacitances ($C_T = C_J + C_C$). The Q-factor for a VVC can be as high as 600 at a 50 MHz frequency. However, since the Q-factor varies with bias voltage and frequency, it is used only as a figure of merit for comparing the performance of different VVCs.

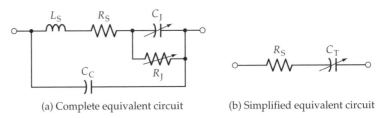

(a) Complete equivalent circuit   (b) Simplified equivalent circuit

Figure 22-3 The complete equivalent circuit of a VVC has five components. The simplified circuit is made up of the semiconductor resistance R_S and the total capacitance C_T.

Specification and Characteristics

A wide selection of VVC capacitances is available, ranging approximately from 6 pF to 700 pF. The *capacitance tuning ratio (TR)* is the ratio of C_T at a small reverse voltage to C_T at a large reverse voltage. In the partial specification for

Typical VVC Specification						
C_T $V_R = 1\ V, f = 1\ MHz$		C_1/C_{10} $f = 1\ MHz$	Q $V_R = 1\ V, f = 1\ MHz$	$V_{R(max)}$	$I_{R(max)}$	$I_{F(max)}$
min	max					
400 pF	600 pF	14	200	15 V	100 nA	200 mA

Figure 22-4 Partial specification for a voltage-variable capacitor diode (VVC).

a VVC shown in Fig. 22-4, the tuning ratio is listed as C_1/C_{10}. This is the ratio of the device capacitance at a 1 V reverse bias to that at a 10 V reverse bias. The 400 pF minimum capacitance (C_T) listed for a 1 V bias is changed to 400 pF/14 when the bias is 10 V. The specification also lists the Q-factor, as well as maximum reverse voltage, reverse leakage current, and the maximum forward current that can be passed when the device is forward-biased.

A typical graph of capacitance (C_T) versus reverse-bias voltage (V_R) for a hyperabrupt junction VVC is reproduced in Fig. 22-5 together with the VVC circuit symbol. It is seen that C_T varies (approximately) from 500 pF to 25 pF when V_R is changed from 1 V to 10 V. It should be noted from the specification in Fig. 22-4 that the nominal capacitance has a large tolerance (400 pF to 600 pF), and this must be taken into account when the C_T/V_R graphs are used.

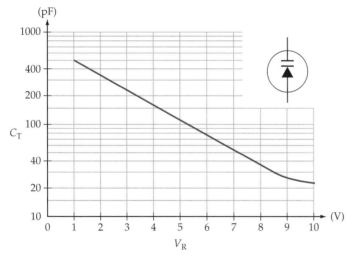

Figure 22-5 Capacitance/voltage characteristic for a hyperabrupt junction VVC, and VVC circuit symbol.

Applications

The major application of VVCs is as tuning capacitors to adjust the frequency of resonance circuits. An example of this is the circuit shown in Fig. 22-6, which is an amplifier with a tuned circuit load. The amplifier produces an output at the resonance frequency of the tuned circuit. The VVC (D_1) provides the capacitance (C_T) of the resonant circuit, and this can be altered by adjusting the diode (reverse) bias voltage (V_D). So the resonance frequency of the circuit can be varied. C_1 is a coupling capacitor with a capacitance much larger than that of the VVC, and R_2 limits the VVC forward current in the event that it becomes forward-biased.

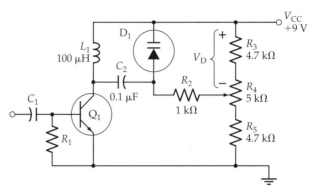

Figure 22-6   Amplifier stage with an *LC* tank circuit load. The resonance frequency of the *LC* circuit can be varied by adjusting the VVC reverse-bias voltage.

Example 22-1

Determine the maximum and minimum resonance frequency for the circuit in Fig. 22-6. Assume that D_1 has the C_T/V_R characteristic in Fig. 22-5.

Solution

$$V_{D(min)} = \frac{V_{CC} \times R_3}{R_3 + R_4 + R_5} = \frac{9\,V \times 4.7\,k\Omega}{4.7\,k\Omega + 5\,k\Omega + 4.7\,k\Omega}$$

$$= 2.9\,V$$

$$V_{D(max)} = \frac{V_{CC}\,(R_3 + R_4)}{R_3 + R_4 + R_5} = \frac{9\,V \times (4.7\,k\Omega + 5\,k\Omega)}{4.7\,k\Omega + 5\,k\Omega + 4.7\,k\Omega}$$

$$\approx 6.1\,V$$

From Fig. 22-5, at $V_D = 2.9\,V$, $C_T \approx 250\,pF$:

$$f_{(min)} = \frac{1}{2\pi\sqrt{(LC_T)}} = \frac{1}{2\pi\sqrt{(100\,\mu H \times 250\,pF)}}$$

$$\approx 1\,MHz$$

From Fig. 22-5, at $V_D = 6.1\,V$, $C_T \approx 70\,pF$:

$$f_{(min)} = \frac{1}{2\pi\sqrt{(LC_T)}} = \frac{1}{2\pi\sqrt{(100\,\mu H \times 70\,pF)}}$$

$$\approx 1.9\,MHz$$

Practice Problem

22-1.1 A tuned amplifier circuit as in Fig. 22-6 is to have a resonance frequency adjustable from 1.5 MHz to 2.5 MHz. A 12 V supply is used, and the inductor (L_1) is 80 µH. The VVC (D_1) has the characteristics in Fig. 22-5 and the specification in Fig. 22-4. Determine suitable resistance values for R_3, R_4, and R_5.

22-2 THERMISTORS

Thermistor Operation

The word *thermistor* is a combination of thermal and resistor. A thermistor is a resistor with definite thermal characteristics. Most thermistors have a negative temperature coefficient (NTC), but positive temperature coefficient (PTC) devices are also available. Thermistors are widely applied for measurement and control of temperature, liquid level, gas flow, and so on.

Silicon and germanium are not normally used for thermistor manufacture, because larger and more predictable temperature coefficients are available with metallic oxides. Various mixtures of manganese, nickel, cobalt, copper, iron, and uranium are pressed into desired shapes and sintered (or baked) at a high temperature to form thermistors. Electrical connections are made either by including fine wires during the shaping process or by silvering the surfaces after sintering (see Fig. 22-7a). Thermistors are made in the shape of beads, probes, discs, washers, and so on (Fig. 22-7b). Beads may be glass-coated or enclosed in evacuated or gas-filled glass envelopes for protection against corrosion.

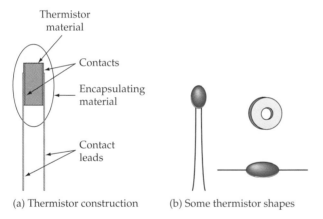

(a) Thermistor construction   (b) Some thermistor shapes

Figure 22-7 Thermistors are resistors that are very sensitive to temperature.

Characteristics and Specifications

The typical thermistor resistance/temperature characteristic in Fig. 22-8 shows that the device resistance (R) decreases substantially when its temperature is raised. At 0°C, $R \approx 1.5$ kΩ; and at 60°C, $R \approx 70$ Ω. Current flow through a thermistor causes power dissipation that can raise its temperature and change its resistance. This could introduce errors in the thermistor application, so device currents are normally kept to a minimum.

Figure 22-9 shows partial specifications for two thermistors with widely different resistance values. Both devices have the resistance specified at 25°C as the *zero power resistance*. This, of course, means that there must be zero power dissipation in the thermistor to give this resistance value. The *dissipation constant* is the device power dissipation that can raise its temperature

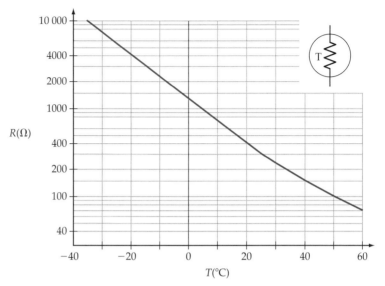

Figure 22-8 Typical resistance/temperature characteristics for a negative temperature co-efficient (NTC) thermistor.

		Typical Thermistor Specifications				
Thermistor	Zero power resistance at 25°C	Resistance ratio 25°C/125°C	β (0 to 50°C)	Maximum working temperature	Dissipation constant	
44002A	300 Ω	15.15	3118	100°C	1 mW/°C in still air, 8 mW/°C in moving liquid	
44008	30 kΩ	29.15	3810	150°C		

Figure 22-9 Partial specifications for two thermistors, one with a 300 Ω 25°C resistance, and the other with a 30 kΩ 25°C resistance.

through 1°C. The dissipation constant in both cases is specified as *1 mW/°C in still air*, and *8 mW/°C in moving liquid*. Thus, a thermistor located in still air conditions could have its temperature increased by 1°C if its current produces 1 mW of power dissipation.

An indication of how much the thermistor resistance changes is given by the *resistance ratio at 25/125°C*. Clearly, with this ratio specified as 15.15, the resistance at 25°C is divided by 15.15 to determine the resistance at 125°C. Note that maximum working temperatures are listed for both devices. The resistance change with temperature is also defined by the constant beta (β), this time for the range 0°C to 50°C. This constant is used in an equation that relates resistance values at different temperatures:

$$\ln \frac{R_1}{R_2} = \beta \left(\frac{1}{T_1} - \frac{1}{T_2} \right)$$

(22-1)

In Equation 22-1, R_1 is the resistance at temperature T_1, and R_2 is the resistance T_2. It is important to note that T_1 and T_2 are *absolute* (or *Kelvin*) temperature values: (°C + 273) K.

Example 22-2

Calculate the resistance of the 300 Ω thermistor specified in Fig. 22-9 at temperatures of 20°C and 30°C.

Solution

For $T = 20°C$: $T_1 = 25°C + 273 = 298$ K

and $T_2 = 20°C + 273 = 293$ K

From Eq. 22-1, $R_2 = \dfrac{R_1}{e^{B(1/T1-1/T2)}} = \dfrac{300\ \Omega}{e^{3118(1/298-1/293)}}$

                      $= 358\ \Omega$

For $T = 30°C$: $T_1 = 25°C + 273 = 298$ K

and $T_2 = 30°C + 273 = 303$ K

From Eq. 22-1, $R_2 = \dfrac{R_1}{e^{B(1/T1-1/T2)}} = \dfrac{300\ \Omega}{e^{3118(1/298-1/303)}}$

                      $= 252\ \Omega$

Applications

Figure 22-10 shows a thermistor connected as a feedback resistor in an inverting amplifier circuit. (Note the device circuit symbol.) In this case, the thermistor is supplied with a constant current determined by R_1 and V_i. The output voltage is directly proportional to the thermistor resistance, and so V_o varies with temperature change.

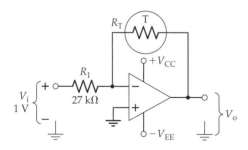

Figure 22-10   Use of an inverting amplifier to produce a constant current through a thermistor.

The circuit in Fig. 22-11 illustrates how a thermistor can be used for triggering a Schmitt circuit at a predetermined temperature. This could be air temperature or the temperature of a liquid or perhaps the temperature of some type of heating appliance. When the thermistor resistance (R_T) is increased by the device temperature decrease, the Schmitt input voltage is raised to the upper trigger point, causing the output to switch negatively.

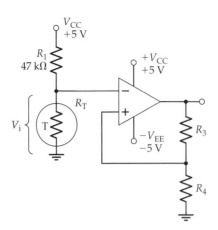

Figure 22-11 Schmitt trigger circuit using a thermistor input stage for temperature level detection.

Example 22-3

Calculate V_i for the Schmitt circuit in Fig. 22-11 at 25°C and at 28°C if the thermistor is the 300 Ω device specified in Fig. 22-9.

Solution

At 25°C, $\qquad\qquad R_T = 300\ \Omega$

$$V_i = \frac{V_{CC} \times R_T}{R_1 + R_T} = \frac{5\ \text{V} \times 300\ \Omega}{47\ \text{k}\Omega + 300\ \Omega}$$

$$= 31.7\ \text{mV}$$

For $T = 28°C$: $\qquad\qquad T_1 = 25°C + 273 = 298\ \text{K}$

and $\qquad\qquad T_2 = 28°C + 273 = 301\ \text{K}$

From Eq. 22-1, $\qquad R_2 = \dfrac{R_1}{e^{\beta(1/T1 - 1/T2)}} = \dfrac{300\ \Omega}{e^{3118(1/298 - 1/301)}}$

$$= 270\ \Omega$$

$$V_i = \frac{V_{CC} \times R_T}{R_1 + R_T} = \frac{5\ \text{V} \times 270\ \Omega}{47\ \text{k}\Omega + 270\ \Omega}$$

$$= 28.6\ \text{mV}$$

Practice Problems

22-2.1 Calculate the output voltage from the circuit in Fig. 22-10 at 25°C and 28°C if the 30 kΩ thermistor specified in Fig. 22-9 is used.

22-2.2 If the Schmitt circuit in Fig. 22-11 has $UTP = 1$ V, calculate a suitable resistance value for R_1 for the circuit to trigger at 18°C. The thermistor used is the 300 Ω device specified in Fig. 22-9.

22-3 TUNNEL DIODES

Tunnel Diode Operation

A *tunnel diode* (sometimes called an *Esaki diode* after its inventor, Leo Esaki) is a two-terminal *negative resistance* device that can be employed as an amplifier, an oscillator, or a switch. Recall from Chapter 1 that the width of the depletion region at a *pn*-junction depends upon the doping density of the semiconductor material. Lightly doped material has a wide depletion region, whereas heavily doped material has a narrow region. A tunnel diode uses very heavily doped semiconductor material, so the depletion region is extremely narrow. This is illustrated in Fig. 22-12 along with three tunnel diode circuit symbols.

The depletion region is an insulator because it lacks charge carriers, and usually charge carriers can cross it only when the external bias is large enough to overcome the barrier potential. However, because the depletion region in a tunnel diode is so narrow, it does not constitute a large barrier to electron flow. Consequently, a small forward or reverse bias (not large enough to overcome the barrier potential) can give charge carriers sufficient energy to cross the depletion region. When this occurs, the charge carriers are said to be *tunnelling* through the barrier.

When a tunnel diode junction is reverse-biased (negative on the *p*-side, positive on the *n*-side), substantial current flow occurs because of the tunnelling effect (electrons moving from the *p*-side to the *n*-side). Increasing levels of reverse-bias voltage produce more tunnelling and a greater reverse current. So, as shown in Fig. 22-13, the reverse characteristic of a tunnel diode is linear, just like that of a resistor.

A forward-biased tunnel diode initially behaves like a reverse-biased device. Electron tunnelling occurs from the *n*-side to the *p*-side, and the forward

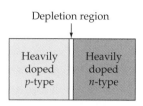

(a) A heavily-doped *pn*-junction has a very narrow depletion region

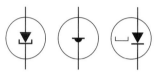

(b) Tunnel diode circuit symbols

Figure 22-12 A tunnel diode has a heavily doped *pn*-junction, which results in a very narrow depletion region.

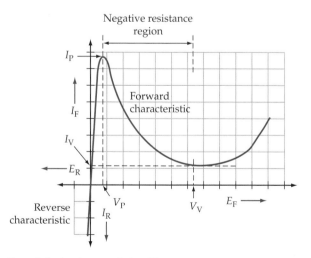

Figure 22-13 Tunnel diode characteristics. The current increases to a peak level (I_P) as the forward bias is increased, and then falls off to a valley current (I_V) with increasing bias voltage.

current (I_F) continues to increase with increasing levels of forward voltage (E_F). Eventually, a peak level of tunnelling is reached, and then further increase in E_F actually causes I_F to decrease. (See the forward characteristic in Fig. 22-13.) The decrease in I_F with increasing E_F continues until the normal process of current flow across a forward-biased junction begins to take over when the bias voltage becomes large enough to overcome the barrier potential. I_F now starts to increase with increasing levels of E_F, so that the final portion of the tunnel diode forward characteristics is similar to that for an ordinary *pn*-junction. The shape of the tunnel diode characteristics can be explained in terms of energy band diagrams for the semiconductor material.

Characteristics and Parameters

Consider the typical tunnel diode forward characteristics shown in Fig. 22-14. The *peak current* (I_P) and *valley current* (I_V) are easily recognized on the forward characteristic as the maximum and minimum levels of I_F before the junction is completely forward-biased. The peak voltage (V_P) is the level of forward bias voltage (E_F) corresponding to I_P, and the *valley voltage* (V_V) is the E_F level at I_V. V_F is the forward voltage drop when the device is completely forward-biased. The dashed line at the bottom of the forward characteristic shows the characteristic for an ordinary forward-biased diode. It is seen that this joins the tunnel diode characteristic as V_F is approached.

When a voltage is applied to a resistance, the current normally increases as the applied voltage is increased. Between I_P and I_V on the tunnel diode characteristic, I_F actually decreases as E_F is increased. So this region of the characteristic is named the *negative resistance region*, and the *negative resistance* (R_D) of the tunnel diode is its most important property.

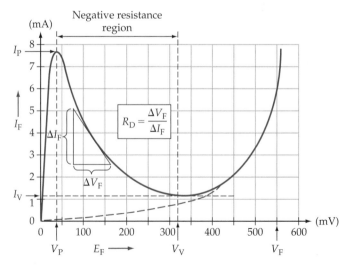

Figure 22-14 Typical forward characteristic for a tunnel diode. Note that the negative resistance region exists between forward-bias voltages of approximately 50 mV and 325 mV.

The negative resistance value can be determined as the reciprocal of the slope of the characteristic in the negative resistance region. From Fig. 22-14, it can be seen that the negative resistance is $R_D = \Delta V_F/\Delta I_F$, and the *negative conductance* is $G_D = \Delta I_F/\Delta V_F$. If R_D is measured at different points on the negative resistance portion of the characteristic, slightly different values will be obtained at each point because the slope is not constant. Therefore, R_D is usually specified at the centre of the negative resistance region. Figure 22-15 lists typical tunnel diode parameters.

It is shown in Chapter 2 that a straight-line approximation of diode characteristics can sometimes be conveniently employed. For a tunnel diode, the *piecewise linear characteristics* can usually be constructed from data provided by the device manufacturer.

Typical Tunnel Diode Parameters					
I_P (mA)	V_P (mV)	I_V (mA)	V_V (mV)	V_F (V)	R_D (Ω)
1 to 100	50 to 200	0.1 to 5	350 to 500	0.5 to 1	-10 to -200

Figure 22-15   Tunnel diode specification data showing the range of parameters.

Example 22-4

Construct the piecewise linear characteristics and determine R_D for a 1N3712 tunnel diode from the following data: $I_P = 1$ mA, $I_V = 0.12$ mA, $V_P = 65$ mV, $V_V = 350$ mV, and $V_F = 500$ mV at $I_F = I_P$.

Solution

Refer to Fig. 22-16.

Plot point 1 at $I_P = 1$ mA   and $V_P = 65$ mV

Plot point 2 at $I_V = 0.12$ mA   and $V_V = 350$ mV

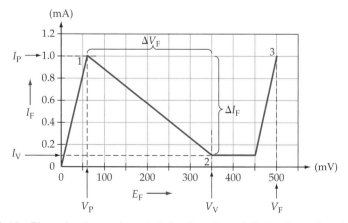

Figure 22-16   Piecewise linear characteristics for a tunnel diode, drawn from information provided on the device specification.

Draw the first portion of the characteristic from the zero point to point 1. Draw the negative resistance portion between points 1 and 2.

Plot point 3 at $\qquad I_F = I_P \quad$ and $\quad V_F = 500\ \text{mV}$

Draw the final portion of the characteristics at the same slope as the line between point 0 and point 1.

Draw the horizontal part of the characteristic from point 2 to the final portion.

$$R_D = \frac{\Delta E_F}{\Delta I_F} = \frac{350\ \text{mV} - 65\ \text{mV}}{-(1\ \text{mA} - 0.12\ \text{mA})}$$

$$= -324\ \Omega$$

Parallel Amplifier

For operation as an amplifier, a tunnel diode must be biased to the centre of its negative resistance region. Figure 22-17a shows the basic circuit of a tunnel diode *parallel amplifier*. Load resistor R_L is connected in parallel with diode D_1 and supplied with current from voltage source E_B and signal source e_s.

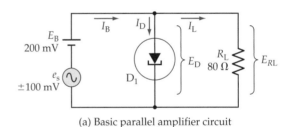

(a) Basic parallel amplifier circuit

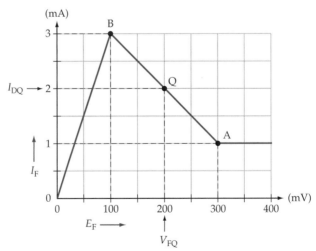

(b) Circuit current and voltage levels

Figure 22-17   A basic tunnel diode parallel amplifier has a load resistance in parallel with the diode and the (series-connected) bias and signal sources applied directly to the diode and load.

Figure 22-17b uses the tunnel diode piecewise linear characteristics to show the dc conditions of the diode when the signal voltage is zero ($e_s = 0$) and when $e_s = \pm100$ mV. The operation of the circuit is explained by the analysis in Example 22-5, which also demonstrates that a parallel amplifier has current gain but no voltage gain.

Example 22-5

Assuming that E_B and e_s have zero source resistance, calculate the current gain and voltage gain for the tunnel diode parallel amplifier in Fig. 22-17a. The device piecewise linear characteristics are given in Fig. 22-17b.

Solution

When $e_s = 0$:

$$E_{DQ} = E_B = 200 \text{ mV (point } Q \text{ on Fig. 22-17b)}$$

At the Q-point, $I_{DQ} = 2$ mA

Also, $E_{RLQ} = E_B = 200$ mV

$$I_{RLQ} = \frac{E_{RL}}{R_L} = \frac{200 \text{ mV}}{80 \text{ } \Omega}$$

$$= 2.5 \text{ mA}$$

$$I_{BQ} = I_{RL} + I_D = 2.5 \text{ mA} + 2 \text{ mA}$$

$$= 4.5 \text{ mA}$$

When $e_s = +100$ mV:

$$E_B + e_s = 200 \text{ mV} + 100 \text{ mV} = 300 \text{ mV}$$

$$E_D = E_{RL(A)} = 300 \text{ mV (point A on Fig. 22-17b)}$$

and $I_{D(A)} = 1$ mA

Also, $I_{RL(A)} = \frac{E_{RL(A)}}{R_L} = \frac{300 \text{ mV}}{80 \text{ } \Omega}$

$$= 3.75 \text{ mA}$$

$$I_{B(A)} = I_{RL(A)} + I_{D(A)} = 3.75 \text{ mA} + 1 \text{ mA}$$

$$= 4.75 \text{ mA}$$

When $e_s = -100$ mV:

$$E_B + e_s = 200 \text{ mV} - 100 \text{ mV} = 100 \text{ mV}$$

$$E_D = E_{RL(B)} = 100 \text{ mV (point B on Fig. 22-17b)}$$

and $I_{D(B)} = 3$ mA

Also,
$$I_{RL(B)} = \frac{E_{RL(B)}}{R_L} = \frac{100 \text{ mV}}{80 \text{ }\Omega}$$

$$= 1.25 \text{ mA}$$

$$I_{B(B)} = I_{RL(B)} + I_{D(B)} = 1.25 \text{ mA} + 3 \text{ mA}$$

$$= 4.25 \text{ mA}$$

Total load current change:
$$\Delta I_{RL} = I_{RL(A)} - I_{RL(B)} = 3.75 \text{ mA} - 1.25 \text{ mA}$$

$$= 2.5 \text{ mA}$$

Total signal current change:
$$\Delta I_B = I_{B(A)} - I_{B(B)} = 4.75 \text{ mA} - 4.25 \text{ mA}$$

$$= 0.5 \text{ mA}$$

Current gain:
$$A_i = \frac{\Delta I_{RL}}{\Delta I_B} = \frac{2.5 \text{ mA}}{0.5 \text{ mA}}$$

$$= 5$$

Voltage gain:
$$A_v = \frac{\Delta E_{RL}}{e_s} = \frac{\pm 100 \text{ mV}}{\pm 100 \text{ mV}}$$

$$= 1$$

The current gain equation for a tunnel diode parallel amplifier can be shown to be

$$A_i = \frac{R_D}{R_D - R_L} \tag{22-2}$$

Note that R_D is already taken as negative in Equation 22-2, so that only the absolute value should be used in calculating A_i. For $R_D = 100 \text{ }\Omega$ and $R_L = 80 \text{ }\Omega$, as in Example 22-5,

$$A_i = \frac{100}{100 \text{ }\Omega - 80 \text{ }\Omega} = 5$$

From Equation 22-2, it can be seen that when $R_L \ll R_D$, $A_i \approx 1$; when $R_L \gg R_D$, $A_i < 1$; and when $R_L = R_D$, $A_i = \infty$. A current gain of infinity means that the circuit is likely to oscillate. For maximum stable current gain, R_L should be selected to be just slightly less than R_D.

Figure 22-18 shows the circuit of a practical tunnel diode parallel amplifier. The signal voltage e_s and load resistor R_L are capacitor-coupled to the diode, while dc bias is provided by source voltage E_B and voltage divider R_1 and R_2. Inductor L_1 and capacitor C_1 isolate the bias supply from ac signals.

A tunnel diode *series amplifier* can be constructed. In this case the device is connected in series with the load, and voltage amplification is obtained instead of current amplification. Oscillators and switching circuits can also be constructed with tunnel diodes.

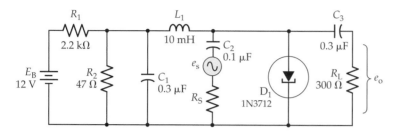

Figure 22-18 In a practical parallel amplifier circuit, the load and signal source are capacitor-coupled to the tunnel diode.

Practice Problem

22-3.1 Draw the dc and ac equivalent circuits for the tunnel diode parallel amplifier in Fig. 22-18. Calculate the current gain.

22-4 SCHOTTKY, PIN, AND CURRENT-LIMITING DIODES

The Schottky Diode

The *Schottky diode* (also known as a *hot-carrier diode*) has a metal-to-semiconductor junction instead of a *pn*-junction. This gives the device a lower forward voltage drop and faster reverse recovery than a *pn*-junction diode. Figure 22-19a illustrates the basic structure of the device. An *n*-type epitaxial layer is created on a low-resistance *n*-type substrate (see Chapter 7), and a metal film (the *barrier metal*) is deposited on the epitaxial layer. Additional layers of metal are deposited on the outer surfaces for anode and cathode connections, as illustrated. The type of metal used at the junction affects the device performance.

The metal side of the junction has an abundance of free electrons, and there are no *holes* in either the metal or the *n*-type material. Thus there are no minority charge carriers and there is virtually no depletion region. This means that when the junction is switched from forward to reverse bias, there are no charge carriers to be withdrawn from a depletion region (see Section 2-6), and there is no charge carrier recombination. Consequently, the reverse recovery time is extremely small—on the order of picoseconds.

The typical forward characteristic in Fig. 22-19b show a voltage drop (V_F) ranging from 0.2 V at a 10 mA forward current to about 0.4 V at 1 A. As illustrated, the reverse current is approximately 0.5 mA at 25 °C, which is substantially greater than for most *pn*-junction diodes. Another disadvantage is that the 50 V typical reverse breakdown voltage is lower than for other types of diodes.

The two important advantages of Schottky diodes over *pn*-junction diodes are the lower forward voltage drop and the much faster turn-off time. Figure 22-20 shows the use of a Schottky diode in an application where both of these advantages are employed: clamping a switched BJT to keep it from saturating. Without the diode, Q_1 saturation voltage would be $V_{CE(sat)} \approx 0.2$ V,

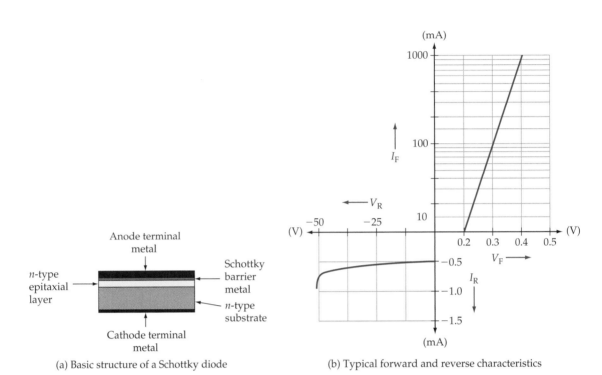

(a) Basic structure of a Schottky diode

(b) Typical forward and reverse characteristics

Figure 22-19 A Schottky diode has a metal-to-semiconductor junction, which gives it a lower forward voltage drop and faster turn-off time than a *pn*-junction diode.

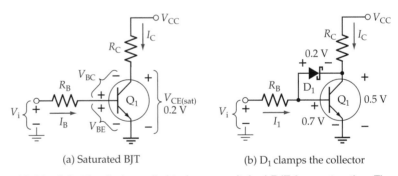

(a) Saturated BJT

(b) D_1 clamps the collector

Figure 22-20 Schottky diode applied to keep a switched BJT from saturating. The diode clamps the BJT collector at 0.2 V below the base voltage.

which gives $V_{BC} \approx 0.5$ V (Fig. 22-20a). If D_1 in Fig. 22-20b has $V_F = 0.2$ V, then the collector is clamped to 0.2 V below the base voltage. Thus, the minimum level of V_{CE} is 0.5 V, which means that Q_1 is *not* saturated, and it can be switched off much faster than a saturated BJT. Note the graphic symbol used for the Schottky diode.

Schottky diodes can be fabricated as high-current devices for use as rectifiers, and in this application the low forward voltage drop means that the (wasted) power dissipation in the device is lower than with *pn*-junction rectifiers. Another important application is as a catch diode in switching regulators (see Chapter 18).

PIN Diode

A *PIN diode* consists of a layer of *intrinsic* (undoped) semiconductor material sandwiched between *p*-type and *n*-type material. This is illustrated in Fig. 22-21a. The intrinsic layer gives the device a very low reverse leakage current and a high reverse breakdown voltage. When forward-biased, the device does not begin to conduct until the bias voltage typically exceeds 0.75 V, and V_F may approach 1 V at normal I_F levels. When reverse-biased, it behaves like a constant-value capacitor.

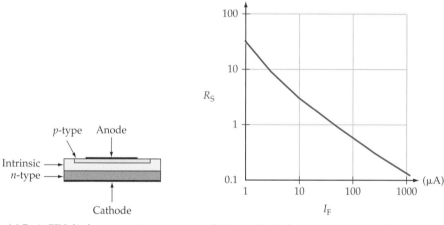

(a) Basic PIN diode construction   (b) Typical high-frequency resistance characteristic

Figure 22-21 A PIN diode has a layer of intrinsic material sandwiched between *p*-type and *n*-type layers. At high frequencies the device has a resistance that is dependent on the dc forward current level.

At dc and low-frequency operation, a PIN diode has similar characteristics to an ordinary *pn*-junction diode, except for the higher forward voltage drop. When operated in forward bias at radio frequencies, the device behaves like a resistor with its resistive value dependent on the dc forward bias current. From the typical R_S/I_F graph in Fig. 22-21b, R_S ranges from 0.1 Ω at $I_F = 1$ mA to approximately 20 Ω at $I_F = 1$ μA.

The PIN diode is applied as a variable resistor and as a switch in RF circuits, and also as a power rectifier in situations where the high reverse voltage is important.

Current-Limiting Diode

Refer to the JFET I_D/V_{DS} characteristic in Fig. 9-8, and note that, with $V_{GS} = 0$, I_D remains constant over the V_{DS} range from V_P to the break-down voltage. So a JFET can be operated as a constant-current device if the gate and source terminals are connected. This is the basis of the *current-limiting diode (CLD)*, also known as a *field-effect diode* and as a *current regulator diode (CRD)*.

A CLD is a field-effect device designed to operate at a particular current level and to maintain the current constant over a wide range of applied voltage. The typical I_F/V_F characteristics in Fig. 22-22 show that there is a lower limit to the CLD operating voltage. Termed the *knee voltage* (V_K), the lowest

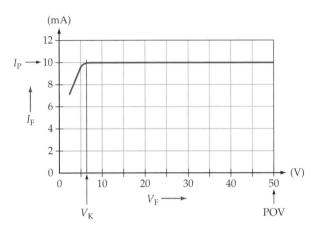

Figure 22-22 Forward characteristics for a current-limiting diode (CLD). The current remains constant over a wide range of applied voltage.

operating voltage is usually about 6 V. The device will, of course, break down when a high enough voltage is applied. So there is an upper voltage limit, known as the *peak operating voltage (POV)*. The constant-current level is identified as the *regulator current* (I_P). Because the current remains substantially constant when the voltage changes, the device has a very high *regulator impedance* ($Z_s = \Delta V_F/\Delta I_F$).

CLDs are available with I_P levels ranging from 35 μA to 15 mA, Z_s values of 200 kΩ to 25 MΩ, and maximum POVs of 50 V to 100 V. The 1N5414 CLD has I_P = 4.7 mA, Z_s = 235 kΩ, and POV = 100 V. Although the CLD is referred to as a diode, it *should not be reverse-biased* because it does not block reverse current flow.

Some CLD applications are illustrated in Fig. 22-23 and 22-24. (Note the the CLD graphic symbol.) The CLD in Fig. 22-23a holds the LED current constant regardless of supply voltage variations. Thus the LED illumination level does not change when V_{CC} increases or decreases. The differential amplifier circuit in Fig. 22-23b has a CLD providing a constant emitter current that will not change when common-mode voltages appear at the amplifier input terminals. The circuit in Fig. 22-24 is a *current multiplier/divider*. The constant level of I_P maintains a constant voltage drop across resistor R_1, and the voltage across R_2

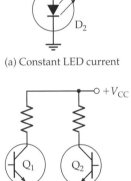

(a) Constant LED current

(b) Differential amplifier constant emitter current

Figure 22-23 Current-limiting diode used for maintaining a constant current level through a LED and in the emitter circuit of a differential amplifier.

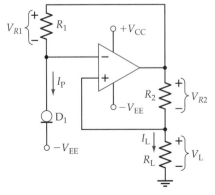

Figure 22-24 Current multiplier/divider using a CLD. The load current is $I_L = I_P \times R_1/R_2$.

equals V_{R1}. So the load current (I_L) (through R_2) is held constant at a higher or lower level than I_P, depending on the ratio R_1/R_2.

Example 22-6

In the circuit in Fig. 22-24, $V_{CC} = \pm15$ V, $R_1 = 6.8$ kΩ, $R_2 = 680$ Ω, and D_1 has $I_P = 1$ mA. Calculate the V_L, V_{D1}, and the op-amp output voltage (V_o) when $R_L = 50$ Ω, and when $R_L = 200$ Ω.

Solution

$$V_{R1} = I_P R_1 = 1 \text{ mA} \times 6.8 \text{ k}\Omega$$

$$= 6.8 \text{ V}$$

$$V_{R2} = V_{R1} = 6.8 \text{ V}$$

$$I_L = \frac{V_{R2}}{R_2} = \frac{6.8 \text{ V}}{680 \text{ }\Omega}$$

$$= 10 \text{ mA}$$

With $R_L = 50$ Ω,
$$V_{RL} = I_L R_L = 10 \text{ mA} \times 50 \text{ }\Omega$$

$$= 0.5 \text{ V}$$

$$V_{(-)} = V_{RL} = 0.5 \text{ V}$$

$$V_{D1} = V_{(-)} - V_{EE} = 0.5 \text{ V} - (-15 \text{ V})$$

$$= 15.5 \text{ V}$$

$$V_o = V_{RL} + V_{R2} = 0.5 \text{ V} + 6.8 \text{ V}$$

$$= 7.3 \text{ V}$$

With $R_L = 200$ Ω,
$$V_{RL} = I_L R_L = 10 \text{ mA} \times 200 \text{ }\Omega$$

$$= 2 \text{ V}$$

$$V_{(-)} = V_{RL} = 2 \text{ V}$$

$$V_{D1} = V_{(-)} - V_{EE} = 2 \text{ V} - (-15 \text{ V})$$

$$= 17 \text{ V}$$

$$V_o = V_{RL} + V_{R2} = 2 \text{ V} + 6.8 \text{ V}$$

$$= 8.8 \text{ V}$$

Practice Problems

22-4.1 The circuit in Fig. 22-20b has the following quantities: $V_{CC} = +5$ V, $R_C = 1$ kΩ, $R_B = 12$ kΩ, and $V_i = +3$ V. Calculate the diode current.

22-4.2 Modify the circuit in Example 22-6 to pass a 100 μA current through a 10 kΩ load. Calculate V_{D1} and V_o if $V_{CC} = \pm18$ V.

Review Questions

Section 22-1

22-1 Using illustrations, explain the operation of a VVC diode. Sketch the doping profile at abrupt and hyperabrupt junctions, and explain the difference between the two.

22-2 Sketch the equivalent circuit for a VVC. Explain the origin of each component and show how the circuit may be simplified.

22-3 List the most important VVC parameters and state typical parameter values.

22-4 Sketch a circuit to show a typical VVC application. Briefly explain.

Section 22-2

22-5 Sketch typical resistance/temperature characteristics for a thermistor, and discuss the thermistor operation.

22-6 List the most important parameters for a thermistor, and state typical parameter values.

22-7 Sketch a circuit diagram to show how a thermistor can be used to control a Schmitt trigger circuit. Explain the circuit operation.

22-8 Draw a diagram to show how a thermistor might be used to compensate for V_{BE} variations (due to temperature change) in a voltage-divider biased BJT circuit.

Section 22-3

22-9 Discuss the difference between a tunnel diode and an ordinary pn-junction diode. Explain what is meant by *tunnelling*.

22-10 Sketch typical forward and reverse characteristics for a tunnel diode. Discuss the shape of the characteristics, and identify the regions and important points on the characteristics.

22-11 List the most important parameters for a tunnel diode, and state typical parameter values.

22-12 Sketch the basic circuit of a tunnel diode parallel amplifier, explain its operation, and write the equation for amplifier current gain.

22-13 Sketch a practical tunnel diode parallel amplifier circuit, and discuss the function of each component.

Section 22-4

22-14 Sketch the basic construction and typical characteristics for a Schottky diode. Briefly explain the device operation, and discuss its most important characteristics.

22-15 Draw a circuit diagram to show how a Schottky diode is used to improve the turn-off time of a switched BJT. Explain the circuit operation.

22-16 Draw a sketch to show the basic construction of a PIN diode, and explain how it operates in comparison to a pn-junction diode. Sketch the typical high frequency resistance/current characteristic of the device, and discuss its operation at high frequencies.

22-17 Draw typical forward characteristics for a current-limiting diode. Explain how the CLD operates, and discuss its limitations.

22-18 Show how a CLD may be applied for LED current stabilization and for differential amplifier emitter current stabilization. Briefly explain.

22-19 Sketch the circuit diagram of a current multiplier/divider. Explain how the load current relates to the CLD current.

Problems

Section 22-1

22-1 A tuned amplifier circuit similar to Fig. 22-6 has $V_{CC} = 15$ V and the following component values: $L_1 = 80$ μH, $R_3 = 1$ kΩ, $R_4 = 10$ kΩ, and $R_5 = 4.7$ kΩ. Assuming that D_1 has the C_T/V_R characteristic in Fig. 22-5, determine the maximum and minimum resonance frequency for the circuit.

22-2 If the VVC in the circuit in Problem 22-1 is replaced with the VVC specified in Fig. 22-4, calculate the highest and lowest possible resonance frequency for the circuit.

22-3 A tuned amplifier circuit as in Fig. 22-6 is to have its resonance frequency adjustable from 0.8 MHz to 1.2 MHz. Determine suitable resistance values for R_3, R_4, and R_5 if $V_{CC} = 18$ V, $L_1 = 100$ μH, and the VVC characteristics are those in Fig. 22-5.

22-4 Determine the bias voltage for the VVC in Problem 22-3 to give a 1 MHz resonance frequency.

22-5 The VVC in the circuit in Problem 22-1 is replaced with another one that gives a resonance frequency ranging from 900 kHz to 3.5 MHz. Specify the new VVC in terms of its capacitance and tuning ratio from 1 V to 10 V.

Section 22-2

22-6 A thermistor with a 1 kΩ resistance at 25°C has β specified as 3395. Calculate the thermistor resistance at 5°C and at 35°C temperatures.

22-7 Calculate the temperature of the 30 kΩ thermistor specified in Fig. 22-9 when its resistance is measured as 24.5 kΩ.

22-8 In a thermistor circuit such as in Fig. 22-10, $V_i = -1$ V and $R_1 = 22$ kΩ. Calculate the output voltage at 25°C and 35°C if the thermistor is the 30 kΩ device specified in Fig. 22-9.

22-9 The 300 Ω thermistor specified in Fig. 22-9 is connected in series with a 1.5 kΩ resistor (R_1) and a 12 V supply. Determine the voltage drop across R_1 at temperatures of 22°C, 28°C, and 31°C.

Section 22-3

22-10 A tunnel diode is specified as having $I_P = 6$ mA, $V_P = 50$ mV, $I_V = 0.5$ mA, $V_V = 400$ mV, and $V_F = 550$ mV at $I_F = I_P$. Construct the piecewise linear characteristics for the device, and determine its negative resistance value.

22-11 Construct the piecewise linear characteristics for a 1N3715 from the following data: $I_P = 2.2$ mA, $I_V = 0.21$ mA, $V_P = 65$ mV, $V_V = 355$ mV, and $V_F = 510$ mV at $I_F = I_P$. Determine R_D for the device.

22-12 A parallel amplifier uses the tunnel diode specified in Problem 22-10 and a load resistance of 47 Ω. Calculate the circuit current gain.

22-13 A 1N3715 is to be connected as a parallel amplifier. Referring to the piecewise linear characteristics drawn for Problem 22-11, calculate the current gain if $R_L = 120$ Ω.

22-14 A practical tunnel diode parallel amplifier circuit such as in Fig. 22-18 uses a device with $I_P = 5$ mA, $V_P = 50$ mV, $I_V = 1$ mA, and $V_V = 400$ mV. Construct the piecewise linear characteristics and calculate the circuit current gain if $R_L = 75$ Ω.

Section 22-4

22-15 Calculate the new peak output current and load voltage for the rectifier circuit in Example 3-1 when a Schottky diode is used instead of a *pn*-junction diode.

22-16 Recalculate the current levels in the AND gate circuit referred in Example 3-24 if D_1, D_2, and D_3 are Schottky diodes.

22-17 A current multiplier/divider circuit as in Fig. 22-24 has $R_L = 1$ kΩ and a IN5309 CLD with $I_P = 3$ mA. Determine suitable resistor values for the circuit to pass a 1 mA load current. The op-amp has $V_{CC} = \pm 12$ V.

22-18 An LED is connected as the load in a current multiplier/divider circuit (as in Fig. 22-24). The supply voltage is $V_{CC} = \pm 12$ V, and the CLD is a IN5302 with $I_P = 1.5$ mA. Determine suitable resistor values if the LED current is to be maintained constant at 20 mA.

Practice Problem Answers

22-1.1 39 kΩ, 35 kΩ, 47 kΩ
22-2.1 -1.1 V, -0.98 V
22-2.2 1.5 kΩ
22-3.1 13.5
22-4.1 100 μA
22-4.2 68 kΩ, 19 V, 7.8 V

APPENDICES

APPENDIX A   DEVICE DATA SHEETS

Data Sheet A-1
1N914 to 1N917 Diodes

National Semiconductor™

Discrete POWER & Signal Technologies

1N/FDLL 914/A/B / 916/A/B / 4148 / 4448

DO-35

LL-34

THE PLACEMENT OF THE EXPANSION GAP HAS NO RELATIONSHIP TO THE LOCATION OF THE CATHODE TERMINAL

COLOR BAND MARKING		
DEVICE	**1ST BAND**	**2ND BAND**
FDLL914	BLACK	BROWN
FDLL914A	BLACK	GRAY
FDLL914B	BROWN	BLACK
FDLL916	BLACK	RED
FDLL916A	BLACK	WHITE
FDLL916B	BROWN	BROWN
FDLL4148	BLACK	BROWN
FDLL4448	BROWN	BLACK

High Conductance Fast Diode

Sourced from Process D3.

Absolute Maximum Ratings*   TA = 25°C unless otherwise noted

Symbol	Parameter	Value	Units
W_{IV}	Working Inverse Voltage	75	V
I_O	Average Rectified Current	200	mA
I_F	DC Forward Current	300	mA
i_f	Recurrent Peak Forward Current	400	mA
$i_{f(surge)}$	Peak Forward Surge Current Pulse width = 1.0 second Pulse width = 1.0 microsecond	 1.0 4.0	 A A
T_{stg}	Storage Temperature Range	-65 to +200	°C
T_J	Operating Junction Temperature	175	°C

*These ratings are limiting values above which the serviceability of any semiconductor device may be impaired.

NOTES:
1) These ratings are based on a maximum junction temperature of 200 degrees C.
2) These are steady state limits. The factory should be consulted on applications involving pulsed or low duty cycle operations.

Thermal Characteristics   TA = 25°C unless otherwise noted

Symbol	Characteristic	Max	Units
		1N/FDLL 914/A/B / 4148 / 4448	
P_D	Total Device Dissipation Derate above 25°C	500 3.33	mW mW/°C
$R_{\theta JA}$	Thermal Resistance, Junction to Ambient	300	°C/W

Data Sheet A-1 *(Continued)*

1N/FDLL 914/A/B / 916/A/B / 4148 / 4448

High Conductance Fast Diode
(continued)

Electrical Characteristics   TA = 25°C unless otherwise noted

Symbol	Parameter	Test Conditions	Min	Max	Units
B_V	Breakdown Voltage	$I_R = 100\ \mu A$	100		V
		$I_R = 5.0\ \mu A$	75		V
I_R	Reverse Current	$V_R = 20\ V$		25	nA
		$V_R = 20\ V$, $T_A = 150°C$		50	μA
		$V_R = 75\ V$		5.0	μA
V_F	Forward Voltage 1N914B / 4448	$I_F = 5.0\ mA$	620	720	mV
	1N916B	$I_F = 5.0\ mA$	630	730	mV
	1N914 / 916 / 4148	$I_F = 10\ mA$		1.0	V
	1N914A / 916A	$I_F = 20\ mA$		1.0	V
	1N916B	$I_F = 30\ mA$		1.0	V
	1N914B / 4448	$I_F = 100\ mA$		1.0	V
C_O	Diode Capacitance				
	1N916/A/B / 4448	$V_R = 0$, $f = 1.0\ MHz$		2.0	pF
	1N914/A/B / 4148	$V_R = 0$, $f = 1.0\ MHz$		4.0	pF
T_{RR}	Reverse Recovery Time	$I_F = 10\ mA$, $V_R = 6.0\ V$ (60 mA), $I_{rr} = 1.0\ mA$, $R_L = 100\ \Omega$		4.0	nS

Typical Characteristics

REVERSE VOLTAGE vs REVERSE CURRENT
BV - 1.0 to 100 uA

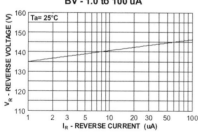

REVERSE CURRENT vs REVERSE VOLTAGE
IR - 10 to 100 V

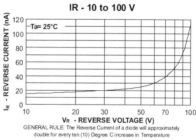

GENERAL RULE: The Reverse Current of a diode will approximately double for every ten (10) Degree C increase in Temperature

FORWARD VOLTAGE vs FORWARD CURRENT
VF - 1 to 100 uA

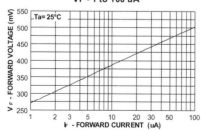

FORWARD VOLTAGE vs FORWARD CURRENT
VF - 0.1 to 100 mA

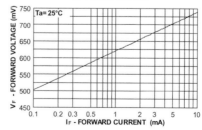

Data Sheet A-2
1N4001 to 1N4007 Rectifier Diodes

MOTOROLA
SEMICONDUCTOR TECHNICAL DATA

Order this document
by 1N4001/D

Axial Lead
Standard Recovery Rectifiers

This data sheet provides information on subminiature size, axial lead mounted rectifiers for general–purpose low–power applications.

Mechanical Characteristics

- Case: Epoxy, Molded
- Weight: 0.4 gram (approximately)
- Finish: All External Surfaces Corrosion Resistant and Terminal Leads are Readily Solderable
- Lead and Mounting Surface Temperature for Soldering Purposes: 220°C Max. for 10 Seconds, 1/16″ from case
- Shipped in plastic bags, 1000 per bag.
- Available Tape and Reeled, 5000 per reel, by adding a "RL" suffix to the part number
- Polarity: Cathode Indicated by Polarity Band
- Marking: 1N4001, 1N4002, 1N4003, 1N4004, 1N4005, 1N4006, 1N4007

**1N4001
thru
1N4007**

1N4004 and 1N4007 are
Motorola Preferred Devices

**LEAD MOUNTED
RECTIFIERS
50–1000 VOLTS
DIFFUSED JUNCTION**

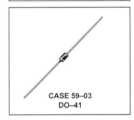

CASE 59–03
DO–41

MAXIMUM RATINGS

Rating	Symbol	1N4001	1N4002	1N4003	1N4004	1N4005	1N4006	1N4007	Unit
*Peak Repetitive Reverse Voltage Working Peak Reverse Voltage DC Blocking Voltage	V_{RRM} V_{RWM} V_R	50	100	200	400	600	800	1000	Volts
*Non–Repetitive Peak Reverse Voltage (halfwave, single phase, 60 Hz)	V_{RSM}	60	120	240	480	720	1000	1200	Volts
*RMS Reverse Voltage	$V_{R(RMS)}$	35	70	140	280	420	560	700	Volts
*Average Rectified Forward Current (single phase, resistive load, 60 Hz, see Figure 8, T_A = 75°C)	I_O	1.0							Amp
*Non–Repetitive Peak Surge Current (surge applied at rated load conditions, see Figure 2)	I_{FSM}	30 (for 1 cycle)							Amp
Operating and Storage Junction Temperature Range	T_J T_{stg}	− 65 to +175							°C

ELECTRICAL CHARACTERISTICS*

Rating	Symbol	Typ	Max	Unit
Maximum Instantaneous Forward Voltage Drop (i_F = 1.0 Amp, T_J = 25°C) Figure 1	v_F	0.93	1.1	Volts
Maximum Full–Cycle Average Forward Voltage Drop (I_O = 1.0 Amp, T_L = 75°C, 1 inch leads)	$V_{F(AV)}$	—	0.8	Volts
Maximum Reverse Current (rated dc voltage) (T_J = 25°C) (T_J = 100°C)	I_R	0.05 1.0	10 50	μA
Maximum Full–Cycle Average Reverse Current (I_O = 1.0 Amp, T_L = 75°C, 1 inch leads)	$I_{R(AV)}$	—	30	μA

*Indicates JEDEC Registered Data

Preferred devices are Motorola recommended choices for future use and best overall value.

Rev 5

 **MOTOROLA**

Data Sheet A-2 (*Continued*)

1N4001 thru 1N4007

PACKAGE DIMENSIONS

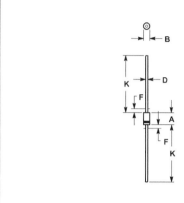

NOTES:
1. ALL RULES AND NOTES ASSOCIATED WITH JEDEC DO–41 OUTLINE SHALL APPLY.
2. POLARITY DENOTED BY CATHODE BAND.
3. LEAD DIAMETER NOT CONTROLLED WITHIN F DIMENSION.

DIM	MILLIMETERS		INCHES	
	MIN	MAX	MIN	MAX
A	4.07	5.20	0.160	0.205
B	2.04	2.71	0.080	0.107
D	0.71	0.86	0.028	0.034
F	—	1.27	—	0.050
K	27.94	—	1.100	—

CASE 59–03
(DO–41)
ISSUE M

Mfax is a trademark of Motorola, Inc.

How to reach us:
USA/EUROPE/Locations Not Listed: Motorola Literature Distribution;
P.O. Box 5405, Denver, Colorado 80217. 303–675–2140 or 1–800–441–2447

Mfax™: RMFAX0@email.sps.mot.com – TOUCHTONE 602–244–6609
                                    – US & Canada ONLY 1–800–774–1848

INTERNET: http://motorola.com/sps

JAPAN: Nippon Motorola Ltd.: SPD, Strategic Planning Office, 4–32–1,
Nishi–Gotanda, Shinagawa–ku, Tokyo 141, Japan. 81–3–5487–8488

ASIA/PACIFIC: Motorola Semiconductors H.K. Ltd.; 8B Tai Ping Industrial Park,
51 Ting Kok Road, Tai Po, N.T., Hong Kong. 852–26629298

 **MOTOROLA**

◊

1N4001/D

Data Sheet A-3
1N5391 to 1N5399 Low-Power Rectifiers

◗ RECTRON
SEMICONDUCTOR
TECHNICAL SPECIFICATION

| 1N5391 |
| THRU |
| 1N5399 |

SILICON RECTIFIER

VOLTAGE RANGE 50 to 1000 Volts   CURRENT 1.5 Amperes

FEATURES
* Low cost
* Low leakage
* Low forward voltage drop
* High current capability

MECHANICAL DATA
* Case: Molded plastic
* Epoxy: Device has UL flammability classification 94V-O
* Lead: MIL-STD-202E method 208C guaranteed
* Mounting position: Any
* Weight: 0.38 gram

MAXIMUM RATINGS AND ELECTRICAL CHARACTERISTICS
Ratings at 25 °C ambient temperature unless otherwise specified.
Single phase, half wave, 60 Hz, resistive or inductive load.
For capacitive load, derate current by 20%.

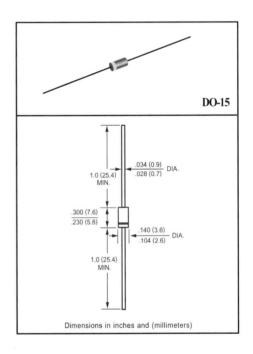

DO-15

Dimensions in inches and (millimeters)

MAXIMUM RATINGS (At TA = 25°C unless otherwise noted)

RATINGS	SYMBOL	1N5391	1N5392	1N5393	1N5394	1N5395	1N5396	1N5397	1N5398	1N5399	UNITS
Maximum Recurrent Peak Reverse Voltage	V$_{RRM}$	50	100	200	300	400	500	600	800	1000	Volts
Maximum RMS Voltage	V$_{RMS}$	35	70	140	210	280	350	420	560	700	Volts
Maximum DC Blocking Voltage	V$_{DC}$	50	100	200	300	400	500	600	800	1000	Volts
Maximum Average Forward Rectified Current .375" (9.5mm) lead length at TL = 70°C	I$_O$					1.5					Amps
Peak Forward Surge Current 8.3 ms single half sine-wave superimposed on rated load (JEDEC method)	I$_{FSM}$					50					Amps
Typical Junction Capacitance (Note)	C$_J$					20					pF
Typical Thermal Resistance	R$_{\theta JA}$					50					°C/W
Operating and Storage Temperature Range	TJ, TSTG					-65 to + 175					°C

ELECTRICAL CHARACTERISTICS (At TA = 25°C unless otherwise noted)

CHARACTERISTICS		SYMBOL	1N5391	1N5392	1N5393	1N5394	1N5395	1N5396	1N5397	1N5398	1N5399	UNITS
Maximum Instantaneous Forward Voltage at 1.5A DC		V$_F$					1.4					Volts
Maximum DC Reverse Current	@TA = 25°C						5.0					uAmps
at Rated DC Blocking Voltage	@TA = 100°C	I$_R$					50					
Maximum Full Load Reverse Current Average, Full Cycle .375" (9.5mm) lead length at TL = 75°C							30					uAmps

NOTES :　Measured at 1 MHz and applied reverse voltage of 4.0 volts

2001-5

Data Sheet A-4a
Zener diodes: 1N746 to 1N759, 1N957A to 1N986A, 1N4370 to 1N4372

- 1N746A-1 THRU 1N759-1 AVAILABLE IN **JAN, JANTX** AND **JANTXV**
  PER MIL-PRF-19500/127
- 1N4370A-1 THRU 1N4372A-1 AVAILABLE IN **JAN, JANTX** AND **JANTXV**
  PER MIL-PRF-19500/127
- DOUBLE PLUG CONSTRUCTION
- METALLURGICALLY BONDED

1N746 thru 1N759A
and
1N746A-1 thru 1N759A-1
and
1N4370 thru 1N4372A
and
1N4370A-1 thru 1N4372A-1

MAXIMUM RATINGS

Operating Temperature: -65°C to +175°C
Storage Temperature: -65°C to +175°C
DC Power Dissipation: 500 mW @ +50°C
Power Derating: 4 mW / °C above +50°C
Forward Voltage @ 200mA: 1.1 volts maximum

ELECTRICAL CHARACTERISTICS @ 25°C

JEDEC TYPE NUMBER (NOTE 1)	NOMINAL ZENER VOLTAGE V_Z @ I_{ZT} (NOTE 2)	ZENER TEST CURRENT I_{ZT}	MAXIMUM ZENER IMPEDANCE (NOTE 3) Z_{ZT}@ I_{ZT}	MAXIMUM REVERSE CURRENT I_R @ V_R		MAXIMUM ZENER CURRENT I_{ZM}
	VOLTS	mA	OHMS	μA	VOLTS	mA
1N4370A	2.4	20	30	100	1.0	155
1N4371A	2.7	20	30	60	1.0	140
1N4372A	3.0	20	29	30	1.0	125
1N746A	3.3	20	28	5	1.0	120
1N747A	3.6	20	24	3	1.0	110
1N748A	3.9	20	23	2	1.0	100
1N749A	4.3	20	22	2	1.0	90
1N750A	4.7	20	19	5	1.5	85
1N751A	5.1	20	17	5	2.0	75
1N752A	5.6	20	11	5	2.5	70
1N753A	6.2	20	7	5	3.5	65
1N754A	6.8	20	5	2	4.0	60
1N755A	7.5	20	6	2	5.0	55
1N756A	8.2	20	8	1	6.0	50
1N757A	9.1	20	10	1	7.0	45
1N758A	10.0	20	17	1	8.0	40
1N759A	12.0	20	30	1	9.0	35

NOTE 1   Zener voltage tolerance on "A" suffix is $\pm$5%. No Suffix denotes $\pm$ 10% tolerance, "C" suffix denotes $\pm$ 2% tollerance and "D" suffix denotes $\pm$ 1% tolerance.

NOTE 2   Zener voltage is measured with the device junction in thermal equilibrium at an ambient temperature of 25°C $\pm$ 3°C.

NOTE 3   Zener impedance is derived by superimposing on I_{ZT} A 60Hz rms a.c. current equal to 10% of I_{ZT}

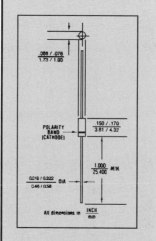

FIGURE 1

DESIGN DATA

CASE: Hermetically sealed glass case. DO – 35 outline.

LEAD MATERIAL: Copper clad steel.

LEAD FINISH: Tin / Lead

THERMAL RESISTANCE: ($R_{\theta JEC}$): 250 °C/W maximum at L = .375 inch

THERMAL IMPEDANCE: ($Z_{\theta JX}$): 35 °C/W maximum

POLARITY: Diode to be operated with the banded (cathode) end positive.

MOUNTING POSITION: Any.

6 LAKE STREET, LAWRENCE, MASSACHUSETTS 01841
PHONE (978) 620-2600   FAX (978) 689-0803
WEBSITE: http://www.microsemi.com

Data Sheet A-4b
Zener Diodes

- 1N962B-1 THRU 1N986B-1 AVAILABLE IN **JAN, JANTX AND JANTXV**
  PER MIL-PRF-19500/117
- METALLURGICALLY BONDED
- DOUBLE PLUG CONSTRUCTION

**1N957 thru 1N986B
and
1N962B-1 thru 1N986B-1**

MAXIMUM RATINGS

Operating Temperature: -65°C to +175°C
Storage Temperature: -65°C to +175°C
DC Power Dissipation: 500 mW @ +50°C
Power Derating: 4 mW / °C above +50°C
Forward Voltage @ 200mA: 1.1volts maximum

ELECTRICAL CHARACTERISTICS @ 25°C

JEDEC TYPE NUMBER (NOTE 1)	NOMINAL ZENER VOLTAGE V_Z (NOTE 2)	ZENER TEST CURRENT I_{ZT}	MAXIMUM ZENER IMPEDANCE (NOTE 3)			MAX. DC ZENER CURRENT I_{ZM}	MAX. REVERSE LEAKAGE CURRENT $I_R @ V_R$	
			$Z_{ZT} @ I_{ZT}$	$Z_{ZK} @ I_{ZK}$				
	VOLTS	mA	OHMS	OHMS	mA	mA	μ A	VOLTS
1N957B	6.8	18.5	4.5	700	1.0	55	5	5.2
1N958B	7.5	16.5	5.5	700	.5	50	5	5.7
1N959B	8.2	15.0	6.5	700	.5	45	5	6.2
1N960B	9.1	14.0	7.5	700	.5	41	5	6.9
1N961B	10	12.5	8.5	700	.25	38	2	7.6
1N962B	11	11.5	9.5	700	.25	32	1	8.4
1N963B	12	10.5	11.5	700	.25	31	1	9.1
1N964B	13	9.5	13	700	.25	28	0.5	9.9
1N965B	15	8.5	16	700	.25	25	0.5	11
1N966B	16	7.8	17	700	.25	24	0.5	12
1N967B	18	7.0	21	750	.25	20	0.5	14
1N968B	20	6.2	25	750	.25	18	0.5	15
1N969B	22	5.6	29	750	.25	16	0.5	17
1N970B	24	5.2	33	750	.25	15	0.5	18
1N971B	27	4.6	41	750	.25	13	0.5	21
1N972B	30	4.2	49	1000	.25	12	0.5	23
1N973B	33	3.8	58	1000	.25	11	0.5	25
1N974B	36	3.4	70	1000	.25	10	0.5	27
1N975B	39	3.2	90	1000	.25	9.5	0.5	30
1N976B	43	3.0	93	1500	.25	8.8	0.5	33
1N977B	47	2.7	105	1500	.25	7.9	0.5	36
1N978B	51	2.5	125	1500	.25	7.4	0.5	39
1N979B	56	2.2	150	2000	.25	6.8	0.5	43
1N980B	62	2.0	185	2000	.25	6.0	0.5	47
1N981B	68	1.8	230	2000	.25	5.5	0.5	52
1N982B	75	1.7	270	2000	.25	5.0	0.5	56
1N983B	82	1.5	330	3000	.25	4.6	0.5	62
1N984B	91	1.4	400	3000	.25	4.1	0.5	69
1N985B	100	1.3	500	3000	.25	3.7	0.5	76
1N986B	110	1.1	750	4000	.25	3.3	0.5	84

NOTE 1 Zener voltage tolerance on "B"suffix is $\pm$ 5%. Suffix letter A denotes +10%. No Suffix denotes $\pm$ 20% tolerance, "C" suffix denotes $\pm$ 2% and "D" suffix denotes $\pm$ 1%.

NOTE 2 Zener voltage is measured with the device junction in thermal equilibrium at an ambient temperature of 25°C $\pm$ 3°C.

NOTE 3 Zener impedance is derived by superimposing on I_{ZT} A 60Hz rms a.c. current equal to 10% of I_{ZT}

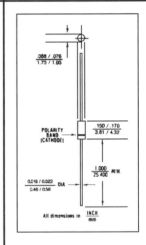

FIGURE 1

DESIGN DATA

CASE: Hermetically sealed glass case. DO – 35 outline.

LEAD MATERIAL: Copper clad steel.

LEAD FINISH: Tin / Lead

THERMAL RESISTANCE: ($R_{\theta JEC}$): 250 ˚C/W maximum at L = .375 inch

THERMAL IMPEDANCE: ($Z_{\theta JX}$): 35 ˚C/W maximum

POLARITY: Diode to be operated with the banded (cathode) end positive.

MOUNTING POSITION: Any.

COMPENSATED DEVICES INCORPORATED
22 COREY STREET, MELROSE, MASSACHUSETTS 02176
PHONE (781) 665-1071   FAX (781) 665-7379
WEBSITE: http://www.cdi-diodes.com   E-mail: mail@cdi-diodes.com

Data Sheet A-5
2N3903 and 2N3904 *npn* BJTs

Micro Commercial Components
21201 Itasca Street Chatsworth
CA 91311
Phone: (818) 701-4933
Fax: (818) 701-4939

2N3904

Features

- Through Hole Package
- Capable of 600mWatts of Power Dissipation

Pin Configuration
Bottom View
C B E

NPN General Purpose Amplifier

TO-92

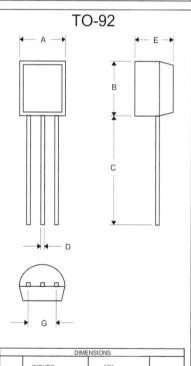

Electrical Characteristics @ 25°C Unless Otherwise Specified

Symbol	Parameter	Min	Max	Units
OFF CHARACTERISTICS				
$V_{(BR)CEO}$	Collector-Emitter Breakdown Voltage* (I_C=1.0mAdc, I_B=0)	40		Vdc
$V_{(BR)CBO}$	Collector-Base Breakdown Voltage (I_C=10µAdc, I_E=0)	60		Vdc
$V_{(BR)EBO}$	Emitter-Base Breakdown Voltage (I_E=10µAdc, I_C=0)	6.0		Vdc
I_{BL}	Base Cutoff Current (V_{CE}=30Vdc, V_{BE}=3.0Vdc)		50	nAdc
I_{CEX}	Collector Cutoff Current (V_{CE}=30Vdc, V_{BE}=3.0Vdc)		50	nAdc
ON CHARACTERISTICS				
h_{FE}	DC Current Gain* (I_C=0.1mAdc, V_{CE}=1.0Vdc) (I_C=1.0mAdc, V_{CE}=1.0Vdc) (I_C=10mAdc, V_{CE}=1.0Vdc) (I_C=50mAdc, V_{CE}=1.0Vdc) (I_C=100mAdc, V_{CE}=1.0Vdc)	40 70 100 60 30	300	
$V_{CE(sat)}$	Collector-Emitter Saturation Voltage (I_C=10mAdc, I_B=1.0mAdc) (I_C=50mAdc, I_B=5.0mAdc)		0.2 0.3	Vdc
$V_{BE(sat)}$	Base-Emitter Saturation Voltage (I_C=10mAdc, I_B=1.0mAdc) (I_C=50mAdc, I_B=5.0mAdc)	0.65	0.85 0.95	Vdc
SMALL-SIGNAL CHARACTERISTICS				
f_T	Current Gain-Bandwidth Product (I_C=10mAdc, V_{CE}=20Vdc, f=100MHz)	300		MHz
C_{obo}	Output Capacitance (V_{CB}=5.0Vdec, I_E=0, f=1.0MHz)		4.0	pF
C_{ibo}	Input Capacitance (V_{BE}=0.5Vdc, I_C=0, f=1.0MHz)		8.0	pF
NF	Noise Figure (I_C=100µAdc, V_{CE}=5.0Vdc, R_S=1.0kΩ f=10Hz to 15.7kHz)		5.0	dB
SWITCHING CHARACTERISTICS				
t_d	Delay Time	(V_{CC}=3.0Vdc, V_{BE}=0.5Vdc	35	ns
t_r	Rise Time	I_C=10mAdc, I_{B1}=1.0mAdc)	35	ns
t_s	Storage Time	(V_{CC}=3.0Vdc, I_C=10mAdc	200	ns
t_f	Fall Time	I_{B1}=I_{B2}=1.0mAdc)	50	ns

*Pulse Width ≤ 300µs, Duty Cycle ≤ 2.0%

DIMENSIONS

DIM	INCHES MIN	MAX	MM MIN	MAX	NOTE
A	.175	.185	4.45	4.70	
B	.175	.185	4.46	4.70	
C	.500	---	12.7	---	
D	.016	.020	0.41	0.63	
E	.135	.145	3.43	3.68	
G	.095	.105	2.42	2.67	

www.mccsemi.com

Data Sheet A-5 (*Continued*)

2N3904

·M·C·C·

DC Current Gain vs Collector Current

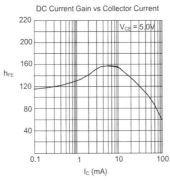

Base-Emitter ON Voltage vs Collector Current

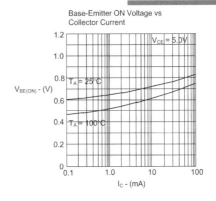

Collector Saturation Volatge vs Collector Current

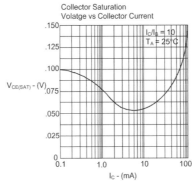

Base Saturation Voltage vs Collector Current

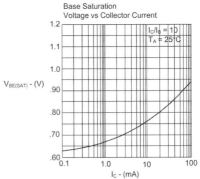

Collector Cutoff Current vs Ambient Temperature

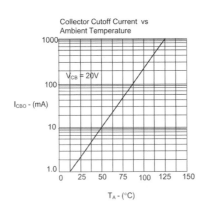

Capacitance vs Reverse Bias Voltage

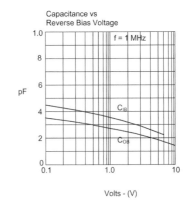

Data Sheet A-6
2N3905 and 2N3906 *pnp* BJTs

2N3905, 2N3906	**R**	Switching Transistors

PNP	Si-Epitaxial PlanarTransistors	PNP

Version 2004-01-20

Standard Pinning
1 = C   2 = B   3 = E

Power dissipation – Verlustleistung   625 mW

Plastic case   TO-92
Kunststoffgehäuse   (10D3)

Weight approx. – Gewicht ca.   0.18 g

Plastic material has UL classification 94V-0
Gehäusematerial UL94V-0 klassifiziert

Standard packaging taped in ammo pack
Standard Lieferform gegurtet in Ammo-Pack

Maximum ratings ($T_A = 25°C$)   **Grenzwerte ($T_A = 25°C$)**

			2N3905, 2N3906
Collector-Emitter-voltage	B open	- V_{CE0}	40 V
Collector-Base-voltage	E open	- V_{CE0}	40 V
Emitter-Base-voltage	C open	- V_{EB0}	5 V
Power dissipation – Verlustleistung		P_{tot}	625 mW [1])
Collector current – Kollektorstrom (dc)		- I_C	100 mA
Peak collector current – Kollektorspitzenstrom		- I_{CM}	200 mA
Junction temp. – Sperrschichttemperatur		T_j	150°C
Storage temperature – Lagerungstemperatur		T_S	- 55…+ 150°C

Characteristics ($T_j = 25°C$)   **Kennwerte ($T_j = 25°C$)**

		Min.	Typ.	Max.
Collector saturation volt. – Kollektor-Sättigungsspannung				
- $I_C = 10$ mA, - $I_B = 1$ mA	- V_{CEsat}	–	–	250 mV
- $I_C = 50$ mA, - $I_B = 5$ mA	- V_{CEsat}	–	–	400 mV
Base saturation voltage – Basis-Sättigungsspannung				
- $I_C = 10$ mA, - $I_B = 1$ mA	- V_{BEsat}	–	–	850 mV
- $I_C = 50$ mA, - $I_B = 5$ mA	- V_{BEsat}	–	–	950 mV
Collector cutoff current – Kollektorreststrom				
- $V_{CE} = 30$ V, - $V_{EB} = 3$ V	- I_{CEV}	–	–	50 nA
Emitter cutoff current – Emitterreststrom				
- $V_{CE} = 30$ V, - $V_{EB} = 3$ V	- I_{EBV}	–	–	50 nA

---

[1]) Valid, if leads are kept at ambient temperature at a distance of 2 mm from case
   Gültig, wenn die Anschlußdrähte in 2 mm Abstand von Gehäuse auf Umgebungstemperatur gehalten werden

Data Sheet A-6 (*Continued*)

| General Purpose Transistors | **R** | 2N3905, 2N3906 |

Characteristics ($T_j = 25°C$)			Kennwerte ($T_j = 25°C$)		
			Min.	**Typ.**	**Max.**
DC current gain – Kollektor-Basis-Stromverhältnis					
- $V_{CE} = 1$ V, - $I_C = 0.1$ mA	2N3905	h_{FE}	30	–	–
	2N3906	h_{FE}	60	–	–
- $V_{CE} = 1$ V, - $I_C = 1$ mA	2N3905	h_{FE}	40	–	–
	2N3906	h_{FE}	80	–	–
- $V_{CE} = 1$ V, - $I_C = 10$ mA	2N3905	h_{FE}	50	–	150
	2N3906	h_{FE}	100	–	300
- $V_{CE} = 1$ V, - $I_C = 50$ mA	2N3905	h_{FE}	30	–	–
	2N3906	h_{FE}	60	–	–
- $V_{CE} = 1$ V, - $I_C = 100$ mA	2N3905	h_{FE}	15	–	–
	2N3906	h_{FE}	30	–	–
Gain-Bandwidth Product – Transitfrequenz					
- $V_{CE} = 20$ V, - $I_C = 10$ mA, f = 100 MHz	2N3905	f_T	200 MHz	–	–
	2N3906	f_T	250 MHz	–	–
Collector-Base Capacitance – Kollektor-Basis-Kapazität					
- $V_{CB} = 5$ V, $I_E = i_e = 0$, f = 100 kHz		C_{CBO}	–	–	4.5 pF
Emitter-Base Capacitance – Emitter-Basis-Kapazität					
- $V_{EB} = 0.5$ V, $I_C = i_c = 0$, f = 100 kHz		C_{EBO}	–	–	10 pF
Noise figure – Rauschzahl					
- $V_{CE} = 5$ V, - $I_C = 100$ μA	2N3905	F	–	–	5 dB
$R_G = 1$ kΩ f = 10 Hz ...15.7 kHz	2N3906	F	–	–	4 dB
Switching times – Schaltzeiten					
turn-on time $I_{Con} = 10$ mA,		t_{on}	–	–	70
turn-off time $I_{Bon} = - I_{Boff} = 1$ mA		t_{off}	–	–	300
Thermal resistance junction to ambient air Wärmewiderstand Sperrschicht – umgebende Luft		R_{thA}			200 K/W [1]
Recommended complementary NPN transistors Empfohlene komplementäre NPN-Transistoren					2N3903, 2N3904

[1]   Valid, if leads are kept at ambient temperature at a distance of 2 mm from case
   Gültig, wenn die Anschlußdrähte in 2 mm Abstand von Gehäuse auf Umgebungstemperatur gehalten werden

Data Sheet A-7
2N3251 *pnp* BJT

MAXIMUM RATINGS

Rating	Symbol	2N3250 2N3251	2N3251A	Unit
Collector-Emitter Voltage	V_{CEO}	−40	−60	Vdc
Collector-Base Voltage	V_{CBO}	−50	−60	Vdc
Emitter-Base Voltage	V_{EBO}	−5.0		Vdc
Collector Current	I_C	−200		mAdc
Total Device Dissipation @ T_A = 25°C Derate above 25°C	P_D	0.36 2.06		Watt mW/°C
Total Device Dissipation @ T_C = 25°C Derate above 25°C	P_D	1.2 6.9		Watts mW/°C
Operating and Storage Temperature Temperature Range	T_J, T_{stg}	−65 to +200		°C

THERMAL CHARACTERISTICS

Characteristic	Symbol	Max	Unit
Thermal Resistance, Junction to Ambient	$R_{\theta JA}$	486	°C/W
Thermal Resistance, Junction to Case	$R_{\theta JC}$	146	°C/W

2N3250
2N3251,A★

CASE 22-03, STYLE 1
TO-18 (TO-206AA)

3 Collector

2
Bas

1 Emitter

3 2 1

**GENERAL PURPOSE
TRANSISTORS**

PNP SILICON

★2N3251A is a Motorola
designated preferred device.

ELECTRICAL CHARACTERISTICS (T_A = 25°C unless otherwise noted.)

Characteristic		Symbol	Min	Max	Unit
OFF CHARACTERISTICS					
Collector-Emitter Breakdown Voltage(1) (I_C = −10 mAdc)	2N3250, 2N3251 2N3251A	$V_{(BR)CEO}$	−40 −60	—	Vdc
Collector-Base Breakdown Voltage (I_C = −10 μAdc)	2N3250, 2N3251 2N3251A	$V_{(BR)CBO}$	−50 −60	—	Vdc
Emitter-Base Breakdown Voltage (I_E = −10 μAdc)		$V_{(BR)EBO}$	−5.0	—	Vdc
Collector Cutoff Current (V_{CE} = −40 Vdc, V_{EB} = −3.0 Vdc)		I_{CEX}	—	−20	nA
Base Cutoff Current (V_{CE} = −40 Vdc, V_{EB} = −3.0 Vdc)		I_{BL}	—	−50	nAdc
ON CHARACTERISTICS					
DC Forward Current Transfer Ratio (I_C = −0.1 mAdc, V_{CE} = −10 Vdc)	2N3250 2N3251, 2N3251A	h_{FE}	40 80	— —	—
(I_C = −1.0 mAdc, V_{CE} = −1.0 Vdc)	2N3250 2N3251, 2N3251A		45 90	— —	
(I_C = −10 mAdc, V_{CE} = −1.0 Vdc)(1)	2N3250 2N3251, 2N3251A		50 100	150 300	
(I_C = −50 mAdc, V_{CE} = −1.0 Vdc)(1)	2N3250 2N3251, 2N3251A		15 30	— —	
Collector-Emitter Saturation Voltage (1) (I_C = −10 mAdc, I_B = −1.0 mAdc) (I_C = −50 mAdc, I_B = −5.0 mAdc		$V_{CE(sat)}$	— —	−0.25 −0.5	Vdc
Base-Emitter Saturation Voltage (1) (I_C = −10 mAdc, I_B = −1.0 mAdc) (I_C = −50 mAdc, I_B = −5.0 mAdc)		$V_{BE(sat)}$	−0.6 —	−0.9 −1.2	Vdc
SMALL-SIGNAL CHARACTERISTICS					
Current-Gain — Bandwidth Product (I_C = −10 mAdc, V_{CE} = −20 Vdc, f = 100 MHz)	2N3250 2N3251, 2N3251A	f_T	250 300	— —	MHz
Output Capacitance (V_{CB} = −10 Vdc, I_E = 0, f = 1.0 MHz)		C_{obo}	—	6.0	pF
Input Capacitance (V_{EB} = −1.0 Vdc, I_C = 0, f = 1.0 MHz)		C_{ibo}	—	8.0	pF

Data Sheet A-7   (*Continued*)

2N3250 2N3251,A

ELECTRICAL CHARACTERISTICS (continued) (T_A = 25°C unless otherwise noted.)

Characteristic		Symbol	Min	Max	Unit
Input Impedance (I_C = −1.0 mA, V_{CE} = −10 V, f = 1.0 kHz)	2N3250 2N3251, 2N3251A	h_{ie}	1.0 2.0	6.0 12	kohms
Voltage Feedback Ratio (I_C = −1.0 mA, V_{CE} = −10 V, f = 1.0 kHz)	2N3250 2N3251, 2N3251A	h_{re}	— —	10 20	X 10^{-4}
Small-Signal Current Gain (I_C = −1.0 mA, V_{CE} = −10 V, f = 1.0 kHz)	2N3250 2N3251, 2N3251A	h_{fe}	50 100	200 400	—
Output Admittance (I_C = −1.0 mA, V_{CE} = −10 V, f = 1.0 kHz)	2N3250 2N3251, 2N3251A	h_{oe}	4.0 10	40 60	μmhos
Collector Base Time Constant (I_C = −10 mA, V_{CE} = −20 V, f = 31.8 MHz)		$rb'C_C$	—	250	ps
Noise Figure (I_C = −100 μA, V_{CE} = −5.0 V, R_S = 1.0 kΩ, f = 100 Hz)		NF	—	6.0	dB

SWITCHING CHARACTERISTICS

Characteristic			Symbol	Max	Unit
Delay Time	(V_{CC} = −3.0 Vdc, V_{BE} = +0.5 Vdc		t_d	35	ns
Rise Time	I_C = −10 mAdc, I_{B1} = −1.0 mA)		t_r	35	ns
Storage Time	I_C = −10 mAdc, I_{B1} = I_{B1} = −1.0 mAdc) (V_{CC} = −3.0 V)	2N3250 2N3251, 2N3251A	t_s	175 200	ns
Fall Time			t_f	50	ns

(1) Pulse Test: PW = 300 μs, Duty Cycle = 2.0%.

SWITCHING TIME CHARACTERISTICS

FIGURE 1 — DELAY AND RISE TIME   FIGURE 2 — STORAGE AND FALL TIME

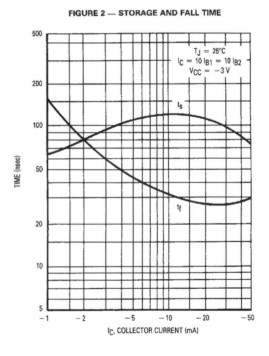

Data Sheet A-8
2N3055 and MJ2955 Complementary Power BJTs

MOTOROLA
SEMICONDUCTOR TECHNICAL DATA

Order this document
by 2N3055/D

Complementary Silicon Power Transistors

. . . designed for general–purpose switching and amplifier applications.

- DC Current Gain — $h_{FE} = 20-70$ @ $I_C = 4$ Adc
- Collector–Emitter Saturation Voltage —
  $V_{CE(sat)} = 1.1$ Vdc (Max) @ $I_C = 4$ Adc
- Excellent Safe Operating Area

**NPN
2N3055 ***
**PNP
MJ2955 ***

*Motorola Preferred Device

**15 AMPERE
POWER TRANSISTORS
COMPLEMENTARY
SILICON
60 VOLTS
115 WATTS**

**CASE 1–07
TO–204AA
(TO–3)**

MAXIMUM RATINGS

Rating	Symbol	Value	Unit
Collector–Emitter Voltage	V_{CEO}	60	Vdc
Collector–Emitter Voltage	V_{CER}	70	Vdc
Collector–Base Voltage	V_{CB}	100	Vdc
Emitter–Base Voltage	V_{EB}	7	Vdc
Collector Current — Continuous	I_C	15	Adc
Base Current	I_B	7	Adc
Total Power Dissipation @ $T_C = 25°C$ Derate above 25°C	P_D	115 0.657	Watts W/°C
Operating and Storage Junction Temperature Range	T_J, T_{stg}	−65 to +200	°C

THERMAL CHARACTERISTICS

Characteristic	Symbol	Max	Unit
Thermal Resistance, Junction to Case	$R_{\theta JC}$	1.52	°C/W

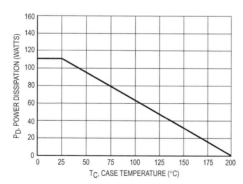

Figure 1. Power Derating

Preferred devices are Motorola recommended choices for future use and best overall value.

MOTOROLA

Data Sheet A-8 (*Continued*)

2N3055 MJ2955

ELECTRICAL CHARACTERISTICS (T_C = 25°C unless otherwise noted)

Characteristic	Symbol	Min	Max	Unit
*OFF CHARACTERISTICS				
Collector–Emitter Sustaining Voltage (1) (I_C = 200 mAdc, I_B = 0)	$V_{CEO(sus)}$	60	—	Vdc
Collector–Emitter Sustaining Voltage (1) (I_C = 200 mAdc, R_{BE} = 100 Ohms)	$V_{CER(sus)}$	70	—	Vdc
Collector Cutoff Current (V_{CE} = 30 Vdc, I_B = 0)	I_{CEO}	—	0.7	mAdc
Collector Cutoff Current (V_{CE} = 100 Vdc, $V_{BE(off)}$ = 1.5 Vdc) (V_{CE} = 100 Vdc, $V_{BE(off)}$ = 1.5 Vdc, T_C = 150°C)	I_{CEX}	— —	1.0 5.0	mAdc
Emitter Cutoff Current (V_{BE} = 7.0 Vdc, I_C = 0)	I_{EBO}	—	5.0	mAdc
*ON CHARACTERISTICS (1)				
DC Current Gain (I_C = 4.0 Adc, V_{CE} = 4.0 Vdc) (I_C = 10 Adc, V_{CE} = 4.0 Vdc)	h_{FE}	20 5.0	70 —	—
Collector–Emitter Saturation Voltage (I_C = 4.0 Adc, I_B = 400 mAdc) (I_C = 10 Adc, I_B = 3.3 Adc)	$V_{CE(sat)}$	—	1.1 3.0	Vdc
Base–Emitter On Voltage (I_C = 4.0 Adc, V_{CE} = 4.0 Vdc)	$V_{BE(on)}$	—	1.5	Vdc
SECOND BREAKDOWN				
Second Breakdown Collector Current with Base Forward Biased (V_{CE} = 40 Vdc, t = 1.0 s, Nonrepetitive)	$I_{s/b}$	2.87	—	Adc
DYNAMIC CHARACTERISTICS				
Current Gain — Bandwidth Product (I_C = 0.5 Adc, V_{CE} = 10 Vdc, f = 1.0 MHz)	f_T	2.5	—	MHz
*Small–Signal Current Gain (I_C = 1.0 Adc, V_{CE} = 4.0 Vdc, f = 1.0 kHz)	h_{fe}	15	120	—
*Small–Signal Current Gain Cutoff Frequency (V_{CE} = 4.0 Vdc, I_C = 1.0 Adc, f = 1.0 kHz)	f_{hfe}	10	—	kHz

* Indicates Within JEDEC Registration. (2N3055)
(1) Pulse Test: Pulse Width ≤ 300 μs, Duty Cycle ≤ 2.0%.

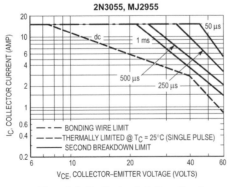

Figure 2. Active Region Safe Operating Area

There are two limitations on the power handling ability of a transistor: average junction temperature and second breakdown. Safe operating area curves indicate I_C – V_{CE} limits of the transistor that must be observed for reliable operation; i.e., the transistor must not be subjected to greater dissipation than the curves indicate.

The data of Figure 2 is based on T_C = 25°C; $T_{J(pk)}$ is variable depending on power level. Second breakdown pulse limits are valid for duty cycles to 10% but must be derated for temperature according to Figure 1.

Data Sheet A-9

2N6121 and 2N6124 Complementary Power BJTs

⚡MOSPEC

COMPLEMENTARY SILICON PLASTIC
POWER TRANSISTORS

... designed for use in power amplifier and switching circuit applications

FEATURES:

* Collector-Emitter Sustaining Voltage-
  $V_{CEO(SUS)}$ = 45 V (Min) - 2N6121, 2N6124
  = 60 V (Min) - 2N6122, 2N6125
  = 80 V (Min) - 2N6123, 2N6126

* Collector-Emitter Saturation Voltage
  $V_{CE(sat)}$ = 0.6V (Max.) @ I_C = 1.5 A, I_B = 0.15 A

NPN	PNP
2N6121	2N6124
2N6122	2N6125
2N6123	2N6126

4 AMPERE
COMPLEMENTARY SILICON
POWER TRANSISTORS
45-80 Volts
40 Watts

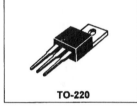

TO-220

MAXIMUM RATINGS

Characteristic	Symbol	2N6121 2N6124	2N6122 2N6125	2N6123 2N6126	Unit
Collector-Emitter Voltage	V_{CEO}	45	60	80	V
Collector-Base Voltage	V_{CBO}	45	60	80	V
Emitter-Base Voltage	V_{EBO}	5.0			V
Collector Current - Continuous - Peak	I_C	4.0 8.0			A
Base Current	I_B	1.0			A
Total Power Dissipation@T_C = 25°C Derate above 25°C	P_D	40 0.32			W W/°C
Operating and Storage Junction Temperature Range	T_J, T_{STG}	-65 to +150			°C

THERMAL CHARACTERISTICS

Characteristic	Symbol	Max	Unit
Thermal Resistance Junction to Case	Rθjc	3.125	°C/W

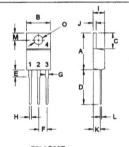

PIN 1.BASE
2.COLLECTOR
3.EMITTER
4.COLLECTO(CASE)

DIM	MILLIMETERS	
	MIN	MAX
A	14.68	15.31
B	9.78	10.42
C	5.01	6.52
D	13.06	14.62
E	3.57	4.07
F	2.42	3.66
G	1.12	1.36
H	0.72	0.96
I	4.22	4.98
J	1.14	1.38
K	2.20	2.97
L	0.33	0.55
M	2.48	2.98
O	3.70	3.90

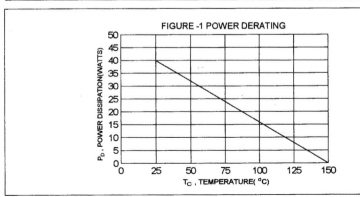

FIGURE -1 POWER DERATING

Data Sheet A-9 *(Continued)*

2N6121, 2N6122,2N 6123 NPN / 2N6124, 2N6125,2N6126 PNP

ELECTRICAL CHARACTERISTICS (T_C = 25°C unless otherwise noted)

Characteristic		Symbol	Min	Max	Unit
OFF CHARACTERISTICS					
Collector - Emitter Sustaining Voltage (1) (I_C = 100 mA, I_B = 0) 2N6121, 2N6124 2N6122, 2N6125 2N6123, 2N6126		$V_{CEO(SUS)}$	45 60 80		V
Collector Cutoff Current (V_{CE} = 45 V, I_B = 0) 2N6121, 2N6124 (V_{CE} = 60 V, I_B = 0) 2N6122, 2N6125 (V_{CE} = 80 V, I_B = 0) 2N6123, 2N6126		I_{CEO}		1.0 1.0 1.0	mA
Collector Cutoff Current (V_{CE} = 45 V, $V_{BE(off)}$ = 1.5 V) 2N6121, 2N6124 (V_{CE} = 60 V, $V_{BE(off)}$ = 1.5 V) 2N6122, 2N6125 (V_{CE} = 80 V, $V_{BE(off)}$ = 1.5 V) 2N6123, 2N6126 (V_{CE} = 45 V, $V_{BE(off)}$ = 1.5 V, T_C = 125°C) 2N6121, 2N6124 (V_{CE} = 60 V, $V_{BE(off)}$ = 1.5 V, T_C = 125°C) 2N6122, 2N6125 (V_{CE} = 80 V, $V_{BE(off)}$ = 1.5 V, T_C = 125°C) 2N6123, 2N6126		I_{CEX}		0.1 0.1 0.1 2.0 2.0 2.0	mA
Emitter Cutoff Current (V_{EB} = 5.0 V , I_C = 0)		I_{EBO}		1.0	mA
ON CHARACTERISTICS (1)					
DC Current Gain (I_C = 1.5 A, V_{CE} = 2.0 V) 2N6121, 2N6124 2N6122, 2N6125 2N6123, 2N6126 (I_C = 4.0 A, V_{CE} = 2.0 V) 2N6121, 2N6124 2N6122, 2N6125 2N6123, 2N6126		h_{FE}	25 25 20 10 10 7.0	100 100 80	
Collector-Emitter Saturation Voltage (I_C = 1.5 A, I_B = 0.15 A) (I_C = 4.0 A, I_B = 1.0 A)		$V_{CE(sat)}$		0.6 1.4	V
Base-Emitter On Voltage (I_C = 1.5 A, V_{CE} = 2.0 V)		$V_{BE(on)}$		1.2	V
DYNAMIC CHARACTERISTICS					
Current-Gain-Bandwidth Product (2) (I_C = 1.0 A, V_{CE} = 4.0 V, f = 1.0 MHz)		f_T	2.5		MHz
Small-Signal Current Gain (I_C = 0.1 A, V_{CE} = 2.0 V, f = 1.0 KHZ)		h_{fe}	25		

(1) Pulse Test: Pulse width = 300 us , Duty Cycle $\leq$ 2.0%
(2) f_T = $\left| h_{fe} \right| \cdot f_{test}$

Data Sheet A-10
Transistor Heat Sinks

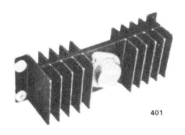

401

403

MODEL 401
NATURAL CONVECTION COOLER

Produces maximum semiconductor cooling per unit volume, often eliminating the need for costly redesigns required to accommodate bulkier, less efficient coolers. Unexcelled where space is critical. Additionally affords excellent forced convection characteristics, with a thermal resistance under 0.5°C/W. at moderate air flows. Uses 3 Teflon mounting washers

MODEL 403
NATURAL CONVECTION COOLER

Affords heat dissipation characteristics which are difficult to equal with less "engineered" products. A thermal resistance of 1.8°C/W under medium power input permits adequate cooling in all but the most stringent applications. Under forced convection conditions, a thermal resistance of only 0.3°C/W is achieved with moderate air flows! Permits safe, reliable operation of the semiconductor under most power requirements where space is limited, and allowable temperature rise is restricted. Uses 4 Teflon mounting washers

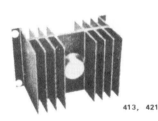

413, 421

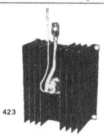

423

MODEL 421
NATURAL CONVECTION COOLER

Permits design engineers to free themselves from the question of how to maintain their power transistors at safe operating temperatures. Requiring no greater chassis base mounting area than that required by Model 403, this unit affords a still lower thermal resistance that is generally adequate for most high-power applications. Natural convection thermal resistance as low as 1.3°C/W; forced convection thermal resistance as low as 0.3°C/W. Uses 4 Teflon mounting washers

MODEL 423
NATURAL CONVECTION COOLER

Will provide semiconductor cooling never before possible under natural convection conditions, with thermal resistance as low as 0.8°C/W, and only 0.25°C/W with moderate air flows! Ideally suited to high-power transistors, rectifiers, and silicon-controlled rectifiers. Where installation area is not greatly restricted, and design conditions do not permit forced convection, Model 423 will achieve the lowest possible temperature rise per unit volume — and at the lowest possible cost.

MODEL 441
NATURAL CONVECTION HEAT SINKS

Model 441 is designed for mounting on a vertical surface (no special bracket required) to take better advantage of natural convection. Thermal resistance of .55°C/W at higher power levels is obtained. With moderate forced convection, thermal resistance of .18°C/W may be achieved. Electrical isolation may be achieved by the use of 4 teflon mounting washers (not included).

441

* Courtesy of EG & G, Wakefield Engineering.

Data Sheet A-11
2N5457 FET

2N5457, 2N5458

Preferred Device

JFETs - General Purpose

N−Channel − Depletion

N−Channel Junction Field Effect Transistors, depletion mode (Type A) designed for audio and switching applications.

Features

- N−Channel for Higher Gain
- Drain and Source Interchangeable
- High AC Input Impedance
- High DC Input Resistance
- Low Transfer and Input Capacitance
- Low Cross−Modulation and Intermodulation Distortion
- Unibloc Plastic Encapsulated Package
- Pb−Free Packages are Available*

MAXIMUM RATINGS

Rating	Symbol	Value	Unit
Drain−Source Voltage	V_{DS}	25	Vdc
Drain−Gate Voltage	V_{DG}	25	Vdc
Reverse Gate−Source Voltage	V_{GSR}	−25	Vdc
Gate Current	I_G	10	mAdc
Total Device Dissipation @ $T_A = 25°C$ Derate above 25°C	P_D	310 2.82	mW mW/°C
Operating Junction Temperature	T_J	135	°C
Storage Temperature Range	T_{stg}	−65 to +150	°C

Stresses exceeding Maximum Ratings may damage the device. Maximum Ratings are stress ratings only. Functional operation above the Recommended Operating Conditions is not implied. Extended exposure to stresses above the Recommended Operating Conditions may affect device reliability.

ON Semiconductor®

http://onsemi.com

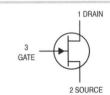

1 DRAIN

3 GATE

2 SOURCE

MARKING DIAGRAM

TO−92
CASE 29
STYLE 5

1 2 3

2N
545x
AYWW▪

2N545x = Device Code
    x = 7 or 8
A   = Assembly Location
Y   = Year
WW = Work Week
▪   = Pb−Free Package
(Note: Microdot may be in either location)

ORDERING INFORMATION

Device	Package	Shipping
2N5457	TO−92	1000 Units/Box
2N5457G	TO−92 (Pb−Free)	1000 Units/Box
2N5458	TO−92	1000 Units/Box
2N5458G	TO−92 (Pb−Free)	1000 Units/Box

Preferred devices are recommended choices for future use and best overall value.

*For additional information on our Pb−Free strategy and soldering details, please download the ON Semiconductor Soldering and Mounting Techniques Reference Manual, SOLDERRM/D.

Data Sheet A-11 (*Continued*)

2N5457, 2N5458

ELECTRICAL CHARACTERISTICS (T_A = 25°C unless otherwise noted)

Characteristic		Symbol	Min	Typ	Max	Unit
OFF CHARACTERISTICS						
Gate–Source Breakdown Voltage (I_G = −10 µAdc, V_{DS} = 0)		$V_{(BR)GSS}$	−25	–	–	Vdc
Gate Reverse Current (V_{GS} = −15 Vdc, V_{DS} = 0) (V_{GS} = −15 Vdc, V_{DS} = 0, T_A = 100°C)		I_{GSS}	– –	– –	−1.0 −200	nAdc
Gate–Source Cutoff Voltage (V_{DS} = 15 Vdc, i_D = 10 nAdc)	2N5457 2N5458	$V_{GS(off)}$	−0.5 −1.0	– –	−6.0 −7.0	Vdc
Gate–Source Voltage (V_{DS} = 15 Vdc, i_D = 100 µAdc) (V_{DS} = 15 Vdc, i_D = 200 µAdc)	2N5457 2N5458	V_{GS}	– –	−2.5 −3.5	– –	Vdc
ON CHARACTERISTICS						
Zero–Gate–Voltage Drain Current (Note 1) · (V_{DS} = 15 Vdc, V_{GS} = 0)	2N5457 2N5458	I_{DSS}	1.0 2.0	3.0 6.0	5.0 9.0	mAdc
DYNAMIC CHARACTERISTICS						
Forward Transfer Admittance (Note 1) (V_{DS} = 15 Vdc, V_{GS} = 0, f = 1 kHz)	2N5457 2N5458	$|Y_{fs}|$	1000 1500	3000 4000	5000 5500	µmhos
Output Admittance Common Source (Note 1) (V_{DS} = 15 Vdc, V_{GS} = 0, f = 1 kHz)		$|Y_{os}|$	–	10	50	µmhos
Input Capacitance (V_{DS} = 15 Vdc, V_{GS} = 0, f = 1 kHz)		C_{iss}	–	4.5	7.0	pF
Reverse Transfer Capacitance (V_{DS} = 15 Vdc, V_{GS} = 0, f = 1 kHz)		C_{rss}	–	1.5	3.0	pF

1. Pulse Width ≤ 630 ms, Duty Cycle ≤ 10%.

Data Sheet A-12
2N4856 FET

TECHNICAL DATA

N-CHANNEL J-FET
Qualified per MIL-PRF-19500/385

Devices

| 2N4856 | 2N4857 | 2N4858 | 2N4859 | 2N4860 | 2N4861 |

Qualified Level

**JAN
JANTX
JANTXV**

ABSOLUTE MAXIMUM RATINGS (T_C = +25^{0}C unless otherwise noted)

Parameters / Test Conditions	Symbol	2N4856 2N4857 2N4858	2N4859 2N4860 2N4861	Unit
Gate-Source Voltage	V_{GS}	-40	-30	V
Drain-Source Voltage	V_{DS}	40	30	V
Drain-Gate Voltage	V_{DG}	40	30	V
Gate Current	I_G	50		mA
Power Dissipation $\quad T_A = +25^0C$ $^{(1)}$ $\quad\quad\quad\quad\quad\quad T_C = +25^0C$ $^{(2)}$	P_T	0.36 1.8		W W
Operating Junction & Storage Temperature Range	T_j, T_{stg}	-65 to +200		^{0}C

(1) Derate linearly 2.06 mW/^{0}C for $T_A > 25^0$C.
(2) Derate linearly 10.3 mW/^{0}C for $T_C > 25^0$C.

TO-18*
(TO-206AA)

*See appendix A for
package outline

ELECTRICAL CHARACTERISTICS (T_C = 25^{0}C unless otherwise noted)

Parameters / Test Conditions		Symbol	Min.	Max.	Units
Gate-Source Breakdown Voltage $V_{DS} = 0$, $I_G = 1.0$ µAdc	2N4856, 2N4857, 2N4858 2N4859, 2N4860, 2N4861	$V_{(BR)GSS}$	-40 -30		Vdc
Gate-Source "Off" State Voltage $V_{DS} = 15$ Vdc, $I_D = 0.5$ ηAdc	2N4856, 2N4859 2N4857, 2N4860 2N4858, 2N4861	$V_{GS(on)}$	-4.0 -2.0 -0.8	-10 -6.0 -4.0	Vdc
Gate Reverse Current $V_{DS} = 0$, $V_{GS} = -20$ Vdc $V_{DS} = 0$, $V_{GS} = -15$ Vdc	2N4856, 2N4857, 2N4858 2N4859, 2N4860, 2N4861	I_{GSS}		-0.25 -0.25	ηA
Drain Current $V_{GS} = -10$ Vds, $V_{DS} = 15$ Vdc		$I_{D(off)}$		0.25	ηA

Data Sheet A-12 (*Continued*)

2N4856, 2N4857, 2N4858, 2N4859, 2N4860, 2N24861 JAN SERIES

ELECTRICAL CHARACTERISTICS ($T_C = 25^0C$ unless otherwise noted) (con't)

Parameters / Test Conditions		Symbol	Min.	Max.	Units
Drain Current		I_{DSS}			mA
$V_{GS} = 0$, $V_{DS} = 15$ Vdc 2N4856, 2N4859			50	175	
2N4857, 2N4860			20	100	
2N4858, 2N4861			8.0	80	
Static Drain - Source "On" State Resistance		$r_{ds(on)}$			Ω
$V_{GS} = 0$, $I_D = 1.0$ mAdc 2N4856, 2N4859				25	
2N4857, 2N4860				40	
2N4858, 2N4861				60	
Drain-Source "On" State Voltage		$V_{DS(on)}$			Vdc
$V_{GS} = 0$, $I_D = 20$ mAdc 2N4856, 2N4859				0.75	
$V_{GS} = 0$, $I_D = 10$ mAdc 2N4857, 2N4860				0.50	
$V_{GS} = 0$, $I_D = 5.0$ mAdc 2N4858, 2N4861				0.50	
Small-Signal, Common-Source Reverse Transfer Capacitance		C_{rss}		8.0	pF
$V_{GS} = -10$ Vdc, $V_{DS} = 0$, f = 1.0 MHz					
$C_1 = 0.1\mu F$, $L_1 = L_2 \geq 500 \mu H$					
Small-Signal, Common-Source Short-Circuit Input Capacitance		C_{iss}		18	pF
$V_{GS} = -10$ Vdc, $V_{DS} = 0$, f = 1.0 MHz					
$C_1 = 0.1\mu F$, $C_2 = 20.1$ m					
$FL_1 = L_2 \geq 500 \mu H$					
Turn-On Delay Time 2N4856, 2N4859		$^t d_{on}$		6	ηs
2N4857, 2N4860	See Figure 3			6	
2N4858, 2N4861	of MIL-PRF-			10	
Rise Time 2N4856, 2N4859	19500/385	$^t r$		3	ηs
2N4857, 2N4860				4	
2N4858, 2N4861				10	
Turn-Off Delay Time 2N4856, 2N4859		$^t d_{off}$		25	ηs
2N4857, 2N4860				50	
2N4858, 2N4861				100	

Data Sheet A-13
741 Op-amp

National *Semiconductor*

November 1994

LM741 Operational Amplifier

General Description

The LM741 series are general purpose operational amplifiers which feature improved performance over industry standards like the LM709. They are direct, plug-in replacements for the 709C, LM201, MC1439 and 748 in most applications.

The amplifiers offer many features which make their application nearly foolproof: overload protection on the input and output, no latch-up when the common mode range is exceeded, as well as freedom from oscillations.

The LM741C/LM741E are identical to the LM741/LM741A except that the LM741C/LM741E have their performance guaranteed over a 0°C to +70°C temperature range, instead of −55°C to +125°C.

Schematic Diagram

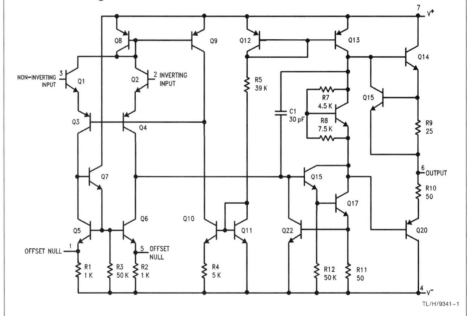

TL/H/9341–1

Offset Nulling Circuit

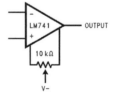

TL/H/9341–7

Data Sheet A-13 (*Continued*)

Absolute Maximum Ratings
If Military/Aerospace specified devices are required, please contact the National Semiconductor Sales Office/Distributors for availability and specifications. (Note 5)

	LM741A	LM741E	LM741	LM741C
Supply Voltage	±22V	±22V	±22V	±18V
Power Dissipation (Note 1)	500 mW	500 mW	500 mW	500 mW
Differential Input Voltage	±30V	±30V	±30V	±30V
Input Voltage (Note 2)	±15V	±15V	±15V	±15V
Output Short Circuit Duration	Continuous	Continuous	Continuous	Continuous
Operating Temperature Range	−55°C to +125°C	0°C to +70°C	−55°C to +125°C	0°C to +70°C
Storage Temperature Range	−65°C to +150°C	−65°C to +150°C	−65°C to +150°C	−65°C to +150°C
Junction Temperature	150°C	100°C	150°C	100°C
Soldering Information				
N-Package (10 seconds)	260°C	260°C	260°C	260°C
J- or H-Package (10 seconds)	300°C	300°C	300°C	300°C
M-Package				
Vapor Phase (60 seconds)	215°C	215°C	215°C	215°C
Infrared (15 seconds)	215°C	215°C	215°C	215°C

See AN-450 "Surface Mounting Methods and Their Effect on Product Reliability" for other methods of soldering surface mount devices.

| ESD Tolerance (Note 6) | 400V | 400V | 400V | 400V |

Electrical Characteristics (Note 3)

Parameter	Conditions	LM741A/LM741E Min	Typ	Max	LM741 Min	Typ	Max	LM741C Min	Typ	Max	Units
Input Offset Voltage	$T_A = 25°C$ $R_S \leq 10\,k\Omega$ $R_S \leq 50\Omega$		0.8	3.0		1.0	5.0		2.0	6.0	mV mV
	$T_{AMIN} \leq T_A \leq T_{AMAX}$ $R_S \leq 50\Omega$ $R_S \leq 10\,k\Omega$			4.0			6.0			7.5	mV mV
Average Input Offset Voltage Drift				15							μV/°C
Input Offset Voltage Adjustment Range	$T_A = 25°C, V_S = ±20V$	±10				±15			±15		mV
Input Offset Current	$T_A = 25°C$		3.0	30		20	200		20	200	nA
	$T_{AMIN} \leq T_A \leq T_{AMAX}$			70		85	500			300	nA
Average Input Offset Current Drift				0.5							nA/°C
Input Bias Current	$T_A = 25°C$		30	80		80	500		80	500	nA
	$T_{AMIN} \leq T_A \leq T_{AMAX}$			0.210			1.5			0.8	μA
Input Resistance	$T_A = 25°C, V_S = ±20V$	1.0	6.0		0.3	2.0		0.3	2.0		MΩ
	$T_{AMIN} \leq T_A \leq T_{AMAX}$, $V_S = ±20V$	0.5									MΩ
Input Voltage Range	$T_A = 25°C$							±12	±13		V
	$T_{AMIN} \leq T_A \leq T_{AMAX}$				±12	±13					V
Large Signal Voltage Gain	$T_A = 25°C, R_L \geq 2\,k\Omega$ $V_S = ±20V, V_O = ±15V$ $V_S = ±15V, V_O = ±10V$	50			50	200		20	200		V/mV V/mV
	$T_{AMIN} \leq T_A \leq T_{AMAX}$, $R_L \geq 2\,k\Omega$, $V_S = ±20V, V_O = ±15V$ $V_S = ±15V, V_O = ±10V$ $V_S = ±5V, V_O = ±2V$	32 10			25			15			V/mV V/mV V/mV

2

Data Sheet A-13 (*Continued*)

Electrical Characteristics (Note 3) (Continued)

Parameter	Conditions	LM741A/LM741E			LM741			LM741C			Units
		Min	Typ	Max	Min	Typ	Max	Min	Typ	Max	
Output Voltage Swing	$V_S = \pm 20V$ $R_L \geq 10\,k\Omega$ $R_L \geq 2\,k\Omega$	± 16 ± 15									V V
	$V_S = \pm 15V$ $R_L \geq 10\,k\Omega$ $R_L \geq 2\,k\Omega$				± 12 ± 10	± 14 ± 13		± 12 ± 10	± 14 ± 13		V V
Output Short Circuit Current	$T_A = 25°C$ $T_{AMIN} \leq T_A \leq T_{AMAX}$	10 10	25	35 40		25			25		mA mA
Common-Mode Rejection Ratio	$T_{AMIN} \leq T_A \leq T_{AMAX}$ $R_S \leq 10\,k\Omega, V_{CM} = \pm 12V$ $R_S \leq 50\Omega, V_{CM} = \pm 12V$	80	95		70	90		70	90		dB dB
Supply Voltage Rejection Ratio	$T_{AMIN} \leq T_A \leq T_{AMAX},$ $V_S = \pm 20V$ to $V_S = \pm 5V$ $R_S \leq 50\Omega$ $R_S \leq 10\,k\Omega$	86	96		77	96		77	96		dB dB
Transient Response Rise Time Overshoot	$T_A = 25°C$, Unity Gain		0.25 6.0	0.8 20		0.3 5			0.3 5		μs %
Bandwidth (Note 4)	$T_A = 25°C$	0.437	1.5								MHz
Slew Rate	$T_A = 25°C$, Unity Gain	0.3	0.7			0.5			0.5		$V/\mu s$
Supply Current	$T_A = 25°C$					1.7	2.8		1.7	2.8	mA
Power Consumption	$T_A = 25°C$ $V_S = \pm 20V$ $V_S = \pm 15V$		80	150		50	85		50	85	mW mW
LM741A	$V_S = \pm 20V$ $T_A = T_{AMIN}$ $T_A = T_{AMAX}$			165 135							mW mW
LM741E	$V_S = \pm 20V$ $T_A = T_{AMIN}$ $T_A = T_{AMAX}$			150 150							mW mW
LM741	$V_S = \pm 15V$ $T_A = T_{AMIN}$ $T_A = T_{AMAX}$					60 45	100 75				mW mW

Note 1: For operation at elevated temperatures, these devices must be derated based on thermal resistance, and T_j max. (listed under "Absolute Maximum Ratings"). $T_j = T_A + (\theta_{jA} P_D)$.

Thermal Resistance	Cerdip (J)	DIP (N)	HO8 (H)	SO-8 (M)
θ_{jA} (Junction to Ambient)	100°C/W	100°C/W	170°C/W	195°C/W
θ_{jC} (Junction to Case)	N/A	N/A	25°C/W	N/A

Note 2: For supply voltages less than $\pm 15V$, the absolute maximum input voltage is equal to the supply voltage.

Note 3: Unless otherwise specified, these specifications apply for $V_S = \pm 15V$, $-55°C \leq T_A \leq +125°C$ (LM741/LM741A). For the LM741C/LM741E, these specifications are limited to $0°C \leq T_A \leq +70°C$.

Note 4: Calculated value from: BW (MHz) = 0.35/Rise Time(μs).

Note 5: For military specifications see RETS741X for LM741 and RETS741AX for LM741A.

Note 6: Human body model, 1.5 $k\Omega$ in series with 100 pF.

Data Sheet A-13 *(Continued)*

Connection Diagrams

Metal Can Package

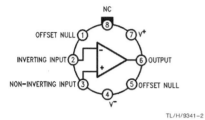

TL/H/9341-2

**Order Number LM741H, LM741H/883*,
LM741AH/883 or LM741CH
See NS Package Number H08C**

Ceramic Dual-In-Line Package

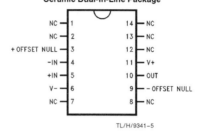

TL/H/9341-5

Order Number LM741J-14/883*, LM741AJ-14/883
See NS Package Number J14A**

*also available per JM38510/10101

**also available per JM38510/10102

Dual-In-Line or S.O. Package

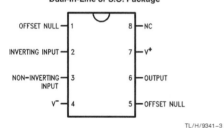

TL/H/9341-3

**Order Number LM741J, LM741J/883,
LM741CM, LM741CN or LM741EN
See NS Package Number J08A, M08A or N08E**

Ceramic Flatpak

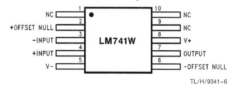

TL/H/9341-6

**Order Number LM741W/883
See NS Package Number W10A**

*LM741H is available per JM38510/10101

4

Data Sheet A-14
108 and 308 Op-amps

 **National** *Semiconductor*

December 1994

LM108/LM208/LM308 Operational Amplifiers

General Description

The LM108 series are precision operational amplifiers having specifications a factor of ten better than FET amplifiers over a $-55°C$ to $+125°C$ temperature range.

The devices operate with supply voltages from $\pm2V$ to $\pm20V$ and have sufficient supply rejection to use unregulated supplies. Although the circuit is interchangeable with and uses the same compensation as the LM101A, an alternate compensation scheme can be used to make it particularly insensitive to power supply noise and to make supply bypass capacitors unnecessary.

The low current error of the LM108 series makes possible many designs that are not practical with conventional amplifiers. In fact, it operates from 10 MΩ source resistances,

introducing less error than devices like the 709 with 10 kΩ sources. Integrators with drifts less than 500 μV/sec and analog time delays in excess of one hour can be made using capacitors no larger than 1 μF.

The LM108 is guaranteed from $-55°C$ to $+125°C$, the LM208 from $-25°C$ to $+85°C$, and the LM308 from 0°C to $+70°C$.

Features

■ Maximum input bias current of 3.0 nA over temperature
■ Offset current less than 400 pA over temperature
■ Supply current of only 300 μA, even in saturation
■ Guaranteed drift characteristics

Compensation Circuits

Standard Compensation Circuit

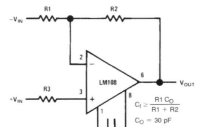

$$C_f \geq \frac{R1\,C_O}{R1 + R2}$$
$$C_O = 30 \text{ pF}$$

TL/H/7758-1

Alternate* Frequency Compensation

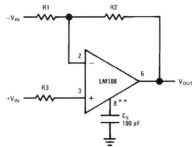

C_S 100 pF

TL/H/7758-2

**Bandwidth and slew rate are proportional to $1/C_f$

*Improves rejection of power supply noise by a factor of ten.

**Bandwidth and slew rate are proportional to $1/C_S$

Feedforward Compensation

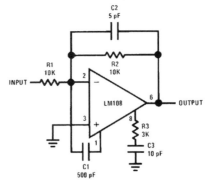

TL/H/7758-3

RRD-B30M115/Printed in U. S. A.

(vertical text, right margin) LM108/LM208/LM308 Operational Amplifiers

Data Sheet A-14 *(Continued)*

Absolute Maximum Ratings

If Military/Aerospace specified devices are required, please contact the National Semiconductor Sales Office/Distributors for availability and specifications. (Note 5)

	LM108/LM208	LM308
Supply Voltage	$\pm$ 20V	$\pm$ 18V
Power Dissipation (Note 1)	500 mW	500 mW
Differential Input Current (Note 2)	$\pm$ 10 mA	$\pm$ 10 mA
Input Voltage (Note 3)	$\pm$ 15V	$\pm$ 15V
Output Short-Circuit Duration	Continuous	Continuous
Operating Temperature Range (LM108)	$-55°C$ to $+125°C$	$0°C$ to $+70°C$
(LM208)	$-25°C$ to $+85°C$	
Storage Temperature Range	$-65°C$ to $+150°C$	$-65°C$ to $+150°C$
Lead Temperature (Soldering, 10 sec)		
DIP	260°C	260°C
H Package Lead Temp		
(Soldering 10 seconds)	300°C	300°C
Soldering Information		
Dual-In-Line Package		
Soldering (10 seconds)	260°C	
Small Outline Package		
Vapor Phase (60 seconds)	215°C	
Infrared (15 seconds)	220°C	

See AN-450 "Surface Mounting Methods and Their Effect on Product Reliability" for other methods of soldering surface mount devices.

ESD Tolerance (Note 6)	2000V

Electrical Characteristics (Note 4)

Parameter	Condition	LM108/LM208			LM308			Units
		Min	Typ	Max	Min	Typ	Max	
Input Offset Voltage	$T_A = 25°C$		0.7	2.0		2.0	7.5	mV
Input Offset Current	$T_A = 25°C$		0.05	0.2		0.2	1	nA
Input Bias Current	$T_A = 25°C$		0.8	2.0		1.5	7	nA
Input Resistance	$T_A = 25°C$	30	70		10	40		MΩ
Supply Current	$T_A = 25°C$		0.3	0.6		0.3	0.8	mA
Large Signal Voltage Gain	$T_A = 25°C$, $V_S = \pm 15V$ $V_{OUT} = \pm 10V$, $R_L \geq 10\,k\Omega$	50	300		25	300		V/mV
Input Offset Voltage				3.0			10	mV
Average Temperature Coefficient of Input Offset Voltage			3.0	15		6.0	30	μV/°C
Input Offset Current				0.4			1.5	nA
Average Temperature Coefficient of Input Offset Current			0.5	2.5		2.0	10	pA/°C
Input Bias Current				3.0			10	nA
Supply Current	$T_A = +125°C$		0.15	0.4				mA
Large Signal Voltage Gain	$V_S = \pm 15V$, $V_{OUT} = \pm 10V$ $R_L \geq 10\,k\Omega$	25			15			V/mV
Output Voltage Swing	$V_S = \pm 15V$, $R_L = 10\,k\Omega$	$\pm$ 13	$\pm$ 14		$\pm$ 13	$\pm$ 14		V

2

Data Sheet A-14   (*Continued*)

Electrical Characteristics (Note 4) (Continued)

Parameter	Condition	LM108/LM208			LM308			Units
		Min	Typ	Max	Min	Typ	Max	
Input Voltage Range	$V_S = \pm 15V$	± 13.5			± 14			V
Common Mode Rejection Ratio		85	100		80	100		dB
Supply Voltage Rejection Ratio		80	96		80	96		dB

Note 1: The maximum junction temperature of the LM108 is 150°C, for the LM208, 100°C and for the LM308, 85°C. For operating at elevated temperatures, devices in the H08 package must be derated based on a thermal resistance of 160°C/W, junction to ambient, or 20°C/W, junction to case. The thermal resistance of the dual-in-line package is 100°C/W, junction to ambient.

Note 2: The inputs are shunted with back-to-back diodes for overvoltage protection. Therefore, excessive current will flow if a differential input voltage in excess of 1V is applied between the inputs unless some limiting resistance is used.

Note 3: For supply voltages less than $\pm 15V$, the absolute maximum input voltage is equal to the supply voltage.

Note 4: These specifications apply for $\pm 5V \leq V_S \leq \pm 20V$ and $-55°C \leq T_A \leq +125°C$, unless otherwise specified. With the LM208, however, all temperature specifications are limited to $-25°C \leq T_A \leq 85°C$, and for the LM308 they are limited to $0°C \leq T_A \leq 70°C$.

Note 5: Refer to RETS108X for LM108 military specifications and RETs 108AX for LM108A military specifications.

Note 6: Human body model, 1.5 kΩ in series with 100 pF.

Schematic Diagram

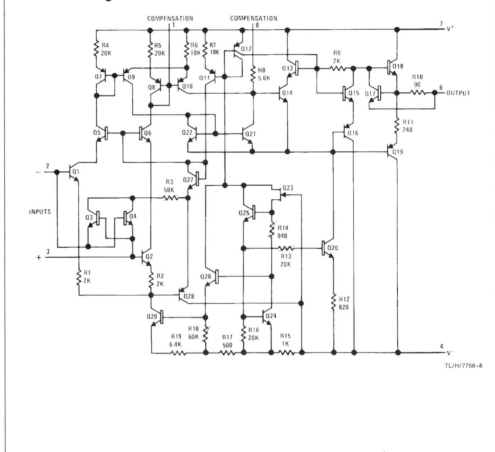

TL/H/7758–8

3

Data Sheet A-14 (*Continued*)

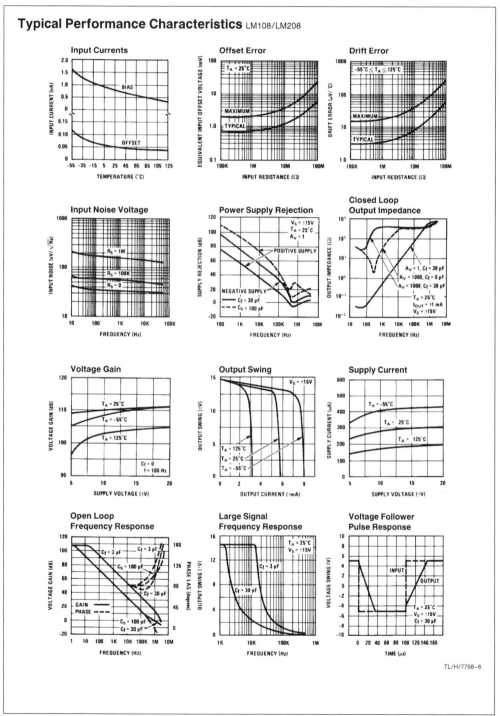

Typical Performance Characteristics LM108/LM208

TL/H/7758–6

4

Data Sheet A-15
353 Op-amp

December 2003

 **National Semiconductor**

LF353
Wide Bandwidth Dual JFET Input Operational Amplifier

General Description

These devices are low cost, high speed, dual JFET input operational amplifiers with an internally trimmed input offset voltage (BI-FET II™ technology). They require low supply current yet maintain a large gain bandwidth product and fast slew rate. In addition, well matched high voltage JFET input devices provide very low input bias and offset currents. The LF353 is pin compatible with the standard LM1558 allowing designers to immediately upgrade the overall performance of existing LM1558 and LM358 designs.

These amplifiers may be used in applications such as high speed integrators, fast D/A converters, sample and hold circuits and many other circuits requiring low input offset voltage, low input bias current, high input impedance, high slew rate and wide bandwidth. The devices also exhibit low noise and offset voltage drift.

Features

- Internally trimmed offset voltage: 10 mV
- Low input bias current: 50pA
- Low input noise voltage: 25 nV/√Hz
- Low input noise current: 0.01 pA/√Hz
- Wide gain bandwidth: 4 MHz
- High slew rate: 13 V/μs
- Low supply current: 3.6 mA
- High input impedance: $10^{12}\Omega$
- Low total harmonic distortion :   ≤0.02%
- Low 1/f noise corner: 50 Hz
- Fast settling time to 0.01%:   2 μs

Typical Connection

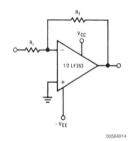

00564914

Connection Diagram

Dual-In-Line Package

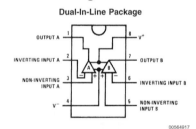

00564917

Top View
Order Number LF353M, LF353MX or LF353N
See NS Package Number M08A or N08E

Simplified Schematic

1/2 Dual

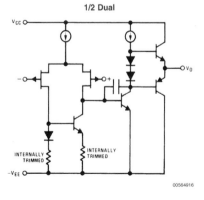

00564916

BI-FET II™ is a trademark of National Semiconductor Corporation.

Data Sheet A-15 (*Continued*)

Absolute Maximum Ratings (Note 1)

If Military/Aerospace specified devices are required, please contact the National Semiconductor Sales Office/Distributors for availability and specifications.

Supply Voltage	±18V
Power Dissipation	(Note 2)
Operating Temperature Range	0°C to +70°C
T_j(MAX)	150°C
Differential Input Voltage	±30V
Input Voltage Range (Note 3)	±15V
Output Short Circuit Duration	Continuous
Storage Temperature Range	−65°C to +150°C
Lead Temp. (Soldering, 10 sec.)	260°C
Soldering Information	
Dual-In-Line Package	
Soldering (10 sec.)	260°C
Small Outline Package	
Vapor Phase (60 sec.)	215°C
Infrared (15 sec.)	220°C

See AN-450 "Surface Mounting Methods and Their Effect on Product Reliability" for other methods of soldering surface mount devices.

ESD Tolerance (Note 8)	1000V
θ_{JA} M Package	TBD

Note 1: Absolute Maximum Ratings indicate limits beyond which damage to the device may occur. Operating ratings indicate conditions for which the device is functional, but do not guarantee specific performance limits. Electrical Characteristics state DC and AC electrical specifications under particular test conditions which guarantee specific performance limits. This assumes that the device is within the Operating Ratings. Specifications are not guaranteed for parameters where no limit is given, however, the typical value is a good indication of device performance.

DC Electrical Characteristics
(Note 5)

Symbol	Parameter	Conditions	LF353 Min	LF353 Typ	LF353 Max	Units
V_{OS}	Input Offset Voltage	R_S=10kΩ, T_A=25°C		5	10	mV
		Over Temperature			13	mV
$\Delta V_{OS}/\Delta T$	Average TC of Input Offset Voltage	R_S=10 kΩ		10		µV/°C
I_{OS}	Input Offset Current	T_j=25°C, (Notes 5, 6)		25	100	pA
		T_j≤70°C			4	nA
I_B	Input Bias Current	T_j=25°C, (Notes 5, 6)		50	200	pA
		T_j≤70°C			8	nA
R_{IN}	Input Resistance	T_j=25°C		10^{12}		Ω
A_{VOL}	Large Signal Voltage Gain	V_S=±15V, T_A=25°C	25	100		V/mV
		V_O=±10V, R_L=2 kΩ				
		Over Temperature	15			V/mV
V_O	Output Voltage Swing	V_S=±15V, R_L=10kΩ	±12	±13.5		V
V_{CM}	Input Common-Mode Voltage	V_S=±15V	±11	+15		V
	Range			−12		V
CMRR	Common-Mode Rejection Ratio	R_S≤ 10kΩ	70	100		dB
PSRR	Supply Voltage Rejection Ratio	(Note 7)	70	100		dB
I_S	Supply Current			3.6	6.5	mA

AC Electrical Characteristics
(Note 5)

Symbol	Parameter	Conditions	LF353 Min	LF353 Typ	LF353 Max	Units
	Amplifier to Amplifier Coupling	T_A=25°C, f=1 Hz–20 kHz (Input Referred)		−120		dB
SR	Slew Rate	V_S=±15V, T_A=25°C	8.0	13		V/µs
GBW	Gain Bandwidth Product	V_S=±15V, T_A=25°C	2.7	4		MHz
e_n	Equivalent Input Noise Voltage	T_A=25°C, R_S=100Ω, f=1000 Hz		16		nV/$\sqrt{Hz}$
i_n	Equivalent Input Noise Current	T_j=25°C, f=1000 Hz		0.01		pA/$\sqrt{Hz}$

Data Sheet A-15 (*Continued*)

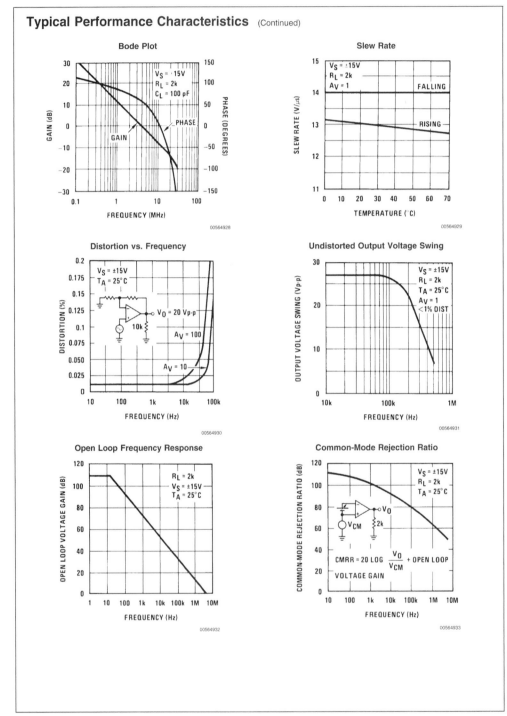

Typical Performance Characteristics (Continued)

Bode Plot

$V_S = \pm 15V$
$R_L = 2k$
$C_L = 100\,pF$

00564928

Slew Rate

$V_S = \pm 15V$
$R_L = 2k$
$A_V = 1$

FALLING

RISING

00564929

Distortion vs. Frequency

$V_S = \pm 15V$
$T_A = 25°C$

$V_O = 20\,Vp\text{-}p$

$A_V = 100$

$A_V = 10$

00564930

Undistorted Output Voltage Swing

$V_S = \pm 15V$
$R_L = 2k$
$T_A = 25°C$
$A_V = 1$
$<1\%$ DIST

00564931

Open Loop Frequency Response

$R_L = 2k$
$V_S = \pm 15V$
$T_A = 25°C$

00564932

Common-Mode Rejection Ratio

$V_S = \pm 15V$
$R_L = 2k$
$T_A = 25°C$

$CMRR = 20\,LOG\,\dfrac{V_O}{V_{CM}} + OPEN\ LOOP$
VOLTAGE GAIN

00564933

Data Sheet A-16
723 Voltage Regulator

	Min.	Type	Max.
Continuous voltage V_{cc}^+ to V_{cc}^-			40 V
Input-output voltage differential			40 V
Input voltage range	9.5 V		37 V
Output voltage range	2 V		37 V
Output current			150 mA
Standby current		2.3 mA	3.5 mA
Reference voltage	6.95 V	7.15 V	7.35 V
Current from V_{ref}			15 mA
Maximum power dissipation at or below 25°C free air temperature:			
Metal can			1000 mW
DIP			12.5 W

Data Sheet A-17
SCRs—Comparison of Typical Parameters

Parameter		Device		
		2N1597	2N5170	2N6400
V_{RRM}	Peak repetitive blocking voltage	200 V	400 V	50 V
$I_{T(rms)}$	RMS on-state current	1.6 A	20 A	16 A
V_{TM}	Peak on-state voltage	1.1 V	1.5 V	1.7 V
I_{GT}	Gate trigger current	2 mA	40 mA	5 mA
V_{GT}	Gate trigger voltage	0.7 V	1.5 V	0.7 V
I_H	Holding current	5 mA	50 mA	6 mA
t_{gt}	Turn-on time	0.8 μs	1 μs	1 μs
t_q	Turn-off time	10 μs	20 μs	15 μs

Data Sheet A-18

UJTs—Comparison of Typical Parameters

Parameter		Device		
		2N4870	2N4949	2N2647
P_D	Maximum power dissipation	300 mW	360 mW	300 mW
	Derate linearly at	3 mW/°C	2.4 mW/°C	3 mW/°C
R_{BB}	Interbase resistance	6 kΩ	7 kΩ	7 kΩ
η	Intrinsic stand-off ratio	0.56 to 0.75	0.55 to 0.82	0.68 to 0.82
$V_{EB1(sat)}$	Emitter–B1 saturation voltage	2.5 V	2.5 V	3.5 V
I_P	Peak point current	1 μA	0.6 μA	1 μA
I_V	Valley point current current	5 mA	4 mA	10 mA
$I_{E(rms)}$	Maximum rms emitter current	50 mA	50 mA	50 mA
i_E	Peak emitter pulse current	1.5 A	1 A	2 A
V_{B1B2}	Maximum B1-B2 voltage	35 V	35 V	35 V
V_{B2E}	Maximum emitter reverse voltage	30 V	30 V	30 V

APPENDIX B   STANDARD-VALUE COMPONENTS

Table B-1
Typical Standard-Value Resistors

10% tolerance resistors

Ω	Ω	Ω	kΩ	kΩ	kΩ	MΩ	MΩ
—	10	100	1	10	100	1	10
—	12	120	1.2	12	120	1.2	12
—	15	150	1.5	15	150	1.5	15
—	18	180	1.8	18	180	1.8	18
—	22	220	2.2	22	220	2.2	22
2.7	27	270	2.7	27	270	2.7	—
3.3	33	330	3.3	33	330	3.3	—
3.9	39	390	3.9	39	390	3.9	—
4.7	47	470	4.7	47	470	4.7	—
5.6	56	560	5.6	56	560	5.6	—
6.8	68	680	6.8	68	680	6.8	—
8.2	82	820	8.2	82	820	8.2	—

5% tolerance resistors

Ω	Ω	Ω	kΩ	kΩ	kΩ	MΩ	MΩ
—	10	100	1	10	100	1	10
—	11	110	1.1	11	110	1.1	11
—	12	120	1.2	12	120	1.2	12
—	13	130	1.3	13	130	1.3	13
—	15	150	1.5	15	150	1.5	15
—	16	160	1.6	16	160	1.6	16
—	18	180	1.8	18	180	1.8	18
—	20	200	2	20	200	2	20
—	22	220	2.2	22	220	2.2	22
—	24	240	2.4	24	240	2.4	—
2.7	27	270	2.7	27	270	2.7	—
3	30	300	3	30	300	3	—
3.3	33	330	3.3	33	330	3.3	—
3.6	36	360	3.6	36	360	3.6	—
3.9	39	390	3.9	39	390	3.9	—
4.3	43	430	4.3	43	430	4.3	—
4.7	47	470	4.7	47	470	4.7	—
5.1	51	510	5.1	51	510	5.1	—
5.6	56	560	5.6	56	560	5.6	—
6.2	62	620	6.2	62	620	6.2	—
6.8	68	680	6.8	68	680	6.8	—
7.5	75	750	7.5	75	750	7.5	—
8.2	82	820	8.2	82	820	8.2	—
9.1	91	910	9.1	91	910	9.1	—

Table B-1 (*Continued*)

1% tolerance resistors (Basic values in Ω)

For values above 10 Ω multiply by 10, 100, etc.							
1.00	6.04	11.0	20.0	36.5	66.5	121	221
1.10	6.19	11.3	20.5	37.4	68.1	124	226
1.21	6.34	11.5	21.0	38.3	69.8	127	232
1.30	6.49	11.8	21.5	39.2	71.5	130	237
1.50	6.65	12.1	22.1	40.2	73.2	133	243
1.62	6.81	12.4	22.6	41.2	75.0	137	249
1.82	6.98	12.7	23.2	42.2	76.8	140	255
2.00	7.15	13.0	23.7	43.2	78.7	143	261
2.21	7.32	13.3	24.3	44.2	80.6	147	267
2.43	7.50	13.7	24.9	45.3	82.5	150	274
2.67	7.68	14.0	25.5	46.4	84.5	154	280
3.01	7.87	14.3	26.1	47.5	86.6	158	287
3.32	8.06	14.7	26.7	48.7	88.7	162	294
3.57	8.25	15.0	27.4	49.9	90.9	165	301
3.92	8.45	15.4	28.0	51.1	93.1	169	309
4.32	8.66	15.8	28.7	52.3	95.3	174	316
4.75	8.87	16.2	29.4	53.6	97.6	178	324
4.99	9.09	16.5	30.1	54.9	100	182	332
5.11	9.31	16.9	30.9	56.2	102	187	340
5.23	9.53	17.4	31.6	57.6	105	191	348
5.36	9.76	17.8	32.4	59.0	107	196	356
5.49	10.0	18.2	33.2	60.4	110	200	365
5.62	10.2	18.7	34.0	61.9	113	205	374
5.76	10.5	19.1	34.8	63.4	115	210	383
5.90	10.7	19.6	35.7	64.9	118	215	392

Potentiometers

Ω	Ω	kΩ	kΩ	kΩ	MΩ
10	100	1	10	100	1
—	150	1.5	15	—	—
20	200	2	20	200	2
—	250	2.5	25	250	2.5
—	350	3.5	35	—	—
50	500	5	50	500	—
—	750	7.5	75	750	—

Table B-2
Typical Standard-Value Capacitors

pF	pF	pF	pF	μF	μF	μF	μF	μF	μF	μF
5	50	500	5000	—	0.05	0.5	5	50	500	5000
—	51	510	5100	—	—	—	—	—	—	—
—	56	560	5600	—	0.056	0.56	5.6	56	—	5600
—	—	—	6000	—	0.06	—	6	—	—	6000
—	62	620	6200	—	—	—	—	—	—	—
—	68	680	6800	—	0.068	0.68	6.8	68	680	6800
—	75	750	7500	—	—	—	—	75	—	—
—	—	—	8000	—	—	—	8	80	—	—
—	82	820	8200	—	0.082	0.82	8.2	82	—	—
—	91	910	9100	—	—	—	—	—	—	—
10	100	1000	—	0.01	0.1	1	10	100	1000	10000
—	110	1100	—	—	—	—	—	—	—	—
12	120	1200	—	0.012	0.12	1.2	—	—	—	—
—	130	1300	—	—	—	—	—	—	—	—
15	150	1500	—	0.015	0.15	1.5	15	150	1500	15000
—	160	1600	—	—	—	—	—	—	—	—
18	180	1800	—	0.018	0.18	1.8	18	180	—	—
20	200	2000	—	0.02	0.2	2	20	200	2000	—
22	220	2200	—	—	0.22	2.2	22	220	2200	22000
24	240	2400	—	—	—	—	—	240	—	—
—	250	2500	—	—	0.25	—	25	250	2500	—
27	270	2700	—	0.027	0.27	2.7	27	270	—	—
30	300	3000	—	0.03	0.3	3	30	300	3000	—
33	330	3300	—	0.033	0.33	3.3	33	330	3300	—
36	360	3600	—	—	—	—	—	—	—	—
39	390	3900	—	0.039	0.39	3.9	39	—	—	—
—	—	4000	—	0.04	—	4	—	400	—	—
43	430	4300	—	—	—	—	—	—	—	—
47	470	4700	—	0.047	0.47	4.7	47	470	4700	—

APPENDIX C ANSWERS FOR ODD-NUMBERED PROBLEMS

2-1   4.25 Ω, 1.5 × 10⁹ Ω

2-3   13 Ω, 1.7 Ω

2-5   6 V

2-7   Pt. A (0 mA, 0.7 V), Pt. B (300 mA, 0.85 V)

2-9   170 mA, 260 mA

2-11   9.1 V

2-13   625 mA

2-15   1.7 A, 1.25 A

2-17   26.7 mA, 28 mA

2-19   20 ns

2-21   80 ns

2-23   800 V, 1.5 A

2-25   2 V/cm, 200 Ω, 0.05 V/cm

2-27   44 mA, 33.4 mA

3-1   34.7 V, 57.8 mA, 35.4 V

3-3   536 mW

3-5   34 V, 56.6 mA, 14 mW, 35.4 V

3-7   3300 μF

3-9   50.4 V, 200 mA, 4.38 A, 1N4002, 0.84 Ω

3-11   1500 μF

3-13   115 V, 11.6 V, 109 mA, 1.08 A

3-15   1500 μF

3-17   25.9 V, 200 mA, 2.2 A, 1N4001, 0.86 Ω

3-19   560 μF

3-21   82 μF

3-23   0.4 V, 0.5 V, 3.3%, 4.2%

3-25   10%, 2.08%

3-27   680 Ω, 0.27 W, ±3.7 mA

3-29   129 mV, 1.16 V, 5.85 × 10⁻²

3-31   4.59 V, 5.61 V, 92.2 Ω

3-33   10 kΩ, 12 V, 1 mA

3-35   23 mA, 7 V

3-37   2.7 kΩ, 6 V, 1.96 mA

3-39   ±3.3 V, 6.8 kΩ

3-41   1N756, 1 kΩ

3-43   1N751, 820 Ω

3-45   132 mV

3-47   2.7 μF, 5.6 kΩ

3-49   +3.6 V, −12.4 V, 4.4 V

3-51   1N759, 1.5 μF, 1.5 kΩ

3-53   1.5 μF, 3 μF

3-55   1.2 kΩ

4-1   1.62 mA, 1.67 mA, 32.3

4-3   1.98 mA, 2 mA

4-5   1 mA

4-7   120

4-9   68

4-11   3.02 mA, 280 μA, 10.8

4-13   5.3 V

4-17   4 mA

4-19   40 μA

4-21   7.7 mA, 3.7 mA, 123

4-23   9.3 V

4-25   1 V/cm, 500 Ω, 0.8 V/cm

5-1   (0.97 mA, 7.7 V), (0.94 mA, 4.5 V)

5-3   ±7.3 V, ±4.5 V

5-5   2.8 V, 2.35 V, 12.8 V

5-7   5.95 mA, 4.3 V

5-9   25 V, 4 V

5-11   5.45 mA, 5.1 V

5-13   33.2, 3.86 V

5-15   4.7 mA, 6.5 V

5-17   10.6 V

5-19   5.93 V

5-21   (1.5 V, 10.72 V), (4.46 V, 8.7 V), (6.98 V, 7.66 V)

5-23 $V_{CE} = 0.45$ V

5-25   4.7 kΩ

5-27   10.7 V, (≈0.2 V)

5-29   270 kΩ, 2.2 kΩ

5-31   270 kΩ, 4.7 kΩ

5-33   220 kΩ, 56 kΩ, 15 kΩ, 5.6 kΩ

5-35   100 kΩ, 33 kΩ, 2.7 kΩ

5-37   7.8 V, 1.4 mA

5-39   680 Ω, 1.2 kΩ, 1.8 kΩ

5-41   2.7 kΩ, 6.8 kΩ

5-43   10 kΩ, 5.6 kΩ, 1.2 kΩ, 470 Ω

5-45   39.5 μA, 12.3 μA, 2.6 μA

5-47   16 μA, 33 μA, 50 μA

5-49   9.9

5-51   1.8 kΩ, 10 kΩ

5-53   5.6 kΩ, 22 kΩ

6-1 7.92 V, 6 V
6-3 point A ($I_C = 0$, $V_{CE} = 18$ V);
point B ($I_C = 1.6$ mA, $V_{CE} = 0$)
point Q ($I_C = 0.9$ mA, $V_{CE} = 7.8$ V);
point C ($I_C = 0$, $V_{CE} = 12.1$ V)
$\Delta V_C = \pm 4.3$ V
6-5 point A ($I_C = 0$, $V_{CE} = 15$ V);
point B ($I_C = 1.39$ mA, $V_{CE} = 0$)
point Q ($I_C = 1$ mA, $V_{CE} = 4$ V);
point C ($I_C = 0$, $V_{CE} = 12.1$ V)
6-7 1 kΩ, 62.5 µS, 135
6-9 123
6-11 50 µS
6-13 15.4 Ω
6-15 893 Ω, 5.54 kΩ, −425
6-17 1.29 kΩ, 3.28 kΩ, −98.3
6-19 1.48 kΩ, 3.29 kΩ, −213
6-21 8.2 kΩ, −0.88
6-23 13.1 kΩ, 18.5 Ω
6-25 62.9 kΩ, 55 Ω
6-27 11.6 Ω, 5.5 kΩ, 425
6-29 31.4 Ω, 1.8 kΩ, 54.9
6-31 7.9 Ω, 3.28 kΩ, 380
6-33 (12.9 kΩ to 13.3 kΩ), (21 Ω to 42.3 Ω)

8-1 60 V, 4 A, 5 V, 25, 100
8-3 −1.25 dB
8-5 1.26 V
8-7 2.76 nF
8-9 6 pF
8-11 4.02 GHz
8-13 36.5 MHz
8-15 79.6 kHz
8-17 70 ns, 250 ns
8-19 50 ns
8-21 390 pF, 975 ns
8-23 2.4 dB
8-25 64°C
8-27 24 W
8-31 0.5°C/W
8-33 1.25°C/W

9-1 5 mA, 5 V
9-3 9 mA, 8 V
9-5 2 mS
9-7 40 V, 360 mW, −6 V, 100 mA, 40 Ω

9-9 540 µS, 1150 µS
9-11 20 mS, 30 mS
9-13 20 mS, 30 mS
9-15 25 V, 310 mW, 6 V, 5 mA
9-17 ±0.5 V, ±0.1 V, 5, 1
9-19 3.3 kΩ
9-21 4 V, 2.5 V, 12.5 W
9-23 5.7 V, 1.5 V, 7.5 W

10-1 −1.5 V
10-3 4 kΩ
10-5 2.3 mA, 8.1 V
10-7 7.8 V, 19.7 V
10-9 11.9 V, 19.1 V
10-11 5.2 V, 19.5 V
10-13 17.3 V
10-15 6.5 V, 9.8 V
10-17 9.5 V, 14.2 V
10-19 −0.9 V, 1 MΩ, 3.3 kΩ
10-21 1 MΩ, 5.6 kΩ, 1 kΩ
10-23 3.9 MΩ, 1 MΩ, 3.3 kΩ, 3.3 kΩ
10-25 (6.8 MΩ + 1.8 MΩ), 1 MΩ, 3.9 kΩ, 3.9 kΩ
10-27 1 MΩ, 3.9 kΩ, 10 kΩ
10-29 1 MΩ, 4.7 kΩ, 10 kΩ
10-31 1.8 MΩ, 1 MΩ, 3.3 kΩ, 3.3 kΩ
10-33 6.8 MΩ, 1 MΩ, 3.9 kΩ, 3.9 kΩ
10-35 (82 kΩ + 1.5 kΩ), 27 kΩ, 3.9 kΩ, 2.7 kΩ
10-37 5.2 V
10-39 8.9 V
10-41 (5.6 MΩ + 2.2 MΩ), 1 MΩ, 1.8 kΩ
10-43 3.9 MΩ, 1 MΩ, 2.2 kΩ, 2.2 kΩ
10-45 1 MΩ, 1 Ω, 5.5 V
10-47 1 MΩ, 10 kΩ, −5 V
10-49 0.6 Ω, 5.7 V

11-1 $R_{L(dc)} = 5.4$ kΩ, $R_{L(ac)} = 2.5$ kΩ
11-3 $R_{L(dc)} = 3.6$ kΩ, $R_{L(ac)} = 1.6$ kΩ
11-5 880 kΩ, 3.7 kΩ, −12.5
11-7 470 kΩ, 4.5 kΩ, −19.3
11-9 9.1 kΩ, −36
11-11 470 kΩ, 4.5 kΩ, −1.8
11-13 629 kΩ, 161 Ω, 0.96
11-15 820 kΩ, 241 Ω, 0.96
11-17 685 kΩ, 318 Ω, 0.95
11-19 160 Ω, 3.7 kΩ, 21.2
11-21 266 Ω, 3.7 kΩ, 12.5

11-23 241 Ω, 6.4 kΩ, 14.1
11-25 0.99 V
11-27 693 mV, 217 mV
11-29 3.2 MHz
11-31 5.1 MHz
11-33 81.7 MHz

12-1 100 kΩ, 56 kΩ, 6.8 kΩ, 5.6 kΩ
12-3 236, 1.23 kΩ, 6.8 kΩ
12-5 15 μF, 150 μF, 0.33 μF, 430 pF
12-7 56 kΩ, 39 kΩ, 3.9 kΩ, 3.9 kΩ
12-9 158, 2.07 kΩ, 3.4 kΩ
12-11 0.039 μF, 15 μF, 0.39 μF, 510 pF
12-13 3.3 MΩ, 1 MΩ, 6.8 kΩ, 6.8 kΩ
12-15 24.3, 767 kΩ, 6.8 kΩ
12-17 0.027 μF, 15 μF, 0.39 μF
12-19 (56 kΩ + 1.5 kΩ), 47 kΩ, 3.9 kΩ, 4.7 kΩ,
(56 kΩ + 1.5 kΩ), 47 kΩ, 3.9 kΩ, 4.7 kΩ
12-21 4995, 1.24 kΩ, 3.9 kΩ
12-23 6.8 μF, 150 μF, 10 μF, 150 μF, 0.2 μF
12-25 (56 kΩ + 33 kΩ), 47 kΩ, 8.2 kΩ, 4.7 kΩ, 6.8 kΩ,
(8.2 kΩ + 560 Ω)
12-27 10 047, 1.25 kΩ, 6.8 kΩ
12-29 20 μF, 500 μF, 240 μF, 0.33 μF
12-31 33 kΩ, (56 kΩ + 2.2 kΩ), 3.3 kΩ, 4.7 kΩ, 6.8 kΩ,
3.9 kΩ, 15 μF, 150 μF, 150 μF, 0.25 μF
12-33 −129
13-35 −223
12-37 14 792
12-39 8.2 kΩ, 150 kΩ, 150 kΩ, 3.3 kΩ, 10 μF,
1.8 μF, 20 μF
12-41 3.3 kΩ, 120 kΩ, 220 kΩ, 3.3 kΩ, 8 μF, 1 μF, 15 μF
12-43 6.8 MΩ, 820 kΩ, 3.9 kΩ, 3.9 kΩ, (120 kΩ + 15 kΩ),
47 kΩ, 12 kΩ, 4.7 kΩ
12-45 3276, 732 kΩ, 12 kΩ
12-47 0.068 μF, 15 μF, 150 μF, 0.18 μF
12-49 1 MΩ, 6.8 kΩ, 2.2 kΩ, 220 kΩ, 68 kΩ, 18 kΩ,
6.8 kΩ, 0.02 μF, 25 μF, 2 μF, 100 μF, 0.1 μF
12-51 1 MΩ, 6.8 kΩ, 2.2 kΩ, 12 kΩ, 3.3 kΩ, 0.02 μF,
25 μF, 150 μF, 0.1 μF
12-53 0.47 μF, 0.15 μF
12-55 (33 kΩ + 3.3 kΩ), 22 kΩ, 6.8 kΩ, 2.2 kΩ, 6.8 kΩ,
(33 kΩ + 3.3 kΩ), 22 kΩ, 1 μF, 0.3 μF
12-57 3.9 kΩ, 27 kΩ, 2.7 kΩ, 2.7 kΩ, 10 μF, 100 μF,
0.82 μF
12-59 (56 kΩ + 3.3 kΩ), 18 kΩ, 33 kΩ, 4.7 kΩ, 3.3 kΩ,
15 μF, 15 μF, 150 μF, 0.39 μF

12-61 15.1 kΩ
12-63 1.91 kΩ

13-1 124.9, 123.9
13-3 44.3 Ω, 629 Ω
13-5 470 Ω, (22 kΩ + 1 kΩ), 2.7 μF, 4.7 μF
13-7 470 Ω, (82 kΩ + 2.2 kΩ), 3.9 μF, 2.2 μF
13-9 15 μF, 220 Ω, (15 kΩ + 1.2 kΩ), 18 μF
13-11 220 Ω, 27 kΩ, 39 μF, 30 μF
13-13 18 kΩ, 220 Ω, 8 μF, 6 μF, 180 pF
13-15 150 Ω, 2.7 kΩ, 2.7 μF, 15 μF
13-17 4.7 kΩ, 10 kΩ, 5.6 kΩ, 10 kΩ, 820 kΩ, 4.7 kΩ,
4.7 kΩ, 12 kΩ
13-19 10 kΩ, 6.8 kΩ, 3.9 kΩ, 6.8 kΩ, (680 kΩ + 68 kΩ),
10 kΩ, 2.7 kΩ, 8.2 kΩ
13-21 100 kΩ, 47 kΩ, 10 kΩ, 180 Ω, 4.7 kΩ, 2.7 μF,
18 μF, 0.2 μF
13-23 (R_{E1} = 270 Ω), 2.2 μF, 10 μF, 0.39 μF
13-25 (R_4 = 82 Ω), (R_9 = 150 Ω), 5 μF, 75 μF, 2 μF,
33 μF, 0.47 μF
13-27 1535, 6.5 kΩ, 6.3 kΩ
13-29 8.2 kΩ, 270 Ω, 270 kΩ, 6.8 kΩ, 270 Ω, 4.7 kΩ,
1 μF, 10 μF, 0.27 μF
13-31 18 kΩ, 12 kΩ, 390 kΩ, 15 kΩ, 560 Ω, 4.7 kΩ,
0.68 μF, 1.5 μF
13-33 12 kΩ, 8.2 kΩ, 330 kΩ, 4.7 kΩ, 270 Ω, 2.7 kΩ,
8 μF, 2.7 μF
13-35 0.006%
13-37 22.5 MHz

14-1 2 MΩ, 80 nA, 75 Ω, 25 mA, 1.7 mA, 50 mW
14-3 3.6 mA, 100 dB, ±13.5 V, 1 mV
14-5 180 kΩ, 180 kΩ, 82 kΩ
14-7 1 MΩ, 1 MΩ, 470 kΩ, 8.75 V, 8.75 V, 8.985 V
14-9 10^{11} Ω, 0.5 × 10^{-3} Ω
14-11 68 kΩ, 0.2 μF, 1.5 μF
14-13 1 kΩ, 120 kΩ, 1 kΩ
14-15 15 kΩ, 1 MΩ, 15 kΩ
14-17 5.9 × 10^9 Ω, 1.5 kΩ
14-19 22 kΩ, 1 MΩ, 22 kΩ
14-21 820 Ω, (150 kΩ + 15 kΩ), 820 Ω, 820 Ω
14-23 (47 kΩ + 3.3 kΩ), 1 MΩ, 47 kΩ
14-25 680 Ω, 33 kΩ, 680 Ω, 6 μF, 3.3 μF
14-27 1.2 kΩ, (56 kΩ + 3.9 kΩ), 120 kΩ, 120 kΩ, 1.8 μF,
4.7 μF
14-29 60 μA, 80 μA, 140 μA, −1.4 V
14-31 −0.528 V

14-33 5.5 mV, −5.92 mV, −0.11 mV

14-35 50 μA, 50 μA, 50 μA, 22.2 μA, 22.2 μA, 88.9 μA, 88.9 μA, +1.8 V, +450 mV, +150 mV, −1.2 V, 2.4 V, −0.6 V, 3 V

14-37 (270 kΩ + 22 kΩ), 39 kΩ

14-39 39 kΩ, 330 kΩ

14-41 4.7 kΩ, (39 kΩ + 6.8 kΩ), (82 kΩ + 10 kΩ)

15-1 $\theta_L = -360°$ (unstable)

15-3 $\theta_m = 75°$ (stable)

15-5 44°, 1.1°

15-7 200 pF

15-9 800 kHz, 20 kHz

15-11 80 kHz, 3.4 MHz

15-13 20 kHz, 5.6 kHz

15-15 31.8 kHz, 159 kHz

15-17 39.8 kHz, 3.97 V (peak)

15-19 0.4 pF, 15.4 pF

15-21 8.2 pF

15-23 4.4 pF, 0.4 pF

15-25 265 pF

15-27 147 pF

15-29 265 pF

16-1 6.8 kΩ, 220 kΩ, 6.8 kΩ, 3300 pF, 6.8 kΩ

16-3 39 kΩ, 22 kΩ, 1.2 kΩ, 2.2 kΩ, 2.2 kΩ, 2.2 kΩ, 3300 pF, 25 μF

16-5 0.5 μF, 0.068 μF

16-7 8.2 kΩ, 82 kΩ, 8.2 kΩ, 3900 pF, 470 pF

16-9 0.06 μF, 5600 pF, 0.082 μF, 220 kΩ, 82 kΩ, 12 kΩ, 8.2 kΩ, 0.68 μF

16-11 17.8 kHz

16-13 1300 pF, 8.2 kΩ, 1300 pF, 8.2 kΩ, 18 kΩ, 8.2 kΩ, 45 kHz, 1.2 V/μs

16-15 10.5 kHz

16-17 220 Ω, 1.5 kΩ, 5.6 kΩ

16-19 0.015 μF, 3.3 kΩ, 0.015 μF, 3.3 kΩ, 3.3 kΩ, 5.6 kΩ, 1.5 kΩ, 9 kHz, 0.2 V/μs

16-21 ±12 V, 1.35 kHz

16-23 (1.5 kΩ + 15 kΩ pot.), 1 μF, 16.5 kΩ, 470 Ω

16-25 893 Hz, 67%

16-27 446 Hz

16-29 10 kΩ, 0.27 μF, 1 kΩ, (10 kΩ + 1 kΩ)

16-31 ±1.95 V, 282 Hz

16-33 crystal in series with R_1 = 270 Ω, 0.24 mW

16-35 1.5 kΩ, 2.2 kΩ, 4.7 kΩ, 2.2 kΩ, 330 pF, 330 pF, 0.85 mW

17-1 8.84 kHz, 15 dB

17-3 2.18 kHz, 0.7 dB

17-5 1.4 dB, 10.4 kHz

17-7 [430 pF, (82 kΩ + 10 kΩ), 100 kΩ], [1000 pF, 39 kΩ, 50 kΩ]

17-9 4.7 kHz, 800 kHz

17-11 1000 pF, 2000 pF, 22.6 kΩ, 22.6 kΩ, 47 kΩ

17-13 1000 pF, 2000 pF, 32.4 kΩ, 32.4 kΩ, 68 kΩ

17-15 4 kHz

17-17 1000 pF, 1000 pF, 187 kΩ, 93.1 kΩ, 100 kΩ

17-19 1000 pF, 5.76 kΩ, 5.6 kΩ, 2000 pF, 1000 pF, 4.99 kΩ, 4.99 kΩ, 10 kΩ

17-21 642 kHz, 39 dB

17-23 0.047 μF, 1000 pF, 3.32 kΩ, 3.32 kΩ, 3.3 kΩ

17-25 9100 pF, 9100 pF, 34.8 kΩ, 69.8 kΩ, 68 kΩ, 402 Ω

17-27 (1000 pF, 1.62 kΩ, 1.5 kΩ), (1000 pF, 16.2 kΩ, 15 kΩ), (20 kΩ, 20 kΩ, 20 kΩ, 6.8 kΩ)

17-29 4000 pF, 1000 pF, (4 × 29.4 kΩ)

17-31 5100 pF, 39 kΩ, 39 kΩ, (18 kΩ + 35 kΩ)

17-33 1000 pF, 69.8 kΩ, 536 kΩ

17-35 6.04 kΩ, 34.8 kΩ, 121 kΩ, 173 kHz

18-1 220 Ω, IN758, 8.2 kΩ

18-3 36 mA, 19.6 mA, 0.77 mA, 21.4 mA

18-5 12.5 mV, 10 mV, 0.07%, 0.06%

18-7 680 Ω, 470 Ω, IN758, 5.6 kΩ, 10 kΩ

18-9 0.3%, 0.15%, 33 dB

18-11 1.2 kΩ, 8.2 kΩ, 2.5 kΩ

18-13 2.7 kΩ, 15 kΩ

18-15 3.3 kΩ, 1N758, 470 Ω, 3.9 kΩ, 8.2 kΩ, 2.5 kΩ, 15 kΩ

18-17 470 Ω, (2 × 1N758)

18-19 1N751, 3.9 kΩ, 1.8 kΩ, 23 V

18-21 2.27 Ω, 4.4 W

18-23 1 A, 313 mA

18-25 220 Ω, 2.2 kΩ, 4.7 kΩ, 2.5 kΩ, 12 kΩ, 1N754

18-27 0.75×10^{-3}%, 0.4×10^{-3}%

18-29 25 V, 6.8 kΩ, 4.7 kΩ, 2.5 kΩ, 37.5 mA

18-31 −17 V, 1 kΩ, 3.9 kΩ, 400 mW

18-33 61.5%, 82.5%

18-35 330 μH, 15 μF

18-37 1.2 kΩ, (12 kΩ + 1.8 kΩ), 0.55 Ω, 620 pF

18-39 1.2 kΩ, (10 kΩ + 820 Ω), 0.39 Ω, 510 pF

19-1 Point A: (0 mA, 20 V), Point Q: (49.7 mA, 15 V), Point B: (49.7 mA and 20 V from Q)

19-3 Point A: (0 mA, 25 V), Point Q: (1.19 mA, 22.3 V), Point B: (1.19 mA and 9.6 V from Q)

19-5 Point Q: (0 mA, 40 V),
Point B: (92.6 mA, 0 V)

19-7 16.1 V, 16.2 V

19-9 386 mW

19-11 90 V, 900 mA, 2.8 W

19-13 Point Q: (183 mA, 19 V), Point A: (0 mA, 24 V),
Point B: (19.03 V and 183 mA from Q)

19-15 12 V, 24 V, 144 mA, 215 mW

19-17 23 kΩ, 1.8 kΩ, 1.5 kΩ, 25 Ω, 180 Ω

19-19 $\pm$15 V, 30 V, 492 mA, 675 mW

19-21 $\pm$21 V, Q_2 Q_3: (2.7 W, 42 V, 778 mA),
Q_4 Q_5: (135 mW, 42 V, 39 mA)

19-23 $\pm$15 V, 3.3 kΩ, 50 Ω, 680 Ω, 1.5 Ω, 1.5 Ω

19-25 5.6 kΩ, 1.2 kΩ, 1 kΩ

19-27 330 Ω, 400 μF, 53 V

19-29 1.8 kΩ, 250 Ω, 1.8 kΩ, 1 kΩ, 1 kΩ, 2.2 Ω, 2.2 Ω

19-31 18 μF, 68 μF, 680 pF

19-33 1.72 W, 30 V, 644 mA

19-35 0.85 mA, 1.01 A

19-37 $\pm$23 V, 2.4 W

19-39 4.7 kΩ, 18 kΩ, 10 kΩ, 330 Ω, 4.7 kΩ, 100 kΩ,
1.2 MΩ, 1.8 MΩ, 100 kΩ, 4.7 μF, 6 μF, 0.2 μF,
0.2 μF

19-41 $\pm$19 V

19-43 4.7 kΩ, 15 kΩ, 8.2 kΩ, 180 Ω, 4.7 kΩ, 100 kΩ,
680 kΩ, 1 MΩ, 100 kΩ, 8.2 μF, 22 μF, 0.47 μF,
0.47 μF

19-45 15 kΩ, (560 Ω + 82 Ω), 15 kΩ, 560 Ω, 560 Ω

19-47 ($R_8 = R_9 = R_{10} = R_{11} = 1.2$ kΩ), 100 μF, 15.3 V

19-49 $\pm$28 V, 14.3 V, $\pm$1.25 V

19-51 $\pm$15 V

19-53 2.7 μF, 4 μF, 1600 pF, 240 pF

19-55 0.53 Ω

20-1

	2N1595	2N6167
V_{DRM}	50 V	100 V
$I_{T(rms)}$	1.6 A	13 A
V_{TM}	1.1 V	1.5 V
I_H	5 mA	3.5 mA
I_{GT}	2 mA	2.1 mA
V_{GT}	0.7 V	0.63 V

20-3 2N1597, 1.27 V

20-5 1.5 kΩ, 100 Ω, 12 Ω

20-7 560 Ω, 22 kΩ, 200 kΩ

20-9 12.8 V, 680 Ω

20-11 56 kΩ, 6 μF, 0.43 V

20-13

	2N6071	2N6343
V_{DRM}	200 V	400 V
$I_{T(rms)}$	4 A	8 A
V_G	1.4 V	0.9 V

20-15 49°

20-17 3.3 kΩ, 180 Ω, 10 kΩ, 3.3 kΩ, 0.82 μF, 0.15 μF

20-19 1 V, 15 V, 830 μA, 2 mA

20-21 165 mW, 34 V

20-23 435 Hz

20-25 6.8 kΩ, 10 kΩ, 100 Ω, 330 Ω, 0.56 μF

20-27 25 kΩ, (12 kΩ + 2.2 kΩ), 3.9 kΩ

20-29 15 kΩ, (8.2 kΩ + 820 Ω), 15.7 V, 0.9 V, 24 kΩ, 5.6 kΩ

21-1 184 W

21-3 4.72×10^{14} Hz, 5.15×10^{14} Hz, 5.31×10^{14} Hz

21-5 $R_E = 270$ Ω

21-7 10 kΩ, 820 Ω

21-9 690 μW

21-11 4.7 kΩ, 10 lx

21-13 2 mA, 26 mA

21-15 120 Ω, -0.12 V, -7 mA

21-17 0.38 V

21-19 102

21-21 1.2 kΩ

21-23 13 V, 5.5 mA

21-25 330 Ω

21-27 5.7 V

21-29 9 mA, 4.4 mA, 49%

22-1 3.6 MHz, 796 kHz

22-3 15 kΩ, 20 kΩ, 150 kΩ

22-5 390 pF, 15

22-7 30°C

22-9 9.8 V, 10.2 V, 10.3 V

22-11 146 Ω

22-13 120 Ω, 225 mV, $\pm$145 mV, 5.6

22-15 30.9 V, 61.8 mA

22-17 1.2 kΩ, 3.57 kΩ

INDEX